Organic Reactions

Organic Reactions

VOLUME 48

JOHN WILEY & SONS, INC.

New York　•　Chichester　•　Brisbane　•　Toronto　•　Singapore

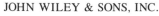

Library of Congress Catalog Card Number 42-20265

ISBN 0-471-14699-4

Printed in the United States of America

10 9 8 7 6 5 4 3 2 1

PREFACE TO THE SERIES

In the course of nearly every program of research in organic chemistry the investigator finds it necessary to use several of the better-known synthetic reactions. To discover the optimum conditions for the application of even the most familiar one to a compound not previously subjected to the reaction often requires an extensive search of the literature; even then a series of experiments may be necessary. When the results of the investigation are published, the synthesis, which may have required months of work, is usually described without comment. The background of knowledge and experience gained in the literature search and experimentation is thus lost to those who subsequently have occasion to apply the general method. The student of preparative organic chemistry faces similar difficulties. The textbooks and laboratory manuals furnish numerous examples of the application of various syntheses, but only rarely do they convey an accurate conception of the scope and usefulness of the processes.

For many years American organic chemists have discussed these problems. The plan of compiling critical discussions of the more important reactions thus was evolved. The volumes of *Organic Reactions* are collections of chapters each devoted to a single reaction, or a definite phase of a reaction, of wide applicability. The authors have had experience with the processes surveyed. The subjects are presented from the preparative viewpoint, and particular attention is given to limitations, interfering influences, effects of structure, and the selection of experimental techniques. Each chapter includes several detailed procedures illustrating the significant modifications of the method. Most of these procedures have been found satisfactory by the author or one of the editors, but unlike those in *Organic Syntheses* they have not been subjected to careful testing in two or more laboratories.

Each chapter contains tables that include all the examples of the reaction under consideration that the author has been able to find. It is inevitable, however, that in the search of the literature some examples will be missed, especially when the reaction is used as one step in an extended synthesis. Nevertheless, the investigator will be able to use the tables and their accompanying bibliographies in place of most or all of the literature search so often required.

Because of the systematic arrangement of the material in the chapters and the entries in the tables, users of the books will be able to find information desired by reference to the table of contents of the appropriate chapter. In the interest of economy the entries in the indices have been kept to a minimum, and, in particular, the compounds listed in the tables are not repeated in the indices.

The success of this publication, which will appear periodically, depends upon the cooperation of organic chemists and their willingness to devote time and effort to the preparation of the chapters. They have manifested their interest already by the almost unanimous acceptance of invitations to contribute to the work. The editors will welcome their continued interest and their suggestions for improvements in *Organic Reactions*.

Chemists who are considering the preparation of a manuscript for submission to *Organic Reactions* are urged to write either secretary before they begin work.

CONTENTS

Organic Reactions

CHAPTER 1

ASYMMETRIC EPOXIDATION OF ALLYLIC ALCOHOLS: THE KATSUKI–SHARPLESS EPOXIDATION REACTION

TSUTOMU KATSUKI

Department of Chemistry, Faculty of Science, Kyushu University, Japan

VICTOR S. MARTIN

*Centro de Productos Naturales Organicos "Antonio Gonzalez,"
Instituto Universitario de Bio-Organica, Universidad de la Laguna,
Tenerife, Spain*

CONTENTS

Organic Reactions, Vol. 48, Edited by Leo A. Paquette et al.
ISBN 0-471-14699-4   © 1996 Organic Reactions, Inc. Published by John Wiley & Sons, Inc.

ACKNOWLEDGMENT

This article is dedicated to Professor K. Barry Sharpless, who discovered and extensively developed the titanium-mediated asymmetric epoxidation reaction, and who introduced the authors to this fertile area of chemistry. The authors are also grateful to him for his generous help with editing our manuscript and for his encouragement throughout this work. This article is also dedicated to the memory of the late Professor Bryant Rossiter, who made a great contribution to epoxidation chemistry and shared an excellent time with us at Stanford and at M. I. T. The authors also thank the editors of *Organic Reactions* and Drs. Soo Y. Ko, Albert W. M. Lee, Janice M. Klunder, Robert M. Hanson, and Roy A. Johnson for their kind help in the preparation of this chapter.

INTRODUCTION

In 1980, Sharpless and Katsuki discovered a system for the asymmetric epoxidation of primary allylic alcohols that utilizes $Ti(OPr-i)_4$, a dialkyl tartrate as a chiral ligand, and *tert*-butyl hydroperoxide as the oxidant.[1] Notably, this reaction exhibits high levels of enantioselectivity (usually > 90% ee). Like other metal-catalyzed epoxidations, this reaction also proceeds under mild conditions with good chemical yield and with high regio- and chemoselectivity. Various aspects of this reaction, including its mechanism,[2] early synthetic applications,[3–6] and further transformations of the epoxy alcohol product,[7] have been reviewed. In the following sections, the full scope and limitations of this reaction, its synthetic applications, and typical experimental conditions are described.

ASYMMETRIC EPOXIDATION WITH THE
TITANIUM(IV)–TARTRATE CATALYST

The combination of Ti(OPr-i)$_4$, a dialkyl tartrate, and *tert*-butyl hydroperoxide (hereafter referred to as TBHP) epoxidizes most allylic alcohols in good chemical yield and with predictably high enantiofacial selectivity according to the empirical rule illustrated in Scheme 1. When an allylic alcohol (R^4, R^5 = H) is

Scheme 1

drawn in a plane with the hydroxymethyl group positioned at the lower right, the delivery of oxygen occurs from the bottom side of the olefin to give the (2S)-epoxide if an (R,R)-dialkyl tartrate is used as the chiral auxiliary. Of course, when an (S,S)-dialkyl tartrate is employed, oxygen is delivered from the top side. The enantiofacial selectivity of the reaction is > 90% ee (usually > 95% ee) for substrates without a Z olefinic substituent (R^3 = H). The degree of facial selectivity for a Z allylic alcohol depends on the nature of the Z substituent R^3. The enantioselectivity for substrates with unbranched R^3 substituents ranges from 80 to 94% ee, but that for substrates with a branched substituent is lower.[2] Representative examples of epoxidations of allylic alcohols with diethyl tartrate (DET) as the chiral auxiliary are shown in Eqs. 1–4.[1,8,9]

(77%) 95% ee

(Eq. 1)

(81%) >95% ee

(Eq. 2)

(Eq. 3)

(90%) 94% ee

(Eq. 4)

(54%) 66% ec

This reagent combination is also effective for the kinetic resolution of racemic secondary allylic alcohols (R^4 = H, R^5 = alkyl and R^4 = alkyl, R^5 = H). When (R,R)-tartrate is used as the chiral auxiliary, the S enantiomer (R^5 = H) of the allylic alcohol reacts faster than the R enantiomer (R^4 = H), and the R enantiomer may be recovered with high enantiomeric purity (Eq. 5).[10] The relative re-

(Eq. 5)

>96% ee

action rates (k_{fast}/k_{slow}) for enantiomeric pairs usually range from 15 to 700,[2,10-16] except for peculiar substrates like allyl-*tert*-butylcarbinol and cyclohexenol. This kinetic resolution is an effective way of obtaining *sec*-allylic alcohols of high optical purity. The relationship between the enantiomeric purity of the unreacted allylic alcohol, the relative reaction rate of a pair of enantiomers, and percent conversion of the starting racemic allylic alcohol is described in the section on the kinetic resolution of secondary allylic alcohols. This reaction is also effective for preparing epoxy alcohols. The epoxy alcohol derived from the fast-reacting enantiomer, except for substrates with Z olefinic substituents, possesses the *erythro* configuration.[10] This is consonant with the empirical rule defining the stereochemical outcome of the reaction presented at the beginning of this section.

By adding molecular sieves to the reaction system, the asymmetric epoxidation and kinetic resolution processes can be carried out with a catalytic amount of the titanium–tartrate complex without impairing the enantioselectivity of the reaction.[17-19]

The titanium–tartrate complex and its modified relatives have also been employed for the asymmetric oxidation of heteroatoms like nitrogen,[20,21] sulfur,[22–25] and selenium.[26]

MECHANISM

The reaction sequence proposed for the metal-catalyzed epoxidation of allylic alcohols is shown in Scheme 2.[27–30]

Scheme 2

Metal alkoxides generally undergo rapid ligand exchange with alcohols. When a metal alkoxide, an allylic alcohol, and an alkyl hydroperoxide are mixed, ligand exchange occurs to afford a mixture of complexes $M(OR)_{n-x-y}$-$(OCH_2CH=CH_2)_x(OOR)_y$. Among them, only species such as **1**, bearing both allylic alkoxide and alkyl hydroperoxide groups, are responsible for the epoxidation. The incorporated alkyl hydroperoxide is thought to be further activated by coordination of the second oxygen atom (O-2) to the metal center (see structures **2** and **3**). That the ensuing transfer of O-1 to the double bond of the allylic alcohol occurs in an intramolecular fashion is supported by comparison of the epoxidation rate of allyl alcohol with that of allyl methyl ether.[31] However, controversy still surrounds the oxygen transfer process (**2**→**5**). One suggestion is that the double bond first coordinates to the metal center and then inserts into the $\mu2$-alkyl hydroperoxide ligand to give an epoxide via the peroxometallocycle intermediate **4**.[32–34] An alternative proposal is that the double bond attacks the distal oxygen along the axis of the O–O bond that is broken.[2,30,35–39] Frontier molecular orbital treatment of peroxometal complexes also suggests that d transition metal complexes of ROO− exhibit electrophilic behavior.[40] Finally, exchange of *tert*-butoxide and the epoxy alkoxide so formed with allylic alcohol and alkyl hydroperoxide completes the reaction cycle.

The titanium–tartrate mediated asymmetric epoxidation of allylic alcohols also follows the same basic reaction pathway of Scheme 2. Therefore, the remaining mechanistic question is how oxygen is transferred enantioselectively to substrates. To answer this question, structures of titanium–dialkyl tartrate complexes,[37-38,41-45] as well as those prepared from Ti(OPr-i)$_4$ and (R,R)-N,N'-dibenzyltartramide and from Ti(OEt)$_4$, (R,R)-diethyl tartrate, and PhC(O)-N(OH)Ph were determined.[46-48] Based on the X-ray analysis of these complexes, the structure of the asymmetric epoxidation catalyst has been proposed as **6**.

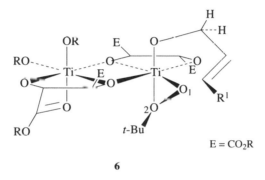

6

When structure **6** is viewed down the distal peroxide oxygen–titanium bond axis (O^1-Ti), the symmetry of the tartrate "windmill arms" becomes apparent. Within this model, conformer **7**, in which the allylic alcohol

7

and the TBHP-ligand align meridionally and the TiO—C—C=C dihedral angle is as small as 30°, has been suggested as a transition state.[2]

This conformer experiences severe steric interactions only when $R^5 \neq$ H. This explains the high efficiency of kinetic resolution of racemic secondary allylic alcohols where one enantiomer (R^4 = alkyl, R^5 = H) reacts much faster than the other isomer (R^4 = H, R^5 = alkyl). The poor reactivity of tertiary allylic alcohols (R^4 and R^5 = alkyl) is rationalized analogously.[49] We also see that the Z olefinic substituent (R^3) is close to the hydroxymethyl group bound to titanium because of the small O—C—C=C dihedral angle. These interactions destabilize conformer **8** and lower the reactivity of this complex. The C-2 substituent (R^2) in

8

7 is also in the vicinity of the titanium complex, and only the E olefinic substituent (R^1) projects toward an open quadrant.

This model explains the following three observations. First, bulky Z olefinic substituents retard epoxidation reactions, and substrates with branched Z substituents exhibit poor reactivity and decreased enantioselectivity. This may be rationalized by the conformational requirement for minimization of allylic strain due to the small C=C—C—OTi dihedral angle.[2] That is, the conformation in which H is in the plane of the olefin (H in-plane conformation) is energetically more accessible than the other two conformations (R and R' in-plane conformations).[50] Thus the disposition of an alkyl group (R') to the bottom side raises the energy of the transition state depicted as **8** [using (R,R)-tartrate], causing retardation of the reaction and decreased enantioselectivity. When R ≠ R', each enantiomer of a racemic substrate has different reactivity and treatment of such a racemic mixture with Ti(OPr-i)$_4$-tartrate effects kinetic resolution.

Second, because the C-2 substituent is near the Ti–tartrate moiety, its chirality also affects substrate reactivity. Thus enantiomers of a racemic substrate bearing a chiral C-2 substituent have different reactivities, and in some cases a good level of kinetic resolution is observed.

Third, except for a few examples, the E substituent, which is located in an open quadrant, has little effect on the stereoselectivity of the reaction. Therefore, the epoxidation of chiral E allylic alcohols should proceed with the same high level of enantioselectivity seen with achiral E allylic alcohols.[51]

Although a proposal that the dimeric titanium tartrate complex is in equilibrium with a monomeric ion pair which catalyzes epoxidation has been presented,[52] kinetic studies of the epoxidation reaction seem to exclude the possibility that the active catalyst species is monomeric.[36]

SCOPE AND LIMITATIONS

Asymmetric Epoxidation of Achiral Primary Allylic Alcohols

There are two general procedures for asymmetric epoxidation: stoichiometric and catalytic. These procedures exhibit similar stereo-, chemo-, and regioselectivities. However, the catalytic procedure provides several practical advantages over the stoichiometric one: a) ease of isolating the product, b) extended scope due to use of a catalytic amount of Lewis acidic titanium catalyst, and c) more economical. General features of asymmetric epoxidation are discussed first, and the advantage of the catalytic procedure is described later in this section.

Substrate Reactivity. The reactivity of allylic alcohols changes with the level of olefinic substitution (Table A).[2,4,43,44] Like epoxidation reactions with

TABLE A. SUBSTRATE REACTIVITY

Entry	Substrate	Relative Reactivity
1		0.048
2		0.600
3		0.220
4		0.420
5		1.00
6		1.20
7		4.39
8		1.48

Reprinted with permission from Sharpless, K. B., et al., *Asymmetric Synthesis,* Morrison, J. D., Ed., Academic Press, Vol. 5, p. 268.

peracids or other metal-catalyzed epoxidations, the reactivity of the substrates generally rises as the olefinic electron density increases.

The reactivity of cinnamyl alcohols changes with the electronic nature of the aromatic substituent. This trend indicates the nucleophilic nature of the olefin (entries 4–7). However, the level of reactivity of the 4-chloro derivative is curious (entry 6).

A Z-olefinic substituent decreases substrate reactivity (Eqs. 6, 7),[18] especially when it is branched at C-4 (see section on mechanism).

$$n\text{-}C_8H_{17} \diagdown \diagup OH \xrightarrow[\text{TBHP, 4 Å MS, -10°, 1.5 h}]{\text{Ti(OPr-}i)_4 \text{ (0.05 eq), (}R,R\text{)-(+)-DET (0.06 eq)}} n\text{-}C_8H_{17} \overset{O}{\diagup\!\!\!\triangle} OH$$

(78%) 94% ee

(Eq. 6)

$$n\text{-}C_8H_{17} \diagup\diagdown_{OH} \xrightarrow[\text{TBHP, 4 Å MS, -12°, 42 h}]{\text{Ti(OPr-}i)_4 \text{ (0.05 eq), (}R,R\text{)-(+)-DET (0.06 eq)}} n\text{-}C_8H_{17} \overset{O}{\triangle}_{OH}$$

(63%) >83% ee

(Eq. 7)

Epoxidations of tertiary allylic alcohols are usually sluggish, though (E)-2,5-dihydroxy-2,5-dimethylhex-3-ene is epoxidized smoothly to give the corresponding epoxide with moderate enantioselectivity (Eq. 8).[53]

$$\xrightarrow[\text{TBHP, 4 Å MS, -20°}]{\text{Ti(OPr-}i)_4, (R,R)\text{-(+)-DIPT}}$$

(96%) 70% ee

(Eq. 8)

However, the epoxidation of dimethyl(2-phenylvinyl)silanol (9) proceeds with high enantioselectivity (85–95% ee), whereas that of diphenyl(2-phenylvinyl)silanol (10) exhibits an enantioselectivity of only 20%.[54]

$$Ph \diagdown\diagup SiR_2OH \xrightarrow[\text{TBHP}]{\text{Ti(OPr-}i)_4, (R,R)\text{-(+)-DET}} Ph \overset{O}{\diagup\!\!\triangle} SiR_2OH$$

9 R = Me
10 R = Ph

(71%) 85-95% ee
(70%) 20% ee

The presence of a coordinating functional group adjacent to the allylic hydroxy group sometimes decreases the reactivity of the allylic alcohol below useful levels (Eq. 9).[55]

$$\text{(Eq. 9)}$$

Stereoselectivity. The stereochemistry of this reaction can be predicted by the empirical rule shown in Scheme 1, except for a few substrates bearing bulky chiral substituents near the site of epoxidation. There have been no exceptions for achiral substrates to date. Thus the reaction has been used to assign or verify absolute configuration. For example, the (−)-(E)-cycloalkenemethanol **11** was assigned the S configuration by chemical correlation to a substance whose stereochemistry had been determined on the basis of an ORD curve. However, the (−)-enantiomer [(−)-**11**] did not react under the titanium-mediated epoxidation conditions using (R,R)-diethyl tartrate, although the (+)-enantiomer [(+)-**11**)] reacted smoothly. The empirical rule therefore indicates that (−)-**11**

(+)-**11**

(60%)

(−)-**11**

should have the R, not the S, configuration. In fact, a different and unambiguous chemical correlation later established that the conclusion based on the empirical rule was correct.[56]

The following examples illustrate the effect of allylic alcohol substitution on stereoselectivity. The asymmetric epoxidation of a substrate with an E tert-butyl group exhibits normal enantioselectivity (Eq. 10).[57] This confirms that E allylic

$$\text{(Eq. 10)}$$

(52%) >95% ee

alcohols are good substrates for this titanium-mediated asymmetric epoxidation, and there are many such examples (Eq. 11).

n-C$_{15}$H$_{31}$ ⟍⟍⟍⟍OH $\xrightarrow[\text{TBHP}]{\text{Ti(OPr-}i)_4\text{, (S,S)-(−)-DET}}$ n-C$_{15}$H$_{31}$ ⟍⟍⟍⟍ epoxide OH   (Eq. 11)

(88%) >95% ee

The reaction with 2-*tert*-butylallyl alcohol gives lower enantioselectivity (Eq. 12). However, the optical purity (85% ee) determined for the product is

⟍⟍⟍OH $\xrightarrow[\text{TBHP}]{\text{Ti(OBu-}t)_4\text{, (R,R)-(+)-DET}}$ ⟍⟍OH   (Eq. 12)

(42%) 85% ee

probably the lower limit, since the major enantiomer of the alcohol is preferentially attacked and consumed by an alcoholic nucleophile under the influence of the chiral titanium catalyst.[57]

The propensity for large C-2 substituents to erode enantioselectivity is also illustrated by the epoxidation of compound 12.[58] This reaction shows diminished

12   (80%) 70% ee

enantioselectivity (70% ee), though that of the parent 2-phenylcinnamyl alcohol exhibits high enantioselectivity (>95% ee). Although these examples are consonant with the mechanistic view that places the C-2 substituent in the vicinity of the titanium–tartrate complex, the effect of an achiral C-2 substituent on enantioselectivity is usually small. Consequently, the epoxidation of most C-2 substituted allylic alcohols proceeds with high enantioselectivity (>90% ee) (Eqs. 13, 14).

n-C$_{14}$H$_{29}$ ⟍⟍OH $\xrightarrow[\text{TBHP}]{\text{Ti(OPr-}i)_4\text{, (R,R)-(+)-DET}}$ n-C$_{14}$H$_{29}$ ⟍⟍OH   (Eq. 13)

(91%) >96% ee

⟍⟍OH $\xrightarrow[\text{TBHP}]{\text{Ti(OPr-}i)_4\text{, (R,R)-(+)-DET}}$ ⟍⟍OH   (Eq. 14)

(86%) >95% ee

In spite of the high enantioselectivity of the titanium–tartrate complex, substandard stereoselectivity is observed for some sterically crowded allylic alcohols. The enantioselectivity of the epoxidation of Z-olefinic substrates with C-4 branched substituents decreases as the substituent become larger (Eqs. 4, 15–18).[8,57,59–61] A substrate with a Z tert-butyl group undergoes epoxidation with very poor asymmetric induction, although face selection for this substrate is still in the normal direction (Eq. 18).[57]

(Eq. 15)

(45%) >95% ee

(Eq. 16)

(68%) 92% ee

(Eq. 17)

(59%) 80% ee

(Eq. 18)

(77%) 25% ee

Asymmetric epoxidation of some allylic alcohols proceeds with diminished enantioselectivity. In many cases, however, the product epoxy alcohols are crystalline compounds that can be recrystallized to optical purity. As an example, alcohol **13** produces epoxy alcohol **14** in 86% ee, and recrystallization of **14** affords

13

cat. Ti(OPr-i)$_4$, (S,S)-(–)-DET

MS 4 Å, TBHP

14   86% ee (separated by chromatography)
(79%) 100% ee (recrystallized)

only a single enantiomer in 79% yield.[62] Noncrystalline epoxy alcohols can often be converted to crystalline derivatives such as **15**,[18] **16**,[63,64] and **17**,[65] which also

15

16

allow enhancement of enantiomeric purity through recrystallization. These derivatives can be obtained by catalytic asymmetric epoxidation and subsequent in situ derivatization (vide infra).[18,66,67] Unfortunately, increases in enantiomeric purity by recrystallization are not universal. Some compounds like **18** decrease in enantiomeric purity upon repeated recrystallization.[67]

17

18

Epoxidation of bisallylic alcohols possessing C_s symmetry provides diepoxy products of high enantiomeric purity together with a small amount of *meso* isomer.[68-72] For example, if epoxidation of each allylic alcohol group in compound **19** proceeded with the usual enantioselectivity (95% ee), the product should be a mixture of (2S,3S,10S,11S)-**20**, (2R,3R,10R,11R)-**20**, and *meso*-**21** isomers in the

19   Ti(OPr-i)₄, (R,R)-(+)-DET   TBHP   (2S,3S,10S,11S)-**20**   +

(2R,3R,10R,11R)-**20**   +   meso-**21**

ratio 361:1:38. Actually, the reaction of **19** gives a mixture of **20** and **21** in a ratio of 9:1, though the enantiomeric purity of **20** has not been determined.[68]

Chemoselectivity. One of the important features of this metal-catalyzed epoxidation is its high chemoselectivity. Allylic alcohols bearing other functional groups such as ether (**22**),[73,74] epoxide (**23**),[75-78] acetal [e.g., acetonide (**22**),[73] tetrahydropyranyl (THP) ether (**24**),[79] 1,3-dioxane (**25**),[80] and ethoxyethyl (EE) ether (**26**)[81]], silyloxy (**23**),[77,78,82-89] carbonyl (**27**),[90] enone (**23**),[77,78] ester (**28**)[1,91] α,β-unsaturated ester (**29**),[92,93] carbonate (**30**),[94] urethane (**31**),[95-97] toluenesulfonamide (**32**),[98] acetylene (**33**),[99-104] 4,5-diphenyloxazole (**34**),[105] nitrile

(**35**),[106–108] nonfunctionalized olefin (**36**),[1,109,110] vinylsilane (**37**),[111] trialkylsilylacetylene (**38**),[112] allylsilane (**29**),[93] *tert*-allylic alcohol (**39**),[113] furan (**40**)[114] (except for furfuryl alcohol)[114], and lactim ether (**41**)[115] units, as exemplified by the structures in Fig. 1, can be successfully epoxidized without interference from the

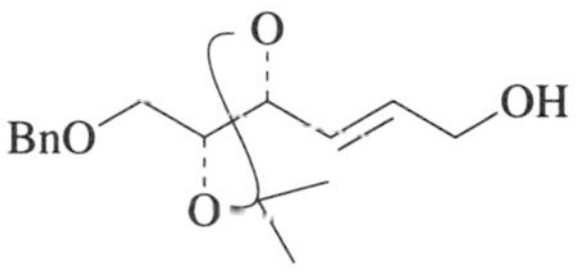

22

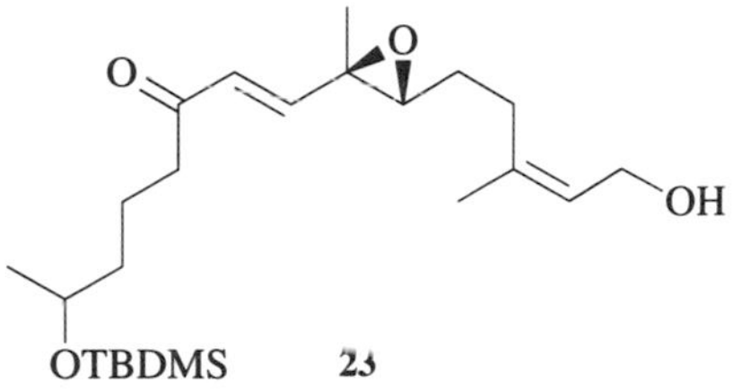

23

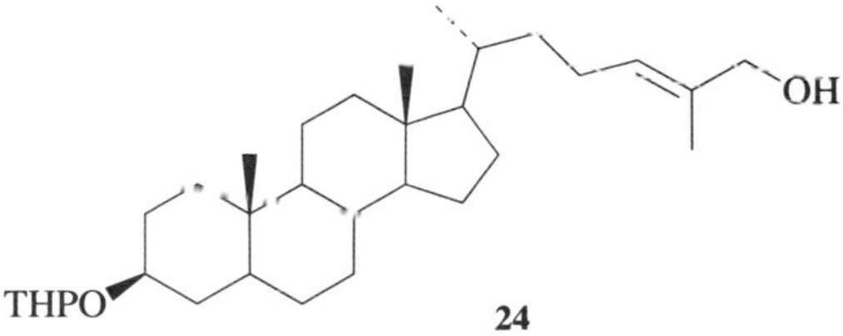

24

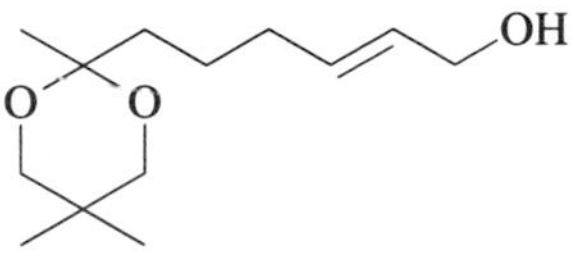

25

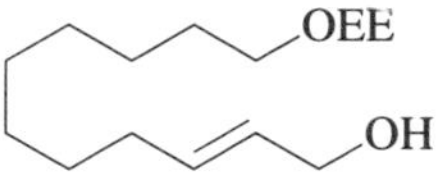

26

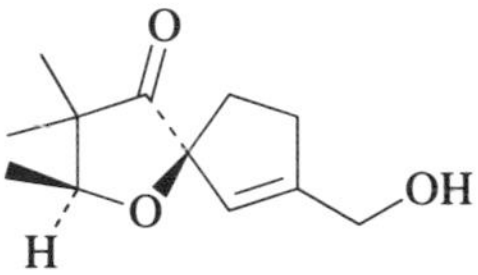

27

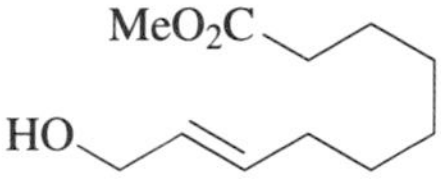

28

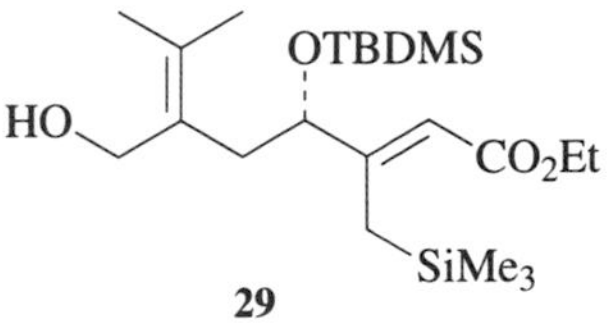

29

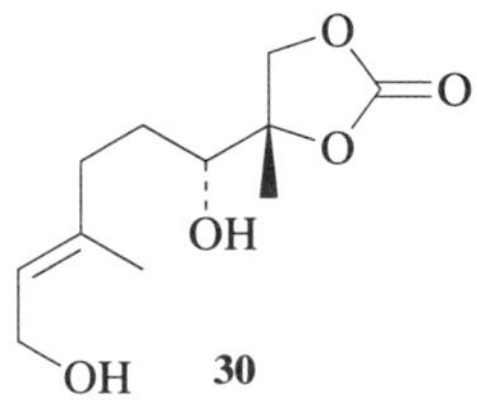

30

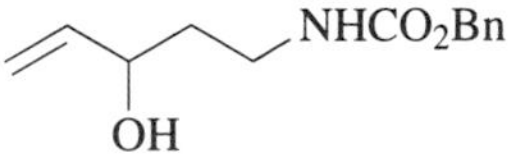

31

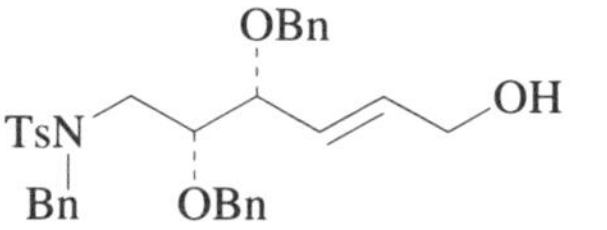

32

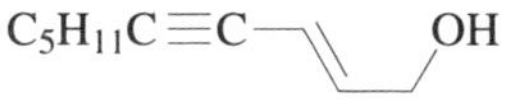

33

Fig. 1

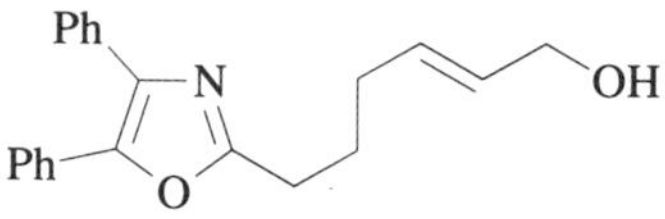

34

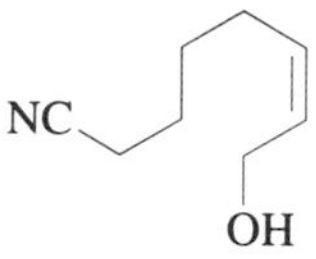

35

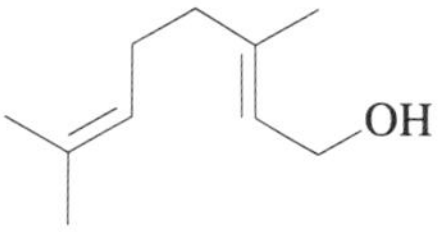

36

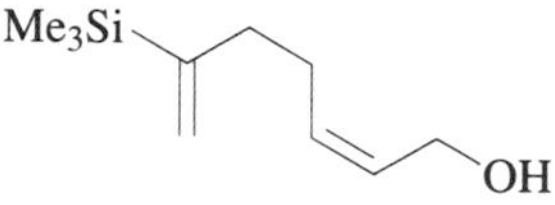

37

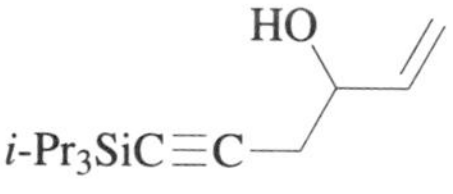

38

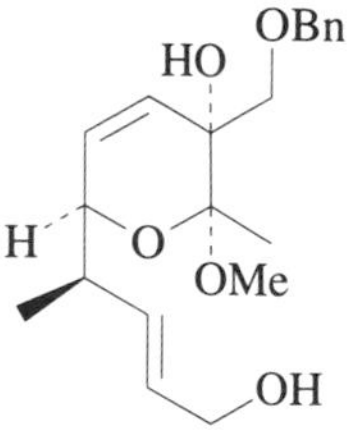

39

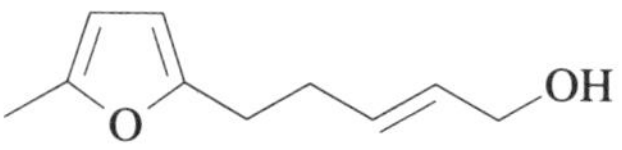

40

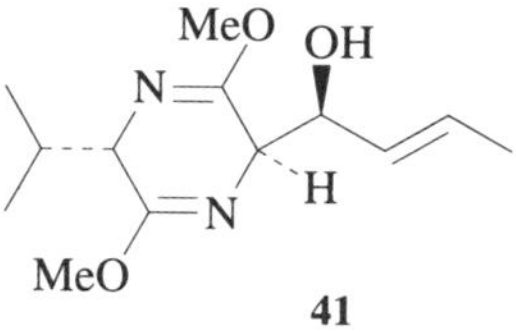

41

Fig. 1 (*cont.*)

other resident functional group. However, when a carbonyl group[116] (including ester[117] and amide carbonyls), a 4,5-diphenyloxazole, or a hydroxy group[117–119] is located in an appropriate position for intramolecular attack on the epoxide that is formed, a subsequent epoxide-opening reaction often occurs in situ or during workup to give a cyclic product (Eq. 19)[116] or a mixture of unidentified products other than the desired epoxide (Eq. 20).[105]

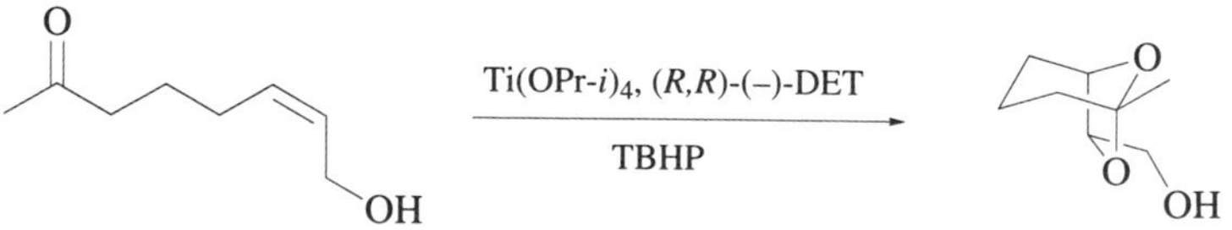

(70%) 95% ee

(Eq. 19)

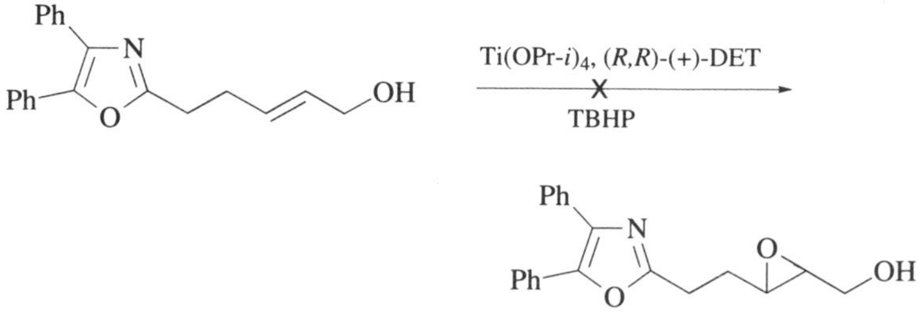

(Eq. 20)

Asymmetric epoxidation of phenol **42** proceeds with diminished enantioselectivity (53% ee), possibly because of the strong coordination of phenols to titanium(IV).[120,121] The epoxidation of compound **43**, in which the phenolic hydroxy groups are protected, proceeds with high enantioselectivity.[122]

Ti(OPr-*i*)₄ (5 eq), (*S,S*)-(–)-DET (5 eq)

TBHP (10 eq)

42

(85%) 53% ee

Ti(OPr-*i*)₄, (*S,S*)-(–)-DET

TBHP, -20°

43

(80%) 96% ee

Certain functional groups in the vicinity of the reacting site may affect the stereoselectivity. For example, asymmetric epoxidation of ether **44** using (*R,R*)- and (*S,S*)-diethyl tartrate gives 3:1 and 1:3 mixtures of **45** and **46**, respec-

Ti(OPr-*i*)₄

(*R,R*)-(+)-DET

45   3:1   + **46**

Ti(OPr-*i*)₄

(*S,S*)-(–)-DET

46   3:1   + **45**

44

tively.[123] The asymmetric epoxidation of alcohol **47**, however, proceeds with high selectivity.[124]

Scheme: $n\text{-}C_{12}H_{25}$ (OMOM) (E)-allylic alcohol **47** $\xrightarrow[\text{TBHP, }-20°]{\text{Ti(OPr-}i)_4,\ (S,S)\text{-}(-)\text{-DET}}$ $n\text{-}C_{12}H_{25}$ (OMOM) epoxy alcohol (63%) 95% de

Regioselectivity. In metal-catalyzed epoxidation reactions, the reactivity of allylic alcohols is superior to that of homoallylic alcohols when both substrates have the same substitution pattern.[125] Although there is no detailed study on the regioselectivity of titanium(IV)-tartrate epoxidation, the above trend in regioselectivity probably holds in this reaction. However, when the degree of substitution of a homoallylic alcohol is greater than that of an allylic alcohol, the homoallylic alcohol (e.g., **48**) can be epoxidized preferentially.[126] Olefins without coordinat-

Scheme: $(CH_2)_{10}$ ... OH **48** $\xrightarrow[\text{TBHP}]{\text{Ti(OPr-}i)_4,\ (S,S)\text{-}(-)\text{-DET}}$ $(CH_2)_{10}$ epoxide, OH (35%) + $(CH_2)_{10}$ OH (26%) 10% ee

ing functional groups are inert to the titanium(IV) mediated asymmetric epoxidation reaction conditions.

Product Stability. The epoxy alcohol product is usually stable to the reaction conditions for titanium-mediated asymmetric epoxidation. However, the stoichiometric titanium-mediated asymmetric epoxidation[1] has some limitations. For example, the unsubstituted conjugated dienol **49** reacts slowly but the desired

Scheme: dienol **49** OH $\xrightarrow[\text{TBHP}]{\underset{\times}{\text{stoich. Ti(OPr-}i)_4,\ (S,S)\text{-}(-)\text{-DET}}}$ epoxide OH (0%)

epoxide is not obtained because of its instability.[2,19,102] However, the C2-substituted dienol **50**[4] gives the desired epoxide in good yields. A related secondary

Scheme: $n\text{-}C_5H_{11}$ ... OH **50** $\xrightarrow[\text{TBHP}]{\text{Ti(OPr-}i)_4,\ (R,R)\text{-}(+)\text{-DET}}$ $n\text{-}C_5H_{11}$ epoxide OH (80%)

dienol **51**[127] also gives the desired epoxides **52** with an *erythro:threo* ratio (>13:1), though product yields are not given.

erythro-**52**   *erythro : threo* = >13 : 1   *threo*-**52**

Epoxidation of *p*-methoxycinnamyl alcohol also does not give the desired epoxide because of product instability due to the presence of the *p*-methoxy group (Eq. 21),[2,128] though *meta*-substituted compound **53** is a good substrate (Eq. 22).[129] Product instability due to the presence of a neighboring nucleophilic

(Eq. 21)

53   (82%) 97% ee

(Eq. 22)

functional group has already been discussed (Eqs. 19 and 20). In addition, epoxy alcohols lacking substituents at C-3 are sensitive to nucleophilic opening,[17,18] and this sensitivity is enhanced by complexation to a metal ion. For example, the reaction of 2-*tert*-butylallyl alcohol **54** with Ti(OPr-*i*)$_4$-tartrate gives predominantly the isopropanol substitution product.[130–132] Use of Ti(OBu-*t*)₄ instead of Ti(OPr-*i*)₄, however, improves the yield of the epoxy alcohol. Epoxidation of 2-alkyl-2-cyclopropylidenethanols provides 2-alkyl-2-hydroxymethylcyclobuta-nones, probably via a labile epoxy alcohol (Eq. 23).[133,134]

(Eq. 23)

Catalytic Asymmetric Epoxidation. Since the principal difficulties (isolation of unstable and/or water-soluble epoxy alcohols) with the stoichiometric reaction are mainly attributed to the mild Lewis acidity of titanium alkoxide and the aqueous workup required for hydrolysis of the stoichiometric catalyst, it is not surprising that these problems are minimized when the reaction is conducted in a catalytic manner. In 1986, it was discovered that addition of molecular sieves to the reaction mixture allows epoxidation to proceed to completion in the presence of only 5–10% of the Ti(IV)–tartrate complex.[17,18] A catalyst with 5 mol% Ti(OPr-*i*)₄ and 6 mol% tartrate has been recommended as the most widely applicable system for asymmetric epoxidation. Below the 5 mol% level, the enantioselectivity of the reaction decreases remarkably. The amount of tartrate ester must be carefully controlled, because a large excess of tartrate (>100% excess) decreases the reaction rate [the titanium–tartrate (1:2) complex is inactive], while with too little tartrate (<10% excess) the enantioselectivity may suffer.

By using the catalytic procedure, the unstable and water-soluble glycidol (Eq. 24) and low molecular weight 2-substituted glycidols are obtained in moder-

(65%) 91% ee
stoich: (17%) 87% ee

(Eq. 24)

ate to good yields.[19,131,132] Other unstable epoxy alcohols like 3-arylglycidols and 3-vinylglycidols (Eq. 25), which decompose under stoichiometric conditions, are

(45%) 95% ee
stoich: (0%)

(Eq. 25)

also isolated in acceptable yields via the catalytic procedure.[19] These examples demonstrate several advantages of the catalytic procedure (economy, mildness of conditions, ease of product isolation, and increased yield) over the stoichiometric procedure, though enantioselectivity is often reduced by 1–5% relative to stoichiometric reactions.

Furthermore, epoxy alcohols produced by the catalytic procedure can be converted in situ into p-nitrobenzoates or p-toluenesulfonates, which are more easily isolated than the parent epoxy alcohols and can serve as versatile intermediates for further transformations (Eq. 26).[18,135–137] The combination of asymmetric

(65%) >98% ee; after recrystallization

(Eq. 26)

epoxidation and in situ titanium-mediated epoxide-opening reactions also provides access to diol derivatives without isolating unstable and/or water-soluble epoxy alcohols. For example, allyl alcohol can be converted to (2S)-propranolol (**55**) without isolating the glycidol.[135,136]

Another advantage of the catalytic process is high substrate concentration. In the stoichiometric reaction, the substrate concentration must be kept low (0.1–0.3 M) to avoid undesired side reactions like epoxide opening, while the catalytic process can be performed at concentrations up to 0.5–1.0 M. Even with the catalytic procedure, the epoxidation of a sensitive substrate like cinnamyl alcohol should be carried out at around 0.1 M concentration.

A heterogeneous catalyst prepared from montmorillonite pillared by Ti(OPr-i)$_4$ and dialkyl tartrate promotes asymmetric epoxidation of allylic alcohols with high enantiomeric selectivity in a catalytic manner (Eq. 27).[138] This procedure provides easy separation of epoxy alcohols by simple filtration without tedious workup.

$$\text{(Eq. 27)}$$

(68-91%) 90-98% ee

Ti-PIL = montmorillonite pillared with Ti(OPr-i)$_4$

Modification of The Titanium–Tartrate Complex. In the standard asymmetric epoxidation, the complex with 2:2 Ti:tartrate stoichiometry is the active catalyst. [Despite the 2:2 stoichiometry, it is strongly recommended that at least 1.2 equivalents of tartrate to Ti(OPr-i)$_4$ be used in both the catalytic and stoichiometric epoxidation of allylic alcohols. See "Experimental Conditions."][18] Modifications of the 2:2 catalyst by varying the chiral auxiliary or the titanium-to-auxiliary stoichiometry lead to new catalyst systems with different enantioselectivity.[2,130] When tartramide 56 is used as a chiral auxiliary, the catalyst with 2:1 [Ti(OPr-i)$_4$:56] stoichiometry exhibits reversed enantiofacial selection from the standard 2:2 asymmetric epoxidation catalyst, though the enantioselectivity decreases to some extent (Eq. 28).

$$\text{(Eq. 28)}$$

Reversed facial selection is also observed when a TiCl$_2$(OPr-i)$_2$ and diisopropyl tartrate (2:1) system is used. In this reaction, chloro diol 57 is obtained instead of epoxy alcohol 58, but 57 is readily cyclized to 58 on alkaline treatment.

Catalysts with other than 2:2 stoichiometry also show high efficiency for the oxidation of heteroatoms such as nitrogen[20,21,139] and sulfur,[22-25] as described in the section on asymmetric oxidations of sulfides, selenides, and amines.

Addition of catalytic amounts of CaH_2 and silica gel to the reaction mixture also enhances the epoxidation rate without affecting the enantioselectivity, though use of a stoichiometric amount of catalyst is required.[140-142] Tertiary allylic alcohols can be epoxidized under these conditions (Eq. 29).

(Eq. 29)

(70%) 100% de

Combinations of $Ti(OPr-i)_4$ and vicinal diols other than tartaric acid derivatives usually provide poor enantioselectivity.[3-6] However, use of $(1S,2S)$-1,2-di(o-methoxyphenyl)ethylene glycol (**59**) as a chiral auxiliary gives high enantioselectivity.[143]

(78%) 95% ee

$$59 = \quad \underset{HO \quad OH}{o\text{-MeOC}_6H_4 \quad C_6H_4OMe\text{-}o}$$

Tartrate esters linked to 1% cross-linked polystyrene resin can be used as chiral auxiliaries, but are less effective than dialkyl tartrate in inducing asymmetry.[144]

Asymmetric Epoxidation of Chiral Primary Allylic Alcohols

While the sense of facial selectivity for asymmetric epoxidation of a chiral allylic alcohol is determined by the chirality of the tartrate, the magnitude is affected by the substrate chirality. The level of influence depends on the location of the stereocenter and on the bulkiness and polarity of its substituents.[51] The reactions of (+)-**11** and (−)-**11** described in the previous section are examples of double asymmetric epoxidations where one side of the olefin is blocked by a methylene chain. The (+) isomer makes a matched pair with the Ti-(R,R)-diethyl tartrate system, while the (−) isomer constitutes a mismatched pair. In the following sections, examples of asymmetric epoxidation of chiral substrates are described. Except for a few isolated cases, the results follow the empirical face-selection rule presented in Scheme 1.

Chiral Z Allylic Alcohols.   The C-4 stereogenic center in this type of allylic alcohol strongly influences the stereochemistry of the reaction. For example, effective kinetic resolution of racemic (Z)-4-phenyl-2-penten-1-ol is observed when it is treated with Ti(OPr-i)$_4$/(R,R)-(+)-diethyl tartrate/TBHP (Eq. 30).[145,146] The mismatched double asymmetric epoxidation of compound **60**

(Eq. 30)

with the titanium-(S,S)-tartrate catalyst is very slow, and the desired epoxy alcohol **61** is not produced, while epoxidation of the same substrate with (R,R)-tartrate as a chiral source (matched pair) proceeds with high stereoselectivity (30:1).[73] A bulky and highly oxygenated Z olefinic substituent sometimes retards epoxidation using either enantiomer of the dialkyl tartrate auxiliary (Eq. 31).[147–149]

(R,R)-(+)-DET   (55%) 30:1
(S,S)-(–)-DET   Products were not isolated

(Eq. 31)

Both reactions are slow and no yield was reported.

Chiral 2-Substituted Allylic Alcohols.   Racemic 2-(1-phenylethyl)allyl alcohol (**62**) can be effectively resolved.[145] Epoxidation of chiral, nonracemic allylic alcohol **63** with (R,R)-(+)-diisopropyl tartrate provides epoxides **64** and **65**

in a 96:4 ratio, whereas a 25:75 mixture of **64** and **65** is obtained with (−)-diisopropyl tartrate.[150-152] However, kinetic resolution of racemic cyclic allylic alcohol **66** is not very efficient. The enantiomeric purity of recovered allylic alcohol is only 70% ee, even after 70% conversion of **66**.[153]

Chiral E Allylic Alcohols.   A stereogenic center in the 3E substituent is expected to have little influence on the facial selectivity of asymmetric epoxidation. Indeed, the kinetic resolution of racemic (E)-4-phenyl-2-penten-1-ol is very poor. Therefore, the high facial selectivity observed for achiral E allylic alcohols can also be expected for the epoxidation of chiral E allylic alcohols,[51] and many successful examples of the epoxidation of substrates of this type are known.[154-160] For example, the stereoselective construction of the ansa chain of rifamycin S,

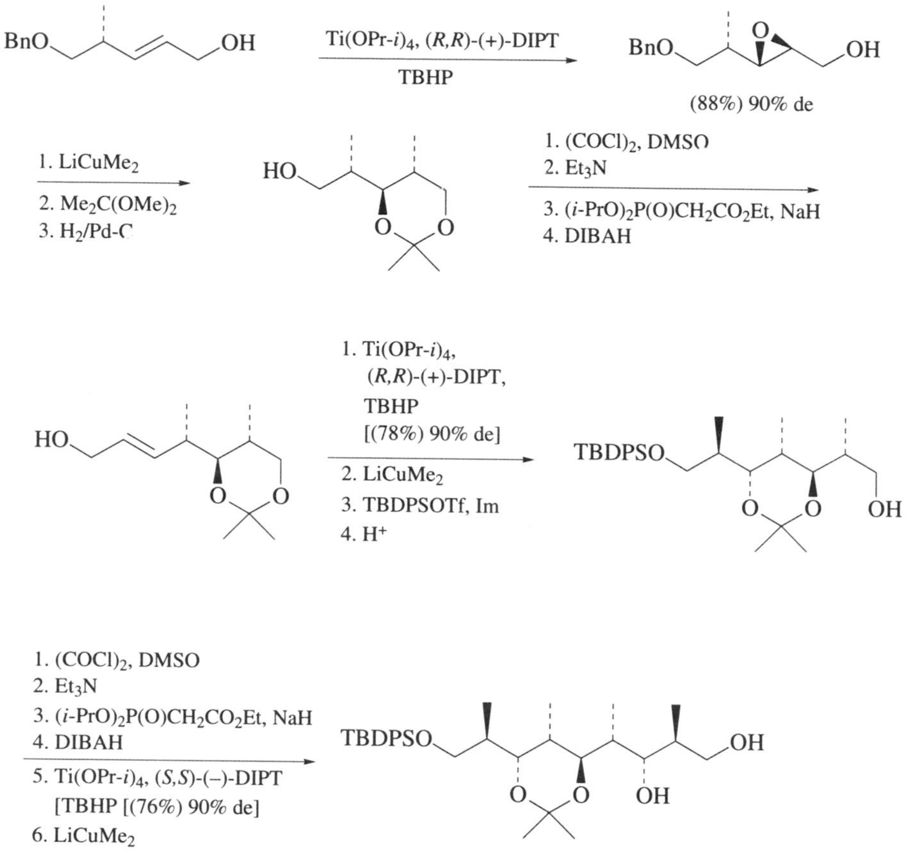

Scheme 3

where the methyl and hydroxy groups are arranged consecutively in an alternating pattern, has been achieved by using asymmetric epoxidation for the key steps (Scheme 3).[154] Other examples are found in sugar synthesis. A combination of titanium-mediated asymmetric epoxidation and regioselective opening of the resulting epoxy alcohols allows the stereoselective construction of consecutively polyhydroxylated compounds.[73,160-162] High asymmetric induction in the expected direction has been attained in all epoxidations whether they are matched or mismatched combinations (Eqs. 32, 33). This carbohydrate chain-extension method has also been applied successfully to syntheses of various segments of palytoxin, a large complex organic molecule.[163-166]

Even in the case of E allylic alcohols, a bulky and highly oxygenated E olefinic substituent affects enantiofacial selectivity of the reaction. Epoxidation of **67** with (R,R)-diisopropyl tartrate gives the single epoxy alcohol **68**, while epoxidation with (S,S)-diisopropyl tartrate gives a mixture of **68** and its

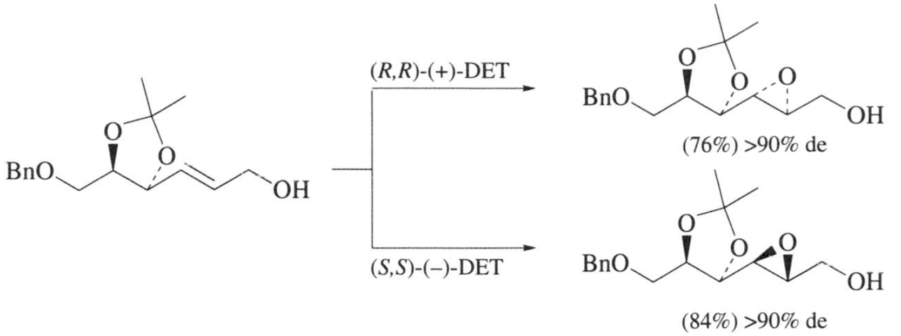

(Eq. 32)

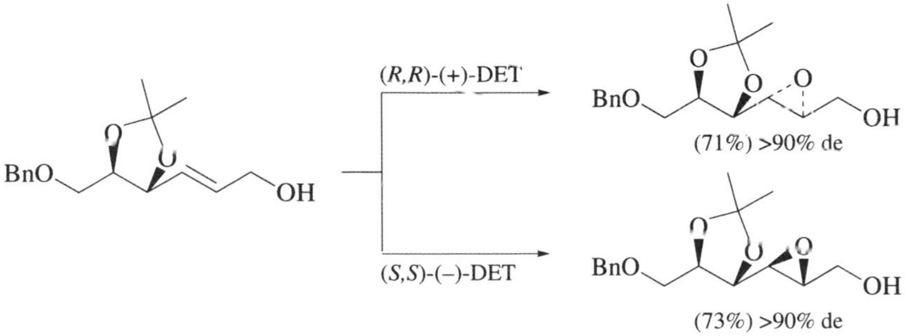

(Eq. 33)

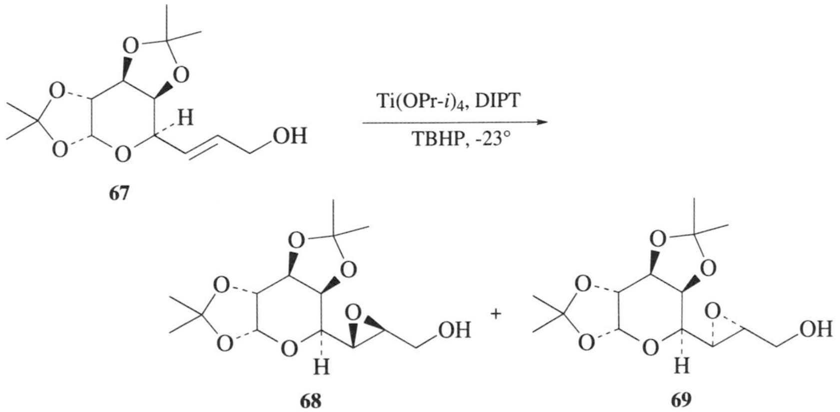

	68	69
(R,R)-(+)-DIPT	(66%)	(0%)
(S,S)-(–)-DIPT	A mixture containing a slight excess of **68**. (The reaction is very slow and yields are not given).	

diastereomer **69**.[147,148] Epoxidation of **70** with both (*R*,*R*)- and (*S*,*S*)-diisopropyl tartrates proceeds with only moderate diastereofacial selectivity.[149]

When a substrate has a bulky and polar *E* substituent, substantial kinetic resolution has also been observed (Eq. 34).[167]

70

Ti(OPr-i)$_4$, DIPT

TBHP, -23°, 5 d

(R,R)-(+)-DIPT   (63%)   5:1
(S,S)-(–)-DIPT   (—)   1:4

i-Pr

Ph$_2$PO

Ti(OPr-i)$_4$, (R,R)-(+)-DIPT

TBHP (47% conversion)

i-Pr

Ph$_2$PO

OH

84% ee   (Eq. 34)

i-Pr

Ph$_2$PO

OH

65% ee

With C-2 substituted E allylic alcohols, the stereogenic center at C-4 has considerable influence on the facial selectivity.[168–172] On treatment with Ti-(R,R)-(+)-diethyl tartrate, racemic allylic alcohol **71** affords two enantiomeric epoxides **72** and **73** having the $(2S,3S,4S)$ and $(2R,3R,4R)$ configurations; no

OTBDMS

HO   R

(±)-**71**

R= CH$_2$C(OMe)$_2$CH(OMe)CH$_2$OAc

Ti(OPr-i)$_4$, (R,R)-(+)-DET

TBHP, -20°, 12 h

OTBDMS

HO   R

72 29%

OTBDMS

HO   R

73 (19%)

OTBDMS

HO   R

$(4R)$-**71** (10%) >95% ee

($2S,3S,4R$) isomer is obtained. The enantiomeric purity of recovered ($4R$)-**71** is >95% ee. This very strong diastereofacial control by the C-4 stereogenic center is attributed to the forced in-plane conformation of the C-4 hydrogen by the C-2 substituent.[168,169] However, less highly substituted chiral E allylic alcohols can be epoxidized with high selectivity [e.g., 99% de using the Ti-(R,R)-(+)-diethyl tartrate catalyst (Eq. 35)].[173-175] In this example, the C-4 stereogenic center and the

(Eq. 35)

titanium-(R,R)-tartrate catalyst may constitute a matched pair in terms of double diastereoselection, although epoxidation of the same substrate with titanium (S,S)-tartrate has not been examined.

Miscellaneous Chiral Substrates.   Other types of racemic substrates, including compounds **74** and **75**, possessing axial chirality can also be resolved. The resulting stereoselectivity can be explained by the empirical rule.[145,146]

Asymmetric Epoxidation of Homo-, Bishomo-, Trishomoallylic Alcohols and Unfunctionalized Olefins

The Ti(OPr-i)$_4$, dialkyl tartrate, and *tert*-butyl hydroperoxide system has been applied to asymmetric epoxidation of homoallylic alcohols with limited success.[176] The reaction is fairly slow and shows rather low enantioselectivities ranging from 23 to 55% ee. The sense of asymmetric induction with homoallylic alcohols is opposite to that with allylic alcohols. That is, epoxidation of homoallylic alcohols with (R,R)-(+)-diethyl tartrate gives products enriched in the 3R enantiomer, while epoxidation of allylic alcohols with the same chiral source gives the 2S isomer with high enantioselectivity (Scheme 4).

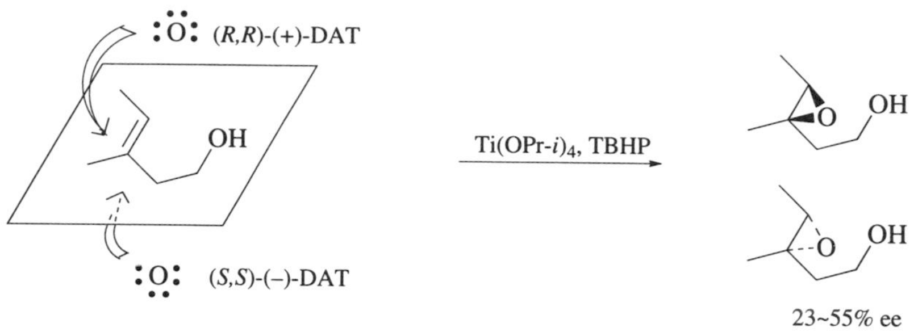

Scheme 4

Treatment of chiral homoallylic alcohol **76** with a Ti(OPr-i)$_4$, (R,R)-(+)-dimethyl tartrate, *tert*-butyl hydroperoxide system provides a 3:1 mixture of diastereomeric tetrahydrofuran derivatives via attack of the intermediate epoxide by the neighboring hydroxy group.[76]

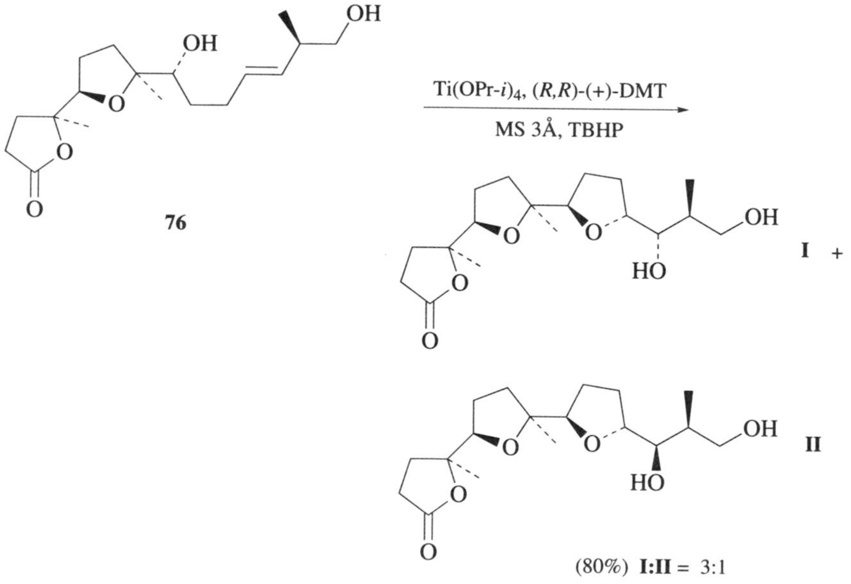

Epoxidation of alkenylethylene glycols which have hydroxy groups at both allylic and homoallylic carbons shows high but reversed diastereofacial selection to that expected for secondary allylic alcohols (vide infra) (Eq. 36).[177] A Zr(OPr-n)$_4$, dicyclohexyltartramide, and *tert*-butyl hydroperoxide system shows slightly better asymmetric induction for the epoxidation of Z homoallylic alcohols,[178] though the chemical yield is only modest (Eq. 37).

(Eq. 36)

R = Me, Et, n-C$_5$H$_{11}$ (23-28%) 50-74% ee

Epoxidation of trishomoallylic alcohol **77** with (S,S)-$(-)$-diethyl tartrate as a chiral source provides epoxide **78** preferentially and does not produce an appreciable amount of its diastereomer when trityl hydroperoxide is used instead of *tert*-butyl hydroperoxide.[179] No diastereoselectivity is observed when TBHP is used as an oxidant. On the other hand, epoxidation of **77** with (R,R)-$(+)$-diethyl tartrate provides a 3:4 mixture of **78** and its diastereomer.

Olefins that have no neighboring hydroxy group are not epoxidized by the present reaction. However, oxidation of olefins with singlet oxygen in the presence of Ti(OBu-t)$_4$ and (R,R)-$(+)$-diethyl tartrate provides optically active epoxy alcohols. The allylic hydroperoxides produced by the ene reaction are converted to allylic alcohols and further epoxidized enantioselectively by the titanium–tartrate catalyst (Eq. 38).[180-182]

$$\text{(Eq. 38)}$$

(79%) 72% ee

Kinetic Resolution of Secondary Allylic Alcohols

The kinetic resolution of secondary allylic alcohols was first reported in 1981 (Scheme 5),[10] wherein some examples were performed with as little as

Scheme 5

13–25% catalyst. Although this catalytic procedure has been used by other researchers,[183-186] only recently has there been reported a way to accomplish kinetic resolution in a truly catalytic manner with selectivity only slightly lower (0–4%) than that achieved in the stoichiometric reaction.[18] The key feature of this catalytic procedure is the use of molecular sieves (zeolites).

With cyclohexyl (E)-1-propenyl carbinol as the model ($R^1 = CH_3$, $R^2 = R^3 = R^4 = H$, $R^5 = C_6H_{11}$, and $R^1 = CH_3$, $R^2 = R^3 = R^5 = H$, $R^4 = C_6H_{11}$ in Scheme 5), it was found that the S enantiomer reacts 74 times faster than the R enantiomer at 0° when (R,R)-(+)-diisopropyl tartrate is used as the chiral auxiliary.[43] As in the epoxidation of primary allylic alcohols,[1] the stereochemical course of the kinetic resolution processes has been highly predictable. When the secondary allylic alcohol is drawn so that the hydroxy group lies in the lower right corner of the plane (Scheme 5), the enantiomer that reacts rapidly with

(R,R)-$(+)$-dialkyl tartrates is the one in which the substituent (R^4) on C-1 is located above the plane. Epoxidation occurs from the underside to give the usual $2S$ epoxide (*erythro* selectivity, 98:2). The slow-reacting enantiomer is the one in which the C-1 substituent (R^5) is located on the underside, interfering with the "normal" delivery of the oxygen atom. This interference reduces the expected *threo* selectivity for the slow-reacting enantiomer (38 *erythro*:62 *threo*, Scheme 6). This enantioselection rule has consistently been observed for all secondary

Scheme 6

allylic alcohols except for those with bulky Z substituents and 1,2-divinylethylene glycols. Kinetic resolution is very poor for allylic alcohols with bulky Z substituents,[10] and reversed but high enantioselectivity is observed in the kinetic resolutions of 1,2-divinylethylene glycols (Eq. 39).[187,188]

(Eq. 39)

The most important parameters in kinetic resolution are the relative rates of reaction of the two allylic alcohol enantiomers. The graph in Fig. 2, which represents solutions of Eq. 40,[189] enables the relative rate difference to be related to the percent ee of unreacted allylic alcohol. Three variables influence solutions to Eq. 40: the percent ee of the remaining substrate, the percent conversion of the racemic material, and the relative rate (k_{rel}) of reaction of the two enantiomers.

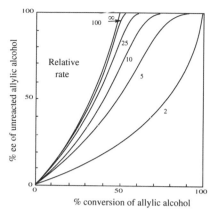

$$K_{rel} = \frac{k_f}{k_s} = \frac{\ln(F/F^\circ)}{\ln(S/S^\circ)} = \frac{\ln(1-c)(1-ee)}{\ln(1-c)(1+ee)} \qquad \text{(Eq. 40)}$$

k_f = rate constant of fast reacting isomer
k_s = rate constant of slow reacting isomer
c = Fraction of consumption of racemate
ee = % ee/100
F = Concentration of the fast reacting isomer
S = Concentration of the slow reacting isomer

Fig. 2

Knowledge of any two allows specification of the third. The maximum effectiveness of a kinetic resolution procedure is, of course, when $k_{rel} = \infty$, but a value of 50–100 is almost as effective. Actual values are in the range 15 to 700.[10,12,190–198]

The kinetic resolution of racemic secondary allylic alcohols has almost the same substituent effects as the normal asymmetric epoxidation of primary allylic alcohols excepting, of course, the substituents labeled R^4 (or R^5) in Scheme 5.[10] Bulky R^1 groups (Scheme 5) increase the rate of epoxidation of the fast-reacting enantiomer and decrease the rate of the slow-reacting enantiomer, thus increasing k_{rel}. It should be recalled that E-substituted primary allylic alcohols, even with very bulky groups, are the best substrates for asymmetric epoxidation in terms of rate, yield, and enantioselectivity. The most efficiently resolved substrates to date are those in which the E substituent R^1 is trimethylsilyl, iodo, or tri-n-butylstannyl; at 50% conversion of the racemic substrate, both the recovered allylic alcohol and the *erythro*-epoxy alcohol have more than 99% ee.[12,190–195] A careful measurement of k_{rel} for (E)-1-trimethylsilyl-1-octen-3-ol shows a value of $k_{rel} = 700$.[16] Optical purities, k_{rel}, and conversion are in complete agreement with Eq. 40 and the graph in Fig. 2.

Bulky Z substituents (R^3) are especially deleterious for achieving good kinetic resolution. A few examples where stereochemistry of the epoxidation does not follow the empirical rule belong to this class.[10] For example, the kinetic resolution of (Z)-2-cyclohexylethenyl cyclohexyl carbinol is inefficient, and the configuration of the recovered allylic alcohol is opposite to that expected from the empirical rule (Eq. 41). In contrast, resolution of the corresponding E isomer proceeds with high enantioselectivity to leave the expected R allylic alcohol (Eq. 42).

$$\underset{OH}{\overset{\begin{array}{c}C_6H_{11}\\ \\C_6H_{11}\end{array}}{\diagup}} \quad \xrightarrow[\text{TBHP (0.6 eq)}]{\text{Ti(OPr-}i)_4,\ (R,R)\text{-}(+)\text{-DIPT}} \quad \underset{OH}{\overset{\begin{array}{c}C_6H_{11}\\ \\C_6H_{11}\end{array}}{\diagup}} \qquad \text{(Eq. 41)}$$

$$(38\%) \approx 10\% \text{ ee}$$

$$C_6H_{11}\diagdown\diagup C_6H_{11} \quad \xrightarrow[\text{TBHP (0.6 eq)}]{\text{Ti(OPr-}i)_4, (R,R)\text{-(+)-DIPT}} \quad C_6H_{11}\diagdown\diagup C_6H_{11}$$

$$\overset{|}{OH} \qquad\qquad\qquad\qquad\qquad\qquad \overset{|}{OH}$$

(32%) >96% ee

(Eq. 42)

Kinetic resolution proceeds smoothly when R^2 is an unbranched alkyl group. The effect of very bulky groups has been tested by introducing sterically demanding *tert*-butyl substituents in all positions labeled R^1–R^5 (Scheme 5).[57] The results observed are consistent with examples in which $R^1 = R^3 = $ *tert*-butyl. When $R^? - $ *tert* butyl, at 60% conversion the recovered allylic alcohol had only 30% ee ($k_{rel} \approx 2$) (Eq. 43), although the epoxy alcohol consisted largely (ca.

(Eq. 43)

≈30% ee

40:1) of one of the two possible diastereoisomers (the absolute configurations of both the remaining allylic alcohol and the epoxy alcohol have not been determined). This is one substitution pattern where there is not complete correlation between normal asymmetric epoxidation of the primary allylic alcohol and effective kinetic resolution of the corresponding secondary allylic alcohol [the epoxidation of the primary allylic alcohol with the same alkene substituents is achieved with 85% ee. (Eq. 12)].

Finally, when R^4 (or R^5) = *tert*-butyl, the substrate is not useful for kinetic resolution, a surprising result considering that branched (aryl and secondary) substituents are among the best substrates [e.g., cyclohexyl propenyl carbinol (Scheme 6)]. Kinetic resolution is not effective in these cases, probably because of steric difficulties associated with the formation and/or reaction of the allylic alkoxide complex.[57]

Another important feature of kinetic resolution is that k_{rel} increases as the temperature is lowered, as is seen by comparing the k_{rel} for cyclohexyl propenyl carbinol with (R,R)-diisopropyl tartrate at 0° ($k_{rel} = 74$) and −20° ($k_{rel} = 104$). The immediate conclusion is that kinetic resolution should be performed at −20° or lower, depending on the reactivity of the substrate. For convenience, most kinetic resolutions are run by storing the reaction mixture in a freezer. Both the reaction rate and the enantiomeric excess improve with stirring.[17,18] Hence, when possible, the resolution should be carried out with constant stirring and appropriate temperature control.

The efficiency of a kinetic resolution decreases when titanium tetra-*tert*-butoxide is used to generate titanium–tartrate complex.[38]

When the epoxy alcohol is the desired product of a kinetic resolution, care should be taken to avoid extending the reaction time unnecessarily. For example, when a 4:6 mixture of (±)-*erythro*- and (±)-*threo*-1,2-epoxynonan-3-ol is exposed to the standard conditions of kinetic resolution, preferential decomposition of the *threo* diastereomer is observed. After several days at 0° (conversion ≈ 70%), the remaining epoxide has an *erythro*:*threo* ratio of 8:2. The remaining *threo*-epoxy alcohol has 90% ee of the 3S enantiomer; the remaining *erythro*-epoxy alcohol has 12% ee of the 3R enantiomer (Eq. 44). Because of its

$$n\text{-}C_6H_{13} \quad \xrightarrow[\text{TBHP, }0°]{\text{Ti(OPr-}i)_4,\ (R,R)\text{-}(+)\text{-DET}} \quad \text{(conversion = 70%)}$$

(±)-*erythro*:(±)-*threo* = 4:6

(Eq. 44)

erythro 12% ee   threo 90% ee

erythro:threo = 8:2

Lewis acidity, the titanium–tartrate system acts here as a catalyst for the ring opening reaction, promoting attack of some alcohols in the reaction mixture (isopropanol, allylic alcohol, epoxy alcohol, or TBHP). The resulting diol ethers are potent inhibitors of kinetic resolution because of their capacity to chelate strongly with the titanium atom, thereby reducing access of TBHP and the allylic alcohol to the metal. Although such ring-opening processes are attenuated at −20°, they retain the potential (above all with slowly reacting substrates) to degrade both the diastereomeric ratio and the enantiomeric purity. To minimize such problems, monitoring of the reaction mixture (GLC with an internal standard) is recommended.

Like all resolutions, this kinetic resolution is limited to a theoretical yield of 50% of one enantiomer from the racemic mixture. If an epoxy alcohol is converted to an allylic alcohol of the same absolute configuration as the unreacted enantiomer, a racemic allylic alcohol is converted to a single enantiomer with theoretically 100% yield. This conversion is effected by a combination of methanesulfonation and telluride reduction, though substrates are limited to terminal epoxides (Eq. 45).[199] The actual combined yields of allylic alcohol from both the kinetic resolution and telluride steps range from 75 to 88%.

$$\xrightarrow[\text{TBHP}]{\text{Ti(OPr-}i)_4,\ (+)\text{-DIPT}}$$

I

II (79%) 92% ee

(MeSO₂)₂O
DMAP

OMs

Te
HOCH₂SO₂Na

II

(Eq. 45)

Kinetic resolution can be extended to other secondary carbinol systems in which a suitably oxidizable group is located in the allylic position. Thus 2-furyl and 2-thienyl carbinols can be resolved efficiently in both the stoichiometric and catalytic manner.[14,15] The expected epoxy alcohol **79** is the precursor of the isolated oxidized materials **80**.

α-Furfuryl toluenesulfonamide is also resolved efficiently by using modified reaction conditions (Eq. 46),[200] although the stereochemistry is opposite to the empirical rule.

(Eq. 46)

Asymmetric Oxidation of Sulfides, Selenides, and Amines

Titanium–tartrate complexes are also applicable to asymmetric oxidation of heteroatoms. Two procedures have been reported for oxidation of sulfides.[22–25] When methyl p-tolyl sulfide is used as a substrate, the standard procedure using Ti(OPr-i)$_4$, dialkyl tartrate, and TBHP (1:1:2) leads to a mixture of the racemic sulfoxide and the sulfone (Eq. 47).[22] Although increasing the tartrate/titanium te-

Me$\diagdown$$_S$$\diagup$C$_6H_4$Me-$p$ $\longrightarrow$ Me$\diagdown$$_S$$\diagup$C$_6H_4$Me-$p$ **I** + Me$\diagdown$$_S$$\diagup$C$_6H_4$Me-$p$ **II**

Ti(OPr-i)$_4$: (R,R)-(+)-DIPT : TBHP = 1:1:2, **I** (41%) 0% ee + **II** (17%)

Ti(OPr-i)$_4$: (R,R)-(+)-DIPT : TBHP = 1:4:2, **I** (60%) 88% ee

Ti(OPr-i)$_4$: (R,R)-(+)-DIPT : H$_2$O : TBHP = 1:2:1:1, **I** (90%) 90% ee

Ti(OPr-i)$_4$: (R,R)-(+)-DET : H$_2$O : CHP = 1:2:1:1, **I** (90%) 93% ee   (Eq. 47)

traisopropoxide ratio improves the enantioselectivity,[24] a combination of titanium tetraisopropoxide/tartrate/water/hydroperoxide (1:2:1:1) further enhances both the chemical yield and enantioselectivity.[22,23] Use of cumene hydroperoxide (CHP) instead of TBHP improves enantioselectivity.[25] Reaction enantioselectivity is also temperature dependent, and an optimal ee is obtained around −21°. Functional groups such as allylic alcohols, isolated double bonds, amines, phenols, and alcohols are compatible with this reaction.[201–207] Although the role of water is not completely understood, it has been suggested that the water hydrolyzes a Ti-OPr-i bond with formation of a new μ-oxo bridge (Ti-O-Ti) between two dimeric species.[23]

Simple chiral selenoxides readily racemize through a tetracoordinated achiral hydrate. However, some selenides bearing bulky substituents are enantiomerically stable. Oxidation of selenide **81** with a modified titanium–tartrate system (2:4) system proceeds with moderate enantioselectivity.[26]

81   (72%) 40% ee

Racemic β-amino alcohols are efficiently resolved by using the complex with 2:1 [Ti(OPr-i)$_4$:tartrate] stoichiometry as a catalyst (Eq. 48), and the standard 2:2 catalyst is less effective.[20,21,139]

(37%) 95% ee   (59%) 63% ee

Synthetic Applications: Transformations of Epoxy Alcohols into Other Functional Units

The great utility of the titanium-mediated asymmetric epoxidation in organic synthesis is attributable to its enantioselectivity and to the numerous applications of epoxy alcohols as precursors to diversely functionalized compounds. However, since epoxy alcohols have three reactive sites (Scheme 7), regio- and stereoselec-

Scheme 7

tive reactions are essential for their use, and many studies have been directed toward developing regioselective transformations of epoxy alcohols. For convenience, these reactions are classified into three categories: (1) direct substitution reactions and other transformations of the hydroxy group at C-1; (2) rearrangement of 2,3-epoxy alcohols into 1,2-epoxy alcohols (Payne rearrangement) which undergo regioselective substitution at C-1; and (3) epoxide ring opening at C-2 or C-3.

Transformations of the Hydroxy Group at C-1.   Epoxy alcohols can be converted directly into the corresponding epoxy ethers by using Mitsunobu procedures.[208,209] Activation of the hydroxy group as the corresponding mesylate or tosylate also provides a useful means of replacing it with various nucleophiles like organolithium and organocopper reagents[109,110] and hydride sources.[1] For example, insect pheromones bearing an isolated epoxide, such as disparlure (**83**) and the saltmarsh caterpillar moth pheromone **85**, have been prepared enantiospecifically by the alkylation of epoxy tosylates **82** and **84**.[109,110]

Alkynyl epoxides can be obtained by treatment of epoxy triflates with alkynyllithiums (Eq. 49).[210]

Treatment of epoxy tosylates with a telluride reagent gives allylic alcohols (Eq. 50).[211]

Reduction of epoxy tosylates with dibutylaluminum hydride provides 2-alkanols directly with high regioselectivity (>99:1) (Eq. 51).[212]

1. Ti(OPr-i)$_4$, (R,R)-(+)-DET,
   TBHP
2. DNBCl, C$_5$H$_5$N

3. K$_2$CO$_3$, MeOH
4. TsCl

82   (55%) 100% ee

(the intermediate dinitrobenzoate
was recrystallized.)

(n-C$_9$H$_{19}$)$_2$CuLi/ether

83

1. Ti(OPr-i)$_4$, (R,R)-(+)-DET,
   TBHP
2. DNBCl, C$_5$H$_5$N

3. K$_2$CO$_3$, MeOH
4. TsCl

84 (36%) 100% ee

(the intermediate dinitrobenzoate
was recrystallized.)

(n-C$_{10}$H$_{21}$)$_2$CuLi/ether

85

1. Tf$_2$O

2. MeCH(OLi)C≡CLi   (Eq. 49)

Te, HOCH$_2$SO$_2$Na

NaOH   (66%)   (Eq. 50)

n-C$_{10}$H$_{21}$ DIBAH, 0°

CH$_2$Cl$_2$ n-C$_{10}$H$_{21}$ (98%) (Eq. 51)

Epoxy mesylates and tosylates can be converted further into epoxy halides,[1,213,214] which also undergo several useful reactions.[215-217] For example, treatment of epoxy bromide **86** with (diisopropoxymethylsilyl)methyl-magnesium

86

87 (38%)

chloride followed by hydrogen peroxide oxidation gives the 3,4-epoxy alcohol **87**.[218] This sequence is a useful alternative to asymmetric epoxidation of the corresponding homoallylic alcohol because only moderate enantioselectivity is usually observed in asymmetric epoxidations of homoallylic alcohols.[176] Epoxy iodides can also be extended by two carbons by treatment with the lithium enolate of *tert*-butyl acetate to give the γ,δ-epoxy esters (Eq. 52).[76]

(Eq. 52)

The reactions of epoxy iodides with vinylmagnesium bromide–cuprous iodide provide allylic alcohols, but ordinary substitution products can be obtained when vinylmagnesium bromide is added to a solution of an epoxy iodide in the presence of hexamethylphosphoric triamide (Eq. 53).[219,220] Treatment of epoxy iodides with *tert*-butyllithium also gives allylic alcohols.[221]

(Eq. 53)

Transformations of epoxy halides to allylic alcohols can also be effected by treatment with zinc–acetic acid or $Bu_3SnAlEt_2$ (Eq. 54).[213,214,222]

(Eq. 54)

(−)-nerolidol

Direct transformation of epoxy alcohols to allylic alcohols is effected by treatment with bis(cyclopentadienyl)titanium chloride (Eq. 55).[223]

(88%)

(Eq. 55)

Treatment of epoxy alcohols with triphenylphosphine–carbon tetrachloride gives epoxy chlorides, which are converted into propargylic alcohols by further treatment with 3 equivalents of an alkyllithium or lithium diisopropylamide (Eq. 56).[224,225] On the other hand, treatment with 1 equivalent of lithium amide or

(Eq. 56)

lithium diisopropylamide gives vinyl chlorides. However, use of 1 equivalent of an alkyllithium instead of lithium amide gives a mixture of propargylic alcohol and vinyl chloride.[226–228]

Epoxy alcohols can be oxidized without epimerization to epoxy aldehydes by various procedures such as Swern, Collins, and Mukaiyama oxidations.[229–232] Epoxy aldehydes can be further converted to vinyl epoxides by Wittig olefination.[233] For example, oxidation of epoxy alcohols **88** with Collins reagent followed by Wittig olefination provides vinyl epoxide **89**, hydrogenation of which gives (−)-disparlure (**83**).[59]

88   1. CrO$_3$, C$_5$H$_5$N   **89**   H$_2$ / Rh/Al$_2$O$_3$ **83**
    2. Ph$_3$P=CHCH$_2$Bu-i

Addition of LiCuMe$_2$ to epoxy allyl alcohol **90** proceeds in an *anti*-S$_{N2'}$ manner to give an *E* allylic alcohol predominantly.[234–238]

90

R = Me$_2$C=CH(CH$_2$)$_2$

(68%)   (13%)

Wittig product **91** is reduced regioselectively by diisobutylaluminum hydride,[85–87] samarium iodide,[239] or [tris(dibenzylideneacetone)-chloroform]-dipalladium-PBu$_3$-HCO$_2$H[240] to give the δ-hydroxy allylic alcohol, δ-hydroxy-β,γ-unsaturated ester, and δ-hydroxy-α,β-unsaturated ester, respectively.

DIBAH

SmI$_2$

Pd(dba)$_3$CHCl$_3$-PBu$_3$-HCO$_2$H

91

Repetition of a sequence of asymmetric epoxidation, oxidation, Wittig olefination, and DIBAH reduction provides easy access to 1,3,5-polyols (Scheme 8).[85]

On the other hand, unsaturated epoxide **92** is converted into carbonate **93** or carbamate **94** with retention of epoxide configuration by respective treatment with carbon dioxide or phenyl isocyanate in the presence of a Pd(0) catalyst.[241–243]

Both *endo*- and *exo*-brevicomins can be synthesized from the same epoxy tosylate **95** by the choice of appropriate epoxide-opening methodology.[244]

The [3 + 3] annulation reaction of epoxy aldehydes with 3-iodo-2-[(trimethylsilyl)methyl]propene in the presence of tin(II) fluoride provides cyclohexanediols in a single step with high stereoselectivity (Eq. 57).[245]

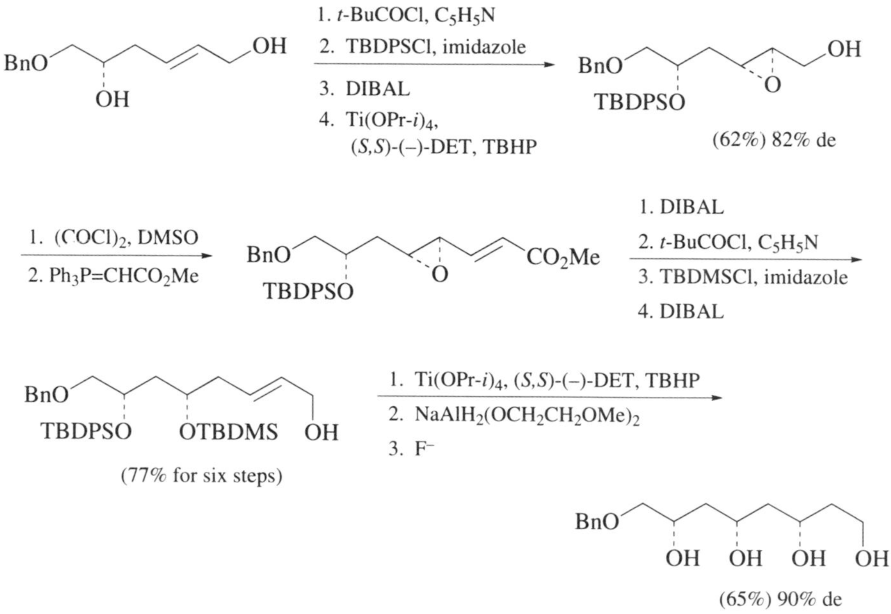

Scheme 8

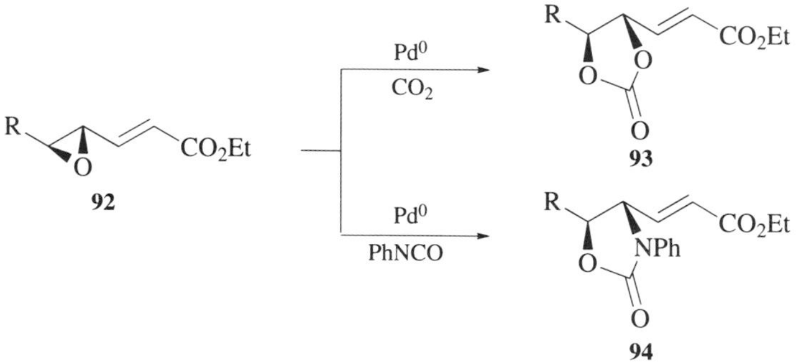

92 93 94

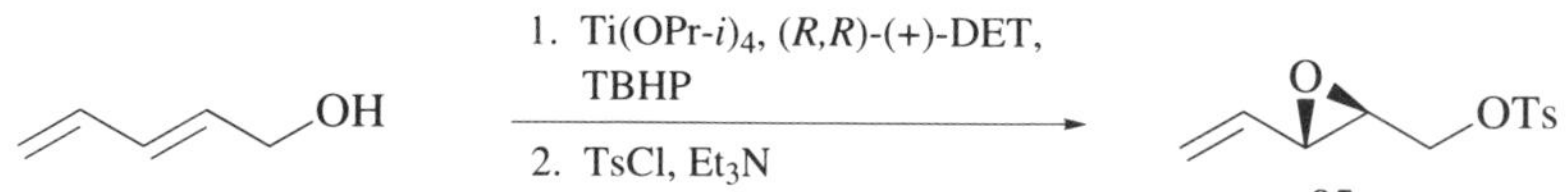

95

Reaction of Grignard reagents with β-trimethylsilyl-α,β-epoxyketones, which are readily available by kinetic resolution of *dl*-β-trimethylsilylvinyl carbinol and subsequent Swern oxidation, gives γ-trimethylsilyl-β,γ-epoxy alcohols with high diastereoselectivity (Eq. 58).[246] The asymmetric allylboration of (*R*,*R*)-epoxy

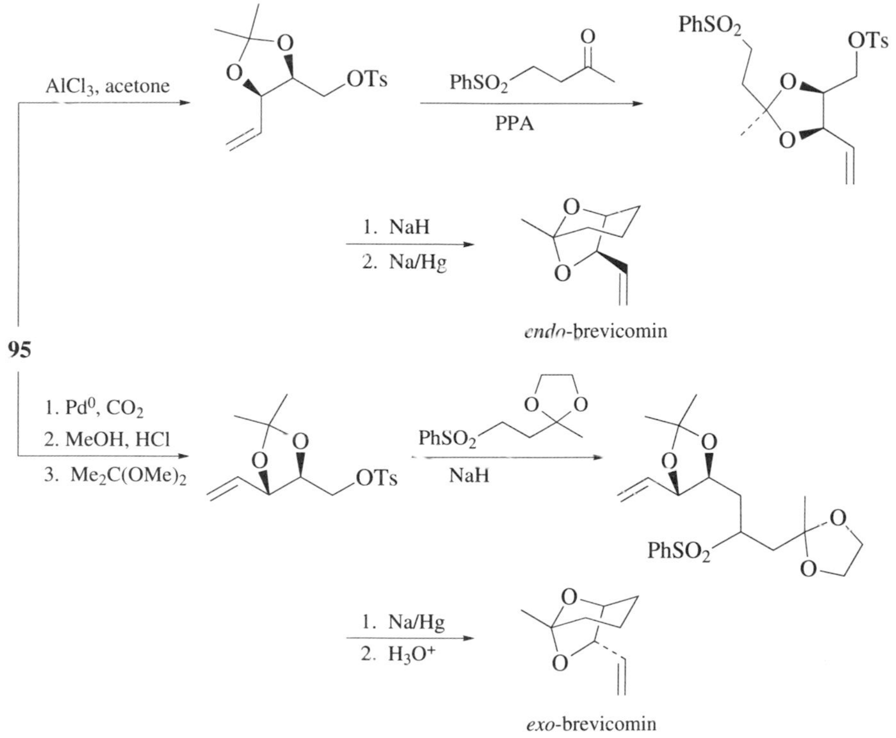

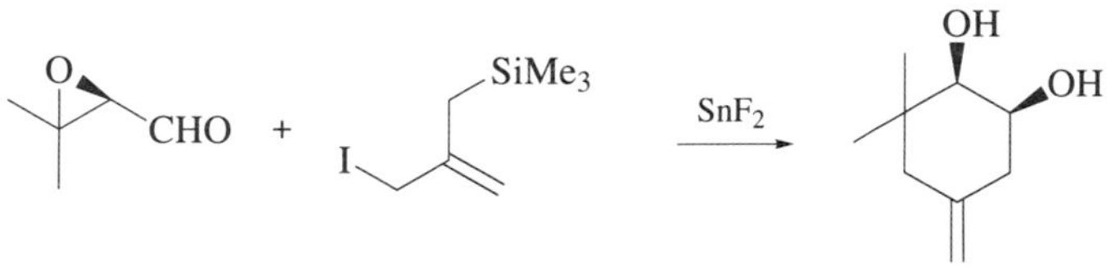

endo-brevicomin

95

exo-brevicomin

(Eq. 57)

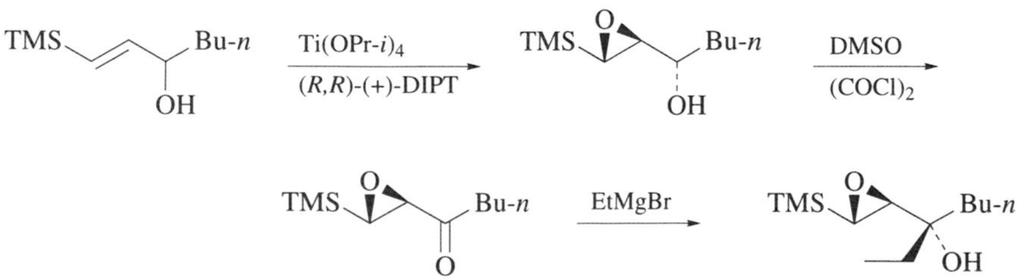

(Eq. 58)

aldehydes with optically active (*S,S*)-(−)-diethyl tartrate as a chiral source provides *erythro* epoxy alcohols exclusively (Eq. 59).[247]

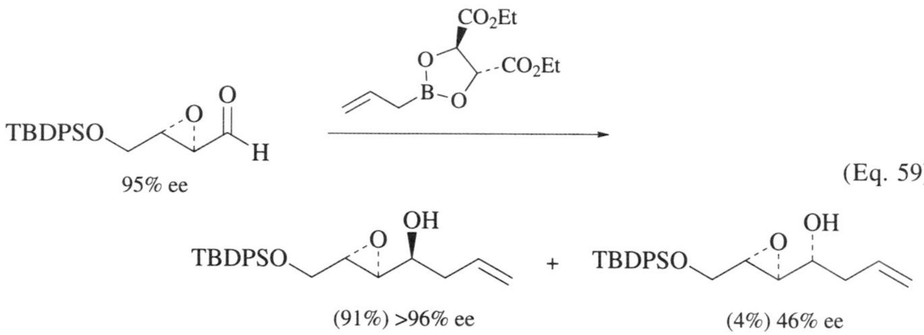

(Eq. 59)

(91%) >96% ee   (4%) 46% ee

Reduction of epoxy ketones by sodium borohydride–diphenyl diselenide[248,249] or samarium iodide[250,251] provides β-hydroxy ketones regioselectively (Eq. 60).

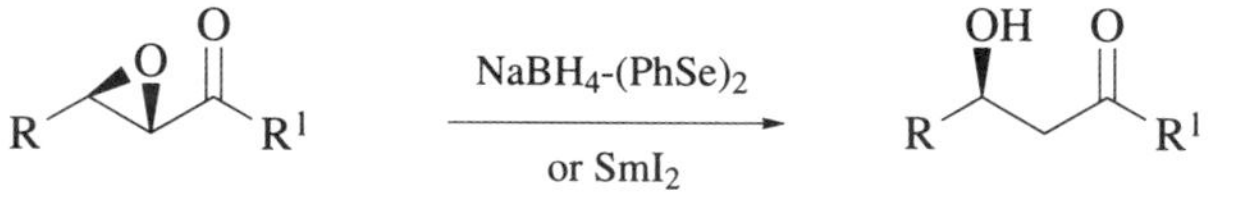

(Eq. 60)

Epoxy aldehydes can be used as chiral sources in imine–ketene cycloadditions, which provide *cis*-substituted 3-amino-4-alkylazetidinones of high enantiomeric purity (Eq. 61).[252]

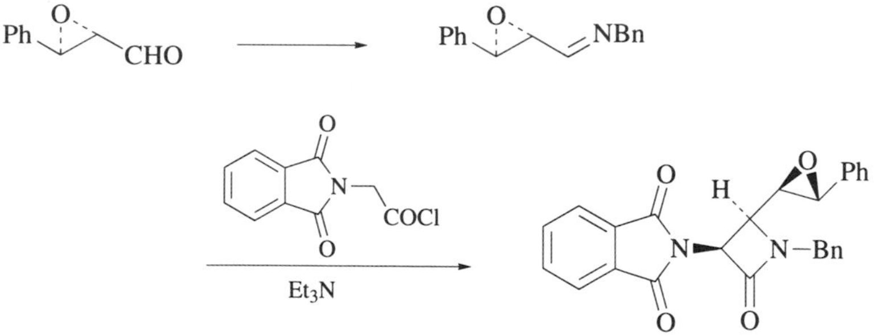

(Eq. 61)

94% de

Epoxy alcohols can be oxidized directly without epimerization to epoxy acids by ruthenium tetroxide or potassium permanganate (Eq. 62).[253–255]

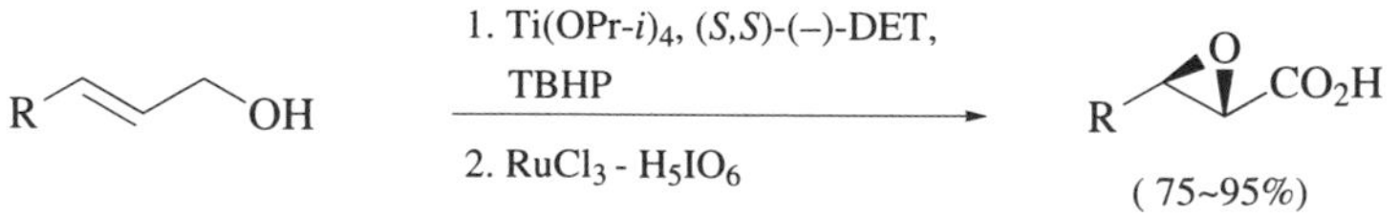

(Eq. 62)

Epoxy acids can be further converted into γ-hydroxy-α,β-unsaturated esters via epoxy diazomethyl ketones (Eq. 63).[256,257]

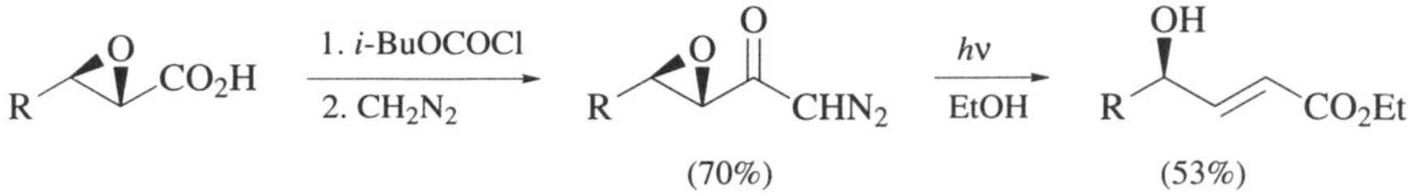

(Eq. 63)

Payne Rearrangement–Epoxide-Opening Sequence. 2,3-Epoxy alcohols are rapidly equilibrated with 1,2-epoxy alcohols under alkaline conditions (Payne rearrangement).[258] The equilibrium ratio of 1,2- to 2,3-epoxy alcohols is remarkably dependent on the substrate. However, as a 1,2-epoxide is considerably more reactive than a 2,3-epoxide, treatment of the equilibrium mixture with a nucleophile provides preferentially the product from the 1,2-epoxide (Eq. 64).[258,259]

(Eq. 64)

Thus the Payne rearrangement–epoxide-opening sequence is a useful alternative for activating C-1 for substitution,[7,73,260,261] although this provides 2,3-diols while direct C-1 substitution provides 2,3-epoxides. For example, the Payne rearrangement–epoxide-opening sequence using phenylthiolate as the nucleophile has permitted the straightforward synthesis of sugars via iterative asymmetric epoxidation cycles (Scheme 9).[73,74] Other nucleophiles including OH⁻,[73,145,148]

Scheme 9

BH_4^-,[145,259] $TsHN^-$,[145] CN^-,[260,262] N_3^-,[261,263] and R_2NH[259] have also been used successfully in Eq. 64.

Use of the Payne rearrangement provides an easy approach to unstable vinyl epoxy alcohol **96**, which can be obtained in only moderate yield via the direct asymmetric epoxidation even under catalytic conditions. Epoxidation of 1,4-pentadien-3-ol (**97**) followed by Payne rearrangement gives **96** in good yield.[264-273] Another advantage of this procedure is that epoxidation of **97** gives **98** (and hence **96**) with extremely high enantiomeric purity.[274-280]

The major limitation of this Payne rearrangement–epoxide-opening strategy is that most organometallic reagents are not compatible with the aqueous conditions that are essential for the Payne rearrangement. This limitation can be overcome by isolating the 1,2-epoxy alcohols and then treating them with organometallic nucleophiles under nonaqueous conditions.[261,281] However, the equilibrium in the Payne reaction usually favors the 2,3-epoxy alcohols. Thus procedures for the synthesis of 1,2-epoxy alcohols have been developed.[261,282,283] For example, enantiomeric *erythro* 1,2-epoxy alcohols **99** and **100** can be derived

stereoselectively from common 2,3-epoxy alcohols via dihydroxy mesylate and dihydroxy sulfide intermediates, respectively.[259,261] These *erythro* epoxides can be converted further to the *threo* 1,2-epoxides by the Mitsunobu reaction (Eq. 65).[208] The 1,2-epoxides thus obtained undergo nucleophilic epoxide opening at the terminal carbon.[259,282]

Recently it has been found that lithium chloride catalyzes the Payne rearrangement in tetrahydrofuran (Eq. 66).[284,285] This enables the use of organometallic reagents like RCu or LiCuCNR as nucleophiles. However, the more reactive LiCuR$_2$ species react predominantly with the unrearranged 2,3-epoxide.

(Eq. 66)

Treatment of epoxy alcohol with alkyllithium in the presence of boron trifluoride etherate gives a mixture of 1,2- and 1,3-diols (Eq. 67).[286,287]

(Eq. 67)

Epoxide Ring Opening at C-2 or C-3. Although the regio- and stereochemistry in epoxide-opening reactions of 2,3-epoxy alcohols depend on the steric and electronic factors in the substrates and on reaction conditions, the following general features are noted. Nucleophilic substitution under neutral and basic conditions occurs preferentially from the less-substituted side in an S_N2 manner, where the configuration of the attacked carbon is inverted.[7,288,289] Nucleophilic attack under acidic conditions occurs at the more-substituted side, also in an S_N2 manner.[7,290] With sterically unbiased epoxy alcohols or their O-protected derivatives, epoxide opening with nucleophiles occurs preferentially at C-3 (Eq. 68).[7,261,291] This regioselectivity is attributed to the presence of the electro-

(Eq. 68)

negative hydroxy group at C-1,[7,261] which retards S_N2 substitution at the vicinal carbon.

The presence of an acetal group at C-1 also promotes epoxide opening at C-3 to give a 2-hydroxy acetal selectively (Eq. 69).[7,261] The substituent at C-4, how-

$$R \overset{}{\diagdown} \underset{O}{\overset{O}{\diagup}} \xrightarrow{\text{Nu}} R \overset{OH}{\underset{Nu}{\diagup}} \underset{O}{\overset{O}{\diagup}} \qquad \text{(Eq. 69)}$$

Nu = H⁻, N₃⁻, PhS⁻, Me⁻

ever, lowers selectivity for C-3 attack by differing degrees, depending on its size and electron-withdrawing ability.[7,261]

Selectivity for nucleophilic opening at C-3 is greatly enhanced by the use of appropriate organometallic reagents. Alkyl, aryl, and alkynyl groups[292-295] as well as hydride[123,292] can be introduced regioselectively at C-3 by using aluminum reagents (Eq. 70). The azide group is also introduced regioselectively

$$R \overset{O}{\diagdown}\diagup OH \xrightarrow{(R^1)_3Al} R \overset{R^1}{\underset{OH}{\diagup}} OH \qquad \text{(Eq. 70)}$$

(98:2) with trimethylsilyl azide–diethylaluminum fluoride to give 3-azido diols (Eq. 71),[296] while reaction with sodium azide–ammonium chloride exhibits only modest C-3 selectivity.[261]

$$R \overset{O}{\diagdown}\diagup OH \xrightarrow{Me_3SiN_3,\ Et_2AlF} R \overset{N_3}{\underset{OH}{\diagup}} OH \qquad \text{(Eq. 71)}$$

Treatment of epoxy alcohol **101** with trimethylaluminum, however, leads to the formation of a complicated mixture. With such a substrate, the sequence in-

$$\underset{\textbf{101}}{\overset{O}{\diagdown}\diagup OH} \xrightarrow[X]{AlMe_3} \overset{OH}{\diagup} OH$$

1. Ph₃CCl, Py
2. LiCuMe₂

TsOH, MeOH, H₂O

OCPh₃

(80%)   (14%)

volving hydroxy protection as a trityl ether, alkylation with lithium dialkyl-cuprate, and deprotection is an effective alternative.[297-299] Although alkylative epoxide-opening with trialkylaluminum usually occurs in an S_N2 manner, reaction of 3-p-tolylglycidol (102) with trimethylaluminum proceeds with retention of the configuration to give 103. Treatment of 102 with lithium dimethylcuprate gives the usual S_N2 reaction product 104.[300]

The formation of chelates of epoxy alcohols with metal ions also enhances C-3 selectivity in the nucleophilic opening as well as the reactivity of the substrate, especially when titanium tetraalkoxide is used as a mediator.[301] For example, the reaction of (2S,3S)-epoxyhexan-1-ol with diethylamine in the presence of Ti(OPr-i)$_4$ proceeds to completion at room temperature, although the same reaction in the absence of Ti(OPr-i)$_4$ is sluggish even under reflux conditions (Eq. 72).

Et$_2$NH (xs), reflux; (4%) 3.7:1
Et$_2$NH (xs), Ti(OPr-i)$_4$, rt; (90%) 20:1

(Eq. 72)

In the presence of Ti(OPr-i)$_4$, nucleophiles such as dialkylamines,[301] monoalkylamines,[301,302] alcohols,[301] borohydrides,[303] and carboxylic acids[301,304] also exhibit high C-3 selectivity (>100:1), while nucleophiles like thiols,[301] azides,[301] chlorides,[301,305] bromides,[305] and cyanides[301] show moderate levels of C3-selectivity (5–15:1). For ring openings with azides, use of Ti(OPr-i)$_2$(N$_3$)$_2$ instead of Ti(OPr-i)$_4$ and NaN$_3$ is recommended.[306] C3-Selective halohydrin formation is also effected by the use of Ti(OPr-i)$_4$-halogen (Br$_2$, I$_2$),[307] TiCl$_4$-

lithium halide,[308] or $ClBH_2 \cdot SMe_2$.[309] When the nucleophiles are chlorides, thio-cyanides, or thiols, C3-selective opening is also promoted by tetrakis(triphenyl-phosphine)palladium.[310]

Treatment of 2,3-epoxy alcohols with lithium iodide at 70° affords *threo*-1-iodo-2,3-diols (Eq. 73).[311]

(Eq. 73)

Since $Ti(OPr-i)_4$ is a weak Lewis acid, it promotes the rearrangement of some epoxy alcohols in the absence of an appropriate nucleophile.[312] For example, ex-posure of 2,3-epoxygeraniol and 2,3-epoxynerol to $Ti(OPr-i)_4$ leads stereoselec-tively to allylic alcohols **105** and **106**.

105   (70-80%)

106   (70-80%)

The titanium chelation effect is also observed in the epoxide opening reactions of epoxy acids (Eq. 74).[254]

	I + II	I:II
$Ti(OPr-i)_4$-Et_2NH (xs), rt	(92%)	>20:1
Et_2NH-H_2O (2:1), reflux	(95%)	1:6

(Eq. 74)

Treatment of epoxy ester **107** with azidotrimethylsilane in the presence of a catalytic amount of zinc chloride gives the 3-azido hydroxy ester exclusively.[313]

Nucleophiles supported on calcium ion-exchanged Y-type zeolite also react with epoxy alcohols regioselectively to give C-3-opened products (Eq. 75).[314–316]

(Eq. 75)

$$Nu = N_3^-, Cl^-, Br^-, PhS^-$$
CaY = calcium ion-exchanged Y-type zeolite

Magnesium iodide reacts regioselectively with epoxy alcohols to give iodohydrins that are further reduced to vicinal diols (Eq. 76).[317,318]

(Eq. 76)

Intramolecular nucleophilic opening of an epoxy alcohol where the nucleophile is anchored to the hydroxy group is effective for selective epoxide opening at C-2. For example, epoxy carbonates **108** and epoxy urethanes **109**, obtained by treatment of epoxy alcohols with alkoxycarbonyl chlorides or isocyanates, undergo intramolecular epoxide opening under acidic conditions by attack on the carbonyl oxygen to give hydroxy carbonates **110** with inversion at

C-2.[105,160,183,184,319–322] Treatment of the carbamates derived from 3,3-disubstituted epoxy alcohols, however, provides a mixture of C-2 and C-3 ring-opened products, the ratio of which is dependent on the Lewis acid used (Eqs. 77,[323] 78[324]).

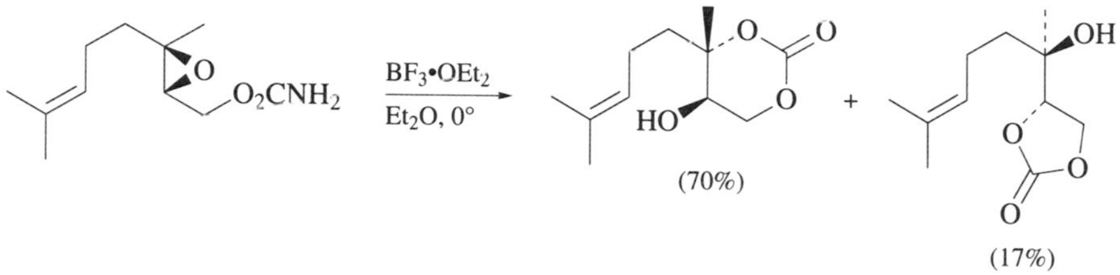

(Eq. 77)

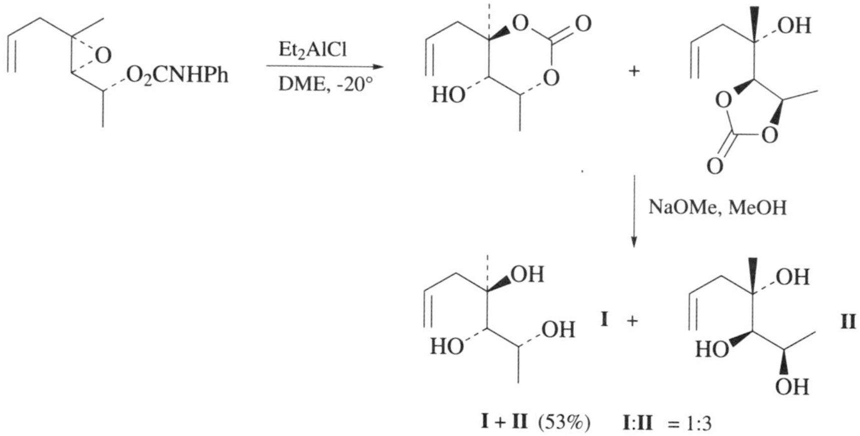

(Eq. 78)

With acid-sensitive substrates such as 111, treatment with cesium carbonate and powdered 3-Å molecular sieves in dimethylformamide under carbon dioxide provides 3-hydroxy carbonate 112 together with a small amount of 1-hydroxy carbonate 113.[325,326]

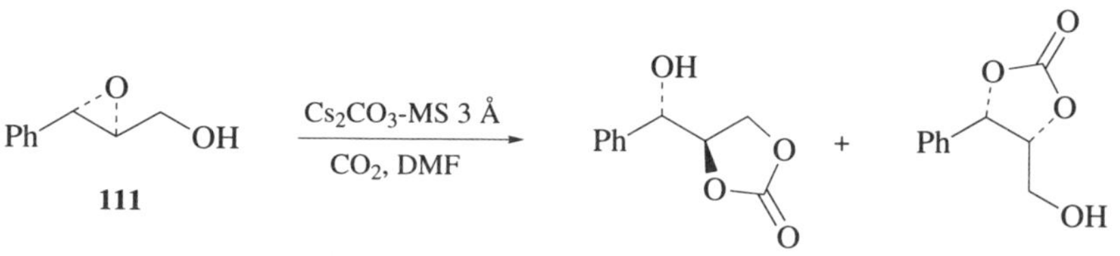

Treatment of epoxy alcohols with cesium carbonate and paraformaldehyde gives hydroxy dioxolanes in good yields (Eq. 79).[327]

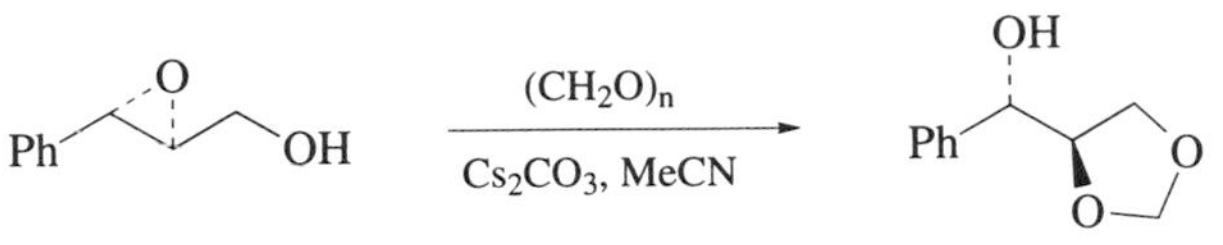

(Eq. 79)

The reaction of epoxy alcohols with carbon disulfide in the presence of sodium hydride gives cyclic xanthates which can be transformed into episulfides (Eq. 80).[328]

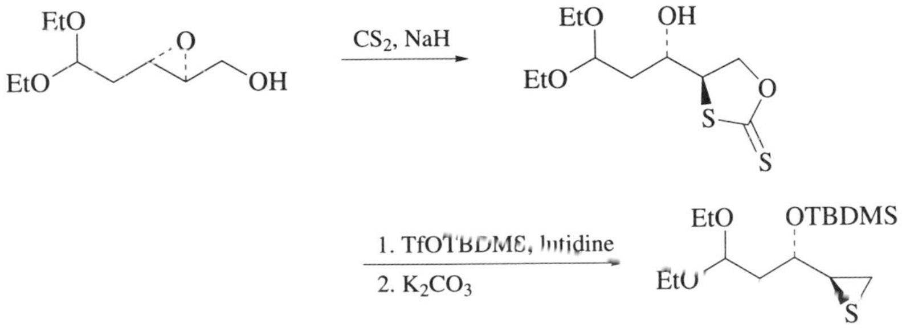

(Eq. 80)

On treatment of **109** with a base such as sodium hydride, the urethane nitrogen attacks the C-2 epoxide carbon to give hydroxy carbamates **114**.[160,319,321]

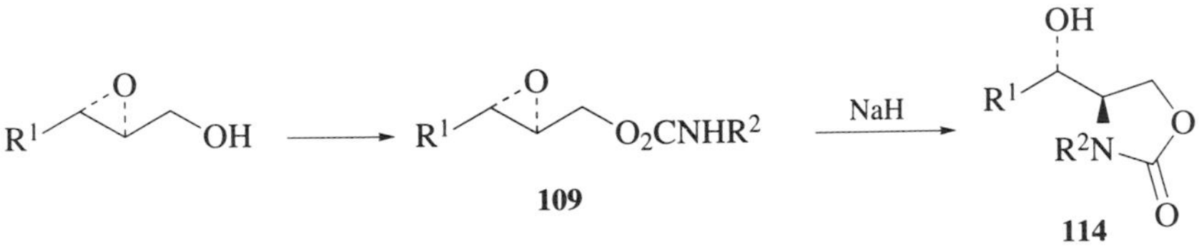

Nitrogen delivery can be effected under mild basic conditions when epoxy N-benzoylcarbamate **115** is used as a substrate (Eq. 81).[329]

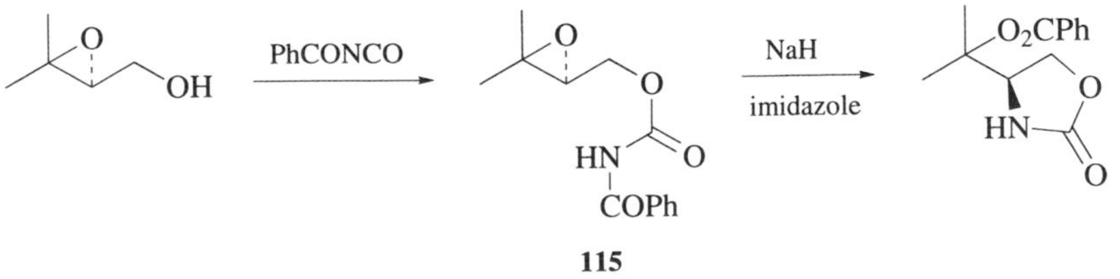

(Eq. 81)

Reduction of epoxy alcohols with sodium bis(methoxyethoxy)aluminum hydride (Red-Al®) in tetrahydrofuran provides 1,3-diols selectively, presumably by a pathway in which hydride is delivered to C-2 from an intermediate epoxy alcohol-Al(OR)$_2$H$^-$ complex (Eq. 82).[160,330,331] This notion is supported by the ob-

$$n\text{-}C_7H_{15} \overset{O}{\diagup\!\!\diagdown}\!\!\diagup\!\!\diagdown OH \quad \xrightarrow[\text{THF}]{\text{Red-Al}} \quad n\text{-}C_7H_{15}\overset{OH}{\diagup\!\!\diagup\!\!\diagdown\!\!\diagup}OH \qquad \text{(Eq. 82)}$$
$$(90\%)$$

servation that reduction of the epoxy ether is slow and exhibits poor regioselectivity (Eq. 83).[331] In contrast, reduction of epoxy alcohols with diisobutylaluminum

$$n\text{-}C_7H_{15}\overset{O}{\diagup\!\!\diagdown}\!\!\diagup\!\!\diagdown OBn \quad \xrightarrow[\text{THF}]{\text{Red-Al}} \quad n\text{-}C_7H_{15}\overset{OH}{\diagup\!\!\diagup\!\!\diagdown\!\!\diagup}OBn \quad +$$
$$(50\%) \; 1{:}1.5$$

$$n\text{-}C_7H_{15}\diagup\!\!\diagdown\!\!\underset{OH}{\diagup}\diagdown OBn$$

$$\text{(Eq. 83)}$$

hydride[123] gives 1,2-diols preferentially (Eq. 70). Compound **116** with a C-2 substituent, however, undergoes opening at C-3 because of steric hindrance at C-2.[331]

$$n\text{-}C_7H_{15}\overset{O}{\diagup\!\!\diagdown}\!\!\underset{|}{\diagup}\!\!\diagup OH \quad \xrightarrow[\text{THF}]{\text{Red-Al}} \quad n\text{-}C_7H_{15}\diagup\!\!\diagdown\!\!\underset{OH}{\diagup}\diagdown OH \quad (70\%)$$
$$\textbf{116}$$

In contrast, C-4 branching causes hydride delivery to occur at C2 (Eq. 84).[332] In the reduction of 2,3-epoxycinnamyl alcohol by sodium bis(methoxyethoxy)-

$$\text{BOMO}\diagup\!\!\diagdown\!\!\diagup\!\!\diagdown\!\!\overset{O}{\diagup\!\!\diagdown}\!\!\diagup OH \quad \xrightarrow[\text{THF}]{\text{Red-Al}} \quad \text{BOMO}\diagup\!\!\diagdown\!\!\diagup\!\!\diagdown\!\!\overset{OH}{\diagup}\!\!\diagdown OH$$
$$(89\%)$$

$$\text{(Eq. 84)}$$

aluminum hydride, higher regioselectivity (22:1) is obtained by using dimethoxyethane instead of tetrahydrofuran as a solvent.[333]

Reaction of epoxy alcohol **117** with $LiCuMe_2$ gives C-2 methylated 1,3-diols selectively.[154,334] With sterically unbiased epoxy alcohols, methylation with

$LiCuMe_2$ is nonregioselective (Eq. 85), but reaction with $Li_2Cu(CN)Me_2$ prepared from methyllithium not contaminated with lithium chloride, in a coordinating solvent system [THF-1,3-dimethyl-2-imidazolidone (DMEU)], shows moderate C-2 selectivity (Eq. 86).[335,336] Use of a Grignard reagent in the presence of a catalytic amount of CuI also shows moderate C-2 selectivity.[337]

(Eq. 85)

(Eq. 86)

Treatment of **118**, derived from epoxy alcohols, with titanium tetrachloride or boron trifluoride etherate promotes 1,2-migration of the vinyl group, providing the *anti*-β-hydroxy ketone stereoselectively.[338–339]

Treatment of epoxy silyl ethers with methylaluminum bis(4-bromo-2,6-di-*tert*-butylphenoxide) (MABR) promotes rearrangement of the *tert*-butyldimethyl-siloxymethyl group to give β-hydroxy aldehyde derivatives (Eq. 87).[340–343]

(Eq. 87)

Cyclization by intramolecular nucleophilic substitution of epoxides follows Baldwin's rule.[344,345] For example, 3-7-*exo-tet* systems are favored and 5-6-*endo-tet* systems disfavored (Eq. 88).

(Eq. 88)

Epoxide opening with a hydroxy group leads to stereospecific formation of tetrahydrofuran[75,76,82,83,94,346] or tetrahydropyran ring systems (Eqs. 89, 90).[82,83,347,348] However, formation of a tetrahydropyran or oxepane ring is ob-

(Eq. 89)

(Eq. 90)

served in the acid treatment of hydroxy vinylepoxides, which are readily derived from the corresponding epoxy alcohols (Eq. 91).[349,350]

(Eq. 91)

Although the vinyl carbanion derived from vinyl sulfone **119** cyclizes smoothly to give dihydrofuran **120**, cyclization of the cyano ester **121** to lactone **122** does not proceed because of the unfavorable stereoelectronic requirement

119

120 (72%)

121

122

that the enolate π system must achieve colinearity with the oxirane C_2-O bond.[351–353]

Intramolecular nucleophiles such as carbonyl groups, alcohols, and carboxylic acids can react with the epoxide with C-3 selectivity if they are located at an appropriate position. For example, alkaline treatment of epoxy ester **123** provides a triol derivative stereoselectively via an intermediate lactone.[354,355]

123

1. DHP, H⁺
2. NaOH

(82%)

Amino groups can also serve as intramolecular nucleophiles (Eq. 92).[356]

1. BnBr, NaH
2. NaC₁₀H₉, -60°

(70%)

(Eq. 92)

Nucleophilic π-bonds will also open epoxides. For example, treatment of epoxy ether **124** with boron trifluoride etherate gives the tricyclic compound **125**

with high stereoselectivity.[357] Baker's yeast also catalyzes this type of cyclization. For example, epoxide **126** provides C-28 hydroxylated sterol **127** on treat-

ment with Baker's yeast, but the enantiomer of **126** does not react under the same conditions.[358]

Treatment of unsaturated epoxy alcohols **128** or **129** with tin tetrachloride gives mixtures of diastereomeric cyclopropyl lactones in a ratio of 1.5:1, regardless of the geometry of the double bonds.[359]

Reactions of alkynyl epoxy alcohols with dialkylbromomagnesium cuprates in the presence of dimethyl sulfide give dihydroxy allenes stereoselectively (Eq. 93).[101]

$$(Eq. 93)$$

2,3-Epoxy alcohols can be converted into the corresponding 2,3-epithio alcohols and 2,3-aziridino alcohols stereoselectively. Treatment of epoxy alcohols with thiourea in the presence of Ti(OPr-i)$_4$ gives 2,3-epithio alcohols stereospecifically (Eq. 94).[360] Transformation of epoxy alcohols into aziridino al-

$$(Eq. 94)$$

cohols is also achieved in five steps by using the modified Blum procedure (Eq. 95).[361,362]

$$(Eq. 95)$$

The selectivity of reactions of 2,3-epithio alcohols and 2,3-aziridino alcohols is similar to that of the corresponding epoxy alcohols. For example, reduction with sodium bis(methoxyethoxy)aluminum hydride provides 3-thio or 3-amino alcohols regioselectively (Eqs. 96, 97).[361,363] Treatment of an aziridino alcohol with LiCuEt$_2$ provides a 3-amino-2-ethyl alcohol exclusively (Eq. 98).[362]

$$(38\%)$$

$$(Eq. 96)$$

$$(92\%)$$

$$(Eq. 97)$$

$$(80\%)$$

$$(Eq. 98)$$

OTHER METHODS FOR SYNTHESIS OF NONRACEMIC EPOXIDES

Epoxides can be prepared in various ways, including epoxidation of olefins, reaction of carbonyl compounds with sulfur ylides or with α-halo esters, and alkaline closure of halohydrins. Among these methodologies, the epoxidation of olefins is the most practical from the viewpoint of easy availability of olefins and mildness of the reaction. Three classes of asymmetric epoxidation of olefins are summarized in this section.

1. Epoxidation with chiral peracids or related compounds.[364-368] This includes the first example of asymmetric epoxidation using an optically active peroxycamphoric acid (Eq. 99), though the enantioselectivity of the reaction is poor (< 10%

$$\text{(Eq. 99)}$$

ee).[364] Several optically active peracids were examined for asymmetric epoxidation but the optical yields did not exceed 5% ee.[365-368] However, the enantiomerically pure N-sulfamyloxaziridine derivatives have proven quite effective for asymmetric epoxidation of unfunctionalized alkenes: good enantioselectivities (up to 65% ee) have been observed (Eq. 100).[369-371]

$$\text{(Eq. 100)}$$

2. Epoxidation of conjugated enones with hydrogen peroxide in the presence of a chiral phase transfer reagent (Eq. 101).[372-376]

$$\text{(Eq. 101)}$$

3. Epoxidation using a metal catalyst bearing chiral ligand(s). In 1970, it was found that the epoxidation of allylic alcohols with an alkyl hydroperoxide under catalysis by $VO(acac)_2$ proceeds smoothly in comparison with that of isolated olefins.[377] These metal-catalyzed epoxidations of allylic alcohols are chemo-, regio-, and stereoselective (Eqs. 102,103).[61,378,379] Reaction occurs in the coordi-

$$V^{5+}, TBHP \qquad (Eq.\ 102)$$

$$V^{5+}, TBHP \qquad (Eq.\ 103)$$

nation sphere of the metal, and the reacting ligand sites are located adjacent to the bystander ligand (L) sites of complex **130**.[30] Therefore, the use of chiral bystander

130

ligands has the effect of placing the reaction site in an asymmetric environment, and several approaches using chiral metal catalysts have been examined.[30,380–383] Although optical yields were unsatisfactory from a practical point of view, these initial results indicated the potential of chiral metal complex-catalyzed asymmetric epoxidation and, in this context, the titanium-mediated epoxidation discussed in this chapter was discovered.

In contrast to the success realized in the asymmetric epoxidation of allylic alcohols, there is still considerable room for improvement in the asymmetric epoxidation of olefins lacking coordinating functional groups, though several precedent studies have been reported.[384,385] However, chiral porphyrin complexes were shown to be effective catalysts for the epoxidation of mono- and Z-disubstituted olefins.[386–392] Furthermore, quite recently the chiral salen complexes **131** were

	R^1	R^2	R^3	R^4
131a	t-Bu	H	H	Ph
131b	t-Bu	H	t-Bu	$-(CH_2)_4-$
131c	(S)-1-phenylpropyl	H	H	Ph
131d	(S)-1-(4-$tert$-butylphenyl)propyl	Me	H	$-(CH_2)_4-$

found to be very effective catalysts for this purpose, and the highest enantioselectivity to date was realized.[393–403] For example, 2,2-dimethylchromene (132) was converted to the corresponding epoxide with 98% ee.

Highly enantioselective dihydroxylation of olefins proceeds in a catalytic manner by using $K_2OsO_2(OH)_4$-phthaladine (133 or 134) complex as a catalyst (Eq. 104).[404] This dihydroxylation reaction can be applied to a wide range of

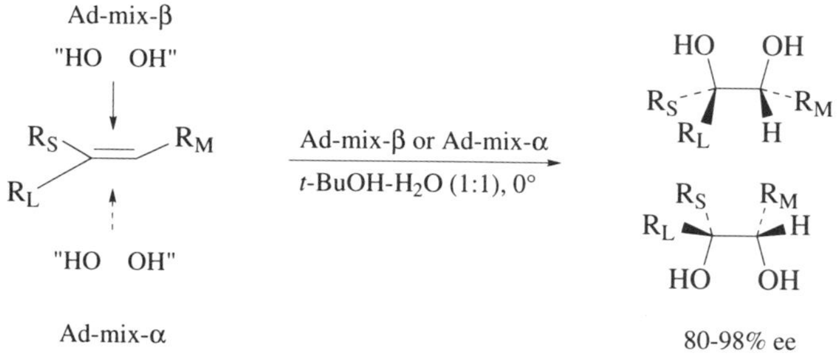

Ad-mix-β: $K_2OsO_2(OH)_4$, 133, K_2CO_3, $K_3Fe(CN)_6$
Ad-mix-α: $K_2OsO_2(OH)_4$, 134, K_2CO_3, $K_3Fe(CN)_6$

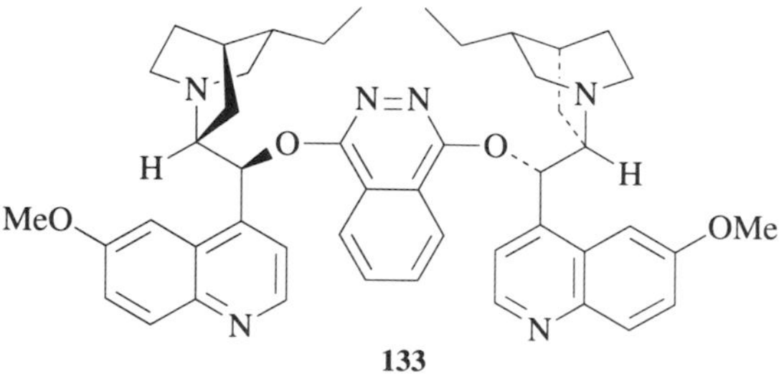

(Eq. 104)

133

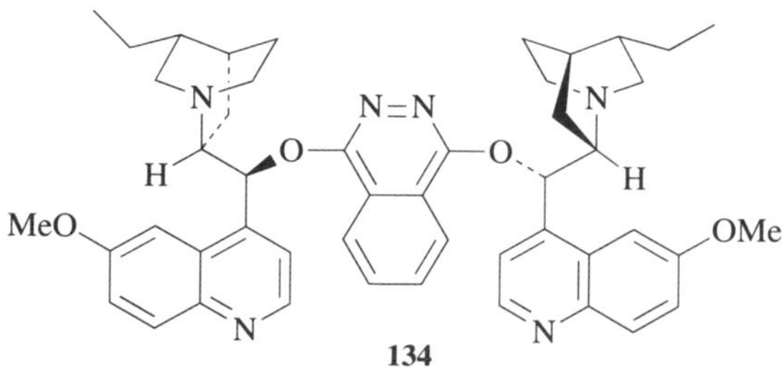

134

olefins irrespective of the presence or absence of coordinating functional groups, except for *cis*-disubstituted and tetrasubstituted olefins which show decreased enantioselectivity to some extent. The resulting diols are stereospecifically converted into epoxides (Eq. 105).[405,406]

(Eq. 105)

Experimental Conditions

Reagents

Titanium Alkoxide. Titanium tetraisopropoxide, $Ti(OPr-i)_4$, has been almost the only metal alkoxide used in asymmetric epoxidation and kinetic resolution reactions. It can be distilled under vacuum (bp 78–79.5°/1.1 mm Hg) and stored under an inert gas for long periods of time (~1 year) without decomposition. This reagent can be used for most any epoxidation as received from commercial sources, but if poor percentages of ee are obtained, especially in the catalytic procedure, purification by distillation is recommended. Titanium tetraisopropoxide can be transferred to the reaction vessel via syringe or cannula, avoiding exposure to atmospheric moisture, which would quickly destroy the metal alkoxide. $Ti(OBu-t)_4$ can be used in the epoxidation of primary allylic alcohols when epoxide opening is problematic (see Experimental Procedures). However, use of $Ti(OBu-t)_4$ is not recommended when the substrate is a secondary allylic alcohol because the relative rate constant obtained with $Ti(OBu-t)_4$ is low relative to that with $Ti(OPr-i)_4$.[38]

Dialkyl Tartrates. Diethyl tartrate (DET) and diisopropyl tartrate (DIPT) are the chiral auxiliaries most often used in either stoichiometric or catalytic asymmetric epoxidation, the choice being irrelevant in terms of the enantioselectivity except in the asymmetric epoxidation of allyl alcohol.[18] Dimethyl tartrate (DMT) is useful when a water-soluble tartrate is needed (e.g., when the epoxy alcohol needs to be purified without tartrate hydrolysis). The choice of tartrate ester is more critical in kinetic resolution since k_{rel} increases in the order

DMT < DET < DIPT. In the catalytic procedure, dicyclohexyl tartrate (DCHT)[18] and dicyclododecyl tartrate (DCDT)[18] have shown greater enantioselectivity than DIPT, although DIPT is usually the first choice because it is most readily available and gives acceptable selectivity.

Like titanium tetraisopropoxide, the tartrate ester usually can be used as received from commercial sources, but if substandard enantioselectivity is observed, distillation under high vacuum followed by storage under an inert gas is recommended (DMT, mp 48–50°, bp 163°/23 mm Hg; DET, bp 89°/0.5 mm Hg; DIPT, bp 76°/0.1 mm Hg). The distillation temperature should be kept below 100° in order to avoid polymerization of the tartrate. Both DCHT (mp 69.5–70.5°) and DCDT (mp 122–123°) can easily be prepared by Fisher esterification of tartaric acid with the corresponding alcohol.[18] Liquid tartrates can be handled by syringe. However, since they are extremely viscous, a more convenient method is to weigh the amount required into a flask, dissolve it in a minimum amount of dichloromethane, and transfer the solution to the reaction vessel via syringe or cannula.

Organic Peroxide.   *tert*-Butyl hydroperoxide is the most often used oxidant in asymmetric epoxidation and kinetic resolution reactions. It is available as a 70% solution in water, but preparation of anhydrous organic solutions is necessary. Anhydrous isooctane solutions[18] are commercially available (Aldrich Chemical Company). However, care must be taken when an isooctane solution is used. Some reactions using TBHP in isooctane with higher substrate concentrations produce lower reaction rates and/or percent ee. Introduction of too much isooctane into the reaction medium (dichloromethane) changes the solvent polarity, with negative effects on the rate and percent ee.[18] Accordingly, use of >5 M solutions of TBHP in isooctane is highly recommended. Other organic solvents such as dichloromethane, dichloroethane, heptane, and toluene can also be used. Anhydrous TBHP solutions can be prepared by extraction of the commercial 70% aqueous solution with the chosen solvent and by azeotropic distillation of the organic phase to ensure the removal of all remaining water.[17,18,30,407] The concentration of hydroperoxide is determined by iodometric titration,[17,18,30] and the solutions are stored with refrigeration (0–5°) over 3 Å molecular sieves (dichloromethane) or at room temperature (isooctane) without zeolites in high-density polyethylene bottles. The latter are recommended instead of glass bottles because of possible pressurization by gas evolution. Although TBHP solutions have proved to be safe during manipulation, peroxides are potentially hazardous with respect to violent decomposition, so special care should be taken when handling them. Thus, never add transition metal salts or strong acid to the concentrated peroxide solutions. In the same sense, never work with pure hydroperoxide, and avoid high concentrations of TBHP whenever possible. *All heating of peroxide solutions should be done behind an adequate blast shield in a well-ventilated fume hood.* The desired amount of TBHP solution should be added to a flask or graduated cylinder containing activated 3 or 4 Å molecular sieves. After standing several minutes in the stoppered flask, the solution can be transferred to the reaction vessel by syringe, by cannula, or simply by pouring. The syringe or cannula should never be inserted directly into the stock solution.

Commercially available cumyl hydroperoxide (Aldrich Chemical Company, 80% solution in cumene) can be used without further treatment. Asymmetric epoxidations using cumyl hydroperoxide are faster than those using TBHP. However, in general, TBHP solutions are recommended because the products are more easily isolated.

Reaction Solvent. Dichloromethane is the solvent used in all asymmetric epoxidation and kinetic resolution reactions. When dichloromethane is methanol-free, it can be prepared for use just by storing over activated 3 Å molecular sieve pellets (4-Å should not be used since pressurization in bottles has been observed). A simple check of the dichloromethane by ^{1}H NMR can show whether it contains methanol. In general, care must be taken to avoid contamination by chelating solvents (alcohols, esters, nitriles, amines, ketones) which decrease the rate, yield, and enantioselectivity of the epoxidation reaction.

Although most allylic alcohols have sufficient solubility in dichloromethane, some long-chain allylic alcohols are poorly soluble. For example, (E)-octadec-2-en-4-yn-1-ol is scarcely soluble in dichloromethane at −20°, but use of a 1:1 mixture of 2,3-dimethyl-2-butene and dichloromethane as a solvent allows smooth epoxidation.[102,103]

Molecular Sieves. The use of 3 or 4 Å molecular sieves (zeolites) is essential to accomplish both asymmetric epoxidation and kinetic resolution in a catalytic manner. Their use in the stoichiometric reaction is also highly recommended. The amount of molecular sieves added to the reaction mixture is not critical, so long as the allylic alcohol and TBHP solutions are predried. Tartrates and Ti(OPr-i)$_4$ should not be stored over molecular sieves. Molecular sieves should be preactivated in a vacuum oven (160° and 0.05 mm Hg pressure for at least 3 hours) and crushed. Preactivated, powdered 4 Å molecular sieves are commercially available (Aldrich Chemical Company). Either 3 or 4 Å molecular sieves can be used, except for epoxidations of low molecular weight substrates (e.g., allyl alcohol) where 3 Å sieves are recommended.

Reaction Conditions. Asymmetric epoxidation and kinetic resolution reactions can be run either stoichiometrically (50% or more catalyst) or catalytically (5–10% catalyst). The stoichiometric procedure is useful when very unreactive substrates are oxidized or when the reaction product is not amenable to further percent ee enrichment by recrystallization. In general, use of the minimum amount of catalyst is recommended because the reaction products are more easily isolated. Reaction conditions are similar in both procedures. The catalyst is unstable if stored over long periods of time, and thus should be prepared in situ. It is prepared by mixing the tartrate with Ti(OPr-i)$_4$ at −20°, in any order of addition, and adding either the allylic alcohol or the TBHP solution in a second step. *The use of at least 10% excess of tartrate over Ti(OPr-i)$_4$ is essential to obtain* *optimum results (a ratio of 1.2 : 1 is recommended).* The three components should be aged at this temperature for 20–30 minutes. Although this aging period is important for the catalytic procedure, most stoichiometric reactions have been run after aging for only 1–5 minutes. After aging, the precooled (especially important in large-scale reactions) fourth component (the allylic alcohol or the TBHP solution) is added to the reaction mixture. It is recommended that TBHP be the last component added to maintain more control over the reaction temperature. This is especially important in large-scale procedures because its addition is slightly exothermic. The use of 2.0 equivalents of TBHP is generally suitable, although 3 equivalents may be used for sluggish substrates. In kinetic resolution reactions, 0.6 equivalent of TBHP is the normal amount used. However, if very unreactive substrates are resolved, more TBHP can be used (about 2 equivalents), as long as substrate conversion is monitored (e.g., GLC with an internal standard). The temperature of the reaction mixture must be adjusted to within the range −40 to 0°, depending on the substitution pattern of the substrate. If the epoxidations are rapid enough, lower temperatures are recommended to avoid side reactions such as transesterification and epoxide opening. Most asymmetric epoxidations and kinetic resolutions are carried out at −20° (carbon tetrachloride/dry ice slurry).

The concentration of the substrate is another factor to be considered. The highest suggested concentration for the stoichiometric reaction and for the catalytic procedure in which the product is very reactive (e.g., cinnamyl alcohol epoxide) is 0.1 M in order to minimize side reactions (mainly epoxide opening). For some substrates the concentration for the catalytic procedure can be up to 1.0 M, although substrate solubility can be a limitation.

Finally, since asymmetric epoxidation and kinetic resolution are sensitive to humidity, these reactions must be run under dry conditions, usually under a dry nitrogen or argon atmosphere.

Workup. Choice of the appropriate isolation procedure is essential to the success of an asymmetric epoxidation or kinetic resolution reaction and depends on the nature of the product. The development of workup procedures has paralleled the development of the asymmetric epoxidation and kinetic resolution procedures.[1,18,41,59,347,408,409] The most important factors to be considered in workup are: (1) hydrolysis of the titanium complex; (2) destruction of excess oxidant; (3) tartrate removal; and (4) purification of the product, having in mind the possibility that the product might be water soluble or react with the reagents used.

Hydrolysis of the titanium complex has been effected with aqueous solutions of hydroxy carboxylic acids,[1,17,18] water/acetone,[10] saturated sodium sulfate/ether,[4,348] sodium fluoride solution,[59] triethanolamine,[347] anhydrous citric acid/ether (or acetone/ether),[18] and sodium hydroxide in saturated brine.[17,18] Hydrolysis of the complex can also be combined with destruction of the tartrate and/or the excess TBHP. Tartrate removal can be carried out either by alkaline hydrolysis of the ester to give a water-soluble tartrate salt or by chromatographic separation at the purification step. Excess TBHP can be reduced with dimethyl sulfide,[59,347] trimethyl phosphite,[18] aqueous sodium thiosulfate,[30,363] aqueous sodium sulfite,[30] sodium borohydride,[408] triphenylphosphine,[409] or aqueous ferrous sulfate.[18]

These steps are followed by a product purification step involving chromatographic separation, distillation, or recrystallization, either alone or in combination. Although various combinations of these procedures have led to an adequate workup method for most epoxy alcohols, an ideal method has not yet been found to isolate very water-soluble or reactive epoxy alcohols. Thus in situ derivatization of the reaction products can sometimes be useful because more stable and easily isolable products are obtained.[18,135,136] Such in situ derivatization is one of the most important advantages of the catalytic procedure, since it is impractical to perform in the stoichiometric reaction because of the large amount of hydroxylic contaminants.[18] The most water-soluble glycidol, however, can be extracted into water and purified by distillation when tributylphosphine is used as a reducing agent for excess TBHP.[410] A few standard workup procedures are presented here that can be used for any epoxy alcohols included in the group indicated in each case.

Aqueous Acidic Workup. This is the most common procedure for water-insoluble products from either asymmetric epoxidation or kinetic resolution.[1] The reaction mixture is poured into a precooled (0°) aqueous solution of tartaric acid (10% w/v) or citric acid (11% w/v),[18] using 5–20 mL of the acid solution per 1 mmol of Ti(OPr-i)$_4$. This mixture is vigorously stirred until a clear organic phase is obtained (10–30 minutes). In small-scale reactions the excess TBHP can be ignored because it can be easily removed by azeotropic distillation with toluene or carbon tetrachloride (up to 0.5 mol has been removed by this procedure) or in the product purification step. However, oxidant destruction may be desired to avoid peroxide hazard and is essential in large-scale reactions. If so, the solution of tartaric or citric acid should include ferrous sulfate heptahydrate (30% w/v), in which case the resulting organic phase contains only the reaction product, the tartrate, and *tert*-butyl alcohol. Further alkaline treatment (30% NaOH w/v in saturated brine, approximately 2 mL for 1 mmol of tartrate) at 0° (approximately 1 hour) with vigorous stirring leads, after solvent removal, to a crude reaction product that contains fairly pure epoxy alcohol, or epoxy and allylic alcohols from kinetic resolution reactions. Although the original workup procedure[1] used aqueous alkali for tartrate hydrolysis, the saturation of this solution with sodium chloride is strongly recommended to avoid losing partially water-soluble products and to suppress Payne rearrangement.[258]

Aqueous Nonacidic Workup. Modified workup procedures have been developed to simplify the acidic procedure and to accommodate the instability of some water-insoluble epoxy alcohols.[18,42,59,347] Thus an alternative procedure entails adding water [20–30 times the weight of Ti(OPr-i)$_4$] to hydrolyze the titanium complex and stirring (30–60 minutes) until room temperature is reached. Hydrolysis of the tartrate is performed in situ by adding an alkaline hydroxide solution (30% NaOH w/v in saturated brine) (1 mL for 1 mmol of tartrate) and stirring vigorously until phase separation occurs (addition of a small amount of methanol, ca. 5% v/v, can help achieve the separation). After drying the organic phase over sodium sulfate or magnesium sulfate, removal of the solvent gives a crude reaction mixture containing the epoxy alcohol, TBHP, and *tert*-butyl alcohol.

When dealing with very acid-sensitive products, the alkaline solution (10% NaOH w/v in saturated brine, 1–2 mL for 1 mmol of tartrate) and diethyl ether (10% v/v) are added directly to the reaction and the mixture is stirred to hydrolyze both the complex and the tartrate ester. The addition of magnesium sulfate and Celite gives a solution containing the epoxy alcohol and TBHP; the latter can be removed as described above.[18]

Nonaqueous Workup. When the product is very water soluble, a nonaqueous workup procedure must be used. A solution of citric acid monohydrate [1 equivalent relative to Ti(OPr-i)$_4$] in acetone-ether (11% w/v in 1:9 mixture) or ether (4% w/v) is added to the reaction mixture. After stirring, solvent removal, and filtration through a pad of Celite, a crude reaction product is obtained that contains the water-soluble epoxy alcohol, tartrate, and TBHP. These last two substances must be removed in the purification step. It is important for acid-sensitive compounds that the molar amount of citric acid used be equal to or slightly less than that of Ti(OPr-i)$_4$ to avoid excess acid in the solution.

An alternative workup useful for both acid-sensitive and water-soluble epoxy alcohols uses triethanolamine [1.5 equivalents relative to Ti(OPr-i)$_4$] to neutralize the acidity of the titanium species, and dimethyl sulfide to reduce the excess TBHP.[347] The cold reaction mixture is then quickly filtered through a pad of silica gel and eluted with ether. Removal of solvents under vacuum leads to a crude mixture containing the product, tartrate, and *tert*-butyl alcohol.

In Situ Derivatization. This procedure is useful only for the products of a catalytic asymmetric epoxidation; the amounts of isopropyl alcohol and tartrate present in the stoichiometric procedure make in situ derivatization impractical. It is also necessary to destroy excess TBHP before doing any in situ derivatization because the hydroperoxide reacts with most alkylating and acylating agents, whereas *tert*-butyl alcohol does not.

Trimethyl phosphite is the recommended reagent for reducing excess TBHP because it reacts completely and rapidly at low temperature, and because both trimethyl phosphite and trimethyl phosphate are volatile and can be removed under high vacuum. Care should be taken not to use excess trimethyl phosphite, particularly for a subsequent sulfonation, because epoxy sulfinates could be formed. The destruction of TBHP should be monitored by TLC eluting with 40% ethyl acetate/hexane using tetramethylphenylenediamine spray (10% w/v in methanol:water:ethyl acetate 128:28:1 mixture).

Two important derivative classes are the p-nitrobenzoates (PNB) and p-toluenesulfonates. Although both derivatives are useful synthetic intermediates, p-nitrobenzoates are especially useful because they are easily isolated and are usually crystalline. Enhancement of optical purity is often possible by recrystallization of p-nitrobenzoates. Furthermore, epoxy alcohol p-nitrobenzoates undergo most of the ring-opening reactions of the free substance.[137]

Derivatization is carried out on the catalytic epoxidation mixture when the reaction is found to be completed (TLC). The mixture is cooled to $-20°$, and trimethyl phosphite is carefully added until all TBHP is destroyed, taking care that the temperature does not rise above $-20°$. The mixture is then treated with triethylamine (1.2 equivalents to the original amount of allylic alcohol), 4-(N,N-dimethylamino)pyridine (0.05 equivalent), and a solution of the derivatizing agent (acyl or sulfonyl chloride, 1 equivalent) in dichloromethane, and allowed to stand (usually at $-20°$) until the reaction is completed. The reaction mixture is filtered through a pad of Celite. The filtrate is successively washed with aqueous tartaric acid (10% w/v), saturated sodium bicarbonate and saturated brine, dried over magnesium sulfate, and concentrated. Chromatographic purification and/or crystallization of the crude mixture yields the pure derivatives.

EXPERIMENTAL PROCEDURES

$$n\text{-}C_{15}H_{31} \overset{\displaystyle\triangle}{\underset{O}{}}\!\!\!\diagup\!\!\diagup\text{OH}$$

(2R,3R)-2,3-Epoxy-1-octadecanol (Stoichiometric Epoxidation of an E Allylic Alcohol).[321] Freshly distilled titanium tetraisopropoxide (3.8 mL, 12.7 mmol) was added to dichloromethane (115 mL), and the resulting solution was cooled to a temperature between -30 and $-20°$ (dry ice/CCl$_4$). Freshly distilled (S,S)-$(-)$-diethyl tartrate (2.89 mL, 16.9 mmol) was then added. The resulting mixture was stirred for 15 minutes, and then a solution of (E)-octadec-2-en-1-ol (2.83 g, 10.6 mmol) in dichloromethane (20 mL) was added. Ten minutes later $tert$-butyl hydroperoxide solution (8.7 mL of a 3.64 M solution in toluene, 31.7 mmol) was added, and the reaction mixture was then stored in a $-20°$ freezer for 24 hours. The reaction was quenched by addition of dimethyl sulfide (3 mL, 41 mmol). The resulting mixture was stirred at $-20°$ for 30 minutes, and then saturated aqueous Na$_2$SO$_4$ (13 mL) was added. This suspension was allowed to warm to room temperature, then filtered through a pad of Celite and washed with diethyl ether. Concentration of the filtrate provided an oil, which was purified by chromatography on a flash silica gel column with solvent, increasing the polarity from hexane to 5:1 hexane-diethyl ether. The appropriate fractions were combined and concentrated under reduced pressure to give 2.65 g (88%) of the crystalline product, mp 77–78°; $[\alpha]_D^{23}$ +22.5° (c 0.79, CHCl$_3$) (>95% ee); IR (CH$_2$Cl$_2$) 3780, 2915, 2860, 1420, 1250, 1110 cm^{-1}; ^{1}H NMR δ 3.90 (ddd, J = 12.6, 5.5, 2.5 Hz, 1 H, H_{1a}), 3.61 (ddd, J = 12.6, 4.2, 3.1 Hz, 1 H, H_{1b}), 2.91

(m, 2 H, H$_{2,3}$), 1.65 (dd, J = 7.3, 5.6 Hz, 2 H), 1.23–1.57 (m, 27 H), 0.86 (t, J = 6.6 Hz, 3 H, CH$_3$); mass spectrum, m/z 284 (M$^+$).

(2R,3R)-2,3-Epoxyoctadec-4-yn-1-ol (Stoichiometric Asymmetric Epoxidation of an E Unsaturated Allylic Alcohol).[103]

A solution of freshly distilled titanium tetra-*tert*-butoxide (41.1 mL, 107.6 mmol) in absolute dichloromethane (100 mL) was cooled to $-25°$. (S,S)-(−)-Diethyl tartrate (35 mL of a 3.14-M solution in absolute dichloromethane, 110 mmol) was added over 15 minutes. After an additional 15 minutes at $-25°$, a solution of (E)-octadec-2-en-4-yn-1-ol (15.0 g, 56.7 mmol) in absolute dichloromethane was added at such a rate (approximately 60 minutes) that the mixture remained homogeneous. This was followed by the addition of *tert*-butyl hydroperoxide (34 mL, 3.79 M in absolute toluene, 129 mmol). After 4–5 hours at $-30°$, 10% aqueous tartaric acid (500 mL) was added and the mixture was warmed to room temperature. Dilution with diethyl ether (1 L), washing with 10% aqueous tartaric acid (2 × 500 mL) and saturated NaCl solution (2 × 750 mL), drying (MgSO$_4$), and concentration under reduced pressure afforded a yellow oil. Separation of the product by flash silica gel column chromatography gave 1.05 g (6.6%) of unreacted (E)-octadec-2-en-4-yn-1-ol and 12.6 g [86%, based on recovered (E)-octadec-2-en-4-yn-1-ol] of the crystalline product. Two recrystallizations from hexane ($-10°$) gave pure (2R,3R)-2,3-epoxyoctadec-4-yn-1-ol, mp 55–56°; [α]$_D^{25}$ $-41.5°$ (c 2.05, CHCl$_3$) (100% ee); IR (KBr) 3300 (br), 3180, 3000, 2960, 2850, 2240, 1460, 1320,1070, 1030, 875, 725 cm^{-1}; ^{1}H NMR (CDCl$_3$) δ 3.94 (ddd, J = 12.9, 4.9, 2.2 Hz); with D$_2$O: dd, J = 12.9, 3.4 Hz, H-C1), 3.70 (ddd, J = 12.9, 7.9, 3.4 Hz; with D$_2$O: dd, J = 12.9, 2.2 Hz, H-C1), 3.43 (q, J = 1.7 Hz, H-C3), 3.27 (ddd, J = 3.4, 2.2, 1.7 Hz, H-C2), 2.20 (td, J = 7.0, 2, 1.7.Hz, H-C6), 1.55 (m, exchangeable with D$_2$O, OH), 1.26 (m, 22 H), 0.88 (t, J = 6.7 Hz, 3 H-C18); ^{13}C NMR δ 85.6, 75.8 (2s, C-4, C-5), 60.4 (t, C-1), 60.0 (d, C-3), 43.1 (d, C-2), 31.9 (t, C-16), 29.6–28.3 (9t, C-7–C-15), 22.6 (t, C-17), 18.7 (t, C-6), 14.0 (q, C-18); CI-mass spectrum, m/z 281 (M+1)$^+$. Anal. Calcd for C$_{18}$H$_{32}$O$_2$: C, 77.09; H, 11.50. Found: C, 76.81; H, 11.44.

4,5-Anhydro-3-deoxy-3-[(phenylmethoxy)methyl]-1,2-bis-O-(phenylmethyl)galactiol (Stoichiometric Asymmetric Epoxidation of a Chiral E Allylic Alcohol).[87]

(2E,4R,5R)-5,6-bis(Phenylmethoxy)-4-[(phenylmethoxy)-methyl]-2-hexen-1-ol (12.5 g, 28.7 mmol) was dissolved in anhydrous dichloromethane (287 mL) and cooled to $-20°$ under argon. To this stirred solution were sequentially added (S,S)-$(-)$-diethyl tartrate (8.88 g, 43.1 mmol), titanium tetraisopropoxide (9.79 g, 34.4 mmol), and *tert*-butyl hydroperoxide (21.0 mL of a 3.4 M solution in dichloromethane, 63.1 mmol). The reaction mixture was kept at $-20°$ for 16 hours, quenched at that temperature with 10% aqueous tartaric acid solution (75 mL), and vigorously stirred for 1 hour at $25°$. The resulting precipitate was filtered (Celite), and the filtrate was dried over Na_2SO_4. Filtration followed by concentration gave an oily residue, which was diluted with diethyl ether (250 mL), cooled to $0°$, and treated with NaOH solution (1 N, 80 mL). The resulting two-phase mixture was vigorously stirred at $0°$ for 30 minutes, and the organic phase was separated and washed with aqueous HCl (1 N, 30 mL), saturated aqueous $NaHCO_3$ (2 × 25 mL), and brine (50 mL). Drying ($MgSO_4$), concentration and flash column chromatography (silica, 30% ethyl acetate in petroleum ether) gave 9.65 g (75%) of the product, $[\alpha]_D^{20}$ +13.6° (c 0.73, MeOH); IR (film) 3460, 3080, 3060, 3030, 2940, 2860, 1600, 1580, 1500, 1480, 1450, 1360, 1200, 1100, 1020, 900, 670 cm^{-1}; ^{1}H NMR (C_6D_6) δ 7.40–7.05 (m, 15 H, aromatic), 4.82 (d, J = 11.6 Hz, 1 H, PhCH$_2$O), 4.56 (d, J = 11.6 Hz, 1 H, PhCH$_2$O), 4.37 (s, 2H, PhCH$_2$O), 4.19 (s, 2 H, PhCH$_2$O), 4.05 (m, 1 H, CHO), 3.75–3.50 (m, 5 H, CHO, CH$_2$O), 3.06 (dd, J = 8.6, 2.1 Hz, 1 H, CH epoxide), 2.83 (m, 1 H, CH epoxide), 1.89–1.81 (m, 1 H, CH), 1.61 (br s, 1 H, OH); CI-mass spectrum, m/z calculated for $C_{28}H_{32}O_5$ + H, 449.2328, found 449.2321 (M + H).

(2S,3S)-2,3-Epoxy-2-methyl-1-butanol (Stoichiometric Asymmetric Epoxidation of a 2,3-Disubstituted E Allylic Alcohol).[347] To a solution of (E)-2-methyl-2-buten-1-ol (2.57 g, 29.8 mmol) in dichloromethane (100 mL) at $-20°$ were added (R,R)-$(+)$-diethyl tartrate (1.43 mL, 1.78 g, 8.35 mmol) and titanium tetraisopropoxide (1.77 mL, 1.69 g, 6.00 mmol). After 15 minutes, the solution was cooled to $-30°$, and *tert*-butyl hydroperoxide (8.2 mL of a 4.0 M solution in benzene) was added in one portion. The solution was stirred at -40 to $-30°$ for 30 minutes and then placed in a $-20°$ freezer. After 10 hours at $-20°$, dimethyl sulfide (3.6 mL) was added, and the reaction mixture was kept at $-20°$ an additional 5 hours. Triethanolamine (9 mL of a 1.0 M solution in dichloromethane) was added, and the solution was stirred for 30 minutes at $0°$. The solution was filtered through silica gel (75 g) in a sintered glass funnel and eluted with diethyl ether (500 mL). Concentration of the filtrate and bulb-to-bulb distillation (oven temperature 120°, 5 mm Hg) of the residue gave 2.33 g (77%) of the product as a colorless oil, $[\alpha]_D^{20}$ $-22.2°$ (c 3.0, CH$_2$Cl$_2$) (94% ee); IR (film) 3420 (br), 3000, 2970, 2930, 2880, 1460, 1380, 1030, 855 cm^{-1}; ^{1}H NMR (CDCl$_3$) δ 3.68 and 3.55 (AB quartet, 2 H, J = 12.3 Hz, CH$_2$OH), 3.16 (q, 1 H, J = 6.6 Hz, CHCH$_3$), 2.64 (br s, 1 H, OH), 1.32 (d, 3 H, J = 6.6 Hz, CHCH$_3$), 1.29 (s, 3 H, CHCH$_3$); ^{13}C NMR (CDCl$_3$) d 65.5, 60.9, 55.9, 13.8, 13.4.

$$\text{TBDPSO}\diagup\diagdown\diagup\overset{\overset{\displaystyle O}{\diagup\!\!\diagdown}}{}\diagup\text{OH}$$

5-[(*tert*-Butyldimethylsilyl)oxy]-3-methyl-(2*R*,3*R*)-2,3-epoxypentan-1-ol (Stoichiometric Asymmetric Epoxidation of a 3,3-Disubstituted Allylic Alcohol).[255] Freshly distilled (*S*,*S*)-(−)-diethyl tartrate (3.6 mL, 20.6 mmol) was added to a −20° solution of titanium tetraisopropoxide (4.13 mL, 13.8 mmol) in dichloromethane (60 mL). The solution was stirred for 20 minutes at −20°. A solution of (2*E*)-5-[(*tert*-butyldimethylsilyl)oxy]-3-methyl-2-penten-1-ol (3.02 g, 13.1 mmol) in dichloromethane (50 mL) followed by *tert*-butyl hydroperoxide (26.2 mmol, 6.6 mL of 3.98 M solution in toluene) was then added dropwise. The mixture was left in a −20° freezer for 17 hours before being quenched with saturated aqueous Na$_2$SO$_4$ (5 mL). The mixture was stirred vigorously and then filtered through Celite and evaporated. The residue was partitioned between diethyl ether (80 mL) and saturated aqueous NaCl (80 mL) and then cooled to 0°. Aqueous NaOH (10 mL of 0.5 N solution) was added and the mixture was stirred vigorously for 1.5 hours at 0° until analytical TLC showed complete disappearance of (*S*,*S*)-(−)-diethyl tartrate. The layers were separated and the aqueous phase was washed with ether (50 mL). The combined organic layers were dried (Na$_2$SO$_4$), filtered, and evaporated. The crude epoxide was purified by flash chromatography (50-mm column, 1:1 hexane–ether) to afford 2.63 g (81%) of the product, $[\alpha]_D^{19}$ +2.1° (*c* 1.59, CHCl$_3$) (>95% ee); IR (film) 3340, 2950, 2926, 2858, 1470, 1256, 1098, 872, 772 cm^{-1}; ^{1}H NMR (CDCl$_3$) δ 3.91–3.63 (m, 4 H, H$_1$ and H$_5$), 3.06 (dd, *J* = 7, 4 Hz, 1 H, H$_2$), 1.90 (dt, *J* = 14, 6 Hz, 1 H, H$_{4a}$), 1.71 (t, *J* = 7 Hz, 1 H, OH), 1.65 (dt, *J* = 14, 6 Hz, 1 H, H$_{4b}$), 1.34 (s, 3H, H$_6$), 0.90 (s, 9H, *t*-Bu), 0.06 (s, 6 H, SiMe$_2$); ^{13}C NMR (CDCl$_3$) δ 63.2, 61.3, 59.8, 59.5, 41.4, 25.8, 18.1, 17.2, −5.5; mass spectrum, m/z 246 (M$^+$). Anal. Calcd for C$_{12}$H$_{26}$O$_3$Si: C, 58.49; H, 10.63. Found: C, 58.28; H, 10.89.

$$\text{Ph}\diagdown\overset{\overset{\displaystyle Ph}{}}{}\overset{\overset{\displaystyle O}{\diagup\!\!\diagdown}}{}\diagup\text{OH}$$

(2*R*,3*R*)-3,4-Diphenyl-2-methyl-2,3-epoxybutan-1-ol (Stoichiometric Asymmetric Epoxidation of a 2,3,3-Trisubstituted Allylic Alcohol)[411] To a stirred solution of titanium tetraisopropoxide (698 mg, 2.45 mmol) in dichloromethane (20 mL) at −20° was added (*S*,*S*)-(−)-diethyl tartrate (698 mg, 3.07 mmol). The pale yellow solution was stirred at −20° for 5 minutes followed by the addition of (*E*)-3,4-diphenyl-2-methyl-2-buten-1-ol (487 mg, 2.04 mmol) and *tert*-butyl hydroperoxide (0.63 mL of 6.54 M solution in dichloromethane). The solution was stirred at −20° for 5 hours and the reaction was stopped by addition of saturated Na$_2$SO$_4$ (0.5 mL) and diethyl ether (1.5 mL). The mixture was warmed to room temperature and stirred for 3 hours, then filtered through Celite, dried over MgSO$_4$, filtered, and evaporated to give a colorless oil. This residue was dissolved in diethyl ether (20 mL) and this solution stirred vigorously for 30 minutes with a 10% solution of NaOH in saturated NaCl at room temperature. The organic phase was separated, washed with NaCl (2 × 10 mL), dried over MgSO$_4$, filtered, and evaporated. The crude product was purified by flash chromatography to afford 469 mg of the product as a colorless oil (94% ee): IR (film) 3420 (br), 3080, 3060, 3030, 2960, 2920, 1600, 1580, 1490, 1450, 1445, 1375, 1265 cm^{-1}; ^{1}H NMR (C$_6$D$_6$) δ 6.87–7.15 (m, 10 H), 3.71 (br s, 2 H), 3.14 (ABq, 2 H), 1.67 (s, 1 H, D$_2$O exchange), 1.00 (s, 3 H).

(2S,3S)-Epoxycinnamyl Alcohol [(2S-*trans*)-3-Phenyloxiranemethanol] (Catalytic Asymmetric Epoxidation of an Aromatic E Allylic Alcohol).[18] A flame-dried, 5-L, three-necked flask was fitted with an overhead mechanical stirrer, thermometer, and dropping funnel, flushed with nitrogen, and charged with (R,R)-(+)-diethyl tartrate (6.55 g, 0.028 mol) and dichloromethane (3.5 L). After the mixture was cooled to −20°, activated powdered 4 Å molecular sieves (20 g), titanium tetraisopropoxide (5.55 mL, 5.30 g, 0.019 mol), and *tert* butyl hydroperoxide (96.9 mL of 7.7 M solution in dichloromethane, 0.746 mol) were added sequentially. The mixture was stirred at −20° for 1 hour and a solution of freshly distilled (E)-3-phenylpropen-1-ol (cinnamyl alcohol) (50.0 g, 0.373 mol) in dichloromethane (70 mL) was added dropwise over a period of 1 hour. After 3 hours at −20°, the reaction was quenched at −20° with 30 mL of a 10% aqueous solution of sodium hydroxide saturated with sodium chloride. After diethyl ether (400 mL) was added, the cold bath was allowed to warm to 10°. Stirring was maintained at 10°, while MgSO$_4$ (30 g) and Celite (4 g) were added. After a final 15 minutes of stirring, the mixture was allowed to settle and the clear solution was filtered through a pad of Celite and washed with diethyl ether. Azeotropic removal of the *tert*-butyl hydroperoxide with toluene at reduced pressure and finally subjection of the residue to high vacuum (0.2 mm Hg) gave a yellow oil. Recrystallization of this material from petroleum ether/ethyl ether at −20° gave 50.0 g (89%) of slightly yellow crystals, mp 51.5–53°, $[\alpha]_D^{25}$ −49.6° (c 2.4, CHCl$_3$) (>98% ee); IR (CHCl$_3$) 3580, 3450, 2980, 2920, 2870, 1600, 1450, 1380, 1100, 1070, 1020, 880, 860, 845 cm^{-1}; ^{1}H NMR (CDCl$_3$) δ 7.2–7.5 (m, 5 H), 4.18 (dd, 1 H, J = 3, 13 Hz), 3.95 (d, 1 H, J = 3 Hz), 3.81 (dd, 1 H, J = 5.13 Hz), 3.25-3.3 (m, 1 H), 2.2 (br s, 1 H, w$_{1/2}$ = 40 Hz).

(S)-2,3-Epoxypropanol (Glycidol) (Catalytic Epoxidation to Obtain a Water-Soluble Epoxy Alcohol).[18,304] To a stirred mixture of powdered acti-

vated 3 Å molecular sieves (3.5 g) and dichloromethane (190 mL) were added
(*R,R*)-(+)-diisopropyl tartrate (1.39 g, 1.25 mL, 5.95 mmol) and allyl alcohol
(5.81 g, 6.8 mL, 100 mmol, stored over 3 Å sieves). The solution was stirred and
cooled to −5°, and titanium tetraisopropoxide (1.4 g, 1.5 mL, 5 mmol) was
added. The reaction mixture was stirred at −5 ± 2° for 10–30 minutes, and then
80% technical grade cumene hydroperoxide (36 mL, ≈200 mmol, dried over 3 Å
sieves prior to use) was added slowly (30 minutes) via a dropping funnel. The
mixture was stirred under N_2 at −5 ± 2° for 5 hours. Triethanolamine (10 mL of
1 M solution in dichloromethane) was added, and the mixture was stirred for
30 minutes at −5°. The cooling bath was removed, and after 15 minutes the mixture was filtered through a pad of Celite over a layer (0.5 cm) of flash silica gel.
The pad was rinsed with diethyl ether (500 mL). After the solvent was removed
on a rotary evaporator, the crude product was distilled quickly through a Bantamware simple distillation apparatus, collecting all material with a boiling point
of about 50° at 7–8 mm Hg of pressure. This distillate was redistilled through a
Bantamware 10-cm Vigreux column, and 3.77 g (51%) of the product was col-
lected at bp 49–50° (7.7 mm Hg). $[\alpha]_D$ −13.02° (neat) (>86% ee). The ^{1}H NMR
($CDCl_3$) spectrum was identical with that of glycidol except that it showed the
presence of small amounts of cumyl alcohol and cumene.

**(*R*)-2,3-Epoxypropyl *p*-Nitrobenzoate (Catalytic Asymmetric Epoxida-
tion with in situ Derivatization).[18]**   Epoxidation of allyl alcohol was carried
out as described above on a 1.0 mol scale with (*R,R*)-(+)-diisopropyl tartrate.
After 6 hours at −5 ± 2°, the mixture was cooled to −20° and carefully treated
with trimethyl phosphite (180 mL, 189 g, 1.5 mol) added over a period of 1 hour,
taking care that the temperature did not rise above −20°. The mixture was then
treated with triethylamine (170 mL, 123 g, 1.2 mol) and a solution of *p*-nitroben-
zoyl chloride (185.6 g, 1 mol) in dichloromethane (250 mL) and stirred for 1 hour
at 0°. After filtration through a pad of Celite, the filtrate was washed with 10%
aqueous tartaric acid (2 × 250 mL), saturated $NaHCO_3$ (3 × 250 mL), and brine
(2 × 250 mL). The organic phase was dried over Na_2SO_4, filtered through a
small pad of silica gel, and concentrated to an oil (first at 12 mm Hg, then at
0.2 mm Hg at 60°) to remove any remaining cumene, 2-phenylpropan-2-ol,
trimethyl phosphite, and trimethyl phosphate. The oil solidified on standing and
was recrystallized twice from diethyl ether to give 135.7 g (61%) of the product,
mp 59.5–60.0°; $[\alpha]_D^{20}$ −38.8° (*c* 3.02, $CHCl_3$) (92–94% ee); IR: 3120, 2970, 2920,
2860, 1730, 1610, 1520, 1460 cm^{-1}; ^{1}H NMR ($CDCl_3$) δ 8.21–8.37 (m, 4 H), 4.76
(dd, 1 H, *J* = 3, 13 Hz), 4.21 (dd, 1 H, *J* = 7, 13 Hz), 3.37 (m, 1 H), 2.96 (t, 1 H,
J = 5 Hz), 2.77 (dd, 1 H, *J* = 5, 3 Hz). Anal. Calcd for $C_{10}H_9NO_5$: C, 53.81; H,
4.06; N, 6.28. Found: C, 53.68; H, 4.20; N, 6.23.

(1S,2S)-(2-Methyl-1,2-epoxycyclotetradecyl)methanol (Stoichiometric Asymmetric Epoxidation of a Cyclic Allylic Alcohol).[412] To dichloromethane (80 mL) at −23° solution under an argon atmosphere was added titanium tetraisopropoxide (2.76 mL, 9.28 mmol), followed by dropwise addition of (R,R)-(+)-diethyl tartrate (2.07 mL, 12.07 mmol). The mixture was stirred for 5 minutes, and then (E)-(2-methyl-1-cyclotetradecenyl)methanol (2.211 g, 9.28 mmol) in dichloromethane (15 mL) was added dropwise followed by anhydrous *tert*-butyl hydroperoxide (3.18 mL of 3.5 M solution in 1,2-dichloroethane, 11.14 mmol) at −23°. The reaction mixture was stirred for 1.5 hours at −23° and then stored in a freezer at −30° for 18.5 hours. The mixture was poured into a −23° solution of water (5.0 mL) and acetone (195 mL) and stirred at −23° for 45 minutes and at room temperature for 2 hours. The clear solution was filtered through Celite, concentrated, and extracted with dichloromethane. The combined extracts were dried over potassium carbonate, filtered, and concentrated under reduced pressure. Flash chromatography on silica gel (7:1 hexane–diethyl ether) afforded 1.815 g (77%) of the product as a white solid, mp 62.5–65.0°; $[\alpha]_D^{25}$ −7.13° (c 3.49, CHCl$_3$) (>90% ee). IR (film) 3350, 2920, 2850, 1470, 1050 cm^{-1}; ^{1}H NMR δ 1.23–1.60 (env, ring CH$_2$), 1.61–2.13 (m), 2.26 (t, J = 6.0 Hz, CH$_2$), 3.80 (ABq, J = 3.6 Hz, CH$_2$O). Anal. Calcd for C$_{16}$H$_{30}$O$_2$: C, 75.54; H, 11.89. Found, C, 75.47; H, 11.94.

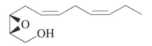

(2R,3S)-(5Z,8Z)-2,3-Epoxy-5,8-undecadien-1-ol (Stoichiometric Asymmetric Epoxidation of a Z-Allylic Alcohol and Derivatization to Increase Optical Purity).[110] Titanium tetraisopropoxide (2.15 mL, 7.23 mmol) was added dropwise to stirred dry dichloromethane (60 mL) at −23° under an argon atmosphere. To this was added (S,S)-(−)-diethyl tartrate (1.74 g, 8.43 mmol). The mixture was stirred for 5 minutes at 23°. A solution of (2Z,5Z,8Z)-2,5,8-undecatrien-1-ol (1.0 g, 6.02 mmol) in dry dichloromethane (2 mL) and 4.4 M *tert*-butyl hydroperoxide (3.42 mL, 15.1 mmol, dichloromethane) were added to the stirred mixture at −23°. The mixture was left to stand for 42 hours at −23°. The reaction was quenched by the addition of 10% aqueous tartaric acid (15 mL) at −23° with stirring. After stirring for 30 minutes at −23°, the mixture was warmed to room temperature and stirred for 1 hour. The organic layer was separated, washed with water, dried (Na$_2$SO$_4$), and concentrated under reduced pressure. The residue was purified by silica gel chromatography and distilled (bp 96–97°/0.15 mm Hg) to afford 886 mg (81%) of the product, $[\alpha]_D^{23}$ +9.27° (c 0.90,

CHCl$_3$) (81% ee). To a solution of this material (852 mg, 4.68 mmol) in dry diethyl ether (20 mL) was added 3,5-dinitrobenzoyl chloride (1.4 g, 6.09 mmol) and dry pyridine with stirring and ice cooling. The stirring was continued overnight at 0°. The mixture was poured into ice water and extracted with diethyl ether. The ether solution was washed with saturated aqueous CuSO$_4$, water and brine, dried (MgSO$_4$), and concentrated under reduced pressure. This residue was repeatedly recrystallized from n-hexane-ether (9:1) until constant optical rotation was reached, giving 793 mg (45.1 mmol) of pure (2R,3S,5Z,8Z)-2,3-epoxy-5,8-undecadienyl 3, 5-dinitrobenzoate as colorless leaflets, mp 38–39°; $[\alpha]_D^{22}$ +5.68° (c 0.37, diethyl ether). K$_2$CO$_3$ (60 mg, 0.43 mmol) was added to a solution of (2R,3S,5Z,8Z)-2,3-epoxy-5,8-undecadienyl 3,5-dinitrobenzoate (610 mg, 1.62 mmol) in methanol (6 mL) with stirring and ice cooling. The stirring was continued for 30 minutes at 0°. The mixture was concentrated under reduced pressure to remove MeOH. The residue was diluted with water and extracted with diethyl ether. The ether solution was washed with water and brine, dried (MgSO$_4$), and concentrated under reduced pressure. The residue was chromatographed over silica gel and distilled (bp 90–91°/0.11 mm Hg) to give 257 mg (87%) of the product, $[\alpha]_D^{22}$ +11.6° (c 0.84, CHCl$_3$) (>99% ee); IR 3450 (br), 3040, 2990, 2960, 2900, 1655, 1460, 1400, 1040, 975, 920 cm^{-1}; ^{1}H NMR (CCl$_4$) δ 0.96 (t, $J = 7$ Hz, 3 H), 1.70–2.50 (m, 4 H), 2.50–3.20 (m 4 H), 3.25–3.80 (m, 3 H), 5.0–5.7 (m, 4 H); Anal. Calcd for C$_{11}$H$_{18}$O$_2$: C, 72.49; H, 9.96. Found: C, 72.22; H, 10.11.

(2R,3S)-1,2-Epoxy-4-penten-3-ol [Catalytic Asymmetric Epoxidation and Kinetic Resolution of a *meso*-Allylic Alcohol].[279]   To a −23° (dry ice–CCl$_4$) cooled mixture of powdered 4 Å molecular sieves (4.3 g) and dichloromethane (250 mL) were added titanium tetraisopropoxide (5.0 mL, 16.8 mmol) and (R,R)-(+)-diisopropyl tartrate (4.50 mL, 21.5 mmol) via syringe. After stirring at −23° for 10 minutes, divinyl carbinol (20 g, 238 mmol) was added via cannula, followed by *tert*-butyl hydroperoxide (160 mL of 3.0 M solution in isooctane, 480 mmol) in several portions via syringe. The reaction vessel was then placed in a −15° freezer and stored for 118 hours. At freeze temperature, aqueous Na$_2$SO$_4$ (17 mL) was added, and the mixture was diluted with diethyl ether (250 mL). After stirring at ambient temperature for 2 hours, the resulting slurry was filtered through a pad of Celite, washing with several portions of ether. The Celite pad was transferred to an Erlenmeyer flask, heated gently with diethyl ether, and the supernatant filtered through a second fresh pad of Celite. Most of the solvent was removed on a rotary evaporator with cooling to avoid undue loss of the somewhat volatile product. The resulting oil was subjected to flash column chromatography followed by distillation at aspirator pressure to afford 23.6 g of the desired epoxide, which was about 50% pure by NMR. A second flash column chromatography (3:1 pentane–ether) in two batches provided 13.1 g (50%) of the product as a clear, colorless oil, $[\alpha]_D^{23}$ +48.8° (c 0.73, CHCl$_3$) (>99% ee), IR (film) 3400 (br), 3019, 1108 cm^{-1}; ^{1}H NMR (CDCl$_3$) δ 5.75-5.89 (ddd, J = 17.2, 10.5, 6.2 Hz, 1 H), 5.34 (dt, J = 17.2, 1.4 Hz, 1 H), 5.22 (dt, J = 10.5, 1.3 Hz, 1 H), 4.26 (br s, 1 H), 3.05 (dd, J = 3.3, 3.1 Hz, 1 H), 2.70–2.78 (m, 2 H), 2.61 (br s, 1 H); ^{13}C NMR (CDCl$_3$) δ 135.6, 117.3, 70.3, 53.9, 43.5; mass spectrum m/z (percent) 99 (M-H, 0.4), 57 (100).

(1E,3R)-1-Trimethylsilyl-1-octen-3-ol and **(1S,2S,3S)-1-Trimethylsilyl-1,2-epoxyoctan-3-ol (Stoichiometric Kinetic Resolution of a Secondary E Allylic Alcohol).**[13] To a solution of titanium tetraisopropoxide (4.6 mL, 15.6 mmol) in dichloromethane (140 mL) was added (R,R)-(+)-diisopropyl tartrate (3.94 mL, 18.72 mmol) and the resulting solution was stirred for 10 minutes at −20°. After addition of (E)-1 trimethylsilyl-1-octen-3-ol (6.29 g, 15.6 mmol), the mixture was stirred for an additional 10 minutes. tert-Butyl hydroperoxide (4.46 mL of 3.5 M solution in dichloromethane, 15.6 mmol, 1 equiv) was added slowly and the solution was stirred at −20° for 7 hours. Dimethyl sulfide (3.94 mL, 46.8 mmol) was added slowly and the mixture was stirred for 30 minutes at −20°. Tartaric acid aqueous solution (10%, 280 mL), diethyl ether (280 mL), NaF (10.9 g), and Celite (6.2 g) were added sequentially. The resulting mixture was stirred for 30 minutes at room temperature, filtered through a pad of Celite with diethyl ether, and concentrated. The residue was chromatographed on triethylamine-deactivated silica gel to afford 2.64 g (42%) of (R)-(E)-1-trimethylsilyl-1-octen-3-ol, $[\alpha]_D^{25}$ −9.8° (c 1.10, CHCl$_3$) (>99% ee), and 2.75 g (42%) of (1S,2S,3S)-1-trimethylsilyl-1,2-epoxyoctan-3-ol, $[\alpha]_D^{25}$ −7.5° (c 1.04, CHCl$_3$) (>99% ee), IR (neat) 3420, 2390, 1247 cm^{-1}; ^{1}H NMR (CCl$_4$, C$_6$H$_6$, D$_2$O) δ 0.03 (s, 9 H), 0.93 (t, 3 H, J = 4.8 Hz), 1.07–1.72 (m, 8 H), 2.26 (d, 1 H, J = 4.0 Hz), 2.71 (t, 1 H, J = 4.0 Hz), 3.30–3.70 (m, 1 H).

(R)-Non-1-en-3-ol (Catalytic Kinetic Resolution of a Secondary Allylic Alcohol).[18] To a room-temperature solution of (±)-1-nonen-3-ol (284 mg, 2 mmol) and (R,R)-(+)-dicyclododecyl tartrate (135.6 mg, 0.3 mmol) in dichloromethane (8 mL) were added powdered and activated 3 Å molecular sieves (60 mg) and a saturated hydrocarbon internal standard (n-decane, 80 μL) for gas chromatographic (GC) monitoring of percent conversion. The stirred mixture, maintained under an inert atmosphere, was cooled to −10 to −20°, treated with titanium tetraisopropoxide (60 μL, 0.2 mmol) and allowed to stir for 20–30 minutes at $-20°$. During this time, a small aliquot (ca. 100 μL) was removed, diluted with 100 μL of ether, and quenched into 200 μL of a freshly prepared aqueous solution of $FeSO_4 \cdot 7H_2O$ (330 mg, 1.2 mmol) and citric acid (110 mg, 0.6 mmol) diluted to 10 mL with deionized water at $\sim 0°$, to provide a T_0 GC reference sample. The reaction was then treated with *tert*-butyl hydroperoxide [190 μL, 5.8 M solution in isooctane (dried with freshly activated 3 Å pellets for 30 minutes prior to use), 1.1 mmol] added by gastight syringe. The reaction was maintained at $-22 \pm 2°$ and periodically monitored by GC. After 13 days (51% conversion), the reaction was quenched with 2 mL of an aqueous solution of $FeSO_4$ and citric acid at $-20°$ and stirred vigorously without cooling for 30 minutes until two clear phases appeared. The phases were separated and the aqueous phase was extracted twice with dichloromethane. The combined organic phases were washed with saturated brine and dried ($MgSO_4$). After precipitation of most of the crystalline dicyclododecyl tartrate, the crude product was then purified by flash chromatography to give 95 mg of (R)-non-1-en-3-ol (99% based on percent conversion, 33% overall), $[\alpha]_D^{25}$ $-19.1°$ (c 6.7, EtOH) ($>98\%$ee).

(R)-1-(2-Furyl)hexan-1-ol (Catalytic Kinetic Resolution of a Secondary Furyl Carbinol).[413] To a mixture of crushed 4 Å molecular sieves (5 g) and 0.2 equiv of titanium tetraisopropoxide (7.35 mL, 24.7 mmol) in dichloromethane (100 mL) at $-21°$ was added 0.24 equiv of (R,R)-(+)-diisopropyl tartrate (6.23 mL, 29.6 mmol). The mixture was stirred for 10 minutes at $-21°$ and cooled to $-30°$. ($\pm$)-1-(2-Furyl)hexan-1-ol (20.7 g, 123 mmol) dissolved in dichloromethane (20 mL) was added and the mixture was stirred at temperatures between -30 and $-20°$ for 30 minutes. The mixture was cooled again to $-30°$, and 0.6 equiv of *tert*-butyl hydroperoxide (17.0 mL, 4.35 M solution in dichloromethane, 74.0 mmol) was added slowly. After stirring for 14 hours at $-21°$, dimethyl sulfide (5.43 mL, 74.0 mmol) was added slowly and the mixture was stirred for 30 minutes at $-21°$. To this mixture were added 10% aqueous tartaric acid (5 mL), diethyl ether (100 mL), and NaF (30 g), and the resulting mixture was vigorously stirred for 2 hours at room temperature. The white precipitate was filtered through a pad of Celite and washed with diethyl ether (100 mL). The filtrate was concentrated to give an oil, which was dissolved in diethyl ether (200 mL) and treated with NaOH (3 N, 100 mL) for 30 minutes at $0°$ with vigorous stirring. The organic layer was washed with brine, dried ($MgSO_4$), and concentrated to give an oil, which was passed through a short silica gel column to afford 7.94 g (38 %) of the product, $[\alpha]_D^{25}$ $+13.8°$ (c 1.07, $CHCl_3$) (>95 % ee).

Ethyl (3R)-3-Hydroxy-4-methyl-4-pentenoate (Stoichiometric Kinetic Resolution of a Secondary 2-Substituted Allylic Alcohol).[414] A solution of (R,R)-(+)-diisopropyl tartrate (42 g, 179 mmol) in anhydrous dichloromethane (1 L) at $-25°$ was treated with freshly distilled titanium tetraisopropoxide (40.8 g, 143 mmol). After 15 minutes, ($\pm$)-ethyl 3-hydroxy-4-methyl-4-pentenoate (18.9 g, 119 mmol) was added, and the mixture was stirred for an additional 15 minutes. tert-Butyl hydroperoxide (358 mmol, 125 mL of a 2.86 M solution in 1,2-dichloroethane) was added slowly, and the reaction temperature was maintained at $-25°$ for 96 hours (at which time the reaction was 57% complete by GC analysis). The reaction mixture was poured into a stirred solution of water (50 mL) in acetone (2 L) at $-50°$ and, after being allowed to reach ambient temperature, was filtered through Celite. Removal of the solvents in vacuo followed by fractional distillation (60–65°, 1 mm Hg) gave a mixture of ethyl (3R)-3-hydroxy-4-methyl-4-pentenoate and the corresponding epoxy ester, which were separated by silica gel flash chromatography to afford 2.95 g (16 %) of the pentenoate as a clear oil, $[\alpha]_D^{25}$ $+14.6°$ (c 2.31, EtOH) (>99% ee).

4-(n-Decyloxy)phenyl 4-[(1S,2S,3S)-2,3-Epoxy-1-hydroxynonyl]-benzoate (Stoichiometric Kinetic Resolution of a Secondary Aromatic Allylic Alcohol).[415] To a cooled solution ($-30°$) of titanium tetraisopropoxide (5.49 g, 18.44 mmol) in dry dichloromethane (100 mL) was added with stirring (R,R)-(+)-diisopropyl tartrate (4.65 g, 22.1 mmol) . The mixture was stirred for 20 minutes. A precooled $-20°$ solution of 4'-(n-decyloxy)phenyl 4-[(E)-1-hydroxy-2-nonenyl]benzoate (9.12 g, 18.44 mmol) in dry dichloromethane (50 mL) was added, and dichloromethane (34 mL) was used to transfer all residual allylic alcohol. The reaction mixture was stirred for 10 minutes before tert-butyl hydroperoxide (3.32 mL of a 2.5 M solution in toluene, 8.3 mmol) was added. The reaction flask was stored in a freezer ($-25°$) for 18 hours. The reaction mixture was then poured over an ice-cold solution of $FeSO_4$ (9.22 g) and water (37 mL). The mixture was then warmed to ambient temperature and stirred for 1 hour. The two layers were separated, and the aqueous layer was washed three times with 50 mL portions of diethyl ether. The combined organic layers were dried over Na_2SO_4, and the solvent was removed in vacuo. The oil obtained was dissolved in a minimum amount of 20% ethyl acetate/hexane solution and was filtered through a pad of silica gel, which was then washed thoroughly with 20% ethyl acetate/hexane. The combined filtrate was concentrated and the crude product was purified by flash chromatography on silica gel with 25% ethyl acetate/hexane as eluant. The product was then recrystallized from hexane at $-20°$. The gel obtained was centrifuged at $-20°$ and the solvent was decanted to afford 3.73 g (79%) of the product, mp 68.9–69.5°; IR (CHCl$_3$) 3350, 3010, 2930, 2850, 1736, 1510, 1270, 1235, 1185 cm^{-1}; ^{1}H NMR (CDCl$_3$) δ 0.87 [m 6 H, (RCH$_3$)$_2$], 1.15–1.60 [m, 24 H, ArOCH$_2$CH$_2$(CH$_2$)$_7$CH$_3$, CHOCH(CH$_2$)$_5$CH$_3$], 1.73 (quint, J = 6.8 Hz, 2 H, ArOCH$_2$CH$_2$), 2.40 (d, J = 1.8 Hz, 1 H, ArCHOH), 3.0 (t, J = 3.0 Hz, ArCHOHCHOCH, 1 H), 3.12 (m, 1 H, ArCHOHCHOCH), 3.94 (t, J = 6.5 Hz, 2 H, ArOCH_2), 4.97 (s, 1 H, ArCHOH), 6.91 (d, J = 9.1 Hz, 2 H, ArH), 7.08 (d, J = 9.0 Hz, 2 H, ArH), 7.50 (d, J = 8.2 Hz, 2 H, ArH), 8.18 (d, J = 8.3 Hz, 2 H, ArH); ^{13}C NMR δ 13.9, 14.0, 22.4, 22.6, 25.8, 26.0, 28.8, 29.3, 29.4, 29.5, 31.3, 31.6, 31.9, 55.3, 61.0, 68.6, 70.7, 115.2, 115.3, 126.2, 129.5, 130.4, 144.4, 145.6, 157.0, 165.2; mass spectrum m/z (percent) 510 (M$^+$, 0.49), 147.1 (92), 133.1 (100). Anal. Calcd for C$_{32}$H$_{46}$O$_5$: C, 75.42; H, 9.10. Found: C, 75.33; H, 9.15.

(R)-2-[1-(Phenyl)ethyl]propenol (Stoichiometric Kinetic Resolution of a 2-Substituted Chiral Primary Allylic Alcohol).[416] To a cooled ($-20°$), stirred solution of titanium tetraisopropoxide (5.95 mL, 20 mmol) in dry dichloromethane (200 mL) were added (R,R)-(+)-diisopropyl tartrate (5.616 g, 24 mmol) and (±)-2-[1-(phenyl)ethyl]propenol (3.22 g, 20 mmol). The mixture was stirred for 10 minutes and treated with *tert*-butyl hydroperoxide (3.75 mL of a 3.73 M solution in dry dichloromethane, 14 mmol). The reaction was stored in a freezer ($-20°$) for 18 hours. Acetone (50 mL) was added at $-20°$ and the mixture was stirred for 5 minutes. Then water (6 mL) was added and the reaction mixture was vigorously stirred at ambient temperature for 1 hour. The resulting slurry was filtered through a pad of Celite, washing with several portions of diethyl ether. The filtrate was concentrated to give an oil, which was dissolved in ether (50 mL) and treated with NaOH (15% in brine, 20 mL) for 30 minutes at 0° with vigorous stirring. The organic layer was washed with brine, dried (MgSO$_4$), and concentrated to give an oil, which was passed through a silica gel column to afford 700 mg (35%) of the product, $[\alpha]_D^{25}$ +98.0° (c 3.20, EtOH) (>95% ee).

(R)-N,N-Dimethyl(2-hydroxydecyl)amine (Kinetic Resolution of Racemic β-Hydroxyamines).[21] To a mixture of (±)-N,N-dimethyl(2-hydroxydecyl)amine (404 mg, 2.01 mmol) and (R,R)-(+)-diisopropyl tartrate (2.43 mmol, 1.21 equiv) was added dichloromethane (20 mL) followed by titanium tetraisopropoxide (1.20 mL, 4.09 mmol, 2.04 equiv). The mixture was stirred for 30 minutes at room temperature. After this aging period, the flask was cooled, while stirring, in a dry ice/CCl$_4$ bath (ca. $-20°$). To this solution was added *tert*-butyl hydroperoxide (365 mL, 1.20 mmol, 3.29 M solution in toluene, 0.6 equiv). After stirring for 2 hours at $-20°$, the reaction was quenched by adding diethyl ether (20 mL), water (0.8 mL), and a 40% NaOH solution (0.8 mL). This mixture was vigorously stirred for 4–5 hours at room temperature, yielding a gelatinous precipitate which was filtered through a pad of Celite. The precipitate was stirred vigorously in refluxing chloroform for 3 minutes before filtering it again through the Celite pad. The combined filtrates were concentrated to leave a pale yellow viscous oil, which was dried under high vacuum. This oil was triturated with *n*-hexane (20 mL). The clear supernatant solution was filtered and the filtrate was washed with *n*-hexane (20 mL). The filtered solid was the optically active *N*-oxide of dimethyl(2-hydroxydecyl)amine (233 mg, 54%). The hexane extracts were diluted with ether (5 mL), washed with water (~ 200 mL $\times$ 2), and dried over anhydrous Na$_2$SO$_4$. The solvent was evaporated to afford 144 mg, (36%) of the product as an oil, $[\alpha]_D^{20}$ $-3.58°$ (c 1.65, EtOH) (91% ee).

(R)-Methyl p-Tolyl Sulfoxide (Asymmetric Oxidation of a Sulfide).[23]

Titanium tetraisopropoxide (1.49 mL, 5 mmol) and (R,R)-(+)-diethyl tartrate (1.71 mL, 10 mmol) were dissolved at room temperature in dichloromethane (50 mL) under nitrogen. Water (90 mL, 5 mmol) was introduced via syringe. Stirring was maintained until the yellow solution became homogeneous (15–20 minutes) and 0.7 g (5 mmol) of methyl p-tolyl sulfide was added. The solution was cooled to $-20°$ and *tert*-butyl hydroperoxide (2.75 mL, 2 M solution in dichloromethane, 5.5 mmol) was then introduced. After the reaction was completed (4 hours), water (0.9 mL, 10 equiv) was added dropwise via microsyringe to the solution at $-20°$. The mixture was stirred vigorously for 1 hour at $-20°$ and for 1 hour at room temperature. The white gel was filtered (a small amount of alumina added to the solution helps the filtration) and thoroughly washed with dichloromethane. The filtrate was kept in the presence of NaOH (5%) and brine for 1 hour and then separated. The organic phase was dried over Na$_2$SO$_4$ and concentrated to give the crude product, which did not contain sulfone. Chromatography (ethyl acetate, cyclohexane 1:1) of the crude product on silica gel afforded 0.70 g (80%) of the product, $[\alpha]_D^{20}$ $+132°$ (c 2, acetone) (90% ee) ($[\alpha]_D^{20}$ $+145.5°$ for enantiomerically pure R isomer).

TABULAR SURVEY

Asymmetric oxidations of allylic alcohols, amines, sulfides, and selenides are grouped in Tables I–XVII. Allylic alcohols are classified as nonchiral or chiral and are further subdivided according to their substitution patterns. Within each table, entries are in the order of increasing number of carbon atoms in the substrate, omitting the carbon atoms in nonreacting groups attached to the substrate through a heteroatom. The literature has been reviewed through April 1992.

Reactions were carried out in dichloromethane using *tert*-butyl hydroperoxide as oxidant, titanium tetraisopropoxide as the titanium tetraalkoxide, and dialkyl tartrate as the chiral auxiliary unless noted otherwise. Workup conditions are not described. A dash in the Conditions column indicates that no conditions were given, and a dash in a Product column of the kinetic resolution tables means that the indicated product was not reported. The symbol (—) after a product means that no yield was reported. Optical rotations were measured with the Na D line in chloroform unless noted otherwise.

The following abbreviations are used in the tables:

Ac	acetyl
Aib	α-aminoisobutyryl
Ala	alanyl
Bn	benzyl
Boc	benzyloxycarbonyl
BOM	benzyloxymethyl
Bz	benzoyl
catalytic	catalytic amounts of Ti(OPr-i)$_4$ and DAT were used
CUHP	cumene hydroperoxide
DAT	dialkyl tartrate
(−)-DAT	(S,S)-(−)-DAT
(+)-DAT	(R,R)-(+)-DAT
DBTA	N,N'-dibenzyltartramide
DCHT	dicyclohexyl tartrate
DCDT	dicyclododecyl tartrate
DET	diethyl tartrate
DIPT	diisopropyl tartrate
DMT	dimethyl tartrate
DNB	3,5-dinitrobenzoyl
EE	1-ethoxyethyl
MEM	methoxyethoxymethyl
MMTr	m-methoxytrityl
MOM	methoxymethyl
MPM	p-methoxyphenylmethyl
MS	molecular sieves
Nap	naphthyl
Nps	naphthalenesulfonyl

Phe	phenylalanyl
PNB	*p*-nitrobenzoyl
Pro	prolyl
Pyr	pyridyl
SEM	(2-trimethylsilylethoxy)methoxy
stoichiometric	stoichiometric amounts of Ti(OPr-*i*)$_4$ and DAT were used
TBDMS	*tert*-butyldimethylsilyl
TBDPS	*tert*-butyldiphenylsilyl
TBHP	*tert*-butyl hydroperoxide
TES	triethylsilyl
THP	tetrahydropyranyl
TI	titanium(IV) isopropoxide
TIPS	triisopropylsilyl
TMS	trimethylsilyl
Tr	trityl
Ts	*p*-toluenesulfonyl

TABLE I. ASYMMETRIC EPOXIDATION OF PRIMARY ALLYLIC ALCOHOLS

Substrate	Conditions TI (eq), DAT (eq), MS	Product(s) and Yield(s) (%)	Refs.
C₃			
⟍OH (allyl alcohol)	TI (0.05), (+)-DIPT (0.06), 5 h[a]	(epoxide)⟍OH (65) 90-92% ee	18, 304
	TI (1.2), (+)-DIPT (1.4), 0°, 24 h	" (15)[b] 73% ee	1
TMS⟍⟍OH	TI (1), (–)-DIPT (1.05), -23°	TMS⟍(epoxide)⟍OH (60) >95% ee	145, 417, 418
	TI (0.03), (+)-DIPT (0.036), -20°, 3.5 d	TMS⟍(epoxide)⟍OH (90) 99% ee, [α] -25.3°	419
TBDMS⟍⟍OH	(+)-DET	TBDMS⟍(epoxide)⟍OH (75) >98% ee, [α] -26.9°	420
TIPS⟍⟍OH	(+)-DET	TIPS⟍(epoxide)⟍OH (70) 98% ee, [α] -12.7°	421
Ph₃Si⟍⟍OH (D,D)	TI (1), (+)-DIPT (1.25), -20°, 16 h	Ph₃Si⟍(epoxide)⟍OH (D, D) (96) 94% ee	422
(Ph₃Si)⟍⟍OH (D)	TI (1), (–)-DIPT (1.25), -20°, 16 h	(Ph₃Si)⟍(epoxide)⟍OH (D) (—) 92% ee	422
C₄			
⟍OH (methallyl)	TI (1), (+)-DET (1), -20°, 3 d	(epoxide)⟍OH (—) 88% ee	423
	Catalytic	" (—) 93% ee	424

TABLE I. ASYMMETRIC EPOXIDATION OF PRIMARY ALLYLIC ALCOHOLS (*Continued*)

Substrate	Conditions: Ti (eq), DAT (eq), MS	Product(s) and Yield(s) (%)	Refs.
	Ti[c] (0.08), (−)-DET (0.1), 4 Å, -40 to -20°	[structure] OH (47) >95% ee, [α] +10.62°	131, 425, 426
	Ti (1), (−)-DET (1), -30°, 40 h	" (32) >94% ee, [α] +9.8°[d]	409, 427, 428
	Ti (0.1), (−)-DIPT (0.12)	" (−) 87% ee	429
	Ti (1.4), (−)-dibutyl tartrate (1.64), SiO2, CaH[a]	" (81) 91% ee	430
CD2OAc OH	—, Stoichiometric	CD2OAc O OH (−) 90% ee	431
CD2OBn OH	Ti (0.08)[c], (+)-DET (0.1), 4 Å, -20°, overnight	CD2OBn O OH (75) [α] -10.9°[d]	132
	Ti (0.08)[c], (−)-DET (0.1), 4 Å, -20°, overnight	CD2OBn O OH (74) [α] +11.0°[d]	131
OH	Ti (0.05), (+)-DIPT (0.06), 3 Å, -20°, 2 h	O OH (70) 91% ee, [α] -50.1°	18
	Ti (1), (+)-DIPT (1), -20°, 20 h	" (58) [α] -49°[e]	432
	(+)-DIPT, -20°, 24 h	(40) 95% ee, [α] -53.1°[e]	433-437
	(+)-DET, (catalytic)	" (−) >91% ee, [α] -55.8°	438
	Ti (0.03)f, (+)-DIPT (0.06), -20 to -15°, 4.2 h	" (91) 95% ee	138

TABLE I. ASYMMETRIC EPOXIDATION OF PRIMARY ALLYLIC ALCOHOLS (*Continued*)

Substrate	Conditions TI (eq), DAT (eq), MS	Product(s) and Yield(s) (%)	Refs.
	TI (1), (–)-DIPT (1), –23°	(45) >95% ee, [α] +55° e	59, 436, 437
	(–)-DET (catalytic, 3 Å, –23°	90%ee	129
	TI (0.05), (+)-DIPT (0.06), 4 Å, –15°, overnight	(52) [α] -54° e	439
	TI (0.05), (+)-DIPT (0.06), 4 Å, –15°, overnight	(43) [α] -51° e	439
	TI (0.05), (–)-DIPT (0.06), 4 Å, –15°, overnight	(43) [α] +48° e	439
	TI (0.05), (+)-DIPT (0.08), –20 to 0°	(85) 88% ee	224, 440
	(+)-DIPT	" (—) >95% ee	441
	(+)-DET (stoichiometric), –20°, 24 h	" (84) 98% ee	73, 161, 162, 171
	(+)-DET (stoichiometric),	" (74) [α] -20.3°	282, 442
	TI (1.1), (–)-DET (1.1), –23°, 5 h	(80) 98% ee [α] +21.8°	85, 87, 295, 443

TABLE I. ASYMMETRIC EPOXIDATION OF PRIMARY ALLYLIC ALCOHOLS (*Continued*)

Substrate	Conditions TI (eq), DAT (eq), MS	Product(s) and Yield(s) (%)	Refs.
MMTrO⌒⌒OH	(+)-DET	MMTrO epoxide OH (34) 97% ee	444
	(−)-DET	MMTrO epoxide OH (79) 93% ee	444
2-(OBn)phenyl-O⌒⌒OH	TI (1), (+)-DET (1), −20°, overnight	epoxide-CH₂O-2-(OBn)phenyl OH (86) 92% ee [α] −15.8°	445, 446
	TI (1), (−)-DET (1), −20°, overnight	epoxide-CH₂O-2-(OBn)phenyl OH (86) >96% ee [α] +16.4°	445, 446
TBDPSO⌒⌒OH	TI (0.14), (+)-DET (0.2), −20°, 3 d	TBDPSO epoxide OH (86) >95% ee [α] −15.3° *d*	247
	Stoichiometric	" (−) >95% ee, [α] −9.6°	438
MeO₂CH⌒⌒OH (MeO, MeO acetal)	TI (1), (+)-DET (1), −23°, 20 h	(MeO)₂CH epoxide OH (81) >97% ee [α] −37.5° *g*	57, 447, 448
Ts(Bn)N⌒⌒OH	TI (1.2), (−)-DIPT (1.5), −20°, 2.5 h	Ts(Bn)N epoxide OH (91) 95% ee [α] +20°	98
Ph₂P(O)⌒⌒OH	TI (1), (+)-DET (1.2), 4 Å, −16°	Ph₂P(O) epoxide OH (76) 82% ee	449

TABLE I. ASYMMETRIC EPOXIDATION OF PRIMARY ALLYLIC ALCOHOLS (Continued)

Substrate	Conditions TI (eq), DAT (eq), MS	Product(s) and Yield(s) (%)	Refs.
NC⌁⌁OH	TI (0.3), (+)-DIPT (0.37), 4 Å, 5°, 10 d	NC⟨O⟩OH (81) 91% ee	106, 108
$C_2H_5O_2C$⌁⌁OH	TI (1), (−)-DIPT (1.2), 3 Å, -25°, 72 h	$C_2H_5O_2C$⟨O⟩OH (85) 90% ee	450
⌁⌁OH	TI (0.05), (+)-DIPT (0.06)	⟨O⟩OH (68) 92% ee	18
BnO⌁⌁OH	TI (0.1), (+)-DET (0.14), 4 Å, -20°, 43 h	BnO⟨O⟩OH (−)[h] 85% ee	18
	TI (1), (−)-DIPT (1), -20°, 40 h	BnO⟨O⟩OH (86) 89% ee [α] +25°	63, 442
	(−)-DET (stoichiometric)	" >95% ee, [α] +25°	451, 73, 353
p-BrC$_6$H$_4$O⌁⌁OH	TI (0.1), (+)-DET (0.13), 4 Å, -20°, 2 h	p-BrC$_6$H$_4$O⟨O⟩OH (73) 100% ee[i], [α] -17.3°[j]	62
p-BrC$_6$H$_4$O⌁⌁OH	TI (0.1), (−)-DET (0.13), 4 Å, -20°, 2 h	p-BrC$_6$H$_4$O⟨O⟩OH (86) [α] +14.7°; (71) 100% ee[i], [α] +17.4°[j]	62

TABLE I. ASYMMETRIC EPOXIDATION OF PRIMARY ALLYLIC ALCOHOLS (*Continued*)

Substrate	Conditions TI (eq), DAT (eq), MS	Product(s) and Yield(s) (%)	Refs.
Ph–CH(Ph)–O–CH₂CH=CHCH₂OH	TI (1.2), (+)-DIPT (1.5), -20°, 18 h	epoxy alcohol (92) [α] -17.6°	162
MMTrO–CH₂CH=CHCH₂OH	(+)-DET	(78) 97% ee	444
MMTrO–CH₂CH=CHCH₂OH	(-)-DET	(73) 94% ee	444
TBDMSO–CH₂CH=CHCH₂OH	(+)-DET, -25°	(90) 98% ee	89
TBDPSO–CH₂CH=CHCH₂OH	(+)-DET	(—) 75% ee [α] -4.4° d	452, 453
Sugar (AcO, AcO, OAc, OAc) glycoside allylic alcohol	TI (0.2), (+)-DET (0.24), 4 Å, -16°, 6 d	(65) 96% de	454
Sugar (AcO, AcO, OAc, OAc) glycoside allylic alcohol	TI (0.2), (-)-DET (0.24), 4 Å, -16°, 6 d	~79) 80% de, [α] -9.8°	454

91

TABLE I. ASYMMETRIC EPOXIDATION OF PRIMARY ALLYLIC ALCOHOLS (*Continued*)

Substrate	Conditions TI (eq), DAT (eq), MS	Product(s) and Yield(s) (%)	Refs.
(per-O-acetyl disaccharide with OAc groups, allylic OH)	TI (0.1), (+)-DET, (0.2), 4 Å	(75) [α] -23.7°	454
	TI (0.1), (−)-DET, (0.2), 4 Å	(65) [α] -17.0°	454
TBDMS allylic alcohol	(+)-DET	(78) >98% ee [α] -24.9°	420
(2-methylene-butanol, OH)	TI (0.048), (+)-DET (0.06), 4 Å, -20 to -15°, 1 h	(97) 86% ee	455-457
	TI (0.048), (−)-DET (0.06), 4 Å, -20 to -15°, 1 h	(97) 86% ee [α] +32°	455-457
C5 (pent-2-en-1-ol, OH)	TI (1), (−)-DIPT (1.2), -23°	(80) >95% ee [α] +32.1°	332

TABLE I. ASYMMETRIC EPOXIDATION OF PRIMARY ALLYLIC ALCOHOLS (*Continued*)

Substrate	Conditions TI (eq), DAT (eq), MS	Product(s) and Yield(s) (%)	Refs.
(structure) OH	Catalytic	(structure) OH (40)[b] 95% ee	19
	(+)-DIPT (catalytic), -25°	" (56) >91% ee, [α] -164°	244
HC≡C (structure) OH	(-)-DET, -28°, 12 h	HC≡C (structure) OH (43) [α] -10.3°	101
BnO (structure) OH	(+)-DIPT, 4 Å	BnO (structure) OH (80) >90% ee [α] -33.3°	458
	TI (1), (-)-DET (1.002), -25°, 12 h	EnO (structure) OH (80) [α] +29.76°	459
	TI (0.1), (-)-DET (0.12), 4 Å, -20°, 12 h	" (89) 91% ee	460-462
	(-)-DIPT, 4 Å	" (72) >90% ee, [α] +32.3°	458
(structure) O (structure) OH	(+)-DET, -20°, 3 h	(structure) OH (72), 98% ee	463
TBDPSO O (structure) OH	(+)-DET (stoichiometric)	TBDPSO O (structure) OH (93) 97% ee, [α] -17.56°[d]	362, 464
	TI (0.06), (+)-DIPT (0.08), 4 Å, -20°, 20 h	" (85) >95% ee	221

93

TABLE I. ASYMMETRIC EPOXIDATION OF PRIMARY ALLYLIC ALCOHOLS (*Continued*)

Substrate	Conditions Ti (eq), DAT (eq), MS	Product(s) and Yield(s) (%)	Refs.
TrO〜OH	(+)-DET (stoichiometric)	TrO—epoxide—OH (—) >95% ee [α] -23.8°	438
Me₂Si(t-Bu)O〜OH	Ti (1.1), (+)-DET (1.3), 4 Å, -23°, 18 h	Me₂Si(t-Bu)O—epoxide—OH (—) 99% ee, [α] -28.5°	465
(same)	Ti (1.1), (—)-DET (1.3), 4 Å, -23°, 18 h	Me₂Si(t-Bu)O—epoxide—OH (78) 96% ee, [α] +28.6°	465
dioxane〜OH	(—)-DIPT (catalytic)[a], 4 Å, -20°, 22 h	dioxane—epoxide—OH (90)	466
MeO/MeO〜OH	Ti (1), (+)-DET (1.2), -20°, 20 h	MeO/MeO—epoxide—OH (81) [α] -37.5°[g]	447
"	(—)-DIPT (stoichiometric), -20°, SiO₂, 2 d	" (74-90) >95% ee	467

94

TABLE I. ASYMMETRIC EPOXIDATION OF PRIMARY ALLYLIC ALCOHOLS (*Continued*)

Substrate	Conditions TI (eq), DAT (eq), MS	Product(s) and Yield(s) (%)	Refs.
BnO, BnO (structure)	(–)-DET, 4 Å, –20°	(structure) (91) >97% ee	468
BnO, BnO, TBDMSO (structure)	TI, (–)-DIPT	(structure) (98) 90% de	228
(structure)	TI (0.05), (+)-DIPT (0.08), –23 to 0°	(structure) (83) 91% de	224
	(+)-DET, –23°, 2 d	" (74) 87% de, [α] -21.5°	160, 2, 73
	TI (0.05), (–)-DIPT (0.08), –23 to 0°	(structure) (73) 92% de	224
	(–)-DIPT	" (–) 98% ee	162
	(–)-DET, –23°, 2 d	" (77) 95% de, [α] +38.6°	160, 2, 73
TBDMSO (structure)	TI (0.05), (+)-DET (0.07), –12°	(structure) (80)	243
(structure)	(+)-DIPT	(structure) (60) 80% ee [α] -11.8°[d]	60, 469
	(+)-DIPT (stoichiometric)	" (42) [α] -3.0°	210

95

TABLE I. ASYMMETRIC EPOXIDATION OF PRIMARY ALLYLIC ALCOHOLS (*Continued*)

Substrate	Conditions Ti (eq), DAT (eq), MS	Product(s) and Yield(s) (%)	Refs.
BnO⌒OH	(–)-DIPT, stoichiometric	[α] +3.1°[k]	210
	" (–)-DIPT, stoichiometric	(67) [α] +8.1°	470
BnO	(+)-DIPT, 4 Å	(88) 92% ee [α] -6.7°	458
BnO	(–)-DIPT, 4 Å	(75) 92% ee [α] +8.5°	458
MPMO	(+)-DIPT, -20°, 48 h	92% ee [α] -9.02°	471
TBDMSO	Ti (1.06), (–)-DIPT (1.06), 4 Å, -30°, 64 h	(90) [α] -7.81°	472
	Ti (1.10), (–)-DET (1.13), 4 Å, -30°, 4 d	" (95)	294
TBDPSO	(–)-DET, -20°	(87)	473
	(+)-DET (stoichiometric), -20°, 14 d	(55) >94% de	73
dioxolane substrate	Ti (1), (+)-DET (1), -23°, 11 d	" (57) 85% de	160

96

TABLE I. ASYMMETRIC EPOXIDATION OF PRIMARY ALLYLIC ALCOHOLS (*Continued*)

Substrate	Conditions TI (eq), DAT (eq), MS	Product(s) and Yield(s) (%)	Refs.
(2-methylbut-2-en-1-ol)	TI (1), (–)-DET (1), -23°, 11 d	(–) 20% de	160, 73
(BnO-substituted allylic alcohol)	TI (0.2), (+)-DET (0.28), -20°	(77)^l 94% ee [α] -22.2° d	347
	TI (0.05), (–)-DET (0.07), -23°	(–)	242
(BnO-substituted allylic alcohol)	(+)-DET (catalytic)	(–) 80-90% ee	235
	(–)-DIPT, -15°	(87) 90% ee [α] +22°	263
(BnO-substituted allylic alcohol)	TI (1), (+)-DET (1.42), -20°, 14 h	(80) 80% ee [α] -10.86°	235
(prenol-type allylic alcohol)	TI(1), (+)-DET (1), -20°, 24 h	(55) >93% ee [α] -20.1°	474
	TI(0.05)^a, (+)-DIPT (0.075), 4 Å, -20°, 0.5 h	" (90) [α] -18°	475, 476

97

TABLE I. ASYMMETRIC EPOXIDATION OF PRIMARY ALLYLIC ALCOHOLS (*Continued*)

Substrate	Conditions TI (eq), DAT (eq), MS	Product(s) and Yield(s) (%)	Refs.
Ph₂P(O)CH₂... —OH	TI (0.05), (–)-DIPT (0.075)	(—) 90% ee	252
	TI(1), (–)-dibutyl tartrate (1), –20°	" (25) 90% ee, [α] +17.1°	477
Ph₂P(O)CH₂... —OH	TI(1), (+)-DET (1.2), 4 Å, -16°	(91) 96% ee	449
	TI(1), (+)-DET (1.2), 4 Å, -16°	(85) 92% ee	449
C₆			
—OH	TI (0.05), (+)-DET (0.06), 3 Å, -12°, 11 h	(88) 95% ee, [α] -25.9°	18
—OH	TI (0.65), (+)-DET (1.2)	(56) 86% ee	478
OTBDMS —OH	TI (0.1), (+)-DET (1.3), 4 Å, -20°, overnight	OTBDMS (79) 98% ee	426

TABLE I. ASYMMETRIC EPOXIDATION OF PRIMARY ALLYLIC ALCOHOLS (*Continued*)

Substrate	Conditions TI (eq), DAT (eq), MS	Product(s) and Yield(s) (%)	Refs.
(structure: OH)	TI (0.05), (+)-DET (0.06), 4 Å, -20°, 2.5 h	(structure: O, OH) (85) 94% ee [α] -46.3°	18, 479
	TI (1.1), (+)-DET (1.1), -20°, 14 h	" (90) 92% ee, [α] -44.6°	480-483, 363
	TI (0.03)f, (+)-DIPT, -20 to -15°, 3.5 h	" (86) 94% ee	138
	TI (1.1), (+)-DIPT (1.11), -20°, 24 h	* (63.5) 93.5% ee, [α] -45°	484
	TI (1), (–)-DET (1)	(structure: O, OH) (—) [α] +40.8°c	363
	(–)-DET	" (63) >96% ee	482
(structure: OH)	TI (0.05), (+)-DET (0.06), 4 Å, -20°, 2.5 h	(structure: O, OH) (82) >95% ee [α] -32.7°	485
	(+)-DET	" (64) >90% ee	336, 483
	(–)-DET, -20°	(structure: O, OH) (48) >90% ee	486
	(–)-DET	" (59) [α] +32.2°	487
(structure: THPO...OH)	(+)-DET, -20°	(structure: THPO...O...OH) (85) >95% ee [α] -15.2°	488

TABLE I. ASYMMETRIC EPOXIDATION OF PRIMARY ALLYLIC ALCOHOLS (*Continued*)

Substrate	Conditions TI (eq), DAT (eq), MS	Product(s) and Yield(s) (%)	Refs.
	(–)-DET	(72)	489
	(+)-DET, -23°	(0)	105
	(+)-DET (stoichiometric)	(61) >95% ee, [α] -10.0°	438, 490
	(+)-DET	[α] -23.3°	491
	TI (1.2), (–)-DIPT (1.2), 4 Å, -20°, 24 h	(94) >98% ee [α] +36.8°	492, 493
	(+)-DET (catalytic), -20°	(92)[m] [α] -34° [d]	494

100

TABLE I. ASYMMETRIC EPOXIDATION OF PRIMARY ALLYLIC ALCOHOLS (*Continued*)

Substrate	Conditions TI (eq), DAT (eq), MS	Product(s) and Yield(s) (%)	Refs.
	(–)-DET (stoichiometric), -20°	 (77)	495
	TI (1.19), (+)-DIPT (1.59), -20°, overnight	 (75) [α] -22.5°	496
	TI (1.19), (–)-DET (1.59), -20°, overnight	 (72) [α] +19.8°	496
	TI (1.1), (+)-DET (1.1), -23°, 0.7 h	 (88) 90% de, [α] -27.7°	154, 158, 159
	TI (1.1), (–)-DET (1.1), -23°, 0.7 h	 (––) 90% de, [α] +32.4°	154, 158, 159
	(–)-DET (stoichiometric)	 (98) 80% de	497

TABLE I. ASYMMETRIC EPOXIDATION OF PRIMARY ALLYLIC ALCOHOLS (*Continued*)

Substrate	Conditions Ti (eq), DAT (eq), MS	Product(s) and Yield(s) (%)	Refs.
BOMO … OH	—	BOMO … O … OH (87)	498
THPO … OH	(+)-DIPT, -25°	THPO … O … OH (—)	499
TBDPSO … OH	Ti (0.1), (–)-DET (0.13), 4 Å, -20°, 32 h	TBDPSO … O … OH (96) >95% de	221
OH … OTBDMS	(+)-DET	O … OH OTBDMS (90)	500
	(–)-DET	O … OH OTBDMS (95)	500
BnO … OH	(–)-DET (stoichiometric), -20°, 18 h	BnO … O … OH (76) >98% de	330
O … OH OBn	(–)-DET (stoichiometric), -20°, 3 h	O … OH OBn (70) >98% de	330
TBDMSO … OH	Ti (1.4), (+)-DET (1.4), 4 Å, -20°, 14 h	O … O … OH (78)[a] [α] -22.4°	501

TABLE I. ASYMMETRIC EPOXIDATION OF PRIMARY ALLYLIC ALCOHOLS (*Continued*)

Substrate	Conditions TI (eq), DAT (eq), MS	Product(s) and Yield(s) (%)	Refs.
	(+)-DIPT (stoichiometric), -20°, 36 h	(70) >90% de	73, 161, 162, 257
	(−)-DIPT (stoichiometric), -20°, 36 h	(70) >90% de	161, 162, 257
	(+)-DET (stoichiometric)	(71) >90% de	161, 162
	(−)-DET (stoichiometric)	(73) >90% de	161, 162
	TI (1.2), (−)-DIPT (1.5), -20°	(84) [α] +14.7°	162

TABLE I. ASYMMETRIC EPOXIDATION OF PRIMARY ALLYLIC ALCOHOLS (*Continued*)

Substrate	Conditions Ti (eq), DAT (eq), MS	Product(s) and Yield(s) (%)	Refs.
BnO‚ ‚ ‚ OH TBDPSO	Ti (1.1), (−)-DET (1.1), −23°, 5 h	BnO‚ ‚ ‚ epoxide ‚ ‚ OH TBDPSO (60) 82% de [α] -1.2°	85, 87
BocNH‚ CO$_2$CH$_2$C$_6$H$_4$OMe ‚ ‚ OH H	Ti (1.1), (+)-DET (1.1), −23°, 18 h	MeOC$_6$H$_4$CO$_2$ ‚ HN, O, OH, H, OH (cyclic carbamate) (55-60)	502
OH	Ti (1.1), (+)-DIPT (1.2), −20°, 3 d	O epoxide OH (57) 87% ee; 97% ee[n] [α] -5.73°	64, 484
	Ti (1.1), (−)-DIPT (1.2), −20°, 3 d	O epoxide OH (56) 80% ee [α] +4.91°; 99% ee[n] [α] +5.69°	64
OH	(+)-DET	O epoxide OH (54) 66% ee, [α] -9.9°	8
TBDPSO‚ ‚ OH	(−)-DET , −20°	TBDPSO‚ ‚ O epoxide OH (88) 89% ee	503
BnO‚ ‚ OH	Ti (1.1), (+)-DET (1.1), −23°, 24 h	BnO‚ ‚ O epoxide OH (79) 67% ee	154
	Ti (1.1), (−)-DET (1.1), −23°, 24 h	(0)	154

TABLE I. ASYMMETRIC EPOXIDATION OF PRIMARY ALLYLIC ALCOHOLS (*Continued*)

Substrate	Conditions TI (eq), DAT (eq), MS	Product(s) and Yield(s) (%)	Refs.
	(+)-DET (stoichiometric)	(—)	73
	TI (1), (+)-DMT (1.05), -20°, 16 h	(79) >95% ee	59
	TI (1), (+)-DET (1), -20°	" [α] -13.5°	504
	TI (0.05), (+)-DET (0.06), 3 Å, -25°, 3 d	" (85) >98% ee	505
	TI (0.1), (+)-DET (0.1), 4 Å, -23°, 2 h	" (91) >98% ee, [α] -21.3°	506, 507
	(–)-DMT, 3 Å, -20°	(78) >98% ee	508
	TI (1), (–)-DET (1), -20°	" (—) [α] +14.3°	504
	(+)-DET (stoichiometric)	(38) >95% ee [α] -23.92°	509
	TI (1), (+)-DIPT (1.26), -23°	(66) 34% de	510

105

TABLE I. ASYMMETRIC EPOXIDATION OF PRIMARY ALLYLIC ALCOHOLS (*Continued*)

Substrate	Conditions Ti (eq), DAT (eq), MS	Product(s) and Yield(s) (%)	Refs.
	Ti (1), (+)-DIPT (1.26), -23°	(64) 90% de	510
	Ti (1), (–)-DIPT (1.26), -23°	" (71)^a 80% de	510
	Ti (1), (+)-DIPT (1.26), -23°	(63) 50% de	510
	Ti (1), (–)-DIPT (1.26), -23°	" (65) 70% de	510
	Ti (0.1), (–)-DIPT (0.15), 4 Å, -20°, 8 h	(90) 95% ee [α] +19°^{g}	511
	Ti (0.1), (+)-DIPT (0.12), 3 Å, -15°, overnight	(89) 80% ee [α] +15.3°	512

TABLE I. ASYMMETRIC EPOXIDATION OF PRIMARY ALLYLIC ALCOHOLS (*Continued*)

Substrate	Conditions TI (eq), DAT (eq), MS	Product(s) and Yield(s) (%)	Refs.
HC≡C— ... —OH	(+)-DET, −20°, 4 h	HC≡C epoxide —OH (80) [α] +11.36°	513
OH ... C≡CH	—	OH (—) 60% ee	104
TBDMSO ... OH	TI(1.05), (−)-DET (1.57), −20°, 17 h	TBDMSO epoxide OH (81) 95% ee [α] +2.1°	255
EEO ... OH	TI (1), (+)-DET (1), −20°, 24 h	EEO epoxide OH (90) [α] −13.3° g	288, 514
EEO ... OH	TI(1), (−)-DET (1.04), −20°	EEO epoxide OH (—)	288
D, EEO ... OH	TI (1), (+)-DET (1)	EEO epoxide D OH (83) 92% ee	288, 514

TABLE I. ASYMMETRIC EPOXIDATION OF PRIMARY ALLYLIC ALCOHOLS (Continued)

Substrate	Conditions TI (eq), DAT (eq), MS	Product(s) and Yield(s) (%)	Refs.
TBDMSO — OH	TI(1), (–)-DET (1.04), -20°	(82) [α] +2.9°	288
OH	(+)-DET (stoichiometric), -30°, 12 h	(98) 89% ee [α] -22°	515
	(–)-DET (stoichiometric), -30°, 12 h	(98) 88% ee [α] +24°	515, 89
OH	TI (1), (+)-DIPT (1.2), -50°, 24 h	(54) 92% ee, [α] +18.2°	133
	TI (1), (–)-DIPT (1.2), -50°, 24 h	(38) 93% ee, [α] -18.5°	133
	TI (1), (–)-DET (1.2), -50°, 24 h	" (53) 89% ee, [α] -17.9°	133
C7 Bu-n OH	TI (1), (+)-DET (1.2), -20°, 24 h	(67) [α] -6.2° k	516

TABLE I. ASYMMETRIC EPOXIDATION OF PRIMARY ALLYLIC ALCOHOLS (*Continued*)

Substrate	Conditions: TI (eq), DAT (eq), MS	Product(s) and Yield(s) (%)	Refs.
Bu-*t* ⟍OH	TI[c] (1.2), (+)-DET (1.5), -20°	(42) 85% ee	57
	TI[c] (1.1), (+)-DET (1.1),	(79)[o] 72% ee	180
n-Bu ⟍OH	Stoichiometric[c]	*n*-Bu ⟍OH (90) >95% ee [α] -49.9° e	298, 299
	TI (1[p], (+)-DET (1.2), -23°, 1 h	" (76) 95% ee. [α] -29.7°	140
⟍OH	(+)-DIPT	⟍OH (90) 90% ee	517
t-Bu ⟍OH	TI (1.2), (+)-DET (1.5), -20°	*t*-Bu ⟍OH (52) >95% ee	57
MeC≡C ⟍OH	TI (0.075), (+)-DET (0.11), 4 Å, 0 to 23°, overnight	MeC≡C ⟍OH (74) [α] -20.3° j	518
TBDMSO ⟍OH	TI (0.1), (–)-DET (0.15), 4 Å, -40°, 8 h	TBDMSO ⟍OH (85) 9?% ee	350
MOMO ⟍OH	(+)-DIPT, -23°	MOMO ⟍OH (57) 96% ee, [α] -24.7°	519

TABLE I. ASYMMETRIC EPOXIDATION OF PRIMARY ALLYLIC ALCOHOLS (*Continued*)

Substrate	Conditions TI (eq), DAT (eq), MS	Product(s) and Yield(s) (%)	Refs.
BnO— ... —OH	TI (0.2), (–)-DET (0.26), 4 Å, 0°, 8 h	(85) 92% ee [α] +8.38°	520, 521
Ph—oxazole(N)—Ph ... —OH	(+)-DET, -23°	(75) 99% ee, [α] -16.8°	105
... —OH	(–)-DET (stoichiometric), -45°	(74)	522, 523
AcO— ... —OH	(+)-DET, MS	(77)	524
AcO— ... —OH	(–)-DET, MS	(83)	525
OTBDPS— ... —OH	(+)-DIPT, -30°	(89) 90% de	526
OTBDPS— ... —OH	(–)-DET, MS, -20°, 24 h	(88) 88% de	527

TABLE I. ASYMMETRIC EPOXIDATION OF PRIMARY ALLYLIC ALCOHOLS (Continued)

Substrate	Conditions TI (eq), DAT (eq), MS	Product(s) and Yield(s) (%)	Refs.
[structure: acetonide furanose, BnO, OH allylic]	TI (1.2), (+)-DIPT (1.5), -23°, overnight	[epoxide structure] $(88)^m$ $[\alpha]$ +43°	528
[structure: TBDPSO, alkyne, OH]	(+)-DET (stoichiometric), -20°, 3 h	[epoxide structure] (83)	346
[structure: BnO, Ts(Bn)N, BnO, OH]	TI (1.2), (+)-DET (1.5), -20°	[epoxide structure] (—) >99.5% de	98
	TI (1.2), (–)-DIPT (1.5), -20°, 21 h	[epoxide structure] (9:) >99.4% de	98
[structure: OBn, BnO, BnO, OH]	TI (1.2), (–)-DET (1.5), -20°, 16 h	[epoxide structure] $(75)^m$ $[\alpha]$ +13.6°g	86, 87

TABLE I. ASYMMETRIC EPOXIDATION OF PRIMARY ALLYLIC ALCOHOLS (*Continued*)

Substrate	Conditions Tl (eq), DAT (eq), MS	Product(s) and Yield(s) (%)	Refs.
	(–)-DET	OH (89)	529
	(+)-DET	(0)	529
	(+)-DET	OH (80) 95% ee [α] -5.2°	530
t-Bu OH	Tl (1.2), (+)-DET (1.5), -20°	*t*-Bu OH (77) 25% ee	57
OH TMS	Tl (0.23), (+)-DET (0.3), 4 Å, 4°, 15 h	OH (78) 81% ee [α] -4.73° TMS	111
BnO OH	(+)-DET	BnO OH (—)	531
BnO OH	Tl (1), (–)-DET (1), -20°, 48 h	BnO OH (—) (73) 86% ee, [α] +1.55°	532
MeO$_2$C OH	Tl (1), (+)-DET (1), -20°, 8 h	MeO$_2$C OH (57) 94% ee, [α] -2.5°	354, 355

112

TABLE I. ASYMMETRIC EPOXIDATION OF PRIMARY ALLYLIC ALCOHOLS (*Continued*)

Substrate	Conditions TI (eq), DAT (eq), MS	Product(s) and Yield(s) (%)	Refs.
(MeO, H, O, O structure with CH=CH-OH)	TI (0.24), (+)-DET (0.34), 3 Å, -15°, overnight[a]	(MeO, H, O, O epoxide structure with OH) (77) [α] -45.9°	533
(cyclohexene with CH2OH)	TI (0.05), (+)-DET (0.075), -12°	(epoxide with OH) (77) 93% ee, [α] -22.8°	18
	(+)-DET	" (−)	534
	(−)-DET, -40°	(O epoxide with OH) (70) 92% ee, [α] +22.4°	535, 536
TBDMSO (chain with CH2OH)	TI (0.75), (−)-DET (1), -20°, 12 h	TBDMSO (chain with O epoxide, OH) (88) 93% ee, [α]_j +10.7°	490
TMSC≡C (chain with TMS, OH)	Catalytic, (−)-DET, -5°, 36 h	TMSC≡C (O epoxide with TMS, OH) (83) 93% ee	537
n-Bu (TMS chain with OH)	(+)-DET	n-Bu (TMS epoxide O with OH) (91)	538

113

TABLE I. ASYMMETRIC EPOXIDATION OF PRIMARY ALLYLIC ALCOHOLS (Continued)

Substrate	Conditions TI (eq), DAT (eq), MS	Product(s) and Yield(s) (%)	Refs.
BnO — OH	TI (1), (+)-DET (1.05), -25°, overnight	BnO — O — OH (93) 99% de, [α] -14.03°	173, 174
TBDMSO — OH	TI (0.045), (+)-DET (0.065), 3 Å, -20°, 10 h	TBDMSO — O — OH (94) 50% de	539
TBDPSO — OH	TI (1), (-)-DET (1.2), -20°, overnight	TBDPSO — O — OH (97) [α] +17.33°	540, 459
(cyclopropyl) — OH	TI (1), (-)-DIPT (1.2), -50°, 24 h	(cyclobutanone, OH) (80) 96% de, [α] +46.0°	133
(dioxolane) — OH	TI (1), (+)-DIPT (1), -23°, 72 h	(dioxolane, O, OH) (80) [α] -10.8°	541
C8			
n-C$_5$H$_{11}$ — OH	TI (0.05), (+)-DET (0.08), -20 to 0°	n-C$_5$H$_{11}$ — O — OH (75) 95% ee	250, 251
	TI^p (1), (+)-DET (1.2), -40°	" (76) 95% ee, [α] -29.7°	542
	Stoichiometric	" >95% ee, [α] -35.6°	282, 438

114

TABLE I. ASYMMETRIC EPOXIDATION OF PRIMARY ALLYLIC ALCOHOLS (*Continued*)

Substrate	Conditions Ti (eq), DAT (eq), MS	Product(s) and Yield(s) (%)	Refs.
	Ti (0.03)f, (+)-DIPT, -20 to -15°, 9 h	" (76) 98% ee	138
	Ti (1), (+)-DIPT (1.04), -23°	" (51) 90% ee	222
	(+)-DIPT	* (89) 98% ee	311
	Ti (1), (−)-DIPT (1.04), -23°	r-C$_5$H$_{11}$ epoxide —OH (51) 80% ee	222
	Ti (1), (+)-DET (1)	epoxide —OH (80) >95% ee	59
	(+)-DET (0.2), -20°, 5 h	epoxide —OH (83) >98% ee	543
	Ti (0.05), (−)-DET (0.06), 4 Å, -20°, 4 h	epoxide —OH (90) [α] +16.27°	227
	(+)-DIPT	epoxide —OH (82)	544
	(+)-DIPT, -20°, 18 h	epoxide —OH (70)	545
	Ti (0.03)f, (+)-DIPT, -20 to -15°, 8 h	" (82) [α] -17.8°	138

115

TABLE I. ASYMMETRIC EPOXIDATION OF PRIMARY ALLYLIC ALCOHOLS (*Continued*)

Substrate	Conditions TI (eq), DAT (eq), MS	Product(s) and Yield(s) (%)	Refs.
[enyne D-labeled allylic alcohol]	TI (1.9), (+)-DET (1.94), -30°, 5 h	[epoxide, D-labeled] OH (55) >97% ee	546
	TI (1.9), (−)-DET (1.94), -30°, 5 h	[epoxide, D-labeled] OH (55) >97% ee	546
NC⌁⌁⌁OH	(+)-DIPT	[epoxide] OH (72) 96% ee	107
	(−)-DIPT	[epoxide] OH (−)	225
MPMO⌁⌁OH	(+)-DET, 4 Å	[epoxide] OH (69) 95% ee	129
THPO⌁⌁OH	TI (0.11), (+)-DIPT (0.14), 3 Å, -40 to -5°	[epoxide] OH (85) 93% ee	80
[dioxane-containing allylic alcohol]	TI (1.13), (+)-DIPT (1.2 5), -30°, 6 h	" (70) 93% ee, [α] -13.4°	80

116

TABLE I. ASYMMETRIC EPOXIDATION OF PRIMARY ALLYLIC ALCOHOLS (*Continued*)

Substrate	Conditions Ti (eq), DAT (eq), MS	Product(s) and Yield(s) (%)	Refs.
(allylic alcohol, dioxane-protected)	Ti (0.11), (–)-DIPT (0.14), 3 Å, –40 to –5°	(73) 93% ee [α] +13.2°	80
	Ti (1.13), (–)-DIPT (1.25), –30°, 6 h	" (70) 91% ee, [α] +13.2[c]	80
NC—, Ph, O, N, Ph (oxazole)	Ti (0.3), (+)-DET (0.3), 4 Å, –8°	NC—, O, OH (—) >94% ee	547, 548
	(+)-DET, –23°	Ph, O, N, Ph, O, OH (66[c] 99% ee, [α] –11.1°	105
BnO—, OH	(+)-DET (stoichiometric), 3 Å, –20°, 2 h	BnO—, O, OH (88) >95% ee [α] –16.5°	549
BnO—, OH, OBn	(+)-DET (stoichiometric), 3 Å, –20°	O, OH, OBn (88) [α] –25.0°	549

117

TABLE I. ASYMMETRIC EPOXIDATION OF PRIMARY ALLYLIC ALCOHOLS (Continued)

Substrate	Conditions TI (eq), DAT (eq), MS	Product(s) and Yield(s) (%)	Refs.
	TI (1.1), (+)-DIPT (1.3), -25°, 48 h	 (95) 60% de, [α] +108.8°	550
	TI (1.1), (−)-DIPT (1.3), -25°, 48 h	 (93), [α] +116.9° j	550
	TI (1), (−)-DET (2.0), -20°	(70) 95% de	157, 520
	TI (1.2), (+)-DIPT (1.5), -23°, overnight	 (66)m [α] -90°	148

118

TABLE I. ASYMMETRIC EPOXIDATION OF PRIMARY ALLYLIC ALCOHOLS (*Continued*)

Substrate	Conditions TI (eq), DAT (eq), MS	Product(s) and Yield(s) (%)	Refs.
	(−)-DIPT, -23°, 8 d	(68)[q]	148
	TI (1.1), (+)-DET (1.1), -20°, overnight	(72)[m]	551
	(+)-DET (stoichiometric), -23°	(−) 50% de	123
	(−)-DET (stoichiometric), -23°	(−) 50% de	123
	TI (3.6), (+)-DET (5), -20°	(85) >94% de	348

TABLE I. ASYMMETRIC EPOXIDATION OF PRIMARY ALLYLIC ALCOHOLS (*Continued*)

Substrate	Conditions TI (eq), DAT (eq), MS	Product(s) and Yield(s) (%)	Refs.
	TI (3.6), (–)-DET (5), –20°	(74) >94% de	348
	TI (3.6), (+)-DET (5), –20°	(88) >94% de	348
	TI (3.6), (–)-DET (5), –20°	(70) >94% de	348
	TI (1.1), (–)-DET (1.1), –23°		85
	(–)-DET	(—)	164

TABLE I. ASYMMETRIC EPOXIDATION OF PRIMARY ALLYLIC ALCOHOLS (*Continued*)

Substrate	Conditions TI (eq), DAT (eq), MS	Product(s) and Yield(s) (%)	Refs.
(structure with OTBDMS, OH, methyluracil, acetonide)	TI (1), (+)-DIPT (1.1), -25°, 36 h	(86) 100% de, [α] +62.5° (structure with OTBDMS, OH epoxide)	552
	TI (1), (−)-DIPT (1.1), -25°, 48 h	(30) >92% de, [α] +57.5° (structure with OTBDMS, OH epoxide)	552
(structure: n-C$_5$H$_9$ allylic alcohol OH)	TI (0.1), (+)-DET (0.13), 4 Å, -25°, 40 h	(62) 74.6% ee [α] -5.42° (n-C$_5$H$_9$ epoxide OH)	553
	TI (1), (−)-DET (1.1), -20°	(78) 90 ee [α] +4.3° (n-C$_5$H$_9$ epoxide OH)	554
	TI (1), (−)-DET (1.2), -20°	" (78)	555, 556

TABLE I. ASYMMETRIC EPOXIDATION OF PRIMARY ALLYLIC ALCOHOLS (*Continued*)

Substrate	Conditions Ti (eq), DAT (eq), MS	Product(s) and Yield(s) (%)	Refs.
	Ti (1), (+)-DIPT (1.1), -29°, 65 h	(75) 88% ee [α] -5.4° r	220
	Ti (1), (–)-DIPT (1.1), -29°, 65 h	(77) 88% ee [α] +6.5° r	220
	Ti (1), (+)-DET (1.2), -20°, 6 d	(78) >94% ee [α] -3.51°	284, 285
	Ti (1), (+)-DET (1.1), -20°, 40 h	(76) 92% ee [α] -10.24°	518
	Ti (1), (+)-DET (1.1), -25°	(73) 95% ee	116
	Ti (1), (–)-DET (1.1), -25°	(82) 95% ee	116
	(+)-DIPT	(72) 96% ee [α] +12.5°	107

122

TABLE I. ASYMMETRIC EPOXIDATION OF PRIMARY ALLYLIC ALCOHOLS (*Continued*)

Substrate	Conditions TI (eq), DAT (eq), MS	Product(s) and Yield(s) (%)	Refs.
	TI (1), (+)-DET (1.1), -25°	(30) 95% ee	116
	TI (1), (−)-DET (1.1), -25°	(30) 95% ee	116
	(+)-DET, -20°, 31 d	(60)^m	557
	(+)-DIPT (stoichiometric), -23°	(0)	148
	(−)-DIPT (stoichiometric), -23°	(0)	148
	(−)-DET, -10°	(70)^m [α] -143°	558

TABLE I. ASYMMETRIC EPOXIDATION OF PRIMARY ALLYLIC ALCOHOLS (*Continued*)

Substrate	Conditions TI (eq), DAT (eq), MS	Product(s) and Yield(s) (%)	Refs.
	TI (1.10). (+)-DET (1.11), -25°, 3.5 h	(71) 96% ee [α] -18.1°	559-561
	TI (1.4), (−)-DIPT (1.5), -30°, 2.5 h	(84) 96% ee [α] +4.3° s	561
	(+)-DET	(—) [α] -10.6° s	240
	(−)-DET (stoichiometric)	(92) >95% ee [α] +9° s	562
	(+)-DET	(81)	538
	TI (1), (−)-DET (1), -20°, overnight	(—)	563, 564
	(−)-DET (−) 95% de	(—)	565

124

TABLE I. ASYMMETRIC EPOXIDATION OF PRIMARY ALLYLIC ALCOHOLS (*Continued*)

Substrate	Conditions TI (eq), DAT (eq), MS	Product(s) and Yield(s) (%)	Refs.
(TBDMSO-substituted cyclohexene methanol)	(+)-DET	(—) 95% de	88
(Bu-*t* substituted allylic alcohol)	TI^c (1.2), (+)-DET (1.5), -20°	(43) 60% ee	57
(cyclohexylidene ethanol)	(+)-DET (catalytic), 4 Å, -25°, 2 h	(80) 82% ee [α] -16.1°	566
(TBDPSO-substituted allylic alcohol)	(+)-DIPT (catalytic)	(—) 86% ee, [α] -2.6°	567
(thymidine-derived allylic alcohol)	TI (1.1), (+)-DIPT (1.3), -25°, 5 d	(75) 100% de [α] -5° ^{g}	568

125

TABLE I. ASYMMETRIC EPOXIDATION OF PRIMARY ALLYLIC ALCOHOLS (*Continued*)

Substrate	Conditions TI (eq), DAT (eq), MS	Product(s) and Yield(s) (%)	Refs.
	TI (1.1), (–)-DIPT (1.3), -25°, 5 d	" 100% de, [α] -2° [g]	568
	TI (1.1), (+)-DIPT (1.3), -25°, 5 d	(70) 100% de [α] -17.5° [g]	568
	TI (1.1), (–)-DIPT (1.3), -25°, 5 d	" 100% de, [α] -15° [g]	568
	TI (1.1), (+)-DIPT (1.3), -25°	(73) 100% de [α] -62.5° [g]	568
	TI (1.1), (–)-DIPT (1.3), -25°	" 100% de, [α] -67.5° [g]	568

TABLE I. ASYMMETRIC EPOXIDATION OF PRIMARY ALLYLIC ALCOHOLS (*Continued*)

Substrate	Conditions Ti (eq), DAT (eq), MS	Product(s) and Yield(s) (%)	Refs.
	Ti (1.1), (+)-DIPT (1.3), –25°	(50) 100% de [α] –50° *g*	568
	Ti (1.1), (–)-DIPT (1.3), –25°	" 100% de, [α] –60° *g*	568
	Ti (1), (–)-DET (1.2), –50°, 24 h	(70) 93% ee, [α] +36.9°	133
	Ti (1), (+)-DET (1.2), –50°, 48 h	(80) 89% ee	134
	Ti (1), (–)-DET (1.2), –50°, 48 h	(73) 89% ee, [α] +101.4°	133, 134

TABLE I. ASYMMETRIC EPOXIDATION OF PRIMARY ALLYLIC ALCOHOLS (*Continued*)

Substrate	Conditions TI (eq), DAT (eq), MS	Product(s) and Yield(s) (%)	Refs.
C₉			
C_6H_{13}-n allylic alcohol (OH)	(+)-Tartrate	C_6H_{13}-n epoxide, OH (87) 97% ee	569
C_6H_{11} allylic alcohol (OH)	TI (1), (+)-DET (1), -23°	C_6H_{11} epoxide, OH (81) >95% ee	1
$(CH_2)_6OC_6H_4Cl$-p allylic alcohol (OH)	TI (0.51), (+)-DET (0.77), 4 Å, -20°, 3 h	$(CH_2)_6OC_6H_4Cl$-p epoxide, OH (49) [α] -30.7°	570
cyclohexene with OMPM and OH (H)	TI (1), (+)-DET (1.2), -23°, 5.5 h	epoxide, OH, OMPM (95) <80% de	571
p-MeOC₆H₄ structure, racemic	(+)-DET[c], -23°, 5.5 h	p-MeOC₆H₄ epoxide (47) [α] -243°[d] + p-MeOC₆H₄ epoxide (47)	572

128

TABLE I. ASYMMETRIC EPOXIDATION OF PRIMARY ALLYLIC ALCOHOLS (*Continued*)

Substrate	Conditions TI (eq), DAT (eq). MS	Product(s) and Yield(s) (%)	Refs.
(spiro dioxolane allylic alcohol, =CH2, CH2OH)	TI (0.21), (–)-DET (0.38), 4 Å, –20°, 5 h	(96) 87% de, [α] -10.4°	573
n-C$_6$H$_{13}$—〜—OH	TI (1), (+)-DET (1), -10°, 24 h	r-C$_6$H$_{13}$—epoxide—OH (—) [α] -38.9°	574
	TI (1), (–)-DET (1), -10°, 24 h	t-C$_6$H$_{13}$—epoxide—OH (—) [α] +38.7°	574
(dienol chain with terminal vinyl, OH)	TI (0.52), (+)-DIPT (0.62), -20°, 3d	epoxide—OH (82) 96% ee	575, 576
MeO$_2$C—〜—OH	(–)-DET (stoichiometric)	MeO$_2$C—epoxide—OH (69) [α] -1.5°	577
(3,6-dihydro-2H-pyran allylic alcohol)	(–)-DIPT, -20°	epoxide—OH (82) 90% ee	578
TMS—〜—〜—OH	TI (0.1), (–)-DIPT, 4 Å, -25°	TMS—〜—epoxide—OH (96) >93% ee	579

129

TABLE I. ASYMMETRIC EPOXIDATION OF PRIMARY ALLYLIC ALCOHOLS (*Continued*)

Substrate	Conditions TI (eq), DAT (eq), MS	Product(s) and Yield(s) (%);	Refs.
Ph⌒⌒OH	(+)-DET (catalytic)	Ph△⌒OH (55) 93% ee	479
	TI (0.05), (+)-DIPT (0.075), 4 Å, -20°, 3h	" (89) >98% ee, [α] -49.6°	224, 438
	Stoichiometric	" (36) >98% ee, [α] -10.7°	282
	TI (1),[p] (+)-DET (1.2), -23°, 0.75 h	" (54) 89% ee, [α] -49.5°	140
	TI (1), (+)-DET (1.2), -23°, 1.2 h	" (44) 95% ee, [α] -51.7°	140
	TI (2), (+)-*N,N'*-dibenzyl tartramide (1)	Ph△⌒OH (—) 65% ee	130
	TI (1), (–)-DIPT (0.075)	" (—) >98% ee	252
m-MeOC₆H₄⌒⌒OH	TI (0.05), (+)-DIPT (0.075), 4 Å, -30°, 7 h	m-MeOC₆H₄△⌒OH (91) 94% ee [α] -66.8°[d]	580
m-BOMOC₆H₄⌒⌒OH	(–)-DIPT, 4 Å, -30°	m-BOMOC₆H₄△⌒OH (—) 96% ee	129
p-O₂NC₆H₄⌒⌒OH	TI (0.05), (+)-DIPT (0.075), 4 Å, -20°, 2 h	p-O₂NC₆H₄△⌒OH (89) >98% ee[i] [α] -37.4°	18
p-BrC₆H₄⌒⌒OH	TI (0.05), (+)-DET (0.075), 4 Å, -20°, 0.75 h	p-BrC₆H₄△⌒OH (69) >98% ee[i] [α] -35.2	18

130

TABLE I. ASYMMETRIC EPOXIDATION OF PRIMARY ALLYLIC ALCOHOLS (*Continued*)

Substrate	Conditions TI (eq), DAT (eq), MS	Product(s) and Yield(s) (%)	Refs.
(structure)	(+)-DET, -20°, 4 h	(—)	513
(structure)	TI (1), (−)-DIPT (1), -20°, 18 h	(85) [α] +22°	581-583
(structure)	TI (1), (+)-DET (1), -20°	(85) 90% ee	584
(structure)	TI (1), (+)-DIPT (1), -20°, 7 h	(88) [α] -28.5°	585
(structure)	TI (1.1), (+)-DET (1.1), -23°	(78) 90% de [α] -44.0°	154
(structure)	TI (0.2), (+)-DET (0.26), 4 Å, 0 41, 12 h	(80) >98% de	521
(structure)	TI (0.1), (+)-DET (0.15), 4 Å	(78) 68% de	349

TABLE I. ASYMMETRIC EPOXIDATION OF PRIMARY ALLYLIC ALCOHOLS (*Continued*)

Substrate	Conditions TI (eq), DAT (eq), MS	Product(s) and Yield(s) (%)	Refs.
(structure: BnO, TBDPSO, OH, vinyl)	TI (0.1), (–)-DET (0.15)	TBDMSO (structure with epoxide, OH) (92) 90% de	349
(structure: TBDPSO, O pyran, OH)	(–)-DET, -20°, 15 h	(structure: BnO, TBDPSO, epoxide, OH) (—)	586
	TI (0.1), (+)-DET (0.15), 4 Å	TBDPSO (structure, epoxide) OH (80) 62% de	349
	TI (0.1), (–)-DET (0.15), 4 Å	TBDPSO (structure, epoxide) OH (87) 89.5% de	349
dl (structure: BnO, TMS, OH)	TI (1), (–)-DET (1.2), 4 Å, -25°	(structure: BnO, O epoxide, OH, TMS) (89) >90% de	587

TABLE I. ASYMMETRIC EPOXIDATION OF PRIMARY ALLYLIC ALCOHOLS (*Continued*)

Substrate	Conditions TI (eq), DAT (eq), MS	Product(s) and Yield(s) (%)	Refs.
dl	TI (1), (–)-DET (1.2), 4 Å, -25°	(92) >90% de	587
	TI (1), (–)-DET (1.2), 4 Å, -25°	(79) 100% de	587
	TI (0.19), (–)-DET (0.3), 4 Å, -25°	(91) 100% de [α] +51°	587
	1. TI (2), (+)-DET (2.1), -20°, 20 h, 2. CSA	(75)	119

TABLE I. ASYMMETRIC EPOXIDATION OF PRIMARY ALLYLIC ALCOHOLS (*Continued*)

Substrate	Conditions TI (eq), DAT (eq), MS	Product(s) and Yield(s) (%)	Refs.
	TI (0.15) (–)-DET (0.2), 4 Å, -20°, 14 h	(89)	588
	(+)-DET	(—)	589
	(–)-DET	(—)	589
	(–)-DET	(—)	590
	TI (1), (+)-DET (1.2), 1.2 h	(56) 85% ee, [α] -50°	140-142, 2, 313, 352
	TI (1)*, (+)-DET (1.2), 48 h	" (50) 85% ee, [α] -30°	140

TABLE I. ASYMMETRIC EPOXIDATION OF PRIMARY ALLYLIC ALCOHOLS (*Continued*)

Substrate	Conditions TI (eq), DAT (eq), MS	Product(s) and Yield(s) (%)	Refs.
$p\text{-}O_2NC_6H_4$	(+)-DET, –20°, 7 d	$t\text{-}O_2NC_6H_4$ (85) 95% ee $[\alpha]$ –98.3°	591
	TI (1), (+)-DET (1.1), –25°, 40 h	(76)	592
	TI (1), (+)-DET (1), –20°, overnight	(64) >90% ee $[\alpha]$ –12.2°	217
	(+)-DMT	" (90) >95% ee	593
	TI (1), (–)-DET (1), –20°, overnight	(59) >91% ee $[\alpha]$ +12.6°	217
TBDMSO BnO	(–)-DET (1), –23°	TBDMSO (92) BnO	594
TBDMSO	TI (1.2), (–)-DET (1.9), –20°, 2 d	TBDMSO (46) 92% de $[\alpha]$ +85.7° [t]	170, 172
	(+)-DET	TBDMSO (—) 68% de	172
	(–)-DET	" (—) 20% de	172

135

TABLE I. ASYMMETRIC EPOXIDATION OF PRIMARY ALLYLIC ALCOHOLS (*Continued*)

Substrate	Conditions Ti (eq), DAT (eq), MS	Product(s) and Yield(s) (%)	Refs.
	Ti (0.1), (+)-DIPT (0.15), 4 Å, -20°, 2.5 h	(88)	595
	Ti (0.1), (−)-DIPT (0.15), 4 Å, -20°, 2.5 h	(95)	595
	(+)-DET	(−) 95% ee	88
	Ti (2), (+)-DET (2), -20°, overnight	(68) 95% ee [α] -0.99°	596
	(−)-DET, -20°	(84) [α] -9.8°	227

136

TABLE I. ASYMMETRIC EPOXIDATION OF PRIMARY ALLYLIC ALCOHOLS (*Continued*)

Substrate	Conditions TI (eq), DAT (eq), MS	Product(s) and Yield(s) (%)	Refs.
	Catalytic (+)-DET, −20°, 4 h	(80) 90-95% ee [α] +3.96°	513
	(+)-Tartrate	(—)	597
	TI (1), (−)-DET (1.2), −50°, 24 h	(70) 94% ee, [α] +37.3°	133
	TI (1), (−)-DET (1.2), −50°, 48 h	(96) 91% ee, [α] +26.2°	133
dl	TI (1), (+)-DET (1), −20°, 24 h	(29) 78% ee + (31) 70% ee	598

137

TABLE I. ASYMMETRIC EPOXIDATION OF PRIMARY ALLYLIC ALCOHOLS (*Continued*)

Substrate	Conditions TI (eq), DAT (eq), MS	Product(s) and Yield(s) (%)	Refs.
(image: cyclohexanone with OH)	(+)-DET, -23°	(image: epoxide product with OH) (—) >90% ee [α] -2.56°	599
C$_{10}$			
(image: Ph-substituted allylic alcohol with OH)	(+)-DAT	(image: epoxide with Ph, OH) (60) 80% ee, [α] -26°	600
(image: n-C$_7$H$_{15}$ allylic alcohol OH)	TI (0.05), (+)-DET (0.075), 4 Å, -23°, 2.5 h	(image: n-C$_7$H$_{15}$ epoxide OH) (99) 96% ee [α] -36.5°	18
(image: n-C$_7$H$_{15}$, F, F allylic alcohol OH)	TI (1), (+)-DIPT (1.2), 4 Å, -38°, 72 h	(image: n-C$_7$H$_{15}$, F, F epoxide OH) (81) 80% ee	601, 602
(image: MeO$_2$C chain OH)	(+)-DET	(image: MeO$_2$C epoxide OH) (69) 93% ee, [α] -15.5°	91
(image: allylic alcohol OH with isopropylidene)	TI (1), (+)-DET (1), -23°, 2.5 h	(image: epoxide OH) (74) 93% ee, [α] -35.4°	603
	TI (0.05), (+)-DET (0.06), -20°, 2.5 h	" (86) [α] -32.2°	227

138

TABLE I. ASYMMETRIC EPOXIDATION OF PRIMARY ALLYLIC ALCOHOLS (*Continued*)

Substrate	Conditions TI (eq), DAT (eq), MS	Product(s) and Yield(s) (%)	Refs.
(structure with CO₂Me, OH) + (structure with CO₂Me, OH)	(–)-DET	(epoxide structure, CO₂Me, OH) + (91) (epoxide structure, CO₂Me, OH)	359
p-MeC₆H₄ —OH	TI (0.1), (+)-DIPT (0.15), –20°, 34 h	*p*-MeC₆H₄ (epoxy)OH (62) >98% ee [α] -36.65°	300
p-MeOC₆H₄O(CH₂)₇ —OH	(+)-DET	*p*-MeOC₆H₄O(CH₂)₇ (epoxy) OH (87) >97% ee	604
	(–)-DET	*p*-MeOC₆H₄O(CH₂)₇ (epoxy) OH (74) >97% ee	604
Ph —OH	(+)-DET	Ph (epoxy) OH (—) >95% ee	605
(furan structure with OH, CH₃)	(+)-DET	(furan epoxide structure) OH (92) [α] -35.23°	114

139

TABLE I. ASYMMETRIC EPOXIDATION OF PRIMARY ALLYLIC ALCOHOLS (*Continued*)

Substrate	Conditions TI (eq), DAT (eq), MS	Product(s) and Yield(s) (%)	Refs.
[structure: OH, OMe allylic alcohol]	TI (1), (–)-DET (1), –20°, 18 h	$(68)^m$ [α] +28.0°	256
[structure: vinyl cyclopentene, OH]	TI (3), (+)-DET (3.5), –30°, 15 h	(73) 97% de [α] +293.5°^{d}	606
[structure: vinyl cyclopentene, OH]	TI (3), (–)-DET (3.5), –30°, 15 h	(71) 96% de [α] +238.3°^{d}	606
[structure: EtO₂C, OH]	(–)-DIPT, 4 Å, –20°	(95)	607
[structure: TBDPSO, OH]	TI (2), (+)-DIPT (2.4), –20°, 6 h	(70)	118

140

TABLE I. ASYMMETRIC EPOXIDATION OF PRIMARY ALLYLIC ALCOHOLS (*Continued*)

Substrate	Conditions TI (eq), DAT (eq), MS	Product(s) and Yield(s) (%)	Refs.
	(+)-DET , 4 Å	 (88)	608
	(+)-DET	 (—)	164
	TI (1.37), (+)-DIPT (1.52), -23°, 5 d	 (63) 67% de, [α] -58° *j*	147, 149
	TI (1.45), (+)-DIPT (1.6), -23°, 4 d	 (80) 60% de	147, 149

141

TABLE I. ASYMMETRIC EPOXIDATION OF PRIMARY ALLYLIC ALCOHOLS (*Continued*)

Substrate	Conditions TI (eq), DAT (eq), MS	Product(s) and Yield(s) (%)	Refs.
	TI (1.45), (–)-DIPT (1.6), -23°, 4 d	 (61) [α] -65°	147
	(+)-DET, -30°, 6 h	 (90) 96% de, [α] +25.8°	163, 166
	TI (2), (+)-DET (2), -30°, 3 h	 (78)	609
	TI (0.1), (+)-DET (0.14), 4 Å, -10°, 29 h	 (74) 86%, ee [α] -4.8°	18
	(+)-DIPT, 4 Å, -20°, 29 h	" (74) 86% ee	610
	TI (0.1), (–)-DET (0.14), 4 Å, -20°, 3 d	 (92) 84%, ee	18

142

TABLE I. ASYMMETRIC EPOXIDATION OF PRIMARY ALLYLIC ALCOHOLS (*Continued*)

Substrate	Conditions TI (eq), DAT (eq), MS	Product(s) and Yield(s) (%)	Refs.
(structure: OH)	TI (1.2), (+)-DET (1.4), -20°, 18 h	(structure: O, OH) (81) 84.2% ee	109, 110, 611
Ph (structure: OH)	(−)-Tartrate	Ph (structure: O, OH) (83) 91% ee [α] -8.40° s	145
(structure: OBn, OMPM, TBDMSO, OH)	(+)-DET, -15°	(structure: OBn, OMFM, TBDMSO, O, OH) (—)	612
(structure: OH)	TI (1), (+)-DET (1), -20°, overnight	(structure: O, OH) (70)	94
Ph (structure: OH)	TI (0.05), (+)-DIPT (0.075), 4 Å, -35°, 2 h	Ph (structure: O, OH) (79) >98% ee [α] -16.9° j	18
AcO (structure: OH)	TI (1), (+)-DET (1), -23°	AcO (structure: O, OH) (70) 95% ee	1
	TI (1), (−)-DET (1), -15°, overnight	AcO (structure: O, OH) (50) [α] +5.86°	94

143

TABLE I. ASYMMETRIC EPOXIDATION OF PRIMARY ALLYLIC ALCOHOLS (Continued)

Substrate	Conditions TI (eq), DAT (eq), MS	Product(s) and Yield(s) (%)	Refs.
TBDPSO...OH	TI (1), (–)-DIPT (1), –20°, 14 h	" (79) [α] +13.2°	83
	(–)-DIPT, 4 Å	" (75) 90% ee	75
	(–)-DMT (catalytic), 3 Å, –23°, 12 h	" (–) 92% ee	76
TBDPSO...coumarin	(–)-DMT (catalytic), 3 Å, –23°, 18 h	TBDPSO...OH (92)	76
coumarin	TI (1), (–)-DET (1), –20°	(72) >95% ee, [α] +15.8°	215, 216
Ts...OH	TI (0.05), (+)-DIPT (0.07), 4 Å, –23°, 3.5 h	Ts...OH (89) >98% ee, [α] -3.56° d	613
BzO...OH	TI (0.05), (+)-DIPT (0.073), 3 Å, –10°, 3.5 h	BzO...OH (83) >95% ee, [α] -5.77°	357

TABLE I. ASYMMETRIC EPOXIDATION OF PRIMARY ALLYLIC ALCOHOLS (*Continued*)

Substrate	Conditions TI (eq), DAT (eq), MS	Product(s) and Yield(s) (%)	Refs.
	(+)-DET	(—)	614
	(−)-DET, -23°	(—) 80% de	615
dl	TI (1), (+)-DIPT (1.2), -23°	(94)[a] 96% de + 96% de	332
	TI (1.1), (+)-DIPT (1.1), -20°, overnight	(41)	616
	TI (1.1), (−)-DIPT (1.1), -20°, overnight	(36)	616

145

TABLE I. ASYMMETRIC EPOXIDATION OF PRIMARY ALLYLIC ALCOHOLS (*Continued*)

Substrate	Conditions TI (eq), DAT (eq), MS	Product(s) and Yield(s) (%)	Refs.
	TI (0.05), (+)-DET (0.074), 3 Å, -20°, 0.75 h	(95) 91% ee [α] -5.3°	18, 238
	TI (1), (+)-DET (1), -20°	" (77) 95% ee	1, 234, 617, 618
	TI (0.03)*f*, (+)-DET, -20 to -15°, 7.5 h	" (79) 98% ee	138
	TI (0.03)*f*, (+)-DET, -20 to -15°, 9 h	" (72) 96% ee	138
	TI (1)*f*, (+)-DET, -40°, 1.5 h	" (77) [α] -5.4°	141
	(–)-DIPT, -20°	(98) 88% ee	619
	TI (1), (–)-DET (1.2)	" (79) [α] +5.5°	618
	TI (1), (+)-DET (1), -20°	(79) 94% ee	1, 620
	TI (0.1), (+)-DET (0.15), 3 Å, -23°, 2.5 h	" (88) 72% ee	238
	TI (0.15), (–)-DET (0.2), 3 Å, -20°, 3 h	(99) >70% ee [α] +15.4°	621, 622

TABLE I. ASYMMETRIC EPOXIDATION OF PRIMARY ALLYLIC ALCOHOLS (*Continued*)

Substrate	Conditions TI (eq), DAT (eq), MS	Product(s) and Yield(s) (%)	Refs.
	TI (0.05), (–)-DET (0.075), 4 Å, –20°, 7 h		623
		(–) 93% ee, [α] +5.0° c	
	TI (1), (–)-DET (1), –15°, overnight	(60) [α] +9.4° j	94
	(–)-DMT (catalytic), 3 Å, –23°		76
		(93)	
	(+)-DIPT (catalytic)	(80) 80–85% ee	624
	(–)-DET, –25°, 20 h	(70) [α] –13.3°	566

147

TABLE I. ASYMMETRIC EPOXIDATION OF PRIMARY ALLYLIC ALCOHOLS (*Continued*)

Substrate	Conditions TI (eq), DAT (eq), MS	Product(s) and Yield(s) (%)	Refs.
	TI (1), (+)-DET (1), -20°, overnight	(80) 95% ee [α] -25.8°	625
	TI (1), (−)-DET (1), -20°, overnight	(−) [α] +25°	625
	TI (0.2), (+)-DET, (0.28), 4 Å, -47 to -7°	(95) 94.8% ee [α] -21.9°	626
	TI (0.2), (−)-DET, (0.28), 4 Å, -47 to -7°	(93) 97.4% ee [α] +22.9°	626
C₁₁ C₈H₁₇-*n*	TI (0.1), (+)-DET (0.13), 4 Å, -20°, overnight	C₈H₁₇-*n* (82) 96% ee	426
Ph	TI (0.1), (+)-DET (0.13), 4 Å, -20°, overnight	Ph (87) 87% ee	426

TABLE I. ASYMMETRIC EPOXIDATION OF PRIMARY ALLYLIC ALCOHOLS (*Continued*)

Substrate	Conditions: TI (eq), DAT (eq), MS	Product(s) and Yield(s) (%)	Refs.
	(+)-DIPT[a] (catalytic)	(81) 92% de	151, 152
	(−)-DIPT[a] (catalytic)	(69) 50% de	150, 152
$n\text{-}C_8H_{17}$ OH	TI (0.05), (+)-DET (0.06), 4 Å, −10°, 1.5 h	$n\text{-}C_8H_{17}$ ⟍OH (78) 94% ee [α] −35.5°	18
	TI (1.15), (+)-DMT (1.5), −23°, 24 h	" (85) [α] −4.2°	627
$n\text{-}C_5H_{11}C{\equiv}C$ OH	TI (1.05), (−)-DMT (1.1), −20°, 24 h	$n\text{-}C_5H_{11}C{\equiv}C$ OH (75) [α] +10.7°	99
OH	—	OH (88) [α] −10.0°	282
OH	(−)-DMT	OH (76) [α] +8.3°	100

149

TABLE I. ASYMMETRIC EPOXIDATION OF PRIMARY ALLYLIC ALCOHOLS (Continued)

Substrate	Conditions TI (eq), DAT (eq), MS	Product(s) and Yield(s) (%)	Refs.
$n\text{-}C_5H_{11}$ —＼＝／— OH	TI (1.05), (−)-DMT (1.1)	$n\text{-}C_5H_{11}$ —＼＜O＞／— OH (−) 93% ee, [α] +17.9°	628
OH (diene substrate)	(+)-DET (stoichiometric), -20°, 20 h	epoxide —OH (82) >95% ee [α] -10.2°	629
OH (diene substrate)	(+)-DET (stoichiometric), -20°	epoxide —OH (85) >95% ee [α] -16.8°	282
	(−)-DET, -20°, 10 h	epoxide —OH (85) >95% ee [α] +16.8°	283
Ph —＼＝／— OH	(+)-DET, -23°	Ph —＼＜O＞／— OH (96) 94% ee	630
	TI (1.24), (+)-DET (1.36), 4 Å, -23°, 3 h	" (87) >95% ee, [α] -32.4°	631
OMEM substrate —OH	(+)-DIPT	OMEM epoxide —OH (−) >95% ee	632

150

TABLE I. ASYMMETRIC EPOXIDATION OF PRIMARY ALLYLIC ALCOHOLS (Continued)

Substrate	Conditions TI (eq), DAT (eq), MS	Product(s) and Yield(s) (%)	Refs.
(allylic alcohol with pyran ring, HO, OMe, methyl substituents)	TI (1.15), (+)-DIPT (1.38), -20°, overnight	(epoxide product) (75) 83% de	113
(Ts-N piperidine with allylic alcohol)	TI (0.3), (+)-DET (0.36), 3 Å, -23°, 6 h	(epoxide product) (90) 92% de, [α] +18.3°	356
(Ts-N piperidine with allylic alcohol)	TI (0.3), (-)-DET (0.36), 3 Å, -23°, 6 h	(epoxide product) (80) 91% de, [α] +54.19°	356
(OBn, C≡C alkyne chain allylic alcohol)	(+)-DET, 4 Å	(epoxide product) (80)	633
(MeO, BnO, MeO pyran with allylic alcohol)	(-)-DET	(epoxide product) (—)	634

TABLE I. ASYMMETRIC EPOXIDATION OF PRIMARY ALLYLIC ALCOHOLS (*Continued*)

Substrate	Conditions Ti (eq), DAT (eq), MS	Product(s) and Yield(s) (%)	Refs.
dl (HO-substituted structure)	(–)-DET, 4 Å	HO (92)^u >95% ee + HO >90% ee	635
dl (SEMO-substituted structure)	(–)-DET, 4 Å	SEMO (78)^u >90% ee + SEMO >86% ee	635
(MEMO-substituted structure)	Ti (0.05), (+)-DET (0.057), 4 Å, –18°, 15 h	MEMO (72) 72% de	636

152

TABLE I. ASYMMETRIC EPOXIDATION OF PRIMARY ALLYLIC ALCOHOLS (*Continued*)

Substrate	Conditions Ti (eq), DAT (eq), MS	Product(s) and Yield(s) (%)	Refs.
(structure: BnO, OAc furanose with OH allylic)	Ti (1.4), (–)-DET (1.4), –20°, 12 h	(structure with OAc, BnO, OH) (99) >98% de	155, 156
(structure: BnO, dioxolane with OH)	Ti (0.15), (–)-DET (0.18), 4 Å, 0°, 7 h	(structure: BnO, epoxide, OH) (80) $[\alpha]$ -9.43°	521
(structure: Ph, H, OTBDMS bicyclic with OH)	Ti (0.15), (–)-DET (0.2), 4 Å, –28°, 14 h	(structure: Ph, H, OT3DMS, epoxide OH) (87)	588
(structure: TBDPSO pyranose with OH)	(+)-DET	(structure: TBDPSO, epoxide, OH) (—)	163
(structure: TBDPSO pyranose with OH)	(–)-DET	(structure: TBDPSO, epoxide, OH) (—)	163

153

TABLE I. ASYMMETRIC EPOXIDATION OF PRIMARY ALLYLIC ALCOHOLS (*Continued*)

Substrate	Conditions TI (eq), DAT (eq), MS	Product(s) and Yield(s) (%)	Refs.
(cyclopentane lactone with OBz, allylic alcohol side chain)	TI (1.1), (+)-DET (1.1), −20°, overnight	(epoxy alcohol product) (92)^m [α] −100.8°	637
(pyranose with OBn, BnO, OBn substituents and allylic alcohol chain)	(−)-DET	(epoxide product) (—)	165
n-C$_8$H$_{17}$ OH	TI (0.05), (+)-DIPT (0.074), 3 Å, −12°, 42 h	n-C$_8$H$_{17}$ (epoxy alcohol) OH (63) >80% ee [α] −3.5°	18
n-C$_5$H$_{11}$ OH	TI (1.1), (+)-DMT (1.1), −25°, 18 h	n-C$_5$H$_{11}$ (epoxy alcohol) OH (71) 94% ee, [α] −10.2°	638
n-C$_5$H$_{11}$ OH	TI (1), (−)-DIPT (1.2), −25°, 7 d	n-C$_5$H$_{11}$ (epoxy alcohol) OH (90) 85.7% ee, [α] +10.6°	553
(diene allylic alcohol, OH)	TI (1.2), (+)-DET (1.4), −20°, 26 h	(epoxy alcohol product) OH (55) 80.6% ee	109, 110

154

TABLE I. ASYMMETRIC EPOXIDATION OF PRIMARY ALLYLIC ALCOHOLS (*Continued*)

Substrate	Conditions TI (eq), DAT (eq), MS	Product(s) and Yield(s) (%)	Refs.
	TI (1)^p, (+)-DET (1.4), −20°, 26 h	" (60) [α] +8.0°	141
	(+)-DET, 4 Å	(—)	639
	TI (1), (−)-DET (1.5), 4 Å, −25°, 15 h	(60) 88% ee [α] +5.53°	640, 641
	(−)-DET, 4 Å	" (60) 87–88% ee	639
	(+)-DET	(80) 95% ee	328
	(−)-DET	(—)	531
	TI (1), (+)-DMT (1.1), −20°, overnight	(72) 87% ee, [α] −10.0°	642

TABLE I. ASYMMETRIC EPOXIDATION OF PRIMARY ALLYLIC ALCOHOLS (*Continued*)

Substrate	Conditions TI (eq), DAT (eq), MS	Product(s) and Yield(s) (%)	Refs.
MeO / Br / MeO / OH	TI (0.2), (+)-DMT (0.2)	MeO / Br / MeO / OH / O (85) 91% ee $[\alpha]$ -103.6°	643
dl OH	(+)-DIPT, -20°	OH / O (84)^{u} 100% de + OH / O 86% de	153
OH / BnO / OH	(+)-DET, -20°	HO / OH / O / BnO (100) 100% de	347
OH / Ph	(−)-DET (stoichiometric), -20°	OH / O / Ph (90) 91%ee $[\alpha]$ -33.05° ^{s}	145

156

TABLE I. ASYMMETRIC EPOXIDATION OF PRIMARY ALLYLIC ALCOHOLS (*Continued*)

Substrate	Conditions Tl (eq), DAT (eq), MS	Product(s) and Yield(s) (%)	Refs.
	(+)-DET	>95% ee	644
	(–)-DIPT, -25°, 3 d	(60) >95% ee [α] -4.1° g,j	645
	(+)-DET (stoichiometric), -20°, 11 h	(—) 84% ee	326
	(–)-DET (stoichiometric), -20°, 11 h	(—) 84% ee	326

157

TABLE I. ASYMMETRIC EPOXIDATION OF PRIMARY ALLYLIC ALCOHOLS (*Continued*)

Substrate	Conditions TI (eq), DAT (eq), MS	Product(s) and Yield(s) (%)	Refs.
	(+)-DET (stoichiometric), −20°, 11 h	 (−) 88% ee	326
	(−)-DET (stoichiometric), −20°, 11 h	 (−) 88% ee	326
	TI (0.1), (−)-DIPT (0.12), 3 Å, −50°	 (83) 82-86% de	646
	TI (1), (−)-DET (1.2), −50°, 24 h	 (89) 83% ee, [α] +56.2°	133
	TI (1), (−)-DET (1.2), −50°, 24 h	 (82) 73% ee, [α] +47.2°	133

TABLE I. ASYMMETRIC EPOXIDATION OF PRIMARY ALLYLIC ALCOHOLS (*Continued*)

Substrate	Conditions TI (eq), DAT (eq), MS	Product(s) and Yield(s) (%)	Refs.
C_{12}	(–)-DET, -25°	(30) 90-95% de [α] +65.3°	93
	TI (1), (+)-DET (1.1), -40°, 48 h	C_9H_{19}-n (53) >96% ee, [α] +11.9°	647
	TI (1), (+)-DIPT (1), -20°, 20 h	" (55) [α] -13.2°	648
	TI (1), (–)-DIPT (1), -40°, 48 h	C_9H_{19}-n (41) >96% ee, [α] -11.0°	647
n-C_9H_{19}	TI (1), (+)-DET (1.2), -45°, 3 d	n-C_9H_{19} (80) 96% ee	1, 649
EEO	(+)-DET	EEO (83) 92% ee [α] -18°	81
	(–)-DET	EEO (82) 92% ee [α] +17°	81

159

TABLE I. ASYMMETRIC EPOXIDATION OF PRIMARY ALLYLIC ALCOHOLS (*Continued*)

Substrate	Conditions Ti (eq), DAT (eq), MS	Product(s) and Yield(s) (%)	Refs.
	Ti (0.05), (+)-DIPT (0.08), -20 to 0°	(76) 79% de	224, 650-652
	Ti (0.05), (-)-DIPT (0.08), -20 to 0°	(92) 75% de	224
	(+)-DIPT, 4 Å, -30°	(97) >90% de [α] -30.67°	653
	Ti (1.19), (+)-DIPT (1.5), -20°, 5 h	(90)	113
	(-)-DET	(—)	531

TABLE I. ASYMMETRIC EPOXIDATION OF PRIMARY ALLYLIC ALCOHOLS (*Continued*)

Substrate	Conditions Ti (eq), DAT (eq), MS	Product(s) and Yield(s) (%)	Refs.
	(–)-DMT (stoichiometric), –23°, 12 h	(59)	76
	Ti (1.1), (–)-DET (1.1), –23°, 0.7 h	 (76) 90% de, [α] -21.0°	154
	Ti (0.05). (+)-DIPT (0.075), –20°, 48 h	(92)	654
	Ti (1). (–)-DET (1.3), –23°, 16 h	(80)	655

161

TABLE I. ASYMMETRIC EPOXIDATION OF PRIMARY ALLYLIC ALCOHOLS (*Continued*)

Substrate	Conditions TI (eq), DAT (eq), MS	Product(s) and Yield(s) (%)	Refs.
	TI (1), (–)-DET (1), -20°, 18 h	(57) [α] +5.43°	650, 652
dl	(DET), -20°	(—)	656
	(–)-DET	(90)^u >95% de + >95% de	90
	TI (0.1), (+) DET, (0.15), 4 Å, -40°, 2.5 h	(97) 93% ee 100% ee^i [α] -21.7°	657, 658

TABLE I. ASYMMETRIC EPOXIDATION OF PRIMARY ALLYLIC ALCOHOLS (*Continued*)

Substrate	Conditions Ti (eq), DAT (eq), MS	Product(s) and Yield(s) (%)	Refs.
MeO$_2$C ⋯ OH	(–)-DIPT, 4 Å, –20°	MeO$_2$C ⋯ OH (72) 99% de	75
t-BuO$_2$C ⋯ OH	(+)-DMT (catalytic), 3 Å, –23°, 20 h	*t*-BuO$_2$C ⋯ OH (80) 90% de	76
C$_6$H$_4$Me-*p* OH	Ti (1), (+)-DET (1.2), 3 Å, –50°, 48 h	O ⋯ OH ⋯ C$_6$H$_4$Me-*p* (75) 78% ee, [α] –48.4°	133
	Ti (1), (–)-DET (1.2), 3 Å, –50°, 48 h	O ⋯ OH ⋯ C$_6$H$_4$Me-*p* (76) 79% ee, [α] +53.5°	133
C$_{13}$ *n*-C$_{10}$H$_{21}$ ⋯ OH	(+)-DET	*n*-C$_{10}$H$_{21}$ O ⋯ OH >98% ee	212
	Ti (1)*p*, (+)-DET (1.2), –40°, 6 h	" (76) 96% ee, [α] +25.9°	141, 649
OH	Ti (1), (–)-DET (1.2), –20°, overnight	O ⋯ OH (71) 87% ee	659

163

TABLE I. ASYMMETRIC EPOXIDATION OF PRIMARY ALLYLIC ALCOHOLS (*Continued*)

Substrate	Conditions TI (eq), DAT (eq), MS	Product(s) and Yield(s) (%)	Refs.
	TI (1), (–)-DET (1.3), –20°, 16 h		655
	TI (3.0), (+)-DET (4.0), –23°	(80) [α] -14.3° (72)	84
	(+)-Tartrate	 (—)	163
	(–)-Tartrate	 (—)	163

TABLE I. ASYMMETRIC EPOXIDATION OF PRIMARY ALLYLIC ALCOHOLS (*Continued*)

Substrate	Conditions Ti (eq), DAT (eq), MS	Product(s) and Yield(s) (%)	Refs.
(structure with OTBDMS, MeO$_2$C, TBDMSO, OMe, OH)	(+)-DET	(structure with OTBDMS, MeO$_2$C, TBDMSO, OMe, O, OH) (—)	660
	(–)-DET	(—)	660
(structure with OTBDMS, MeO$_2$C, TBDMSO, OMe, OH)	(+)-DET	(—)	660
	(–)-DET	(—)	660
n-C$_{10}$H$_{21}$ ⌒ OH	Ti (1), (+)-DET (1.2), –45°, 3 d	n-C$_{10}$H$_{21}$ (epoxide) OH (80) 96% ee	649
	Ti (1.03)^p, (+)-DET (1.15), –40°, 10 h	" (43) >95% ee^z, [α] +7.8^{os}	141, 661

165

TABLE I. ASYMMETRIC EPOXIDATION OF PRIMARY ALLYLIC ALCOHOLS (*Continued*)

Substrate	Conditions: TI (eq), DAT (eq), MS	Product(s) and Yield(s) (%)	Refs.
	TI (1)^p, (−)-DET (1.2), −40°, 8 h	$n\text{-}C_{10}H_{21}$ (66) >95% ee^i [α] -7.8° ^{s}	649
	TI (1), (−)-DET (1.05), −40°, 4 d	" (80) 91% ee	59, 661
	TI (1), (+)-DET (1.7), −23°, 24 h	(95) [α] +124.7°	175
	1. TI (1.05), (+)-DIPT (2), 4 Å, −20°, 3 h 2. $n\text{-}Bu_4NF$	(85) [α] +26°	83
	(−)-DET	(—)	662

166

TABLE I. ASYMMETRIC EPOXIDATION OF PRIMARY ALLYLIC ALCOHOLS (*Continued*)

Substrate	Conditions TI (eq), DAT (eq), MS	Product(s) and Yield(s) (%)	Refs.
C_{14}			
n-C$_{11}$H$_{23}$ $\diagup$ OH	TI (1), (+)-DET (1.2), 4 Å, -20°, 15 h	n-C$_{11}$H$_{23}$ $\diagup$ OH (78) 94% ee [α] -25.5°	663
n-C$_5$H$_{11}$ OH	TI (1), (+)-DET (1), -23°, overnight	n-C$_5$H$_{11}$ OH (63) >90% ee, [α] -23.6°	664
"	TI (1), (+)-DET (1), -23°, overnight	(70) >90% ee, [α] -2.0°	665
OH C≡C TMS	TI (1.1), (−)-DET (1.2), -40°	OH C≡C TMS (−) 100% ee [α] -20.4°	666
n-C$_{11}$H$_{23}$ OH	TI (1.2), (+)-DIPT (1.3), -23°, 36 h	n-C$_{11}$H$_{23}$ OH (83.8) [α] +8.3°^{j,s}	667
"	TI (1)^p, (+)-DET (1.2), -40°, 8 h	(75), [α] +7.4°	141
n-C$_{11}$H$_{23}$ OH	TI (1.2), (−)-DIPT (1.3), -23°, 36 h	n-C$_{11}$H$_{23}$ OH (−) [α] -8.1°^{j,s}	667

TABLE I. ASYMMETRIC EPOXIDATION OF PRIMARY ALLYLIC ALCOHOLS (*Continued*)

Substrate	Conditions TI (eq), DAT (eq), MS	Product(s) and Yield(s) (%)	Refs.
	TI (1,1), (+)-DET (1.2), -40°	(—) >90% ee [α] -1.2°	666
	(+)-DET, -20°, 24 h	(8i) [α] -25.2°	283
	(–)-DET (stoichiometric), -20°	(87) >95% ee [α] +17.69°	668
	TI (1.2), (+)-DIPT (1.4), -20°	(79) 96% ee	92

TBDPSO, AcO, TMS, MeO, OMe, MeO₂C

OH

TABLE I. ASYMMETRIC EPOXIDATION OF PRIMARY ALLYLIC ALCOHOLS (*Continued*)

Substrate	Conditions TI (eq), DAT (eq), MS	Product(s) and Yield(s) (%)	Refs.
	(−)-DMT, 3 Å, −23°, 4 h	(−) 88% de	77, 78
C$_{15}$	TI (1), (+)-DET (1.3), −23°, 2 h	(59) 94% ee, [α] −9.56°	412, 126
TBDMSO	TI (1), (−)-DIPT (1.25), 4 Å, −20°, 16 h	(60) 90% de	669
n-C$_{12}$H$_{25}$	(+)-DET	(−)	670
HO	TI (0.05), (−)-DIPT (0.075), 4 Å, −20°, 5 h	(60)	671

169

TABLE I. ASYMMETRIC EPOXIDATION OF PRIMARY ALLYLIC ALCOHOLS (*Continued*)

Substrate	Conditions TI (eq), DAT (eq), MS	Product(s) and Yield(s) (%)	Refs.
	TI (0.05), (+)-DET (0.075), -20°, 12 h	(92)	654
	TI (1), (+)-DET (1.4), 4 Å, -23°, 40 h	(72) 85.4% ee [α] -7.2°	672
	TI (1), (−)-DET (1.4), 4 Å, -23°, 40 h	(65) 85.4% ee [α] +7.4°	672
	(+)-DET	(—)	531

170

TABLE I. ASYMMETRIC EPOXIDATION OF PRIMARY ALLYLIC ALCOHOLS (Continued)

Substrate	Conditions TI (eq), DAT (eq), MS	Product(s) and Yield(s) (%)	Refs.
	TI (1.07), (−)-DET (1.12), −50 to −45°, 1 h	(84) 96% ee, [α] +6.53°	213, 214, 179
	(+)-DIPT	(82) 80% ee	673
	TI (0.05), (+)-DET (0.074), 3 Å, −20°, 1.5 h	(−) 91% ee	18
	TI (1), (+)-DET (1), −20°	" (87) >95% ee	1
	TI (1), (+)-DBTA (1.2), −20°	" (−) 96% ee	130
	TI (2), (+)-DET (1)	" (−) 78% ee	130
	TI (2),(+)-DBTA (1), −20°	(−) 82% ee	130
	Ti (1), (+)-DIPT (1), −20°, 15 h	(81) >90% ee [α] +60.9°ˢ	58

TABLE I. ASYMMETRIC EPOXIDATION OF PRIMARY ALLYLIC ALCOHOLS (*Continued*)

Substrate	Conditions TI (eq), DAT (eq), MS	Product(s) and Yield(s) (%)	Refs.
	Ti (1), (+)-DIPT (1), -20°, 15 h	(81) >90% ee [α] +39.7° s	58
	Ti (1), (+)-DIPT (1), -20°, 15 h	(81) >70% ee [α] +148.2° s	58
	TI (1), (+)-DET (1), -20°, 14 h	(80.6), [α] -6.5° g	674
	TI (0.1), (−)-DIPT (1.17), 4 Å, -45°, overnight	(76) >95% ee	675

172

TABLE I. ASYMMETRIC EPOXIDATION OF PRIMARY ALLYLIC ALCOHOLS (*Continued*)

Substrate	Conditions TI (eq), DAT (eq), MS	Product(s) and Yield(s) (%)	Refs.
	(–)-DET (catalytic), 4 Å, –20°	(92)	179
C_{16}	TI (1.02), (–)-DET (1.11), –45°, 1 h	" (84) 96% ee, [α] +6.53°	213, 214
MOMO, $n\text{-}C_{12}H_{25}$	TI (1.07), (–)-DET (1.29), –20°, 15 h	MOMO OH (63) 95% de [α] +21.7°	124
MeO_2C	TI, (+)-DMT	MeO_2C OH (—) 95% ee	676
$(CH_2)_{12}$ OH	TI (1), (+)-DET (1.3), –23°, 21 h	$(CH_2)_{12}$ OH (77), [α] –7.13°	412
C_{17} $C_{14}H_{29}\text{-}n$ OH	TI (0.1), (+)-DET (0.13), 3 Å, –12°, 11 h	$C_{14}H_{29}\text{-}n$ OH (91) 96% ee, [α] –10.9°	18
	(+)-DET (stoichiometric)[c] –20°	" (51)	130

TABLE I. ASYMMETRIC EPOXIDATION OF PRIMARY ALLYLIC ALCOHOLS (*Continued*)

Substrate	Conditions TI (eq), DAT (eq), MS	Product(s) and Yield(s) (%)	Refs.
n-$C_{14}H_{29}$ —OH	TI (1), (+)-DIPT (1), -27°, 18 h	" (26) >94% ee, [α] -9.45°	677
	TI (2)^y, (+)-DET (1), 0°	$C_{14}H_{29}$-n —OH (73) 68% ee	130
n-$C_{12}H_{25}$ (OAc) —OH	(+)-DIPT (stoichiometric)	n-$C_{14}H_{29}$ —OH (77) >95% ee [α] -27.0°^{e}	299, 678
(AcO, O-ring structure with OH)	TI (1), (+)-DET (1), -25°, overnight	n-$C_{12}H_{25}$ —OH (70) [α] -14°	679
	TI (3.6), (+)-DET (5), -20°, overnight	(OAc, O-ring, AcO, H structure) (87) [α] -28°	680
C_8H_{17}-n (polyene) —OH	TI (1), (+)-DIPT (1.2), -30°, 9 d	C_8H_{17}-n —OH (50) 58% ee [α] -2.9°^{d}	681

TABLE I. ASYMMETRIC EPOXIDATION OF PRIMARY ALLYLIC ALCOHOLS (*Continued*)

Substrate	Conditions TI (eq), DAT (eq), MS	Product(s) and Yield(s) (%)	Refs.
	(+)-DET	 (—)	531
	TI (1), (+)-DET (1.1), -20°, 1.5 h	 (98) 90% ee, [α] -14.9°	682
	(-)-DMT, 3 Å, -23°, 4 h	 (80) 88% de	77, 78

175

TABLE I. ASYMMETRIC EPOXIDATION OF PRIMARY ALLYLIC ALCOHOLS (*Continued*)

Substrate	Conditions TI (eq), DAT (eq), MS	Product(s) and Yield(s) (%)	Refs.
(Ph, Ph, —OH substrate)	TI (1.2), (–)-DET (1.5), -20°, 5 h	(90) 94% ee	411
C$_{18}$			
n-C$_{15}$H$_{31}$ —OH	TI (1.2), (–)-DET (1.6), -20°, 38 h	(75) [α] +22.5°	650-652, 321
n-C$_{13}$H$_{27}$C≡C —OH	TI (1)^c, (–)-DET (1.2), -25°, 4-5 h^w	(86)^w >98% ee [α] -2.2°	102, 103
(—OH, C$_9$H$_{19}$-n substrate)	TI (1), (+)-DIPT (1.2), -30°, 9 d	71% ee [α] -3.4°^{d}	681
	TI (1.2), (+)-DMT (1.4), -23 to -30°, 69 h	" (77) 68% ee	683, 684
(—OH, C$_9$H$_{19}$-n substrate)	TI (1), (–)-DIPT (1.2), -30°, 9 d	60% ee [α] +3.3°^{d}	681
	TI (1), (–)-DMT (1), -20°, 2 d	" (70) 65% ee	683, 684

TABLE I. ASYMMETRIC EPOXIDATION OF PRIMARY ALLYLIC ALCOHOLS (*Continued*)

Substrate	Conditions Tl (eq), DAT (eq), MS	Product(s) and Yield(s) (%)	Refs.
BOMO / OTMS structure with OH	Tl (0.05), (–)-DET (0.08), 4 Å, -23°, 16 h	(94) 48% ee, [α] +6.8°	685
cyclic alkenol with OH	Tl (1), (+)-DET (1.3), -23°	(—) >90% ee	686
"	Tl (1), (+)-DET (1.3), -23°, 16.5 h	" (74), [α] -5.82°	412
C$_{19}$ Ph(CH$_2$)$_{10}$ D epoxide OH	Tl (0.05), (+)-DET (0.05), 4 Å, -22°, 2.5 h	Ph(CH$_2$)$_{10}$ D epoxide OH (65) >97% ee [α] -19.15° [d]	687
	Tl (0.05), (–)-DET (0.05), 4 Å, -22°, 2.5 h	Ph(CH$_2$)$_{10}$ D epoxide OH (60) >97% ee [α] +19.3° [d]	687

TABLE I. ASYMMETRIC EPOXIDATION OF PRIMARY ALLYLIC ALCOHOLS (*Continued*)

Substrate	Conditions Ti (eq), DAT (eq), MS	Product(s) and Yield(s) (%)	Refs.
MOMO i-Pr(CH$_2$)$_9$ ⋯ OH	Ti (0.4), (−)-DIPT (0.5), 4 Å, −20°, 15 h	MOMO i-Pr(CH$_2$)$_9$ ⋯ O OH (90) 50% de	688
OH C$_{10}$H$_{21}$-n	Ti (1), (+)-DIPT (1.2), −30°, 9 d	O OH C$_{10}$H$_{21}$-n (−) 90% ee [α] -3.4° ^{d}	681
	Ti (1), (−)-DIPT (1.2), −30°, 9 d	O OH C$_{10}$H$_{21}$-n (−) 81% ee [α] +4.0° ^{d}	681
OAc MeO O ... O	Ti (1), (+)-DET (1.2), 4 Å, −20°, 15 h	OAc MeO O ... O OH (80) 96% ee	122
C$_{20}$ OH C$_{11}$H$_{23}$-n	Ti (1), (+)-DIPT (1.2), −30°, 9 d	O OH C$_{11}$H$_{23}$-n (−) 87% ee [α] -3.4° ^{d}	681

TABLE I. ASYMMETRIC EPOXIDATION OF PRIMARY ALLYLIC ALCOHOLS (Continued)

Substrate	Conditions TI (eq), DAT (eq), MS	Product(s) and Yield(s) (%)	Refs.
	TI (1), (–)-DIPT (1.2), -30°, 9 d	(–) 81% ee [α] +3.8° d	681
	TI (0.106), (+)-DET (0.127), -20°, 4 h	(51) 95% ee	689
	TI (0.106), (–)-DET (0.127), -20°, 4 h	(57) 95% ee [α] +7.88°	689
	1. TI (1.1), (+)-DIPT (1.5), 4 Å, -20°, 4 h 2. Bu₄NF	(98) [α]+3.0°	82, 83

179

TABLE I. ASYMMETRIC EPOXIDATION OF PRIMARY ALLYLIC ALCOHOLS (*Continued*)

Substrate	Conditions TI (eq), DAT (eq), MS	Product(s) and Yield(s) (%)	Refs.
	TI (1), (+)-DET (1), -20°, 18 h	(57)	690
	TI (1), (−)-DET (1), -20°, 18 h	(36)	690
	TI (0.4), (−)-DET (0.6), 4 Å, -25°, 2 h	(93) 80% de	690
	TI (1), (+)-DET (1), -20°, 18 h	(74)	691

TABLE I. ASYMMETRIC EPOXIDATION OF PRIMARY ALLYLIC ALCOHOLS (*Continued*)

Substrate	Conditions TI (eq), DAT (eq), MS	Product(s) and Yield (s) (%)	Refs.
	TI (1), (–)-DET (1), –20°, 18 h	 (65)	691
	TI (4.6, (–)-DET (4.7), –50°, 2 h	 (65) 50% de, [α] +28°	692
	(–)-DMT, –20°	 (85) 95% ee [α] +4.39° s	693-695
	TI (1), (–)-DIPT (1), 4 Å, –30°, 16 h	" (95), [α] +4.88°	653, 696

181

TABLE I. ASYMMETRIC EPOXIDATION OF PRIMARY ALLYLIC ALCOHOLS (Continued)

Substrate	Conditions TI (eq), DAT (eq), MS	Product(s) and Yield(s) (%)	Refs.
C21	TI (1), (+)-DET (1), -20°, 20 h	(80) [α] -8.90°	650-652
	TI (5.8), (-)-DET (5.8), -10°, 40 h	(85) 53% ee	120, 121
C23	(-)-DET	(—)	697
C25	(+)-DET (stoichiometric)	(—) >95% ee, [α] -23.8°	438

182

TABLE I. ASYMMETRIC EPOXIDATION OF PRIMARY ALLYLIC ALCOHOLS (Continued)

Substrate	Conditions Ti (eq), DAT (eq), MS	Product(s) and Yield(s) (%)	Refs.
C_{26} 	Ti (1), (−)-DET (1), −20°, overnight	 (72) >95% de, [α] +42° [g]	698
C_{27} 	Ti (0.8), (+)-DET (0.8), −25°, 18 h	 (86) >98% de, [α] −38.4°	79

183

TABLE I. ASYMMETRIC EPOXIDATION OF PRIMARY ALLYLIC ALCOHOLS (*Continued*)

Substrate	Conditions TI (eq), DAT (eq), MS	Product(s) and Yield(s) (%)	Refs.
	TI (0.8), (–)-DET (0.8), -25°, 18 h	 (79) >98% de, [α] -4.8°	79
	TI (1), (–)-DET (1), -25°, 24 h	 (82) 90% de	699
	TI (2.4), (+)-DET (2.4), 3 Å, -25°, 16 h	 (74) >95% ee	700

TABLE I. ASYMMETRIC EPOXIDATION OF PRIMARY ALLYLIC ALCOHOLS (*Continued*)

Substrate	Conditions TI (eq), DAT (eq), MS	Product(s) and Yield(s) (%)	Refs.
	TI (1), (+)-DET (1), −20°, 18 h	(86) 98% de, [α] -38.4°	79
	TI (1), (−)-DET (1), −20°, 18 h	(70) 98% de, [α] -4.8°	79
C$_{30}$	TI (1), (+)-DET (1), −20°, 1 h	(70) 100% de, [α] +43°	701

185

TABLE I. ASYMMETRIC EPOXIDATION OF PRIMARY ALLYLIC ALCOHOLS (*Continued*)

Substrate	Conditions TI (eq), DAT (eq), MS	Product(s) and Yield(s) (%)	Refs.
	TI (1), (−)-DET (1), −20°, 1 h	(82) 100% de, [α] +64°	701
	TI (0.2), (+)-DET (0.2), 4 Å, −20°, 1 h	(87) 87% ee, [α] -3.31°	358, 702
	TI (1), (+)-DMT (1), −20°, 3 h	" (70) 80% ee, [α] -3.05°	703
	TI (0.2), (−)-DET (0.2), 4 Å, −20°, 1 h	(93) 78% ee, [α] -5.8°	358

186

TABLE I. ASYMMETRIC EPOXIDATION OF PRIMARY ALLYLIC ALCOHOLS (*Continued*)

Substrate	Conditions TI (eq), DAT (eq), MS	Product(s) and Yield(s) (%)	Refs.

a The oxidant was cumene hydroperoxide.

b The product was isolated as the mesylate.

c The titanium alkoxide was Ti(OBu-*t*)$_4$.

d The rotation was measured in dichloromethane.

e The rotation was measured in benzene.

f The titanium alkoxide was titanium-pillared montmorillonite.

g The rotation was measured in methanol.

h The reaction was slow.

i The % ee was reported for recrystallized material.

j The rotation was reported for recrystallized material.

k The rotation was measured in diethyl ether.

l The reaction was quenched with triethanolamine.

m A single isomer was obtained.

n The optical purity was enhanced by the sequence: 3,5-dinitrobenzoylation, recrystallization, and hydrolysis.

o The intermediate allylic alcohol produced in situ by the ene reaction of singlet oxygen with 2-*tert*-butylpropene was oxidized with Ti(OBu-*t*)$_4$ and (+)-DET.

p The reaction was conducted in the presence of catalytic amounts of calcium hydride and silica gel.

q The reaction furnished a diastereomeric mixture containing a marginal excess of 2*R*,3*R* isomer.

r The rotation was reported for neat material.

s The rotation was measured in ethanol.

t The rotation was reported for chromatographically purified material.

u The total yield of diastereomeric products was reported.

v The titanium alkoxide was TiCl$_2$(OPr-*i*)$_2$. The resulting 2-chloromethylhexane-1,3-diol was converted into the epoxy alcohol by treatment with base.

w The reaction was carried out in a 1:1 mixture of 2,3-dimethyl-2-butene and dichloromethane.

TABLE II. ASYMMETRIC EPOXIDATION OF BISALLYLIC ALCOHOLS

Substrate	Conditions TI (eq), DAT (eq), MS	Product(s) and Yield(s) (%)	Refs.
C$_8$	(+)-DIPT (0.6), 4 Å, -20°, 84 h	(96) 70% ee, [α] -18.4°	53
C$_{10}$	(+)-Tartrate (catalytic)	(65)	70
C$_{11}$	(+)-DET (0.6), 4 Å, -40°, 13 h	(81)	72
C$_{12}$	TI (2.5), (+)-DET (3), -15°, 18 h	(27) 99.45% ee[a]	68, 69
	TI (2.5), (+)-DET (3), -15°, 18 h	(—)	68
C$_{13}$	TI (0.6), (+)-DET (0.6), -20°, 4 d	(58.4) [α] -33.8°	71

[a] The ee was calculated (see text, p. 14).

188

TABLE III. ASYMMETRIC EPOXIDATION OF ALKENYLSILANOLS

Substrate	Conditions TI (eq), DAT (eq), MS	Product(s) and Yield(s) (%)	Refs.
C$_8$			
Ph Si OH / Me Me	(+)-DET (stoichiometric), -20°, 40 h	O Ph Si OH / Me Me (≤0) 85-90% ee	54
Ph Si OH / Ph Ph	(+)-DET (catalytic), MS^a, -20°, 40 h	O Ph Si OH / Ph Ph 20% ee	54
	(+)-DET (stoichiometric), -20°, 40 h	" (70) 7% ee	54

^{a} The type of molecular sieves used was not reported.

TABLE IV. KINETIC RESOLUTION OF RACEMIC PRIMARY ALLYLIC ALCOHOLS

Substrate	Conditions	Allylic Alcohol(s) and Yield(s) (%)	Epoxy Alcohol(s) and Yield(s) (%)	Refs.
C$_9$ — allene: H, Me, C=C=C, Bu-n, OH	(–)-DIPT, -20°	allene: H, Me, C=C=C, Bu-n, OH — (55) 40% ee. [α] +1.0° (—)		146
C$_{10}$ — cyclohexene with CH$_2$OH and isopropenyl group	Ti (1.0), (+)-DIPT (1.2), TBHP (0.6), -20° 50-60% conversion	cyclohexene with CH$_2$OH, isopropenyl, H — (—) 48% ee		145
C$_{11}$ — Ph-CH(CH$_3$)-CH=CH-CH$_2$OH	"	Ph, OH (—) 6% ee		145
Ph-CH(CH$_3$)-C(=CH$_2$)-CH$_2$OH	"	Ph, OH (—) >95% ee		145
Ph-CH(CH$_3$)-CH=CH-CH$_2$OH (cis)	"	Ph, OH (—) 80% ee		145
H, C=C=C, n-C$_7$H$_{15}$, H, CH$_2$OH	"	H, C=C=C, n-C$_7$H$_{15}$, H, OH (—) 40% ee		145

190

TABLE IV. KINETIC RESOLUTION OF RACEMIC PRIMARY ALLYLIC ALCOHOLS (*Continued*)

Substrate	Conditions	Allylic Alcohol(s) and Yield(s) (%)	Epoxy Alcohol(s) and Yield(s) (%)	Refs.
+ enantiomer	TI (1.0), (+)-DIPT (1.2), TBHP (0.6), -20° 50-60% conversion	(—) 85% ee (—)		145
+ enantiomer R =	TI (1.5), (+)-DET (1.52), TBHP (3.0), -20°, 12 h	(15) >95% ee	(29) >95% ee + (19) >95% ee	168, 169
C₁₂	TI (1.0), (+)-DIPT (1.2), TBHP (0.6), -20° 50-60% conversion	(—) 70% ee	(—)	145, 146

TABLE IV. KINETIC RESOLUTION OF RACEMIC PRIMARY ALLYLIC ALCOHOLS (Continued)

Substrate	Conditions	Allylic Alcohol(s) and Yield(s) (%)	Epoxy Alcohol(s) and Yield(s) (%)	Refs.
C13 structure + enantiomer	(−)-DIPT	(−)	total (90) >95% ee	90
C13 structure	TI (1.0), (+)-DET (1.2), TBHP (0.5), −23°, 50 min	(62) >95% ee $[\alpha]$ -64.7°	(30) >95% ee	56
C17–C22 structure	TI (0.5), (+)-DIPT (0.6), TBHP (0.5), 4 Å, −16°, 5 d	OH	OH I + II	449

	R		anti (I)	anti:syn	syn (II)
C17	Me	(−) 10% ee	(−)	54:46	(−)
C18	Et	(−) 31% ee	(−) 82% ee	65:35	(−)
C19	Pr-i	(−) 65% ee	(−) 85% ee	93:7	(−)
C21	C5H11-n	(−) 36% ee	(−)	68:32	(−)
C22	C6H11	(−) 65% ee	(−) 75% ee	90:10	(−)

TABLE IV. KINETIC RESOLUTION OF RACEMIC PRIMARY ALLYLIC ALCOHOLS (Continued)

Substrate	Conditions	Allylic Alcohol(s) and Yield(s) (%)	Epoxy Alcohol(s) and Yield(s) (%)	Refs.
C_{17} + enantiomer	Ti (1.0), (+)-DET (1.2), TBHP (0.58), −23°, 2.5 h	(45) 70% ee, $[\alpha]$ +39.4° [a]	(—)	704
C_{18} + enantiomer	Ti (1.0), (+)-DET (1.2), TBHP (0.6), −23°, 10 min	(89) 95% ee	(—)	705
n-Bu + enantiomer	Ti (1.0), (+)-DIPT (1.2), −20°		n-Bu	412

	n	$\dfrac{\text{TBHP}}{}$	t	Allylic Alcohol(s)	Epoxy Alcohol(s)	
C_{18}	10	0.6	42 min	(35) 98% ee, $[\alpha]$ +67.8°	(58) 56% ee	
C_{20}	12	0.6	20 min	(40) 99% ee, $[\alpha]$ +133°	(31) 76% ee	
	12	1.5	6.8 h	(—)	(62) 0% ee	
C_{21}	13	0.6	1.5 h	(41) 90% ee, $[\text{c}]$ +121.1°	(4) 66% ee	
	13	1.5	5 h	(—)	(55) 0% ee	
C_{22}	14	0.6	16 min	(33) 0% ee	(29) 80% ee	
	14	1.5	3.5 h	(—)	(62) 0% ee	

Substrate	Conditions	Allylic Alcohol(s) and Yield(s) (%)	Epoxy Alcohol(s) and Yield(s) (%)	Refs.
C_{25} + enantiomer	Ti (1.0), (+)-DIPT (1.2), TBHP (0.6), −23°, 10 min	(42) >90% ee, $[\alpha]$ +46.9°	(—)	686

[a] The type of solvent was not reported.

TABLE V. ASYMMETRIC EPOXIDATION OF HOMO-, BISHOMO-, AND TRISHOMOALLYLIC ALCOHOLS

Substrate	Conditions TI (or ZR^a) (eq), DRT (eq), MS	Product(s) and Yield(s) (%)	Refs.
C$_4$	(+)-DET^b, 0°, 4 d	(11-25) 55% ee [α] +12.48° ^{c}	176
	Zr (1), (+)-DCTA^d (1.3), 0°,12 d	" (4) 40% ee	178
C$_5$	Zr (1), (+)-DCTA^d (1.3), 0°, 9 d	(25) 77% ee	178
C$_6$	(+)-DET^b, -20°	(35-40) 41% ee [α] +17.69° ^{e}	176
	Zr (1), (+)-DCTA^d (1.3), 0°,14 d	" (38) 43% ee	178
	(+)-DET^b, 0°, 5 d	(50) 36% ee [α] +5.35° ^{e}	176
	(+)-DET^b, -20°	" (30) 50% ee	176
	Zr (1), (+)-DCTA^d (1.3), 0°, 8 d	" (23) 72% ee	178
	(+)-DET^b, -20°	(41) 27% ee	176
	(+)-DET^b, -20°	^{f} (60) 23% ee [α] +1.46° ^{e}	176

TABLE V. ASYMMETRIC EPOXIDATION OF HOMO-, BISHOMO-, AND TRISHOMOALLYLIC ALCOHOLS (*Continued*)

Substrate	Conditions: Tl (or ZR[a]) (eq), DRT (eq), MS	Product(s) and Yield(s) (%)	Refs.
C$_7$	(+)-DET[b], −20°	(15) [c] +1.63°[e]	176
	Zr (1), (+)-DCTA[d] (1.3), 0°, 8 d	(28) 74% ee	178
	(+)-DET[b], −20°	(62) 48% ee [c] +2.70°[e]	176
C$_8$	Tl (1), (+)-DET (1.2), 144 h	(50) 60% ee [α] +4.8°	140
	Tl (1)[g], (+)-DET (1.2), 24 h	" (60) 60% ee, [α] +4.6°	140
C$_9$	Zr (1), (+)-DCTA[d] (1.3), 0°, 16 d	(21) 53% ee	178
C$_{10}$	Tl (1), (+)-DET (1.17), 0°, 48 h	(22) 29% ee	706

Substrate	Conditions TI (or ZR[a]) (eq), DRT (eq), MS	Product(s) and Yield(s) (%)	Refs.
C$_{11}$	TI (1), (+)-DET (1.17), 0°, 48 h	f (<10)	706
C$_{12}$	TI (1), (+)-DET (1.17), 0°, 48 h	f (49) 56% ee	706
	(–)-DET, (catalytic), 3 Å, –23°, 4 d	(80) 50% de	76
	TI (1.1), (–)-DET (1.3), (MS)[h,i], 0–23°, 15 h	(74j)	179

196

TABLE V. ASYMMETRIC EPOXIDATION OF HOMO-, BISHOMO-, AND TRISHOMOALLYLIC ALCOHOLS (Continued)

Substrate	Conditions TI (or ZR[a]) (eq), DRT (eq), MS	Product(s) and Yield(s) (%)	Refs.
	(+)-DET, (MS)[h,i]	(—) 14% de	179
dl	TI (1)[k], (+)-DET (1.5)	(35) [α] -1.3°	126

[a] ZR represents Zr(OPr-n)₄.
[b] The Ti(OPr-i)₄:DRT ratio was 1:1.1–1.2.
[c] The rotation was measured in dichloromethane.
[d] DCTA represents (R,R)-(+)-N,N'-dicyclohexyltartramide.
[e] The rotation was measured in ethanol.
[f] The stereochemistry of the product was not reported.
[g] The reaction was conducted in the presence of catalytic amounts of calcium hydride and silica gel.
[h] The type of molecular sieves used was not reported.
[i] The oxidant was trityl hydroperoxide.
[j] The reaction did not produce an appreciable amount of the diastereomeric 6R,7R-epoxide.
[k] The reaction was carried out in dichloroethane.

TABLE VI. KINETIC RESOLUTION OF RACEMIC SECONDARY ALLYLIC ALCOHOLS

Substrate	Conditions	Allylic Alcohol(s) and Yield(s) (%)	Epoxy Alcohol(s) and Yield(s) (%)	Refs.
		A. Secondary Allylic Alcohols		
C$_4$	Ti (0.1), (–)-DIPT (0.12), 41.5 h	(–) 88% ee	(—)	38
	Ti[a] (0.1), (–)-DIPT (0.12), 51 h	" (—) 72% ee	(—)	38
	(–)-DIPT	(—)	(51) 91% ee [α] -16.3° [b]	707, 708
	Ti (1.0), (–)-DIPT (1.2), TBHP (0.5), -20°	(—)	" (27) >95% ee	709
	Ti (1.0), (–)-DIPT(1.2), TBHP (0.6), -20°, 26 h	OBn (37) 95% ee [α] +5.9°	OBn (41) 94% ee [α] -10.4°	581, 582
	Ti (0.1), (–)-DIPT (0.12), 3 Å, TBHP (0.6), -15°, 14 d	(—)	(41) >95% ee, [α] -11.23°	710, 711
	(–)-DIPT, 4 Å, -20°, 30 h	(—)	(31) >95% ee [α] -29.9°	712
	Ti (1.0), (+)-DIPT (1.2), TBHP (0.6), -20°, 6 d	(–) 91% ee [α] -7.7° [c]	(—)	10
C$_5$	—	TMS (—)	(—)	713

198

TABLE VI. KINETIC RESOLUTION OF RACEMIC SECONDARY ALLYLIC ALCOHOLS (*Continued*)

Substrate	Conditions	Allylic Alcohol(s) and Yield(s) (%)	Epoxy Alcohol(s) and Yield(s) (%)	Refs.
	Ti (1.0), (+)-DIPT (1.2), TBHP (0.6), -20°, 15 h	(—)	(24)	714
	Ti (1.0), (–)-DIPT (1.2), TBHP (0.6), -20°, 15 h	(—)	(23)	714
	(+)-DIPT, 70% conversion	(—) 100% ee	(—)	715
	Ti (0.04), (+)-DIPT (0.06), cumene hydroperoxide (1.95), 4 Å, -5°	" 100% ee, [α] -24°	(—)	716
	—	(—)ᵈ	(—)	717
	Ti (1.0), (+)-DIPT (1.2), TBHP (0.6), -20°, 8 d	(32) 92% ee, [α] -2.60°	(24), [α] +5.93°	95, 96
	Ti (1.0), (–)-DIPT (1.2), TBHP (0.6), -20°, 8 d	(44) 92% ee, [α] +2.64°	(34), [α] -5.93°	96

199

TABLE VI. KINETIC RESOLUTION OF RACEMIC SECONDARY ALLYLIC ALCOHOLS (*Continued*)

Substrate	Conditions	Allylic Alcohol(s) and Yield(s) (%)	Epoxy Alcohol(s) and Yield(s) (%)	Refs.
(chloro allylic alcohol, Cl–CH₂CH₂–CH(OH)–CH=CH₂)	(+)-DIPT	[structure: Cl] (−) 94% ee [α] -7.9°	(−)	718
(OMPM-substituted diol)	(−)-DIPT, -20°	(−)	[structure: OMPM epoxy alcohol] (45) 88% ee	719, 720
(methyl-substituted allylic alcohol)	(+)-Tartrate	[structure, OH] (−)	(−)	721, 722
	TI (0.1), (−)-DIPT (0.12), 10 h	[structure, OH] (−) 89% ee	(−)	38
	TI[a] (0.1), (−)-DIPT (0.12), 10 h	" (−) 92% ee	(−)	38
(two allylic alcohols +)	—	[structures] OH + OH (−) 60% ee	(−)	723
(OBn-substituted allylic alcohol)	TI (1.0), (+)-DIPT (1.2), -20°, overnight	[structure] OBn (30) 79% ee	(−)	724
(Ph₂P(O)-substituted allylic alcohol)	TI (0.5), (+)-DET (0.6), TBHP (0.5), 4 Å, -16°, 5 d	[structure] Ph₂P(O) 95% ee	[structure: epoxy alcohol with Ph₂P(O)] >95% ee	449

200

TABLE VI. KINETIC RESOLUTION OF RACEMIC SECONDARY ALLYLIC ALCOHOLS (Continued)

Substrate	Conditions	Allylic Alcohol(s) and Yield(s) (%)	Epoxy Alcohol(s) and Yield(s) (%)	Refs.
(cyclopentenol, OH)	—	(—)	(—)OH (—) 60% ee	725
C6 (C≡CTIPS, OH)	TI (1.0), (–)-DIPT (1.1) TBHP (1.5), -20°, 16 h	(—)	C≡CTIPS OH (40) 90% ee, [α] +26.2°	112
Pr-i (OH)	TI (1.0), (–)-DMT TBHP (0.5), -20°	(—)	Pr-i OH (27) 93% ee	709
(OH)	—	(–) 98% ee [α] -4.924	(—)	726
	(+)-DIPT	(39) [α] +4.8°	(—)	727
(CO2Et, OH)	TI (1.2), (+)-DIPT (1.5) TBHP (3.0), -25°, 96 h	CO_2Et OH (16) >95% ee [α] +14.6°cc	(—)	414

201

TABLE VI. KINETIC RESOLUTION OF RACEMIC SECONDARY ALLYLIC ALCOHOLS (Continued)

Substrate	Conditions	Allylic Alcohol(s) and Yield(s) (%)	Epoxy Alcohol(s) and Yield(s) (%)	Refs.
	(−)-DIPT	96% ee	(−)	728
	TI (0.1), (−)-DIPT (0.12), 12 h	" (−) 92% ee	(−)	38
	TI[a] (0.1), (−)-DIPT (0.12), 20 h	" (−) 82% ee	(−)	38
	TI (1.0), (−)-DIPT (1.2), TBHP (0.4), −20°	(−)	(35) >95% ee	145
	(+)-Tartrate	(−)	(35) >95% ee	729
	TI (1.0), (+)-DIPT (1.2), TBHP (0.6), −20°, 4 d	(−) 30% ee [α] +37.1°	(−)	10
	TI (1.0), (+)-DIPT, CaH$_2$ (cat.), SiO$_2$ (cat.), TBHP (0.6), −20°, 6 h	(25) 94% ee [α] −4.3°	(24) 96% ee [α] −2.6°	140
	TI (0.5), (+)-DET (0.6), 4 Å, TBHP (0.5), −16°, 5 d	91% ee	>95% ee	449

TABLE VI. KINETIC RESOLUTION OF RACEMIC SECONDARY ALLYLIC ALCOHOLS (*Continued*)

Substrate	Conditions	Allylic Alcohol(s) and Yield(s) (%)	Epoxy Alcohol(s) and Yield(s) (%)	Refs.
Bu-*n* structure, OH	—	(—) >95% ee	(—)	730
	TI (1.0), (+)-DET (1.1), TBHP (1.0), -20°, overnight	" (43.5) >90% ee, [α] -6.0°	(—)	731
	TI (1.0), (–)-DIPT (1.17), TBHP (1.39), -20 41, 40 h	(—)	(37) 94% ee [α] -24.5°	732
	TI (0.5), (+)-DIPT (0.6), TBHP (2), -20°, 4.5 h	(—)	(49) [α] -11.2° [b]	733
	TI (0.5), (–)-DIPT (0.6), TBHP (2), -20°, 4.5 h	(—)	(48) [α] +11° [b]	733
	(–)-DIPT, -10°		" [α] +8.7° [b]	734
Bu-*t* structure, OH	TI (1.0), (+)-DET (1.0)	(—)	(60) 10% ee	57
structure, OH	TI (0.1), (+)-DIPT (0.15)[e], 3 Å, -20°, 12 h	(28) [α] +3.4°		735, 736
structure, OH	TI (0.1), (+)-DIPT (0.15)[e], 3 Å, -20°	(36) [α] +2.0°		735, 736

203

TABLE VI. KINETIC RESOLUTION OF RACEMIC SECONDARY ALLYLIC ALCOHOLS (*Continued*)

Substrate	Conditions	Allylic Alcohol(s) and Yield(s) (%)	Epoxy Alcohol(s) and Yield(s) (%)	Refs.
(structure: 1,3-dioxolane with allylic OH)	(+)-DIPT	(55) 48% ee	(—)	737
(structure: C≡CTMS allylic OH)	TI (0.1), (+)-DCHT (0.15), 3 Å, -20°, 44 h	(35) 98% ee	(—)	738
(structure: dioxinone with allylic OH)	TI (1.1), (+)-DIPT (1.2), 4 Å, TBHP (1.5), -20°, 44 h	(46) 95% ee	(44) 98% ee, [α] -11.6°	493
(structure: methylenecyclohexanol)	TI (0.13), (-)-DIPT (0.17), 4 Å, TBHP (0.67), -25°, 20 h	(46) 80% ee [α] +6.7°	(—)	739
(structure: Pr-*i* allylic OH)	—	(—) 90% ee	(—)	721, 740-742
(structure: Pr-*i* allylic OH)	TI (0.1), (-)-DIPT (0.12), 18 h	(—) 90% ee	(—)	38
	TI[a] (0.1), (-)-DIPT (0.12), 16 h	" (—) 47% ee	(—)	38

Substrate	Conditions	Allylic Alcohol(s) and Yield(s) (%)	Epoxy Alcohol(s) and Yield(s) (%)	Refs.
(structure) OH	Ti (0.13), (+)-DIPT (0.21), TBHP (0.42), –25°	(35) 90% ee	(43.5) 90% ee	743
	Ti (0.13), (–)-DIPT (0.21), TBHP (0.45), –20°, 5 d	(39) 90% ee	(40) 90% ee	183
	(–)-DCHT	(—)	(—) >96% ee	744
(structure) OH	Ti (1.0), (–)-DIPT (1.5), TBHP (0.4), –20°, 18 h	(30) 72% ee, [α] –8.9°	(27) >95% ee, [α] +3.0°	183, 247
	Ti (0.1), (+)-DIPT (0.16), TBHP (0.42), –20°, 43 h	(46) 68% ee	(37) >95% ee	183
	Ti (0.1), (–)-DIPT (0.15), 4 Å, TBHP (0.5), –25°, 20 h	(—)	" (43) >96% ee	322
BnO (structure) OH	(+)-DET, –20°	(—)	BnO (structure) OH (46)	745
I $\langle\ \rangle_3$ CO$_2$Me OH	Ti (0.31), (–)-DIPT (0.37), 4 Å, TBHP (1.5), –20° 48 h	I $\langle\ \rangle_3$ CO$_2$Me OH (38) >98% ee	(—)	746

205

TABLE VI. KINETIC RESOLUTION OF RACEMIC SECONDARY ALLYLIC ALCOHOLS (*Continued*)

Substrate	Conditions	Allylic Alcohol(s) and Yield(s) (%)	Epoxy Alcohol(s) and Yield(s) (%)	Refs.
(cycloheptenol) OH	TI (1.0), (+)-DIPT (1.1), TBHP (0.6), -20°, 4 d	(—) 80% ee [α] +25° b	(—)	10
(methylcyclohexenol) OH 69% ee	(+)-DIPT	(31) 69% ee OH	(41) 50% ee OH	747
	(—)-DIPT	(10) 38% ee OH	(59) 86% ee OH	747
C8 $C_5H_{11}\text{-}n$ OH	TI (1.0), (—)-DIPT (1.2) TBHP (2.0), -20°, 14.5 h	(—)	$C_5H_{11}\text{-}n$ (42) 93% ee [α] -22.3°	748,749
	TI (0.56), (+)-DIPT (0.67) TBHP (1.13), -20°, 16 h	(—)	$C_5H_{11}\text{-}n$ (41) [α] +24.7°	749
(tetrahydropyranyl) OH	(—)-DIPT, -20°, 6 d	(—) >98% ee [α] +14.9° b	(—)	750
(dienol) OH	—	(—) 78% ee, [α] +7.3°	(—)	751

TABLE VI. KINETIC RESOLUTION OF RACEMIC SECONDARY[v] ALLYLIC ALCOHOLS (*Continued*)

Substrate	Conditions	Allylic Alcohol(s) and Yield(s) (%)	Epoxy Alcohol(s) and Yield(s) (%)	Refs.
structure with (CH₂)₃C≡CTMS, OH	Ti (0.1), (+)-DCHT (0.15), 3 Å, −20°, 44 h	(—)	(—)	738
structure with C≡CH, OTBDPS, OH	(−)-DCHT, −20°	(46) 98% ee	(—)	752
Bu-n, OH structure	Ti (1.0), (+)-DIPT (1.2), TBHP (0.6), −20°, 15 h	Bu-n (—) >96% ee [α] +3.2° [c]	(—) >96% ee	10
	Ti(0.1), (+)-DIPT (0.15) TBHP (0.7), −20°, 27 h	" (44) 94% ee, [α] +3.6°[f]	(—)	18
	Ti(0.1), (+)-DCHT (0.15) TBHP (0.7), −20°, 29 h	" (44) 97% ee, [c] +3.8°[f]	(—)	18
	Ti(0.1), (+)-DCDT (0.15) TBHP (0.7), −20°, 24 h	" (42) >98% ee, [α] +4.2°[f]	(—)	18
C≡CTMS, OH structure (Bu-t)	(−)-DIPT, TBHP (0.6), −22°	C≡CTMS, OH structure (—)	epoxy structure C≡CTMS, OH (—) 93% ee	753
OH structure (Bu-t)	Ti (1.2)[a], (+)-DIPT (1.5)	(—)	epoxy structure, Bu-t, OH (60) 30% ee	57

207

TABLE VI. KINETIC RESOLUTION OF RACEMIC SECONDARY ALLYLIC ALCOHOLS (*Continued*)

Substrate	Conditions	Allylic Alcohol(s) and Yield(s) (%)	Epoxy Alcohol(s) and Yield(s) (%)	Refs.
TMS, C$_5$H$_{11}$-n, OH (allylic alcohol)	Ti (1.0), (+)-DIPT (1.2), TBHP (0.6), -20°, 24 h	TMS, C$_5$H$_{11}$-n, OH (44) 87% ee [α] +6.8°	TMS, O, C$_5$H$_{11}$-n, OH (8) 79% ee [α] -7.0°	191, 13
carbamate O–C(O)N(Pr-i)$_2$ structure, *dll*	Ti (1), (+)-DIPT (1.2), TBHP (0.55), -16°, 24 h	N(Pr-i)$_2$ carbamate, OH (59) 41% ee, [α] +8.6° f	N(Pr-i)$_2$ carbamate, O epoxide, OH (26) 57% ee, [α] -14.6° d	754
n-Bu, OH	—	n-Bu, OH (—)	(—)	730
OBn, OH (+ enantiomer)	(+)-DET	(—)	OBn, O epoxide, OH (35)	755
	(–)-DET	(—)	OBn, O epoxide, OH (35)	755

208

TABLE VI. KINETIC RESOLUTION OF RACEMIC SECONDARY ALLYLIC ALCOHOLS (*Continued*)

Substrate	Conditions	Allylic Alcohol(s) and Yield(s) (%)	Epoxy Alcohol(s) and Yield(s) (%)	Refs.
(Bu-*n* allylic alcohol)	—	(—)	(—)	721, 722
(Bu-*t* allylic alcohol)	TI (1.2), (+)-DMT (1.5) (—)		(—)	57
(Pr-*i* allylic alcohol)	(+)-DIPT	Pr-*i* (—) 67% ee (—) [α] +7.2^c		756
	—	" (—) 78% ee	(—)	757
	(+)-DIPT	(—)	(—) 96% ee, [α] -28.3^{oc}	758
(CO$_2$Et allylic alcohol)	TI (1.0), (—)-DET (1.5) TBHP (0.7), -25°, 25 min (—)		(—), [α] +17.6°	414

209

TABLE VI. KINETIC RESOLUTION OF RACEMIC SECONDARY ALLYLIC ALCOHOLS (*Continued*)

Substrate	Conditions	Allylic Alcohol(s) and Yield(s) (%)	Epoxy Alcohol(s) and Yield(s) (%)	Refs.
(structure: OBn, C≡C, OH)	TI (1.0), (–)-DET (1.0), TBHP (0.6), -20°	(—)	(structure: O epoxide, C≡C, OBn, OH) (—)	157
(structure: Ph₂P=O, OH) + enantiomer	TI (0.5), (+)-DET (0.6), 4 Å, TBHP (0.5) -16°, 5 d	(structure: Ph₂P=O, OH) 89% ee	(structure: Ph₂P=O, O epoxide, OH) >98% ee	449
(structure: Ph₂P=O, OH) + enantiomer	TI (0.5), (+)-DET (0.6), 4 Å, TBHP (0.5) -16°, 5 d	(structure: Ph₂P=O, OH) 37% ee	(structure: Ph₂P=O, O epoxide, OH) 52% de	449
(cyclohexenyl-CH(CH₃)OH)	TI (1.0), (+)-DIPT (1.2), TBHP (0.6), -20°, 15 h	(—) >96% ee	(—)	10
"	TI (0.1), (+)-DIPT (0.15), 3 Å, TBHP (0.7), -20°, 3.5 h	(46) >98% ee, [α] +3.29° c	(—)	18
(cyclohexenyl-CH(CH₃)OH)	TI (0.1), (+)-DCHT (0.15), 3 Å, TBHP (0.7), -20°, 3.5 h	(43) >98% ee, [α] +3.59° c	(—)	18

TABLE VI. KINETIC RESOLUTION OF RACEMIC SECONDARY ALLYLIC ALCOHOLS (Continued)

Substrate	Conditions	Allylic Alcohol_s) and Yield(s) (%)	Epoxy Alcohol(s) and Yield(s) (%)	Refs.
	Tl (0.1), (+)-DCDT (0.15), 3 Å, TBHP (0.7), -20°, 4 h	" (43) >98% ee, [α] +3.16° ^{c}	(—)	18
	Tl (0.5), (+)-DET (0.6), 4 Å, TBHP (0.5), -16°, 5 d	80% ee	1:1	449
	(–)-DET, -15°, 50% conversion	(—) 50% ee	(—) 78% ee	759
	(–)-DET, -15°, 75% conversion	" (—) 88% ee	" (—) 19% ee	759
	Tl (1.0), (+)-DIPT (1.2) TBHP (0.6), -20°, 1 h	(41) >99% ee, [α] +17.5°	(42) >99% ee, [α] -4.86°	191, 13

Substrate	Conditions	Allylic Alcohol(s) and Yield(s) (%)	Epoxy Alcohol(s) and Yield(s) (%)	Refs.
C9				
C_6H_{13}-n, OH (structure)	Ti (1.0), (+)-DIPT (1.2), TBHP (0.6), -20°, 12 d	C_6H_{13}-n (—) >96% ee [α] -19.1°c	C_6H_{13}-n (—) OH [α] -4.0°	10, 408
	Ti (0.1), (+)-DIPT (0.15), 3 Å, TBHP (0.6), -20°, 13 d	" (45) 86% ee, [α] -14.9°c	(—)	18
	Ti (0.1), (+)-DCHT (0.15), 3 Å, TBHP (0.7), -20°, 7.5 d	" (41) >98% ee, [α] -17°c	(—)	18
	Ti (0.1), (+)-DCHT (0.15), 3 Å, TBHP (1.5), -20°, 63 h	" (28) 95% ee	(—)	18
	Ti (0.1), (+)-DCDT (0.15), 3 Å, TBHP (0.7), -20°, 11 d	" (34) >98% ee	(—)	18
(structure), OH	—	(—)	(structure) O, OH	760
Ph, OH (structure)	(+)-DIPT	Ph (31) 95% ee [α] -7.8° g, OH	(—)	761, 762
Ph, OH (structure)	Ti (1.0), (−)-DMT (1.5), TBHP (0.5), -20°	(—)	Ph (23) 90% ee O, OH	709

TABLE VI. KINETIC RESOLUTION OF RACEMIC SECONDARY ALLYLIC ALCOHOLS (Continued)

Substrate	Conditions	Allylic Alcohol(s) and Yield(s) (%)	Epoxy Alcohol(s) and Yield(s) (%)	Refs.
C_6F_5 allylic alcohol (OH)	Ti (0.1), (+)-DIPT (0.16), TBHP (0.68), 4 Å, -20°, 27 d	C_6F_5 OH (47) 97% ee	C_6F_5 OH (45) 97% ee	763
C_6H_{11} allylic alcohol (OH)	Ti (1.0), (+)-DET (1.1), TBHP (1.0), -20°, overnight	C_6H_{11} OH (—) >90% ee [α] -9.5°	(—)	731
	Ti (1), (+)-DIPT (1.2), TBHP (0.6), -10°, 2 d	" (42) >99% ee, [α] -92° b	(—)	764
	Ti (1.1), (-)-DIPT (1.2), TBHP (1.6), -20°, 3 d	C_6H_{11} OH (32) >98% ee [α] +26° c	(—)	765
(structure, OH)	Ti (1.0), (+)-DIPT (1.2), TBHP (0.6), -20°, 15 h	(42) 63% ee, [α] +26.9° h	(—)	766, 727
(structure, OH)	Ti (0.1), (+)-DIPT (0.12), TBHP (0.4), -20°, 3 d	(50)	(30) >96% ee	186
(structure, OH)	Ti (0.4), (-)-DIPT (0.45), TBHP (0.45), -20°, 24 h	(—)	(40) 92% ee, [α] +8.6° f	184

TABLE VI. KINETIC RESOLUTION OF RACEMIC SECONDARY ALLYLIC ALCOHOLS (Continued)

Substrate	Conditions	Allylic Alcohol(s) and Yield(s) (%)	Epoxy Alcohol(s) and Yield(s) (%)	Refs.
n-C$_5$H$_{11}$ —CF$_2$H, OH	Ti (1.0), (+)-DIPT (1.2), TBHP (0.55), -20°, 4 d	n-C$_5$H$_{11}$ —CF$_2$H, OH (31) 97% ee, [α] -11.7° [b]	n-C$_5$H$_{11}$ —CF$_2$H, OH (59)	767
	Ti (0.13), (+)-DIPT (0.19), TBHP (0.55), -20°, 7 d	(39) 98% ee, [α] -11.76° [b]	" (52)	767
n-C$_5$H$_{11}$ —CH$_2$F, OH	Ti (1.0), (+)-DIPT (1.2), TBHP (0.55), -20°, 24 h	n-C$_5$H$_{11}$ —CH$_2$F, OH (43) 98% ee, [α] -0.39° [b]	n-C$_5$H$_{11}$ —CH$_2$F, OH (56)	767
	Ti (0.13), (+)-DIPT (0.19), TBHP (0.55), -20°, 5 d	" (41) 93% ee, [α] -0.35° [b]	" (46)	767
n-C$_5$H$_{11}$ —, OH	Ti (1.0), (+)-DIPT (1.2), TBHP (0.55), -20°, 24 h	n-C$_5$H$_{11}$ —, OH (40) 98% ee	n-C$_5$H$_{11}$ — (—)	767
	Ti (0.13), (+)-DIPT (0.19), TBHP (0.55), -20°, 5 d	" (34) 97% ee	" (—)	767
NC —, C$_5$H$_{11}$-n, OH	Ti (1.0), (+)-DIPT (1.2), 5°, 3d	NC —, C$_5$H$_{11}$-n, OH (40) 95% ee	NC —, C$_5$H$_{11}$-n, OH (—)	106

214

TABLE VI. KINETIC RESOLUTION OF RACEMIC SECONDARY ALLYLIC ALCOHOLS (Continued)

Substrate	Conditions	Allylic Alcohol(s) and Yield(s) (%)	Epoxy Alcohol(s) and Yield(s) (%)	Refs.
	Ti (1.0), (+)-DAT (1.2), –20° DAT / TBHP / Time DCHT 0.7 / 24 h DCHT 0.4 / 24 h DET 0.4 / 24 h DET 0.7 / 7 d DCHT 0.7 / 2 d DIPT 2.0 / 36 h DCHT 1.5 / 10 h	 (—) (—) (—) (30) 69% ee (24) 84% ee (15) 90% ee (13) >99% ee, [α] -145.6°	 (33) 32% ee (25) 67% ee (17) 78% ee (—) (—) (—) (—)	768
	Ti (0.2), (+)-DET (0.24), TBHP (0.7), –20°, 14 d	(—)	(—)	
	—	(—)	(—) 90% ee	769
	Ti (0.1), (+)-DIPT (0.12), TBHP (0.45), –30°, 15 h	(50) 65% ee, [α] +10.8° [c]	(37) 95% ee, [α] -50.0° [c]	770

215

TABLE VI. KINETIC RESOLUTION OF RACEMIC SECONDARY ALLYLIC ALCOHOLS (Continued)

Substrate	Conditions	Allylic Alcohol(s) and Yield(s) (%)	Epoxy Alcohol(s) and Yield(s) (%)	Refs.
(structure)	TI (0.1), (–)-DIPT (0.12), TBHP (0.45), -30°, 15 h	(structure) OH (38) 60% ee, [α] -9.0° c	(structure) OH (45) >99% ee, [α] +53° c	770
(structure, cycloheptenone with OH)	(–)-DET, -15°	(structure) OH 75% ee	(structure) OH 20% ee	759
(structure, trimethylcyclohexenol with OH)	(+)-DET	(structure) OH (—) 52% ee [α] +49.8°	(—)	771
C$_{10}$ (structure, NHBoc, CO$_2$Me, OH)	(+)-DIPT	(—)	(structure, NHBoc CO$_2$Me, OH) (20) 95% de	772
(structure, NHCbz, CO$_2$Me, OH)	(–)-DIPT	(—)	(structure, NHCbz CO$_2$Me, OH) (—) 94% de, [α] -15.8°	773

216

TABLE VI. KINETIC RESOLUTION OF RACEMIC SECONDARY ALLYLIC ALCOHOLS (*Continued*)

Substrate	Conditions	Allylic Alcohol(s) and Yield(s) (%)	Epoxy Alcohol(s) and Yield(s) (%)	Refs.
[structure] X = Aib-L-Phe-D-Pro	Ti (1.1), (–)-DIPT (1.3), TBHP (0.6)	[structure] (48)	[structure] (27) 95% de	774
[structure] X = D-Pro-L-Ala-D-Ala	Ti (1.0), (–)-DIPT (1.2), 4 Å, TBHP (2.0), –20°, 5 d	[structure] (38)	[structure] (20)	775
[structure OTBDMS]	(+)-DIPT	(—)	[structure] (25) 92% de, [α] +2.7°	776
[structure] C_6H_{11} OH	Ti (1.0), (+)-DIPT (1.2), TBHP (0.6), –20°	C_6H_{11} (—) 93% ee [α] +6.2°	(—)	777
[structure] C_6H_{11} OH	Ti (1.0), (+) DIPT (1.2), TBHP (0.6), –20°, 15 h	C_6H_{11} (—) >96% ee, [α] -14.6°c	C_6H_{11} OH (—)	10
	Ti (0.1), (+)-DIPT (0.15), TBHP (0.7), –20°, 15 h	" (44.2) 94% ee, [α] -13.3°c	(—)	18

TABLE VI. KINETIC RESOLUTION OF RACEMIC SECONDARY ALLYLIC ALCOHOLS (Continued)

Substrate	Conditions	Allylic Alcohol(s) and Yield(s) (%)	Epoxy Alcohol(s) and Yield(s) (%)	Refs.
	Ti (0.1), (+)-DCHT (0.15) TBHP (0.55), -20°, 16 h	" (44.2) 95% ee, [α] -13.2°c	(—)	18
	Ti (0.1), (+)-DCDT (0.15) TBHP (0.55), -20°, 16 h	" (44.2) >98% ee, [α] -13.6°c	(—)	18
	Ti (0.1), (−)-DIPT (0.12), 17 h	C6H11 (—) 97% ee, OH	(—)	38
	Ti^a (0.1), (−)-DIPT (0.12), 17 h	" (—) 20% ee	(—)	38
	(+)-DIPT	Ph (—) 72% ee, OH	(—)	730
	(+)-DIPT	" (37) >99% ee, [α] +24.5°	(—)	778, 779
Ph OH	Ti (1), (+)-DIPT (1.2), TBHP (0.5)	(—)	Ph epoxy, OH (—) >98% ee	340, 343
m-MeOC6H4 OH	Ti (0.1), (+)-DIPT (0.15), TBHP (0.7), -20°, 3 h	(—)	m-MeOC6H4 epoxy, OH (48) [α] -40.4° f	580
t-Bu OH (cyclopropyl)	(−)-DIPT, TBHP, -20°, 48 h	t-Bu (cyclopropyl), OH (42) 95% ee, [α] +31.1°	(—)	780

218

TABLE VI. KINETIC RESOLUTION OF RACEMIC SECONDARY ALLYLIC ALCOHOLS (Continued)

Substrate	Conditions	Allylic Alcohol(s) and Yield(s) (%)	Epoxy Alcohol(s) and Yield(s) (%)	Refs.
$n\text{-}C_6H_{13}$ (CF$_3$, OH) + enantiomer	Ti (1.0), (+)-DIPT (1.2) TBHP (0.55), -20°, 5 d	$n\text{-}C_6H_{13}$ (CF$_3$, OH) (46) 60% ee, [α] -1.9° [b]	$n\text{-}C_6H_{13}$ (O, CF$_3$, OH) (50)	767
	Ti (0.13), (+)-DIPT (0.19) TBHP (0.55), -20°, 21 d	" (51) 47% ee, [α] -1.51° [b]	" (47)	767
(tricyclic structure, HO, H, H) + enantiomer	Ti (1.0), (+)-DIPT (1.15) TBHP (0.6), -20°, 4 h	(54) 68% ee [α] +78°	(38) 39% ee [α] -32.6°	781
(bicyclic structure, OH)	(+)-DIPT	(—)	(—)	145
(diene structure, OH)	(+)-DIPT, 90% conversion	(—) >99% ee	(—)	145
	(-)-DIPT, 90% conversion	(—) >99% ee	(—)	145

TABLE VI. KINETIC RESOLUTION OF RACEMIC SECONDARY ALLYLIC ALCOHOLS (*Continued*)

Substrate	Conditions	Allylic Alcohol(s) and Yield(s) (%)	Epoxy Alcohol(s) and Yield(s) (%)	Refs.
(cyclohexylidene alcohol, C_{11})	TI (1.0), (+)-DIPT (1.2) TBHP (0.6), -20°	(—) 95% ee	(—)	777
(Ph alcohol)	TI (1.0), (+)-DIPT (1.2) TBHP (0.6), -20°	(—) 98% ee [α] -0.6°	(—)	777, 727
(alkynyl allylic alcohol)	TI (1.0), (+)-DIPT (1.2) TBHP (0.6), -20°	(35) 95% ee [α] +26°	(—) [α] +22°	782
(NHBoc, C_6H_{11} alcohol)	TI (1.2), (+)-DIPT (1.2), -20°, 48 h	" (35)	" (33)	783
(dienyl CO_2Et alcohol)	(−)-DIPT, -25°	(—)[b]	(—)	55
(trienyl CO_2Et alcohol)	TI (0.1), (−)-DIPT (0.15), 3 Å, -20°	(—)	(29) [α] +13.6° [c]	784, 785

TABLE VI. KINETIC RESOLUTION OF RACEMIC SECONDARY ALLYLIC ALCOHOLS (Continued)

Substrate	Conditions	Allylic Alcohol(s) and Yield(s) (%)	Epoxy Alcohol(s) and Yield(s) (%)	Refs.
	Ti (0.1), (–)-DIPT (0.15), 3 Å, –20°	(—)	 (57) [α] +72.0° c	784, 785
	Ti (0.1), (–)-DIPT (0.15), 3 Å, –20°	(—)	 (29) [α] +51.1° c	784, 785
	Ti (1), (+)-DIPT (1.2), TBHP (0.6), –25°, 15 h	 C6H11 (34) 87% ee [α] -5.8°	 C6H11 (36) 75% ee [α] -24.7°	786
	(+)-DIPT, TBHP (0.6)	 Ph (40) 94% ee	(—)	787
	(–)-DIPT	 (30) 98% ee	(—)	788

Substrate	Conditions	Allylic Alcohol(s) and Yield(s) (%)	Epoxy Alcohol(s) and Yield(s) (%)	Refs.
F F C_5H_{11}-*n* OH	TI (1), (+)-DIPT (1.2) TBHP (0.55), -20°, 4 d	C_5H_{11}-*n* [i] OH (—)	(—)	767
OH *n*-C_5H_{11}	TI (1), (+)-DIPT (1.2) TBHP (0.6), -20°, 2 d	(45) 14% ee, [α] -2.77° [b] OH *n*-C_5H_{11} (—) 82% ee, [α] -17.1° [c]	O *n*-C_5H_{11} OH (—)	10
7 OH + enantiomer 3 OH + enantiomer	TI (0.3), (+)-DET (0.36) TBHP (0.6), -35°, 7.5 h	(<43) 55% ee OH O (4)	OH O (52) 61% ee [α] -0.7° [c]	56
C_{12} TBDPSO O OBn HO HO	TI (1.4), (+)-DET (1.4) TBHP (2.0), -20°, 12 h	TBDPSO O OBn HO HO (47) [α] +35.5° [f]	TBDPSO O OBn HO HO O (42) [α] +20.5° [f]	789

222

TABLE VI. KINETIC RESOLUTION OF RACEMIC SECONDARY ALLYLIC ALCOHOLS (*Continued*)

Substrate	Conditions	Allylic Alcohol(s) and Yield(s) (%)	Epoxy Alcohol(s) and Yield(s) (%)	Refs.
	TI (1), (+)-DIPT (1.2), TBHP (0.6), -25°, 3 h	(45) 61% ee [α] +26.5°	(34) 89% ee [α] -74.0°	786
	TI (1), (+)-DIPT (1.2), TBHP (0.6), -25°, 14 h	" (35) 97% ee [α] -42.4°	" (33) 83% ee [α] -69.5°	786
	(−)-DIPT	(—) 99% ee	(—)	728
	(+)-DIPT	(—)		790
	TI (1.0), (+)-DIPT (1.2), TBHP (0.6), -23°	(—) 95% ee, [α] +38.7°	(—)	782
	TI (1.0), (+)-DIPT (1.0), TBHP (0.6), -50°, 10 h	(39) [α] +26.8° [c]	(—) 84% ee, [α] +2.5°	791, 792

TABLE VI. KINETIC RESOLUTION OF RACEMIC SECONDARY ALLYLIC ALCOHOLS (Continued)

Substrate	Conditions	Allylic Alcohol(s) and Yield(s) (%)	Epoxy Alcohol(s) and Yield(s) (%)	Refs.
(OMe-substituted tetrahydronaphthalenyl ethanol structure)	Ti (1.0), (–)-DIPT (1.0), TBHP (0.6), –50°, 7 h	(47) [α] -128.3°	(46) 98% ee [α] +81.6°	793
	(+)-DIPT	(—)	(41) 98% ee [α] -81.9°	793
(di-OMe tetrahydronaphthalenyl ethanol structure)	Ti (1.0), (–)-DIPT (1.0), TBHP (0.53), –75°, 1 h	(48) >95% ee, [α] -19.2°	(43) >95% ee, [α] +51.6°	794
C_{13} (C₁₀H₂₁-n allylic alcohol structure)	Ti (1.0), (+)-DIPT (1.2), TBHP (2.0), –28°, 15 d	(—)	$C_{10}H_{21}$-n (40) 91% ee [α] +16.2°	141
	Ti (1.0), (+)-DIPT (1.2), CaH₂, SiO₂, TBHP (2.0), –40°, 25 h	(—)	" (42) 85% ee, [α] +15.2°	141
	Ti (1.0), (+)-DIPT (1.2), TBHP (2.0), –20°, 17 h	(—)	" (50) 86-94% ee, [α] +16.2°	795

TABLE VI. KINETIC RESOLUTION OF RACEMIC SECONDARY ALLYLIC ALCOHOLS (Continued)

Substrate	Conditions	Allylic Alcohol(s) and Yield(s) (%)	Epoxy Alcohol(s) and Yield(s) (%)	Refs.
(cyclopropyl–C≡CC$_5$H$_{11}$-n, vinyl, OH)	Ti (1.0), (–)-DIPT (1.2), TBHP (2.0), -20°, 17 h	(—)	C$_{10}$H$_{21}$-n (40) 91% ee [α] -16.6°	795
—	—	(cyclopropyl–C≡CC$_5$H$_{11}$-n, vinyl, OH) (—) (—)		796
(Ph, vinyl, methylene, OH)	Ti (1), (+)-DIPT (1.2), TBHP (0.6), -25°, 11 h	Ph (37) 94% ee [α] +48.2°	Ph (33) 82% ee [α] -45.2°	786
HC≡C...C$_5$H$_{11}$-n, OH	Ti (0.2), (–)-DIPT (0.24), 4 Å, TBHP (0.6), -20°, 20 h	C$_5$H$_{11}$-n (42) 98.6% ee OH	(—)	746
C$_{14}$ (pinene, vinyl, OH)	Ti (0.05), (+)-DIPT (0.06), 3 Å, CUHP (0.55), -8°, 12 h	(35)	(—)	736
(pinene, vinyl, OH)	Ti (0.05), (–)-DIPT (0.06), 3 Å, CUHP (0.55), -8°, 12 h	(35)	(—)	736

225

TABLE VI. KINETIC RESOLUTION OF RACEMIC SECONDARY ALLYLIC ALCOHOLS (*Continued*)

Substrate	Conditions	Allylic Alcohol(s) and Yield(s) (%)	Epoxy Alcohol(s) and Yield(s) (%)	Refs.
(cyclohexenone with CH₂Ph, OH side chain)	TI (0.1), (+)-DIPT (0.15), 3 Å, TBHP (0.7), -20°, 9 h	(35) Ph, OH	(39) 75% ee Ph, OH	797
C₁₅ (7 and 3 terpene vinyl alcohols)	TI (1.0), (+)-DIPT (1.2), TBHP (2.8), -18°, 3 wks	(—)	(46) 7 and 3 epoxy alcohols	798
(2,4-dichlorophenyl substituted alcohol with Cl, Cl)	TI (1.0), (+)-DIPT (1.0), TBHP (0.5), -20°, 15 h	(42) 90% ee, [α] -8.2°	(38) 84% ee, [α] -76.3°	799

226

TABLE VI. KINETIC RESOLUTION OF RACEMIC SECONDARY ALLYLIC ALCOHOLS (*Continued*)

Substrate	Conditions	Allylic Alcohol(s) and Yield(s) (%)	Epoxy Alcohol(s) and Yield(s) (%)	Refs.

Row 1:

Substrate: C$_6$H$_{11}$, OH (allylic alcohol with C$_6$H$_{11}$ substituent)

Conditions: Ti (1.0), (−)-DIPT (1.0) TBHP (0.5), −20°, 15 h

Allylic Alcohol(s): (dichlorophenyl/dichlorophenyl substituted alcohol, Cl, Cl, Cl, OH) (—)

Epoxy Alcohol(s): (dichlorophenyl substituted epoxy alcohol, Cl, Cl, Cl, OH, O) 799

Row 2:

Conditions: Ti (1.0), (+)-DIPT (1.2) TBHP (0.6), −20°, 15 h

Allylic Alcohol(s): C$_6$H$_{11}$, OH (—) >96% ee, [α] −19.8° ^{c}

Epoxy Alcohol(s): C$_6$H$_{11}$, O, OH (41) 84% ee, [α] +69.7° 10

Row 3:

Substrate: Ph, OH, OH (ketone with Ph chain)

Conditions: (+)-DIPT

Allylic Alcohol(s): (—)

Epoxy Alcohol(s): C$_6$H$_{11}$, O, OH (27) 91% ee " 800

Row 4:

Conditions: Ti (0.1), (+)-DIPT (0.15), 3 Å, TBHP (0.7), −20°, 9 h

Allylic Alcohol(s): Ph, O, OH (31)

Epoxy Alcohol(s): Ph, O, O (20) 60% ee 797

Row 5:

C$_{16}$

Substrate: CO$_2$R, *n*-C$_6$H$_{13}$, OH

Conditions: Ti (1.0), (+)-DIPT (0.45), −25°, 18 h

Allylic Alcohol(s): (—)

Epoxy Alcohol(s): *n*-C$_6$H$_{13}$, O, CO$_2$R, OH (30–49) 415

R = *p*-[*n*-C$_{10}$H$_{21}$O]C$_6$H$_4$

TABLE VI. KINETIC RESOLUTION OF RACEMIC SECONDARY ALLYLIC ALCOHOLS (Continued)

Substrate	Conditions	Allylic Alcohol(s) and Yield(s) (%)	Epoxy Alcohol(s) and Yield(s) (%)	Refs.
(structure: $C_{12}H_{15}$-n allylic alcohol with OH)	TI (1.0), (+)-DIPT (1.2) TBHP (0.6), $-20°$, 15 h	(structure: $C_{12}H_{15}$-n, OH) (44) >96% ee, $[\alpha]$ +3.6°	(structure: epoxide $C_{12}H_{15}$-n, OH) (49)	124
(structure: C_6H_{11}, OH)	TI (1.0), (+)-DIPT (1.2) TBHP (0.6), $-20°$, 15 h	(structure: C_6H_{11}, OH) (—) 10% ee, $[\alpha]$ -0.7° [c]	(—)	10
(structure: n-C_6H_{13}, (CH$_2$)$_5$, OH, CO_2Me)	(+)-DET, 3 Å, 0°	(structure: n-C_6H_{13}, OH, CO_2Me) (10) 93–94% ee	(structure: n-C_6H_{13} epoxide, OH, CO_2Me) (66) + (structure: n-C_6H_{13} epoxide, OH, CO_2Me) (23)	801
C_{17} (structure: bicyclic with OH)	TI (1.1), (+)-DET (1.4), TBHP (0.66)	(structure, OH) (26) 15% ee, $[\alpha]$ +2.9°	(structure with O epoxide) (35), $[\alpha]$ -1.3°	126

Substrate	Conditions	Allylic Alcohol(s) and Yield(s) (%)	Epoxy Alcohol(s) and Yield(s) (%)	Refs.
C$_{18}$	Ti (1.2), (+)-DIPT (1.2), TBHP (0.6), –10°, 48 h	(37)	(40)	783
C$_{20}$	Ti (1.0), (+)-DIPT (1.2), TBHP (0.6), –20°, 15 h	(32)	(47)	802

TABLE VI. KINETIC RESOLUTION OF RACEMIC SECONDARY ALLYLIC ALCOHOLS (Continued)

B. Secondary E-Trimethylsilylvinyl Carbinols

	Substrate	Conditions	Allylic Alcohol(s) and Yield(s) (%)	Epoxy Alcohol(s) and Yield(s) (%)	Refs.
C_4	TMS〜OH	(–)-DIPT, TBHP (0.6)	(—) >95% ee	(—)	803
	TMS〜R OH	TI (1.0), (+)-DIPT (1.2), TBHP (1.5), –20°	TMS〜R OH	TMS△R OH	

	R	Time (h)			
C_4	CH_2OPh	13	(47) >99% ee, [α] +8.0°	(46) >99% ee, [α] -17.0°	13
	CH_2OBn	9.5	(43) >99% ee, [α] -1.9°	(48) >99% ee, [α] -2.2°	13, 194
	CH_2OTBS	-	(46) >99% ee	(48) >99% ee	194
C_5	$(CH_2)_2OBn$	9	(43) >99% ee, [α] -3.2°	(45) >99% ee, [α] -10.1°	13, 194
C_6	Pr-i	6	(40) >99% ee, [α] -21.8°	(41) >99% ee, [α] -1.07°	13
C_7	$(CH_2)_3CO_2Me$	20	(43) >99% ee, [α] +6.78°	(45) >99% ee, [α] +6.7°	13, 193

	Substrate	Conditions	Allylic Alcohol(s) and Yield(s) (%)	Epoxy Alcohol(s) and Yield(s) (%)	Refs.
C_7	TMS〜OH〜CO_2Me	TI (1.0), (–)-DIPT (1.2), TBHP (1.5), –21°, 21 h	(43) >99% ee	(45) >99% ee	193
	TMS〜R OH	TI (0.2), (+)-DIPT (0.24), 3 Å, TBHP (1.5), –20°	TMS〜R OH	TMS△R OH	

	R	Time (h)			
C_7	Bu-t	40	(34) 13% ee	(64)	13
C_8	C_5H_{11}-n	2	(42) >99% ee	(44) 94.9% ee	13

TABLE VI. KINETIC RESOLUTION OF RACEMIC SECONDARY ALLYLIC ALCOHOLS (Continued)

Substrate	Conditions	Allylic Alcohol(s) and Yield(s) (%)	Epoxy Alcohol(s) and Yield(s) (%)	Refs.
TMS—R with OH (C8)	Ti (1.0), (+)-DIPT (1.2) TBHP (1.5), −20°	TMS—R with OH	TMS—epoxide—R with OH	
R: C_5H_{11}-n	Time (h): 7	(42) >99% ee, [α] −9.8°	(42) >99% ee, [α] −7.5°	13, 194
"	18	(—) >99% ee	(—) >97.6% ee	13, 194
"	2	(42) >99% ee	(—) >94.9% ee	13
(cis chain)	2	(44) >99% ee	(42) 97.3% ee	804
C_9 Ph	13.5	(44) >99% ee, [α] −10.8°	(42) 97.3% ee, [α] +25.7°	13
C_{11} —C_5H_{11}	3.5	(44) >99% ee, [α] +7.59°	(43) >99% ee, [α] +4.23°	13, 194
				805
$CH_2C\equiv CC_5H_{11}$-n	4	(41) >99% ee, [α] −55.3°	(47) >99% ee, [α] +16.1°	805

C_{11}

TBDMSO (chain) TMS—OH	Ti (1.0), (−)-DIPT (1.2) TBHP (1.5), −20°, 16 h	TBDMSO (chain) TMS—OH	TBDMSO (chain) TMS—epoxide—OH	805
		(43) >99% ee, [α] −4.7°	(42) >99% ee, [α] +6.2° j	

C. Secondary E-Tributylstanny vinyl Carbinols

| n-Bu$_3$Sn—R with OH | Ti (0.3), (+)-DIPT (0.36) TBHP (1.5), −21° | n-Bu$_3$Sn—R with OH | n-Bu$_3$Sn—epoxide—R with OH | |

231

TABLE VI. KINETIC RESOLUTION OF RACEMIC SECONDARY ALLYLIC ALCOHOLS (*Continued*)

Substrate	Conditions	Time (h)	Allylic Alcohol(s) and Yield(s) (%)	Epoxy Alcohol(s) and Yield(s) (%)	Refs.
R					
C_4 CH$_2$OPh		4	(40) >99% ee	(—)	11
C_7 (CH$_2$)$_3$CO$_2$Me		—	(>44) >99% ee	(—)	803
C_8 C$_5$H$_{11}$-n		40	(—) >99% ee	(—) 80% ee	11
C_9 C$_6$H$_{11}$		4	(41) >99% ee	(—)	11
C_{11} (CH$_2$)$_7$CO$_2$Me		36	(>40) >99% ee	(—)	803
C_7 n-Bu$_3$Sn⎯⎯Bu-n, OH	Ti (0.1), (+)-DIPT (0.18), 3 Å, TBHP (0.7), -20°		(—)	n-Bu$_3$Sn⎯⎯Bu-n, O, OH (45) >90% ee, [α] -24.3°	198
C_8 n-Bu$_3$Sn⎯⎯C$_5$H$_{11}$-n, OH	Ti (1.0), (–)-DIPT (1.2) TBHP (1.5), -20°, 4 h		n-Bu$_3$Sn⎯⎯C$_5$H$_{11}$-n, OH (38-42) 95% ee, [α] +3.1°	n-Bu$_3$Sn⎯⎯C$_5$H$_{11}$-n, O, OH (—) 92% ee	11

D. Secondary E-Iodovinyl Carbinols

Substrate	Conditions	Time (h)	Allylic Alcohol(s) and Yield(s) (%)	Epoxy Alcohol(s) and Yield(s) (%)	Refs.
C_8 I⎯⎯C$_5$H$_{11}$-n, OH	(–)-DIPT		I⎯⎯C$_5$H$_{11}$-n, OH (—) >99% ee	(—)	197
	Ti (1.0), (–)-DIPT (1.2), TBHP (1.5), -20°, 42 h		" (>49) >99% ee, [α] +9.87° [b]	I⎯⎯C$_5$H$_{11}$-n, O, OH (—) >99% ee	12
	Ti (1.0), (–)-DIPT (1.2), TBHP (1.5), -20°, 36 h		" (—) 96.2% ee	" (—) >98% ee	12

232

TABLE VI. KINETIC RESOLUTION OF RACEMIC SECONDARY ALLYLIC ALCOHOLS (*Continued*)

Substrate	Conditions	Allylic Alcohol(s) and Yield(s) (%)	Epoxy Alcohol(s) and Yield(s) (%)	Refs.
C_9 ⟍⟍C_6H_{11} OH	Ti (0.2), (–)-DIPT (0.24), TBHP (1.5), 4 Å, -20°, 36 h	" (45) >99% ee	" (–) 96.9% ee	12
	Ti (0.2), (–)-DIPT (0.24), TBHP (1.5), 4 Å, 20°, 36 h	" (–) 98% ee	(–)	12
C_{11} I⟍⟍C_6H_{11} OH	Ti (0.23), (+)-DIPT (0.33), TBHP, (1.5), 4 Å, -21°	I⟍⟍C_6H_{11} OH (42) >99% ee, [α] -11.4[c]	(–)	198, 805
I⟍⟍C_6H_{11} OH	Ti (0.3), (+)-DIPT (0.36), TBHP (1.5), 4 Å, -21°, 40 h	I⟍⟍C_6H_{11} OH (41) >99% ee, [α] 6.1°	(–)	805
I⟍⟍C_8H_{17}-n OH	Ti (0.3), (+)-DIPT (0.36), TBHP (1.5), 4 Å, -21°, 40 h	I⟍⟍C_8H_{17}-n OH (44) >99% ee, [c] -7.8°	(–)	198, 805

[a] Ti(OBu-t)$_4$ was used instead of Ti(OPr-i)$_4$.
[b] The rotation was measured in methanol.
[c] The rotation was measured in ethanol.
[d] The reaction was too slow to be practical.
[e] 2-Phenyl-2-propyl hydroperoxide was used instead of *tert*-butyl hydroperoxide.
[f] The rotation was measured in dichloromethane.
[g] The rotation was measured in benzene.
[h] The rotation was measured in *n*-hexane.
[i] The configuration was not determined.
[j] The rotation was measured in acetone.

233

TABLE VII. ASYMMETRIC EPOXIDATION OF CHIRAL SECONDARY ALLYLIC ALCOHOLS

Substrate	Conditions TI (eq), DAT (eq), MS	Product(s) and Yield(s) (%)	Refs.
C$_4$	TI (1.05), (+)-DIPT (1.05), TBHP (2), -20°, 18 h	(86), [α] +9.2°	582
C$_5$	TI (1.05), (+)-DIPT (1.2), 4 Å, TBHP (2), -20°, 72 h	(92) [α] -11.0°	177
	TI (1.05), (+)-DIPT (1.2), 4 Å, TBHP (2), -20°, 72 h	(88) [α] -22.6°	177
	(+)-DIPT	(60) >95% de	145
C$_6$	TI (1), (−)-DIPT (1.2), 4 Å, TBHP (1.2), 10 h	(42) ~100% de	187
	(+)-DET (1.2)	(—)	806

TABLE VII. ASYMMETRIC EPOXIDATION OF CHIRAL SECONDARY ALLYLIC ALCOHOLS (*Continued*)

Substrate	Conditions TI (eq), DAT (eq), MS	Product(s) and Yield(s) (%)	Refs.
C$_7$	(+)-DIPT (0.13), TBHP (0.5)	(76) 90% ee	807
	TI (1.1), (+)-DIPT (1.2), 4 Å, TBHP (1.5), −20°	OBn (82)	188
	TI (1.1), (−)-DIPT (1.2), 4 Å, TBHP (1.5), −20°	OBn (54) 13% de	188
	TI (1.1), (+)-DIPT (1.2), 4 Å, TBHP (1.5), −20°	OBn (89)	188
	TI (1.1), (−)-DIPT (1.2), 4 Å, TBHP (1.5), −20°	OBn (54) 13% de	188
	TI (1.1), (+)-DIPT (1.2), 4 Å, TBHP (1.5), −20°	OBn (70)	188

TABLE VII. ASYMMETRIC EPOXIDATION OF CHIRAL SECONDARY ALLYLIC ALCOHOLS (*Continued*)

Substrate	Conditions TI (eq), DAT (eq), MS	Product(s) and Yield(s) (%)	Refs.
	TI (1.1), (–)-DIPT (1.2), 4 Å, TBHP (1.5), –20°	(70)	188
	TI (1.1), (+)-DIPT (1.2), 4 Å, TBHP (1.5), –20°	(29) 23% de	188
	TI (1.1), (–)-DIPT (1.2), 4 Å, TBHP (1.5), –20°	(28) 51% de	188
	(+)-DIPT, –20°	(90)	808
C$_8$	TI (2), (–)-DET (2), TBHP (1.5), –20°	(80)	157

Substrate	Conditions TI (eq), DAT (eq), MS	Product(s) and Yield(s) (%)	Refs.
C₉ 84% ee	(+)-DET, 4 Å, -20°, 60 h	(5)) 98% ee + (5)	809
	TI (0.2), (+)-DET (0.24), TBHP (2), -20°, 20 h	(75), [α] +6.7°	197
	TI (1.05), (+)-DIPT (1.2), 4 Å, TBHP (1.1), -20°, 72 h	(8)	177
	TI (1), (+)-DIPT (1.2), 5°, 6 d	(73) [α] -37.9° + (85)	106
64-72% ee	TI (1), (–)-DIPT (1.2), TBHP (0.8), -20°	(71) 88% ee [α] +9.5° ᵃ	810
	(–)-DIPT, 4 Å, -20°	(85) >97% de	811

237

TABLE VII. ASYMMETRIC EPOXIDATION OF CHIRAL SECONDARY ALLYLIC ALCOHOLS (*Continued*)

Substrate	Conditions TI (eq), DAT (eq), MS	Product(s) and Yield(s) (%)	Refs.
	(+)-DIPT, 4 Å, -20°	(83) >97% de	811
	(−)-DIPT, 4 Å, -20°	(97) >97% de	811
	(−)-DET	(83)	768
C₁₀	(−)-DIPT, 4 Å, -20°, 3 d	(97)	812
	TI (1.05), (+)-DIPT (1.2), 4 Å, TBHP (1.1), -20°, 72 h	(56) [α] -32.7°	177
	(+)-DET	(—) 40% de	512

TABLE VII. ASYMMETRIC EPOXIDATION OF CHIRAL SECONDARY ALLYLIC ALCOHOLS (*Continued*)

Substrate	Conditions TI (eq), DAT (eq), MS	Product(s) and Yield(s) (%)	Refs.
	(–)-DET	TBDMSO — (—) 100% de, OH, OBn	512
C11	(+)-DET (catalytic), MS	Ph, OH, O (—)	44
(OH, Ph)	(–)-DET (catalytic), MS	Ph, OH, O (—) 6% de	44
TMS— C5H11-*n*, OH	TI (0.3), (–)-DIPT (0.37), 4 Å, -21°, 4 h	TMS— C5H11-*n*, O, OH (86)	198
TDMSO, TMS— OH	TI (1), (+)-DIPT (1.2), -21°	TDMSO, TMS— O, OH (89)	198
Ph— OH, OH	TI (1.05), (+)-DIPT (1.2), TBHP (1.1), 4 Å, -20°, 72 h	Ph— O, OH, OH (87) [α] -33.4°	177

239

TABLE VII. ASYMMETRIC EPOXIDATION OF CHIRAL SECONDARY ALLYLIC ALCOHOLS (*Continued*)

Substrate	Conditions TI (eq), DAT (eq), MS	Product(s) and Yield(s) (%)	Refs.
C$_{12}$	(+)-DET, −23°	(—) 100% de	512
	(−)-DET, −23°	(—) 100% de	512
	TI (1), (+)-DET (1), TBHP (0.6), −50°, 10 h	(>74)	791, 792
C$_{13}$	(+)-DET, 0°	(—) 99% de	813
	(−)-DET, 0°	" (—) 20% de	813

240

TABLE VII. ASYMMETRIC EPOXIDATION OF CHIRAL SECONDARY ALLYLIC ALCOHOLS (*Continued*)

Substrate	Conditions TI (eq), DAT (eq), MS TI (eq), DAT (eq), MS	Product(s) and Yield(s) (%)	Refs.
C₁₄			
	TI (1), (+)-DIPT (1.2), −20°, 22 h	(96) 90% de	814
C₁₉			
	TI (1.05), (+)-DIPT (1.2), 4 Å, TBHP (1.1), −20°, 72 h	(63) [α] −22.5°	177
	TI (2.1), (+)-DIPT (2.5), CaH₂ (0.4), SiO₂ (0.5), −20°, 96 h	(71) 75% de	140
	TI (2.1), (+)-DIPT (2.5) TBHP (1.0), −20°, 30 d	" (0)	140
	TI (2.1), (−)-DIPT (2.5), CaH₂ (0.4), SiO₂ (0.5), −20°, 30 h	" (81) 80% de	140
	TI (2.1), (−)-DIPT (2.5) TBHP (1.0), −20°, 30 d	" (5)	140

241

TABLE VII. ASYMMETRIC EPOXIDATION OF CHIRAL SECONDARY ALLYLIC ALCOHOLS (*Continued*)

Substrate	Conditions TI (eq), DAT (eq), MS	Product(s) and Yield(s) (%)	Refs.
C_{25}	(+)-DET	I + II (95) I:II = 70:30	815
	(−)-DET	I + II (96) I:II = 44:56	815
C_{27}	(+)-DET	I:II = 33:67	816
	(−)-DET	I:II = 96:4	816

TABLE VII. ASYMMETRIC EPOXIDATION OF CHIRAL SECONDARY ALLYLIC ALCOHOLS (*Continued*)

Substrate	Conditions TI (eq), DAT (eq), MS	Product(s) and Yield(s) (%)	Refs.
	(+)-DET (–)-DET	 **I:II = 98:2** **I:II = 50:50**	816 816
	(+)-DET (100% conversion) (–)-DET (40% conversion)		817 817

^{a} The rotation was measured in dichloromethane.

243

TABLE VIII. EPOXIDATION AND KINETIC RESOLUTION OF *MESO*-SECONDARY ALLYLIC ALCOHOLS

Substrate	Conditions TI (eq), DAT (eq), MS	Product(s) and Yield(s) (%)	Refs.
C₅	TI (1.36), (+)-DIPT (1.8), 4 Å TBHP (4.8), -25° $\dfrac{t\,(h)}{3}$ 24 140	 (40-48) 84% ee, 92% de (40-48) 93% ee, 99.7% de (40-48) >97% ee, >99.7% de	274
	TI (0.07), (+)-DIPT (0.09), 4 Å, TBHP (2.0), -15°, 118 h	" (55) >99% ee, [α] +48.8°	279
	TI (0.1), (+)-DIPT (0.12), 4 Å, TBHP (1.5), -20°, 90 h	" (60) 99% ee, 97% de	274
	TI (1.0), (+)-DET (1.2), TBHP (1.2), -20°, 3 d	" (50-60) >90% ee, [α] +46.7°	274, 275
	(–)-DET	(50), [α] -50.0°	274
	(–)-DIPT	" (—) >97% ee, 99% de	275
	TI (0.2), (–)-DIPT (0.24), 4 Å, TBHP (1.5), -21°, 35 h	(92) >98% ee, >99% de [α] -24.0°	276
	TI (0.2), (+)-DIPT (0.24), 4 Å, TBHP (1.5), -21°, 35 h	(92) >98% ee, >99% de	276

244

	Conditions		
Substrate	TI (eq), DAT (eq), MS	Product(s) and Yield(s) (%)	Refs.

C_6

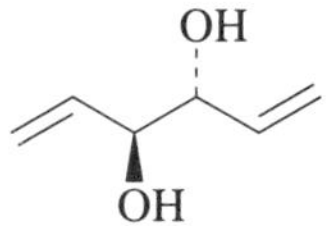

TI (0.4),
(–)-DIPT (0.8),
TBHP (10), 114 h

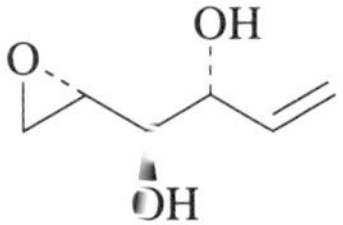

(60) 90% ee 75% de

187

C_7

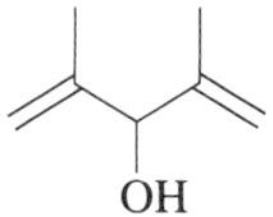

TI (1.1),
(–)-DIPT (1.3),
TBHP (2), -25°

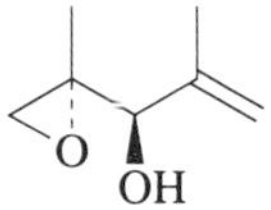

t (h)
0.5
1.0
1.5

(80–85) 88% ee, >99% de
(80–85) 94% ee, >99% de
(80–35) >99.3% ee. >99% de

274, 277

TI (1.15),
(+)-DIPT (1.5),
TBHP (2.6). -25°

t (h)
1
3
44

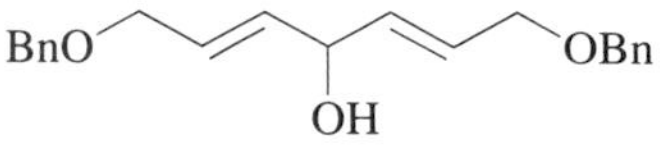

(70–78) 93% ee, 97% de
(70–78) 95% ee, 97% de
(70–78) 93% ee, 97% de

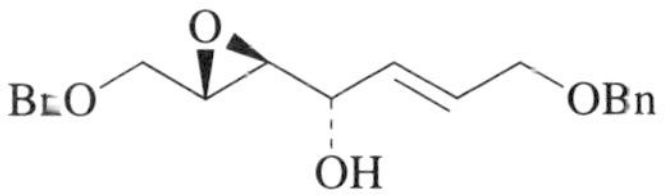

274

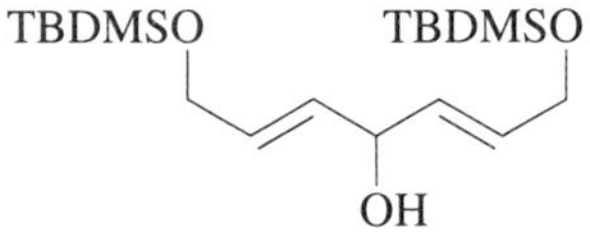

(+)-DET

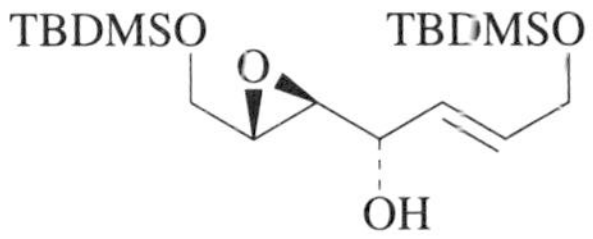

(71) [α] +6.8°

529

245

Substrate	Conditions TI (eq), DAT (eq), MS	Product(s) and Yield(s) (%)	Refs.
	(–)-DET	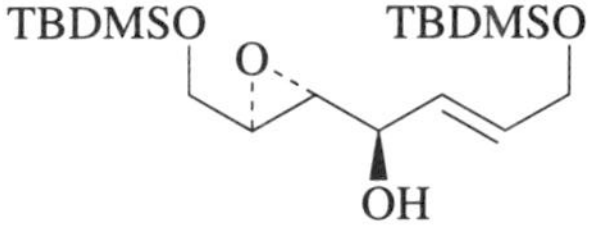 (93) [α] -8.2°	529
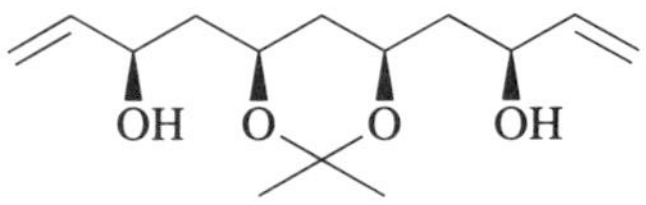	TI (1.36), (+)-DIPT (1.8), 4 Å, TBHP (4.8), -25°	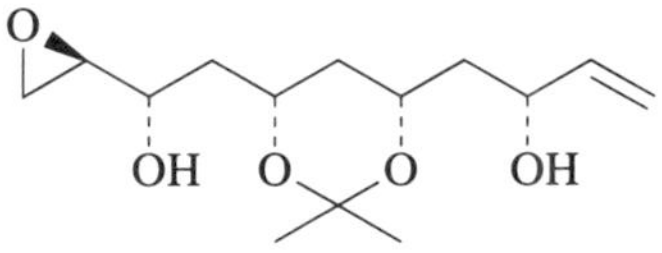 (68)	278
	TI (1.36), (–)-DIPT (1.8), 4 Å, TBHP (4.8), -25°	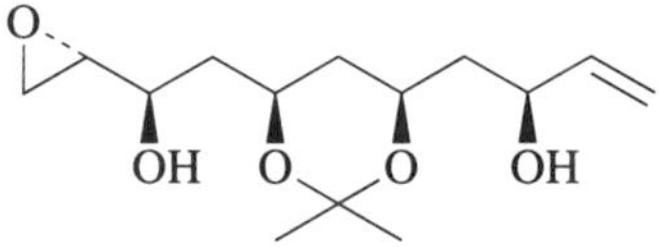 (72)	278
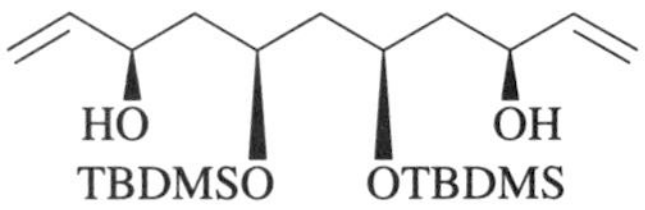	(+)-DIPT, 4 Å, -20°	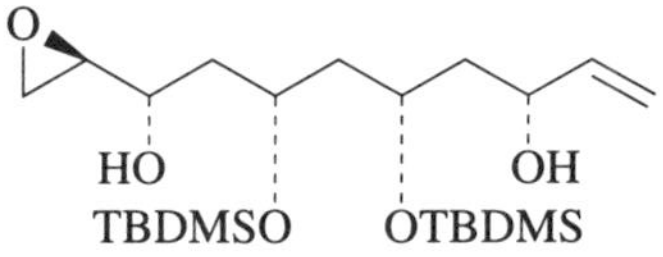 (82)	280
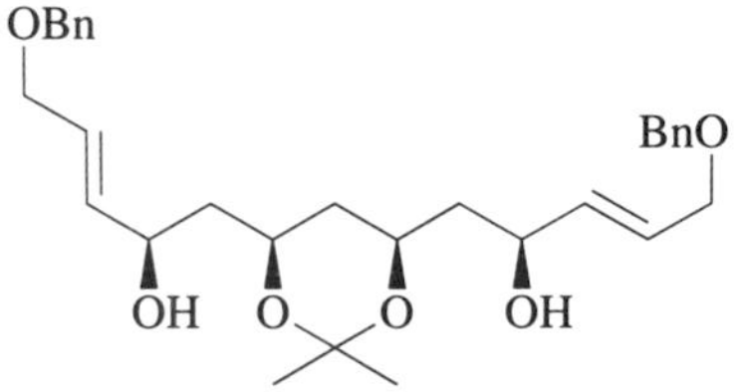	TI (1.36), (+)-DIPT (1.8), 4 Å, TBHP (4.8), -25°	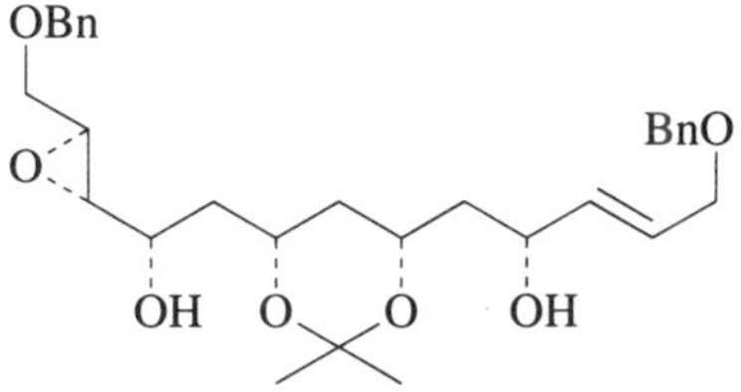 (89)	278

C$_{11}$

C$_{13}$

| | Conditions | | |
| Substrate | TI (eq), DAT (eq), MS | Product(s) and Yield(s) (%) | Refs. |

C_{15}

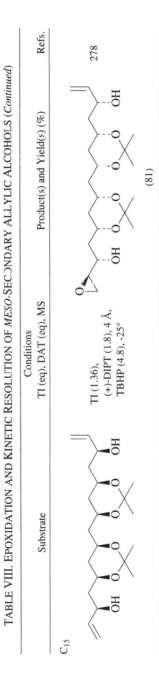

TI (1.36),
(+)-DIPT (1.8), 4 Å,
TBHP (4.8), -25°

(81)

278

TABLE IX. KINETIC RESOLUTION OF α-HYDROXY-FURANS, -THIOPHENES, AND -PYRROLES

Substrate (I): furan with R^1, R^2, R^3, R^4, CH(R^4)OH
Remaining Alcohol I: furan with R^1, R^2, R^3, CH(R^4)OH
Oxidation Product II: R^3, R^2, R^1, R^4, H, HO, O

Conditions: TI (1), (+)-DIPT (1.2), TBHP (0.6), -20°

	R^1	R^2	R^3	R^4	t (h)	Remaining Alcohol(s) and Yield(s) (%) **I**	Oxidation Product(s) and Yield(s) (%) **II**	Refs.
C$_6$	H	H	H	Me	24	(32) >95% ee, [α] +20.8°	(53)	413
C$_7$	H	H	H	CH=CH$_2$	24	(32) >95% ee, [α] -1.74°	(—)	14, 413
	H	H	H	C≡CTMS	20	(38) 88% ee, [α] -16.5°	(—)	14, 413
C$_8$	H	H	H	CH$_2$CH=CH$_2$	36	(42) >95% ee, [α] +39.9°	(—)	413
	H	H	H	Pr-i	25	(39) >95% ee, [α] +18.1°	(55)	14, 413
C$_9$	H	H	H	Bu-t	40	(41) 6% ee	(55)	14, 413
C$_{10}$	H	H	H	C$_5$H$_{11}$-n	25	(42) >95% ee, [α] +13.8°	(53)	14, 413
C$_{11}$	Me	H	H	C$_5$H$_{11}$$n$	6	(40) >95% ee, [α] +7.8°	(55)	14, 413
	H	H	Me	C$_5$H$_{11}$-n	4	(39) >95% ee, [α] +8.9°	(—)	14, 413
	H	H	H	Ph	40	(42) 99% ee, [α] +6.9°	(44)	413

Conditions: TI (0.2), (+)-DIPT (0.24), 4 Å, TBHP (1.5), -20°

	R^1	R^2	R^3	R^4	t (h)	Remaining Alcohol(s) and Yield(s) (%) **I**	Oxidation Product(s) and Yield(s) (%) **II**	Refs.
C$_6$	H	H	H	Me	12	(33) >95% ee	(—)	413
C$_7$	H	H	H	CH$_2$CO$_2$Et	22	(40) >95% ee	(—)	413
C$_9$	H	H	H	Bu-n	50	(38) 99% ee, [α] +16.2°	(—)	413
C$_{10}$	H	H	H	C$_5$H$_{11}$-n	14	(38) >95% ee	(—)	413
C$_{11}$	(ring)	H	H	Me	45	(43) >99% ee, [α] +19.9°	(—)	413
	Me	H	H	C$_5$H$_{11}$-n	6	(40) >95% ee	(—)	413
	H	H	H	Ph	48	(38) >95% ee	(—)	413

248

TABLE IX. KINETIC RESOLUTION OF α-HYDROXY-FURANS, -THIOPHENES, AND -PYRROLES (*Continued*)

Substrate	Conditions TI (eq), DAT (eq), MS	Remaining Alcohol(s) and Yield(s) (%)	Oxidation Product(s) and Yield(s) (%)	Refs.
C_6 (furan-CH(OH)-R)	TI (0.1), (+)-DIPT (0.15), TBHP (0.7), 3 Å	(furan-CH(OH)-R)	(dihydropyranone with R, H)	818
	R Temp Time (h)			
C_6 Me	−20° 5	(36) 80% ee, [α] +17.4°	(38)	
C_7 Et	−20° 3.5	(32) 95% ee, [α] +12.6°	(42)	
C_8 CH=CHMe	−30° 7	(32) 82% ee, [α] −40.4°	(52)	
C_9 Bu-n	−35° 6	(43) 94% ee, [α] +9.2°	(46)	
C_{11} C_6H_{11}	−25° 7	(44) >98% ee, [c] +20.0°	(52)	
C_{15} $C_{10}H_{21}$-n	−25° 4	(44) >98% ee, [c] −9.6°	(51)	
C_8 (furan-CH(OH)-CH=CHMe) + enantiomer	TI (0.11), (+)-DIPT (0.16), 3 Å, TBHP (0.5), −20°, 6 h	(—)	(—)	819
(furan-CH(OH)-CH(CO₂Me with wedges)) + enantiomer	TI (0.22), (+)-DIPT (0.24), 3 Å, TBHP (0.6), −21°, 30 min	(45) >99% ee, [α] +14.7°	(48)	820
(furan-CH(OH)-CH(CO₂Me dashed)) + enantiomer	TI (0.22), (+)-DIPT (0.24), 3 Å, TBHP (0.6), −21°, 30 min	(49) >99% ee [α] +9.52°	(—)	820

249

TABLE IX. KINETIC RESOLUTION OF α-HYDROXY-FURANS, -THIOPHENES, AND -PYRROLES (*Continued*)

Substrate	Conditions TI (eq), DAT (eq), MS	Remaining Alcohol(s) and Yield(s) (%)	Oxidation Product(s) and Yield(s) (%)	Refs.
C8 — furan, TBDMSO, Bu-*n*, OH	TI (0.22), (+)-DIPT (0.24), TBHP (0.6), -25°	furan TBDMSO Bu-*n* OH (41) >95% ee	(—)	821
thiophene R1–S–R2, OH	TI (1), (+)-DIPT (1.2), TBHP (3), 0°		Polymer	822

	R1	R2	h	Remaining Alcohol
C8	H	Pr-*i*	18	(30) 91% ee, [α] +14.2°
C9	H	Bu-*n*	18	(39) >95% ee
	H	Bu-*t*	18	(25) 47% ee
C11	Me	C5H11-*n*	20	(35) >95% ee, [α] +14.9°
	H	Ph	45	(30) >95% ee, [α] -9.8°

Oxidation Product(s): ( ) ( ) ( ) ( ) ( )

Substrate (pyrrole, NTs, R, OH)	Conditions	R	Remaining Alcohol	Oxidation Product	Refs.
NTs pyrrole R OH	TI (1), (–)-DIPT (1.2), TBHP (1.5), CaH2, SiO2, -10°, 16 h			TsNH / O= ... O R H	823
C7		Et	(40) 90% ee, [α] -36.2°	(—)	
C11		C6H13-*n*	(36) >95% ee, [α] -45.5°	(—)	

TABLE IX. KINETIC RESOLUTION OF α-HYDROXY-FURANS, -THIOPHENES, AND -PYRROLES (Continued)

Substrate	Conditions	Remaining Alcohol(s) and Yield(s) (%)	Oxidation Product(s) and Yield(s) (%)	Refs.
(structure: NTs-pyrrole with CH(R)OH)	TI (eq), DAT (eq), MS TI (1), (+)-DIPT (1.2), TBHP (1.5), CaH$_2$, SiO$_2$, -10°, 16 h	(structure: NTs-pyrrole with CH(R)OH)	(structure with O, R, H, TsNH, O)	823

R

	R		
C$_6$	Me	(43) >95% ee, [α] +21.5°	(—)
C$_7$	Et	(38) 92% ee, [α] +37.1°	(—)
C$_8$	CH$_2$CH=CH$_2$	(38) >95% ee, [α] +65.6°	(—)
C$_9$	Bu-n	(35) 90% ee, [α] +42.6°	(—)
C$_{11}$	C$_6$H$_{13}$-n	(33) 94% ee, [α] +42.9°	(—)
C$_{15}$	C$_{10}$H$_{21}$-n	(30) 90% ee, [α] +38.0°	(—)

TABLE X. KINETIC RESOLUTION OF α-TOSYLAMINOFURANS

Substrate	Conditions	Remaining Alcohol(s) and Yield(s) (%)	Oxidation Product(s) and Yield(s) (%)	Refs.
	TI (1.0), (+)-DIPT (1.2) TBHP (2.5), CaH$_2$, SiO$_2$, rt			200

	R	Time (h)		
C$_6$	Me	2	(50) 90% ee, [α] -7.6° [a]	(8)
C$_7$	Et	2	(47) 93.3% ee, [α] -5.0° [a]	(45)
C$_8$	Pr-n	2	(46) 94.7% ee, [α] -5.3° [a]	(47)
C$_9$	Bu-n	2	(46) 90% ee, [α] -5.0° [a]	(47)
	Bu-i	2	(47) 90.7% ee, [α] -7.4° [a]	(41)
C$_{11}$	C$_6$H$_{13}$-n	3	(45) 100% ee, [α] -4.3° [a]	(43)
				(46)

Substrate	Conditions	Remaining Alcohol(s) and Yield(s) (%)	Oxidation Product(s) and Yield(s) (%)	Refs.
	TI (1.0), (–)-DIPT (1.2) TBHP (2.5), CaH$_2$, SiO$_2$, rt			200

	R			
C$_7$	Et	3	(50) 93.5% ee, [α] +5.0° [a]	(46)
C$_{11}$	C$_6$H$_{13}$-n	3.5	(49.5) 100% ee, [α] +4.4° [a]	(48)

[a] The rotation was measured in ethanol.

252

TABLE XI. EPOXIDATION OF ALLYLIC ALCOHOLS WITH IN SITU HYDROXY DERIVATIZATION

Substrate	Conditions TI (eq), DAT (eq), MS	Product(s) and Yield(s) (%)	Refs.
C₃ OH	1. TI (0.05), (–)-DIPT, (0.06), 3 Å[a], 0°, 6 h 2. P(OMe)₃, -20° 3. TsCl, Et₃N, -20°, 10 h	(structure)–OTs (40) 94% ee[b], [α] +17.5° / 97% ee[c], [α] +18.1°	18, 135, 66, 824
	1. TI (0.05), (+)-DIPT, (0.06), 3 Å[a], -5°, 5 h 2. P(OMe)₃, -20° 3. PNBCl, Et₃N, 0°, 1 h	(structure)–OPNB (51) 92-94% ee[b], [α] -38.7°	18
	1. TI (0.05), (–)-DIPT, (0.06), 3 Å[a], 0°, 6 h 2. P(OMe)₃, -20° 3. TBDPSCl, DMAP, -20°, 10 h	(structure)–OTBDPS (45) 91% ee, [α] -2.28°	18, 10b, 86a
	1. TI (0.05), (+)-DIPT, (0.06), 3 Å[a], -20° 2. P(OMe)₃, -20° 3. m-O₂NC₆H₄SO₂Cl, Et₃N	(structure)–OSO₂C₆H₄NO₂-m (57) 96% ee[b], 99% ee, [α] +23°[c]	66, 43, 584a
	1. TI (0.05), (+)-DIPT, (0.06), 3Å[a] 2. P(OMe)₃ 3. p-ClC₆H₄SO₂Cl	(structure)–OSO₂C₆H₄Cl-p (38) 95% ee[b] [α] +22.6°	66
	1. TI (0.05), (–)-DIPT, (0.06), 3 Å[a], -3° 2. P(OMe)₃, -30° 3. TrCl, Et₃N	(structure)–OTr (53)	67
C₄ OH	1. TI (0.05), (–)-DIPT, (0.06), 3 Å, -20°, 4.5 h 2. P(OMe)₃, -20° 3. TsCl, DMAP, -10°	(structure)–OTs (69) 95% ee, [α] +4.84°	18
	1. TI (0.05), (+)-DIPT, (0.06), 3 Å, -20°, 4.5 h 2. P(OMe)₃, -20° 3. PNBCl, Et₃N, 0°	(structure)–OPNB (78) 98% ee[b], [α] -5.87°	18

253

TABLE XI. EPOXIDATION OF ALLYLIC ALCOHOLS WITH IN SITU HYDROXY DERIVATIZATION (*Continued*)

Substrate	Conditions TI (eq), DAT (eq), MS	Product(s) and Yield(s) (%)	Refs.
(allylic alcohol, trans) —OH	1. TI (0.05), (–)-DIPT, (0.06), 3 Å, -20°, 5 h 2. P(OMe)$_3$, -20° 3. 2-C$_{10}$H$_7$SO$_2$Cl, Et$_3$N, -10°	—OSO$_2$C$_{10}$H$_7$-2 (60) 92% ee[b], [α] +5.94°	18
	1. TI (0.05), (–)-DIPT, (0.06), 3 Å, -20°, 2 h 2. P(OMe)$_3$, -20° 3. TsCl, Et$_3$N, -20°, 10 h	OTs (70) 98% ee[b], [α] +34.22°	18
	1. TI (0.05), (+)-DIPT, (0.06), 3 Å, -20°, 2 h 2. P(OMe)$_3$, -20° 3. PNBCl, Et$_3$N, 0°, 1 h	OPNB (65) 98% ee[b], [α] -48.5°	18
	1. TI (0.05), (–)-DIPT, (0.06), 3 Å, -20°, 2 h 2. P(OMe)$_3$, -20° 3. TBDMSCl, DMAP	OTBDMS (68) 92% ee, [α] +13.12°	18
	1. (–)-DET 2. P(OMe)$_3$, -20° 3. TBDPSCl, DMAP	OTBDPS (76)	825
	1. TI (0.05), (–)-DIPT, (0.06), 3 Å[a], -3°, 8 h 2. P(OMe)$_3$, -20° 3. TrCl, Et$_3$N, 2°, overnight	OTr (53)	67
	1. (+)-DET 2. *in situ* tosylation	OTs (—) 98% ee, [α] -33.6°[b]	826
(allylic alcohol, cis) —OH	1. TI (0.05), (+)-DIPT, (0.06), 3 Å, -20°, 20 h 2. P(OMe)$_3$, -20° 3. PNBCl, Et$_3$N, 0°, 1 h	OPNB (68) 92% ee[b], [α] -28.4°	18

254

TABLE XI. EPOXIDATION OF ALLYLIC ALCOHOLS WITH IN SITU HYDROXY DERIVATIZATION (*Continued*)

Substrate	Conditions TI (eq), DAT (eq), MS	Product(s) and Yield(s) (%)	Refs.
C$_5$			
OH	1. TI (0.1), (+)-DIPT (0.12), 4 Å 2. 3,5-(O$_2$N)$_2$C$_6$H$_3$COCl, Et$_3$N	(52) 98% ee[b] [α] -31.3°	827
OH	1. TI (0.05), (+)-DIPT (0.06), 3 Å, -40°, 2 h 2. P(OMe)$_3$, -20° 3. PNBCl, Et$_3$N, 0°, 1 h	OPNB (70) 98% ee[b], [α] -36.09°	18
	1. TI (0.05), (–)-DIPT (0.06), 3 Å, -40°, 2 h 2. P(OMe)$_3$, -20° 3. 2-C$_{10}$H$_7$SO$_2$Cl, DMAP, -10°, 5 h	OSO$_2$C$_{10}$H$_7$-2 (40) [α] +22.43°	18
	1. TI (0.05), (–)-DIPT (0.06), 3 Å, -40°, 2 h 2. P(OMe)$_3$, -20° 3. TsCl, Et$_3$N, -20°, 10 h	OTs (55) 93% ee, [α] +20.15°	18
C$_{10}$ OTBDPS OH	1. TI (0.05), (–)-DET (0.075), 4 Å, -40 to -20°, 4 h 2. P(OMe)$_3$, -20° 3. TsCl, Et$_3$N, -20°	OTBDPS OTs (57) [α] +6.85°[d]	623
OH	1. TI (0.05), (+)-DIPT (0.06), 3 Å, -10°, 15 min 2. P(OMe)$_3$, -20° 3. Ac$_2$O, Et$_3$N, rt	OAc (98) 86% ee, [α] -26.9°	18

255

TABLE XI. EPOXIDATION OF ALLYLIC ALCOHOLS WITH IN SITU HYDROXY DERIVATIZATION (*Continued*)

Substrate	Conditions TI (eq), DAT (eq), MS	Product(s) and Yield(s) (%)	Refs.
	1. TI (0.05), (–)-DET (0.075), 4 Å, -20°, 7 h 2. TsCl, Et$_3$N, 0-5°, 36 h	 (67)^e 93% ee, [α] +17.1° ^{d}	623

^{a} The oxidant was cumene hydroperoxide.
^{b} The ee and rotation were reported for the recrystallized material.
^{c} The rotation was reported for twice recrystallized material.
^{d} The rotation was measured in dichloromethane.
^{e} The corresponding epoxy alcohol (11%) was also obtained.

TABLE XII. EPOXIDATION OF ALLYLIC ALCOHOLS WITH IN SITU EPOXIDE OPENING

Substrate	Conditions TI (eq), DAT (eq), MS	Product(s) and Yield(s) (%)	Refs.
OH	1. TI (0.05), (+)-DIPT[a], (0.06), 3 Å, 0°, 5 h 2. P(OMe)$_3$, -30 to -25° 3. PhSH, Ti(OPr-i)$_4$, rt	PhS OH OH (88) 90% ee [α] +21.3°[b]	136
C$_3$	1. TI (0.05), (+)-DIPT[a], (0.06), 3 Å, 0°, 5 h 2. P(OMe)$_3$, -30 to -25° 3. 1-C$_{10}$H$_7$ONa, Ti(OPr-i)$_4$, rt, overnight	(naphthyl-O) OH CH (54) 90% ee [α] +7.3°[b]	135
	1. TI (0.05), (+)-DIPT[a], (0.06), 3 Å, 0°, 5 h 2. P(OMe)$_3$, -30 to -25° 3. BnNHPr-i, Ti(OPr-i)$_4$, rt, overnight	(i-Pr)BnN OH OH (68) 90% ee [α] -39.8°[b]	136
C$_4$ OH	1. TI (0.05), (+)-DIPT[a], (0.06), 3 Å, -20°, 4 h 2. P(OMe)$_3$, -30 to -25°, 0.5 h 3. PhSH, Ti(OPr-i)$_4$, rt	PhS OH OH (100) 92% ee [α] +2.27°[c]	136

[a] The oxidant was cumene hydroperoxide.
[b] The rotation was measured in ethanol.
[c] The rotation was measured in methanol.

257

TABLE XIII. ASYMMETRIC OXIDATION OF SULFIDES AND DISULFIDES

Substrate	Conditions TI (eq), DAT (eq)	Product(s) and Yield(s) (%)	Refs.
C₁			
Me–S–S–Me	TI (1), (+)-DET (2), H₂O (1), -20°, 16 d	Me–S(=O)–S–Me (60) 41% ee, [α] -4.8°	204
Me–S–S–Me	TI (1), (+)-DET (2), H₂O (1), -20°, 24 h	Me–S(=O)–S–Me (89) 40% ee, [α] -29.7°[a]	202
(benzimidazole, H–N, N, S–Me)	TI (1), (+)-DET (2), H₂O (1), -20°, 24 h	(benzimidazole, H–N, N, O=S–Me) (42) 61% ee, [α] -20.8°	202
C₂			
Me–S–CO₂Et	TI (1), (+)-DET (2), H₂O (1), -20°, 5.5 h	Me–S(=O)–CO₂Et (84) 63% ee, [α] -31.3°[a]	202
C₃			
Me–S–CO₂Me	TI (1), (+)-DET (2), H₂O (1), -20°, 5 h	Me–S(=O)–CO₂Me (85) 64% ee, [α] -50.2°	202
iPr–S–S–iPr	TI (1), (+)-DET (2), H₂O (1), -20°, 6 h	iPr–S(=O)–S–iPr (43) 52% ee, [α] -58.4°	204
NHCOCF₃ / Me–S–CO₂Me	TI (1), (+)-DET (2), H₂O (1), -20°, 24 h	NHCOCF₃ / Me–S(=O)–CO₂Me (73) 92% de[b], [α] -13.6°[a]	203

258

TABLE XIII. ASYMMETRIC OXIDATION OF SULFIDES AND DISULFIDES (*Continued*)

Substrate	Conditions TI (eq), DAT (eq)	Product(s) and Yield(s) (%)	Refs.
C$_4$			
Me—S—Bu-*t*	TI (1), (+)-DET (2), H$_2$O (1), -20°, 22 h	(72) 53% ee, [α] -2.1°[a]	23. 205
n-Bu—S—S—Bu-*n*	TI (1), (+)-DET (2), H$_2$O (1), -20°, 3 d	(62), [α] +0.7°	204
t-Bu—S—S—Bu-*t*	TI (1), (+)-DET (2), H$_2$O (1), -20°, 5 d	(34) 41% ee, [α] -63.5°	204
C$_5$			
Me—S (2-methylbutyl)	TI (1), (+)-DET (2), H$_2$O (1), -21°, 24 h	(93)[b,c]	23
Me—S-(pyridin-2-yl)	TI (1), (+)-DET (2), H$_2$O (1), -20°, 16 h	(63) 77% ee	202
Me—S-(pyridin-3-yl)	TI (1), (+)-DET (2), H$_2$O (1), -50°, 72 h	(55)	207
Me—S-(5-chloropyridin-3-yl)	TI (1), (+)-DET (2), H$_2$O (1), -40°, 20 h	(75)	207

259

TABLE XIII. ASYMMETRIC OXIDATION OF SULFIDES AND DISULFIDES (*Continued*)

Substrate	Conditions TI (eq), DAT (eq)	Product(s) and Yield(s) (%)	Refs.
Me–S–(pyridine)–OMe	TI (1), (+)-DET (2), H$_2$O (1), -40°, 20 h	(70)	207
C$_6$			
Me–S–C≡CBu-*n*	TI (1), (+)-DET (2), H$_2$O (1), -20°	(83) 75% ee, [α] -59.1° [a]	201
Ph–S– (chain, epoxide)	TI[d] (1), (+)-DET (2), H$_2$O (1), -40°	(86) 29% de	482
Ph–S– (chain, epoxide)	TI[d] (1), (+)-DET (2), H$_2$O (1), -40°	(90) 17% de	482
Me–S– (chain, epoxide)	TI[d] (1), (+)-DET (2), H$_2$O (1), -40°	(61) 67% de	482
Me–S– (chain, epoxide)	TI[d] (1), (–)-DET (2), H$_2$O (1), -40°	(72) 64% de	482
Me–S–C$_6$H$_{11}$	TI (1), (+)-DET (2), H$_2$O (1), -20°, 18 h	(67) 54% ee, [α] -44.3° [a]	23, 204

260

TABLE XIII. ASYMMETRIC OXIDATION OF SULFIDES AND DISULFIDES (*Continued*)

Substrate	Conditions TI (eq), DAT (eq)	Product(s) and Yield(s) (%)	Refs.
Me–S–Ph	Tl (1), (+)-DET (2), H$_2$O (1), -20°, 12 h	Me–S(=O)–Ph (81) 89% ee	22
	Tl^d (1), (+)-DET (2), H$_2$O (1), -23°, 20 h	" (93) 93% ee, [α] +135.4° [a]	201
Me–S–C$_6$H$_4$Cl-p	Tl (1), (+)-DET (2), H$_2$O (1), -23°	Me–S(=O)–C$_6$H$_4$Cl-p (85) 78% ee	201
	Tl^d (1), (+)-DET (2), H$_2$O (1), -23°, 20 h	" (85) 91% ee, [α] +114° [a]	201
Cl–CH$_2$–S–Ph	Tl (1), (+)-DET (2), H$_2$O (1), -20°, 16 h	Cl–CH$_2$–S(=O)–Ph (60) 47% ee, [α] +88.1° [a]	203
MeO–S–Ph	Tl (1), (+)-DET (2), H$_2$O (1), -20°, 3 d	MeO–S(=O)–Ph (86) 29% ee	204
CH$_2$=CH–S–Ph	Tl (1), (+)-DET (2), H$_2$O (1), -20°, 16 h	S(=O)–Ph (80) 70% ee, [α] +215° [a]	203
NC–CH$_2$–S–Ph	Tl (1), (+)-DET (2), H$_2$O (1), -20°, 24 h	NC–CH$_2$–S(=O)–Ph (85) 34% ee, [α] +59° [a]	203

TABLE XIII. ASYMMETRIC OXIDATION OF SULFIDES AND DISULFIDES (*Continued*)

Substrate	Conditions TI (eq), DAT (eq)	Product(s) and Yield(s) (%)	Refs.
MeO₂C–CH₂–S–Ph	TI (1), (+)-DET (2), H₂O (1), -20°, 16 h	MeO₂C–CH₂–S(=O)–Ph (81) 64% ee, [α] +98° [a]	203
CH₃C(=O)CH₂–S–Ph	TI (1), (+)-DET (2), H₂O (1), -20°, 24 h	CH₃C(=O)CH₂–S(=O)–Ph (68) 60% ee, [α] +131.3° [a]	203
cyclopropyl–S–Ph	TI (1), (+)-DET (2), H₂O (1), -20°, 24 h	cyclopropyl–S(=O)–Ph (73) 95% ee, [α] +136.7° [a]	204
t-Bu–S–Ph	TI[e] (1), (+)-DET (4), -20°, 30 h	t-Bu–S(=O)–Ph (99) 34.5% ee[b], [α] +62.1° [a]	24
Ph–S–S–Ph	TI (1), (+)-DET (2), H₂O (1), -20°, 1.7 d	Ph–S(=O)–S–Ph (20)[b], [α] +20.2°	204
HO–CH₂CH₂–S–C₆H₄Cl-p	TI[e] (1), (+)-DET (4), -20°, 24 h	HO–CH₂CH₂–S(=O)–C₆H₄Cl-p (41) 14% ee[b] [α] -23.4° [a]	24
Me–S–C₆H₄OMe-p	TI (1), (+)-DET (2), H₂O (1), -20°, 65 h	Me–S(=O)–C₆H₄OMe-p (70) 84% ee	202

TABLE XIII. ASYMMETRIC OXIDATION OF SULFIDES AND DISULFIDES (*Continued*)

Substrate	Conditions TI (eq), DAT (eq)	Product(s) and Yield(s) (%)	Refs.
Me–S–C$_6$H$_4$OMe-o	TI^d (1), (+)-DET (2), H$_2$O (1), −23°, 20 h	(97) 95% ee, [α] +313° ^{a} Me–S(O)–C$_6$H$_4$OMe-o	201
	TI (1), (+)-DET (2), H$_2$O (1), −20°, 4 h	" (77) 89.4% ee, [α] +18.3° ^{a}	203
i-PrO–S–Ph	TI (1), (+)-DET (2), H$_2$O (1), −20°, 3 d	(91) 8.5% ee, [α] −16° ^{a} i-PrO–S(O)–Ph	204
MeO–S–C$_6$H$_3$(NO$_2$)$_2$	TI (1), (+)-DET (2), H$_2$O (1), −20°, 8 d	(0) MeO–S(O)–C$_6$H$_3$(NO$_2$)$_2$	204
C$_7$ Me–S–CH$_2$Ph	TI^e (1), (+)-DET (4), −77°, 24 h	(70) 46% ee, [α] +25.3° ^{a} Me–S(O)–CH$_2$Ph	24
	TI (1), (+)-DET (2), H$_2$O (1), −20°	" (88) 35% ee, [α] −33.6° ^{f,g}	201
	TI (1), (+)-DET (2), H$_2$O (1), −20°	" (84) 62% ee, [α] −59° ^{c}	201

263

TABLE XIII. ASYMMETRIC OXIDATION OF SULFIDES AND DISULFIDES (*Continued*)

Substrate	Conditions TI (eq), DAT (eq)	Product(s) and Yield(s) (%)	Refs.
Me–S–C6H4Me-*p*	TI (1), (+)-DET (2), H2O (1), -40°, 12 h	(image: Me–S(=O)–C6H4Me-*p*) (95) 93% ee, [α] +132° [a]	22, 23
	TI[d] (1), (+)-DET (2), H2O (1), -23°	" (93) 96% ee, [α] +139° [a]	201
	TI[e] (1), (+)-DET (4), -20°, 14 h	" (60) 88.3% ee, [α] +128.5° [a]	24
Me–S–C6H4Me-*o*	TI (1), (+)-DET (2), H2O (1), -20°	(image: Me–S(=O)–C6H4Me-*o*) (77) 89% ee, [α] +183° [a]	202
(image: Et–S–C6H4Me-*p*)	TI (1), (+)-DET (2), H2O (1), -20°, 3 h	(image: Et–S(=O)–C6H4Me-*p*) (71) 74% ee, [α] +139.4° [a]	22, 23
(image: *i*-Pr–S–C6H4Me-*p*)	TI (1), (+)-DET (2), H2O (1), -20°, 3 h	(image: *i*-Pr–S(=O)–C6H4Me-*p*) (56) 63% ee[b], [α] +111° [a]	22
n-Bu–S–C6H4Me-*p*	TI (1), (+)-DET (2), H2O (1), -20°, 3 h	(image: *n*-Bu–S(=O)–C6H4Me-*p*) (28) 20% ee[b], [α] +38° [a]	22, 23
	TI (1), (+)-DET (2), H2O (1), -20°	(image: *n*-Bu–S(=O)–C6H4Me-*p*) (75) 20% ee	202

TABLE XIII. ASYMMETRIC OXIDATION OF SULFIDES AND DISULFIDES (*Continued*).

Substrate	Conditions TI (eq), DAT (eq)	Product(s) and Yield(s) (%)	Refs.
t-BuO$_2$C—S—C$_6$H$_4$Me-p	TI (1), (+)-DET (2), H$_2$O (1), -30°, 60 h	t-BuO$_2$C—CH$_2$—S(=O)—C$_6$H$_4$Me-p (50) 9% ee [α] +15.1° [a]	202, 22
n-BuC≡C—S—C$_6$H$_4$Me-p	TI (1), (+)-DET (2), H$_2$O (1), -20°, 18 h	n-BuC≡C—S(=O)—C$_6$H$_4$Me-p (97) 29% ee [α] +22.2°	201
Ph—CH$_2$—S—C$_6$H$_4$Me-p	TI (1), (+)-DET (2), H$_2$O (1), -20°, 12 h	Ph—CH$_2$—S(=O)—C$_6$H$_4$Me-p (41) 7% ee	22
Me—S—C$_6$H$_4$CH$_2$OH	TI (1), (+)-DET (2), H$_2$O (1), -20°, 48 h	Me—S(=O)—C$_6$H$_4$CH$_2$OH (71) 76% ee	201
Me—S—C$_6$H$_4$CO$_2$Me-p	TI (1), (+)-DET (2), H$_2$O (1), 60 h	Me—S(=O)—C$_6$H$_4$CO$_2$Me-p (50) 91% ee [b]	205
Me—S—C$_6$H$_4$CO$_2$Me-o	TI (1), (+)-DET (2), H$_2$O (1), 7 h	Me—S(=O)—C$_6$H$_4$CO$_2$Me-o (50) 60% ee [b]	205
i-PrNH—S—C$_6$H$_4$Me-p	TI (1), (+)-DET (2), H$_2$O (1), -20°, 3 d	i-PrNH—S(=O)—C$_6$H$_4$Me-p (28) 24% ee [α] +42.7°	203

TABLE XIII. ASYMMETRIC OXIDATION OF SULFIDES AND DISULFIDES (*Continued*)

Substrate	Conditions TI (eq), DAT (eq)	Product(s) and Yield(s) (%)	Refs.
Et_2N–S–C_6H_4Me-p	TI (1), (+)-DET (2), H_2O (1), -20°, 7 d	Et_2N–S(=O)–C_6H_4Me-p (60) 35% ee [α] +42.4°	203
MeO–S–C_6H_4Me-p	TI (1), (+)-DET (2), H_2O (1), -20°, 1 d	MeO–S(=O)–C_6H_4Me-p (88) 36% ee [α] +79.7°	203
p-MeC_6H_4–S–S–C_6H_4Me-p	TI (1), (+)-DET (2), H_2O (1), -20°, 5 d	p-MeC_6H_4–S(=O)–S–C_6H_4Me-p (10) 13% ee [α] +70.0°	203
C₈			
Me–S–C_8H_{17}-n	TI (1), (+)-DET (2), H_2O (1), -20°, 64 h	Me–S(=O)–C_8H_{17}-n (77) 71% ee	23, 205
	TI (1)[d], (+)-DET (2), H_2O (1), -23°	" (71) 80% ee, [α] -66.6° [a]	201
C₉			
(2,4,6-trimethylphenyl) (2,4-dimethylphenyl) sulfide	TI (1), (+)-DET (2), H_2O (1), -20°, 24 h	(0)	22
3-(methylthio)quinoline	TI (1), (+)-DET (2), H_2O (1), -45°, 48 h	3-(methylsulfinyl)quinoline (80)	207

TABLE XIII. ASYMMETRIC OXIDATION OF SULFIDES AND DISULFIDES (*Continued*)

Substrate	Conditions TI (eq), DAT (eq)	Product(s) and Yield(s) (%)	Refs.
C₁₀			
	TI (1), (+)-DET (2), H₂O (1), -20°, 5 h	(98) 89% ee [α] +402° [a]	205
	TI (1), (+)-DET (2), H₂O (1), -20°, 4 h	(88) 90% ee [b] [α] +120° [a]	23, 205
	TI (1), (+)-DET (2), H₂O (1), -21°, 12 h	(78) 24% ee [b] [α] +39° [a]	23, 205
C₁₂			
	TI (1) [e], (+)-DET (2), H₂O (0.5), -15°, 1 h	(52) 78.1% ee [α] +40.4° [g]	828
	TI (1) [d], (–)-DET (2), H₂O (0.5), -15°, 1 h	(47) [α] -40.5° [g]	828

267

TABLE XIII. ASYMMETRIC OXIDATION OF SULFIDES AND DISULFIDES (*Continued*)

Substrate	Conditions TI (eq), DAT (eq)	Product(s) and Yield(s) (%)	Refs.
C$_{14}$			
	TI (1), (+)-DET (2), H$_2$O (1), –20°, 4 h	 (33) 86% ee [α] +97° [a]	205
	TI (1), (+)-DET (2), H$_2$O (1)	 30% ee, [α] +185.4° [g,h]	829

[a] The rotation was measured in acetone.
[b] The configuration of sulfoxide moiety was not determined.
[c] The product contained two diastereomers of 42 and 38% ee, respectively.
[d] Cumene hydroperoxide was used instead of TBHP.
[e] The reaction was carried out in 1,2-dichloroethane.
[f] The rotation was measured in EtOH.
[g] The rotation was reported for recrystallized material.
[h] The rotation was measured in MeCN-MeOH (1:1).

TABLE XIV. ASYMMETRIC OXIDATION OF β-HYDROXY SULFIDES

Substrate	Conditions[a] TI (eq), DAT (eq)	Diastereomeric Product(s)[b] and Yield(s) (%)[c]	Refs.
C₆			
O₂CC₆H₄NO₂-o *dl* (Me—S, Et)	TI (0.25), (+)-DET (1)[d], -20°, 14-16 h	O₂CC₆H₄NO₂-o (O=S, Me, Et) **a**: (58) 68% ee **b**: (22) 65% ee	830
O₂CC₆H₄NO₂-p *dl* (Me—S, Et)	TI (0.25), (+)-DET (1)[d], -20°, 14-16 h	O₂CC₆H₄NO₂-p (O=S, Me, Et) **a**: (46) 60% ee **b**: (13) 38% ee	830
O₂CC₅H₃(NO₂)₂-3,5 *dl* (Me—S, Et)	TI (0.25), (+)-DET (1)[d], -20°, 14-16 h	O₂CC₆H₃(NO₂)₂-3,5 (O=S, Me, Et) **a**: (52) 75% ee **b**: (25) 50% ee	830
OSiPh₃ *dl* (Me—S, Et)	TI (0.25), (+)-DET (1)[d], -20°, 14-16 h	OSiPh₃ (O=S, Me, Et) **a**: (67) 65% ee **b**: (20)	830
C₈			
OH Ph *dl* (Me—S)	TI (0.25), (+)-DET (1), -20°, 14-16 h	OH, Ph (O=S, Me) **a**: (13.6) 3% ee **b**: (6.4) 5% ee	830, 831
O₂CC₆H₄NO₂-o Ph *dl* (Me—S)	TI (0.25), (+)-DET (1)[e], -20°, 14-16 h	O₂CC₆H₄NO₂-o, Ph (O=S, Me) **a**: (38) 54% ee **b**: (33) 50% ee	830, 831

TABLE XIV. ASYMMETRIC OXIDATION OF β–HYDROXY SULFIDES (*Continued*)

Substrate	Conditions[a] TI (eq), DAT (eq)	Diastereomeric Product(s)[b] and Yield(s) (%)[c]	Refs.
OTBDPS, Me–S⋯Ph, *dl*	TI (0.25), (+)-DET (1), -20°, 14-16 h	OTBDPS structure — **a:** (50) 75% ee **b:** (41) 71% ee	830, 831
OSiPh₃, Me–S⋯Ph, *dl*	TI (0.25), (+)-DET (1), -20°, 14-16 h	OSiPh₃ structure — **a:** (44) 70% ee **b:** (34) 64% ee	830, 831
"	TI (0.25), (+)-DET (1), -20°, 14-16 h	**a:** (43) 80% ee; **b:** (43) 75% ee	830, 831
C₉ OH, Me–S⋯Ph, *dl*	TI (0.25), (+)-DET (1), -20°, 14-16 h	OH structure — **a:** (22) 18% ee **b:** (<0.2)	830, 831
OAc, Me–S⋯Ph, *dl*	TI (0.25), (+)-DET (1)[d], -20°, 14-16 h	OAc structure — **a:** (74) 71% ee **b:** (12)	830
OTMS, Me–S⋯Ph, *dl*	TI (0.25), (+)-DET (1)[d], -20°, 14-16 h	OTMS structure — **a:** (73) 66% ee **b:** (11)	830
OSiPh₃, Me–S⋯Ph, *dl*	TI (0.25), (+)-DET (1), -20°, 14-16 h	OSiPh₃ structure — **a:** (79) 70% ee **b:** (11)	830, 831

TABLE XIV. ASYMMETRIC OXIDATION OF β–HYDROXY SULFIDES (*Continued*)

Substrate	Conditions[a] TI (eq), DAT (eq)	Diastereomeric Product(s)[b] and Yield(s) (%)[c]	Refs.
	TI (0.25), (+)-DET (1)[d], -20°, 14-16 h	**a** (83) 78% ee **b**: (8) 70% ee	830, 831

[a] The ratio of *tert*-butyl hydroperoxide to sulfide was 0.5.

[b] The absolute and relative configurations of the two diastereomeric products **a** and **b** were not determined.

[c] The yield was based on the oxidant.

[d] Cumene hydroperoxide was used instead of *tert*-butyl hydroperoxide.

TABLE XV. ASYMMETRIC OXIDATION OF DITHIOACETALS

Substrate	Conditions TI (eq), DAT (eq)	Product(s) and Yield(s) (%)^a	Refs.
C_1	TI (1), (+)-DET (2), H$_2$O (1) $-20°$, 16 h	(78) 20% ee, [α] +41.7°^{b}	203
C_2	TI (1), (+)-DET (2), H$_2$O (1) $-40°$, 50 h	(65) 60:40^c, 80% ee^d	206
C_3	TI (1), (+)-DET (2), H$_2$O (1) $-38°$, 50 h	(65) 100:0^c, 80% ee^d [α] +18.9°^{e}	206

TABLE XV. ASYMMETRIC OXIDATION OF DITHIOACETALS (Continued)

Substrate	Conditions TI (eq), DAT (eq)	Product(s) and Yield(s) (%)[a]	Refs.
C5			
t-Bu (1,3-dithiolane)	TI (1), (+)-DET (4), -20°, 8 h[f]	(82) 99:1[c] 70% ee[d]	832
t-Bu (1,3-dithiane)	TI (1), (+)-DET (2), H$_2$O (1) -78°, 50 h	(49) 90:10[c], 0% ee[d]	206
C6			
t-Bu (2-methyl-1,3-dithiane)	TI (1), (+)-DET (4), -20°, 14 h[f]	(61) 99:1[c], 68% ee[d]	832
C7			
Ph (1,3-dithiolane)	TI (0.19), (+)-DET (0.77), -20°, 15 h[f]	(76) 94:6[c], 76% ee[d]	832
Ph (1,3-dithiane)	TI (0.19), (+)-DET (0.77), -20°, 15 h[f]	(88) 90:10[c], 14% ee[d]	832
C8			
Ph (2-methyl-1,3-dithiane)	TI (0.19), (+)-DET (0.77), -20°, 15 h[f]	(66) 97:3[c], 83% ee[d]	832

TABLE XV. ASYMMETRIC OXIDATION OF DITHIOACETALS (*Continued*)

Substrate	Conditions TI (eq), DAT (eq)	Product(s) and Yield(s) (%)[a]	Refs.
 Ph ⟍ (2-Ph-1,3-dithiane)	TI (0.19), (+)-DET (0.77), -20°, 15 h[f]	(87) 85:15[c], 39% ee[d]	832
	TI (1), (+)-DET (2), H₂O (1), -40°, 50 h	" (65) 100:0[d], 78% ee, [α] +18.9°[g]	206
p-MeOC₆H₄ ⟍ (2-(p-MeOC₆H₄)-1,3-dithiane)	(+)-DET	p-MeOC₆H₄ ⟍ (—) 100:0, 76% ee	833

[a] The absolute configurations of the sulfoxides were not determined.
[b] The rotation was measured in acetone.
[c] The ratio is that of the diastereomers (*trans:cis*).
[d] The value refers to the major *trans* isomer.
[e] The rotation was measured in dichloromethane.
[f] The reaction was carried out in 1,2-dichloroethane.
[g] The rotation was measured in ethanol.

274

TABLE XVI. KINETIC RESOLUTION OF A RACEMIC SULFIDE

Substrate	Conditions TI (eq), DAT (eq)	Sulfide and Yield (%)	Sulfoxide and Yield (%)	Refs.
C17	TI (0.6), (+)-DIPT (1.2 H2O (0.6), TBHP (0.6) 4 Å, –20°, 6 h	(36) 67% ee[a], [α] -56.99°[b]	(—)	834
	TI (0.6), (–)-DIPT (1.2 H2O (0.6), TBHP (0.7) 4 Å, –20°, 6 h	(—) 84% ee[a], [α] +70.41°[b]	(—)	834

[a] The absolute configuration was not determined.
[b] The rotation was measured in methanol.

TABLE XVII. ASYMMETRIC OXIDATION OF SELENIDES

Substrate	Conditions TI (eq), DAT (eq)	Product(s) and Yield(s) (%)	Refs.
C$_{14}$			
Ph–Se–CH$_2$–C(OMe)(Ph)(Ph)	TI (1), (+)-DIPT (4), -5°	Ph–Se(=O)–CH$_2$–C(OMe)(Ph)(Ph) (82) 18% ee, [α] -16°	26
	TI (1), (−)-DIPT (4), -5°	Ph–Se(=O)–CH$_2$–C(OMe)(Ph)(Ph) (75) 20% ee, [α] +14°	26
p-MeOC$_6$H$_4$–Se–CH$_2$–C(OMe)(Ph)(Ph)	TI (1), (+)-DIPT (4), -5°	p-MeOC$_6$H$_4$–Se(=O)–CH$_2$–C(OMe)(Ph)(Ph) (72) 40% ee, [α] -101°	26
	TI (1), (−)-DIPT (4), -5°	p-MeOC$_6$H$_4$–Se(=O)–CH$_2$–C(OMe)(Ph)(Ph) (70) 28% ee, [α] +85°	26

TABLE XVIII. KINETIC RESOLUTION OF β-HYDROXYAMINES[?]

Substrate			Conditions TI (eq), DAT (eq)	Remaining Alcohol(s) and Yield(s) (%)	Oxidation Product(s) and Yield(s) (%)	Refs.
OH / R¹ / N-R²R³			1. TI(2), (+)-DIPT (1.2), rt, 30 min 2. TBHP (0.6), −20°, 2 h	OH / R¹ (bold) / N-R²R³	OH / R¹ / O⁻ N⁺ R²R³	21
R¹	R²	R³				
C₃						
CH_2OBn	Me	Me		(34) 91% ee, $[\alpha]$ −9.75°	(—)	
$CH_2OC_{10}H_7$-1	Me	Me		(40) 92% ee, $[\alpha]$ −2.6°	(56)	
$CH_2OC_{10}H_7$-1	Bn	i-Pr		(36) 32% ee, $[\alpha]$ +9.2°	(56)	
$CH_2OC_6H_4Me$-m	Bn	(CH₂)₂—C₆H₃(OMe)(OMe)		(35) 85% ee, $[\alpha]$ +13.7°	(—)	
C₈						
Ph	Me	Me		(35) 95% ee, $[\alpha]$ −47.7°	(50)	
C_6H_{11}	(CH₂)₂	(CH₂)₂		(36) 92% ee, $[\alpha]$ −20.7°	(50)	
Ph	Bn	i-Pr		(35) 15% ee, $[\alpha]$ −0.74°	(61)	
Ph	(CH₂)₂	(CH₂)₂		(37) 95% ee, $[\alpha]$ −40.3°	(59)	
Ph	Bn	Bn		(19) 10% ee, $[\alpha]$ −0.06°	(—)	
Ph	Me	Bn		(33) 86% ee, $[\alpha]$ −49.2°	(62)	
Ph	(CH₂)₂	(CH₂)₃		(37) 97% ee, $[\alpha]$ −51.2°	(54)	
C₁₀						
n-C₈H₁₇	Me	Me		(36) 91% ee, $[\alpha]$ −3.58°	(53)	
n-C₈H₁₇	(CH₂)₂	(CH₂)₂		(37) 94% ee, $[\alpha]$ −1.20°	(54)	
n-C₈H₁₇	Bn	Bn		(34) 0% ee	(60)	

277

TABLE XVIII. KINETIC RESOLUTION OF β-HYDROXY AMINES[a] (*Continued*)

Substrate	Conditions TI (eq), DAT (eq)	Remaining Alcohol(s) and Yield(s) (%)	Oxidation Product(s) and Yield(s) (%)	Refs.
(structure) OH / R¹—•N(Me)—CH(R²)—Me + enantiomer R¹ R² C₆ (CH₂)₄ C₉ Ph Me	1. TI(2), (+)-DIPT (1.2), rt, 30 min 2. TBHP (0.6), −20°, 2 h	(structure) OH / R¹—•N(Me)—CH(R²)—Me (25) 95% ee, [α] −2.95° (40) 95% ee, [α] +26.3°	(structure) OH / R¹—•N⁺(O⁻)(Me)—CH(R²)—Me (55) (53)	21
(structure) OH / R¹—•N(R³)(R⁴)—CH(R²)— + enantiomer R¹ R² R³ R⁴ C₆ (CH₂)₄ Me Me (CH₂)₄ Bn Bn C₉ Ph Me Me Me	1. TI(2), (+)-DIPT (1.2), rt, 30 min 2. TBHP (0.6), −20°, 2 h	(structure) OH / R¹—•N(R³)(R⁴)—CH(R²)— (40) 92% ee, [α] −25.4° (36) 0% ee (42) 93% ee, [α] −40.2°	(structure) OH / R¹—•N⁺(O⁻)(R³)(R⁴)—CH(R²)— (55) (50) (53)	21
(structure) OH / t-Bu—•N(R)(R)— + enantiomer R C₆ (CH₂)₄ (CH₂)₅	1. TI(2), (+)-DIPT (1.4), rt, 30 min 2. TBHP (0.6), −15°, 4 h	(structure) OH / t-Bu—•N(R)(R)— (39) 93% ee, [α] −62.3° (38) 96% ee, [α] −66.5°	(structure) OH / t-Bu—•N⁺(O⁻)(R)(R)— (46) 59% ee, [α] +36.7° (60) 46.4% ee, [α] +13.0°	139

278

TABLE XVIII. KINETIC RESOLUTION OF β-HYDROXYAMINES[a] (Continued)

Substrate	Conditions TI (eq), DAT (eq)	Remaining Alcohol(s) and Yield(s) (%)	Oxidation Product(s) and Yield(s) (%)	Refs.
C$_8$	1. TI (2), (+)-DIPT, rt, 30 min 2. TBHP (0.6), -20°, 2 h	(37) 95% ee, [α] -40.3°	(59)	21
	(+)-DIPT (1.2) (+)-DIPT (2.4)	(−) 71% ee	(—)	
	1. TI (2), (+)-DIPT (1), rt, 0.5 h 2. TBHP (0.6), -20°, 2 h	(37) 97% ee	(54)	21
C$_9$ + enantiomer	1. TI (2), (+)-DIPT, rt, 30 min 2. TBHP (0.6), -20°, 2 h			21
	(+)-DIPT (1.2) (+)-DIPT (1.3) (+)-DIPT (1.4) (+)-DIPT (1.5) (+)-DIPT (2.4)	(36-39) 58% ee (36-39) 85% ee (36-39) 96% ee (36-39) 95% ee (36-39) 50% ee	(—) (—) (—) (—) (—)	

279

TABLE XVIII. KINETIC RESOLUTION OF β-HYDROXYAMINES[a] (*Continued*)

Substrate	Conditions TI (eq), DAT (eq)	Remaining Alcohol(s) and Yield(s) (%)	Oxidation Product(s) and Yield(s) (%)	Refs.
	1. (+)-DIPT (1.2), H₂O 2. TI (2), rt, 30 min 3. TBHP (0.6), -20°, 2 h			21
+ enantiomer	H₂O (0.5) H₂O (1.0) H₂O (2.0)	(—) 71% ee, 60% conversion (—) 53% ee, 58% conversion (—) 0% ee, 32% conversion	(—) (—) (—)	
	1. TI (2), (+)-DIPT (1.2), rt, 30 min 2. TBHP (0.6), -20°, 2 h	 (52) 0% ee	 (37)	21
C₁₀ 	1. TI (2), (+)-DIPT, rt, 30 min 2. TBHP (0.6), -20°, 2 h			21
	(+)-DIPT (1.2) (+)-DIPT (2.4)	(27) 94% ee, [α] -1.2° (—) 0% ee	(54) (—)	

[a] All optical rotations in this table were measured in ethanol.

280

REFERENCES

[1] Katsuki, T.; Sharpless, K. B. *J. Am. Chem. Soc.*, **1980**, *102*, 5974.

[2] Finn, M. G.; Sharpless, K. B. in *Asymmetric Synthesis*, Morrison, J. D., Ed., Academic Press, Orlando, FL, 1985, Vol. 5, pp. 247–348.

[3] Pfenninger, A. *Synthesis*, **1986**, 89.

[4] Rossiter, B. E. in *Asymmetric Synthesis*, Morrison, J. D., Ed., Academic Press, Orlando, FL, 1985, Vol. 5, pp. 193–246.

[5] Katsuki, T. *J. Syn. Org. Chem. Jpn.*, **1987**, *45*, 90; *Chem. Abstr.*, **1986**, *107*, 134136k.

[6] Johnson, R.; Sharpless, K. B. in *Comprehensive Organic Synthesis*, Trost, B. M., Ed., Pergamon, Oxford, 1991, Vol. 7, Ch. 3, 2.

[7] Behrens, C. H.; Sharpless, K. B. *Aldrichim. Acta*, **1983**, *16*, 67.

[8] Wood, R. D.; Ganem, B. *Tetrahedron Lett.*, **1982**, *23*, 707.

[9] Erickson, T. J. *J. Org. Chem.*, **1986**, *51*, 934.

[10] Martin, V. S.; Woodard, S. S.; Katsuki, T.; Yamada, Y.; Ikeda, M.; Sharpless, K. B. *J. Am. Chem. Soc.*, **1981**, *103*, 6237.

[11] Kitano, Y.; Matsumoto, T.; Okamoto, S.; Shimazaki, T.; Kobayashi, Y.; Sato, F. *Chem. Lett.*, **1987**, 1523.

[12] Kitano, Y.; Matsumoto, T.; Wakasa, T.; Okamoto, S.; Shimazaki, T.; Kobayashi, Y.; Sato, F. *Tetrahedron Lett.*, **1987**, *28*, 6351.

[13] Kitano, Y.; Matsumoto, T.; Sato, F. *Tetrahedron*, **1988**, *44*, 4073.

[14] Kobayashi, Y.; Kusakabe, M.; Kitano, Y.; Sato, F. *J. Org. Chem.*, **1988**, *53*, 1586.

[15] Kitano, Y.; Kusakabe, M.; Kobayashi, Y.; Sato, F. *J. Org. Chem.*, **1989**, *54*, 994.

[16] Carlier, P. R.; Mungall, W. S.; Schroder, G.; Sharpless, K. B. *J. Am. Chem. Soc.*, **1988**, *110*, 2978.

[17] Hanson, R. M.; Sharpless, K. B. *J. Org. Chem.*, **1986**, *51*, 1922.

[18] Gao, Y.; Hanson, R. M.; Kluder, J. M.; Ko, S. Y.; Masamune, H.; Sharpless, K. B. *J. Am. Chem. Soc.*, **1987**, *109*, 5765.

[19] Sharpless, K. B. *Janssen Chim. Acta*, **1988**, *6*, 3; *Chem. Abstr.*, **1988**, *109*, 128034a.

[20] Miyano, S.; Lu, L. D.-L.; Viti, S. M.; Sharpless, K. B. *J. Org. Chem.*, **1983**, *48*, 3608.

[21] Miyano, S.; Lu, L. D.-L.; Viti, S. M.; Sharpless, K. B. *J. Org. Chem.*, **1985**, *50*, 4350.

[22] Pitchen, P.; Kagan, H. B. *Tetrahedron Lett.*, **1984**, *25*, 1049.

[23] Pitchen, P.; Dunach, E.; Deshmukh, M. N.; Kagan, H. B. *J. Am. Chem. Soc.*, **1984**, *106*, 8188.

[24] Di Furia, F.; Modena, G.; Seraglia, R. *Synthesis*, **1984**, 325.

[25] Kagan, H. B.; Rebiere, F. *Synlett*, **1990**, 643.

[26] Tiecco, M.; Tingoli, M.; Testaferri, L.; Bartoli, D. *Tetrahedron Lett.*, **1987**, *28*, 3849.

[27] Parker, R. E.; Isaacs, N. S. *Chem. Rev.*, **1959**, *59*, 737.

[28] Swern, D. in *Organic Peroxides*, D. Swern, Ed., Wiley–Interscience, New York, 1971, Vol. 2, Ch. 5.

[29] Rao, A. S.; Paknikar, S. K.; Kirtane, J. G. *Tetrahedron*, **1983**, *39*, 2323.

[30] Sharpless, K. B.; Verhoeven, T. R. *Aldrichim. Acta*, **1979**, *12*, 63.

[31] Sharpless, K. B. *Proceedings of the Robert A. Welch Foundation Conferences on Chemical Research XXVII*, Houston, Texas, 1983, pp. 59–89.

[32] Chaumette, P.; Mimoun, H.; Saussine, L.; Fischer, J.; Mitschler, A. *J. Organomet. Chem.*, **1983**, *250*, 291.

[33] Mimoun, H.; Charpentier, R.; Mitschler, A.; Fisher, J.; Weiss, R. *J. Am. Chem. Soc.*, **1980**, *102*, 1047.

[34] Mimoun, H. *Angew. Chem. Int. Ed. Engl.*, **1982**, *21*, 734.

[35] Sharpless, K. B.; Woodard, S. S.; and Finn, M. G. *Pure & Appl. Chem.*, **1983** 55, 1823.

[36] Woodard, S. S.; Finn, M. G.; Sharpless, K. B. *J. Am. Chem. Soc.*, **1991**, *113*, 106.

[37] Finn, M. G.; Sharpless, K. B. *J. Am. Chem. Soc.*, **1991**, *113*, 113.

[38] McKee, B. H.; Kalantar, T. H.; Sharpless, K. B. *J. Org. Chem.*, **1991**, *56*, 6966.

[39] Narula, A. S. *Tetrahedron Lett.*, **1982**, *23*, 5579.

[40] Bach, R. D.; Wolber, G. J.; Coddens, B. A. *J. Am. Chem. Soc.*, **1984**, *106*, 6098.

[41] Rossiter, B. E.; Sharpless, K. B. The Scripps Research Institute, La Jolla, CA, unpublished results.

[42] Puchot, C.; Samuel, O.; Dunach, E.; Zhao, S.; Agami, C.; Kagan, H. B. *J. Am. Chem. Soc.*, **1986**, *108*, 2353.

[43] Woodard, S. S. Ph.D Dissertation, Stanford University, Stanford, California, 1981.

[44] Burgess, K.; Jennings, L. D. *J. Am. Chem. Soc.*, **1990**, *112*, 7434.

[45] Carlier, P. R.; Sharpless, K. B. *J. Org. Chem.*, **1989**, *54*, 4016.

[46] Williams, I. D.; Pedersen, S. F.; Sharpless, K. B.; Lippard, S. J. *J. Am. Chem. Soc.*, **1984**, *106*, 6430.

[47] Sharpless, K. B. *Chem. Scripta*, **1987**, *27*, 521.

[48] Pedersen, S. F.; Dewan, J. C.; Eckman, R. R.; Sharpless, K. B. *J. Am. Chem. Soc.*, **1987**, *109*, 1279.

[49] Katsuki, T.; Sharpless, K. B.; Kyushu University, unpublished result.

[50] Chamberlin, A. R.; Mulholland, Jr., R. L.; Kahn, S. D.; Hehre, W. J. *J. Am. Chem. Soc.*, **1987**, *109*, 672.

[51] Masamune, S.; Choy, W.; Petersen, J. S.; Sita, L. R. *Angew. Chem., Int. Ed. Engl.*, **1985**, *24*, 1.

[52] Corey, E. J. *J. Org. Chem.*, **1990**, *55*, 1693.

[53] Takano, S.; Iwabuchi, Y.; Ogasawara, K. *Tetrahedron Lett.*, **1991**, *32*, 3527.

[54] Chan, T. H.; Chen, L. M.; Wang, D. *J. Chem. Soc., Chem. Commun.*, **1988**, 1280.

[55] Luly, J. R.; Hsiao, C.-N.; BaMaung. N.; Plattner, J. J. *J. Org. Chem.*, **1988**, *53*, 6109.

[56] Marshall, J. A.; Flynn, K. E. *J. Am. Chem. Soc.*, **1982**, *104*, 7430.

[57] Schweiter, M. J.; Sharpless, K. B. *Tetrahedron Lett.*, **1985**, *26*, 2543.

[58] Takahashi, K.; Ogata, M. *J. Org. Chem*, **1987**, *52*, 1877.

[59] Rossiter, B. E.; Katsuki, T.; Sharpless, K. B. *J. Am. Chem. Soc.*, **1981**, *103*, 464.

[60] Baker, R.; Swain, C. J.; Head, J. C. *J. Chem. Soc., Chem. Commun.*, **1986**, 874.

[61] Rossiter, B. E.; Verhoeven, T. R.; Sharpless, K. B. *Tetrahedron Lett.*, **1979**, 4733.

[62] Chong, J. M.; Wong, S. *J. Org. Chem.*, **1987**, *52*, 2596.

[63] Mori, K.; Seu, Y.-B. *Tetrahedron*, **1988**, *44*, 1035.

[64] Mori, K.; Nakazono, Y. *Tetrahedron*, **1986**, *42*, 6459.

[65] Hoagland, S.; Morita, Y.; Bai, D.-L.; Marki, H.-P.; Kees, K.; Brown, L.; Heathcock, C. H. *J. Org. Chem.*, **1988**, *53*, 4730.

[66] Klunder, J. M.; Onami, T.; Sharpless, K. B. *J. Org. Chem.*, **1989**, *54*, 1295.

[67] Hendrickson, H. S.; Hendrickson, E. K. *Chem. & Phys. Lipids*, **1990**, *53*, 115.

[68] Hoye, T. R.; Suhadolnik, J. C. *J. Am. Chem. Soc.*, **1985**, *107*, 5312.

[69] Hoye, T. R.; Suhadolnik, J. C. *Tetrahedron*, **1986**, *42*, 2855.

[70] Hoye, T. R.; Jenkins, S. A. *J. Am. Chem. Soc.*, **1987**, *109*, 6196.

[71] Nakahara, Y.; Fujita, A.; Ogawa, T. *Agric. Biol. Chem.*, **1987**, *51*, 1009.

[72] Burke, S. D.; Buchanan, J. L.; Rovin, J. D. *Tetrahedron Lett.*, **1991**, *32*, 3961.

[73] Katsuki, T.; Lee, A. W. M.; Ma, P.; Martin, V. S.; Masamune, S.; Sharpless, K. B.; Tuddenham, D.; Walker, F. J. *J. Org. Chem*, **1982**, *47*, 1373.

[74] Lee, A. W. M.; Martin, V. S.; Masamune, S.; Sharpless, K. B.; Walker, F. J. *J. Am. Chem. Soc.*, **1982**, *104*, 3515.

[75] Russell, S. T.; Robinson, J. A.; Williams, D. J. *J. Chem. Soc., Chem. Commun.*, **1987**, 351.

[76] Paterson, I.; Boddy, I.; Mason, I. *Tetrahedron Lett.*, **1987**, *28*, 5205.

[77] Paterson, I.; Boddy, I. *Tetrahedron Lett.*, **1988**, *29*, 5301.

[78] Paterson I.; Craw, P. A. *Tetrahedron Lett.*, **1989**, *30*, 5799.

[79] Koizumi, N.; Ishiguro, M.; Yasuda, M.; Ikekawa, N. *J. Chem. Soc., Perkin Trans. 1*, **1983**, 1401.

[80] Oehlschlager, A. C.; Johnston, B. D. *J. Org. Chem.*, **1987**, *52*, 940.

[81] Shibuya, H.; Kawashima, K.; Baek, N. I.; Narita, N.; Yoshikawa, M.; Kitagawa, I. *Chem. Pharm. Bull.*, **1989**, *37*, 260.

[82] Hashimoto, M.; Kan, T.; Yanagiya, M.; Shirahama, H.; Matsumoto, T. *Tetrahedron Lett.*, **1987**, *28*, 5665.

[83] Hashimoto, M.; Kan, T.; Nozaki, K.; Yanagiya, M.; Shirahama, H.; Matsumoto, T. *J. Org. Chem.*, **1990**, *55*, 5088.

[84] Nicolaou, K. C.; Duggan, M. E.; Hwang, C.-K.; Somers, P. K. *J. Chem. Soc., Chem. Commun.*, **1985**, 1359.

[85] Nicolaou, K. C.; Uenishi, J. *J. Chem. Soc., Chem. Commun.*, **1982**, 1292.

[86] Nicolaou, K. C.; Daines, R. A.; Uenishi, J.; Li, W. S.; Papahatjis, D. P.; Chakraborty, T. K. *J. Am. Chem. Soc.*, **1987**, *109*, 2205.

[87] Nicolaou, K. C.; Daines, R. A.; Uenishi, J.; Li, W. S.; Papahatjis, D. P.; Chakraborty, T. K. *J. Am. Chem. Soc.*, **1988**, *110*, 4672.

[88] Crimmins, M. T.; Lever, J. G. *Tetrahedron Lett.*, **1986**, *27*, 291.

[89] Bonadies, F.; Fabio, R. D.; Gubiotti, A.; Mecozzi, S.; Bonini, C. *Tetrahedron Lett.*, **1987**, *28*, 703.

[90] Rastetter, W. H.; Adams, J.; Bordner, J. *Tetrahedron Lett.*, **1982**, *23*, 1319.

[91] Baker, S. R.; Boot, J. R.; Morgan, S. E.; Osborne, D. J.; Ross, W. J.; Schrubsall, P. R. *Tetrahedron Lett.*, **1983**, *24*, 4469.

[92] Prestwich, G. D.; Wawrzénczyk, C. *Proc. Natl. Acad. Sci. USA*, **1985**, *82*, 5290.

[93] Petterson, L.; Frejd, T.; Magnusson, G. *Tetrahedron Lett.*, **1987**, *28*, 2753.

[94] Wuts, P. G. M.; D'Costa, R.; Butler, W. *J. Org. Chem.*, **1984**, *49*, 2582.

[95] Takahata, H.; Banba, Y.; Momose, T. *Tetrahedron: Asymmetry*, **1991**, *2*, 445.

[96] Takahata, H.; Banba, Y.; Momose, T. *Tetrahedron*, **1991**, *47*, 7635.

[97] Takahata, H.; Banba, Y.; Tajima, M.; Momose, T. *J. Org. Chem.*, **1991**, *56*, 240.

[98] Adams, C. E.; Walker, F. J.; Sharpless, K. B. *J. Org. Chem.*, **1985**, *50*, 420.

[99] Corey, E. J.; Hopkins, P. B.; Munroe, J. E.; Marfat, A., Hashimoto, S. *J. Am. Chem. Soc.*, **1980**, *102*, 7986.

[100] Corey, E. J.; Pyne, S. G.; Su, W.-G. *Tetrahedron Lett.*, **1983**, *24*, 4883.

[101] Oehlschlager, A. C.; Czyzewska, E. *Tetrahedron Lett.*, **1983**, *24*, 5587.

[102] Bernet, B.; Vasella, A. *Tetrahedron Lett.*, **1983**, *24*, 5491.

[103] Julina, R.; Herzig, T.; Bernet, B.; Vasella, A. *Helv. Chim. Acta*, **1986**, *69*, 368.

[104] Pale, P.; Chuche, J. *Tetrahedron Lett.*, **1987**, *28*, 6447.

[105] Pridgen, L. N.; Shilcrat, S. C.; Lantos, I. *Tetrahedron Lett.*, **1984**, *25*, 2835.

[106] Yamakawa, I.; Urabe, H.; Kobayashi, Y., Sato, F. *Tetrahedron Lett.*, **1991**, *32*, 2045.

[107] Levine, S. G.; Bonner, M. P. *Tetrahedron Lett.*, **1989**, *30*, 4767.

[108] Urabe, H.; Aoyama, Y.; Sato, F. *J. Org. Chem.*, **1992**, *57*, 5056.

[109] Mori, K.; Ebata, T. *Tetrahedron Lett.*, **1981**, *22*, 4281.

[110] Mori, K.; Ebata, T. *Tetrahedron*, **1986**, *42*, 3471.

[111] Overman, L. E.; Thompson, A. S. *J. Am. Chem. Soc.*, **1988**, *110*, 2248.

[112] Corey, E. J.; Tramontano, A. *J. Am. Chem. Soc.*, **1984**, *106*, 462.

[113] Ireland, R. E.; Smith, M. G. *J. Am. Chem. Soc.*, **1988**, *110*, 854.

[114] Takano, S.; Otaki, S.; Ogasawara, K. *J. Chem. Soc., Chem. Commun.*, **1983**, 1172.

[115] Schollkopf, U.; Tiller, T.; and Bardenhagen, J. *Tetrahedron*, **1988**, *44*, 5293.

[116] Johnston, B. D.; Oehlschlager, A. C. *J. Org. Chem.*, **1982**, *47*, 5384.

[117] Abad, A.; Agullo, C.; Arno, M.; Cunat, A. C.; Zaragoza, R. J. *J. Org. Chem*, **1992**, *57*, 50.

[118] Doherty, A. M.; Ley, S. V. *Tetrahedron Lett.*, **1986**, *27*, 105.

[119] de Laszlo, S. E.; Ford, M. J.; Ley, S. V.; Maw, G. N. *Tetrahedron Lett.*, **1990**, *31*, 5525.

[120] Kende, A. S.; Rizzi, J. P. *J. Am. Chem. Soc.*, **1981**, *103*, 4247.

[121] Rizzi, J. P.; Kende, A. S. *Tetrahedron*, **1984**, *40*, 4693.

[122] Naruta, Y.; Nishigaichi, Y.; Maruyama, K. *Tetrahedron Lett.*, **1989**, *30*, 3319.

[123] Finan, J. M.; Kishi, Y. *Tetrahedron Lett.*, **1982**, *23*, 2719.

[124] Sugiyama, S.; Honda, M.; Komori, T. *Justus Liebigs Ann. Chem.*, **1988**, 619.

[125] Kanemoto, S.; Nonaka, T.; Oshima, K.; Utimoto, K.; Nozaki, H. *Tetrahedron Lett.*, **1986**, *27*, 3387.

[126] Marshall, J. A.; Jenson, T. M. *J. Org. Chem.*, **1984**, *49*, 1707.

[127] Falck, J. R.; Manna, S.; Siddhanta, A. K.; Capdevila, J; Buynak, J. D. *Tetrahedron Lett.*, **1983**, *24*, 5715.

[128] Martin, V. S.; Katsuki, T.; Tuddenham, D.; Sharpless, K. B. Kyushu University, unpublished results.

[129] Okamura, H.; Kuroda, S.; Tomita, K.; Ikegami, S.; Sugimoto, Y.; Sakaguchi, S.; Katsuki, T.; Yamaguchi, M. *Tetrahedron Lett.*, **1991**, *32*, 5137.

[130] Lu, L. D.-L.; Johnson, R. A.; Finn, M. G.; Sharpless, K. B. *J. Org. Chem.*, **1984**, *49*, 728.

[131] Tanner, D.; Somfai, P. *Tetrahedron*, **1986**, *42*, 5985.

[132] Tanner, D.; Somfai, P. *Tetrahedron*, **1987**, *43*, 4395.

[133] Nemoto, H.; Ishibashi, H.; Nagamochi, M.; Fukumoto, K. *J. Org. Chem.*, **1992**, *57*, 1707.

[134] Nemoto, H.; Ishibashi, H.; Fukumoto, K. *Heterocycles*, **1992**, *33*, 549.

[135] Klunder, J. M.; Ko, S. Y.; Sharpless, K. B. *J. Org. Chem.*, **1986**, *51*, 3710.

[136] Ko, S. Y.; Sharpless, K. B. *J. Org. Chem.*, **1986**, *51*, 5413.

[137] Ko, S. Y.; Masamune, H.; Sharpless, K. B. *J. Org. Chem.*, **1987**, *52*, 667.

[138] Choudary, B. M; Valli, V. L. K.; Prasad, A. D. *J. Chem. Soc., Chem. Commun.*, **1990**, 1186.

[139] Hayashi, M.; Okamura, F.; Toba, T.; Oguni, N.; Sharpless, K. B. *Chem. Lett.*, **1990**, 547.

[140] Wang, Z. M.; Zhou, W. S. *Tetrahedron*, **1987**, *43*, 2935.

[141] Wang, Z. M.; Zhou, W. S.; Lin, G. Q. *Tetrahedron Lett.*, **1985**, *26*, 6221.

[142] Wang, Z. M.; Zhou, W. S. *Synth. Commun.*, **1989**, *19*, 2627.

[143] Yamamoto, K.; Ando, H.; Shuetake, T.; Chikamatsu, H. *J. Chem. Soc., Chem. Commun.*, **1989**, 754.

[144] Farrall, M. J.; Alexis, M.; Trecarten, M. *Nouv. J. Chim.*, **1983**, *7*, 449.

[145] Sharpless, K. B.; Behrens, C. H.; Katsuki, T.; Lee, A. W. M.; Martin, V. S.; Takatani, M.; Viti, S. M.; Walker, F. J.; Woodard, S. S. *Pure & Appl. Chem.*, **1983**, *55*, 589.

[146] Marshall, J. A.; Robinson, E. D.; Zapata, A. *J. Org. Chem.*, **1989**, *54*, 5854.

[147] Brimacombe, J. S.; Hanna, R.; Kabir, A. K. M. S. *J. Chem. Soc., Perkin Trans. 1*, **1987**, 2421.

[148] Brimacombe, J. S.; Kabir, A. K. M. S.; Bennett, F. *J. Chem. Soc., Perkin Trans. 1*, **1986**, 1677.

[149] Brimacombe, J. S.; Roderick, H.; Kabir, A. K. M. S. *Carbohydr. Res.*, **1986**, *153*, C7.

[150] White, J. D.; Jayasinghe, L. R. *Tetrahedron Lett.*, **1988**, *29*, 2139.

[151] White, J. D.; Amedio, Jr., J. C.; Gut, S.; Jayasinghe, L. *J. Org. Chem.*, **1989**, *54*, 4268.

[152] White, J. D.; Amedio, Jr., J. C.; Gut, S.; Ohira, S.; Jayasinghe, L. R. *J. Org. Chem.*, **1992**, *57*, 2270.

[153] Levine, S. G.; Heard, N. E. *Synth. Commun.*, **1991**, *21*, 549.

[154] Nagaoka, H.; Kishi, Y. *Tetrahedron*, **1981**, *37*, 3873.

[155] Nishiyama, S.; Toshima, H.; Kanai, H.; Yamamura, S. *Tetrahedron Lett.*, **1986**, *27*, 3643.

[156] Nishiyama, S.; Toshima, H.; Kanai, H.; Yamamura, S. *Tetrahedron*, **1988**, *44*, 6315.

[157] Dolle, R. E.; Nicolaou, K. C. *J. Am. Chem. Soc.*, **1985**, *107*, 1691.

[158] Mori, K.; Itou, M. *Justus Liebigs Ann. Chem.*, **1992**, 87.

[159] Smith, III, A. B.; Sarvatore, B. A.; Hull, K. G.; Duan, J. J.-W. *Tetrahedron Lett.*, **1991**, *32*, 4859.

[160] Minami, N.; Ko, S. S.; Kishi, Y. *J. Am. Chem. Soc.*, **1982**, *104*, 1109.

[161] Ko, S. Y.; Lee, A. W. M.; Masamune, S.; Reed, III, L. A.; Sharpless, K. B.; Walker, F. J. *Science*, **1983**, *220*, 949.

[162] Ko, S. Y.; Lee, A. W. M.; Masamune, S.; Reed, III, L. A.; Sharpless, K. B.; Walker, F. J. *Tetrahedron*, **1990**, *46*, 245.

[163] Klein, L. L.; McWhorter Jr., W. W.; Ko, S. S.; Pfaff, K.-P.; Kishi, Y.; Uemura, D.; Hirata, Y. *J. Am. Chem. Soc.*, **1982**, *104*, 7362.

[164] Ko, S. S.; Finan, J. M.; Yonaga, M.; Kishi, Y.; Uemura, D.; Hirata, Y. *J. Am. Chem. Soc.*, **1982**, *104*, 7364.

[165] Fujioka, H.; Christ, W. J.; Cha, J. K.; Leder, J.; Kishi, Y.; Uemura, D.; Hirata, Y. *J. Am. Chem. Soc.*, **1982**, *104*, 7367.

[166] McWhorter, Jr., W. W.; Kang, S. H.; Kishi, Y. *Tetrahedron Lett.*, **1983**, *24*, 2243.

[167] Clayden, J.; Collington, E. W.; Warren, S. *Tetrahedron Lett.*, **1992**, *33*, 7043.

[168] Isobe, M.; Kitamura, M.; Mio, S.; Goto, T. *Tetrahedron Lett.*, **1982**, *23*, 221.

[169] Kitamura, M.; Isobe, M.; Ichikawa, Y.; Goto, T. *J. Org. Chem.*, **1984**, *49*, 3517.

[170] Ichikawa, Y.; Isobe, M.; Bai, D.-L.; Goto, T. *Tetrahedron*, **1987**, *43*, 4737.

[171] Ichikawa, Y.; Isobe, M.; Goto, T. *Tetrahedron*, **1987**, *43*, 4749.

[172] Isobe, M.; Ichikawa, Y.; Goto, T. *Tetrahedron Lett.*, **1985**, *26*, 5199.

[173] Meyers, A. I.; Hudspeth, J. P. *Tetrahedron Lett.*, **1981**, 22, 3925.
[174] Meyers, A. I.; Babiak, K. A.; Campbell, A. L.; Comins, D. L.; Fleming, M. P.; Henning, R.; Heuschmann, M.; Hudspeth, J. P.; Kane, J. M.; Reider, P. J.; Roland, D. M.; Shimizu, K.; Tomioka, K.; Walkup, R. D. *J. Am. Chem. Soc.*, **1983**, 105, 5015.
[175] Nakajima, N.; Tanaka, T.; Hamada, T.; Oikawa, Y.; Yonemitsu, O. *Chem. Pharm. Bull.*, **1987**, 35, 2228.
[176] Rossiter, B. E.; Sharpless, K. B. *J. Org. Chem.*, **1984**, 49, 3707.
[177] Takano, S.; Iwabuchi, Y.; Ogasawara, K. *Synlett*, **1991**, 548.
[178] Ikegami, S.; Katsuki, T.; Yamaguchi, M. *Chem. Lett.*, **1987**, 83.
[179] Corey, E. J.; Ha, D.-C. *Tetrahedron Lett.*, **1988**, 29, 3171.
[180] Adam, W.; Griesbeck, A.; Staab, E. *Tetrahedron Lett.*, **1986**, 27, 2839.
[181] Adam, W.; Griesbeck, A.; Staab, E. *Angew. Chem., Int. Ed. Engl.*, **1986**, 25, 269.
[182] Adam, W.; Braun, M.; Griesbeck, A.; Lucchini, V.; Staab, E.; Will, B. *J. Am. Chem. Soc.*, **1989**, 111, 203
[183] Roush, W. R.; Brown, R. J. *J. Org. Chem.*, **1983**, 48, 5093
[184] Roush, W. R.; Spada, A. P. *Tetrahedron Lett.*, **1983**, 24, 3693.
[185] Page, P. C. B.; Rayner, C. M.; Sutherland, I. O. *Tetrahedron Lett.*, **1986**, 27, 3535.
[186] Page, P. C. B.; Rayner, C. M.; Sutherland, I. O. *J. Chem. Soc., Perkin Trans. 1*, **1990**, 2403.
[187] Takano, S.; Iwabuchi, Y.; Ogasawara, K. *J. Am. Chem. Soc.*, **1991**, 113, 2786.
[188] Takano, S.; Setoh, M.; Takahashi, M.; Ogasawara, K. *Tetrahedron Lett.*, **1992**, 33, 5365.
[189] For precedents to this equation, see footnote 7 in ref. 10.
[190] Kitano, Y.; Matsumoto, T.; Sato, F. *J. Chem. Soc., Chem. Commun.*, **1986**, 1323.
[191] Kitano, Y.; Matsumoto, T.; Takeda, Y.; Sato, F. *J. Chem. Soc., Chem. Commun.*, **1986**, 1732.
[192] Okamoto, S.; Shimazaki, T.; Kobayashi, Y.; Sato, F. *Tetrahedron Lett.*, **1987**, 28, 2033.
[193] Kobayashi, Y.; Shimazaki, T.; Sato. F. *Tetrahedron Lett.*, **1987**, 28, 5849.
[194] Kusakabe, M.; Kato, H.; Sato, F. *Chem. Lett.*, **1987**, 2163.
[195] Shimazaki, T.; Kobayashi, Y.; Sato, F. *Chem. Lett.*, **1988**, 1785.
[196] Kobayashi, Y.; Kusakaba, M.; Kitano, Y.; Sato, F. *J. Org. Chem.*, **1988**, 53, 1587.
[197] Matsumoto, T.; Kitano, Y.; Sato, F. *Tetrahedron Lett.*, **1988**, 29, 5685.
[198] Lohse, P.; Loner, H.; Acklin, P.; Sternfeld, F.; Pfaltz, A. *Tetrahedron Lett.*, **1991**, 32, 615.
[199] Discordia, R. P.; Dittmer, D. C. *J. Org. Chem.*, **1990**, 55, 1414.
[200] Zhou, W.-S.; Lu, Z.-H.; Wang, Z.-M. *Tetrahedron Lett.*, **1991**, 32, 1467.
[201] Zhao, S. H.; Samuel, O.; Kagan, H. B. *Tetrahedron*, **1987**, 43, 5135.
[202] Kagan, H. B.; Dunach, E.; Nemecek, C.; Pitchen, P.; Samuel, O.; Zhao, S. H. *Pure & Appl. Chem.*, **1985**, 57, 1911.
[203] Dunach, E.; Kagan, H. B. *Nouv. J. Chem.*, **1985**, 9, 1.
[204] Nemecek, C.; Dunach, E.; Kagan, H. B. *Nouv. J. Chem.*, **1986**, 10, 761.
[205] Kagan, H. B.; Dunach, E.; Deshmukh, M.; Pitchen, P. *Chem. Scripta*, **1985**, 25, 101.
[206] Samuel, O.; Ronan, B.; Kagan, H. B. *J. Organomet. Chem.*, **1989**, 53, 1587.
[207] Boussad, N.; Trefouel, T.; Dupas, G.; Bourguignon, J.; Queguiner, G. *Phosphorus, Sulfur, Silicon*, **1992**, 66, 127.
[208] Mitsunobu, O. *Synthesis*, **1981**, 1.
[209] Walba, D. M.; Vohra, R. T; Clark, N. A.; Handschy, M. A.; Xue, J.; Parmar, D. S.; Langerwall, S. T.; Skarp, K. *J. Am. Chem. Soc.*, **1986**, 108, 7424.
[210] Wasserman, H. H.; Oku, T. *Tetrahedron Lett.*, **1986**, 27, 4913.
[211] Discordia, R. P.; Murphy, C. K.; Dittmer, D. C. *Tetrahedron Lett.*, **1990**, 31, 5603.
[212] Chong, J. M. *Tetrahedron Lett.*, **1992**, 33, 33.
[213] Kigoshi, H.; Ojima, M.; Shizuri, Y.; Niwa, H.; Yamada, K. *Tetrahedron Lett.*, **1982**, 23, 5413.
[214] Kigoshi, H.; Ojika, M.; Shizuri, Y.; Niwa, H.; Yamada, K. *Tetrahedron*, **1986**, 42, 3789.
[215] Aziz, M.; Rouessac, F. *Tetrahedron Lett.*, **1987**, 28, 2579.
[216] Aziz, M.; Rouessac, F. *Tetrahedron*, **1988**, 44, 101.
[217] Mori, K.; Ueda, H. *Tetrahedron*, **1981**, 37, 2581.
[218] Tius, M. A.; Fauq, A. *J. Am. Chem. Soc.*, **1986**, 108, 6389.
[219] Nicolaou, K. C.; Duggan, M. E.; Ladduwahetty, T. *Tetrahedron Lett.*, **1984**, 25, 2069.

[220] Millar, J. G.; Underhill, E. W. *J. Org. Chem.*, **1986**, *51*, 4726.
[221] Williams, D. R.; Jass, P. A.; Tse, H.-L. A.; Gaston, R. D. *J. Am. Chem. Soc.*, **1990**, *112*, 4552.
[222] Corey, L. D.; Singh, S. M.; Oehlschlager, A. C. *Can. J. Chem.*, **1987**, *65*, 1821.
[223] Yadav, J. S.; Shekharam, T.; Gadgil, V. R. *J. Chem. Soc., Chem. Commun.*, **1990**, 843.
[224] Takano, S.; Samizu, K.; Sugihara, T.; Ogasawara, K. *J. Chem. Soc., Chem. Commun.*, **1989**, 1344.
[225] Mhaskar, S. Y.; Lakshminarayana, G. *Tetrahedron Lett.*, **1990**, *31*, 7227.
[226] Yadav, J. S.; Deshpande, P. K.; Sharma, G. V. M. *Pure & Appl. Chem.*, **1990**, *62*, 1333.
[227] Yadav, J. S.; Deshpande, P. K.; Sharma, G. V. M. *Tetrahedron*, **1990**, *46*, 7033.
[228] Marshall, J. A.; Luke, G. P. *Synlett*, **1992**, 1007.
[229] Omura, K.; Swern, D. *Tetrahedron*, **1977**, *33*, 1651.
[230] Collins, J. C.; Hess, W. W.; Frank, F. J. *Tetrahedron Lett.*, **1968**, 3363.
[231] Collins, J. C.; Hess, W. W. *Org. Synth.*, **1972**, *52*, 5.
[232] Narasaka, K.; Morikawa, A.; Saigo, K.; Mukaiyama, T. *Bull. Chem. Soc. Jpn.*, **1977**, *50*, 2773.
[233] Noyori, R. in *Organic Synthesis, Today and Tomorrow*, Trost, B. M.; Hutchinson, C. R., Eds., Pergamon Press, Oxford, 1980, pp. 273–284.
[234] Marshall, J. A.; Trometer, J. D. *Tetrahedron Lett.*, **1987**, *28*, 4985.
[235] Marshall, J. A.; Trometer, J. D.; Blough, B. E.; Crute, T. D. *J. Org. Chem.*, **1988**, *53*, 4274.
[236] Marshall, J. A.; Blough, B. E. *J. Org. Chem.*, **1991**, *56*, 2225.
[237] Marshall, J. A.; Trometer, J. D.; Blough, B. E.; Crute, T. D. *Tetrahedron Lett.*, **1988**, *29*, 913.
[238] Marshall, J. A.; Trometer, J. D.; Cleary, D. G. *Tetrahedron*, **1989**, *45*, 391.
[239] Molander, G. A.; La Belle, B. E.; Hahn, G. *J. Org. Chem.*, **1986**, *51*, 5259.
[240] Oshima, M.; Yamazaki, H.; Shimizu, I.; Nisar, M.; Tsuji, J. *J. Am. Chem. Soc.*, **1989**, *111*, 6280.
[241] Trost, B. M.; Angle, S. R. *J. Am. Chem. Soc.*, **1985**, *107*, 6123.
[242] Trost, B. M.; Lynch, J. K.; Angle, S. R. *Tetrahedron Lett.*, **1987**, *28*, 375.
[243] Trost, B. M.; Sudhakar, A. R. *J. Am. Chem. Soc.*, **1987**, *109*, 3792.
[244] Wershofen, S.; Scharf, H.-D. *Synthesis*, **1988**, 854.
[245] Molander G. A.; Shubert, D. C. *J. Am. Chem. Soc.*, **1987**, *109*, 576.
[246] Okamoto, S.; Tsujiyama, H.; Yoshio, T.; Sato, F. *Tetrahedron Lett.*, **1991**, *32*, 5789.
[247] Roush, W. R.; Straub, J. A.; VanNiuewenhze, M. S. *J. Org. Chem.*, **1991**, *56*, 1636.
[248] Miyashita, M.; Suzuki, T.; Yoshikoshi, A. *Tetrahedron Lett.*, **1987**, *28*, 4293.
[249] Miyashita, M.; Hoshino, H.; Yoshikoshi, A. *Tetrahedron Lett.*, **1988**, *29*, 347.
[250] Molander, G. A.; Hahn, G. *J. Org. Chem.*, **1986**, *51*, 2596.
[251] Molander, G. A.; Belle, B. E. L.; Hahn, G. *J. Org. Chem.*, **1986**, *51*, 5259.
[252] Evans, D. A.; Williams, J. M. *Tetrahedron Lett.*, **1988**, *29*, 5065.
[253] Carlsen, P. H. J.; Katsuki, T.; Martin, V. S.; Sharpless, K. B. *J. Org. Chem.*, **1981**, *46*, 3936.
[254] Chong, J. M.; Sharpless, K. B. *J. Org. Chem.*, **1985**, *50*, 1560.
[255] Roush, W. R; Blizzard, T. A. *J. Org. Chem.*, **1984**, *49*, 4332.
[256] Thijs, L.; Stokkingreef, E. H. M.; Lemmens, J. M.; Zwanenburg, B. *Tetrahedron*, **1985**, *41*, 2949.
[257] Waanders, P. P.; Thijs, L.; Zwanenburg, B. *Tetrahedron Lett.*, **1987**, *28*, 2409.
[258] Payne, G. B. *J. Org. Chem.*, **1962**, *27*, 3819.
[259] Behrens, C. H.; Ko, S. Y.; Sharpless, K. B.; Walker, F. J. *J. Org. Chem.*, **1985**, *50*, 5687.
[260] Wrobel, J. E.; Ganem, B. *J. Org. Chem.*, **1983**, *48*, 3761.
[261] Behrens, C. H.; Sharpless, K. B. *J. Org. Chem.*, **1985**, *50*, 5696.
[262] Takano, S.; Morimoto, M.; Ogasawara, K. *Synthesis*, **1984**, 834.
[263] Garner, P.; Park, J. M.; Rotello, V. *Tetrahedron Lett.*, **1985**, *26*, 3299.
[264] Hatakeyama, S.; Sakurai, K.; Takano, S. *J. Chem. Soc., Chem. Commun.*, **1985**, 1759.
[265] Hatakeyama, S.; Sakurai, K.; Takano, S. *Tetrahedron Lett.*, **1986**, *27*, 4485.
[266] Hafele, B.; Schroter, D.; Jager, V. *Angew. Chem., Int. Ed. Engl.*, **1986**, *25*, 87.
[267] Koppenhoefer, B.; Walser, M.; Schroter, D.; Hafele, B.; Jager, V. *Tetrahedron*, **1987**, *43*, 2059.
[268] Jager, V.; Hummer, W.; Stahl, U.; Gracza, T. *Synthesis*, **1991**, 769.
[269] Jager, V.; Schroter, D.; Koppenhoefer, B. *Tetrahedron*, **1991**, *47*, 2195.

[270] Babine, R. E. *Tetrahedron Lett.*, **1986**, *27*, 5791.
[271] Askin, D.; Volante, R. P.; Remer, R. A.; Ryan, K. M.; Shinkai, I. *Tetrahedron Lett.*, **1988**, *29*, 277.
[272] Hummer, W.; Gracza, T.; Jager, V. *Tetrahedron Lett.*, **1989**, *30*, 1517.
[273] Schmidt, R. R.; Frische, K. *Justus Liebigs Ann. Chem.*, **1988**, 209.
[274] Schreiber, S. L.; Schreiber, T. S.; Smith, D. B. *J. Am. Chem. Soc.*, **1987**, *109*, 1525.
[275] Nakatsuka, M.; Ragan, J. A.; Sammakia, T.; Smith, D. B.; Uehling, D. E.; Schreiber, S. L. *J. Am. Chem. Soc.*, **1990**, *112*, 5583.
[276] Kobayashi, Y.; Kato, N.; Shimazaki, T.; Sato, F. *Tetrahedron Lett.*, **1988**, *29*, 6297.
[277] Schreiber, S. L.; Smith, D. B. *J. Org. Chem.*, **1989**, *54*, 9.
[278] Schreiber, S. L.; Goulet, M. T.; Schulte, G. *J. Am. Chem. Soc.*, **1987**, *109*, 4718.
[279] Nakatsuka, M.; Ragan, J. A.; Sammakia, T.; Smith, D. B.; Uehling, D. E.; Schreiber, S. L. *J. Am. Chem. Soc.*, **1990**, *112*, 5583.
[280] Schreiber, S. L.; Goulet, M. T. *J. Am. Chem. Soc.*, **1987**, *109*, 8120.
[281] Soulie, J.; Lampilas, M.; Lallemand, J. Y. *Tetrahedron*, **1987**, *43*, 2701.
[282] Palazon, J. M.; Anorbe, B.; Martin, V. S. *Tetrahedron Lett.*, **1986**, *27*, 4987.
[283] Palazon, J. M.; Martin, V. S. *Tetrahedron Lett.*, **1988**, *29*, 681.
[284] Page, P. C. B.; Rayner, C. M.; Sutherland, I. O. *J. Chem. Soc., Chem. Commun.*, **1988**, 356.
[285] Page, P. C. B.; Rayner, C. M.; Sutherland, I. O. *J. Chem. Soc., Perkin Trans. 1*, **1990**, 1375.
[286] Soulie, J.; Ta, C.; Lallemand, J.-Y. *Tetrahedron*, **1992**, *48*, 443.
[287] Yamaguchi, M.; Hirao, I. *J. Chem. Soc., Chem. Commun.*, **1984**, 202.
[288] Schneider, J. A.; Yoshihara, K. *J. Org. Chem.*, **1986**, *51*, 1077.
[289] Sundararaman, P.; Barth, G.; Djerassi, C. *J. Am. Chem. Soc.*, **1981**, *103*, 5004.
[290] Hanson, R. M. *Tetrahedron Lett.*, **1984**, *25*, 3783.
[291] Tucker, H. *J. Org. Chem.*, **1979**, *44*, 2943.
[292] Suzuki, T.; Saimoto, H.; Tomioka, H.; Oshima, K.; Nozaki, H. *Tetrahedron Lett.*, **1982**, *23*, 3597.
[293] Roush, W. R.; Adam, M. A.; Peseckis, S. M. *Tetrahedron Lett.*, **1983**, *24*, 1377.
[294] Herold, P.; Mohr, P.; Tamm, C. *Helv. Chim. Acta*, **1983**, *66*, 744.
[295] Ahn, K. H.; Kim, J. S.; Jin, C. S.; Kang, D. H.; Han, D. S.; Shin, Y. S.; Kim, D. H. *Synlett*, **1992**, 306.
[296] Maruoka, K.; Sano, H.; Yamamoto, H. *Chem. Lett.*, **1985**, 599.
[297] Tung, R. D.; Rich, D. H. *Tetrahedron Lett.*, **1987**, *28*, 1139.
[298] Uchiyama, H.; Kobayashi, Y,; Sato, F. *Chem. Lett.*, **1985**, 467.
[299] Kobayashi, Y.; Kitano, Y.; Takeda, Y.; Sato, F. *Tetrahedron*, **1986**, *42*, 2937.
[300] Takano, S.; Yanase, M.; Sugihara, T.; Ogasawara, K. *J. Chem. Soc., Chem. Commun.*, **1988**, 1538.
[301] Caron, M.; Sharpless, K. B. *J. Org. Chem.*, **1985**, *50*, 1557.
[302] Canas, M.; Poch, M.; Verdaguer, X.; Moyano, A.; Pericas, M. A.; Riera, A. *Tetrahedron Lett.*, **1991**, *32*, 6931.
[303] Dai, L.; Lou, B.; Zhang, Y.; Guo, G. *Tetrahedron Lett.*, **1986**, *27*, 4343.
[304] Burgos, C. E.; Ayer, D. E.; Johnson, R. A. *J. Org. Chem.*, **1987**, *52*, 4973.
[305] Gao. L.; Murai, A. *Chem. Lett.*, **1989**, 357.
[306] Caron, M.; Carlier, P. R.; Sharpless, K. B. *J. Org. Chem.*, **1988**, *53*, 5185.
[307] Alvarez, E.; Nunez, M. T.; Martin, V. S. *J. Org. Chem.*, **1990**, *55*, 3429.
[308] Shimizu, M.; Yoshida, A.; Fujisawa, T. *Synlett*, **1992**, 204.
[309] Bovicelli, P.; Lupattelli, P.; Bersani, M. T. *Tetrahedron Lett.*, **1992**, *33*, 6181.
[310] Choudary, B. M.; Rani, S. S.; Kantam, M. L. *Synth. Commun.*, **1990**, *20*, 2313.
[311] Bonini, C.; Giuliano, C.; Righi, G.; Rossi, L. *Tetrahedron Lett.*, **1992**, *33*, 7429.
[312] Morgan, Jr., D. J.; Sharpless, K. B.; Traynor, S. G. *J. Am. Chem. Soc.*, **1981**, *103*, 462.
[313] Denis, J. N.; Greene, A. E.; Serra, A. A.; Luche, M.-J. *J. Org. Chem.*, **1986**, *51*, 46.
[314] Onaka, M.; Sugita, K.; Izumi, Y. *Chem. Lett.*, **1986**, 1327.
[315] Onaka, M.; Sugita, K.; Takeuchi, H.; Izumi, Y. *J. Chem. Soc., Chem. Commun.*, **1988**, 1173.
[316] Onaka, M.; Sugita, K.; Izumi, Y. *J. Org. Chem.*, **1989**, *54*, 1116.

[317] Bonini, C.; Righi, G.; Sotgiu, G. *J. Org. Chem.*, **1991**, *56*, 6206.

[318] Bonini, C.; Righi, G. *Tetrahedron*, **1992**, *48*, 1531.

[319] Roush, W. R; Brown, R. J. *J. Org. Chem.*, **1982**, *47*, 1371.

[320] Roush, W. R.; Brown, R. J.; DiMare, M. *J. Org. Chem.*, **1983**, *48*, 5083.

[321] Roush, W. R.; Adam, M. A. *J. Org. Chem.*, **1985**, *50*, 3752.

[322] Paterson, I.; Cumming, J. *Tetrahedron Lett.*, **1992**, *33*, 2847.

[323] Kocovsky, P. *Tetrahedron Lett.*, **1986**, *27*, 5521.

[324] Roush, W. R.; Hagadorn, S. M. *Carbohydr. Res.*, **1985**, *136*, 187.

[325] Myers, A. G.; Widdowson, K. L. *Tetrahedron Lett.*, **1988**, *29*, 6389.

[326] Myers, A. G.; Proteau, P. J.; Handel, T. M. *J. Am. Chem. Soc.*, **1988**, *110*, 7212.

[327] McCombie, S. W.; Metz, W. A. *Tetrahedron Lett.*, **1987**, *28*, 383.

[328] Uenishi, J.; Motoyama, M.; Nishiyama, Y.; Wakabayashi, S. *J. Chem. Soc., Chem. Commun.*, **1991**, 1421.

[329] McCombie, S. W.; Nagabhushan, T. L. *Tetrahedron Lett.*, **1987**, *28*, 5395.

[330] Ma, P.; Martin, V. S.; Masamune, S.; Sharpless, K. B.; Viti, S. M. *J. Org. Chem.*, **1982**, *47*, 1378.

[331] Viti, S. M. *Tetrahedron Lett.*, **1982**, *23*, 4541.

[332] Honda, M.; Katsuki, T.; Yamaguchi, M. *Tetrahedron Lett.*, **1984**, *25*, 3857.

[333] Gao, Y.; Sharpless, K. B. *J. Org. Chem.*, **1988**, *53*, 4081.

[334] Johnson, M. R.; Nakata, T.; Kishi, Y. *Tetrahedron Lett.*, **1979**, 4343.

[335] Chong, J. M.; Cyr, D. R.; Mar, E. K. *Tetrahedron Lett.*, **1987**, *28*, 5009.

[336] Lipshutz, B. H.; Kotsuki, H.; Lew, W. *Tetrahedron Lett.*, **1986**, *27*, 4825.

[337] Tius, M. A.; Fauq, A. H. *J. Org. Chem.*, **1983**, *48*, 4131.

[338] Maruoka, K.; Hasegawa, M.; Yamamoto, H.; Suzuki, K.; Shimazaki, M.; Tsuchihashi, G. *J. Am. Chem. Soc.*, **1986**, *108*, 3827.

[339] Suzuki, K.; Miyazawa, M.; Shimazaki, M.; Tsuchihashi, G. *Tetrahedron Lett.*, **1986**, *27*, 6237.

[340] Maruoka, K.; Ooi, T.; Nagahara, S.; Yamamoto, H. *Tetrahedron*, **1991**, *47*, 6983.

[341] Maruoka, K.; Sato, J.; Yamamoto, H. *Tetrahedron*, **1992**, *48*, 3749.

[342] Maruoka, K.; Sato, K.; Yamamoto, H. *J. Am. Chem. Soc.*, **1991**, *113*, 5449.

[343] Maruoka, K.; Ooi, T.; Yamamoto, H. *J. Am. Chem. Soc.*, **1989**, *111*, 6431.

[344] Baldwin, J. E. *J. Chem. Soc., Chem. Commun.*, **1976**, 734.

[345] Stork, G.; Cama, L. D.; Coulson, D. R. *J. Am. Chem. Soc.*, **1974**, *96*, 5268.

[346] Nunez, M. T.; Rodriguez, M. L.; Martin, V. S. *Tetrahedron Lett.*, **1988**, *29*, 1979.

[347] Evans, D. A.; Bender, S. L.; Morris, J. *J. Am. Chem. Soc.*, **1988**, *110*, 2506.

[348] Reed, III, L. A.; Ito, Y.; Masamune, S.; Sharpless, K. B. *J. Am. Chem. Soc.*, **1982**, *104*, 6468.

[349] Nicolaou, K. C.; Prasad, C. V. C.; Somers, P. K.; Hwang, C.-K. *J. Am. Chem. Soc.*, **1989**, *111*, 5330.

[350] Nicolaou, K. C.; Prasad, C. V. C.; Somers, P. K.; Hwang, C.-K. *J. Am. Chem. Soc.*, **1989**, *111*, 5335.

[351] McCombie, S. W.; Shankar, B. B.; Ganguly, A. K. *Tetrahedron Lett.*, **1985**, *26*, 6301.

[352] Pelter, A.; Ward, R. S.; Little, G. M. *J. Chem. Soc., Perkin Trans. 1*, **1990**, 2775.

[353] McCombie, S. W.; Shankar, B. B.; Ganguly, A. K. *Tetrahedron Lett.*, **1989**, *30*, 7029.

[354] Suzuki, M.; Morita, Y.; Yanagisawa, A.; Noyori, R. *J. Am. Chem. Soc.*, **1986**, *108*, 5021.

[355] Suzuki, M.; Morita, Y.; Yanagisawa, A.; Baker, B. J.; Scheuer, P. J.; Noyori, R. *J. Org. Chem.*, **1988**, *53*, 286.

[356] McIntosh, J. M.; Matassa, L. C. *J. Org. Chem.*, **1988**, *53*, 4452.

[357] Tanis, S. P.; Chuang, Y.-H.; Head, D. B. *J. Org. Chem.*, **1988**, *53*, 4929.

[358] Medina, J. C.; Kyler, K. S. *J. Am. Chem. Soc.*, **1988**, *110*, 4818.

[359] White, J. D.; Jensen, M. S. *J. Am. Chem. Soc.*, **1993**, *115*, 2970.

[360] Gao, Y.; Sharpless, K. B. *J. Org. Chem.*, **1988**, *53*, 4114.

[361] Tanner, D.; Somfai, P. *Tetrahedron Lett.*, **1987**, *28*, 1211.

[362] Tanner, D.; Somfai, P. *Tetrahedron*, **1988**, *44*, 619.

[363] Pickenhagen, W.; Bronner–Schindler, H. *Helv. Chim. Acta*, **1984**, *67*, 947.

[364] Ewins, R. C.; Henbest, H. B.; McKervey, M. A. *J. Chem. Soc., Chem. Commun.*, **1967**, 1085.

[365] Montanari, F.; Moretti, I.; Torre, G. *J. Chem. Soc., Chem. Commun.*, **1969**, 135.

[366] Pirkle, W. H.; Rinaldi, P. L. *J. Org. Chem.*, **1977**, *42*, 2080.

[367] Rebek, Jr., J.; McCready, R. *J. Am. Chem. Soc.*, **1980**, *102*, 5602.

[368] Curci, R.; Fiorentino, M.; Serio, M. R. *J. Chem. Soc., Chem. Commun.*, **1984**, 155.

[369] Davis, F. A.; Chattopadhyay, S. *Tetrahedron Lett.*, **1986**, *27*, 5079.

[370] Davis, F. A.; Sheppard, A. C. *Tetrahedron*, **1989**, *45*, 5703.

[371] Davis, F. A.; Harakal, M. E.; Awad, S. B. *J. Am. Chem. Soc.*, **1983**, *105*, 3123.

[372] Julia, S.; Guixer, J.; Masana, J.; Rocas, J. *J. Chem. Soc., Perkin Trans. 1*, **1982**, 1317.

[373] Julia, S.; Masana, J.; Vega, J. C. *Angew. Chem., Int. Ed. Engl.*, **1980**, *19*, 929.

[374] Banfi, S.; Colonna, S.; Molinari, H.; Julia, S.; Guixer, J. *Tetrahedron*, **1984**, *40*, 5207.

[375] Baures, P. W.; Eggleston, D. S.; Flisak, J. R.; Gombatz, K.; Lantos, I.; Mendelson, W.; Remich, J. J. *Tetrahedron Lett.*, **1990**, *31*, 6501.

[376] Wynberg, H.; Greijdanus, B. *J. Chem. Soc., Chem. Commun.*, **1978**, 427.

[377] Sheng, M. N.; Zajacek, J. G. *J. Org. Chem.*, **1970**, *35*, 1839

[378] Sharpless K. B.; Michaelson, R. C. *J. Am. Chem. Soc.*, **1973**, *95*, 6136.

[379] Mihelich, E. D. *Tetrahedron Lett.*, **1979**, 4729.

[380] Yamada, S.; Mashiko, M.; Terashima, S. *J. Am. Chem. Soc.*, **1977**, *99*, 1988.

[381] Michaelson, R. C; Palermo, R. E.; Sharpless, K. B. *J. Am. Chem. Soc.*, **1977**, *99*, 1990.

[382] Takai, K.; Oshima, K.; Nozaki, H. *Tetrahedron Lett.*, **1980**, *21*, 1657.

[383] Takai, K.; Oshima, K.; Nozaki, H. *Bull. Chem. Soc. Jpn.*, **1983**, *56*, 3791.

[384] Kagan, H. B.; Mimoun, H.; Mark, C.; Shurig, V. *Angew. Chem., Int. Ed. Engl.*, **1979**, *18*, 485.

[385] Tani, K.; Hanafusa, M.; Otsuka, S. *Tetrahedron Lett.*, **1979**, 3017.

[386] Groves, J. T.; Myers, R. S. *J. Am. Chem. Soc.*, **1983**, *105*, 5791.

[387] Groves, J. T.; Viski, P. *J. Org. Chem.*, **1990**, *55*, 3628.

[388] Mansuy, D.; Battioni, P.; Renaud, J.-P.; Guerin, P. *J. Chem. Soc., Chem. Commun.*, **1985**, 155.

[389] Naruta, Y.; Tani, F.; Ishihara, N.; Maruyama, K. *J. Am. Chem. Soc.*, **1991**, *113*, 6865.

[390] Konishi, K.; Oda, K.; Nishida, K.; Aida, T.; Inoue, S. *J. Am. Chem. Soc.*, **1992**, *114*, 1313.

[391] Collman, J. P.; Lee, V. J.; Zhang, X.; Ibers, J. A.; Brauman, J. I. *J. Am. Chem. Soc.*, **1993**, *115*, 3834.

[392] Kaku, Y.; Otsuka, M.; Ohno, M. *Chem. Lett.*, **1989**, 611.

[393] Zhang, W.; Loebach, J. L.; Wilson, S. R.; Jacobsen, E. N. *J. Am. Chem. Soc.*, **1990**, *112*, 2801.

[394] Jacobsen, E. N.; Zhang, W.; Guller, M. L. *J. Am. Chem. Soc.*, **1991**, *113*, 6703.

[395] Jacobsen, E. N.; Zhang, W.; Muci, A. R.; Ecker, J. R.; Deng, L. *J. Am. Chem. Soc.*, **1991**, *113*, 7063.

[396] Lee, N. H.; Jacobsen, E. N. *Tetrahedron Lett.*, **1991**, *32*, 6533.

[397] Irie, R.; Noda, K.; Ito, Y.; Matsumoto, N.; Katsuki, T. *Tetrahedron Lett.*, **1990**, *31*, 7345,

[398] Irie, R.; Noda, K.; Ito, Y.; Katsuki, T. *Tetrahedron Lett.*, **1991**, *32*, 1055.

[399] Irie, R.; Ito, Y.; Katsuki, T. *Synlett*, **1991**, *2*, 265,

[400] Irie, R.; Noda, K.; Ito, Y.; Matsumoto, N.; Katsuki, T. *Tetrahedron Asymmetry*, **1991**, *2*, 481.

[401] Hatayama, A.; Hosoya, N.; Irie, R.; Ito, Y.; Katsuki, T. *Synlett*, **1992**, 407.

[402] Hosoya, N.; Irie, R.; Katsuiki, T. *Synlett*, **1993**, 261.

[403] Sasaki, H.; Irie, R.; Katsuki, T. *Synlett*, **1993**, 300.

[404] Sharpless, K. B.; Amberg, W.; Bennani, Y. L.; Crispino, G. A.; Hartung, J.; Jeong, K.-S.; Kwong, H.-L.; Morikawa, K.; Wang, Z.-M.; Xu, D.; Zhang, X.-L. *J. Org. Chem.*, **1992**, *57*, 2768.

[405] Keinan, E.; Sinha, S. C.; Sinha-Bagchi, A.; Wang, Z.-M.; Zhang, X.-L.; Sharpless, K. B. *Tetrahedron Lett.*, **1992**, *33*, 6411.

[406] Kolb, H. C.; Sharpless, K. B. *Tetrahedron* **1992**, *48*, 10515.

[407] Hill, J. G.; Rossiter, B. E.; Sharpless, K. B. *J. Org. Chem.*, **1983**, *48*, 3607.

[408] Bessodes, M.; Abushanab, E.; Antonakis, K. *Tetrahedron Lett.* **1984**, *25*, 5899.

[409] Meister, C.; Scharf, H.-D. *Justus Liebigs Ann. Chem.*, **1983**, 913.

[410] Ko, S. Y. Ph. D Dissertation, Massachusetts Institute of Technology, 1986.

[411] Erickson, T. J. *J. Org. Chem.*, **1986**, *51*, 934.

[412] Marshall, J. A.; Audia, V. H. *J. Org. Chem.*, **1987**, *52*, 1106.

[413] Kusakabe, M.; Kitano, Y.; Kobayashi, Y.; Sato, F. *J. Org. Chem.*, **1989**, *54*, 2085.

[414] Camberlin, A. R.; Dezube, M.; Reich, S. H.; Sall, D. J. *J. Am. Chem. Soc.*, **1989**, *111*, 6247.

[415] Walba, D. M.; Razavi, H. A.; Clark. N. A.; Parmar, D. S. *J. Am. Chem. Soc.*, **1988**, *110*, 8686.

[416] Sharpless, K. B.; Martin, V. S. The Scripps Research Institute, La Jolla, CA, unpublished results.

[417] Katsuki, T. *Tetrahedron Lett.*, **1984**, 25, 2821.

[418] Yoshida, J.; Maekawa, T.; Morita, Y.; Isoe, S. *J. Org. Chem.*, **1992**, *57*, 1321.

[419] Kobayashi, Y.; Ito, T.; Yamakawa, I.; Urabe, H.; Sato, F. *Synlett*, **1991**, 811.

[420] Gilloir, F.; Malacria, M. *Tetrahedron Lett.*, **1992**, *33*, 3859.

[421] Muchowski, J. M.; Naef, R.; Maddox, M. L. *Tetrahedron Lett.*, **1985**, *26*, 5375.

[422] Schwab, J. M.; Ray, T.; Ho, C.-K. *J. Am. Chem. Soc.*, **1989**, *111*, 1057.

[423] Dung, J.-S.; Armstrong, R. W.; Anderson, O. P.; Williams, R. M. *J. Org. Chem.*, **1983**, *48*, 3592.

[424] Scherkenbeck, J.; Barth, M.; Thiel, U.; Metten, K.-H.; Heinemann, F.; Welzel, P. *Tetrahedron*, **1988**, *44*, 6325.

[425] Meister, C.; Scharf, H. D. *Justus Liebigs Ann. Chem.*, **1983**, 913.

[426] Lipshutz, B. H.; Sharma, S.; Dimock, S. H.; Behling, J. R. *Synthesis*, **1992**, 191.

[427] Hosokawa, T.; Makabe, Y.; Shinohara, T.; Murahashi, S. *Chem. Lett.*, **1985**, 1529.

[428] Wershofen, S.; Claben, A.; Scharf, H.-D. *Justus Liebigs Ann. Chem.*, **1989**, 9.

[429] Shimazaki, M.; Hara, H.; Suzuki, K.; Tsuchihashi, G. *Tetrahedron Lett.*, **1987**, *28*, 5891.

[430] Hubscher, J.; Barner, R. *Helv. Chim. Acta*, **1990**, *73*, 1068.

[431] Hill, R. K.; Prakash, S. R. *J. Am. Chem. Soc.*, **1984**, *106*, 795.

[432] White, J. D.; Theramongkol, P.; Kuroda, C.; Engebrecht, J. R. *J. Org. Chem.*, **1988**, *53*, 5909.

[433] Baker, R.; Cummings, W. J.; Hayes, J. F.; and Kumar, A. *J. Chem. Soc., Chem. Commun.*, **1986**, 1237.

[434] Kobayashi, Y.; Kitano, Y.; Sato, F. *J. Chem. Soc., Chem. Commun.*, **1984**, 1329.

[435] Kuroda, C.; Theramongkol, P.; Engebrecht, J. R.; White, J. D. *J. Org. Chem.*, **1986**, *51*, 956.

[436] Gani, D.; O'Hagan, D.; Reynolds, K.; Robinson, J. A. *J. Chem. Soc., Chem. Commun.*, **1985**, 1002.

[437] Reynolds, K. A.; O'Hagan, D.; Gani, D.; Robinson, J. A. *J. Chem. Soc., Perkin Trans. 1*, **1988**, 3195.

[438] Martin, V. S.; Nunez, M. T.; Tonn, C. E. *Tetrahedron Lett.*, **1988**, *29*, 2701.

[439] Pickard, S. T.; Smith, H. E.; Polavarapu, P. L.; Black, T. M.; Rauk, A.; Yang, D. *J. Am. Chem. Soc.*, **1992**, *114*, 6850.

[440] Takano, S.; Sugihara, T.; Samizu, K.; Akiyama, M.; Ogasawara, K. *Chem. Lett.*, **1989**, 1781.

[441] Suzuki, K.; Miyazawa, M.; Shimazaki, M.; Tsuchihashi, G. *Tetrahedron*, **1988**, *44*, 4061.

[442] Takano, S.; Kasahara, C.; Ogasawara, K. *Chem. Lett.*, **1983**, 175.

[443] Yabe, Y.; Guillaume, D.; Rich, D. H. *J. Am. Chem. Soc.*, **1988**, *110*, 4043.

[444] Shibuya, H.; Kawashima, K.: Ikeda, M.; Kitagawa, I. *Tetrahedron Lett.*, **1989**, *30*, 7205.

[445] Clark, R. D.; Kurz, J. *Heterocycles*, **1985**, *23*, 2005.

[446] Nikam, S. S.; Martin, A. R.; Nelson, D. L. *J. Med. Chem.*, **1988**, *31*, 1965.

[447] Hughes, P.; Clardy, J. *J. Org. Chem.*, **1989**, *54*, 3260.

[448] Izawa, T.; Wang, Z.-q.; Nishimura, Y.; Kondo, S.; Umezawa, H. *Chem. Lett.*, **1987**, 1655.

[449] Clayden, J.; Collington, E. W.; Warren, S. *Tetrahedron Lett.*, **1992**, *33*, 7043.

[450] Dunigan, J.; Weigel, L. O. *J. Org. Chem.*, **1991**, *56*, 6225.

[451] Tius, M. A.; Fauq, A. H. *J. Am. Chem. Soc.*, **1986**, *108*, 1035.

[452] Grandjean, D.; Pale, P.; Chuche, J. *Tetrahedron Lett.*, **1991**, *32*, 3043.

[453] Bonini, C.; Fabio, R. D. *Tetrahedron Lett.*, **1988**, *29*, 815.

[454] Rodriguez, E. B.; Scally, G. D.; Stick, R. V. *Aust. J. Chem.*, **1990**, *43*, 1391.

[455] Kuehne, M. E.; Matson, P. A.; Bornmann, W. G. *J. Org. Chem.*, **1991**, *56*, 513.

[456] Magnus, P.; Mendoza, J. S. *Tetrahedron Lett.*, **1992**, *33*, 899.

[457] Magnus, P.; Mendoza, J. S.; Stamford, A.; Ladlow, M.; Willis, P. *J. Am. Chem. Soc.*, **1992**, *114*, 10232.

[458] Hirai, Y.; Chintani, M.; Yamazaki, T.; Momose, T. *Chem. Lett.*, **1989**, 1449.

[459] Diez–Martin, D.; Kotecha, N. R.; Ley, S. V.; Mantegani, S.; Menendez, J. C.; Organ, H. M.; White, A. D.; Banks, B. J. *Tetrahedron*, **1992**, *48*, 7899.

[460] Nicolaou, K. C.; Ahn, K. H. *Tetrahedron Lett.*, **1989**, *30*, 1217.

[461] Kotecha, N. R.; Ley, S. V.; Mantegani, S. *Synlett*, **1992**, 395.

[462] Kozikowski, A. P.; Stein, P. D. *J. Org. Chem.*, **1984**, *49*, 2301.

[463] Schmidt, U.; Mundinger, K.; Mangold, R.; Lieberknecht, A. *J. Chem. Soc., Chem. Commun.*, **1990**, 1216.

[464] Vaccaro, H. A.; Levy, D. E.; Sawabe, A.; Jaetsch, T.; Masamune, S. *Tetrahedron Lett.*, **1992**, *33*, 1937.

[465] Ramaswamy, S.; Prasad, K.; Repic, O. *J. Org. Chem.*, **1992**, *57*, 6344.

[466] White, J. D.; Bolton, G. L. *J. Am. Chem. Soc.*, **1990**, *112*, 1626.

[467] Jung, M. E.; Gardiner, J. M. *J. Org. Chem.*, **1991**, *56*, 2614.

[468] Hager, M. W.; Liotta, D. C. *J. Am. Chem. Soc.*, **1991**, *113*, 5117.

[469] Baker, R.; Head, J. C.; Swain, C. J. *J. Chem. Soc., Perkin Trans. 1*, **1988**, 85.

[470] Szurdoki, F.; Novak, L.; Baitz–Gacs, E.; Szantay, C. *Acta Chim. Hungarica–Models Chem.*, **1992**, *129*, 303.

[471] Rao, A. V. R.; Sharma, G. V. M.; Bhanu, M. N. *Tetrahedron Lett.*, **1992**, *33*, 3907.

[472] Takano, S.; Sugihara, T.; Ogasawara, K. *Tetrahedron Lett.*, **1991**, *32*, 2797.

[473] Still, W. C.; Ohmizu, H. *J. Org. Chem.*, **1981**, *46*, 5242.

[474] Suga, T.; Ohta, S.; Ohmoto, T. *J. Chem. Soc., Perkin Trans. 1*, **1987**, 2845.

[475] Smith, III, A. B.; Sunazuka, T.; Leenay, T. L.; Kingery–Wood, J. *J. Am. Chem. Soc.*, **1990**, *112*, 8197.

[476] Smith, III, A. B.; Kingery–Wood, J.; Leenay, T. L.; Nolen, E. G.; Sunazuka, T. *J. Am. Chem. Soc.*, **1992**, *114*, 1438.

[477] Yamada, S.; Shiraishi, M.; Ohmori, M.; Takayama, H. *Tetrahedron Lett.*, **1984**, *25*, 3347.

[478] Mori. K.; Ebata, T.; Takechi, S. *Tetrahedron*, **1984**, *40*, 1761.

[479] Brunner, H.; Sicheneder, A. *Angew. Chem., Int. Ed. Engl.*, **1988**, *27*, 718.

[480] Mulzer, J.; Lammer, O. *Chem. Ber.*, **1986**, *119*, 2178.

[481] Rayner, C. M.; Westwell, A. D. *Tetrahedron Lett.*, **1992**, *33*, 2409.

[482] Rayner, C. M.; Sin, M. S.; Westwell, A. D. *Tetrahedron Lett.*, **1992**, *33*, 7237.

[483] Gorthey, L. A.; Vairamani, M.; Djerassi, C. *J. Org. Chem.*, **1984**, *49*, 1511.

[484] Nakagawa, N.; Mori, K. *Agric. Biol. Chem.*, **1984**, *48*, 2505.

[485] Caldwell, C. G.; Bondy, S. S. *Synthesis*, **1990**, 34.

[486] Honda, M.; Komori, T. *Tetrahedron Lett.*, **1986**, *27*, 3369.

[487] Baker, R.; Swain, C. J.; Head, J. C. *J. Chem. Soc., Chem. Commun.*, **1985**, 309.

[488] Martin, V. S.; Nunez, M. T.; Ramirez, M. A.; Soler, M. A. *Tetrahedron Lett.*, **1990**, *31*, 763.

[489] Kiefel, M. J.; Maddock, J.; Pattenden, G. *Tetrahedron Lett.*, **1992**, *33*, 3227.

[490] Nicolaou, K. C.; Prasad, C. V. C.; Hwang, C.-K.; Duggan, M. E.; Veale, C. A. *J. Am. Chem. Soc.*, **1989**, *111*, 5321.

[491] Stork, G.; Kobayashi, Y.; Suzuki, T.; Zhao, K. *J. Am. Chem. Soc.*, **1990**, *112*, 1661.

[492] Sakaki, J.; Sugita, Y.; Sato, M.; Kaneko, C. *J. Chem. Soc., Chem. Commun.*, **1991**, 434.

[493] Sugita, Y.; Sakaki, J.; Sato, M.; Kaneko, C. *J. Chem. Soc., Perkin Trans. 1*, **1992**, 2855.

[494] Tanner, D.; Somfai, P. *Tetrahedron Lett.*, **1988**, *29*, 2373.

[495] Masamune, S.; Ma, P.; Okumoto, H.; Ellingboe, J. W.; Ito, Y. *J. Org. Chem.*, **1984**, *49*, 2834.

[496] Roush, W. R.; Adam, M. A.; Walts, A. E.; Harris, D. J. *J. Am. Chem. Soc.*, **1986**, *108*, 3422.

[497] Ireland, R. E.; Thaisrivongs, S.; Dussault, P. H. *J. Am. Chem. Soc.*, **1988**, *110*, 5768.

[498] Barrett, A. G. M.; Edmunds, J. J.; Horita, K.; Parkinson, C. J. *J. Chem. Soc., Chem. Commun.*, **1992**, 1236.

[499] Marshall, J. A.; Yashunsky, D. V. *J. Org. Chem.*, **1991**, *56*, 5493.

[500] Lampilas, M.; Lett, R. *Tetrahedron Lett.*, **1992**, *33*, 773.

[501] Iida, H.; Yamazaki, N.; Kibayashi, C. *J. Org. Chem.*, **1987**, *52*, 3337.

[502] Baldwin, J. E.; Flinn, A. *Tetrahedron Lett.*, **1987**, *28*, 3605.

[503] Suzuki, T.; Sato, O.; Hirama, M.; Yamamoto, Y.; Murata, M.; Yasumoto, T.; Harada, N. *Tetrahedron Lett.*, **1991**, *32*, 4505.

[504] Colvin, E. W.; Robertson, A. D.; Wakharkar, S. *J. Chem. Soc., Chem. Commun.*, **1983**, 312.

[505] Sturmer, R. *Justus Liebigs Ann. Chem.*, **1991**, 311.

[506] Shimizu, I.; Hayashi, K.; Ide, N.; Oshima, M. *Tetrahedron*, **1991**, *47*, 2991.

[507] Shimizu, I.; Hayashi, K.; Oshima, M. *Tetrahedron Lett.*, **1990**, *31*, 4757.

[508] Boeckman, Jr., R. K.; Pruitt, J. R. *J. Am. Chem. Soc.*, **1989**, *111*, 8286.

[509] Sandararaman, P.; Barth, G.; Djerassi, C. *J. Am. Chem. Soc.*, **1981**, *103*, 5004.

[510] Boschetti, A.; Panza, L.; Ronchetti, F.; Russo, G.; Toma, L. *J. Chem. Soc., Perkin Trans. 1*, **1988**, 3353.

[511] Raifeld, Y. E.; Vid, G. Y.; Mikerin, I. E.; Arshava, B. M.; Nikitenko, A. A. *Carbohydr. Res.*, **1992**, *224*, 103.

[512] Marshall, J. A.; Crute III, T. D.; Hsi, J. D. *J. Org. Chem.*, **1992**, 57, 115.

[513] Schmidt, U.; Respondek, M.; Lieberknecht, A.; Werner, J.; Fischer, P. *Synthesis*, **1989**, 256.

[514] Ohta, S.; Nakai, A.; Aoki, T.; Suga, T. *J. Sci. Hiroshima Univ., Ser. S: Phys. Chem.*, **1989**, 49.

[515] Bonadies, F.; Rossi, G.; Bonini, C. *Tetrahedron Lett.*, **1984**, *25*, 5431.

[516] Yadav, J. S.; Reddy, P. S.; Jolly, R. S. *Indian J. Chem.*, **1986**, *25B*, 294.

[517] Wood, J. L.; Jones, D. R.; Hirschmann, R.; Smith, III, A. B. *Tetrahedron Lett.*, **1990**, *31*, 6329.

[518] Aebi, J. D.; Deyo, D. T.; Sun, C. Q.; Guillaume, D.; Dunlap, B.; Rich, D. H. *J. Med. Chem.*, **1990**, *33*, 999.

[519] Hirai, Y.; Terada, T.; Amemiya, Y.; Momose, T. *Tetrahedron Lett.*, **1992**, *33*, 7893.

[520] Masamune, S. *Pure & Appl. Chem.*, **1988**, *60*, 1587.

[521] Blanchette, M. A.; Malamas, M. S.; Nantz, M. H.; Roberts, J. C.; Somfai, P.; Whritenour, D. C.; Masamune, S.; Kageyama, M.; Tamura, T. *J. Org. Chem.*, **1989**, *54*, 2817.

[522] Hanessian, S.; Ugolini, A.; Hodges, P. J.; Beaulieu, P.; Dube, D.; Andre, C. *Pure & Appl. Chem.*, **1987**, *59*, 299.

[523] Hanessian, S.; Ugolini, A.; Dube, D.; Hodges, P. J.; Andre, C. *J. Am. Chem. Soc.*, **1986**, *108*, 2776.

[524] Dommerholt, F. J.; Thijs, L.; Zwanenburg, B. *Tetrahedron Lett.*, **1991**, *32*, 1499.

[525] Dommerholt, F. J.; Thijs, L.; Zwanenburg, B. *Tetrahedron Lett.*, **1991**, *32*, 1495.

[526] Marshall, J. A.; Sedrani, R. *J. Org. Chem.*, **1991**, *56*, 5496.

[527] Boeckmann, Jr., R. K.; Barta, T. E.; Nelson, S. G. *Tetrahedron Lett.*, **1991**, *32*, 4091.

[528] Brimacombe, J. S.; Kabir, K. M. *Carbohydr. Res.*, **1986**, *152*, 329.

[529] Herunsalee, A.; Isobe, M.; Pikul, S.; Goto, T. *Synlett*, **1991**, 199.

[530] Takahashi, T.; Miyazawa, M.; Ueno, H.; Tsuji, J. *Tetrahedron Lett.*, **1986**, *27*, 3881.

[531] Iimori, T.; Still, W. C.; Rheingold, A. L.; Staley, D. L. *J. Am. Chem. Soc.*, **1989**, *111*, 3439.

[532] Nicolaou, K. C.; Veale, C. A.; Webber, S. E.; Katerinopoulos, H. *J. Am. Chem. Soc.*, **1985**, *107*, 7515.

[533] Spada, M. R.; Ubukata, M.; Isono, K. *Heterocycles*, **1992**, *34*, 1147.

[534] Hamon, D. P. G.; Massy–Westropp, R. A.; Newton, J. L. *Tetrahedron: Asymmetry*, **1990**, *1*, 771.

[535] Tanner, D.; He, H. M.; Bergdahl, M. *Tetrahedron Lett.*, **1988**, *29*, 6493.

[536] Tanner, D.; He, H. M. *Tetrahedron*, **1989**, *45*, 4309.

[537] Myers, A. G.; Harrington, P. M.; Kuo, E. Y. *J. Am. Chem. Soc.*, **1991**, *113*, 694.

[538] Takeda, Y.; Matsumoto, T.; Sato, F. *J. Org. Chem.*, **1986**, *51*, 4728.

[539] Marshall, J. A.; Blough, B. E. *J. Org. Chem.*, **1990**, *55*, 1540.

[540] Diez–Martin, D.; Kotecha, N. R.; Ley, S. V.; Menendez, J. C. *Synlett*, **1992**, 399.

[541] Yadav, J. S.; Joshi, B. V.; Sahasrabudhe, A. B. *Synth. Commun.*, **1985**, *15*, 797.

[542] Boeckman, Jr., R. K.; Tagat, J. R.; Johnston, B. H. *Heterocycles*, **1987**, *25*, 33.

[543] Maruoka, K.; Saito, S.; Ooi, T.; Yamamoto, H. *Synlett*, **1991**, 579.

[544] Yadav, J. S.; Deshpande, P. K.; Sharma, G. V. M. *Tetrahedron*, **1992**, *48*, 4465.

[545] Gurjar, M. K.; Reddy, A. S. *Tetrahedron Lett.*, **1990**, *31*, 1783.

[546] Neumann, C.; Boland, W. *Helv. Chim. Acta*, **1990**, *73*, 754.

[547] Lambs, L.; Singh, N. P.; Biellmann, J.–F. *J. Org. Chem.*, **1992**, *57*, 6301.

[548] Lambs, L.; Singh, N. P.; Biellmann, J.–F. *Tetrahedron Lett.*, **1991**, *32*, 2637.

[549] Tonn, C. E.; Palazon, J. M.; Ruiz–Perez, C.; Rodriguez, M. L.; Martin, V. S. *Tetrahedron Lett.*, **1988**, *29*, 3149.

[550] Bessodes, M.; Komiotis, D.; Antonakis, K. *J. Chem. Soc., Perkin Trans. 1*, **1989**, 41.

[551] Valverde, S.; Herradon, B.; Rabanal, R. M.; Martin–Lomas, M. *Can. J. Chem.*, **1987**, *65*, 339.

[552] Komiotis, D.; Bessodes, M.; Antonakis, K. *Carbohydr. Res.*, **1989**, *190*, 153.

[553] Millar, J. G.; Giblin, M.; Barton, D.; Underhill, E. W. *J. Chem. Ecol.*, **1991**, *17*, 911.

[554] Nicolaou, K. C.; Webber, S. E. *Synthesis*, **1986**, 453.

[555] Nicolaou, K. C.; Webber, S. E. *J. Chem. Soc., Chem. Commun.*, **1985**, 297.

[556] Barchi, Jr. J. J.; Moore, R. E.; Patterson, G. M. L. *J. Am. Chem. Soc.* **1984**, *106*, 8193.

[557] Valverde, S.; Herradon, B.; Rabanal, R. M.; Martin–Lomas, M. *Can. J. Chem.*, **1987**, *65*, 332.

[558] Valverde, S.; Martin–Lomas, M.; Herradon, B. *J. Carbohydr. Chem.*, **1987**, *6*, 685.

[559] Niwa, H.; Miyachi, Y.; Uosaki, Y.; Yamada, K. *Tetrahedron Lett.*, **1986**, *27*, 4601.

[560] Niwa, H.; Miyachi, Y.; Okamoto, O.; Uosaki, Y.; Kuroda, A.; Ishikawa, H.; Yamada, K. *Tetrahedron*, **1992**, *48*, 393.

[561] Johnston, B. D.; Oehlschlager, A. C. *Can. J. Chem.*, **1984**, *62*, 2148.

[562] Lee, A. W. M. *J. Chem. Soc., Chem. Commun.*, **1984**, 578.

[563] Diez–Martin, D.; Grice, P.; Kolb, H. C.; Ley, S. V.; Madin, A. *Synlett*, **1990**, 326.

[564] Ley, S. V.; Armstrong, A.; Diez–Martin, D.; Ford, M. J.; Grice, P.; Knight, J. G.; Kolb, H. C.; Madin, A.; Marby, C. A.; Mukherjee, S.; Shaw, A. N.; Slawin, A. M. Z.; Vile, S.; White, A. D.; Williams, D. J.; Woods, M. *J. Chem. Soc., Perkin Trans. 1*, **1991**, 667.

[565] Hatakeyama, S.; Matsui, Y.; Suzuki, M.; Sakurai, K.; Takano, S. *Tetrahedron Lett.*, **1985**, *26*, 6485.

[566] Sabol, J. S.; Cregge, R. J. *Tetrahedron Lett.*, **1990**, *31*, 27.

[567] Hedtmann, U.; Hobert, K.; Klintz, R.; Welzel, P.; Frelek, J.; Strangmann Diekmann, M. *Angew. Chem., Int. Ed. Engl.*, **1989**, *28*, 1515.

[568] Bessodes, M.; Egron, M.–J.; Antonakis, K. *J. Chem. Soc., Perkin Trans. 1*, **1989**, 2099.

[569] Otera, J.; Niibo, Y.; Nozaki, H. *Tetrahedron*, **1991**, *47*, 7625.

[570] Crilley, M. M. L.; Edmunds, A. J. F.; Eistetter, K.; Golding, B. T. *Tetrahedron Lett.*, **1989**, *30*, 885.

[571] Burke, S. D.; Cobb, J. E.; Takeuchi, K. *J. Org. Chem.*, **1990**, *55*, 2138.

[572] Burke, S. D.; Cobb, J. E.; Takeuchi, K. *J. Org. Chem.*, **1985**, *50*, 3420.

[573] Nishikimi, Y.; Iimori, T.; Sodeoka, M.; Shibasaki, M. *J. Org. Chem.*, **1989**, *54*, 3354.

[574] Kitching, W.; Lewis, J. A.; Perkins, M. V.; Drew, R.; Moore, C. J.; Schurig, V.; Konig, W. A.; Francke, W. *J. Org. Chem.*, **1989**, *54*, 3893.

[575] Page, P. C. B.; Rayner, C. M.; Sutherland, I. O. *J. Chem. Soc., Chem. Commun.*, **1986**, 1408.

[576] Page, P. C. B.; Rayner, C. M.; Sutherland, I. O. *J. Chem. Soc., Perkin Trans. 1*, **1990**, 1615

[577] Baker, S. R.; Boot, J. R.; Morgan, S. E. *Tetrahedron Lett.*, **1983**, *24*, 4469.

[578] Kabat, M. M.; Lange, M.; Wovklich, P. M.; Uskokovic, M. R. *ibid.*, **1992**, *33*, 7701.

[579] Hatakeyama, S.; Osanai, K.; Numata, H.; Takao, S. *Tetrahedron Lett.*, **1989**, *30*, 4845.

[580] Evans, D. A.; Gauchet–Prunet, J. A.; Carreira, E. M.; Charette, A. B. *J. Org. Chem.*, **1991**, *56*, 741.

[581] Rao, A. V. R.; Dhar, T. G. M.; Bose, D. S.; Chakraborty, T. K.; Gurjar, M. K. *Tetrahedron*, **1989**, *45*, 7361.

[582] Rao, A. V. R.; Bose, D. S.; Gurjar, M. K.; Ravindranathan, T. *Tetrahedron*, **1989**, *45*, 7031.

[583] Rao, A. V. R.; Dhar, T. G. M.; Chakraborty, T. K.; Gurjar, M. K. *Tetrahedron Lett.*, **1988**, *29*, 2069.

[584] Genet, J. P.; Durand, J. O.; Savignac, M.; Pons, D. *Tetrahedron Lett.*, **1992**, *33*, 2497.

[585] Hori, K.; Hikage, N.; Inagaki, A.; Mori, S.; Nomura, K.; Yoshii, E. *J. Org. Chem.*, **1992**, *57*, 2888.

[586] Masamune, S.; Kaiho, T.; Garvey, D. S. *J. Am. Chem. Soc.*, **1982**, *104*, 5521.

[587] Hatakeyama, S.; Sugawara, K.; Kawamura, M.; Takano, S., *Synlett*, **1990**, 691.

[588] Nicolaou, K. C.; Nugiel, D. A.; Couladouros, E.; Hwang, C.–K. *Tetrahedron*, **1990**, *46*, 4517.

[589] Wang, Y.; Babirad, S. A.; Kishi, Y. *J. Org. Chem.*, **1992**, *57*, 468.

[590] Libing, Y.; Ziquin, W. *J. Chem. Soc., Chem. Commun.*, **1993**, 232.

[591] Rao, A. V. R.; Rao, S. P.; Bhanu, M. N. *J. Chem. Soc. Chem. Commun.*, **1992**, 859.

[592] Sun, C.–Q.; Rich, D. H. *Tetrahedron Lett.*, **1988**, *29*, 5205.

[593] Jung, M. E.; Lew, W. *J. Org. Chem.*, **1991**, *56*, 1347.

[594] Shimizu, I.; Yamazaki, H. *Chem. Lett.*, **1990**, 777.

[595] Marshall, J. A.; Andersen, M. W. *J. Org. Chem.*, **1992**, *57*, 5851.

[596] Mori, K.; Okada, K. *Tetrahedron*, **1985**, *41*, 557.

[597] Abram, T. S.; Biddlecom, W. G.; Jennings, M. A.; Norman, P.; Tudhope, S. R. *Eur. Pat. Appl.*, EP410244; *Chem. Abstr.*, **1991**, *115*, 135803c.

[598] Cregge, R. J.; Lentz, N. L.; Sabol, J. S. *J. Org. Chem.*, **1991**, *56*, 1758.

[599] Hamon, D. P. G.; Shirley, N. J. *J. Chem. Soc., Chem. Commun.*, **1988**, 425.

[600] Ferraboschi, P.; Brembilla, D.; Grisenti, P.; Santaniello, E. *J. Org. Chem.*, **1991**, *56*, 5478.

[601] Gosmini, C.; Dubuffet, T.; Sauvetre, R.; Normant, J.-F. *Tetrahedron: Asymmetry*, **1991**, *2*, 223.

[602] Alexakis, A.; Mutti, S.; Mangeney, P. *J. Org. Chem.* **1992**, *57*, 1224.

[603] Corey, E. J.; Hashimoto, S.; Barton, A. E. *J. Am. Chem. Soc.* **1981**, *103*, 721.

[604] Ernst, B.; Wagner, B. *Helv. Chim. Acta*, **1989**, *72*, 165.

[605] Lentz, N. L.; Peet, N. P. *Tetrahedron Lett.*, **1990**, *31*, 811.

[606] Wirth, D.; Boland, W.; Muller, D. G. *Helv. Chim. Acta*, **1992**, *75*, 751

[607] Takeda, K.; Kawanishi, E.; Nakamura, H.; Yoshii, E. *Tetrahedron Lett.*, **1991**, *32*, 4925.

[608] Romo, D.; Johnson, D. D.; Plamondon, L.; Miwa, T.; Schreiber, S. L. *J. Org. Chem.*, **1992**, *57*, 5060.

[609] Ishibashi, Y.; Nishiyama, S.; Shizuri, Y.; Yamamura, S. *Tetrahedron Lett.*, **1992**, *33*, 521.

[610] Chung, C. B.; Chang, S.-K.; Shim, S. C. *Bull. Korean Chem. Soc.*, **1991**, *12*, 122; *Chem. Abstr.*, **1991**, *115*, 49223g.

[611] Marczak, S.; Masnyk, M.; Wicha, J. *Tetrahedron Lett.*, **1989**, *30*, 2845.

[612] Buszek, K. R.; Fang, F. G.; Forsyth, C. J.; Jung, S. H.; Kishi, Y.; Scola, P. M.; Yoon, K. S. *Tetrahedron Lett.*, **1992**, *33*, 1553.

[613] Yee, N. K. N.; Coates, R. M. *J. Org. Chem.*, **1992**, *57*, 4598.

[614] Ghisalberti, E. L.; Twiss, E.; Rea, P. E. *J. Chem. Res.(S)*, **1991**, 202.

[615] Mori, K.; Harashima, S. *Tetrahedron Lett.*, **1991**, *32*, 5995.

[616] Ando, T.; Jacobsen, N. E.; Toia, R. F.; Casida, J. E. *J. Agric. Food Chem.*, **1991**, *39*, 600.

[617] Rickards, R. W.; Thomas, R. D. *Tetrahedron Lett.*, **1992**, *33*, 8137.

[618] Gill, M.; Smrdel, A. F. *Tetrahedron: Asymmetry*, **1990**, *1*, 453.

[619] Hashimoto, M.; Yanagiya, M.; Shirahama, H. *Chem. Lett.*, **1988**, 645.

[620] Ray, N. C.; Raveendranath, P. C.; Spencer, T. A. *Tetrahedron*, **1992**, *48*, 9427.

[621] Kolb, M.; Van Hijfte, L.; Ireland, R. E. *Tetrahedron Lett.*, **1988**, *29*, 6769.

[622] Van Hijfte, L.; Kolb, M. *Tetrahedron*, **1992**, *31*, 6393.

[623] Davies, M. J.; Heslin, J. C.; Moody, C. J. *J. Chem. Soc., Perkin Trans. 1*, **1989**, 2473.

[624] Coghlan, D. R.; Hamon, D. P. G.; Massy–Westropp, R. A.; Slobedman, D. *Tetrahedron: Asymmetry*, **1990**, *1*, 299.

[625] Oritani, T.; Yamashita, K. *Phytochemistry*, **1983**, *22*, 1909.

[626] Acemoglu, M.; Uebelhart, P.; Rey, M.; Eugster, C. H. *Helv. Chim. Acta*, **1988**, *71*, 931.

[627] Spur, B.; Crea, A.; Peters, W. *Arch. Pharm.*, **1985**, *318*, 225.

[628] Corey, E. J.; Marfat, A.; Munroe, J.; Kim, K. S.; Hopkins, P. B.; Brion, F. *Tetrahedron Lett.*, **1981**, *22*, 1077.

[629] Anorbe, B.; Martin, V. S.; Palazon, J. M.; Trujillo, J. M. *Tetrahedron Lett.*, **1986**, *27*, 4991.

[630] Miyashita, M.; Hoshino, M.; Yoshikoshi, A. *Chem. Lett.*, **1990**, 791.

[631] Nunez, M. T.; Martin, V. S. *J. Org. Chem.*, **1990**, *55*, 1928.

[632] Scherowsky, G.; Gruneberg, K.; Kuhnpast, K. *Ferroelectrics*, **1991**, *122*, 159.

[633] Thijs, L.; Egenberger, D. M.; Zwanenburg, B. *Tetrahedron Lett.*, **1989**, *30*, 2153.

[634] Hong, C. Y.; Kishi, Y. *J. Org. Chem.*, **1990**, *55*, 4242.

[635] Freezou, J. P.; Julia, M.; Li, Y.; Liu, L. W.; Pancrazi, A. *Synlett*, **1990**, 766.

[636] Taber, D. F.; Silverberg, L. J.; Robinson, E. D. *J. Am. Chem. Soc.*, **1991**, *113*, 6639.

[637] Bansal, R.; Cooper, G. F.; Corey, E. J. *J. Org. Chem.*, **1991**, *56*, 1329.

[638] Mills, L. S.; North, P. C. *Tetrahedron Lett.*, **1983**, *24*, 409.

[639] Toth, M.; Buser, H. R.; Pena, A.; Arn, H.; Mori, K.; Takeuchi, T.; Nikolaeva, L. N.; Kovalev, B. G. *Tetrahedron Lett.*, **1989**, *30*, 3405.

[640] Mori, K.; Takeuchi, T. *Justus Liebigs Ann. Chem.*, **1989**, 453.

[641] Toto, M.; Arun, H.; Mori, K.; Ninomiya, Y.; Komata, T.; Senda, S.; Takeuchi, T.; Yuya, M. *Jpn. Kokai Tokkyo Koho* JP 02 262 575 [90 262 575]; *Chem. Abstr.*, **1991**, *114*, 185136n.

[642] Carvalho, C.; Cullen, W. R.; Fryzuk, M. D.; Jacobs, H.; James, B. R.; Kutney, J. P.; Piotrowska, K.; Singh, V. K. *Helv. Chim. Acta*, **1989**, *72*, 205.

[643] Izawa, T.; Wang, Z.–Q.; Nishimura, Y.; Kondo, S.; Umezawa, H. *Chem. Lett.*, **1987**, 1655.

[644] Ko, O. H. *Yakhak Hoechi*, **1986**, *30*, 329; *Chem. Abstr.*, **1986**, *107*, 175711n.

[645] Reddy, K. S.; Ko, O. H.; Ho, D.; Persons, P. E.; Cassady, J. M. *Tetrahedron Lett.*, **1987**, *28*, 3075.

[646] Williams, D. R.; Brown, D. L.; Benbow, J. W. *J. Am. Chem. Soc.*, **1989**, *111*, 1923.

[647] Giese, B.; Rupaner, R. *Justus Liebigs Ann. Chem.*, **1987**, 231.

[648] Noda, Y.; Kikuchi, M. *Synth. Commun.*, **1985**, *15*, 1245.

[649] Guo–qiang, L.; Hai–jian, X.; Bi–chi, W.; Guong–zhong, G.; Wei–Shan, Z. *Tetrahedron Lett.*, **1985**, *26*, 1233.

[650] Mori, K.; Umemura, T. *Tetrahedron Lett.*, **1981**, *22*, 4433.

[651] Mori, K.; Umemura, T. *Tetrahedron Lett.*, **1982**, *23*, 3391.

[652] Umemura, T.; Mori, K. *Agric. Biol. Chem.*, **1987**, *51*, 1973.

[653] Takano, S.; Sugihara, T.; Ogasawara, K. *Synlett*, **1991**, 279.

[654] Nicolaou, K. C.; Veale, C. A.; Hwang, C.–K.; Hutchinson, J.; Prasad, C. V. C.; Ogilvie, W. W. *Angew. Chem. Int. Ed. Engl.*, **1991**, *30*, 299.

[655] Nicolaou, K. C.; Duggan, M. E.; Hwang, C.–K. *J. Am. Chem. Soc.*, **1989**, *111*, 6666.

[656] Fang, F. G.; Kishi, Y.; Matelich, M. C.; Scola, P. M. *Tetrahedron Lett.*, **1992**, *33*, 1557.

[657] Sodeoka, M.; Iimori, T.; Shibasaki, M. *Tetrahedron Lett.*, **1985**, *26*, 6497.

[658] Sodeoka, M.; Iimori, T.; Shibasaki, M. *Chem. Pharm. Bull.*, **1991**, *39*, 323.

[659] Lee, E.; Shin, I.–J.; Kim, T.–S. *J. Am. Chem. Soc.*, **1990**, *112*, 260.

[660] Dulplantier, A. J.; Masamune, S. *J. Am. Chem. Soc.*, **1990**, *112*, 7079.

[661] Prestwich, G. D.; McG. Graham, S.; Kuo, J.–W.; Vogt, R. G. *J. Am. Chem. Soc.*, **1989**, *111*, 636.

[662] Magnus, P.; Davies, M. *J. Chem. Soc., Chem. Commun.*, **1991**, 1522.

[663] Sugiyama, S.; Honda, M.; Komori, T. *Justus Liebigs Ann. Chem.*, **1990**, 1069.

[664] Scherowsky, G.; Gay, J. *Liquid Crystals*, **1989**, *5*, 1253.

[665] Scherowsky, G.; Gay, J.; Sharma, N. K. *Mol. Cryst. Liq. Cryst.*, **1990**, *178*, 179.

[666] Hatakeyama, S.; Numata, H.; Osanai, K.; Takano, S. *J. Chem. Soc., Chem. Commun.*, **1989**, 1893.

[667] Ebata, T.; Mori, K. *Agri. Biol. Chem.*, **1989**, *53*, 801.

[668] Takabe, K.; Okisaka, K.; Uchiyama, Y.; Katagiri, T.; Yoda, H. *Chem. Lett.*, **1985**, 561.

[669] Kitahara, T.; Kiyota, H.; Kurata, H.; Mori, K. *Tetrahedron*, **1991**, *47*, 1649.

[670] Furukawa, J.; Iwasaki, S.; Okuda, S. *Tetrahedron Lett.*, **1983**, *24*, 5257.

[671] Holoboski, M. A.; Koft, E. *J. Org. Chem.*, **1992**, *57*, 965.

[672] Mori, K.; Brevet, J.–L. *Synthesis*, **1991**, 1125.

[673] Tius, M. A.; Cullingham, J. M. *Tetrahedron Lett.*, **1989**, *30*, 3749.

[674] Arm, C.; Pfander, H. *Helv. Chim. Acta*, **1984**, *67*, 1540.

[675] Latli, B.; Prestwich, G. D. *J. Labelled Comp. Radiopharm.*, **1991**, *29*, 1167.

[676] Prestwich, G. D.; Eng, W.–s.; Robles, S.; Vogt, R. G.; Wisniewski, J. R.; Wawrzenczyk, C. *J. Biol. Chem.*, **1988**, *263*, 1398.

[677] Ho, W.; Tarkan, O.; Kiorpes, T. C.; Tutwiler, G. F.; Mohrbacher, R. J. *J. Med. Chem.*, **1987**, *30*, 1094.

[678] Kitano, Y.; Kobayashi, Y.; Sato, F. *J. Chem. Soc., Chem. Commun.*, **1985**, 498.

[679] Moon, S.–S.; Chen, J. L.; Moore, R. E.; Patterson, G. M. L. *J. Org. Chem.*, **1992**, *57*, 1097.

[680] Horita, K.; Nagato, S.; Oikawa, Y.; Yonemitsu, O. *Chem. Pharm. Bull.*, **1989**, *37*, 1705.

[681] Millar, J. G.; Giblin, M.; Barton. D.; Morrison, A.; Underhill, E. W. *J. Chem. Ecol.*, **1990**, *16*, 2317.

[682] Morimoto, Y.; Oda, K.; Shirahama, H.; Matsumoto, T.; Omura, S. *Chem. Lett.*, **1988**, 909.

[683] Cosse, A. A.; Cyjon, R.; Moore, I.; Wysoki, M.; Becker, D. *J. Chem. Ecol.*, **1992**, *18*, 165.

[684] Becker, D.; Cyjon, R.; Cosse, A.; Moore, I.; Kimmel, T.; Wysoki, M. *Tetrahedron Lett.*, **1990**, *31*, 4923.

[685] Nicolaou, K. C.; Duggan, M. E.; Hwang, C.–K. *J. Am. Chem. Soc.*, **1989**, *111*, 6676.

[686] Marshall, J. A.; Flynn, K. E. *J. Am. Chem. Soc.*, **1984**, *106*, 723.

[687] Gorgen, G.; Boland, W.; Preiss, U.; Simon, H. *Helv. Chim. Acta,* **1989**, *72*, 917.

[688] Honda, M.; Ueda, Y.; Sugiyama, S.; Komori, T. *Chem. Pharm. Bull.,* **1991**, *39*, 1385.

[689] Farkas, I.; Pfander, H. *Helv. Chim. Acta,* **1990**, *73*, 1980.

[690] Urones, J. G.; Marcos, I. S.; Cuadrado, J. S.; Basabe, P.; Lithgow, A. M. *Phytochemistry,* **1990**, *29*, 1247.

[691] Urones, J. G.; Marcos, I. S.; Basabe, P.; Sexmero, M.-J.; Diez, D.; Garrido, N. M.; Prieto, J. E. S. *Tetrahedron,* **1990**, *46*, 2495.

[692] Ojika, M.; Kigoshi, H.; Yoshikawa, K.; Nakayama, Y.; Yamada, K. *Bull. Chem. Soc. Jpn.,* **1992**, *65*, 2300.

[693] Inoue, S.; Ikeda, H.; Sato, S.; Horie, K.; Ota, T.; Miyamoto, O.; Sato, K. *J. Org. Chem.,* **1987**, *52*, 5495.

[694] Eisai Co., Ltd. Jpn. Kokai Tokkyo Koho JP 59 29 678 [84 29 678]; C. A., **1984**, *101*, 111227.

[695] Eisai Co., Ltd. Jpn. Kokai Tokkyo Koho JP 59 31 726 [84 31 726]; C. A., **1984**, *101*, 91290.

[696] Takano, S.; Sugihara, T.; Ogasawara, K. *Synlett,* **1990**, 451.

[697] Gleason, J. G.; Hall, R. F.; Perchonock, C. D.; Erhard, K. F.; Frazee, J. S.; Ku, T. W.; Kondrad, K.; McCarthy, M. E.; Mong, S.; Crooke, S. T.; Chi-Rosso, G.; Wasserman, M. A.; Torphy, T. J.; Muccitelli, R. M.; Hay, D. W.; Tucker, S. S.; Vickery-Clark, L. *J. Med. Chem.,* **1987**, *30*, 959.

[698] Lai, C. K.; Gut, M. *J. Org. Chem.,* **1987**, *52*, 685.

[699] Kabat, M. M. *J. Fluorine Chem.,* **1991**, *53*, 249.

[700] Xiao, X.-y.; Prestwich, G. D. *J. Labelled Comp. Radiopharm.,* **1991**, *29*, 883.

[701] Nishitoba, T.; Sato, H.; Oda, K.; Sakamura, S. *Agric. Biol. Chem.,* **1988**, *52*, 211.

[702] Xiao, X. Y.; Sen, S. E.; Prestwich, G. D. *Tetrahedron Lett.,* **1990**, *31*, 2097.

[703] Yamada, Y.; Seo, C. H.; Okada, H. *Agric. Biol. Chem.,* **1981**, *45*, 1741.

[704] Marshall, J. A.; Peterson, J. C.; Lebioda, L. *J. Am. Chem. Soc.,* **1984**, *106*, 6006.

[705] Marshall, J. A.; Flynn, K. E. *J. Am. Chem. Soc.,* **1983**, *105*, 3360.

[706] Hosokawa, T.; Kono, T.; Shinohara, T.; Murahashi, S. *J. Organometal. Chem.,* **1989**, *370*, C13.

[707] White, J. D.; Avery, M. A.; Choudhry, S. C.; Dhingra, O. P.; Kang, M.-c; Whittle, A. J. *J. Am. Chem. Soc.,* **1983**, *105*, 6517.

[708] White, J. D.; Kang, M.-c; Sheldon, B. G. *Tetrahedron Lett.,* **1983**, *24*, 4539.

[709] Jung, M. E.; Jung, Y. H. *Tetrahedron Lett.,* **1989**, *30*, 6637.

[710] Trost, B. M.; Nubling, C. *Carbohydr. Res.,* **1990**, *202*, 1.

[711] Walkup, R. D.; Cunningham, R. T. *Tetrahedron Lett.,* **1987**, *28*, 4019.

[712] Hanessian, S.; Cooke, N. G.; DeHoff, B.; Sakito, Y. *J. Am. Chem. Soc.,* **1990**, *112*, 5276.

[713] Wilson, S. R.; Haque, M. S. *Tetrahedron Lett.,* **1984**, *25*, 3147.

[714] Mori, K.; Seu, Y-B. *Tetrahedron,* **1985**, *41*, 3429.

[715] Paquette, L. A.; Sweeney, T. J. *J. Org. Chem.,* **1990**, *55*, 1703.

[716] Paquette, L. A.; Sweeney, T. J. *Tetrahedron,* **1990**, *46*, 4487.

[717] Ewing, W. R.; Harris, B. D.; Bhat, K. L.; Joullie, M. M. *Tetrahedron,* **1986**, *42*, 2421.

[718] Murakami, M.; Matsuura, M.; Aoki, T.; Nagata, W. *Synlett,* **1990**, 681.

[719] England, P.; Chun, K. H.; Moran, E. J.; Armstrong, R. W. *Tetrahedron Lett.,* **1990**, *31*, 2669.

[720] England, P.; Chun, K. H.; Moran, E. J.; Armstrong, R. W. *Tetrahedron Lett.,* **1990**, *31*, 2669.

[721] Ziegler, F. E.; Kneisley, A.; Wester, R. T. *Tetrahedron Lett.,* **1986**, 27, 1221.

[722] Ziegler, F. E.; Kneisky, A. *Tetrahedron Lett.,* **1985**, *26*, 263.

[723] Owens, K. A.; Berson, J. A. *J. Am. Chem. Soc.,* **1990**, *112*, 5973.

[724] Ikegami, S.; Okamura, H.; Kuroda, S.; Katsuki, T.; Yamaguchi, M. *Bull. Chem. Soc. Jpn.,* **1992**, *65*, 1841.

[725] Mihelich, E. D. Procter & Gamble Co., Miami Valley Laboratories, unpublished results.

[726] Overman, L. E.; Lin, N. H. *J. Org. Chem.,* **1985**, *50*, 3669.

[727] Discordia, R,P.; Dittmer, D. C. *J. Org. Chem.* **1990**, *55*, 1414.

[728] Burke, S. D.; Pacofsky, G. J.; Piscopio, A. D. *J. Org. Chem.,* **1992**, *57*, 2228.

[729] Kim, Y. G.; Cha, J. K. *Tetrahedron Lett.,* **1988**, *29*, 2011.

[730] Zhou, B.; Xu, Y. *J. Org. Chem.,* **1988**, *53*, 4419.

[731] Aggarwal, S. K.; Bradshaw, J. S.; Eguchi, M.; Parry, S.; Rossiter, B. E.; Markides, K. E.; Lee, M. L. *Tetrahedron,* **1987**, *43*, 451.

[732] Mori, K.; Otsuka, T.; Oda, M. *Tetrahedron*, **1984**, *40*, 2929.

[733] Bessodes, M.; Saiah, M.; Antonakis, K. *J. Org. Chem.*, **1992**, *57*, 4441.

[734] Saiah, M.; Bessodes, M.; Antonakis, K. *Tetrahedron: Asymmetry*, **1991**, *2*, 111.

[735] Paquette, L. A.; Maynard, G. D. *Angew. Chem., Int. Ed. Engl.*, **1991**, *30*, 1368.

[736] Paquette, L. A.; Maynard, G. D. *J. Am. Chem. Soc.*, **1992**, *114*, 5018.

[737] Verner, E. J.; Cohen, T. *J. Am. Chem. Soc.*, **1992**, *114*, 375.

[738] Agnel, G.; Negishi, E. *J. Am. Chem. Soc.*, **1991**, *113*, 7424.

[739] Ronald, R. C.; Ruder, S. M.; Lillie, T. S. *Tetrahedron Lett.*, **1987**, *28*, 131.

[740] Ziegler, F. E.; Wester, R. T. *Tetrahedron Lett.*, **1986**, *27*, 1225.

[741] Ziegler, F. E.; Shirchack, E. P.; Wester, R. T. *Tetrahedron Lett.*, **1986**, *27*, 1229.

[742] Ziegler, F. E.; Kneisley, A.; Thottathil, J. K.; Wester, R. T. *J. Am. Chem. Soc.*, **1988**, *110*, 5434.

[743] Dai, L.-x; Lou, B.-l; Zhang, Y.-z. *J. Am. Chem. Soc.* **1988**, *110*, 5195.

[744] Evans, D. A; Kaldor, S. W.: Jones, T. K.; Clardy, J.; Stout, T. J. *J. Am. Chem. Soc.*, **1990**, *112*, 7001.

[745] Roush, W. R.; Hoong, L. K.; Palmer, M. A. J.; Straub, J. A.; Palkowitz, A. D. *J. Org. Chem.*, **1990**, *55*, 4117.

[746] Chemin, D.; Alami, M.; Linstrumelle, G. *Tetrahedron Lett.*, **1992**, *33*, 2681.

[747] Brown, S. M.; Davies, S. G.; de Sousa, J. A. A. *Tetrahedron: Asymmetry*, **1991**, *2*, 511.

[748] Mori, K.; Otsuka, T. *Tetrahedron*, **1985**, *41*, 553.

[749] Mori, K.; Otsuka, T. *Tetrahedron*, **1985**, *41*, 3253.

[750] Alexander, D. L.; Lin, C. H. *Prostaglandins*, **1986**, *32*, 647.

[751] Yadav, J. S.; Gadgil, V. R. *J. Chem. Soc., Chem. Commun.*, **1989**, 1824.

[752] Trost, B. M.; Dumas, J. *J. Am. Chem. Soc.*, **1992**, *114*, 1924.

[753] Marshall, J. A.; Markwalder, J. A. *Tetrahedron Lett.*, **1988**, *29*, 4811.

[754] Peschke, B.; Lubmann, J.; Drybusch, M.; Hoppe, D. *Chem. Ber.*, **1992**, *125*, 1121.

[755] Kufner, U.; Schmidt, R. R., *Angew. Chem., Int. Ed. Engl.*, **1986**, *25*, 89.

[756] Hoppe, D.; Kramer, T. *Angew. Chem., Int. Ed. Engl.*, **1986**, *25*, 160.

[757] Nonoshita, K.; Banno, H.; Maruoka, K.; Yamamoto, H. *J. Am. Chem. Soc.*, **1990**, *112*, 316.

[758] Page, P. C. B.; Carefull, J. F.; Powell, L. H.; Sutherland, I. O., *J. Chem. Soc., Chem. Commun.*, **1985**, 822.

[759] Bailey, M.; Staton, I.; Ashton, P. R.; Marko, I. E.; Ollis, W. D. *Tetrahedron: Asymmetry*, **1991**, *2*, 495.

[760] Bonini, C. *Tetrahedron Lett.*, **1990**, *31*, 5369.

[761] Claremon, D. A.; Lumma, P. K.; Phillips, B. T. *J. Am. Chem. Soc.*, **1986**, *108*, 8265.

[762] Davis, F. A.; Reddy, R. T. *J. Org. Chem.*, **1992**, *57*, 2599.

[763] Jung, M. E.; Jung, Y. H.; Miyazawa, Y. *Tetrahedron Lett.*, **1990**, *31*, 6983.

[764] Mitch, C. H.; Zimmerman, D. M.; Snoddy, J. D.; Reel, J. K.; Cantrell, B. E. *J. Org. Chem.*, **1991**, *56*, 1660.

[765] Aristoff, P. A.; Johnson, P. D.; Harrison, A. W. *J. Am. Chem. Soc.*, **1985**, *107*, 7967.

[766] Nakagawa, N.; Mori, K. *Agric. Biol. Chem.*, **1984**, *48*, 2799.

[767] Hanzawa, Y.; Kawagoe, K.; Ito, M.; Kobayashi, Y. *Chem. Pharm. Bull.*, **1987**, *35*, 1633.

[768] Morales, E. Q.; Vazquez, J. T.; Martin, J. D. *Tetrahedron: Asymmetry*, **1990**, *1*, 319.

[769] Alvarez, E.; Manta, E.; Martin, J. D. *Tetrahedron Lett.*, **1988**, *29*, 2093.

[770] Frater, G.; Müller, U. *Helv. Chim. Acta*, **1989**, *72*, 653.

[771] Mori, K.; Puapoomchareon, P. *Justus Liebigs Ann. Chem.*, **1991**, 1053.

[772] Jacquier, R.; Lazaro, R.; Raniriseheno, H.; Viallefont, P. *Tetrahedron Lett.*, **1984**, *25*, 5525.

[773] Ikegami, S.; Hayama, T.; Katsuki, T.; Yamaguchi, M. *Tetrahedron Lett.*, **1986**, *27*, 3403.

[774] Kawai, M.; Gardner, J. H.; Rich, D. H. *Tetrahedron Lett.*, **1986**, *27*, 1877.

[775] Jacquier, R.; Lazaro, R.; Raniriseheno, H.; Viallefont, P. *Tetrahedron Lett.*, **1986**, *27*, 4735.

[776] Maemoto, S.; Mori, K. *Chem. Lett.*, **1987**, 109.

[777] Gonnella, N. C.; Nakanishi, K.; Martin, V. S.; Sharpless, K. B. *J. Am. Chem. Soc.*, **1982**, *104*, 3775.

[778] Stary, I.; Kocovsky, P. *J. Am. Chem. Soc.* **1989**, *111*, 4981.

[779] Stary, I.; Zajicek, J.; Kocovsky, P. *Tetrahedron*, **1992**, *48*, 7229.

[780] Calverley, M. J. *Synlett*, **1990**, 157.

[781] Ito, T.; Okamoto, Y.; Matsumoto, T. *Bull. Chem. Soc. Jpn.*, **1985**, *58*, 3631.

[782] Rao, A. V. R.; Khrimian, A. P.; Krishna, P. R.; Yadagiri, P.; Yadav, J. S. *Synth. Commun.*, **1988**, *18*, 2325.

[783] Yadav, J. S.; Radhakrishna, P. *Tetrahedron*, **1990**, *46*, 5825.

[784] Crombie, L.; Jarrett, S. R. M. *Tetrahedron Lett.*, **1989**, *30*, 4303.

[785] Crombie, L.; Horsham, M. A.; Jarrett, S. R. M. *J. Chem. Soc., Perkin Trans. 1*, **1991**, 1511.

[786] Hatakeyama, S,; Sugawara, K.; Kawamura, M.; Takano, S. *Tetrahedron Lett.*, **1991**, *32*, 4509.

[787] Yanagisawa, A.; Nomura, N.; Noritake, Y.; Yamamoto, H. *Synthesis*, **1991**, 1130.

[788] Mehmandoust, M.; Petit, Y.; Larcheveque, M. *Tetrahedron Lett.*, **1992**, *33*, 4313.

[789] Nicolaou, K. C.; Hwang, C.-K.; Duggan, M. E. *J. Am. Chem. Soc.*, **1989**, *111*, 6682.

[790] Marples, B. A.; Rogers–Evans, M. *Tetrahedron Lett.*, **1989**, *30*, 261.

[791] Rao, A. V. R.; Yadav, J. S.; Reddy, K. B.; Mehendale, A. R. *J. Chem. Soc., Chem. Commun.*, **1984**, 453.

[792] Rao, A. V. R.: Yadav, J. S.; Reddy, K. B.; Mehendale, A. R. *Tetrahedron*, **1984**, *40*, 4643.

[793] Rao, A. V. R.; Chanda, B.; Borate, H. B.; Gupta, M. *Indian J. Chem.*, **1986**, *25B*, 9.

[794] Holland, H. L.; Viski, P. *J. Org. Chem.*, **1991**, *56*, 5226.

[795] Mori, K.; Otsuka, T. *Tetrahedron*, **1983**, *39*, 3267.

[796] Nicolaou, K. C.; Zipkin, R. E.; Dolle, R. E.; Harris, B. D. *J. Am. Chem. Soc.*, **1984**, *106*, 3548.

[797] Marson, C. M.; Benzies, D. W. M.; Hobson, A. D. *Tetrahedron*, **1991**, *47*, 5491.

[798] Keegan, D. S.; Midland, M. M.; Werley, R. T.; McLoughlin, J. I. *J. Org. Chem.*, **1991**, *56*, 1185.

[799] Ogata, M.; Matsumoto, H.; Takahashi, K.; Shimizu, S.; Kida, S.; Murabayashi, A.; Shiro, M.; Tawara, K. *J. Med. Chem.*, **1987**, *30*, 1054.

[800] Discordia, R. P.; Murphy, C. K.; Dittmer, D. C. *Tetrahedron Lett.*, **1990**, *31*, 5603.

[801] Lewis, M. D.; Menes, R. *Tetrahedron Lett.*, **1987**, *28*, 5129.

[802] Dominguez, D.; Cava, M. P. *J. Org. Chem.*, **1983**, *48*, 2820.

[803] Russell, A. T.; Procter, G. *Tetrahedron Lett.*, **1987**, *28*, 2041.

[804] Okamoto, S.; Kobayashi, Y.; Sato, F. *Tetrahedron Lett.*, **1989**, *30*, 4379.

[805] Kobayashi, Y.; Shimazaki, T.; Taguchi, H.; Sato, F. *J. Org. Chem.*, **1990**, *55*, 5324.

[806] Vanhessche, K.; der Eycken, E. V.; Vandewalle, M. *Tetrahedron Lett.*, **1990**, *31*, 2337.

[807] Evans, D. A.; Polniaszek, R. P.; DeVries, K. M.; Guinn, D. E.; Mathre, D. J. *J. Am. Chem. Soc.*, **1991**, *113*, 7613.

[808] Wang, Z. *Tetrahedron Lett.*, **1989**, *30*, 6611.

[809] Thomas, E. J.; Watts, J. P. *J. Chem. Soc., Chem. Commun.*, **1990**, 467.

[810] Roush,W. R.; Spada, A. P. *Tetrahedron Lett.*, **1982**, *23*, 3773.

[811] Ibuka, T.; Tanaka, M.; Yamamoto, Y. *J. Chem. Soc., Chem. Commun.*, **1989**, 967.

[812] Gao, L.-x.; Murai, A. *Tetrahedron Lett.*, **1992**, *33*, 4349.

[813] Dyer, U. C.; Kishi, Y. *J. Org. Chem.*, **1988**, *53*, 3383.

[814] Baker, R.; Castro, J. *J. Chem. Soc., Perkin Trans. 1*, **1990**, 47.

[815] Back, T. G.; Blazecka, P. G.; Krishna, M. V. *Tetrahedron Lett.*, **1991**, *32*, 4817.

[816] Calverley, M. J. *Tetrahedron*, **1987**, *43*, 4609.

[817] Ekhato, I. V.; Silverton, J. V.; Robinson, C. H. *J. Org. Chem.*, **1988**, *53*, 2180.

[818] Kametani, T.; Tsubuki, M.; Tatsuzaki, Y.; Honda, T. *J. Chem. Soc., Perkin Trans. 1*, **1990**, 639.

[819] Kawatani, T.; Tatsuzaki, Y.; Tsukubi, M.; Honda, T. *Heterocycles*, **1989**, *29*, 1247.

[820] Kusakabe, M.; Sato, F. *J. Org. Chem.*, **1989**, *54*, 3486.

[821] Honda, T.; Kobayashi, Y.; Tsubuki, M. *Tetrahedron Lett.*, **1990**, *31*, 4891.

[822] Kitano, Y.; Kusakabe, M.; Kobayashi, Y.; Sato, F. *J. Org. Chem.*, **1988**, *54*, 994.

[823] Zhou, W.-S.; Wei, D. *Tetrahedron: Asymmetry*, **1991**, *2*, 767.

[824] Baldwin, J. E.; Adlington, R. M.; Bebbington, D.; Russell, A. *J. Chem. Soc., Chem. Commun.*, **1992**, 1249.

[825] Smith, III, A. B.; Hale, K. J.; Laakso, L. M.; Chen, K.; Riera, A. *Tetrahedron Lett.*, **1989**, *30*, 6963.

[826] Benechie, M.; Khuong–Huu, F. *Synlett*, **1992**, 266.

[827] Chong, J. M. *Tetrahedron*, **1989**, *45*, 623.

[828] Nakano, T.; Obata, M.; Yamaguchi, Y.; Marubayashi, N.; Ikeda, K.; Morimoto, Y. *Chem. Pharm. Bull.*, **1992**, *40*, 117.

[829] Sigrist–Nelson, K.; Krasso, A.; Muller, R. K. M.; Fischli, A. E. *Eur. J. Biochem.*, **1987**, *166*, 453.

[830] Conte, V.; Furia, F. D.; Licini, G.; Modena, G.; Sbampato, G.; Valle, G. *Tetrahedron: Asymmetry*, **1991**, *2*, 257.

[831] Conte, V.; Furia, F. D.; Licini, G.; Modena, G. *Tetrahedron Lett.*, **1989**, *30*, 4859.

[832] Bortolini, O.; Furia, F. D.; Licini, G.; Modena, G.; Rossi, M. *Tetrahedron Lett.*, **1986**, *27*, 6257.

[833] Cashman, J. R.; Olsen, L. D.; Bornheim, L. M. *J. Am. Chem. Soc.*, **1990**, *112*, 3191.

[834] Phillips, M. L.; Berry, D. M.; Panetta, J. A. *J. Org. Chem.*, **1992**, *57*, 4047.

CHAPTER 2

RADICAL CYCLIZATION REACTIONS

B. Giese, B. Kopping, T. Göbel, J. Dickhaut, G. Thoma, K. J. Kulicke, and F. Trach

Department of Chemistry, University of Basel, Basel, Switzerland

CONTENTS

Organic Reactions, Vol. 48, Edited by Leo A. Paquette et al.
ISBN 0-471-14699-4   © 1996 Organic Reactions, Inc. Published by John Wiley & Sons, Inc.

INTRODUCTION

Radical cyclization reactions are among the most powerful and versatile methods for the construction of mono- and polycyclic systems. The advantages these reactions offer to the synthetic organic chemist include high functional group tolerance and mild reaction conditions combined with high levels of regio- and stereochemistry. Furthermore, the recent progress in radical chemistry has led to the development of a broad range of very useful practical methods to conduct radical cyclization reactions. In general, radical cyclization reactions comprise three basic steps: selective radical generation, radical cyclization, and conversion of the cyclized radical to the product (Eq. 1).

$$(Eq. 1)$$

For the generation of the initial radical a broad variety of suitable precursors can be employed, such as halides, thio- and selenoethers, alcohols, aldehydes and hydrocarbons. The cyclization step usually involves the intramolecular addition of a radical to a multiple bond. Most often carbon–carbon multiple bonds are employed; however, there are also examples known for the addition to carbon–oxygen and carbon–nitrogen bonds. Depending on the method employed, the cyclized radical is converted to the desired product by trapping with a radical scavenger, by a fragmentation reaction, or by an electron transfer reaction.

The section Mechanism, Regio- and Stereochemistry provides an introduction to the key features of radical cyclization with a special emphasis on the factors controlling the regio- and stereochemistry. The section Scope and Limitations covers the different methods used to conduct radical cyclization. The basic principles of radical chemistry and general practical considerations when conducting radical cyclizations are not discussed in detail. Several excellent review articles[1-5] and books[6-8] dealing with these topics are available. The study of one of these reviews or books is highly recommended, especially for readers who are not familiar with radical chemistry.

MECHANISM, REGIO- AND STEREOCHEMISTRY

Mechanism and Regiochemistry

To achieve synthetically useful radical cyclization, several basic requirements have to be fulfilled:

1. Methods must be available that allow the selective generation of the initial radical from suitable precursors and that effect the transformation of the cyclized radicals to the final products (see Scope and Limitations).
2. The rate constant of ring closure is of special importance because cyclization of the initial radical must be faster than its reaction with the trapping reagent.
3. Each of the reaction steps must be faster than the unwanted side reactions of radicals such as reaction with the solvent or radical recombination.

From these requirements a lower rate constant limit for the cyclization step can be estimated with $k_c \approx 10^2 - 10^3$ s^{-1}, although much larger rate constants ($k_c > 10^5$ s^{-1}) are usually better suited for synthetic applications. Because of its central importance, the rate of cyclization is included in the following discussion.

The regiochemistry of radical cyclization is important because it determines the ring size of the product. In principle, two competing pathways are possible — attack of the radical at the terminal end of the multiple bond (endo cyclization) or attack at the "inner" atom (exo cyclization). Fortunately, radical cyclizations are usually highly regioselective, and exo cyclization (formation of the smaller ring) is often strongly favored over endo cyclization (formation of the larger ring).

Formation of Small Rings.   Radical cyclizations to form 3- and 4-membered ring systems are of limited value in synthesis, and only a few examples are known. Because of the large ring strain, the rate of 3-exo cyclization of the butenyl radical **1** is rather low, and the cyclopentylcarbinyl radical **2** so formed rapidly reopens.[9] The equilibrium usually lies far to the side of the acyclic radical. The formation of cyclobutane radicals **3** by 4-exo cyclization of pentenyl radical **4** is even slower and is also reversible (Eq. 2).[10]

k_c 2×10^3 s^{-1}
$k_{c'}$ 1×10^8 s^{-1}

k_c 1 s^{-1}
$k_{c'}$ 5×10^8 s^{-1}

(Eq. 2)

To achieve synthetically useful small ring formation, the cyclized radicals must be trapped selectively prior to reopening or substituents must be introduced that accelerate the cyclization. The first concept is illustrated in the reaction of thioallyl esters **5**, where the intermediate cyclobutylcarbinyl radicals **6** are trapped by rapid β fragmentation (Eq. 3).[11] The second strategy is illustrated by

R^1	R^2	
PhCH$_2$	H	(47%)
PhCH$_2$	Me	(43%)
n-C$_{10}$H$_{21}$	H	(60%)

(Eq. 3)

the cyclization of acrylate ester **7** that is activated by the ethoxycarbonyl group at the double bond and the *gem*-diethoxy substituents at the 2 position (Eq. 4).[12]

Formation of 5- and 6-Membered Rings. Cyclopentyl rings are almost always formed by 5-exo cyclization of 5-hexenyl radicals, whereas cyclohexyl rings are formed either by 6-endo cyclizations of 5-hexenyl radicals or by 6-exo cyclization of 6-heptenyl radicals. However, the discussion concentrates on cyclizations of 5-hexenyl radicals because they represent the most useful class of radical cyclizations in organic synthesis.

The cyclization reactions of 5-hexenyl and 6-heptenyl radicals are exothermic and irreversible reactions with a preference for formation of the smaller ring size by cyclization in the exo mode.[13,14] For the parent 5-hexenyl radical **8**, this preference results in a ratio of 98:2 in favor of the 5-membered over the 6-membered ring (Eq. 5).

$$k_{5\text{-exo}} \quad 2 \times 10^5 \text{ s}^{-1}$$
$$k_{6\text{-endo}} \quad 4 \times 10^3 \text{ s}^{-1}$$

(Eq. 5)

8   98 : 2

Although the exo mode is again favored, cyclization of 6-heptenyl radical **9** is about 40 times slower than cyclization of the hexenyl radical. Therefore, reactions competing with the cyclization (e.g., reduction of the initial radical) are of much greater importance. Another potential problem of the heptenyl system is caused by a 1,5-hydrogen shift, yielding the resonance-stabilized allyl radical **10** (Eq. 6). This intramolecular H shift is thermodynamically favored in many sys-

$$k_{6\text{-exo}} \quad 5 \times 10^3 \text{ s}^{-1}$$
$$k_{7\text{-endo}} \quad 7 \times 10^2 \text{ s}^{-1}$$

9   88 : 12

(Eq. 6)

$$k_{\text{allyl}} \quad k_{6\text{-exo}}$$

10

tems, but is kinetically disfavored in smaller rings for stereoelectronic reasons and in larger rings for entropic reasons. Nevertheless, the 6-exo cyclization of heptenyl radicals is still a very useful reaction, especially if the exo cyclization is accelerated by the introduction of appropriate substituents.

The preferred formation of the thermodynamically disfavored exo product from the 5-hexenyl radical is best rationalized by a stereoelectronically controlled cyclization with the chair-like transition state **11**.[13,15] This arrangement reflects the early transition state of the reaction with a favorable overlap between the SOMO of the radical and the LUMO of the alkene. The forming C—C bond is very long (ca. 2.3–2.4 Å), and the angle of attack of the radical on the alkene (106°) is close to the angle for the unstrained bimolecular reaction (109°). The corresponding 6-endo transition state **12** is energetically less favored because of

11
5-exo

12
6-endo

the poorer overlap of the orbitals and the higher degree of ring strain. A comparable model has also been derived for the 6-heptenyl radical.[13]

Although hexenyl and heptenyl radicals have an intrinsic preference for exo ring closure, the regioselectivity can be altered or even reversed by the character of the radical (alkyl, vinyl, aryl), the substituents on the radical, or the nature of the radical acceptor (e.g., double or triple bond).

Introduction of alkyl substituents at the 2, 3, 4, or 6 positions of the hexenyl chain usually enhances the rate and improves the regioselectivity in favor of 5-ring formation.[16] Alkyl substituents in the 1 position usually have little effect, whereas substituents in the 5 position sufficiently retard exo cyclization to let 6-ring cyclization become an effective competing reaction.[10]

	k_{rel} (exo)		exo : endo
	1.0	0.02	98 : 2
	1.4	0.02	99 : 1
	16	< 0.5	> 99 : 1
	??	< 0.5	> 99 : 1
	14	< 0.5	> 99 : 1
	0.022	0.04	36 : 64
	2.4	< 0.01	> 99 : 1

Electron-withdrawing substituents at the terminal end of the double bond usually accelerate radical cyclizations owing to favorable FMO interactions. They can be employed to overcome the influence of deactivating substituents and often prove essential for obtaining reasonable yields in the 6-exo closure of heptenyl radicals. In the example shown in Eq. 7, an electron-withdrawing alkoxycarbonyl group is used to accelerate the 6-exo cyclization of the nucleophilic alkyl radical.[17]

t-BuO$_2$C

Br

$$\xrightarrow[\text{C}_6\text{H}_6,\ 80°]{\text{Bu}_3\text{SnH, AIBN}}$$

t-BuO$_2$C

(75%)   (Eq. 7)

t-Bu t-Bu

Radicals bearing endocyclic α-carbonyl substituents often yield preferentially products from 6-endo closure by either kinetic or thermodynamic control. Thus, the incorporation of a ketone (but not an ester or amide) group inside the ring causes a strong preference for 6-ring formation.[18,19] Even in 6-substituted hexenyl radicals 6-endo cyclization still prevails (Eq. 8). Presumably, 5-exo cyclization of

$$R = H \quad (56\%) \quad >97:<3$$
$$R = Me \quad (73\%) \quad 75:25$$

the ketone-substituted radical is retarded because the geometry that allows a stabilizing interaction between the radical and the carbonyl group is distorted in the 5-exo transition state.[18]

It has further been shown that the ring closure of radicals that are stabilized by two nitrile or alkoxycarbonyl groups is reversible.[20,21] By application of suitable reaction conditions it is possible to obtain selectively the thermodynamically favored 6-endo product **13** (Eq. 9).[22]

13

Radical cyclizations onto carbon–carbon triple bonds (5-hexynyl radicals) are somewhat slower than 5-hexenyl cyclizations ($k_c \approx 10^4 \text{ s}^{-1}$).[13] Nevertheless, they are of great synthetic utility because of the good selectivity in favor of 5-ring formation. In addition, the newly formed exocyclic double bond can further be functionalized. Substitution of the alkyne moiety with a trialkylsilyl group seems to accelerate the cyclization, and through the influence of the activating silyl group, even 6-exo cyclizations are possible (Eq. 10).[23]

Vinyl radicals are very reactive and undergo regiospecific exo ring closure under kinetically controlled conditions.[24] However, if the intermediate radicals have a sufficient lifetime, for example, when a low concentration of radical scavenger is present, varying amounts of 6-membered ring products are formed (Eq. 11).[25]

[Bu₃SnH]

0.02 M   75 : 25

1.7 M   >97 : <3

Formation of the more stable cyclohexane derivative is usually not due to a competing endo cyclization. Radical **15** is formed by a rapid rearrangement of the intermediate methylenecyclopentyl radical **14** via a reversible 3-exo cyclization (Eq 12).[26]

As shown in the example in Eq. 13, the increased cyclization rate of vinyl radicals can be used to conduct 6-exo cyclizations.[27] The stereochemistry of the initial precursor **16** is not retained in the product **17** owing to the facile inversion of vinyl radicals.

Vinyl radicals are generated either by halogen abstraction from vinyl halides or by reversible addition of stannyl radicals to triple bonds. An illustrative example of the latter reaction is shown in Eq. 14.[28]

Aryl radicals **18** are widely employed for the formation of benzo-fused ring systems because of their high cyclization rates and excellent exo selectivities.[13] This preference for 5-exo cyclization is maintained even with the 5,5-disubstituted radical **18c** (Eq. 15).[29]

	R^1	X	k_c (s^{-1}, 25°)
a	H	CH$_2$	3.1 x 10^8
b	H	O	5.3 x 10^9
c	Me	O	1.7 x 10^9

(Eq. 15)

As illustrated in Eq. 16, the endo product **19** can be obtained preferentially by inclusion of a radical stabilizing group at the internal alkene atom.[30]

1. Bu$_3$SnH, AIBN, C$_6$H$_6$, 80°

2. HCl, MeOH

(72%)   (Eq. 16)

As in vinyl radical cyclizations, ring expansion of the intermediate 5-exo radical is possible, and mixtures of 5- and 6-membered ring products are sometimes observed. This isomerization is promoted by activating substituents like a carbonyl group (Eq. 17).[31,32]

Bu$_3$SnH, AIBN
C$_6$H$_6$, 80°

+

(Eq. 17)

56 : 44

The cyclization rate constants ($k_c \approx 10^5$–10^6 s^{-1}) and regioselectivities of acyl radicals are very similar to those of alkyl radicals. Examples of the formation of 5-[33] and 6-membered ring ketones[34] (Eq. 18) by exo cyclizations of hexenyl or

Bu$_3$SnH, AIBN
C$_6$H$_6$, 80°

(85%)   (Eq. 18)

heptenyl radicals are known, and synthesis of the bridged ring system **20** is even possible (Eq. 19).[35]

Bu$_3$SnH, AIBN
C$_6$H$_6$, 80°

(69%)   (Eq. 19)

20

A serious side reaction in these types of reactions is decarbonylation of the intermediate acyl radicals prior to cyclization. The rate of decarbonylation depends mainly on the stabilization of the formed alkyl radical. Therefore, systems that form stabilized radicals or have low cyclization rates yield larger amounts of decarbonylated products.[33]

High regioselectivities and yields are also obtained in the cyclization of alkoxycarbonyl radicals to cyclopentanones, although only a few examples are known (Eq. 20).[36]

$$\text{(Eq. 20)}$$

In general, radical cyclization reactions are not restricted to additions to carbon–carbon multiple bonds, and other multiple bonds can also be employed. Cyclizations onto carbon–nitrogen multiple bonds of oximes[37] and nitriles[38] resemble the corresponding additions to carbon–carbon multiple bonds. These cyclizations are irreversible and occur exclusively in the exo mode (attack at the carbon atom). Nitriles are valuable precursors because the resulting imines can be hydrolyzed to the corresponding ketones (Eq. 21).

$$\text{1. Bu}_3\text{SnH, AIBN,}\quad \text{C}_6\text{H}_6, 80^\circ$$
$$\text{2. HOAc}$$
$$(72\%)\qquad\text{(Eq. 21)}$$

Radical additions onto carbon–oxygen double bonds are restricted to aldehydes and some ketones, because ester and amide carbonyl groups are usually not reactive enough. The ring closure is fast, yielding exclusively the exo products. However, the cyclization is reversible, and the equilibrium often lies on the side of the open-chain radical.[39] Because of the increased strain of the cyclopentane ring, fragmentation of the alkoxy radical is much faster when compared to the cyclohexane system. Therefore, the application of this kind of reaction is restricted mainly to the formation of 6-membered rings.[40]

$$k_c \quad 9 \times 10^5 \text{ s}^{-1}$$
$$k_{c'} \quad 5 \times 10^8 \text{ s}^{-1}$$

$$k_c \quad 1 \times 10^6 \text{ s}^{-1}$$
$$k_{c'} \quad 1 \times 10^7 \text{ s}^{-1}$$

One way to overcome these unfavorable equilibria is selective trapping of the cyclic alkoxy radical with a stannane. It has been shown that hydrogen abstraction from the stannane by an alkoxy radical is about 100 times faster than abstraction by the initial carbon–centered radical. It is therefore possible to obtain useful yields of the cyclic product provided that the equilibrium constant is not too small. An example of radical cyclization onto an aldehyde is shown in Eq. 22. This reaction shows furthermore that in a competing system 6-exo cyclization to form cyclohexanol **21** dominates over the possible 5-exo cyclization onto a carbon–carbon double bond.[41]

(Eq. 22)

However, the formation of cyclopentanols is more difficult because ring opening of the five-membered ring is extremely rapid. To obtain useful yields, the equilibrium has to be shifted to the side of the cyclized radical by incorporation of suitable substituents.[42]

If the hexenyl chain contains a nitrogen[43] or oxygen[44] atom in the 3 or 4 position, the rate and the regioselectivity of the radical cyclization reaction are enhanced, owing to a better orbital overlap in the 5-exo transition state. In contrast to the carbocyclic radical **22a**,[45] the 5-substituted hexenyl radicals **22b** and **22c** show a strong preference for 5-ring products (Eq. 23).

	X	23 : 24
a	CH$_2$	36 : 64
b	O	100 : 0
c	NSO$_2$Ph	100 : 0

(Eq. 23)

Radical-stabilizing oxygen or nitrogen atoms in the α position to the radical (2 position) retard the cyclization, and byproduct formation often dominates.[46,47] The poor behavior of these radicals is attributed to increased geometric constraint imposed by delocalization of the radical.[46] Nevertheless, a few examples are known in the literature, even for cyclization of 6-heptenyl radicals (Eq. 24).[48]

(Eq. 24)

Cyclization reactions of systems with an ester or amide group in the radical chain provide a synthetically useful access to lactams and lactones. The regioselectivity is high and favors 5-exo cyclization; the rate of cyclization is usually in a synthetically useful range ($k_c \approx 10^5$ s^{-1}). An example of the formation of a γ-lactone is shown in Eq. 25.[49]

(66%)   (Eq. 25)

However, if the initial radical is in the α position to the carbonyl group, cyclization is strongly decelerated. Presumably the retardation is due to the preferred *trans* configuration of esters and amides and to the barriers to rotation about the ester/amide bond.[50] Therefore, the radical cyclization reaction of allyl esters and allyl amides gives either relatively poor yields or methods have to be employed that tolerate low cyclization rates. A representative example is shown in Eq. 26.[18,51]

(83%)   (Eq. 26)

Several procedures have been developed to overcome these problems. The most popular is the so-called bromoacetal method, in which the ester carbonyl is introduced after the cyclization of an easily accessible bromoacetal precursor (Eq. 27).[52,53]

(81%)

(Eq. 27)

Furthermore, amide cyclization can be accelerated by substitution of the amide nitrogen with strong electron-withdrawing groups like tosyl or trifluoroacetyl (Eq. 28).[54]

(85%)

(Eq. 28)

If silicon atoms are part of the radical chain, the regioselectivity of cyclization strongly depends on the substitution site. In simple 2- and 3-(dimethylsila)hexenyl radicals, the cyclization rate is reduced ($k_c \approx 10^4$ s^{-1}) and cyclizations occur mainly in an endo fashion. The corresponding 4-substituted radical, however, cyclizes mainly in the exo mode.[55]

Of greater synthetic interest are the 5-hexenyl cyclizations of bromomethyl-silyl ethers of allylic alcohols.[56,57] Although the silicon atom is in the position α to the radical center, products of exo cyclization are formed (Eq. 29). Introduc-

$$\text{(Eq. 29)}$$

tion of a 5-substituent can reverse the regioselectivity, and six-membered rings are obtained in good yields (Eq. 30).[58]

$$\text{(Eq. 30)}$$

After cyclization, the carbon–silicon bond can either be cleaved reductively to give the hydrocarbon[57] or oxidized to yield the alcohol.[59] Thus, the heterocycles formed in this type of reaction usually serve the temporary purpose of stereose-lectively introducing an alkyl or hydroxyalkyl group adjacent to an alcohol.

In related reactions, heterocyclic 5- and 6-membered rings are formed by cy-clization of heteroatom-centered radicals. Radicals on oxygen and nitrogen have been employed in synthesis. In comparison to their carbon counterparts, oxygen radicals show enhanced reactivity and high cyclization rates ($k_c \approx 10^8 \text{ s}^{-1}$). The cyclizations are irreversible and strictly exo selective even if steric hindrance is involved (Eq. 31).[60]

$$\text{(Eq. 31)}$$

However, there are only a few examples of oxygen radical cyclizations in the literature[61,62] owing to fast-competing reactions such as 1,5-hydrogen abstraction or β-fragmentation and to the limited methods for generation of these radicals.

With nitrogen-centered radicals one has to distinguish between aminyl and iminyl radicals. Alkyl-substituted aminyl radicals like **25** cyclize very slowly ($k_c \approx 10^3 \text{ s}^{-1}$), and reopening of the cyclized radical **26** has a similar rate con-stant (Eq. 32).[63,64]

$$\text{(Eq. 32)}$$

25

k_c \quad $3 \times 10^3 \text{ s}^{-1}$
$k_{c'}$ \quad $7 \times 10^3 \text{ s}^{-1}$

26

Increasing the electrophilicity at nitrogen accelerates ring closure and thereby shifts the equilibrium toward the cyclized radical. This can be done either by protonation or complexation to a metal center.[65] The 5-hexenyl cyclizations yield 5-ring products regioselectively from exo cyclization. A metal-complexed aminyl radical can be formed from amine **27** (Eq. 33).[66] One way of generating an

(81%)

(Eq. 33)

27

aminium radical cation such as **29** is photolysis of the corresponding *N*-hydroxypyridine-2-thione carbamate **28** under acidic conditions (Eq. 34).[67,68]

28   **29**

(92%)

(Eq. 34)

Iminyl radicals were introduced to synthesis very recently. They behave more like carbon-centered radicals and show a preference for cyclizations in the 5-exo mode.[69]

Formation of Medium-Sized Rings.   Like many other methods, the formation of medium-sized rings by radical cyclizations is usually difficult. Only a few examples of the construction of 7- and 8-membered rings are known. The rates of 7-ring ($k_c \approx 7 \times 10^2$ s^{-1})[13] and 8-ring cyclizations are at the lower limit of synthetic utility. Furthermore, 7-octenyl radicals show reversed regioselectivity, preferentially cyclizing in the 8-endo mode. Therefore, 7- and 8-membered rings are usually constructed by endo ring closure of 6-heptenyl and 7-octenyl radicals, respectively.

Radicals derived from α-keto esters by oxidative methods are especially useful for the formation of medium-sized rings (Eq. 35).[70] This is probably due

(69%) 2.5 : 1

(Eq. 35)

to the increased rate of endo cyclization of α-carbonyl substituted radicals and the absence of fast radical traps for these electron-poor radicals under oxidizing conditions.

The rate of 7-ring closure can be enhanced further if reactive acyl or aryl radicals are employed for the cyclization. Examples of this kind of reaction are shown in Eqs. 36[71] and 37.[35] Whereas the first cyclization proceeds in a 7-endo mode, the second is an example of a rare 7-exo ring closure.

(65%)

(Eq. 36)

(71%)

(Eq. 37)

In cyclizations of heptenyl radicals with α-silyl ether groups in the chain, the product **30** formed from 7-endo ring closure is preferred over 6-exo (Eq. 38).

(92%)

(Eq. 38)

30

However, the regiochemistry can be reversed if a substituent is introduced at the terminal end of the alkene.[72]

In a different approach, medium-sized rings are often easily accessed by radical ring-expansion methods. This type of sequential radical reaction involves a cyclization that is followed by fragmentation of an unstable radical intermediate. Usually a carbon–oxygen double bond is employed as radical acceptor, and the regiochemistry of the fragmentation step is directed by a radical-stabilizing substituent like an alkoxycarbonyl group.[73] The general principle is illustrated by expansion of the 6-membered β-keto ester **31** to the 7-membered homolog **32**, shown in Eq. 39.[74,75]

(Eq. 39)

This method is also suitable for ring expansions by three or four carbon atoms, but not by two atoms because of the low rate of 4-exo closure. A radical-stabilizing substituent is essential to obtain useful yields.[73]

The problem of reversibility of the rearrangement can be overcome by introduction of the trialkylstannyl substituent in the position β to the rearranged radical. The fast β elimination of the stannyl group in radical **33** shifts the equilibrium to the side of ring-expanded product **34**. An example is shown in Eq. 40.[76]

(Eq. 40)

Macrocyclizations.   Radical cyclization reactions can be employed for the construction of 10- to 20-membered macrocycles.[77,78] These cyclizations resemble intermolecular radical additions and consequently occur preferentially in the endo mode. They are controlled by steric and polar effects, and to obtain reasonable yields it is usually essential to accelerate the cyclization by substitution of the alkene with an electron-withdrawing substituent (Eq. 41).

(Eq. 41)

Stereochemistry

Radical cyclization reactions often proceed with high levels of stereoselectivity. Intensive theoretical and empirical studies have resulted in guidelines for rationalization and prediction of the configuration at new stereogenic centers.[13,15] Because radical reactions have early transition states, the interpretation of selectivity usually focuses on the conformational bias of the radical and not on steric interactions in the final product.

Formation of Monocycles. The stereochemistry of 5-hexenyl radical cyclizations can often be rationalized with the guidelines originally proposed by Beckwith.[16] According to this model, the main diastereomer is formed via transition state **35**, which resembles a cyclohexane in the chair form with the

35

chair substituents adopting pseudoequatorial orientations. Thus, cyclizations of 3-substituted radicals yield mainly 3,5-*cis* disubstituted 5-rings, whereas 2- and 4-substituted radicals preferentially give the corresponding *trans* products.

The stereoselectivity of 1-substituted hexenyl radicals depends strongly on the size and electronic properties of the substituent. 5-Hexenyl radicals with simple alkyl groups react in accordance with the Beckwith guidelines, yielding preferentially the 1,2-*cis* product 36 (Eq. 42).[79]

36
cis : trans = 5 : 1

Preferred formation of *trans* disubstituted products is sometimes observed with precursors bearing polar substituents. This is illustrated by the example in Eq. 43, where the silyloxy substituent in the 1-position leads to complete inversion of the stereoselectivity in favor of the *trans* product.[80]

cis : trans = 1 : 7

With the more polar sulfone substituent, exclusive formation of *trans* product **37** is observed (Eq. 44).[81] The high selectivity is attributed to steric and elec-

(94%)

(Eq. 44)

37

trans : *cis* = > 95 : 5

tronic repulsions between the oxygen atoms of the sulfone and the methoxy group of the enol ether. 5-Hexenyl cyclizations of simple radicals with ester or keto substituents in the 1 position show virtually no stereoselectivity.[18]

Good stereoselectivities are frequently observed in the cyclizations of complex precursors. This is illustrated by the reaction of a complex tetrasubstituted precursor **38** that can be derived from a carbohydrate substrate (Eq. 45). Of the four

+ other isomers

38

39   (63%) 78 : 22

(Eq. 45)

possible diastereomers, **39** is formed as the main product in accordance with the Beckwith rules.[82] However, in complex systems, additional factors like allylic strain and boat-like transition states have to be considered for the derivation of transition state models.[83]

Chiral auxiliaries can be also employed to direct addition of the radical to the alkene. The cyclization of a hexenyl radical bearing a chiral 8-phenylmenthyl ester at the 1 position gives the four isomeric cyclopentane isomers in a modest *cis* : *trans* ratio, but a considerably higher $S : R$ selectivity of 4 : 1 (Eq. 46).[84]

+

40   **41**

cis : *trans* = 1.6 : 1   *cis* : *trans* = 1 : 1.5

40 : 41 = 4 : 1 (90%)

(Eq. 46)

Cyclization of a hexenyl radical substituted with a chiral camphor sultam derivative yields the two diastereomeric methylenecyclopentanes **42** and **43** in a ratio of 9 : 1 (Eq. 47).[85]

42 : 43 = 9 : 1 (78%)

(Eq. 47)

Rationalization of the stereoselectivities in the formation of heterocyclic rings is more difficult than in carbocyclic systems. Detailed theoretical and mechanistic studies of the influence of heteroatoms on the cyclization are not available. However, some trends emerge from examples in the literature.

If one carbon atom in the radical is substituted by oxygen, the Beckwith rules still apply. This is illustrated in the cyclization of the 2-substituted iodoethyl allyl ether **44** that preferentially yields the *trans* product (Eq. 48).[86]

cis : trans = 21 : 79 (77%)

(Eq. 48)

On the contrary, introduction of a nitrogen atom in the 3 position of a 2-substituted radical gives mainly the 2,4-*cis*-substituted pyrrolidine derivative (Eq. 49).[87]

R	cis : trans	
H	62 : 38	(90%)
Me	60 : 40	(85%)
Pr	60 : 40	(75%)
Ph	55 : 45	(85%)

(Eq. 49)

Hexenyl radicals with an ester or amide linkage in the chain show similar stereoselectivities. As in carbocyclic systems, 1-substituted radicals preferentially yield the 1,5-*trans*-disubstituted lactone **45**[88] (Eq. 50) or lactam **46**[89] (Eq. 51). The reasons for the selectivities are unclear, but as the amides and esters behave simi-

(47%)

(Eq. 50)

45

trans : cis = > 99 : 1

(90%)

46

trans : cis = 83 : 17

(Eq. 51)

larly, the stereoselectivity is probably determined by the carbonyl substituent in the ring.

Interestingly, the stereoselectivities in these systems seem to depend on the nitrogen substituent of the amide. This is demonstrated by cyclization of the hexenyl-type radicals with substituents in the 4 position. Whereas the benzyl substituted radical preferentially gives the 4,5-*trans*-substituted product, introduction of the tosyl group on nitrogen results in complete reversal of the selectivity (Eq. 52).[90,91]

R	cis : trans	
CH$_2$Ph	10 : 90	(98%)
Ts	90 : 10	(98%)

(Eq. 52)

For the stereoselectivity of 6-exo cyclization of heptenyl radicals, a model can be applied that resembles the Beckwith model. Again, the preferred transition state adopts a chair-like conformation with the substituents in pseudoequatorial orientation. Thus, 2- and 4-substituted radicals should give preferentially *cis* products, and 1-, 3-, and 5-substituted radicals *trans* products. However, cyclizations of 2-, 3-, and 4-methyl-substituted radicals all exhibit a very small preference for the *trans* product (Eqs. 53–55).[17]

cis : trans = 43 : 57 (92%)

(Eq. 53)

cis : trans = 35 : 65 (91%)

(Eq. 54)

cis : trans = 37 : 63 (80%)

(Eq. 55)

Therefore, the cyclohexane model seems to be of only limited value for heptenyl cyclization, and other possible transition states such as cycloheptane-like conformers, should also be considered. Nevertheless, in accordance with the model, an example is known of selective formation of *cis*-**48** from 4-substituted precursor **47** (Eq. 56)[92] and *trans*-**50** from 5-substituted precursor **49** (Eq. 57).[93]

(Eq. 56)

(97%)   (Eq. 57)

Furthermore, the *trans* product is also strongly favored in the cyclization of a heptenyl radical with a phenyl substituent in the 3 position.[48]

Heptenyl radicals with a substituent in the 1 position seem to cyclize preferentially to 1,6-*trans*-substituted 6-membered rings. Two representative examples for carbocyclic[37] and heterocyclic systems[94] are shown in Eqs. 58 and 59.

(71%)   (Eq. 58)

cis : *trans* = 33 :67

(74%)   (Eq. 59)

No data are available for the stereoselectivities in the formation of 7- and 8-membered rings. The stereoselectivities of radical macrocyclizations are comparable to the results in intermolecular radical reactions owing to the small influence of ring constraints. Stereoselectivity is observed in the cyclization onto acrylamide **51** bearing the 2,5-dimethylpyrrolidine chiral auxiliary. The endo:exo ratio of the four isomeric products is about 8:1; the two exo products

53a/b are formed with no stereoselectivity, whereas the products from the endo cyclization **52a/b** are a 93:7 mixture of diastereomers (Eq. 60).[95]

Bu₃SnH
AIBN
C₆H₆, 80°

(Eq. 60)

51

52a/b
$R:S = 14:1$

53a/b
$R:S = 1:1$

52:53 – 8:1 (10-45%)

$R* =$

These results indicate that asymmetric induction is effective only if attack occurs at the α position of the chiral amide.

Formation of Bi- and Polycycles. The formation of bi- and polycyclic systems by radical cyclization constitutes a powerful synthetic method.[5] As in monocyclic systems, 5- and 6-ring cyclizations are the most frequently applied reactions. Annulations are usually achieved either by cyclization of a radical onto a cyclic alkene or by addition of a cyclic radical to a multiple bond in the side chain. Both reaction types afford preferentially the *cis* ring-fused products (5,5- 5,6- and 6,6-bicycles) owing to steric constraint imposed by the ring system. Two examples of these complementary methods are shown in Eqs. 61[96] and 62.[97]

(Eq. 61)

95:5

(Eq. 62)

In addition to the geometry of ring fusion, cyclization also controls the stereochemistry at the 5- or 6-position of the newly formed ring. As in open-chain systems, the 1,5- or 1,6-*cis* configuration is usually preferred (Eq. 63).[49]

Ph₃SnH, AIBN
C₆H₆, 80°

$n = 1$, 70:30 (61%)

$n = 2$, 80:20 (63%)

(Eq. 63)

The stereoselectivities are rationalized by competing cyclohexane-like transition states preferentially adopting chair- or boat-like conformations (Beckwith model). However, in more complex systems, the 1,5- or 1,6-*trans* configuration is sometimes favored. In these cases, more detailed insight is necessary to rationalize the results.[83]

Cyclizations of highly functionalized precursors **54** derived from carbohydrates are intriguing (Eq. 64). In all cases, ring closure yields *cis*-annulated car-

	R^1	R^2	
a	H	H	77 : 23
b	OBn	H	99 : 1
c	H	OBn	>1 : 99

(Eq. 64)

bocycles as mixtures of diastereomers. The stereochemistry of the bicycles is controlled primarily by the configuration at the C-4 carbon of the radical. The C-4 α-configurated radical gives 1,5-*trans*-**55b**, and the 4β-substituted radical gives 1,5-*cis*-**56c**, both with high selectivity. The C-4 deoxy precursor **54a** shows diminished selectivity in favor of 1,5-*trans*-**55a**.[82]

In constrained systems, the formation of *trans*-annulated polycycles is sometimes observed. A prominent example is the synthesis of *trans*-perhydroindanes via radical cyclization of bicyclic lactone **57** (Eq. 65).[98,99] The tricyclic products

58 : 59 = 7 : 1 (93%)

(Eq. 65)

58 and **59** contain 5,5-*cis* and 5,6-*trans* ring fusions. The smaller, less flexible 5-ring lactone takes precedence in the formation of the *cis*-fused bicycle. In accordance with the Beckwith guidelines, the main isomer **58** possesses a 1,5-*cis* configuration.

Cyclization onto cyclic alkenes is the second common strategy used to build up stereoselectively *cis*-fused bi- and polycyclic systems. In these reactions, ring closure generates a cyclic prochiral radical in the α position to the ring junction.

The consecutive reaction of this center is often stereoselective, and the attack of a radical trap occurs preferentially *anti* to the neighboring substituent. The synthetic potency of this method is illustrated in the elegant synthesis of (+)-prostaglandin $F_{2\alpha}$.[100] In the key step, two new chiral centers are stereoselectively generated by 5-ring annulation and consecutive trapping of bicyclic radical **60** with alkene **61** (Eq. 66).

(Eq. 66)

The high stereoselectivities of radical annulations are often used for the regio- and stereoselective introduction of alkyl substituents in cyclic systems. A prominent example of this type of reaction is the silylmethyl radical cyclization. This reaction is employed for stereoselective introduction of a methyl or hydroxymethyl group adjacent to a hydroxy group. The radical precursor is prepared by reaction of the allylic alcohol with (bromomethyl)chlorodimethylsilane. Radical cyclization gives the siloxane ring, which can be converted either into the *cis*-dihydroxy[57] or the α-methylated derivative.[101] In the total synthesis of talaromycin A (**62**), this method was used to introduce stereoselectively the 1,3-diol unit (Eq. 67).[59]

(Eq. 67)

The 5-ring radical annulation of 7- and 8-membered rings preferentially yields *trans*-fused ring systems.[102,103] The altered stereochemistry is rationalized by the different conformational bias compared to the 5- and 6-membered analogs. Introduction of ring substituents has only a slight influence on the *trans:cis* ratio (Eq. 68).[104,105]

$R^1, R^2 = H$	87 : 13 (51%)
$R^1 = OH, R^2 = H$	>99 : <1 (65%)
$R^1 = H, R^2 = OH$	83 : 17 (42%)
$R^1, R^2 = O$	95 : 5 (65%)

(Eq. 68)

Other valuable methods for the construction of bi- and polycyclic systems are tandem reaction sequences (see Scope and Limitations). In these transformations, two or more consecutive radical cyclizations are connected in a reaction series. Contrary to simple cyclizations, the first-formed cyclized radical is not immediately trapped to the product, but instead serves as the initial radical of the second cyclization step. Although the overall transformations can be quite complex, the stereochemistry of the reaction can be rationalized if each step is treated separately.

This concept is exemplified by the synthesis of linear and angular triquinanes by a radical tandem cyclization.[106] In Eq. 69, two consecutive 5-exo cyclizations build up a tricyclic system **63** where all rings are exclusively *cis* fused.[107]

(Eq. 69)

SCOPE AND LIMITATIONS

The following section provides a discussion of the most popular methods for conducting radical cyclization reactions with special emphasis on the scope and limitations of every method.

Metal Hydride Methods

In this type of radical cyclization, metal hydrides are used as precursors and/or promoters of radical chain reactions.

Tin Hydride Method. The use of organotin hydrides is by far the most popular and general method for conducting radical cyclizations.[108] In a chain reaction, the initial radical **65** is generated from a suitable precursor **64** by atom or group abstraction by the trialkylstannyl radical **66**. After cyclization has taken place, trapping of cyclic radical **67** by tin hydride gives the reduced cyclic product together with the chain-propagating trialkylstannyl radical **66** (Eq. 70).

$$R^1X \ + \ R_3Sn\cdot \ \xrightarrow{\ -R_3SnX\ } \ R^1\cdot \ \xrightarrow{\ cyclization\ } \ R^2\cdot \ \xrightarrow{\ +R_3SnH\ } \ R^2H \ + \ R_3Sn\cdot$$

$$\text{64} \qquad \text{66} \qquad\qquad \text{65} \qquad\qquad \text{67} \qquad\qquad\qquad \text{66}$$

(Eq. 70)

The chain reaction is usually initiated by thermally or photolytically induced homolytic cleavage of a chemical initiator, azobisisobutyronitrile (AIBN) or benzoyl peroxide (BPO). A correct match of initiator and reaction temperature is essential for a successful cyclization.[109] For initiation at low temperatures (as low as $-78°$) a mixture of triethylborane/oxygen can be used.[110]

A broad range of functional groups can serve as radical precursors. In order of decreasing reactivity, these include iodides, bromides, phenyl selenides, nitro groups,[111] chlorides, and phenyl sulfides. Iodides and bromides are the most common precursors and they can be used to generate all kinds of radicals including the reactive aryl and vinyl radicals as well as nucleophilic and electrophilic alkyl, stabilized benzyl, and allyl radicals. Numerous examples of these reactions can be found in the tables.

The use of phenyl selenides as precursors can be advantageous for different reasons. They can readily be synthesized by a number of methods and are usually more stable than halides (e.g., against solvolysis). Nevertheless, phenyl selenides are reactive enough for the generation of simple alkyl radicals from alkyl phenyl selenides. An illustrative example is shown in Eq. 71.[112]

(80%)   (Eq. 71)

In addition, acyl phenyl selenides and phenyl selenocarbonates are the precursors of choice for generation of acyl (Eq. 72)[33] and alkoxycarbonyl radicals (Eq. 73).[36]

$$
\text{PhSe} \diagdown \text{O} \quad \xrightarrow[\text{C}_6\text{H}_6,\ 80°]{\text{Bu}_3\text{SnH, AIBN}} \quad \text{(product)} \qquad (86\%) \qquad (\text{Eq. 72})
$$

$$
\text{PhSe} \diagup \text{O} \quad \xrightarrow[\text{C}_6\text{H}_6,\ 80°]{\text{Bu}_3\text{SnH, AIBN}} \quad \text{(product)} \qquad (99\%) \qquad (\text{Eq. 73})
$$

Use of the less reactive phenylthio group is usually restricted to the generation of stabilized radicals. For example, phenyl sulfides are preferred precursors for acylamino radicals because the corresponding halides are prone to solvolysis (Eq. 74).[113]

$$
\xrightarrow[\text{C}_6\text{H}_6,\ 80°]{\text{Bu}_3\text{SnH, AIBN}} \qquad (89\%) \qquad (\text{Eq. 74})
$$

Alcohols **68** can serve as precursors for alkyl radicals via the corresponding thioxanthates **69**, thiocarbamates **70**, or thiocarbonates **71**.[114] The deoxygenated radical is formed via the addition/fragmentation mechanism illustrated in Eq. 75.

$$
\text{ROH} \longrightarrow \underset{\text{RO}}{\overset{S}{\|}} \diagdown \text{Y} \quad \xrightarrow{+\ \text{R}_3\text{Sn}\bullet} \quad \underset{\text{RO}}{\overset{\text{SnR}_3}{S}}\diagdown\text{Y} \quad \longrightarrow \quad \text{R}\bullet \ + \ \underset{O}{\overset{\text{SnR}_3}{S}}\diagdown\text{Y}
$$

68

$$
\text{Y} = \ \text{SMe},\ -\text{N}\diagdown\!\!\!\diagup\text{N},\ \text{OPh}
$$

69   **70**   **71**   (Eq. 75)

This method is best suited for the formation of secondary radicals because the tertiary precursors are difficult to synthesize and fragmentation of the primary radical is often too slow. The utility of the method is demonstrated by the cyclization of the highly functionalized carbohydrate derivative **72** shown in Eq. 76.[82]

(Eq. 76)

Aldehydes and ketones can be employed as precursors for α-oxysubstituted radicals. The carbonyl group can be reacted with tin hydride either directly[115] or via the corresponding mixed thio-[81] or selenoacetals.[116,117] An example of the latter method is shown in Eq. 77.

(Eq. 77)

Alkynes can serve as precursors for vinyl radicals by reversible addition of a stannyl radical to the triple bond. The reversibility of the addition is important because the stannyl radical adds unselectively to all multiple bonds of a system and only the radicals with a favorable pathway will cyclize faster than they revert to the alkyne.[118] This method is of great synthetic utility because the precursors are readily prepared and the stannyl group is easily removed after the cyclization by protodestannylation.[119] A representative example is provided in Eq. 78.[120]

(Eq. 78)

The most crucial point in tin hydride mediated cyclizations is competition between cyclization of the initial radical and its reduction by tin hydride. Since the cyclization is a unimolecular reaction and hydride abstraction is bimolecular, the undesired direct reduction can be suppressed by employing low concentrations of tin hydride. However, in these cases, precautions should be taken to ensure that the chain is efficiently propagated. Therefore, the most reactive precursor available should be employed. The two most popular methods for minimizing tin hydride concentration without the need for large solvent volumes are syringe pump addition of tin hydride and procedures that involve the use of catalytic amounts (0.1 equiv) of trialkyltin hydride or chloride in the presence of a coreductant like sodium borohydride[121] or sodium cyanoborohydride.[122] The borohydride recycles tin hydride by reduction of the tin halide formed during the reaction. This method is restricted to halide precursors because only tin halides will be reduced to the tin hydride. Equation 79 provides an example in which this method has been used to conduct a slow 7-endo cyclization reaction.[72]

The catalytic procedure also simplifies isolation of the desired product because only catalytic amounts of tin compounds have to be removed after the reaction. The removal of tin residues often constitutes an important practical problem, and several special workup procedures have been developed.[18,123]

An alternative approach to overcoming the problem of direct reduction of the initial radical is the use of metal hydrides that are slower hydrogen donors. One possible substitute for tributyltin hydride is the germanium analog. Hydrogen transfer from tributylgermanium hydride to alkyl radicals is 10–20 times slower than from tin hydride, and the germanyl radical is a powerful halogen atom abstractor.[124] In Eq. 80, tributylgermanium hydride is used to promote a slow 6-exo cyclization.[50]

Another suitable tin hydride substitute is tris(trimethylsilyl)silane, a ~10 times slower hydrogen atom donor than tin hydride.[125] In addition, tris(trimethylsilyl)silane is nontoxic, and workup of the reaction mixture is usually simple.[126] Equation 81 shows an example in which tris(trimethylsilyl)silane is employed for a macrocyclization reaction.[127]

Mercury Hydride Method. The mercury hydride method is closely related to the tin method because metal hydrides are propagating the radical chain in both procedures. Precursors for the mercury hydride method are organomercury(II) halides or acetates. They are reduced by sodium borohydride, sodium cyanoborohydride, or tin hydride in an appropriate solvent (methanol, tetrahydrofuran, or dichloromethane). The so-formed transient alkylmercury(II) hydride **73** partly decomposes, thereby initiating the radical chain (Eq. 82).[128] The chain is

$$R^1HgX \xrightarrow{+ "H^-"} \begin{bmatrix} R^1HgH \end{bmatrix} \xrightarrow{- Hg^0} R^1\bullet \qquad \text{(Eq. 82)}$$
$$\phantom{R^1HgX \xrightarrow{+ "H^-"}} \begin{matrix} \mathbf{73} \end{matrix} \qquad \begin{matrix} \mathbf{75} \end{matrix}$$

propagated by hydrogen abstraction by cyclized radical **74** from the alkylmercury hydride followed by spontaneous cleavage of the alkylmercury bond (Eq. 83).

$$R^1\bullet \xrightarrow{\text{cyclization}} R^2\bullet \xrightarrow[- Hg^0]{+ R^1HgH} R^2H + R^1\bullet \qquad \text{(Eq. 83)}$$
$$\phantom{R^1\bullet \xrightarrow{\text{cyclization}}} \begin{matrix} \mathbf{74} \end{matrix} \qquad\qquad\qquad \begin{matrix} \mathbf{75} \end{matrix}$$

As in the tin method, direct reduction of the initial radical **75** always competes with cyclization. As mercury hydrides are very good hydrogen donors (~10 times faster than tin hydrides), only fast cyclizations can be conducted successfully.[129] However, the mercury hydride method has several advantages, including convenient reaction conditions (low temperature, no initiator needed), easy workup (metallic mercury is removed by filtration), and easy preparation of the alkylmercury halides. Methods for the preparation of the precursors include transmetalation of boranes, oxymercuration of cyclopropanes, mercuration of ketone hydrazones, and oxy- or amidomercuration of alkenes.[6,130] A representative example of the most important method (oxymercuration of alkenes) is shown in Eq. 84.[131]

$$\text{(Eq. 84)}$$

Fragmentation Methods

As an alternative to metal hydride based reactions, the chain-transfer agent Y· (**76**) is not generated by hydrogen abstraction from a stannane, but by a fragmentation reaction of cyclic radical **77** or **78**. As shown in Eqs. 85 and 86, the

$$\text{(Eq. 85)}$$

$$\text{(Eq. 86)}$$

chain carrier can be included in either an allylic or vinylic radical acceptor. The use of allylic systems is more common because the alkene substituent sterically and sometimes also electronically biases the addition toward the desired regioselectivity. Vinylation usually needs an activating substituent that directs the addition to the carbon bearing the radical leaving group.

The most important advantage of the fragmentation method is the absence of any metal hydride and therefore there is no competing direct reduction of the initial radical. As usual, all reactions of the sequence have to be faster than unwanted side reactions of the intermediate radicals (recombination, reaction with the solvent). Hence relatively slow cyclizations can be conducted and less reactive precursors can be employed. Furthermore, the cyclizations do not yield reduced products (tin method) and the double bond is retained during the reaction.

Vinyl- and allylstannanes have proven to be especially useful radical accepting groups.[132] As in the tin hydride method, trialkylstannyl radicals carry the chain and therefore similar conditions (initiation, solvent, etc.) and precursors can be employed. Representative examples of both types of reactions are provided in Eqs. 87 and 88.[133,134]

$$\text{(Eq. 87)}$$

$$\text{(Eq. 88)}$$

Generation of the initial radical in the fragmentation method is not restricted to atom or group abstraction methods; it can also be generated by reversible addition to a multiple bond.[135] The example in Eq. 89 shows a radical isomerization of an allyl sulfone 79 that involves a cyclization step.[136]

$$\text{(Eq. 89)}$$

In a method that is closely related to the fragmentation method, the chain-carrying radical is generated via homolytic substitution rather than a fragmentation reaction. Radicals are formed from alkylcobalt derivatives, which can readily be prepared from the corresponding alkyl halides.[137] The cobalt–carbon bonds are very weak, and homolytic substitution at carbon can take place.[138] The released cobalt(II) radical functions as the chain carrier and abstracts a halogen atom from the starting halide. To obtain reasonable rates for the abstraction reaction, only activated precursors like polyhalogenated alkanes or sulfonyl iodides can be employed. However, this method constitutes one of the few radical reactions by which three-membered rings can be formed. An example of this reaction is shown in Eq. 90.[139,140]

$$+ \ ArSO_2I \quad \xrightarrow{CH_2Cl_2, \text{ reflux}} \quad (80\%)$$

(Eq. 90)

Radical cyclizations via homolytic substitution reactions are not restricted to reactions at carbon atoms. There are examples in the literature in which the substitution takes place at selenium[141] or silicon.[142]

Thiohydroxamate Method (Barton Method)

One of the most important radical chain methods where alkyltin radicals are not part of the reaction is the thiohydroxamate or Barton method. This is a useful method that can be employed for a broad variety of transformations that proceed via radicals (e.g., inter- and intramolecular bond formation or functional group interconversion).[143,144] The radical precursors are thiohydroxamate esters **80** that can easily be prepared from carboxylic acids.[145-147] Although thiohydroxamate esters **80** can often be isolated, they are usually prepared and reacted in situ. The radical chain is initiated by thermolysis or by visible light photolysis of thiohydroxamate esters **80**.[148] The propagating steps involve addition of the cyclic radical to the thiohydroxamate unit followed by a rapid fragmentation of the ensuing intermediate **81** to generate the initial alkyl radical **82** (Eq. 91).[149]

(Eq. 91)

In contrast to the metal hydride based reactions, the chain-terminating step is not hydrogen abstraction but transfer of a thiopyridyl group that can be employed for further transformations. The lower rate limit for radical cyclizations that can be conducted successfully by this method is determined by the rate of addition of the uncyclized radical to the starting thiohydroxamate. For primary radicals, the rate of this reaction is very similar to the rate of hydrogen abstraction from tin hydride. As in the metal hydride methods, the formation of uncyclized products can be suppressed by lowering the concentration of the precursor. An example of a cyclization that is conducted by the thiohydroxamate method is shown in Eq. 92.[150]

(Eq. 92)

Use of the thiohydroxamate method is not restricted to the generation of carbon-centered radicals. If carbamate derivatives are used as precursors, aminyl radicals can be generated.[151] High yields of cyclic products are obtained in acidic reaction media.[68,152] Presumably, the nitrogen is protonated under these conditions and the resulting aminyl radical cation cyclizes more rapidly. If a good hydrogen donor like a thiol is present, the cyclic radicals abstract hydrogen to form the reduced products, and the radical chain is propagated by the liberated thiol radical (Eq. 93).[153]

(Eq. 93)

Atom Transfer Methods

The introduction of two new substituents to a multiple bond via a radical chain addition is one of the basic transformations in radical chemistry. A prominent example is the anti-Markownikov addition of hydrogen bromide to alkenes (Kharasch reaction). This type of reaction constitutes the shortest possible radical chain because the initial radical is formed by direct reaction of the substrate with the final radical of the chain. For cyclization reactions, this usually means that the net transformation is isomerization of the starting material. The atom transfer method is ideally suited for conducting slow radical cyclizations because there is no fast radical trap (like tin hydride) present in the reaction medium. The lower rate limits for the cyclization and the atom-transfer step are determined by the unwanted chain termination reactions (radical–radical recombination, reaction with solvent). The cyclization step is usually fast enough because the ring closure is an exothermic reaction (a π bond is converted to a σ bond). However, to obtain suitable rates for the atom-transfer step, the initial radical should be thermodynamically more stable than the cyclic radical.

Depending on the atom that is transferred during the cyclization, the reactions can be divided into hydrogen or halogen transfer methods.

Hydrogen Atom Transfer Method.   Because of the low rates of hydrogen abstraction, application of the hydrogen transfer method is restricted to cyclizations of precursors with activated C H bonds.[1] Useful precursors (e.g., malonates or acetoacetates) bear multiple radical stabilizing substituents like carbonyl or nitrile groups. Cyclization of these electrophilic radicals is often reversible, and because of the slow hydrogen abstraction of the cyclic radical, thermodynamically controlled products can be obtained.[1,20,21]

An illustrative example is shown in Eq. 94.[22] This reaction is initiated by benzoyl peroxide (BPO), a common initiator for hydrogen transfer reactions. Because

$$\text{(76\%)} \qquad \text{(Eq. 94)}$$

of the short reaction chain length resulting from the inefficiency of hydrogen transfer, a large amount of peroxide is necessary. From mechanistic studies it is known that the initial radical cyclizes almost exclusively to the 5-ring.[18] However, because of the long lifetime of the intermediate radicals, thermodynamic equilibrium is established via reopening and 6-endo cyclization to the thermodynamically favored 6-membered ring.

However, because of the slow and unselective nature of bimolecular hydrogen transfer and the limited variety of suitable precursors, hydrogen atom transfer cyclizations are of only limited value for organic synthesis. Furthermore, equivalent radical cyclizations can often be conducted more conveniently by the tin hydride or halogen atom transfer methods.

Halogen Atom Transfer Method.   The halogen atom transfer method is synthetically more useful than the hydrogen atom transfer method. The increased rate of the halogen transfer step results in a more efficient chain process. In addition, the halogen atom is retained in the final product. Suitable precursors for these cyclizations are iodides and polyhaloalkanes. With the latter, low-valent metals are often employed to promote the reaction. Copper(I) salts and ruthenium(II) complexes efficiently catalyze the cyclization of perchlorocarbonyl compounds such as trichloroacetates, trichloroacetamides, and α,α-dichloro esters. A recent review provides an extensive summary of the work in this area.[154] A representative example is shown in Eq. 95.[91]

(98%)   (Eq. 95)

Because there is no good hydrogen donor present, metal-promoted reactions have proven to be especially useful for conducting slow cyclizations. An illustrative example is provided in Eq. 96 where an unfavored bridged system is formed.[155]

(88%)   (Eq. 96)

Metal catalysis is not necessary if α-iodo carbonyl compounds are employed as radical precursors.[156] These iodo compounds are excellent substrates for halogen transfer reactions because abstraction of an iodine atom by the intermediate cyclic radical is exceptionally fast ($k > 10^7$ s^{-1}).[157] These reactions can be run at high concentrations, and they are usually initiated by photolysis in the presence of catalytic amounts of a hexaalkylditin. The cyclic products are formed in high yields under kinetic control and can be isolated or further transformed in situ. Again, this method is suited for conducting slow cyclizations, such as formation of large rings, formation of bridged ring systems, or cyclization of α-iodo esters. Equation 97 provides an example of a relatively slow 6-endo cyclization.[18] The intermediate iodide **83** is not isolated but is directly reduced with tin hydride.

(84%)

(Eq. 97)

However, cyclization of unstabilized radicals by the iodine transfer method is more restricted. In these cases, the intermediate radical and the initial radical are often of similar stability and therefore iodine atom transfer is rather slow. Nevertheless, iodine transfer cyclizations are possible if the initial radical is tertiary and the intermediate is primary or secondary. The reaction chains are usually short and the yields mediocre.[158] The isomerization reactions of alkynyl iodides to cyclic vinyl iodides are high-yielding and fast cyclizations (Eq. 98).[159,160] This

(65%)   (Eq. 98)

is due to the fast and exothermic abstraction of iodine atoms by the intermediate vinyl radical.

The atom transfer method is not restricted to the generation of carbon–centered radicals. Because oxygen–halogen and nitrogen–halogen bonds are relatively weak, they can be used for heteroatom radical cyclizations via the atom transfer method. N-Halo amines and amides can serve as precursors for aminyl and amidyl radicals.[64] The halogen transfer cyclizations are usually promoted by protonation or addition of low valent metals ($TiCl_3$, $CuCl/CuCl_2$).[161] A representative example of a copper-catalyzed cyclization is presented in Eq. 99.[162]

(66%)   (Eq. 99)

The products of these cyclization reactions are nitrogen mustards that are valuable precursors for further transformations via aziridinium ions.[163] Atom transfer cyclizations can also be conducted with N-nitrosoamines and N-nitroso-amides. These cyclizations are initiated by photolysis, and a nitroso group is transferred instead of halogen.[164] Atom transfer cyclizations involving oxygen-centered radicals employ hypohalites as precursors. These are generated in situ from the corresponding alcohols.[165,166] For example, this method has been used in the synthesis of the spiroketal 84 (Eq. 100).[61]

(53%)

84

(Eq. 100)

Radical/Radical Coupling Methods

In contrast to the methods discussed so far, radical/radical coupling methods are nonchain processes. Therefore, it is not sufficient to generate a small amount of radical to start the reaction; the intermediate radical must be generated in stoichiometric amounts. Control of selectivity is the basic problem in these methods. Radical recombinations usually have rates that are close to diffusion control. However, there are methods available that allow the preferred formation of cross-coupled products.

One example is the mixed Kolbe coupling reaction where one precursor is employed in excess (Eq. 101).[167] Because the rate of radical formation from both precursors is similar, the concentration of the radical derived from the excess acid will always be higher than the concentration of the cyclic radical. Therefore, the

(Eq. 101)

cyclic radical will preferentially give the cross-coupled products. However, this method is of only limited value and is probably restricted to examples where one of the acids is inexpensive.

A more general method for selective radical/radical coupling is the cyclization of organocobalt(III) complexes.

RCo(salen)py **85**   RCo(salophen)py **86**   RCo(dmgH)$_2$py **87**
Alkylcobalt salen   Alkylcobalt salophen   Alkylcobaloxime

The alkyl- or acylcobalt(III) salen **85**, salophen **86**, or dimethylglyoximato complexes **87** are air-stable compounds that are readily prepared by substitution reactions of halides or tosylates with the strongly nucleophilic cobalt(I) anions. The carbon–metal bond in these complexes is weak and easily cleaves upon irradiation or thermolysis. Bond homolysis yields a carbon-centered radical **88** and a paramagnetic cobalt(II) complex **89**.[137] The bond cleavage is reversible and the substrate complex can be reformed by rapid radical recombination. However, the lifetime of the alkyl radical is usually long enough so that radical cyclization can take place prior to recombination of the alkyl radical and the cobalt(II) fragment. The cyclization reaction is usually terminated by formation of a double bond by β-hydride elimination of a cobalt hydride complex **90** (Eq.102).

$$R^1Co(III)L \underset{}{\overset{h\nu \text{ or heat}}{\rightleftarrows}} R^1\bullet + Co(II)L \xrightarrow{\text{cyclization}} R^2\bullet + Co(II)L$$

85-87   **88**   **89**

$$R^2\bullet + Co(II)L \rightleftarrows R^2\text{-}Co(III)L \xrightarrow[\text{elimination}]{\beta\text{-hydride}} \text{alkene}$$

90

(Eq. 102)

An example of a 6-exo cyclization of the alkylcobaloxime **91** that was prepared by nucleophilic opening of an epoxide is shown in Eq. 103.[168]

(94%)   (Eq. 103)

With substrates that preferentially react via a $S_{RN}1$ mechanism (e.g., aryl iodides), cyclization occurs in the course of the synthesis of the complex, and the cyclic alkylcobalt(III) complex is the sole reaction product.[169] The cobalt group can either be eliminated or converted into a wide variety of functional groups, like hydroxy, halogen, oxime, phenylthio, and phenylseleno (Eq. 104).[170]

X = Se, S

(Eq. 104)

The retention of functionality in the product, either as a double bond or as the alkyl metal bond, is the main advantage of this method when compared to reductive cyclization methods.

A key feature for the success of the cobalt method is the relatively high concentration of the cobalt(II) complex **89** in the reaction mixture. Thus the desired cross-coupling of the alkyl radicals with cobalt(II) complex **89** is strongly favored over the unwanted cross-coupling of the alkyl radicals with themselves.[171]

In addition to the stoichiometric procedures, methods are available in which only catalytic amounts of cobalt complexes are used in the presence of a reducing agent (see Redox Methods).

Redox Methods

In redox methods, the initial radicals are generated and the cyclic radicals are trapped by electron-transfer processes. For a successful cyclization, both steps have to be selective. Therefore, substrates should be chosen so that the electronic properties of the initial and cyclic radical are different (e.g., the cyclic radical is more easily oxidized than the initial radical). This is usually achieved by incorporation of suitable substituents in the precursors.[154]

Reductive Methods.   In reductive methods, the initial radical is generated by a one-electron transfer to the precursor. This method is suited for the cyclization of nucleophilic radicals because they are not easily reduced. However, a second electron transfer occurs after the cyclization, thereby removing the cyclic radical. Depending on the method employed, this either results in the formation of protonated (reduced) products or organometallic intermediates which can be trapped with electrophiles.

An example of this type of reaction is the titanium-promoted cyclization of epoxides. The reaction proceeds via reductive opening of the epoxide by titanium(III) chloride and subsequent cyclization of the β-hydroxy radical. The cyclic radical is then reduced to an organometallic intermediate that can be trapped with electrophiles like iodine or a proton (Eq.105).[172,173]

(Eq. 105)

Similar transformations are possible with samarium(II) iodide. Precursors are usually aryl iodides or bromides from which the aryl radicals are generated by a single-electron transfer followed by loss of iodide. Again, the cyclic radical is reduced to an organometallic intermediate that can react with electrophiles.[174,175] In Eq. 106 the samarium complex is trapped by addition to a ketone.[176]

(Eq. 106)

The reductive method can be employed further for the cyclization of unsaturated aldehydes and ketones to cycloalkanols. The reaction proceeds via reduction of the carbonyl group to a ketyl radical anion followed by intramolecular addition to a carbon–carbon multiple bond. Potential side reactions include overreduction of the carbonyl group to the alcohol and dimerization of two ketyl radicals to a pinacol. Furthermore, many functional groups have to be protected because they are not compatible with the strongly reductive conditions. For electron transfer, chemical reductants [e.g., samarium(II) iodide, metals in ammonia] or electrochemical techniques are usually employed.[5] Samarium(II) iodide is popular because it combines practical advantages (solubility in organic solvents) with high reducing power and excellent chemoselectivity (nitriles, esters, alkynes, and many other functional groups are tolerated).[177] In addition, high stereoselectivities are often observed.[178] This is illustrated by the example in Eq. 107 in which the organometallic intermediate is trapped by acetone.[179]

(Eq. 107)

SmI$_2$, acetone

THF, -30 to 25°

(75%)

Furthermore, electron transfer can be induced photochemically under very mild reaction conditions.[180] Either hexamethylphosphoric triamide (HMPA) or triethylamine/acetonitrile can be employed as the electron donor. In a recent example, this method is used for the reduction of an α,β-unsaturated ketone (Eq. 108). Cyclization occurs exclusively at the β position of the carbonyl group.[181]

MeCN, Et$_3$N, $h\nu$

(53%)

(Eq. 108)

Another important class of reductive cyclizations is mediated by catalytic amounts of cobalt complexes in the presence of chemical or electrochemical coreductants. This type of reaction is closely related to the cyclization of organocobalt(III) complexes discussed in the preceding section on Radical/Radical Coupling Methods. The coreductant serves to recycle the nucleophilic Co(I) complexes which promote formation of the initial radical. The most common catalysts are vitamin B$_{12}$[182] or vitamin B$_{12}$-model systems like chlorocobaloxime **92**.[170] Either zinc or sodium borohydride can be employed as a chemical coreductant. Suitable precursors are alkyl or aryl bromides and iodides. The cobalt-catalyzed cyclization usually provides reduced products (Eq. 109).[183]

vitamin B$_{12}$

MeOH, LiClO$_4$, -1.8 V

(70%)

(Eq. 109)

However, if suitable substrates and reaction conditions[184] are used, alkenes are formed by hydridocobalt elimination from an intermediate alkylcobalt(III) complex (Eq. 110).[185,186]

(Eq. 110)

The possibility of retaining functionality in the product is an important advantage over other reductive radical cyclization methods. Furthermore, workup is often simplified because only catalytic amounts of metal byproducts have to be removed.

Oxidative Methods.   Oxidative cyclization methods are usually applied to the cyclization of electrophilic radicals because these radicals are relatively inert to further oxidation. The cyclic radicals are oxidized to the cation followed by quenching with a nucleophile. Therefore, the product often possesses a higher degree of functionality than the starting material. The long lifetime of the initial radical makes oxidative methods ideally suited for conducting slow cyclizations. Although the oxidation can be conducted by electrochemical techniques[187] or by photoinduced electron transfer,[188] metal salts are usually applied in synthesis. The manganese(III)-promoted oxidation of enolizable dicarbonyl compounds is one of the most useful methods. The initial radical is chemoselectively generated by oxidation of the most acidic carbon–hydrogen bond. The cation formed by oxidation of the cyclic radical can be trapped intramolecularly by electrophilic substitution[189] or lactonization.[190] An example of the latter reaction is shown in Eq. 111.[191]

(Eq. 111)

By addition of copper(II) salts to the reaction mixture, it is possible to obtain products that contain a double bond. The alkenes are probably formed by metal hydride elimination of an intermediate alkylcopper complex (Eq. 112).[192]

(Eq. 112)

The ability to conduct slow 8-endo cyclizations with the manganese(III)-promoted reaction is demonstrated by the example in Eq. 113.[193,194]

(Eq. 113)

The main disadvantages of oxidative cyclizations are the limited range of possible substrates, the use of large amounts of metal salts (2–3 equiv), and the relatively low yields.

Sequential Reactions

In sequential radical reactions, two or more radical reactions are connected in a reaction series. This is possible because most of the radical reactions, like fragmentation, cyclization, or addition, generate a new radical that can serve as a precursor for a second radical transformation. Actually, all radical cyclization methods are sequences of radical reactions, but here "sequential reactions" are defined as transformations in which a substrate undergoes two or more subsequent radical reactions, excluding steps that involve radical generation and radical removal.[4] The most interesting feature of these reactions is the possibility of constructing complicated structures in a single synthetic step. In general, all the restrictions that apply to single radical transformations are also valid for sequential reactions. Hence, all reactions have to be faster than radical recombination or reaction of the intermediate radicals with the solvent. In addition, the final radical, but not the intermediate radicals, must selectively be converted to a stable product. So far, inter- and intramolecular additions, intramolecular 1,5-hydrogen transfer reactions, and fragmentation reactions have been included in sequential radical reactions. The combination of two or more cyclization steps is the most prominent type of sequential reaction, and can be employed for the ready construction of polycyclic compounds. In principle, every method that is suitable for conducting radical cyclizations can be employed for tandem cyclization as well. Equations 114 and 115 present examples of tandem cyclizations conducted by the tin method[195] and by oxidation with manganese(III).[196]

(Eq. 114)

$$\text{(Eq. 115)} \quad (68\%)$$

In another type of sequential reaction, 1,5-hydrogen abstractions are coupled with radical cyclizations.[197] The hydrogen shift makes it possible to use a carbon–hydrogen bond as a radical precursor for a cyclization. This can be important if the usual precursor in this position is difficult to prepare or is unstable. For a successful transformation the hydrogen transfer step should be rapid to compete with unwanted side reactions. Therefore, reactive radicals like aryl, vinyl, or alkoxyl[198] radicals are generated in the initial step. In Eq. 116 an example is shown of the radical translocation reaction of an aryl iodide.[199]

$$(83\%) \quad \text{(Eq. 116)}$$

A useful class of sequential reactions is the combination of cyclization and fragmentation.[73] Depending on the structure of the substrate, the net result of this sequence is either a group migration[200] or a ring expansion. The latter reaction is especially useful for the formation of medium-sized rings which are not accessible by direct radical cyclizations. The ring expansion proceeds via radical cyclization to a carbonyl double bond followed by rapid β-bond cleavage of the alkoxy radical. The direction of the fragmentation is often controlled by substituents that make the final radical more stable than the starting radical. It is possible to expand rings by one, three, or four atoms. However, expansion by two atoms is not possible because the initial 4-exo cyclization is too slow. As the initial cyclization is often slow and reversible, syringe pump addition of tin hydride and fragmentation methods are suited for these sequences. Equation 117 shows an

$$(73\%) \quad \text{(Eq. 117)}$$

example of a one-carbon expansion reaction conducted by the tin method.[75] The ester group not only stabilizes the final radical but also accelerates the initial cyclization.

The example in Eq. 118 shows the formation of a 10-membered ring by a four-atom expansion of cyclohexenone derivative **93**.[76] Problems in this type of se-

(89%) (Eq. 118)

quence are often associated with direct reduction of the initial radical either by tin hydride or by an intramolecular 1,5-hydrogen transfer.

Sequences are also known where fragmentation precedes cyclization. In the most prominent method, the initial radical is formed by fragmentation of a cyclopropyl carbinyl radical. This sequence can be used for the generation of chiral radicals because the precursors can often be prepared in optically pure form by standard cyclopropanation methods. The cyclopropylcarbinyl radicals can be generated from α-cyclopropyl alcohols either by oxidation[103] or by the tin hydride method via thiohydroxamate **94** (Eq. 119).[201,202] Further suitable precursors are

(89%)

(Eq. 119)

cyclopropyl ketones **95** that are reduced by samarium(II) iodide. An example of this strategy is shown in Eq. 120.[203]

(77%)

(Eq. 120)

If an epoxide is used, this method allows the generation and cyclization of oxygen-centered radicals.[204]

The combination of inter- and intramolecular reaction steps has a great synthetic potential. For sequences where cyclization precedes addition, it is important that cyclization of the initial radical is faster than the competing intermolecular addition. Therefore, fast cyclization steps should be included in such a sequence. Furthermore, the addition step of the intermediate radical to the alkene acceptor has to be optimized, for example, by adjusting the concentration and the electronic properties of the alkene. An excess of activated alkene is often used to accelerate the intermolecular addition and thereby suppress reduction of the cyclic radical. Equation 121 shows an example of a cyclization/addition se-

(Eq. 121)

quence conducted by the fragmentation method.[205] In addition, there are examples that have been conducted by tin hydride[206,207] or reductive methods.[184]

The reverse sequence, addition followed by cyclization, assembles a new ring from two isolated precursors. In this type of reaction, the main problem is differentiation of the initial and final radicals of the sequence. For a successful annulation it is necessary that the initial (but not the final) radical adds to the alkene and that the final (but not the initial) radical selectively abstracts hydrogen from tin hydride. Selectivity is usually achieved by the introduction of substituents or by generating different types of initial and final radicals. This concept is demonstrated by the example presented in Eq. 122. In this reaction, the electrophilic initial radical preferentially adds to the electron-rich alkene **96**, whereas iodine abstraction by the final vinyl radical is faster than the competing addition to another alkene molecule. The intermediate vinyl iodide **97** was not isolated but directly reduced with tin hydride.[208]

(Eq. 122)

Addition/cyclization sequences can also be carried out by the tin hydride method,[69,209] by oxidative methods,[210,211] and by the thiohydroxamate method.[34] An example of the last reaction is shown in Eq. 123.[212]

In addition to sequences consisting of two radical reactions, it is possible to combine three radical reactions in one transformation. The sequence of the steps can differ, and combinations are known of fragmentation/hydrogen–transfer/cyclization,[213] addition/cyclization/addition,[214] cyclization/addition/cyclization,[215]

(Eq. 123)

and fragmentation/ addition/cyclization reactions. An example of one of these sophisticated sequences is shown in Eq. 124.[216] The reaction is promoted by reversible addition of a thiyl radical **99** to the double bond of **98** followed by addition to alkene **100** and cyclization. The reaction is terminated by rapid β fragmentation of **101** that liberates the chain-carrying butylthio radical **99**.

(Eq. 124)

COMPARISON WITH OTHER METHODS

There are too many cyclization methods in the literature for a complete comparison. For an overview of some methods see the review of C. and Y. Thebtaranonth.[216a]

Cationic cyclizations[216b] often lead to the thermodynamically controlled products. Sometimes Wagner–Meerwein rearrangements can occur. 5-Hexenyl cations cyclize in a 6-endo mode. Acidic conditions are used.

Radical reactions lead mainly to products of kinetic control (5-exo cyclization). The reaction conditions are neutral for several methods and various functional groups are tolerated. If organotin compounds are used, the removal of all tin residues can be a problem.

Anionic cyclizations occur under basic conditions and β elimination can be a competing reaction. 5-Hexenyl anions cyclize in a 5-exo mode,[216c,d] approximately 10^8 times slower than the corresponding radical cyclization.[216c] The intramolecular Michael addition was previously reviewed[216e] and is an alternative to radical cyclizations with activated olefins. Thermodynamically and kinetically controlled reactions are possible.

The metal-catalyzed cyclization reactions (mostly with palladium)[216a,f] often occur with high yield and under mild basic conditions. Substrates must usually be chosen to avoid β-hydride elimination.

These methods are compared in Table A. Included are some functional groups which can be used as precursors and functional groups which are tolerated under the reaction conditions.

Table A.   Comparison of Cyclization Methods

	Cationic	Radical	Anionic	Palladium
Precursor	Cl, Br, I, OH, OR, OSO$_2$R, C$=$C, C$=$O, C(OR)$_2$	[Cl]^a, Br, I, OH^b, OOR, SPh, SePh, NH$_2^b$, N$=$NR, N$=$C, NO$_2$, C$=$C, C$\equiv$C, C$=$O, CO$_2$H^b, anions, SnR$_3$, BR$_2$, HgX, CoL$_n$, H	Cl, Br, I, SPh, SePh, SnR$_3$, C$=$C, H	[Cl]^a, Br, I, OH, OAr, O$_2$CR, O$_2$COR, SO$_2$R, OP(O)(OR)$_2$, NO$_2$, C$=$C, cyclopropyl, epoxide, H
Tolerated functional groups		F, [Cl]^a, OH, OR, SR, SO$_2$R, NH$_2$, NR$_2$, CN, [C$=$O]^a, C(OR)$_2$, CO$_2$R, SiR$_3$	OR^c, C(OR)$_2$, SO$_2$R^c, NR$_2$, C$=$O^d	[OH]^a, OR, SO$_2$R, [NH$_2$]^a, NR$_2$, C$=$O, C(OR)$_2$, CO$_2$R
Side reactions	Wagner–Meerwein rearrangement; hydride transfer	Dimerization; hydrogen abstraction	β-Elimination; addition to C$=$O	β-Elimination; isomerization
pH	Acidic	Neutral	Strongly basic	Mildly basic
Preferred regiochemistry of 5-hexenyl cyclization	6-endo	5-exo	5-exo	5-exo
Product control	Thermodynamic	Kinetic	Both possible	Both possible

^{a}This group is sometimes useful.
^{b}Two steps are needed.
^{c}β-Elimination is possible in the precursor or the product.
^{d}Aldol condensation is possible.

EXPERIMENTAL CONDITIONS

Radical reactions require an inert atmosphere (nitrogen or argon). Normal bench/syringe technique is sufficient. Since there is little or no solvent dependence in radical reactions, all solvents (including water) can be used, as long as the reactants are soluble. As mentioned before attention must be paid to the half life time of the initiator. The concentrations of the reactants have to be adjusted well to optimize the expected reaction.[6,7]

EXPERIMENTAL PROCEDURES

2-(Phenylmethyl)cyclopentanone (103) (Formation of a Carbocycle by the Tin Method).[34] A solution of **102** (84.6 mg, 0.316 mmol), tributyltin hydride (190 μL, 0.706 mmol), and AIBN (7.5 mg, 0.045 mmol) in benzene [CAUTION, SUSPECTED CARCINOGEN] (8 mL) was heated at reflux for 1.5 hours. After cooling, the solvent was evaporated, and the residue was purified by flash chromatography (10% EtOAc–hexanes) to provide 49.5 mg (90%) of **103** and 6-phenyl-5-hexenal ($\geq$18:1 by ^{1}H NMR integration). IR (thin film) 2961, 1737, 1160, 690 cm^{-1}; ^{1}H NMR (CDCl$_3$) δ 7.28–7.13 (m, 5 H), 3.13 (dd, J = 13.7, 3.9 Hz, 1 H), 2.51 (dd, J = 13.7, 9.4 Hz, 1 H), 2.37–2.27 (m, 2 H), 2.15–2.02 (m, 2 H), 1.99–1.89 (m, 1 H), 1.75–1.67 (m, 1 H), 1.59–1.46 (m, 1 H); high-resolution mass spectrum, calculated for C$_{12}$H$_{14}$O (M$^+$) m/z 173.0966, found m/z 173.0996; mass spectrum m/z 174 (M$^+$), 156, 146, 130, 117.

(–)-(4R,5R)-2-Ethoxy-5-ethyl-4-methoxycarbonylmethyl-3,4,5,6-tetrahydro-2H-pyran (105) (Formation of a Heterocycle by the Tin Method).[93] A mixture of bromoacetal **104** (549 mg, 1.78 mmol), AIBN (30.3 mg, 0.185 mmol), and Bu$_3$SnH (0.65 mL, 2.42 mmol) in dry benzene (12 mL) was heated under reflux for 1 hour and then evaporated under reduced pressure. Silica gel column chromatography of the crude product with hexane–EtOAc (92:8) as eluant gave tetrahydropyran **105** (395 mg, 97 %) as an oily mixture of two diastereomers. $[\alpha]_D^{24}$ −7.98° (c 0.43, CHCl$_3$); IR (CHCl$_3$): 1732 cm^{-1}; ^{1}H NMR (CDCl$_3$) δ 4.77–4.89 (m, 0.6 H), 4.66–4.69 (m, 0.4 H), 3.40–4.44 (m,

4 H), 3.68 (s, 3 H), 2.61 (dd, J = 15.2, 4.2 Hz, 1 H), 2.51 (dd, J = 15.2, 3.8 Hz, 1 H), 0.73–1.06 (m, 3 H); mass spectrum: m/z 215 (M^+ – Me); Anal. Calcd for $C_{12}H_{22}O_4$: C, 62.6; H, 9.65. Found: C, 62.6; H, 9.7.

106   **107a/b**

83 : 17 *ratio*

Radical Cyclization of Carbohydrate Derivatives (Tin Method).[217] Bromide **106** (1.3 g, 2.5 mmol) and tributyltin hydride (2.4 equiv) were heated by dropwise addition of AIBN (cat.) in dry toluene at 110°. The reaction mixture was cooled and the solvent evaporated. The residue was dissolved in ether and 10% aqueous potassium fluoride solution was added, and the mixture was stirred for 18 hours. The organic phase was separated, dried, and evaporated. After chromatography (hexanes:EtOAc = 4:1) product **107a/b** was obtained in 52% yield (570 mg) as an 83:17 diastereomeric mixture. Recrystallization from hexane gave pure **107a**: mp 116–118° $[\alpha]_D^{24}$ S(24,D) –72° (c 4.5, $CHCl_3$); IR ($CHCl_3$) 3200, 3080, 1755 cm^{-1}; ^{1}H NMR ($CDCl_3$) δ 7.34–7.26 (m, 5 H), 5.44 (q, J = 3.1, 1 H), 5.35 (t, J = 10 Hz, 1 H), 5.21 (t, J = 10 Hz, 1 H), 4.92 (dd, J = 10 and 3.2 Hz, 1 H), 4.61 (s, 2 H), 3.25 (ddd, J = 10.1, 12.4 and 4.2 Hz, 1 H), 2.10, 2.03, 2.01, 1.98 (s, s, s, s, 4 × 3 H), 1.88 (ddd, J = 3.0, 12.4 and 14.8 Hz, 2 H); ^{13}C NMR ($CDCl_3$) δ 170.23, 170.05, 169.97, 169.91, 139.14, 128.67, 128.55, 128.14, 76.99, 71.87, 71.72, 71.20, 67.82, 56.87, 29.78, 21.01, 20.84, 20.75, 20.66; mass spectrum: m/z 438 (M^+); Anal. Calcd for $C_{21}H_{27}NO_9$: C, 57.66; H, 6.17; N, 3.20. Found: C, 57.27; H, 6.12; N, 3.50.

108   **109**

(11E)-3,4,5,6,9,10-Hexahydro-8H-2-benzooxacyclotetradecyne-1,7-dione (109) (Macrocyclization Mediated by Tin Hydride).[127] A solution of tributyltin hydride (157 mg, 0.54 mmol) and AIBN (5 mg, 0.03 mmol)) in dry, degassed benzene (10 mL) was added over 8 hours via syringe pump to a solution of **108** (98 mg, 0.28 mmol) in dry, degassed benzene (100 mL) heated under reflux under an atmosphere of nitrogen. The cooled mixture was evaporated under reduced pressure and the residue was then dissolved in diethyl ether. The ether solution was stirred with 20% aq. KF for 24 hours and filtered, and the organic layer was separated, dried, and evaporated to leave a yellow oil. Purification by chromatography on silica gel using 25% Et$_2$O–hexanes as eluent gave 47 mg (61%) macrolide **109** as white needles, mp 78–80° (pentane); IR (CHCl$_3$) 2920, 1720, 1600, 1460, 1250, and 1075 cm^{-1}; ^{1}H NMR (CDCl$_3$) δ 7.78 (m, 1 H), 7.60–7.15 (m, 3 H), 6.88 (d, J = 16 Hz, 1 H), 5.88 (dt, J = 16, 7 Hz, 1 H), 4.50–4.20 (m, 2 H), 2.65–2.10 (m, 6 H), 2.04–1.50 (m, 6 H); ^{13}C NMR (CDCl$_3$) δ 211.2 (s), 168.7 (s), 138.2 (s) 132.3 (d), 132.0 (d), 131.4 (d), 130.9 (d), 129.1 (s), 127.4 (d), 127.0 (d), 65.7 (t), 43.2 (t), 37.1 (t), 31.3 (t), 27.4 (t), 22.7 (t), 21.5(t); mass spectrum: m/z 272 (M$^+$); Anal. Calcd for C$_{17}$H$_{20}$O$_3$: C, 75.0; H, 7.4. Found: C, 75.1; H, 7.7.

(11E,3S)-14,16-Dimethoxy-3-methyl-3,4,5,6,9,10-hexahydro-8H-2-benzooxacyclotetradecyne-1,7-dione (Zearalenone) (111) (Macrocyclization Mediated by a Silane Hydride).[127]   A solution of tris(trimethylsilyl)silane (35 mg, 0.14 mmol) and AIBN (5 mg, 0.03 mmol) in dry toluene (2 mL) was added over 8 hours via syringe pump to a stirred solution of **110** (41 mg, 0.10 mmol) in dry, degassed toluene (35 mL) at 85° under nitrogen. The solution was heated at 85° for 2 hours and then cooled and evaporated to leave a residue which was purified by chromatography on silica gel using 50% Et$_2$O–hexanes as eluent to give 18 mg (55%) of zearalenone (**111**) as a white crystalline solid: mp 110–111° (Et$_2$O and hexane).

[α]$_D^{24}$ +47.8° (c 1.0, CHCl$_3$); IR 2930, 1710, 1600, 1575 cm^{-1}; ^{1}H NMR (CDCl$_3$) δ 6.55 (d, 1 H, J = 2 Hz), 6.30 (d, 1 H, J = 2 Hz), 6.28 (d, 1 H, J = 16 Hz), 6.05–5.90 (m, 1 H), 5.46–5.06 (m, 1 H), 3.82 (s, 3 H), 3.79 (s, 3 H), 2.90–1.90 (m, 12 H), 1.33 (d, 3 H, J = 7 Hz); ^{13}C NMR δ 211.2 (s), 167.4 (s), 161.1 (s), 157.4 (s), 136.5 (s), 133.0 (d), 128.8 (d), 101.0 (d), 97.5 (d), 71.0 (d), 55.7 (q), 55.2 (q), 43.8 (t), 38.3 (t), 34.9 (t), 31.0 (t), 21.5 (t), 21.1 (t), 19.8 (q); high-resolution mass spectrum, calculated for C$_{20}$H$_{26}$O$_5$ (M$^+$) m/z 346.1780, found m/z 346.1776.

cis/trans-(1,1-Diethoxycarbonyl)-4-(trimethylstannylmethyl)-3-benzyl-cyclopentane (113) (Carbocyclization of a Diene with Tin Hydride).[218]   To a solution of **112** (1.3 g, 5.1 mmol) in 20 mL of anhydrous tert-butyl alcohol under argon, was added trimethyltin chloride (4 g, 10.0 mmol), sodium cyanoborohydride (0.95 g, 15.0 mmol) and AIBN (15 mg, 1.0 mmol). The solution was refluxed for 1 hour (until the diene had disappeared as monitored by TLC using KMnO$_4$ spraying agent). The solution was cooled to room temperature, quenched with 5% aqueous ammonia solution, stirred, and then concentrated under reduced pressure. The residue was dissolved in ether, washed three times with brine, dried (MgSO$_4$), and concentrated. Flash chromatography (3% EtOAc–hexanes) afforded 2.3 g (95%) of stannylated product **113** as an inseparable mixture in the form of a colorless oil (*cis*:*trans* = 4:1 by NMR). IR 3030, 3010, 2980, 2930, 1740, 1607, 1500, 1205, 760, 745, 695 cm^{-1}; ^{1}H NMR (CDCl$_3$) δ 7.14–7.32 (m, 5 H), 4.09–4.26 (m, 4 H), 2.97 (dd, J_{trans} = 13.9, 3.2 Hz, 2 H), 2.79 (dd, J_{cis} = 12.7, 3.8 Hz, 2 H), 2.21–2.46 (m, 2 H), 1.97 (m, 4 H), 1.17–1.27 m, 6 H), 0.99 (d, J = 8.7 Hz, 2 H); ^{13}C NMR δ 172.8 (C = O *cis*), 172.5 (C = O *trans*), 141.3, 140.9, 128.9, 128.8, 128.1, 125.6, 125.5 (arom C's *cis* and *trans*), 61.1 (CH$_2$ from Et *cis* and *trans*), 58.5 (C1 *cis*), 57.7 (C1 *trans*), 50.3 (C3 *trans*), 45.7 (C3 *cis*), 43.8 (C4 *trans*), 43.0 (benzylic C *trans*), 41.6 (benzylic C *cis*), 40.7 (C4 *cis*), 39.7 (C2 *trans*), 39.4 (C5 *trans*), 37.3 (C5 *cis*), 34.8 (C2 *cis*), 14.9 (CH$_2$Sn *trans*), 13.9 (Me from Et *cis* and *trans*), 11.4 (CH$_2$Sn *cis*), −9.6 (MeSn *trans*), −9.8 (MeSn *cis*); high-resolution mass spectra, calculated for C$_{22}$H$_{34}$O$_4$Sn (M$^+$) *m/z* 482.1478, found *m/z* 482.1463.

114   **115a/b** 52:48 ratio

Diethyl 4-(Bromomethyl)-3-(methoxycarbonyl)-3-(tosylmethyl)-cyclopentane-1,1-dicarboxylate (115a/b) (Cyclization of a Diene with Tosyl Bromide).[219] Irradiation (high-pressure mercury lamp) of a solution of diene **114** (1.5 g, 5 mmol) and tosyl bromide (1.18 g, 5 mmol) in 140 mL of acetonitrile for 24 hours, purification on silica gel using EtOAc–petroleum ether (15:85 to 40:60), gave 1.79 g (67%) of **115a/b**. The 52:48 ratio was determined by analytical HPLC (EtOAc:isooctane, 25:75; 2 mL/min). Anal. Calcd for C$_{22}$H$_{29}$BrO$_2$8S: C, 49.54; H, 5.48; S, 6.01 Found: C, 49.49; H, 5.50; S, 5.90.

Product **115a**: ^{1}H NMR (CDCl$_3$) δ 7.75 (d, J = 8.0 Hz, 2 H), 7.36 (d, J = 8.0 Hz, 2 H), 4.30–4.14 (m, 4 H), 3.73 (s, 3 H), 3.60 (d, J = 13.7 Hz, 1 H), 3.57 (dd, J = 9.8, 3.6 Hz, 1 H), 3.38 (d, J = 15.2 Hz, 1 H), 3.35 (d, J = 13.7, 1 H), 3.11 (t, J = 9.8, 1 H), 3.04 (d, J = 15.2, 1 H), 2.73–2.60 (m, 2 H), 2.45 (s, 3 H), 2.48–2.39 (m, 1 H), 1.27 (t, J = 7.1, 6 H); ^{13}C NMR (CDCl$_3$) δ 172.6, 171.7, 171.0, 144.8, 137.6, 129.8, 127.6, 62.1, 61.7, 57.4, 56.7, 53.3, 52.8, 49.7, 39.5, 37.5, 31.0, 21.3, 13.7.

Product **115b**: ^{1}H NMR (CDCl$_3$) δ: 7.77 (d, J = 8.0 Hz, 2 H), 7.36 (d, J = 8.0 Hz, 2 H), 4.31–4.14 (m, 4 H), 4.06 (d, J = 13.8 Hz, 1 H), 3.68 (s, 3 H), 3.40 (dd, J = 10.3, 4.0 Hz, 1 H), 3.34 (d, J = 14.9 Hz, 2 H), 3.29 (d, J = 13.8, 1 H), 3.03 (t, J = 10.3, 9.4 Hz, 1 H), 2.89 (d, J = 14.9, 1 H), 2.45 (s, 3 H), 2.60–2.39 (m, 2 H), 2.24 (t, J = 11.3, 1 H), 1.29 (t, J = 7.1, 3 H), 1.25 (t, J = 7.1 Hz,

3 H); ^{13}C NMR (CDCl$_3$) δ 172.6, 171.6, 170.8, 144.9, 137.8, 129.9, 127.8, 63.3, 62.1, 61.7, 57.5, 53.8, 52.6, 51.7, 40.8, 37.9, 30.9, 21.6, 14.1, 13.9.

Photolysis of (R)-(+)-3-Citronoyloxy-4-methylthiazolin-2-(3H)-thione (116) [Cyclization by the Barton Method].[220] A solution of ester **116** (1.55 g, 5.18 mmol) in ether (20 mL) was irradiated under nitrogen at room temperature with a 100-W medium pressure mercury lamp for 45 minutes. The solvent was then evaporated and the residue chromatographed on silica with pentane–ether (95:5) to give the cyclization product **117** as a 1:1 mixture of diastereomers. The mixture was a colorless analytically pure oil (1.08 g, 82%). IR (film) 3070, 1510, 1445, 1365, 1295, 1140, 1020 cm^{-1}; ^{1}H NMR (CDCl$_3$) δ 7.05 (m, 1 H), 2.55 (m, 3 H), 2.40–1.20 (m, 8 H), 1.40 (s, 6 H), 1.02, and 1.00 (2 d, J = 7.0 Hz ca. 1:1, 3 H); mass spectrum: m/z 255 (M$^+$), 132, 131; Anal. Calcd for C$_{13}$H$_{21}$NS$_2$: C, 61.42; H, 8.28; N, 5.48; S, 25.10 Found: C, 61.1; H, 8.3; N, 5.65; S, 24.95%.

N-Butyl-2-[(2-pyridylthio)methyl]pyrrolidine (119) (Cyclization of an Aminyl Radical).[152] Compound **118** (0.30 g, 1.0 mmol) and malonic acid (0.31 g, 3.0 mmol) were weighed into a round-bottomed flask containing a small stirring bar. The flask was sealed with a septum, wrapped with aluminium foil to exclude light, and purged with nitrogen. Acetonitrile (20 mL) was then added via syringe. The shield was removed, and the mixture was stirred and irradiated with a 100- or 150-W tungsten filament bulb from a distance of about 0.5 m. The reaction was monitored for disappearance of the PTOC carbamate **118** by TLC. When the reaction was judged to be complete, solvent was removed at reduced pressure. The residue was partitioned between ether and 10% aqueous HCl. The aqueous portion was basified and extracted with several portions of ether. The combined ethereal extract was washed with saturated NaCl solution and dried (K$_2$CO$_3$). Purification by chromatography (silica gel, ether elution) gave 0.23 g (0.92 mmol, 92 %) of **119** as a heavy oil. ^{1}H NMR (CDCl$_3$) δ 8.37 (dd, 1 H), 7.40 (dt, 1 H), 7.13 (dd, 1 H), 6.90 (dt, 1 H), 3.59 (dd, 1 H), 3.13 (dt, 1 H), 3.0–2.8 (m, 2 H), 2.60 (m, 1 H), 2.14 (m, 2 H), 2.0–1.2 (m, 8 H), 0.88 (t, 3 H); ^{13}C NMR (CDCl$_3$) δ 160.0, 149.8, 136.2, 122.7, 119.6, 63.8, 54.8, 54.6, 34.4, 31.1, 30.4, 22.7, 21.0, 14.3; mass spectrum: m/z 139, 126, 96.

120   **121** *trans:cis* = 4.7:1

Ethyl *cis/trans*-1-Chloro-2-(chloromethyl)cyclopentane-1-carboxylate (121) (Ruthenium-Catalyzed Cyclization; Atom Transfer Method).[221] General Procedure: Dichloro compound **120** and a transition metal catalyst (8.3 mol%) were placed in a resealable Pyrex™ tube, and 1.0 mL of benzene was added. The mixture was degassed via three freeze/thaw cycles, and the tube was sealed under vacuum, placed in an oil bath, and heated at 155–160° for 22 hours. After cooling, the vessel was opened under argon. A small aliquot was passed through a plug of Florisil™ and was analyzed by GLC to determine if the reaction was complete. If the starting dichloro compound **121** remained, the solution was degassed again and heated as before. Once the reaction was complete, hexane (3 mL) and benzene (2 mL) were added to the solution, and the mixture was filtered through a 3-cm plug of Florisil, which was washed with benzene (3 mL), and the solvent was removed in vacuo. Crude product mixtures were analyzed by GLC. Pure cyclic esters were isolated by preparative TLC, eluting once with 9:1 hexanes–EtOAc and once with 9:1 hexanes–methylene chloride.

trans-**121a**: (61%), bp 50–55° (< 0.1 torr, bulb-to-bulb); IR (film) 2975, 2870, 1735, 1445, 1370, 1200, 1095, 1040, 920, 755 cm^{-1}; ^{1}H NMR (CDCl$_3$) δ 4.26 (q, J = 7.1 Hz, 2 H), 3.82 (dd, J = 11.0, 6.4 Hz, 1 H), 3.54 (dd, J = 10.8, 7.8 Hz, 1 H), 2.87 (m, 1 H), 2.49–1.59 (m, 6 H), 1.32 (t, J = 7.1 Hz, 3 H); ^{13}C NMR (CDCl$_3$) δ 170.5, 77.2, 62.4, 51.6, 44.8, 41.9, 28.2, 21.0, 13.9; mass spectrum: *m/z* 226, 224, 189, 156, 153, 135, 115, 79, 41, 28; high-resolution mass spectrum, calculated for C$_9$H$_{14}$O$_2$Cl *m/z* 224.0371, found *m/z* 224.0372.

cis-**121b**: (13%), bp 50–55° (<0.1 torr, bulb-to-bulb); IR (film) 2975, 2870, 1735, 1445, 1370, 1325, 1265, 1200, 1080, 1035, 1020, 915, 860, 750 cm^{-1}; ^{1}H NMR (CDCl$_3$) δ 4.25 (q, J = 7.1 Hz, 2H), 3.71 (dd, J = 10.9, 4.3 Hz, 1H), 3.43 (dd, J = 10.9, 9.1 Hz, 1H), 2.77 (m, 1H), 2.57 (m, 1H), 2.32–1.61 (m, 5H) 1.33 (t, J = 7.1 Hz, 3 H); ^{13}C NMR (CDCl$_3$) δ 169.5, 74.4, 62.2, 55.0, 44.6, 40.0, 28.1, 21.1, 13.9; mass spectrum: *m/z* 226, 224, 189, 156, 153, 135, 115, 79, 41, 28; high resolution mass spectrum, calculated for C$_9$H$_{14}$O$_2$Cl *m/z* 224.0371, found *m/z* 224.0374.

122   **123**

(1*R*,8*S*)-1-Chloromethyl-2,2-dichloro-3-oxohexahydropyrrolizidine (123) (Copper-Catalyzed Cyclization; Atom Transfer Method).[222] A suspension of **122** (500 mg, 2.06 mmol) and recently recrystallized copper(I) chloride (194 mg, 1.96 mmol) in 14 mL of deoxygenated acetonitrile was heated in a sealed tube with a Teflon tap at 160° for 2 hours. After cooling, solvent was removed under reduced pressure and the residue purified by flash chromatography (hexanes–EtOAc = 4:1) to afford 467 mg (93%) of product **123** as a white solid mp 60–62°. $[\alpha]_D^{24}$ S(24,D) $-24°$ ($c = 1.0$, CH_2Cl_2); IR 2972, 1918, 1413, 843 cm^{-1}; ^{1}H NMR ($CDCl_3$) δ 3.98 (dd, $J = 11.3$, 4.2 Hz, 1 H), 3.86 (dd, $J = 11.3$, 10.3 Hz, 1 H), 3.68 (m, 1 H), 3.55 (m, 1 H), 3.28 (ddd, $J = 12.1$, 8.9, 3.2 Hz, 1 H), 2.76 (m, 1 H), 2.17 (m, 2 H), 1.60 (m, 1 H); mass spectrum: m/z 247, 245, 243, 241, 206. Anal. Calcd for $C_8H_{10}Cl_3NO$: C, 39.62; H, 4.16; N, 5.78. Found: C, 39.67; H, 3.98; 5.53.

2-Methyl-2-phenyl-4-methylenetetrahydrofuran (125) (Cobalt-Mediated Cyclization; Radical/Radical Coupling).[223] To a solution of bromide **124** (2.53g, 10 mmol) in ethanol (50 mL) were added aqueous sodium hydroxide (10 N, 1 mL) and sodium borohydride (380 mg, 10 mmol). The solution was warmed to 50° under nitrogen, and powdered chlorocobaloxime(III) (240 mg, 0.6 mmol) was added in portions over 1 hour. The temperature of the mixture was kept at 50–60°. After the addition, the mixture was stirred for 30 minutes at the same temperature. Most of the ethanol was removed under reduced pressure, and saturated NaCl (50 mL) was added. The mixture was extracted with pentane–ether (4:1) several times. The extracts were washed with saturated NaCl and dried (Na_2SO_4). After evaporation of the solvents, the residue was distilled under reduced pressure to give **125** (1.27 g, 73%); bp 62–63° (0.5 mm Hg). IR (CCl_4) 1671, 1044, 884, 700 cm^{-1}; ^{1}H NMR (CCl_4) δ 7.50–7.20 (m, 5 H), 4.97 (t, $J = 2$ Hz, 1 H), 4.87 (t, $J = 2$ Hz, 1 H), 4.42 (broad t, $J = 15$ Hz, 2 H), 2.91 (broad d, $J = 16$ Hz, 1 H), 2.70 (broad d, $J = 16$ Hz, 1 H), 1.48 (s, 3 H); HRMS, calcd for $C_{12}H_{14}O$ M$^+$): m/z 174.1043, found m/z 174.1041.

trans-1-[(2*R*/*S*,3*aR*,4*R*,5*R*,6*aS*)-5-([(1,1-Dimethylethyl)-dimethylsilyl]oxy)-2-ethoxyhexahydro-2*H*-cyclopenta[*b*]furan-4-yl]octen-3-one (127) (Reductive Cyclization Catalyzed by Vitamin B$_{12}$).[224]** The cathode compartment of an electrochemical cell containing C-felt (2.77 g) as cathode material was charged with a solution of vitamin B_{12} (250 mg, 0.18 mmol) in 0.3 M $LiClO_4$/DMF (250 mL). The anode compartment was charged with 0.3 M $LiClO_4$/DMF (ca. 50 mL). The B_{12} was reduced to Co(I) at a constant cathode potential of -1.4 V (vs. SCE) until the initial current of about 50 mA had diminished to a stable background level and the color had changed from red to dark green. On reducing the potential to -0.9 V a stable background level of approximately 3.5 mA was observed. Keeping the potential at -0.9 V, 1-octyn-3-one (3.72 g, 30.0 mmol) was added, followed by bromoacetal **126** (3.65 g, 10.0 mmol) after 15 minutes. Thereby the color changed to dark red and the current remained at a low level. After switching on the 250-W halogen lamp the current rose to 30 mA and the temperature to $40-41°$, which was maintained throughout the reaction. Additional alkyne was added after 5 hours (2.48 g, 20.0 mmol) and again after 25 hours (1.24 g, 10.0 mmol). After 44 hours the current had diminished to a constant background level of 9 mA, and 1543 Cb e$^-$ (16.0 mmol) had been consumed. The solution was poured into ice water (750 mL) and extracted with Et_2O (6 × 100 mL). The organic solution was washed with brine (3 × 100 mL), dried (Na_2SO_4), and the solvent removed. Excess alkyne was removed by bulb-to-bulb distillation (45°oven temperature/5 × 10^{-2} mbar) and the residual dark brown liquid (8.86 g) was filtered through silica gel (70 g; low-boiling petroleum ether/Et_2O 8:2). Flash chromatography of the crude product (219 g silica gel; low-boiling petroleum ether/Et_2O 19:1 to 8:1) afforded two main fractions. The latter was the desired product **127**: 3.5 g of a yellow liquid comprising four isomers of **127** (containing at least 65 % product by GC). For *cis-trans* isomerization the latter fraction was dissolved in hexanes (50 mL) containing I_2 (90 mg), washed with brine containing a small amount of sodium thiosulfate (2 × 25 mL), and brine (25 mL), dried (Na_2SO_4), and the solvent removed. Flash chromatography (150 g; low-boiling petroleum ether–Et_2O 19:1 to 8:2) of the remaining dark liquid (3.36 g) afforded essentially pure product [2 diastereomers, 1.07 g (26 %) and 0.85 g (21 %)] as slightly yellow oils (assignment of configuration was based on NOE experiments). For elemental analysis a sample was distilled in a glass tube at 150° (oven temperature/5 × 10^{-3} mbar). Anal. Calcd for $C_{23}H_{42}O_4Si$: C, 67.27; H, 10.31. Found: C, 67.27; H, 10.23; IR (neat) 3450, 2960, 2930, 2860, 2250, 1695, 1680, 1630, 1465, 1405, 1380, 1360, 1250, 1115, 1050, 1005, 910, 865, 835, 775, 730, 670, 645 cm^{-1}; ^{1}H NMR (CDCl$_3$) δ 6.64 (dd, $J = 15.83, 8.05$ Hz, 1 H), 6.09 (dd, $J = 15.84, 0.96$, 1 H), 5.19 (d, $J = 4.91$ Hz, 1 H), 4.42 (ddd, $J = 7.02, 7.02, 4.20$ Hz, 1 H), 3.89 (ddd, $J = 8.15, 8.15, 8.15$ Hz, 1 H), 3.67 (dqd, $J = 9.61, 7.11, 0.76$, 2 H), 3.48–3.34 (m, 2 H), 2.48 (t, $J = 7.45$, 1 H), 2.45–2.25 (m, 3 H), 2.01 (ddd, $J = 13.53, 8.82, <1$ Hz, 1 H), 1.84 (ddd, $J = 13.48, 4.83, 4.83$ Hz, 1 H), 1.72 (ddd, $J = 13.48, 8.86, 4.34$ Hz, 1 H), 1.65–1.50 (m, 1 H), 1.35–1.10 (m, 2 H), 1.15 (td, $J = 7.07, 0.77$, 3 H), 0.95–0.70 (m, 1 H), 0.82 (s, 9 H), -0.02 (s, 6 H); ^{13}C NMR (CDCl$_3$) δ 146.2, 130.8, 105.1, 79.5, 78.1, 62.6, 57.0, 44.3, 41.0, 40.4, 38.4, 31.5, 25.7, 24.0, 22.5, 18.0, 15.2, 13.9, -4.5, -4.7; mass spectrum: m/z 365, 354, 353, 309, 308, 307, 281, 234, 233, 199, 189, 187, 178, 161, 153, 151, 136, 135, 133, 131, 129, 117, 107, 105, 101, 99, 91, 81, 79, 75, 73, 72, 71, 59, 55, 44, 43, 41.

128   **129**

***trans*-2-Ethenyl-1,4,4-trimethylcyclopentanol (129) (Reductive Cycliza-
tion with NaC$_{10}$H$_8$).**[225]   General Procedure: To a solution of dry recrystallized
naphthalene (1.92 g, 15.0 mmol) in dry tetrahydrofuran (20 mL), stirred under
nitrogen at room temperature, was added freshly cut sodium (580 mg,
25.2 mmol) in small pieces. The resulting green solution was then stirred at room
temperature for 3 hours. This routinely gave a 0.6 M solution of the reagent. The
solution of sodium naphthalenide (0.6 M, 2.8 mL, 1.68 mmol) was added dropwise under nitrogen to a well-stirred solution of **128** (304.2 mg, 2.0 mmol) in dry
tetrahydrofuran (11 mL) at room temperature until a faint green end point was
reached. The coloration discharged in about 3 minutes after turning off the nitrogen. The mixture was poured into water (30 mL) and then extracted with ether
(3 × 25 mL). The combined ether extracts were washed with dilute hydrochloric
acid (30 mL), followed by water until they were neutral. Evaporation of the dried
extracts gave the crude product, which was purified by column chromatography
on silica gel using ether hexane (1:1) as eluant, giving 97.1 mg (55.8%) of the
desired alcohol **129** (eluted second) and 132.6 mg (22.1%) of recovered starting
material (eluted first). IR 3370, 1540 cm^{-1}; ^{1}H NMR (CDCl$_3$) δ 5.84 (m, 1 H),
5.08 (m, 2 H), 2.69 (m, 1 H), 2.20–0.69 (m, 5 H), 1.16 (s, 3 H), 1.14 (s, 3 H), 1.06
(s, 3 H); ^{13}C NMR (CDCl$_3$) δ 138.1 (d), 115.6 (t), 81.01, 56.04 (t), 55.0 (d), 45.4
(t), 37.8, 32.1 (q), 31.6 (q), 25.1 (q); high resolution mass spectrum, calculated for
C$_{10}$H$_{18}$O m/z 154.1357, found m/z 154.1351.

130   **131** diastereomeric ratio >150:1

**(1R^*,2R^*)-Dimethylcyclopentan-1-ol (131) (Reductive Cyclization with
Samarium(II) Iodide).**[178]   *General Procedure for Preparation of SmI$_2$/THF/
HMPA:* Samarium metal (0.30 g, 2.0 mmol) was added under a flow of argon to
an oven-dried round-bottomed flask containing a magnetic stirring bar and septum inlet. The flask and samarium were gently flame dried and cooled under argon. To the samarium was added 13 mL of tetrahydrofuran followed by
methylene iodide (0.482 g, 1.8 mmol), and the mixture was stirred at room
temperature for 1.5 hours. HMPA [CAUTION, SUSPECTED CARCINOGEN]
(2.51 mL, 14.5 mmol) was added, and the resulting purple solution was stirred 10
minutes before addition of the olefinic ketone.
   General Procedure for Cyclization of Olefinic Ketones: To the SmI$_2$ solution
described above was added olefinic ketone **130** (0.116 g, 1.04 mmol) and *t*-BuOH

(0.160 g, 2.16 mmol) in 14 mL of THF over a 15-minute period. Upon completion, the reaction was quenched with saturated aqueous NaHCO$_3$ and the aqueous layer was extracted with Et$_2$O. The combined organic layers were washed with water and brine, dried over MgSO$_4$, and the solvent was removed under reduced pressure. Final purification involved filtering through a short column of Florisil to remove residual HMPA. Kugelrohr distillation gave 0.102 g (86%) of desired product **131**: bp 60° (25 mm Hg). IR (CCl$_4$) 3419, 2940, 2856 cm^{-1}; ^{1}H NMR (CDCl$_3$) δ 1.96–1.48 (m, 6H), 1.82 (br s, 1H), 1.20–1.12 (m, 1 H), 1.09 (s, 3 H), 0.82 (d, J = 7.1 Hz, 3 H); ^{13}C NMR (CDCl$_3$) δ 80.90, 44.67, 40.06, 31.73, 22.71, 20.45, 15.40. High resolution mass spectrum, calculated for C$_7$H$_{14}$O m/z 114.1045, found m/z 114.1051; Low resolution mass spectrum m/z 114, 85, 71, 58.

Photosensitized Reductive Cyclization of Aminoethylcyclohexenones.[226] Solutions of the (aminoethyl)cyclohexenones (2 mmol) in MeOH (with and without added MeCN) containing 9,10-dicyanoanthracene (DCA) (ca. 1 × 10^{-4} M) were irradiated with uranium glass-filtered light. Irradiations were monitored by GLC and UV and terminated when >95% of starting material was consumed. The photolysates were concentrated in vacuo and filtered to remove DCA. The filtrates were subjected to an acid–base extraction procedure to separate the amine products. The amine-containing fractions were subjected to chromatographic (flash column) separation to provide the photoproducts.

Irradiation of N-Benzyl-N-trimethylsilylmethylaminoethylcyclohexenone (**132**): A solution of **132** (1 mmol) in MeCN containing DCA (6.6 × 10^{-2} mmol) was irradiated for 4 hours (78% conversion). Product yields were determined by GC with triphenylene as internal standard. Workup followed by preparative TLC (silica gel, 1:10 EtOAc–hexanes) separation afforded cyclized products. **133**: IR 2930, 2870, 2800, 2760, 1710, 1495, 1450, 1440, 1465, 1310, 1270, 1165, 1070, 1025, 985 cm^{-1}; ^{1}H NMR (CDCl$_3$) δ 7.33–7.21 (m, 5 H), 3.52 and 3.45 (q, J = 13.2 Hz, 2 H), 2.97 (br d, J = 11.3 Hz, 1 H), 2.89 (br d, J = 10.5 Hz), 2.36 (m, 1 H), 2.29 (ddd, J = 13.8, 13.5, 6.6 Hz, 1 H), 2.05 (m, 1 H), 1.93 (ddd, J = 12.5, 11.3, 2.7 Hz, 1 H), 1.91 (ddd, J = 11.7, 11.6, 3.4 Hz, 1 H), 1.80 (m, 2 H), 1.78 (dd, J = 10.4, 10.5 Hz, 1 H), 1.74–1.64 (m, 3 H), 1.60 (dddd, J = 13.1, 12.5, 11.7, 3.9 Hz, 1 H), 1.38 (dddd, J = 13.7, 13.6, 13.5, 3.8 Hz, 1 H); ^{13}C NMR (CDCl$_3$) δ 221.3, 138.3, 129.0, 128.2, 127.0, 63.0, 60.0, 53.4, 53.3, 43.3, 41.5, 29.6, 26.2, 24.7. MS m/z: 243, 201, 159, 146, 134, 113, 91; HRMS m/z 243.1620 (C$_{16}$H$_{21}$NO requires 243.1623).

Methyl *trans*-2-Oxo-6-[1*E*-propenyl]cyclohexanecarboxylate (135) [Oxidative Cyclization with Manganese).[192] To a solution of Mn(OAc)$_3$·2H$_2$O (1.376 g, 5.10 mmol) and Cu(OAc)$_2$ (0.510g, 2.55 mmol) in 18 mL of glacial acetic acid was added a solution of **134** (0.505 g, 2.55 mmol) in 7 mL of glacial acetic acid to give an opaque brownish green solution containing some undissolved Mn(OAc)$_3$·2H$_2$O. The mixture was stirred for 1 hour at 50° at which time the solution was light blue and contained a white precipitate. Water was added to give a single cloudy phase in which the white precipitate had dissolved. The solution was extracted with five 15-mL portions of methylene chloride. The combined organic layers were washed with saturated aqueous sodium bicarbonate solution until neutral and then with water. The aqueous layer was back extracted with two 15 mL-portions of methylene chloride. The combined organic layers were dried over MgSO$_4$, and the solvent was removed in vacuo to provide 0.512 g of crude cyclized product **135**. Flash chromatography on silica gel (3:1 hexanes–Et$_2$O) gave 0.365 g (71%) of **135** as a 1.3:1 mixture of keto and enol tautomers. The keto and enol tautomers were partially separated by flash chromatography but equilibrated at 25° after 15 days: IR (neat) 1745, 1715 cm^{-1}; Anal. Calcd for C$_{11}$H$_{16}$O$_3$: C, 67.32; H, 8.22. Found: C, 66.90; H, 8.33%.

Keto tautomer: ^{1}H NMR (CDCl$_3$) δ: 5.47–5.08 (m, 2 H), 3.63 (s, 3 H), 3.12 (d, J = 12.0 Hz, 1 H), 2.75 (dddd, J = 12.0, 12.0, 8.0, 4.0 Hz, 1 H), 2.35–2.32 (m, 1 H), 2.29–1.57 (m, 5 H), 1.54 (d, J = 3.7 Hz, 3 H); ^{13}C NMR (CDCl$_3$) δ 205.2, 169.5, 131.7, 126.2, 62.9, 51.6, 44.3, 40.6, 34.1, 24.6, 17.7. Enol tautomer: ^{1}H NMR (CDCl$_3$) δ: 12.32 (enolic H), 5.47–5.08 (m, 2 H), 3.63 (s, 3 H), 3.15–3.05 (m, 1 H), 2.41–2.36 (m, 1 H), 2.29–1.57 (m, 5 H), 1.54 (d, J = 3.7 Hz, 3 H); ^{13}C NMR (CDCl$_3$) δ 172.8, 134.1, 124.5, 99.7, 51.0, 39.6, 30.4, 28.8, 28.0, 16.8, one carbon was not observed.

cis/trans **Methyl 3,3-Dibromo-5-ethenyl-*r*-1-cyclopentanecarboxylate (137) (Sequential Reaction).**[227] A solution containing **136** (50 mg, 0.22 mmol), methyl acrylate (0.3 mL, 3.32 mmol), phenyl disulfide (8 mg, 0.04 mmol), and AIBN (36 mg, 0.22 mmol) was irradiated for 4 hours with a 300-W sun lamp. Purification of the residue by flash chromatography using Et$_2$O–hexanes (5:95) as eluent yielded 51 mg (74%) of cyclopentane **137** (diastereomer ratio = 9:1) as a clear oil. Partial separation of the two diastereomers was achieved by careful flash chromatography using Et$_2$O–hexanes (3:97) as eluent.

  cis: IR (neat) 1715 cm^{-1}; ^{1}H NMR (CDCl$_3$) δ 5.68 (ddd, J = 17.1, 10.1, 8.1 Hz, 1 H), 5.12 (d, J = 17.0 Hz, 1 H), 5.07 (d, J = 10.0 Hz, 1 H), 3.65 (s, 3 H), 3.36 (m, 2 H), 3.14 (dd, J = 14.5, 9.0 Hz, 1 H), 3.01 (m, 1 H), 2.93 (ddd, J = 14.2, 4.3, 2.3 Hz, 1 H), 2.63 (dd, J = 14.2, 9.9 Hz, 1 H); ^{13}C NMR (CDCl$_3$) δ 172.5, 135.3, 117.5, 62.2, 56.0, 53.5, 51.8, 46.3, 44.2; MS *m/z* 312, 233, 173; HRMS calcd for C$_9$H$_{12}$O$_2$Br$_2$, 311.9184; found, 311.9156.

  trans: IR (neat) 1715 cm^{-1}; ^{1}H NMR (CDCl$_3$) δ 5.59 (ddd, J = 17.4, 10.2, 7.6 Hz, 1 H), 4.91 (dt, J = 17.0, 1.2 Hz, 1 H), 4.83 (dd, J = 10.1, 1.2 Hz, 1 H), 3.25 (s, 3H), 3.16 (ddd, J = 17.4, 8.6, 7.7 Hz,1 H), 3.02 (ddd, J = 14.6, 7.9, 1.4 Hz, 1 H), 2.72 (dt, J = 14.0, 1.3 Hz, 1H), 2.70 (dt, J = 14.0, 1.3 Hz, 1 H), 2.60 (heptet, J = 8.6 Hz, 1 H), 2.32 (ddd, J = 14.5, 8.6, 1.1 Hz, 1 H); ^{13}C NMR (CDCl$_3$) δ 173.3, 138.6, 116.0, 61.4, 56.9, 54.4, 52.2, 48.5, 45.5; MS *m/z* 281, 233, 173; HRMS calcd for C$_8$H$_9$O$_2$Br$_2$ (M$^+$ −OMe) *m/z* 280.9000; found, *m/z* 280.8993.

138   **139**

7,8-Dimethyl-8-phenyl-5-(2-propenyl)-3-oxabicyclo[3.3.0]octane-2-thione (139) (Tandem Radical Cyclization of Homoallylic Xanthates; Sequential Reaction).[652] A mixture of **138** (0.225 g, 0.68 mmol), tributyltin hydride (0.224 g, 0.77 mmol), and AIBN (0.011 g, 0.066 mmol) in 35 mL of thiophene-free degassed dry toluene was heated at 80° with stirring for 1 hour under an argon atmosphere. The solvent was removed under reduced pressure. The pale yellow oil was purified by flash chromatography on silica gel (benzene) to give 0.138 g (71%) of the product **139**; mp 87–88°; IR (CHCl$_3$) 3050, 1640, 1270, 1180, 1030 cm^{-1}; ^{1}H NMR (CDCl$_3$) δ 7.43–7.23 (m, 5 H), 5.86–5.77 (m, 1 H), 5.26–5.21 (m, 2 H), 4.60 (d, J = 10.0 Hz, 1 H), 4.44 (d, J = 10.0 Hz, 1 H), 3.58 (s, 1 H), 2.39 (d, J = 7.2 Hz, 2 H), (ddq, J = 12.9, 12.9 and 7.0 Hz, 1 H), 2.10 (dd, J = 12.9, 7.0 Hz, 1 H), 1.74 (dd, J = 12.9 and 12.9 Hz, 1 H), 1.29 (s, 3 H), 0.77 (d, J = 6.7 Hz, 3 H); ^{13}C NMR (CDCl$_3$) INEPT δ 221.8 (C), 144.5 (C), 132.7 (C), 128.0 (CH), 126.8 (CH), 126.5 (CH), 126.4 (CH), 120.2 (CH$_2$), 88.0 (CH$_2$), 78.6 (CH), 53.6 (C), 50.4 (C), 47.8 (CH), 45.0 (CH$_2$), 43.8 (CH$_2$), 14.0 (CH$_3$), 12.7 (CH$_3$); high resolution mass spectra for C$_{18}$H$_{22}$OS (M$^+$), calculated for *m/z* 346.1776, found *m/z* 346.1780.

TABULAR SURVEY

  The computer search of Chemical Abstracts covers the literature from 1964 to the end of 1993.

The tabular organization of the radical cyclization reactions starts with monocyclic rings followed by bicyclic and oligocyclic ring systems. Furthermore, each of these tables is divided into ring systems containing only carbon atoms and rings containing one or more heteroatoms. Within each table the compounds are listed according to ring size. Order of entry is determined by total carbon number of the substrate(s) and then total hydrogen number.

The reaction conditions include the type of mediator, solvent, and temperature (°C) if provided in the literature. Yields are given in parentheses and are based on either isolation or GC analyses. A dash (—) indicates that no yield was reported. Numbers not in parentheses are product ratios. When a reaction has been reported in more than one publication, the conditions producing the highest yield are given, and the reference to that paper is listed first.

The following abbreviations are used in the tables:

Ac	acetyl
ACN	azobiscyclohexylnitrile
AIBN	azobisisobutyronitrile
B_{12}	Vitamin B_{12}
BOC	*tert*-butoxycarbonyl
BOM	benzyloxymethyl
Bn	benzyl
BPO	benzoyl peroxide
bpy	bipyridine
Bu	butyl
Bz	benzoyl
Cbz	benzyloxycarbonyl
Cp	cylopentadienyl
CHD	cyclohexadiene
DBA	di-*tert*-butylamino
DCA	dicyanoanthracene
DCN	dicyanonaphthalene
DDQ	2,3-dichloro-5,6-dicyano-1,4-benzoquinone
DIPHOS	1,2-bis(diphenyphosphino)ethane
DMF	*N,N*-dimethylformamide
dmgH	dimethyl glyoximate
DMP	dimethylpyrrolidinium
DMPU	1,3-dimethyl-3,4,5,6-tetrahydro-2(1*H*)-pyrimidone
DPDC	diisopropylperoxy dicarbonate
DPPE	2-(diphenylphosphino)ethyl
Glu	glucose
HMPA	hexamethylphosphoric triamide
Im	imidazole
IBDA	iodosobenzene diacetate
MEM	2-methoxyethoxymethyl
Mes	mesityl
MOM	methoxymethyl

Ms	methanesulfonyl (mesylate)
NBS	*N*-bromosuccinimide
NIS	*N*-iodosuccinimide
pic	2-pyridinecarboxylate
Piv	pivaloyl
PMB	*p*-methoxybenzyl
PTOC	[(1*H*)-pyridine-2-thione]oxycarbonyl
py	pyridine
pytos	pyridinium tosylate
salen	*N,N'*-1,2-ethylenebis(salicylidimine)
salophen	*N,N'*-1,2-phenylenebis(salicylidimine)
TBAF	tetrabutylammonium fluoride
TBAI	tetrabutylammonium iodide
TBDMS	*tert*-butyldimethylsilyl
TBDPS	*tert*-butyldiphenylsilyl
thexyl	dimethyl-(2,3-dimethyl-2-butyl)
TMS	trimethylsilyl
TMTHF	tetramethyltetrahydrofuran
Tf	trifluoromethanesulfonate (triflate)
THF	tetrahydrofuran
THP	tetrahydropyran
TMP	2,2,6,6-tetramethylpiperidin-1-yl
Tol	toluene
Tr	trityl = triphenylmethyl
Ts	*p*-toluenesulfonyl
p-TSA	*p*-toluenesulfonic acid

TABLE I. MONOCYCLIC RINGS CONTAINING ONLY CARBON

Reactant	Conditions	Product(s) and Yield(s) (%)	Refs.
A. 3-Membered Rings			
C$_7$			
[structure] Br + Cl$_3$CSO$_2$Cl	Zn-Cu, CCl$_4$, 150°	[structure] CN (65)	229
C$_{12}$			
[structure] Fe(CO)$_2$(Cp)	CBr$_4$, CH$_2$Cl$_2$ ArSO$_2$I, CH$_2$Cl$_2$ Cl$_3$CSO$_2$Cl, CH$_2$Cl$_2$	R–[structure] R = CBr$_3$ (65) R = ArSO$_2$ (70) R = CCl$_3$ (65)	230
C$_{17}$			
[structure] Co(dmgH)$_2$py	Cl$_3$CSO$_2$Cl, CH$_2$Cl$_2$, hv	Cl$_3$C–[structure] (75)	138
	TsI, CH$_2$Cl$_2$, 40°	Ts–[structure] (60)	139
C$_{18}$			
R^1 CN [structure] Co(dmgH)$_2$py	Cl$_3$CBr, CH$_2$Cl$_2$, 55°	Cl$_3$C–[structure] R^1 CN R^1: H (26), Me (66), Ph (43)	231
[structure] Co(dmgH)$_2$py	Cl$_3$CBr, CH$_2$Cl$_2$, hv	Cl$_3$C–[structure] (85)	138
[structure] Co(dmgH)$_2$py	Cl$_3$CBr, CH$_2$Cl$_2$, hv	Cl$_3$C–[structure] (82)	138

TABLE I. MONOCYCLIC RINGS CONTAINING ONLY CARBON (*Continued*)

Reactant	Conditions	Product(s) and Yield(s) (%)	Refs.
C_{23} R^1, R^2, Co(dmgH)$_2$py, vinyl	Br$_2$CHCN, CH$_2$Cl$_2$, 55°	cyclopropane product (CN, R^1, R^2, Br) R^1 R^2: Me H (61); Ph CN (87)	231
Co(dmgH)$_2$py, Ph, vinyl	p-MeC$_6$H$_4$SO$_2$I, CH$_2$Cl$_2$, 40°	p-MeC$_6$H$_4$SO$_2$–cyclopropane–Ph (72)	139
thiazole-2-thione N-O-acyl with R^1, R^2	PhMe, heat	cyclopropane (R^2, R^1, vinyl) R^1 R^2: PhCH$_2$ H (47); PhCH$_2$ Me (43); n-C$_{10}$H$_{21}$ H (60)	11

B. 4-Membered Rings

Reactant	Conditions	Product(s) and Yield(s) (%)	Refs.
C_8 bromo-CN reactant	Bu$_3$SnH, AIBN, C$_6$H$_6$, 80°	cyclobutane–CN (45)	232
C_{12} EtO, OEt, Br, CO$_2$Et reactant	Bu$_3$SnH, AIBN, C$_6$H$_6$, 80°	cyclobutane (CO$_2$Et, EtO, EtO) (72)	12
C_{13} SMe, Ts, I reactant	Bu$_3$SnH, AIBN, PhMe, 110°	MeS, Ts cyclobutane (74)	233

TABLE I. MONOCYCLIC RINGS CONTAIN NG ONLY CARBON (*Continued*)

Reactant	Conditions	Product(s) and Yield(s) (%)	Refs.

C. 5-Membered Rings

Reactant	Conditions	Product(s) and Yield(s) (%)	Refs.
C_6			
I (structure)	1. SmI_2, HMPA 2. ArCHO	(structure) OH, Ar (30) Ar = p-MeOC$_6$H$_4$	234
(acid chloride structure)	AIBN, C_6H_6, 80°	**I** + **II** (structures) **I** (34) + **II** (6) **I** (41) + **II** (11)	235 236
(alkyne/iodide structure)	Bu_3SnH, $(TMS)_3SiH$ $(Bu_3Sn)_2$, C_6H_6, 70–75°, hv	(structure I) <table><tr><td>R^1</td><td>R^2</td><td></td><td></td></tr><tr><td>H</td><td>H</td><td>(77)</td><td></td></tr><tr><td>Me</td><td>H</td><td>(87)</td><td>$E{:}Z = 3.3{:}1$</td></tr><tr><td>Me</td><td>Me</td><td>(95)</td><td>$E{:}Z = 15{:}1$</td></tr></table>	237
(epoxide structure)	1. MgI_2 2. Bu_3SnH, AIBN, C_6H_6, 80°	(cyclopentane structure) OH <table><tr><td>R^1</td><td>R^2</td><td></td></tr><tr><td>H</td><td>H</td><td>(30)</td></tr><tr><td>H</td><td>OH</td><td>(53)</td></tr><tr><td>OH</td><td>H</td><td>(35)</td></tr></table>	238
(bromide structure)	Bu_3SnH, AIBN, xylene, 130°	(methylcyclopentane) (92) + (cyclohexane) (1)	239

365

TABLE I. MONOCYCLIC RINGS CONTAINING ONLY CARBON (*Continued*)

Reactant	Conditions	Product(s) and Yield(s) (%)	Refs.
(HgBr-substituted heptenyl structure)	I_2, dioxane	(62)	240
C_7			
OHC—(chain)—CN, *E:Z* = 65:35	Bu_3SnH, AIBN, C_6H_6, 80°	**I** + **II** (73) **I:II** = 52:48	115
(Cl, Cl, CF_3, HO substituted cyclohexene with allyl)	Bu_3SnH, AIBN, 65°, 8 h	(78)	241
O=CH—(chain)—CN	1. Bn_2NH, benzo-triazole 2. SmI_2, THF, 25°	(63) *cis:trans* = 67:33	242
(R, R, Z, X alkene) + (Z—vinyl)	Bu_3SnH, AIBN, C_6H_6, 80°		243

TABLE I. MONOCYCLIC RINGS CONTAINING ONLY CARBON (*Continued*)

Reactant	Conditions	Product(s) and Yield(s) (%)	Refs.

Z	X	R		time (h)	Yield
CN	Br	H		5	(58)
CN	I	Me		4	(65)
CO$_2$Et	I	Me		3	(75)
COMe	I	Me		4	(75)
CHO	I	Me		1.5	(60)

Reactant	Conditions	Product(s) and Yield(s) (%)	Refs.
	SmI$_2$, THF, HMPA	(86)	178
	Hg cathode, DMF, Bu$_4$N$^+$BF$_4^-$, DMP$^+$BF$_4^-$	" (100)	244, 245
	Carbon rod cathode, MeOH, dioxane, Et$_4$NOTs	" (98)	246, 245
	Hg cathode, DMF, Bu$_4$N$^+$BF$_4^-$, DMP$^+$BF$_4^-$	(85)	244
	RuCl$_2$(PPh$_3$)$_3$, C$_6$H$_6$, sealed tube, 150°, 22 h	(61) + (13)	221

367

TABLE I. MONOCYCLIC RINGS CONTAINING ONLY CARBON (*Continued*)

Reactant	Conditions	Product(s) and Yield(s) (%)	Refs.
	$K_3Fe(CN)_6$, N_2, NaOMe, MeOH	 **I** + **II** (71) **I** : **II** = 60 : 40	247
	$(Bu_3Sn)_2$, C_6H_6, 70–80°, hv		248

R^1	R^2	R^3		
H	H	Me	(71)	
H	H	TMS	(41)	
H	Me	H	(65)	*E:Z* = 77:23
Me	Me	H	(60)	*E:Z* = 93:7
Me	Me	Me	(54)	*E:Z* = 79:21
Me	Me	TMS	(45)	*E:Z* = 19:81

Reactant	Conditions	Product(s) and Yield(s) (%)	Refs.
+ CO	Bu_3SnH, AIBN, CO (75–90 atm), C_6H_6, 80°		249

R^1	R^2	R^3	X	
Me	Me	H	Br	(65)
Me	Me	Me	Br	(77)
H	Ph	H	Br	(62)
H	CO_2Et	H	I	(60)

368

TABLE I. MONOCYCLIC RINGS CONTAINING ONLY CARBON (*Continued*)

Reactant	Conditions	Product(s) and Yield(s) (%)	Refs.
	Bu₃SnH, AIBN, C₆H₆, 80°	(27) + (46)	16
	Bu₃SnH, AIBN, C₆H₆, 80°	(48) + (18)	16
	Bu₃SnH, AIBN, C₆H₆, 80°	(16) + (53)	16
	Bu₃SnH, AIBN, C₆H₆, 65°	+ 40:60	249a
	Bu₃SnH, AIBN, C₆H₆, 70°	(88) + (3)	239
	AIBN, n-C₆H₁₄, 74°	(75)	250
	(Bu₃Sn)₂, C₆H₆, sun lamp, 80°	" (72)	106

TABLE I. MONOCYCLIC RINGS CONTAINING ONLY CARBON (Continued)

Reactant	Conditions	Product(s) and Yield(s) (%)	Refs.

C_8

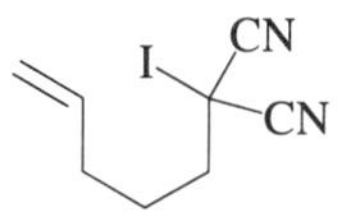

C_6H_6, 80°, 10 h

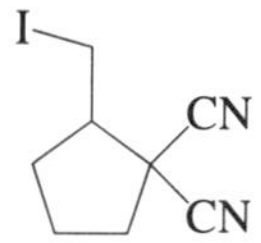

$I : II = 80 : 20$ $I + II$ (84)

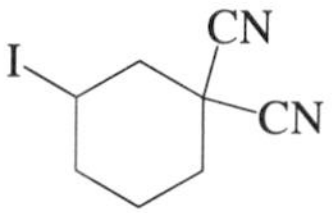

251

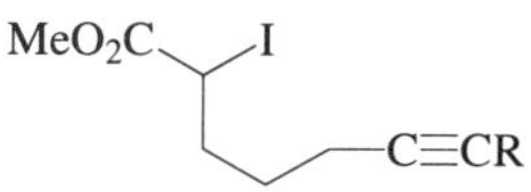

(Me$_3$Sn)$_2$, hv

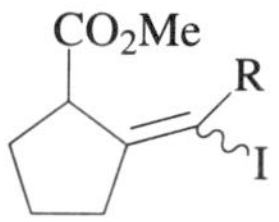

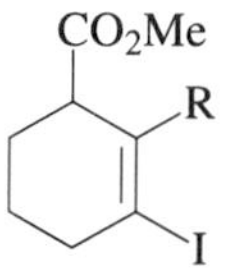

$R = H$ $I + II$ (85) $I : II = 95 : 6$
$R = TMS$ $I + II$ (81) $I : II = 97 : 3$

156

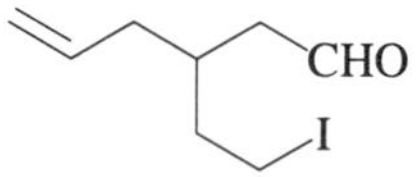

Bu$_3$SnH, AIBN, C_6H_6, 80°

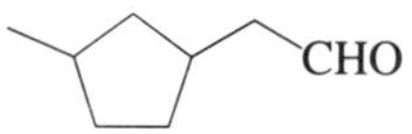

$I + II$ (84) $I:II = 3:1$

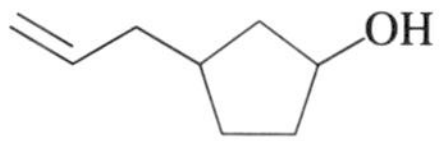

41, 40

Ph$_3$SnH, AIBN, C_6H_6, 80°

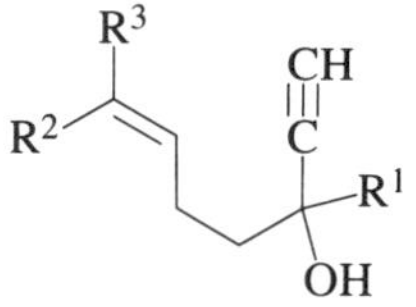

27

R^1	R^2	R^3	$I+II$	I:II
Me	Me	Me	(75)	78:22
Me	H	H	(50)	80:20
Me	Me	(CH$_2$)$_2$CH=CMe$_2$	(78)	79:21
n-C$_5$H$_{11}$	H	Ph	(87)	63:37

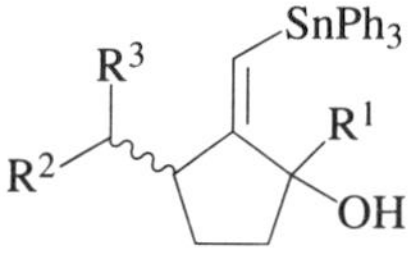

TABLE I. MONOCYCLIC RINGS CONTAINING ONLY CARBON (*Continued*)

Reactant	Conditions	Product(s) and Yield(s) (%)	Refs.
OHC〜〜CO₂Me	VCl₃(THF)₃, Zn, CH₂Cl₂, 25°	(structure I, —OH, CO₂Me) **I** + (structure II, —OH, CO₂Me) **II** **I + II** (68) **I:II = 24:1**	252
(Cl-substituted diene structure)	Bu₃SnH, AIBN, C₆H₆, 80°	**I** + (bicyclic lactone) **III** **I + III** (81) **I:III = 52:48**	115
(O=, I, R diene structure)	(PhSe)₂, DCN, MeCN, *hν*, (>280 nm)	(PhSe cyclopentane vinyl) (70)	253
	(Me₃Sn)₂, AIBN, C₆H₆	(O=, R, I cyclopentane) **I** + (O=, R, I cyclohexane) **II**	156

R		**I** *cis:trans*	**II** *cis:trans*	
MeO		57:43	34:66	(83)
t-BuO		47:53	31:69	(74)
Ph		28:72	30:70	(68)
t-Bu		35:65	33:67	(63)

371

TABLE I. MONOCYCLIC RINGS CONTAINING ONLY CARBON (*Continued*)

Reactant	Conditions	Product(s) and Yield(s) (%)	Refs.

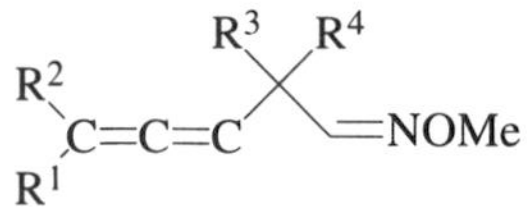

R^2 $\underset{R^1}{\diagup}$ C=C=C=NOMe (R^3, R^4)

R^1	R^2	R^3	R^4
Me	Me	Me	Me
H	Me	Me	Me
Et	Me	Me	Me
Me	Me	H	H

Bu₃SnH, AIBN
C₆H₆, 80°

Bu₃Sn product with R³, R⁴, R¹, R², NHOMe

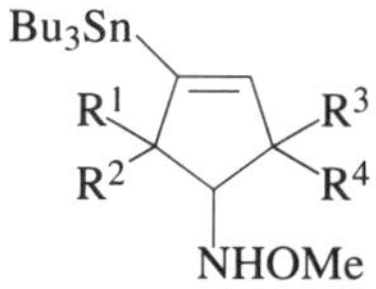

(77)
(74)
(91)
(37)

254

S—C(=S)—OMe

(t-BuO)₂, PhCl,
132°

MeO—C(=S)—S—CH₂—cyclopentyl (61)

255

epoxide with allyl

Cp₂TiCl, H⁺

HO— cyclopentyl product (94)

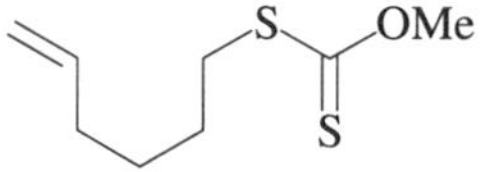

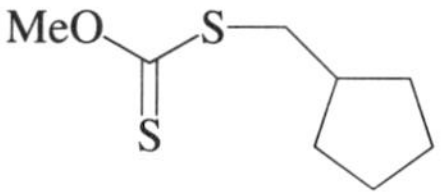

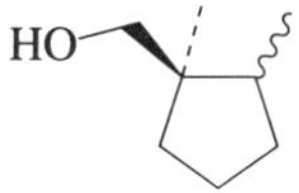

isomeric ratio = 1:1

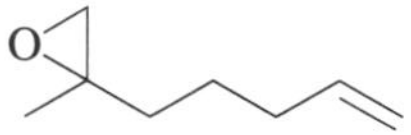

172

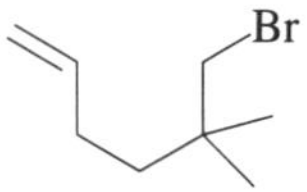

Br, dimethyl, allyl

Bu₃SnH, AIBN
C₆H₆, 80°

dimethylcyclopentane (83)

16

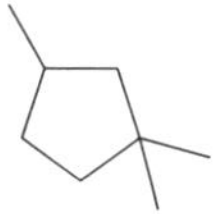

Br, dimethyl, allyl

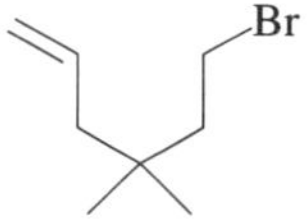

Bu₃SnH, AIBN
C₆H₆, 80°

" (83)

16

TABLE I. MONOCYCLIC RINGS CONTAINING ONLY CARBON (*Continued*)

Reactant	Conditions	Product(s) and Yield(s) (%)	Refs.
	Bu$_3$SnH, AIBN PhMe, 100°	(74) + (4)	239
	Bu$_3$SnH, AIBN PhMe, 100°	(92) + (4)	239
	(Bu$_3$Sn)$_2$, AIBN, C$_6$H$_6$, 80°, *hv*	(75)	106
	K$_3$Fe(CN)$_6$, NaOMe, MeOH	I + II (93) I:II = 40:60	247
	K$_3$Fe(CN)$_6$, NaOMe, MeOH	I + II (93) I:II = 50:50	247

TABLE I. MONOCYCLIC RINGS CONTAINING ONLY CARBON (*Continued*)

Reactant	Conditions	Product(s) and Yield(s) (%)	Refs.

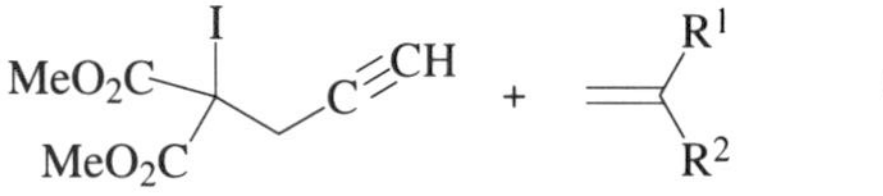

(Bu$_3$Sn)$_2$, C$_6$H$_6$, *hv*

R^1	R^2
H	Ph
H	CH$_2$TMS
Et	Et

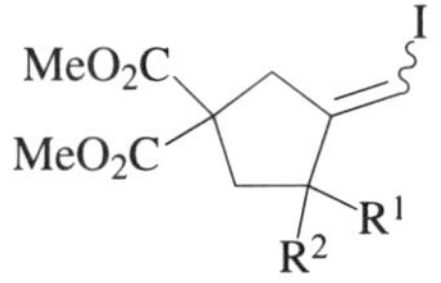

(59) *E:Z* = 3:1
(74) *E:Z* = 3:1
(77) *E:Z* = 4.4:1

256

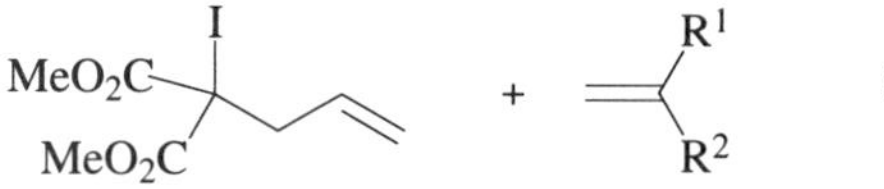

(Bu$_3$Sn)$_2$, C$_6$H$_6$, *hv*

R^1	R^2
H	Ph
H	CH$_2$TMS
Et	Et

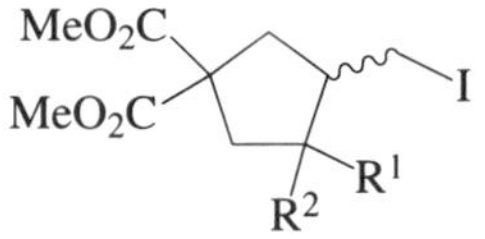

(70) *E:Z* = 1:1.4
(71) *E:Z* = 4.6:1
(72)

256

Bu$_3$SnH, AIBN
C$_6$H$_6$, 80°

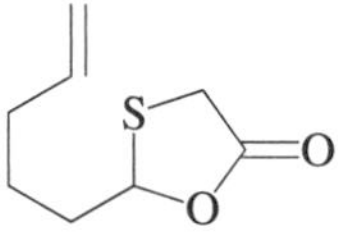

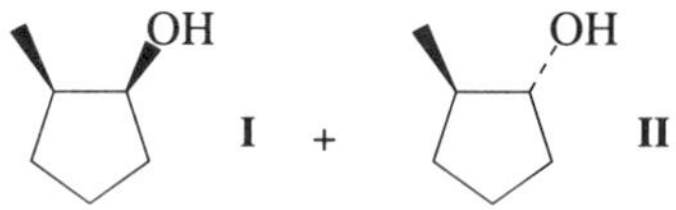

I + II (74)

I:II = 3:2

116

(PhS)$_2$, AIBN, *hv*

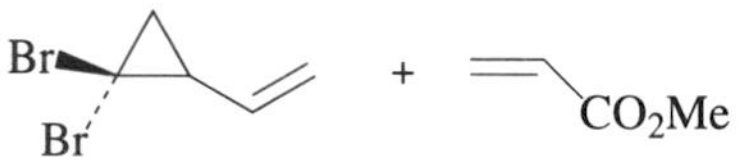

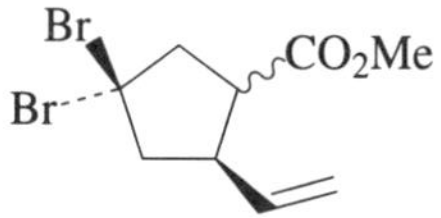

(74)
cis:trans = 9:1

227

C$_9$

TABLE I. MONOCYCLIC RINGS CONTAIN NG ONLY CARBON (*Continued*)

Reactant	Conditions	Product(s) and Yield(s) (%)	Refs.

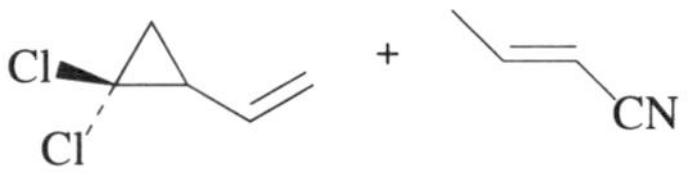

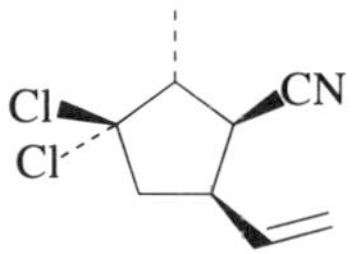

(73)
19:5:5:1
major isomer is shown

227

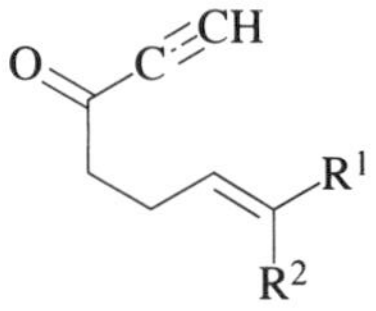

(PhS)₂, AIBN, *hv*

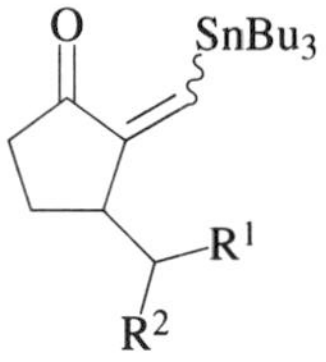

R^1 R^2
Me Me (91)
Ph H (92)
CO₂Et H (88)

257

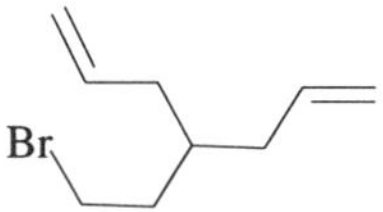

Bu₃SnH, AIBN
C₆H₆, 80°

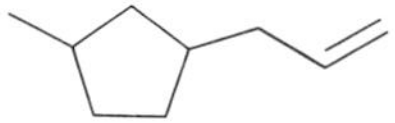

(80 - 100)

258

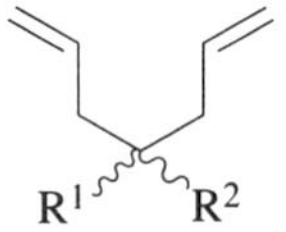

Bu₃SnH, AIBN
C₆H₆, 80°

I:II:III = 69:19:12 (70)
I:II:III = 73:27:0 (76)

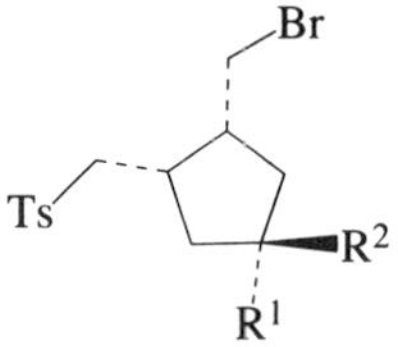

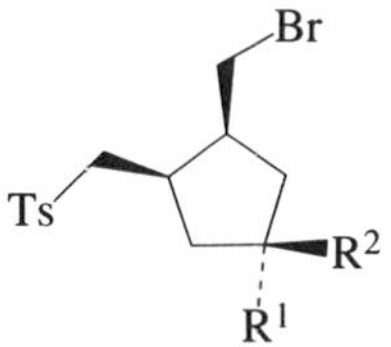

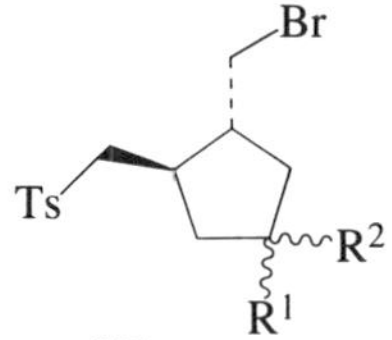

259

TsBr, MeCN, *hv*, 20°

R^1	R^2
CO₂Me	H
Ph	CO₂Me

TABLE I. MONOCYCLIC RINGS CONTAINING ONLY CARBON (*Continued*)

Reactant	Conditions	Product(s) and Yield(s) (%)	Refs.
$X = Br$ $X = OC(S)SMe$	Bu₃SnH, AIBN C₆H₆, 80°	(75) (68)	17
	Bu₃SnH, O₂, PhMe	(38) + (24)	260
	1. Benzotriazole, 2. SmI₂, THF, 25°	(70)	242
	Na, THF, ultrasound NaC₁₀H₈, THF	**I** + **II** **I:II** = 70:30 (44) **I:II** = 74:26 (44)	261

376

TABLE I. MONOCYCLIC RINGS CONTAINING ONLY CARBON (*Continued*)

Reactant	Conditions	Product(s) and Yield(s) (%)	Refs.

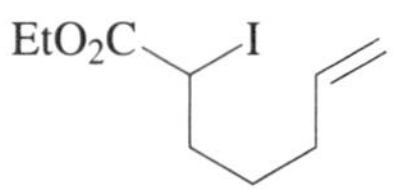

Cr(OAc)$_2$, THF, rt

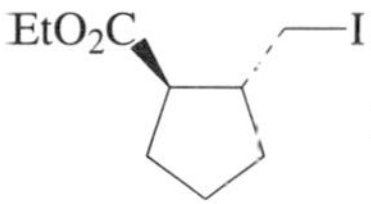

I + **II** + **III** (83)

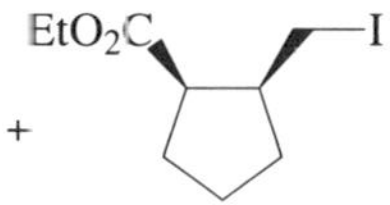

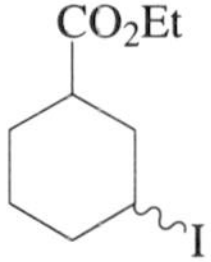

I:II:III = 63:32:5

262

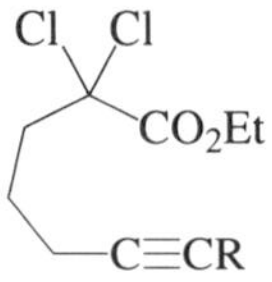

RuCl$_2$(PPh$_3$)$_3$,
cumene, 160°

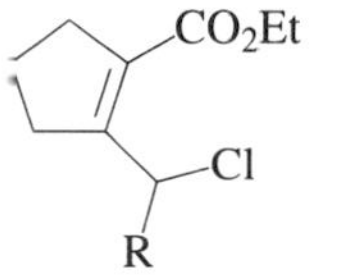

R	
H	(70)
Me	(83)

221

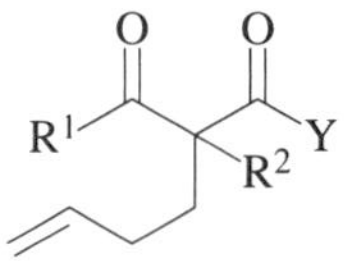

SmI$_2$, THF, 25°

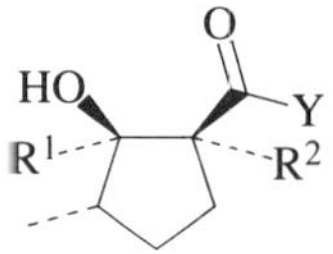

main isomer

Diastereoselectivity

R^1	R^2	Y		
Me	Me	OEt	25:1	(75)
Et	Me	OEt	30:1	(66)
i-Pr	Me	OEt	30:1	(63)
Me	Et	OEt	200:1	(51)
Me	H	OMe	20:1	(60)

263

TABLE I. MONOCYCLIC RINGS CONTAINING ONLY CARBON (*Continued*)

Reactant	Conditions	Product(s) and Yield(s) (%)	Refs.
C_{10}			
[structure: CO_2Me, ketone]	$Mn(OAc)_3$, $Cu(OAc)_2$, HOAc, 50°	[structure: CO_2Me] (36)	192
[structure: CO_2Me, iodocyclohexene]	Bu_3SnH, O_2, PhMe	[structure: CO_2Me, OH] (69)	260
EtO–[xanthate structure]–O	C_7H_{16}, reflux, $h\nu$	[structure: S, OEt] (87)	264
[epoxide structure with I]	Ph_3SnH, Et_3B, C_6H_{14}, C_6H_6, 25°	[structure: OH] **I** + [structure: OH] **II** (48) **I:II = 3:2**	265
[bromoallene structure, Br]	Bu_3SnH, AIBN, C_6H_6, 80°	[structure] (43)	266

378

TABLE I. MONOCYCLIC RINGS CONTAINING ONLY CARBON (*Continued*)

Reactant	Conditions	Product(s) and Yield(s) (%)	Refs.

Reactant 1:

EtO$_2$C / NC with alkene chain

Conditions: CuCl$_2$, DMF, 80°

Product:

EtO$_2$C / NC cyclopentane with CH$_2$Cl (67)

Refs. 267

Reactant 2:

R^{2}C≡C / R^3 silyl ether with Br, Me, Si, Me, O, R^1

R^1	R^2	R^3
H	H	Me
H	n-C$_5$H$_{11}$	Me
H	H	H
H	n-C$_5$H$_{11}$	H
n-C$_5$H$_{11}$	H	Me
H	Ph	Me

Conditions:
1. Ph$_3$SnH, AIBN, C$_6$H$_6$, 80°
2. H$_2$O$_2$

Products:

I + II

	I:II
(65)	100:0
(79)	90:10
(60)	100:0
(67)	100:0
(75)	100:0
(80)	42:58

Refs. 215

Reactant 3:

O with SMe, SMe, Br, R^2, R^1, R^1

Conditions: Bu$_3$SnH, AIBN, C$_6$H$_6$, 80°

Product:

cyclopentanone with SMe, SMe, O, R^2, R^1, R^1

R^1	R^2	
Me	H	(42)
H	Ph	(54)
H	Me	(95)

Refs. 268

379

TABLE I. MONOCYCLIC RINGS CONTAINING ONLY CARBON (*Continued*)

Reactant	Conditions	Product(s) and Yield(s) (%)	Refs.
	Bu₃SnH, AIBN C₆H₆, 80°	 **I** + Bicycle **II** **Bicycle**	269

$$\text{Bu}_3\text{SnH, AIBN}$$
$$\text{C}_6\text{H}_6, 80°$$

Y	Bicycle
CO₂Me	 **I + II** (85) **I:II** = 2.1:1
CN	 **I + II** (94) **I:II** = 3:1
COMe	 **I + II** (93) **I:II** = 3.5:1

380

TABLE I. MONOCYCLIC RINGS CONTAINING ONLY CARBON (*Continued*)

Reactant	Conditions	Product(s) and Yield(s) (%)	Refs.

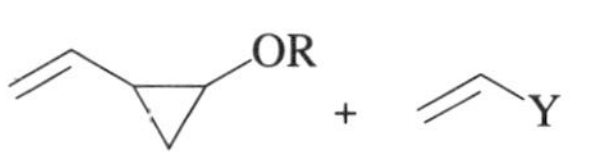

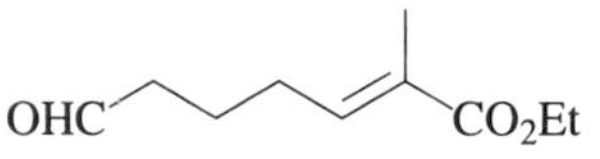

(PhS)$_2$, AIBN, C$_6$H$_6$, 80°

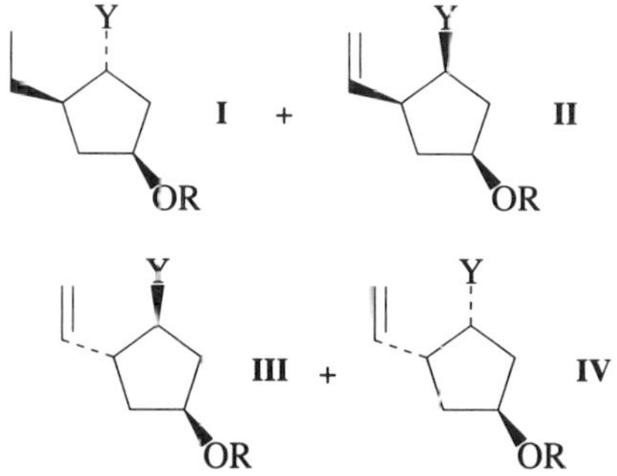

I + II

III + IV

R	Y	I : II : III : IV
Et	CO$_2$Me	3.3 : 2.0 : 1.0 : 4.2 (80)
Bn	CO$_2$Me	2.8 : 1.6 : 1.0 : 3.5 (63)
Bn	CO$_2$Bu-t	4.2 : 1.1 : 1.0 : 2.4 (60)
Bn	O$_2$CBu-t	2.0 : 1.3 : 1.0 : 2.1 (43)
CH$_2$C$_6$H$_4$OMe-p	CO$_2$Me	1.8 : 1.3 : 1.0 : 2.6 (74)
CH$_2$C$_6$H$_4$CF$_3$-p	CO$_2$Me	3.1 : 1.8 : 1.0 : 3.7 (80)
TBDMS	CO$_2$Me	3.8 : 2.3 : 5.1 : 1.0 (78)

270

VCl$_3$(THF)$_3$, Zn, CH$_2$Cl$_2$, 25°

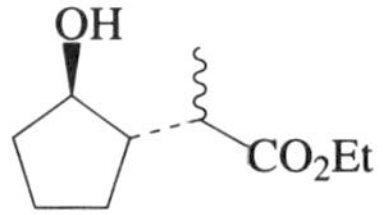

(77)

252

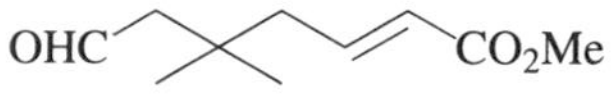

VCl$_3$(THF)$_3$, Zn, CH$_2$Cl$_2$, 25°

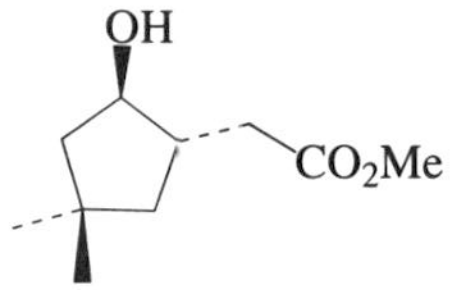

(67)

252

381

TABLE I. MONOCYCLIC RINGS CONTAINING ONLY CARBON (*Continued*)

Reactant	Conditions	Product(s) and Yield(s) (%)	Refs.
(structure: CH₃C(O)CH₂C(CH₃)₂CH₂CH=C=CH₂)	Hg cathode, DMF NaC₁₀H₈	(structure: cyclopentane with HO, vinyl, and methyl substituents) (55) (56)	225
(structure: cyclopentanone with CO₂Me and propenyl chain)	Mn(OAc)₃, Cu(OAc)₂, AcOH, 60°, 1 h	(structure: cyclopentanone with CO₂Me and propenyl) (36) + (structure: cyclopentenone with CO₂Me and propenyl) (10)	271
(structure: Br-CH₂CH₂C(CO₂Me)₂ with allyl)	Ph₂SnH₂, BPO, C₈H₁₈, 110°	(structure: cyclopentane with two CO₂Me groups, MeO₂C and CO₂Me) (62)	272
(structure: Cl(CH₂)₄CH=CH(CH₂)₄Cl)	Bu₃SnH, AIBN C₆H₆, 80°	(structure: cyclopentane with n-butyl chain) (63)	273
(structure: I(CH₂)₃CH₂C≡C-C₄H₉-n)	CrCl₂, DMF, 25°	(structure: cyclopentane with =CH-C₄H₉-n) (85)	274

382

TABLE I. MONOCYCLIC RINGS CONTAINING ONLY CARBON (*Continued*)

Reactant	Conditions	Product(s) and Yield(s) (%)	Refs.
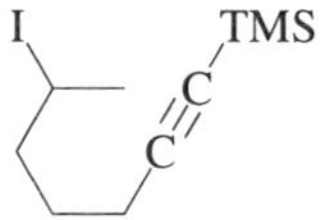	Et_3B, C_6H_{14}, 25°	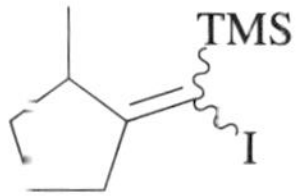 (68)	275
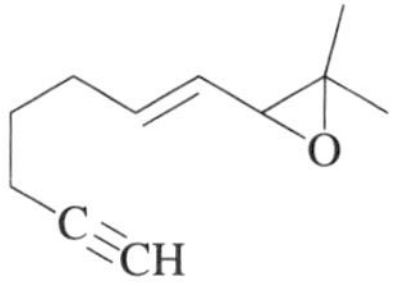	Ph_3SnH, Et_3B, C_6H_{14}, 25°	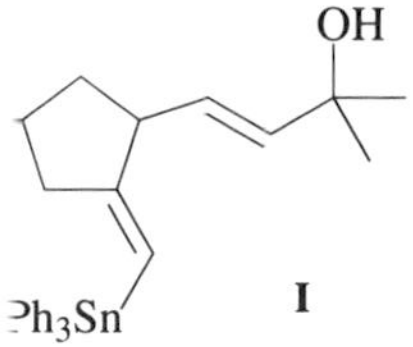 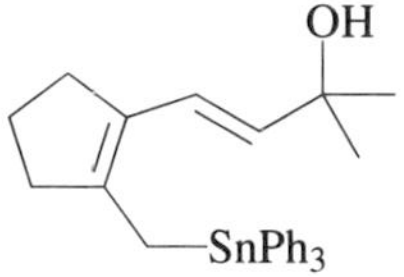 I + II (62) I:II = 18:82	265
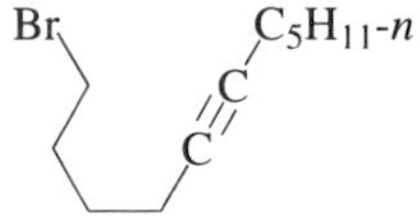	Bu_3SnH, AIBN C_6H_6, 80°	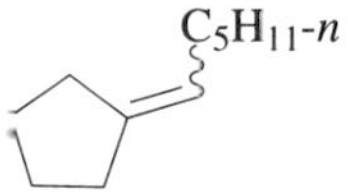 (99)	276
	Zn, AcOH Bu_3SnH, AIBN	(77) (68)	277
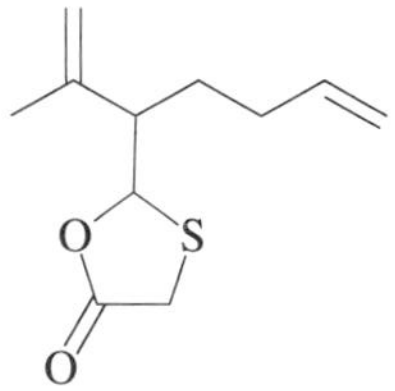	Bu_3SnH, AIBN C_6H_6, 80°	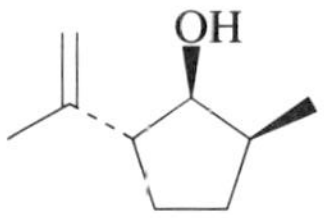 I + II (90) I:II = 8:1	116

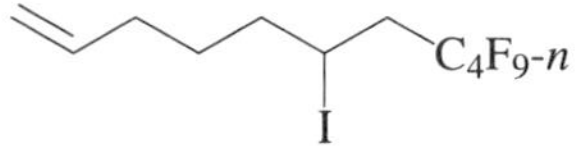

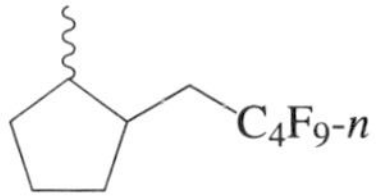

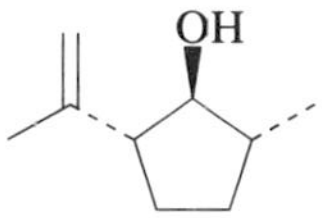

C_{11}

383

TABLE I. MONOCYCLIC RINGS CONTAINING ONLY CARBON (*Continued*)

Reactant	Conditions	Product(s) and Yield(s) (%)	Refs.
[structure: RO$_2$C, CO$_2$R with two allyl groups]	**A**: TsBr, MeCN, $h\nu$, 25° **B**: TsBr, PhCl, 135°	[cyclopentane structure **I**: Ts–CH$_2$, RO$_2$C, CO$_2$R, CH$_2$Br] + [cyclopentane structure **II**: Ts–CH$_2$, RO$_2$C, CO$_2$R, CH$_2$Br]	278

R		I + II	I:II
Me	A	(85)	90:10
	B	(76)	81:19
Et	A	(85)	93:7
	B	(85)	79:21
t-Bu	A	(87)	93:7

Reactant	Conditions	Product(s) and Yield(s) (%)	Refs.
[structure: MeO$_2$C, CO$_2$Me with two allyl groups]	EtSH, (PhS)$_2$, C$_6$H$_6$, $h\nu$	[cyclopentane structure: PhS, R, MeO$_2$C, CO$_2$Me] R = H or SPh *cis:trans* = 6:1 (92)	279
	TsNa, Cu(OAc)$_2$, AcOH, 90°	[cyclopentane structure: Ts, MeO$_2$C, CO$_2$Me] (51)	280
[structure: diacetyl with two allyl groups]	TsSePh, AIBN, CHCl$_3$, 61°	[cyclopentane structure: SePh, Ts, two acetyl groups] (95) 8.3:1	281

384

TABLE I. MONOCYCLIC RINGS CONTAINING ONLY CARBON (*Continued*)

Reactant	Conditions	Product(s) and Yield(s) (%)	Refs.

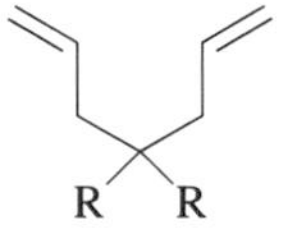

A: (PPh₃)₂ReCl₃, CCl₄, reflux
B: (PPh₃)₃RuCl₂, CCl₄, reflux

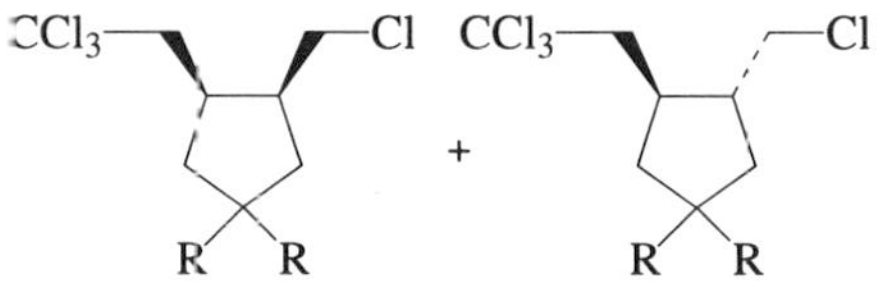

$$I + II \quad I:II$$

R		I + II	I:II
CO₂Et	A	(74)	5:1
	B	(86)	—
COPh	A	(73)	4.2:1
	B	(84)	3.2:1
COMe	B	(65)	—

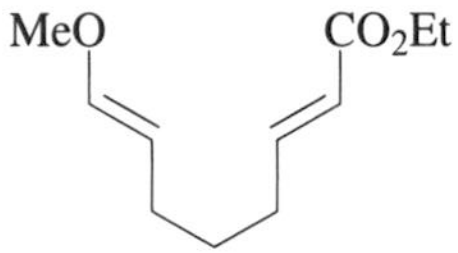

282

1. Hg(OAc)₂, MeOH
2. NaBH₄

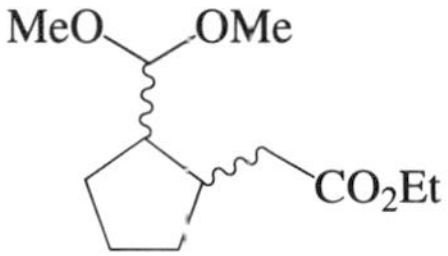

(90) *cis:trans* = 4:1

131, 283

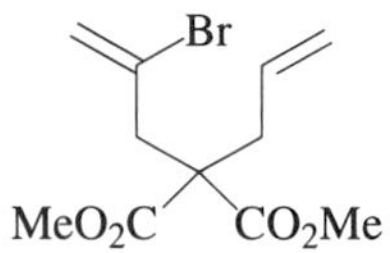

Bu₃SnH, AIBN
C₆H₆, *hv*, heat

I + II

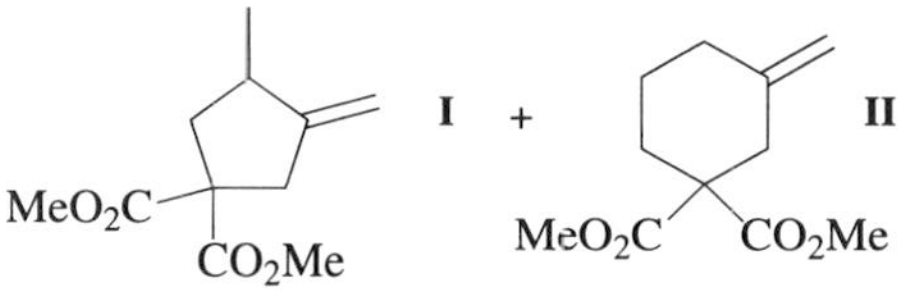

I + II (87) I:II = 3:1

24

1. (Me₃Sn)₂, *hv*
2. Bu₃SnH

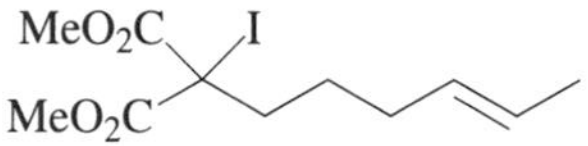

(84)

18

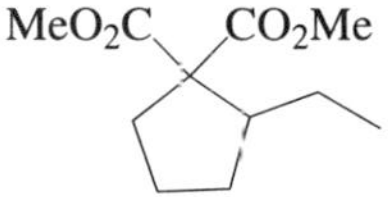

385

TABLE I. MONOCYCLIC RINGS CONTAINING ONLY CARBON (*Continued*)

Reactant	Conditions	Product(s) and Yield(s) (%)	Refs.
R—OOH (cyclohexane with allyl side chain)	FeSO$_4$, AcOH	**I** + **II** $\quad$ $\dfrac{R}{Me}$ $\dfrac{I}{(38)}$ $\dfrac{II}{(30)}$; Et (28) (18) ; Ph (32) (7)	284
EtO$_2$C—CCl$_2$— (diene chain)	(PPh$_3$)$_3$RuCl$_2$, *t*-BuPh, 150–155°	CO$_2$Et, Cl (53) + CO$_2$Et, Cl (22)	221
(isopropenyl dithiolane chain)	Bu$_3$SnH, AIBN, C$_6$H$_6$, 80°, 24 h	**I** + **II** $\quad$ **I:II** = 65:35, (—)	285
(keto ester with pentenyl chain)	Mn(OAc)$_3$, Cu(OAc)$_2$, AcOH, 50°	CO$_2$Et (39)	192

386

TABLE I. MONOCYCLIC RINGS CONTAINING ONLY CARBON (Continued)

Reactant	Conditions	Product(s) and Yield(s) (%)	Refs.
(structure: cyclohexanone with CO_2Me and prenyl side chain)	Mn(OAc)$_3$, Cu(OAc)$_2$, AcOH, 60°	(structure with CO_2Me, O, isopropenyl) (27) + (structure with CO_2Me, O, OAc) (20)	271
(structure: Br, CO_2Et, Cl chain)	Bu$_3$SnH, AIBN, C$_6$H$_6$, 80°	(cyclopentane with CO_2Et and Cl chain) (66) 70:30 stereochemistry unknown	17
(structure: C$_5$H$_{11}$, OHC chain)	Bu$_3$SnH, AIBN, C$_6$H$_6$, 80°	C_5H_{11} ...OH **I** + C_5H_{11} **II** OH **I + II** (32) **I:II** = 66:34	115
(structure with R^1, R^2, Y, allyl)	SmI$_2$, t-BuOH, THF, -78°	(cyclopentane with R^1 OH, R^2, Y)	286

R^1	R^2	Y		Diastereoselectivity
Me	Me	OEt	(75)	25:1
Et	Me	OEt	(66)	30:1
i-Pr	Me	OEt	(63)	30:1
H	H	NEt$_2$	(78)	>200:1

TABLE I. MONOCYCLIC RINGS CONTAINING ONLY CARBON (*Continued*)

Reactant	Conditions	Product(s) and Yield(s) (%)	Refs.
C$_{12}$			
	Bu$_3$SnH, AIBN, C$_6$H$_6$, 80°	(93)	276
	CrCl$_2$, DMF, 25°	(96)	274
	Bu$_3$SnH, AIBN, C$_6$H$_6$, 80°		287
		Z: S(O)Ph (69), SO$_2$Ph (70), POPh$_2$ (79)	
	(PhS)$_2$, AIBN, hv	I + II (72) I:II = 4.1:1	227
	Bu$_3$SnH, AIBN, C$_6$H$_6$, 80°, 10 h	(59)	288

TABLE I. MONOCYCLIC RINGS CONTAINING ONLY CARBON (*Continued*)

Reactant	Conditions	Product(s) and Yield(s) (%)	Refs.

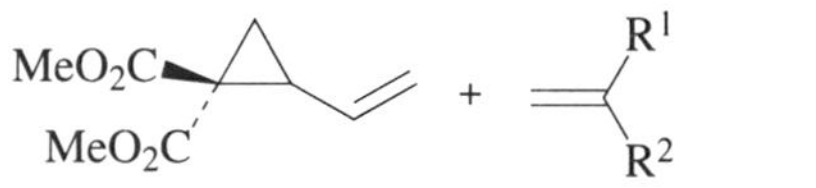

PhSH, 60°

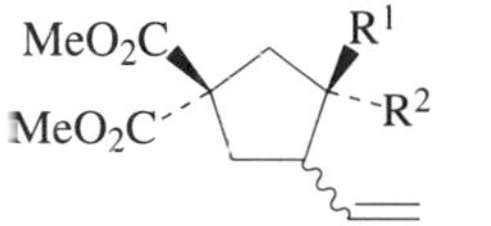

R¹	R²	
H	OBu-*n*	(82)
H	OTMS	(55)
OEt	OEt	(79)
Me	OAc	(74)
H	CH₂TMS	(53)

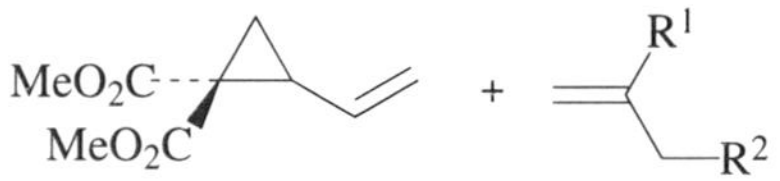

289

(PhS)₂, C₆H₆, *hv*

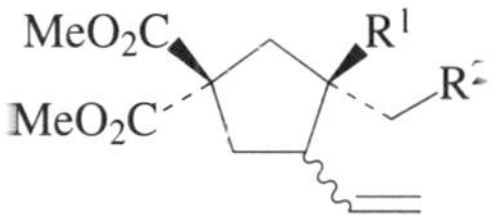

R¹	R²	
Me	Ts	(73)
Me	Cl	(73)
Cl	Cl	(47)

290

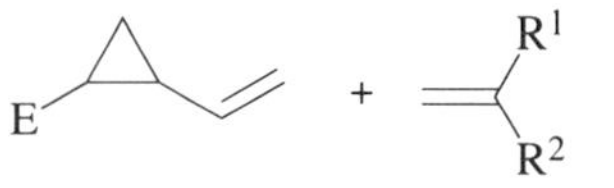

PhSH, 60°

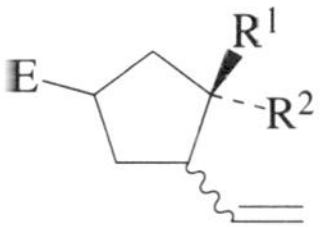

E	R¹	R²	
CO₂Et	H	CO₂Me	(76)
CO₂Et	Me	OAc	(77)
COPh	H	CO₂Me	(63)
COPh	Me	OAc	(71)

289

389

TABLE I. MONOCYCLIC RINGS CONTAINING ONLY CARBON (*Continued*)

Reactant	Conditions	Product(s) and Yield(s) (%)	Refs.

Row 1:

Reactant: structure with OHC and Ph (cinnamaldehyde-type chain)

Conditions: Bu$_3$SnH, AIBN C$_6$H$_6$, 80°

Product(s):

I + II   I:II = 53:47

I + II (80)

Refs.: 115

Row 2:

Reactant: HC≡C, R^1, R^2, NC, CN structure

R^1	R^2	
Bu	H	(59)
n-Pr	n-Pr	(86)
Ph	Me	(82)
i-Pr	Me	(76)

Conditions: Bu$_3$SnH, AIBN C$_6$H$_6$, 80°

Product(s):

Refs.: 291, 208

Row 3:

Reactant: p-O$_2$NC$_6$H$_4$ with Cl and vinyl chain

Conditions: CuCl, DMSO, 80°, ligand

Product(s):

p-O$_2$NC$_6$H$_4$ ... Cl (I) + p-O$_2$NC$_6$H$_4$ ... Cl (II)

ligand	I	II	I + II
bipyridyl	*trans:cis* = 71:29	*trans:cis* = 74:26	(90)
phenanthroline	*trans:cis* = 70:20	*trans:cis* = 71:29	(95)

Refs.: 292

390

TABLE I. MONOCYCLIC RINGS CONTAINING ONLY CARBON (*Continued*)

Reactant	Conditions	Product(s) and Yield(s) (%)	Refs.
	Bu_3SnH, AIBN C_6H_6, 80°	(87)	197
	Bu_3SnH, AIBN C_6H_6, 80°	(62)	293
	Bu_3SnH, AIBN C_6H_6, 80°	**I, II** + **III** **I, II + III** (80)	293
	Bu_3SnH, AIBN C_6H_6, 80°	**I** + **II** **I + II** (52) **I:II** = 20:1	294

391

TABLE I. MONOCYCLIC RINGS CONTAINING ONLY CARBON (*Continued*)

Reactant	Conditions	Product(s) and Yield(s) (%)	Refs.

Reactant 1:

Bu, I, NC CN (allyl)

Conditions: Bu$_3$SnH, AIBN, C$_6$H$_6$, 80°

Products:

I + II: structures with CN, Bu, NC substituents on cyclopentane

I + II (82)   **I:II** = 2.9:1

Ref. 294

Reactant 2:

RO_2C, CO_2R, methylene/allyl

R
Me
Et

Conditions: TsBr, MeCN, *hv*

Products:

I + **II**: Ts, Br, RO$_2$C, CO$_2$R cyclopentane structures

	I + II	**I:II**
	(70–80)	57:43
	(70–80)	63:37

Ref. 278

Reactant 3:

MeO_2C, CO_2Me (diallyl)

Conditions: TsNa, Cu(OAc)$_2$, AcOH, 90°

Product:

Ts, MeO$_2$C, CO$_2$Me, vinyl cyclopentane (73)

Ref. 280

Reactant 4:

MeO_2C, Br, Cl (chain)

Conditions: Bu$_3$SnH, AIBN, C$_6$H$_6$, 80°

Products:

I + II: CO$_2$Me, Cl cyclopentane structures

I + II (60)   **I:II** = 95:5

Ref. 17

392

TABLE I. MONOCYCLIC RINGS CONTAINING ONLY CARBON (*Continued*)

Reactant	Conditions	Product(s) and Yield(s) (%)	Refs.

Row 1

Reactant: structure with R^1, Br, R^2, MeO_2C, CO_2Me

R^1	R^2
Me	Me
OMe	Me
OTBDMS	H
CO_2Me	H
$S(CH_2)_3S$	
$O(CH_2)_3O$	

Conditions: Bu_3SnCl, AIBN, Na(CN)BH₃, *t*-BuOH, 80°

Products: **I** (R^1 R^2, MeO_2C, CO_2Me) + **II** (R^1 R^2, MeO_2C, CO_2Me)

$I + II$	I:II
(65)	—
(82)	14:86
(66)	40:60
(60)	82:18
(96)	—
(56)	—

Refs.: 197, 295

Row 2

Reactant: structure EtO_2C, EtO_2C

Conditions: $Mn(OAc)_3$, EtOH

Product: structure EtO_2C, EtO_2C (40)

Refs.: 296

Row 3

Reactant: structure MeO_2C, O

Conditions: $Mn(OAc)_3$, $Cu(OAc)_2$, AcOH, 50°

Product: **I** (MeO_2C, EtO_2C, O) + **II** (MeO_2C, O)

$I + II$ (67)   I:II = 7:3

Refs.: 192

TABLE I. MONOCYCLIC RINGS CONTAINING ONLY CARBON (Continued)

Reactant	Conditions	Product(s) and Yield(s) (%)	Refs.
	Bu₃SnH, AIBN C_6H_6, 80°	(42)	266
HC≡CR³	(PhS)₂, AIBN, hv	I + II	297

R¹	R²	R³		I:II	I+II
CO₂Bu-t	H	H	C_6H_6, 80°	1.9:1	(50)
CO₂Bu-t	Me	Me	AlMe₃, PhMe, –30°	3.8:1	(53)
OBn	H	H	C_6H_6, 80°	1.7:1	(50)

C₁₃

Reactant	Conditions	Product(s) and Yield(s) (%)	Refs.
	Bu₃SnH, AIBN C_6H_6, 80°	(91)	298

TABLE I. MONOCYCLIC RINGS CONTAINING ONLY CARBON (*Continued*)

Reactant	Conditions	Product(s) and Yield(s) (%)	Refs.
	1. Hg(OAc)$_2$, AcOH 2. NaBH(OMe)$_3$	(70)	299
	Bu$_3$SnH, AIBN,	(95)	300
	Ph$_2$PH, AIBN, C$_6$H$_6$, 80°	(36)	301
	TsSePh, AIBN, C$_6$H$_6$, 80°	(96)	302
	A. TsSePh, AIBN, CHCl$_3$, 61° **B.** TsSePh, CHCl$_3$, *hv*	**A.** (98) 7.2:1 ratio **B.** (89) 8.7:1 ratio	281

TABLE I. MONOCYCLIC RINGS CONTAINING ONLY CARBON (*Continued*)

Reactant	Conditions	Product(s) and Yield(s) (%)	Refs.
(structure: EtO₂C, CO₂Et diallyl)	I—CO₂Et, Cr(OAc)₂, THF, 25°	(88)	262
(structure with R¹, R²)	(R³)₃SiH, (t-BuO)₂, 140°	R¹ R² R³: CO₂Et CO₂Et Cl (84); Ph OMe Et (60)	303
(structure with CH≡C, R, R=H, Me)	(PhSe)₂, C₆H₆, hv, 40°	(70), (84)	304
(structure with CH≡C, CO₂Et)	Bu₃SnH, AIBN, C₆H₆, 80°	(—)	25

396

TABLE I. MONOCYCLIC RINGS CONTAINING ONLY CARBON (*Continued*)

Reactant	Conditions	Product(s) and Yield(s) (%)	Refs.
	Bu₃SnH, C₆H₆, *hv*, 35° → Bu₃SnH, C₆H₆, hv, 35°	(62) + (11)	305
2S, 3R, R_S → $2S, 3R, R_S$	Bu₃SnH, AIBN C₆H₆, *hv* (350 nm), 35° → Bu₃SnH, AIBN, C₆H₆, hv (350 nm), 35°	(75) 5.8:5.4:1:1 ratio of isomers	306
	Bu₃SnH, AIBN C₆H₆, 80° → Bu₃SnH, AIBN, C₆H₆, 80°	(90)	34
	Ph₂PH, AIBN, C₆H₆, 80° → Ph₂PH, AIBN, C₆H₆, 80°	(74)	301
	B(SePh)₃, AIBN → B(SePh)₃, AIBN	(40) + (21)	307

397

TABLE I. MONOCYCLIC RINGS CONTAINING ONLY CARBON (*Continued*)

Reactant	Conditions	Product(s) and Yield(s) (%)	Refs.
	Hg cathode, $CH_2(CO_2Me)_2$ n-Bu_4NBr, MeCN	*trans*:*cis* = 7.5:1 (66)	308
	Hg cathode, $CeCl_3$ $CH_2(CO_2Me)_2$ n-Bu_4NBr, MeCN	*trans*:*cis* = 14.8:1 (73)	
	Bu_3SnH, AIBN, C_6H_6, 80°		81
	Bu_3SnH, AIBN, C_6H_6, 80°	(>80)	309
	Cp_2TiCl, H^+	*cis*:*trans* = 85:15 (68)	172

For the sulfone reactant:

I:II	I + II
84:16	(72)
86:14	(40)
> 95:5	(82)
> 95:5	(94)

n	R
2	H
1	H
2	OMe
1	OMe

TABLE I. MONOCYCLIC RINGS CONTAINING ONLY CARBON (*Continued*)

Reactant	Conditions	Product(s) and Yield(s) (%)	Refs.
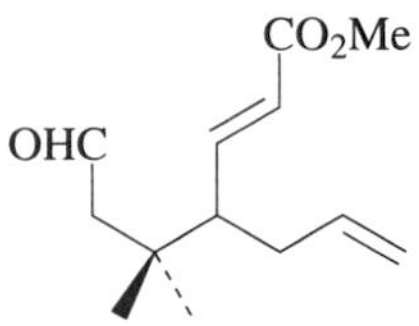	Hg cathode, $CH_2(CO_2Me)_2$ $n\text{-}Bu_4NBr$, MeCN	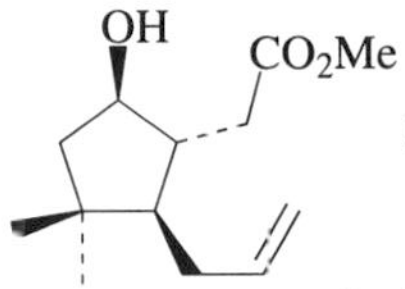 I 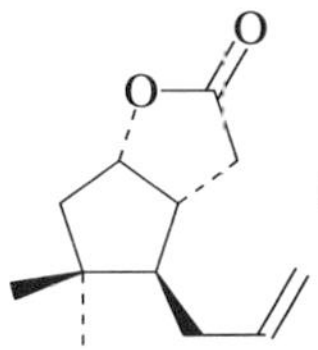 + II (89) I + II (89)	310
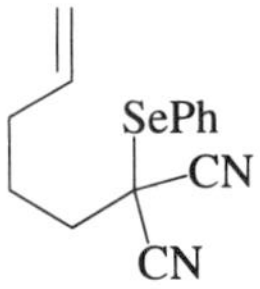	AIBN, C_6H_6, 80°	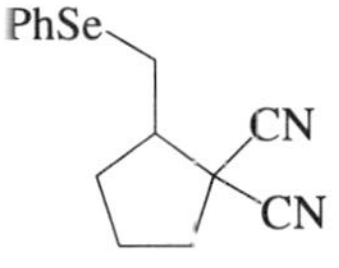 (69) + 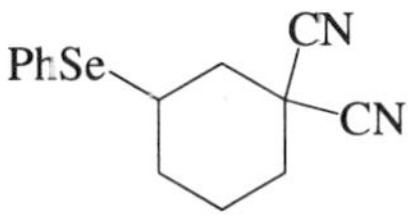 (19)	251
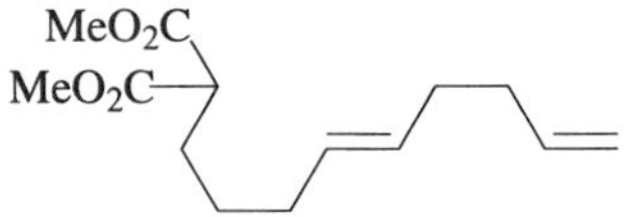	Mn(OAc)$_3$, Cu(OAc)$_2$, AcOH, 55°, 3 d	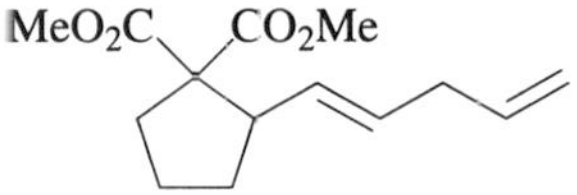 (65)	311
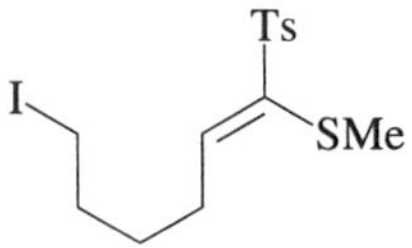	Bu$_3$SnH, AIBN, PhMe, 110°	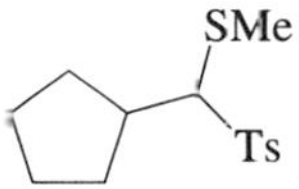 (89)	233
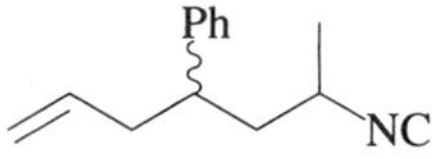	(TMS)$_3$SiH, C_6H_6, AIBN, 70°	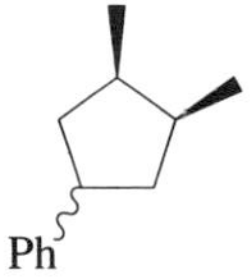 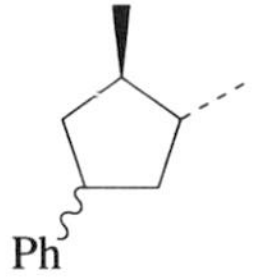 I + II (78) I:II = 4.6:1	312

C 14

399

TABLE I. MONOCYCLIC RINGS CONTAINING ONLY CARBON (*Continued*)

Reactant	Conditions	Product(s) and Yield(s) (%)	Refs.
	Et$_2$O, *hv*	(82)	220
	Bu$_3$SnH, AIBN, C$_6$H$_6$, heat	(75)	313
	A. (PhS)$_2$, AIBN, *hv* B. Ph$_3$SnH, AIBN	I + II + III	314

R		I	II	III	
A	H	(82)	(10)	(—)	Double bond
B	H	(81)	(8)	(3)	isomerizes
A	Me	(83)	(<2)	(—)	
B	Me	(85)	(<2)	(5)	

TABLE I. MONOCYCLIC RINGS CONTAINING ONLY CARBON (*Continued*)

Reactant	Conditions	Product(s) and Yield(s) (%)	Refs.
(structure: I...CF₃, OBz)	Bu₃SnH, AIBN, C₆H₆, 80°	(structure: CF₃, OBz) (88)	313
(structure: Tol, CHO)	NaC₁₀H₈, THF, rt	(structure: OH, Tol) (50) 6 : 1 mixture	315
(structures: Ph, R, X + Z)	Bu₃SnH, AIBN, C₆H₆, heat	(structure: R, Z, Ph)	243

R	X	Z	
Me	I	CN	(67)
Ph	I	CN	(75)
Ph	Br	CN	(70)
Ph	I	CO₂Et	(53)
Ph	Br	CO₂Et	(45)

Reactant	Conditions	Product(s) and Yield(s) (%)	Refs.
(structure: Ts)	BPO, CCl₄ TsNa, AcOH, H₂O	(structure: Ts) (93) (95)	316, 316a

TABLE I. MONOCYCLIC RINGS CONTAINING ONLY CARBON (*Continued*)

Reactant	Conditions	Product(s) and Yield(s) (%)	Refs.

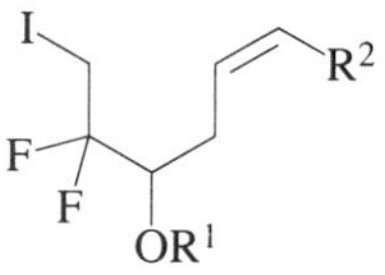

Reactant 1:

R^1	R^2
MOM	n-C$_6$H$_{13}$
MOM	Ph
MOM	BzOCH$_2$
TBDMS	BnCH$_2$

Conditions: Bu$_3$SnH, AIBN, C$_6$H$_6$, heat

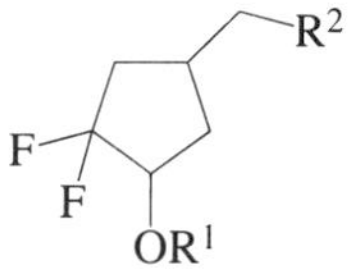

Product:

(96)
(91)
(84)
(60)

Refs. 317

Reactant 2:

R
n-Bu
OAc
CH$_2$TMS

Conditions: Mn(OAc)$_3$, Cu(OAc)$_2$, AcOH

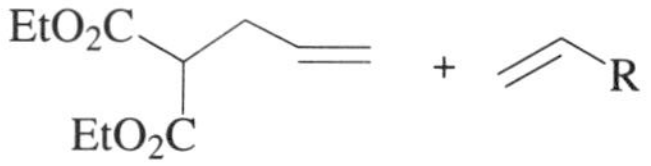

Product:

(35)
(40)
(89)

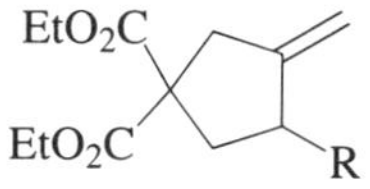

Refs. 211

Reactant 3:

Conditions: Bu$_3$SnH, AIBN, C$_6$H$_6$, 80°

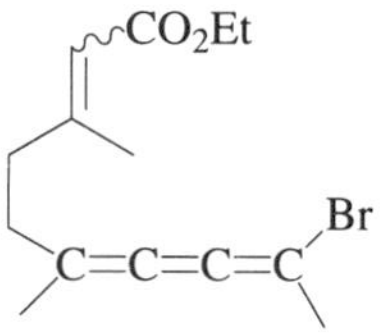

Product:

(45)

Refs. 266

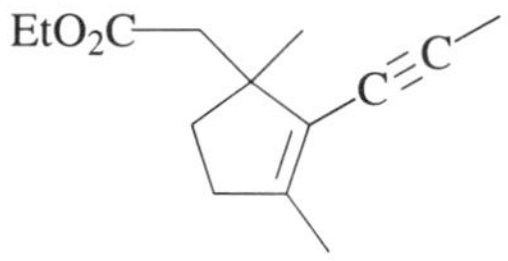

402

TABLE I. MONOCYCLIC RINGS CONTAINING ONLY CARBON (*Continued*)

Reactant	Conditions	Product(s) and Yield(s) (%)	Refs.

Bu$_3$SnH, AIBN, C$_6$H$_6$, 80°

R^1	R^2	R^3		**I**	**II**
H	H	H		(32)	(27)
Ph	H	H		(37)	(20)
Me	Me	H		(30)	(22)
H	H	Me		(31)	(—)

318

SmI$_2$, THF

R^1	R^2		diastereomeric ratio
Me	Me	(79)	31:1
Et	Et	(73)	65:1
i-Pr	i-Pr	(32)	>200:1
	(CH$_2$)$_5$	(58)	200:1
	(CH$_2$)$_4$	(60)	60:1
	CH(Me)(CH$_2$)$_4$	(75)	1:1
	(CH$_2$)$_2$CH(Bu-t)(CH$_2$)$_2$	(61)	10:1
Me	(CH$_2$)$_3$Cl	(65)	1:1
Me	(CH$_2$)$_3$NEt$_2$	(33)	1:1
H	n-Pr	(55)	17:17:1:1

179

TABLE I. MONOCYCLIC RINGS CONTAINING ONLY CARBON (*Continued*)

Reactant	Conditions	Product(s) and Yield(s) (%)	Refs.
(structure: X, Cl, F, F, OAc)	Bu₃SnH, AIBN, C₆H₆, 80°, 10 h	(structure) $\dfrac{X}{\text{OTBDMS (1:1)} \quad (81)}{\text{NBn}_2 \quad (28)}$	288
(structure: C≡CH, OTBDMS, O, S)	Bu₃SnH, AIBN, PhMe, 110°	(structure, OTBDMS, AcO, HO) (60)	319
(structure: EtO₂C, CO₂Et, HC≡C, NNHSO₂Me)	Bu₃SnH, AIBN, C₆H₆, 80°	(structure: EtO₂C, CO₂Et, Bu₃Sn) (67)	320
(structure: CO₂Me, O, Ph)	Mn(OAc)₃, Cu(OAc)₂, AcOH, 50°	(structure: O, CO₂Me, OAc, Ph) (70)	321
(structure: S, N, O, CO₂Me; + CO₂Me, CO₂Me, Z)	Heat	(structure: S, N, CO₂Me, MeO₂C, CO₂Me, MeO₂C) (55)	243

TABLE I. MONOCYCLIC RINGS CONTAINING ONLY CARBON (*Continued*)

Reactant	Conditions	Product(s) and Yield(s) (%)	Refs.
(cyclopropane: CO₂Me, CO₂Me, vinyl) + isopropenyl with R¹, R²	TsCl, BPO, PhMe, 100° *or* (Bu₃Sn)₂, C₆H₆, hv, 25°	cyclopentane with R¹, R², vinyl, CO₂Me, MeO₂C–	322

R¹	R²		isomer ratio	
OAc	Me	(77)	3.5:1	
OBn	Me	(73)	—	
CH₂OBn	Me	(81)	1.2:1	
CH₂OBn	H	(61)	2.4:1	
(CH₂)₂OBn	Me	(73)	1.1:1	
CH₂Cl	Me	(78)	1.2:1	
C₆H₁₃	H	(75)	2.5:1	

(Refs. 323)

Reactant	Conditions	Product(s) and Yield(s) (%)	Refs.
C₁₅ (I, NC, CN allyl) + Ph-CH=CH₂ (styrene)	Bu₃SnH, AIBN, C₆H₆, 80°	I + II I + II (97) I : II = 2.9 : 1	294
(Ts, isopropenyl, C≡CH)	BPO, CCl₄ TsNa, AcOH, H₂O	(Ts, isopropenyl cyclopentane) (80) (80)	316, 316a

TABLE I. MONOCYCLIC RINGS CONTAINING ONLY CARBON (*Continued*)

Reactant	Conditions	Product(s) and Yield(s) (%)	Refs.

Reactant 1: X, R², R², R¹, PhSO$_n$

Bu$_3$SnH, AIBN, C$_6$H$_6$, 80°

Products **I** + **II** (SO$_n$Ph substituted cyclopentanes)

n	X	R¹	R²		I:II	I + II
2	Cl	CO$_2$Et	H	E-alkene	30:70	(81)
2	Cl	CO$_2$Et	H	Z-alkene	50:50	(73)
1	Cl	CO$_2$Et	H	E-alkene	25:75	(70)
1	Cl	CO$_2$Et	H	Z-alkene	30:70	(88)
2	Br	H	CO$_2$Et		30:70	(70)
1	Br	H	CO$_2$Et		35:65	(60)

Refs. 324

Reactant 2: Ts (vinyl-substituted chain)

BPO, CCl$_4$, 77°

Product: Ts-vinyl cyclopentane (93) 3:1 ratio

Refs. 136

Reactant 3: EtO$_2$C, CO$_2$Et, CO$_2$Me, Ts (diene)

TsBr, MeCN, $h\nu$, 18°

Products **I, II** and **III, IV**

I + II (66)
I:II = 58:42

III + IV (34)
III:IV = 45:55

Refs. 219

TABLE I. MONOCYCLIC RINGS CONTAINING ONLY CARBON (*Continued*)

Reactant	Conditions	Product(s) and Yield(s) (%)	Refs.
	TsBr, MeCN, hv	 **I** — **II** **I + II** (67) **I:II** = 52:48	219
	Bu$_3$SnH, AIBN, C$_6$H$_6$, heat	 (60) *cis:trans* = 3:1	320
	Bu$_3$SnH, AIBN, C$_6$H$_6$, 80°, 3 h	 **I** + **II** **I + II** (93)	79
	(PhS)$_2$, AIBN, C$_6$H$_6$, 80° *or* (BuS)$_2$, hv	 **I,II**	324a

R^1	R^2		**I,II**	*cis:trans*
i-Bu	H		(81)	56:54
CH$_2$OH	H		(57)	81:19
PhS	H		(88)	80:20
n-Bu	H		(47)	76:24
OTMS	Me		(64)	33:67

TABLE I. MONOCYCLIC RINGS CONTAINING ONLY CARBON (*Continued*)

Reactant	Conditions	Product(s) and Yield(s) (%)	Refs.
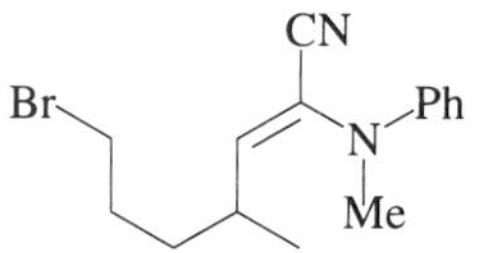	Bu_3SnH, AIBN, C_6H_6, 80°	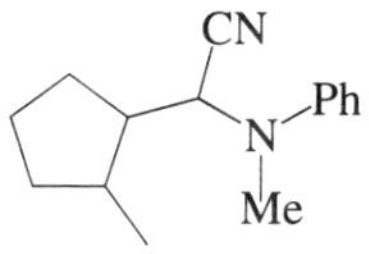 (83) two isomers 2.5:1	325
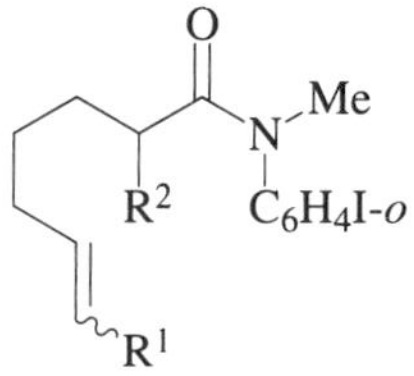 R^1 / R^1: CO_2Et H CO_2Et Me Ph H Me H	Bu_3SnH, AIBN, C_6H_6, 80°	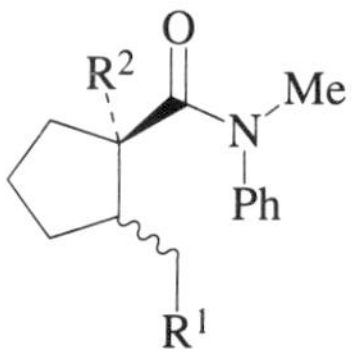 (94) *cis:trans* 2.1:1 (80) 1.4:1 (83) 1.3:1 (67) 1.3:1	199
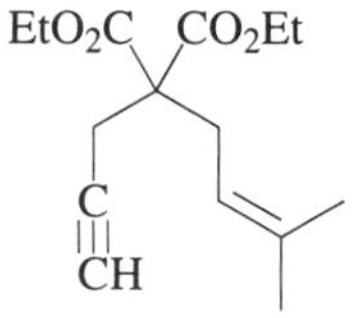	Bu_3SnH, AIBN, C_6H_6, 80°	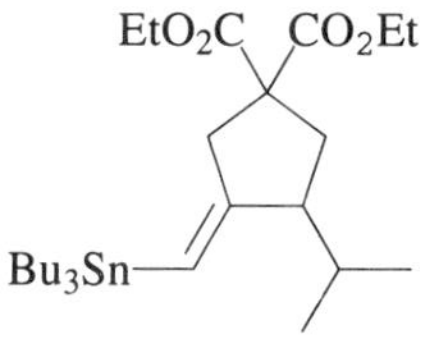 (85)	118
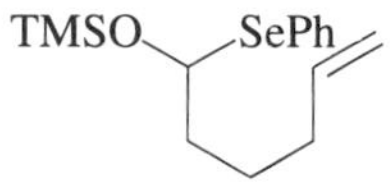	Bu_3SnH, AIBN, C_6H_6, 80°	 (65) *trans:cis* = 7:1	80

TABLE I. MONOCYCLIC RINGS CONTAINING ONLY CARBON (*Continued*)

Reactant	Conditions	Product(s) and Yield(s) (%)	Refs.
	Bu$_3$SnH, AIBN, C$_6$H$_6$, 80°	(67) *trans:cis* = 45:55 (67) *trans:cis* = 35:65	326
	Bu$_3$SnH, AIBN, C$_6$H$_6$, 80°	(45)	79
	Bu$_3$SnH, AIBN, C$_6$H$_6$, 80°, 10 h	(81) (60)	288
	Ph$_3$SnH, Et$_3$B, C$_6$H$_6$	(84)	27

TABLE I. MONOCYCLIC RINGS CONTAINING ONLY CARBON (*Continued*)

Reactant	Conditions	Product(s) and Yield(s) (%)	Refs.
C$_{16}$			
MeO$_2$C, CO$_2$Me, Br, Ph	Bu$_3$SnCl, Na(CN)BH$_3$, AIBN, t-BuOH, 80°	MeO$_2$C, CO$_2$Me, Ph (53)	197
Ts	BPO, CCl$_4$, 77°	(51) 3:1 ratio, Ts	135
Ts	BPO, CCl$_4$ TsNa, AcOH, H$_2$O	Ts (80) (53)	316, 316a
Ts	BPO, CCl$_4$ TsNa, AcOH, H$_2$O	Ts (90) (95)	316, 316a
S, Im, O, CN	Bu$_3$SnH, AIBN, C$_6$H$_6$, 80°	O (72)	38

TABLE I. MONOCYCLIC RINGS CONTAINING ONLY CARBON (*Continued*)

Reactant	Conditions	Product(s) and Yield(s) (%)	Refs.

The reactant is a cyclopropane structure with R¹, R² substituents and CO₂Bu-t group plus an alkene.

Conditions:
A: (PhS)₂, AIBN, C₆H₆, 80°
B: (PhS)₂, AIBN, AlMe₃, C₆H₆, 80°

Products **I**, **II**, **III**, **IV** (cyclopentane derivatives with Y, R², R¹ and CO₂Bu-t)

R¹	R²	Y		I : II : III : IV	
H	H	OBu-n	A	6.1 : 1.4 : 1.3 : 1	(94)
			B	14.2 : 2.0 : 2.2 : 1	(75)
H	H	O₂CBu-t	A	4.0 : 2.2 : — : 1	(55)
			B	12.8 : 10.9 : 1 : 4.4	(52)
H	H	CO₂Bu-t	A	4.2 : 1.2 : 1 : 1.9	(53)
			B	10.3 : 4.0 : 1 : 1.3	(52)
Me	H	OBu-n	A	4.2 : 1.4 : 1 : 1.1	(96)
			B	7.0 : 2.6 : 1 : 1.2	(70)
H	Me	OBu-n	A	1.0 : 1.0 : — : —	(66)
			B	1.8 : 1.0 : — : —	(69)

The second reactant is an epoxide structure with X, OBn substituents and an allyl group.

Conditions: Cp₂TiCl, H⁺

Products: cyclopentane with HO, OBn, X substituents (**I, II, III, IV**)

X = OAc, 4 isomers (74)
X = OBn, 45:30:15:19 (44)

Refs. 327, 270

Refs. 172

TABLE I. MONOCYCLIC RINGS CONTAINING ONLY CARBON (*Continued*)

Reactant	Conditions	Product(s) and Yield(s) (%)	Refs.

Reactant 1:

Structure with CO_2Et, CO_2Et + Y (vinyl)

Y
OBu-*i*
Bu-*n*
SPh

Conditions: $(BuS)_2$, hv, 25°

Product: cyclopentane structure with EtO_2C, EtO_2C, Y, isopropylidene

(81)
(47)
(47)

Refs: 328

Reactant 2:

Aromatic structure with R, O, X, CO_2Et

R	X
H	I
H	Br
Me	I

Conditions: Bu_3SnCl, AIBN, *t*-BuOH, $Na(CN)BH_3$

Product: cyclopentane with R, OBn, CO_2Et

	cis : *trans*
(56)	1 : 2.5
(46)	1 : 2.5
(45)	1 : 4

Refs: 197

Reactant 3:

TBDMS, SnMe₃ structure + R (vinyl), I

R
CN
CO_2Me
COMe
SO_2Ph

Conditions: Bu_3SnH, AIBN, C_6H_6, 80°

Product: TBDMS cyclopentane with R

		E : *Z*
(56)		89 : 11
(56)		98 : 2
(53)		97 : 3
(57)		97 : 3

Refs: 329

TABLE I. MONOCYCLIC RINGS CONTAINING ONLY CARBON (Continued)

Reactant	Conditions	Product(s) and Yield(s) (%)	Refs.
	Bu₃SnH, AIBN, C₆H₆, 80°	R *cis* : *trans* H 1.2 : 1 (81) Me 1 : 1 (84)	199
	Me₃SnCl, Na(CN)BH₃, AIBN, *t*-BuOH, 80°	**I, II** **I + II (92) I : II = 2 : 1**	330
	Bu₃SnH, AIBN, C₆H₆, 80°	(12) + (61)	331
	Mn(OAc)₃, Cu(OAc)₂, AcOH	(36)	211
	Bu₃SnCl, Na(CN)BH₃, AIBN, *t*-BuOH, 80°	(66) *cis* : *trans* = 40 : 60	197

413

TABLE I. MONOCYCLIC RINGS CONTAINING ONLY CARBON (*Continued*)

Reactant	Conditions	Product(s) and Yield(s) (%)	Refs.
	Bu₃SnH, AIBN, C₆H₆, 80°	(78)	332
C₁₇	Bu₃SnH, AIBN, C₆H₆, 80°	(70) 1 : 1 mixture	333, 334
	TsNa, AcOH, 100°	(68) 3 : 1 *ratio*	135
	BPO, CCl₄ TsNa, AcOH, H₂O	(50) (60)	316, 316a

414

TABLE I. MONOCYCLIC RINGS CONTAINING ONLY CARBON (*Continued*)

Reactant	Conditions	Product(s) and Yield(s) (%)	Refs.
	Bu$_3$SnH, AIBN, C$_6$H$_{14}$, 60°	(22) + (35)	335
	BPO, CCl$_4$ TsNa, AcOH, H$_2$O	(60) (90)	316, 316a
	Bu$_3$SnH, AIBN, Na(CN)BH$_3$, *t*-BuOH, 80°	(61) *cis* : *trans* = 1.1 : 1	197
R: *n*-C$_6$H$_{13}$, *n*-C$_9$H$_{19}$, CH$_2$OCOPh, Ph	Bu$_3$SnH, AIBN, C$_6$H$_6$, 80°	(66) (83) (81) (69)	336

TABLE I. MONOCYCLIC RINGS CONTAINING ONLY CARBON (*Continued*)

Reactant	Conditions	Product(s) and Yield(s) (%)	Refs.
MeO$_2$C$\sim\!\sim$CO$_2$Me + R^{1}R^{2}C=CH$_2$ 	R^1 / R^2: H / C$_6$H$_{13}$ Me / (CH$_2$)$_2$OC$_6$H$_{13}$ H / CH$_2$Ts Mn(OAc)$_3$, Cu(OAc)$_2$, AcOH	(81) 2.5:1 (60) 1:1 (56)	210
C$_{18}$			
(cyclopentyl SiPh$_2$Me ketone, with Br chain) py(dmgH)$_2$Co	Bu$_3$SnH, AIBN, C$_6$H$_6$, 80°	OSiMePh$_2$ (cyclopentyl) (81)	337
(HO-substituted alkene SiPh$_2$Me)	C$_6$H$_6$, hv, 25°	(methylenecyclohexanol) (63)	168
(ketone SiPh$_2$Me, I chain)	$\sim\!\sim$SnBu$_3$ AIBN, C$_6$H$_6$, 80°	MePh$_2$SiO (allyl cyclopentane) (61)	338
(divinyl/phenyl dicyclopropyl) + $\sim\!\sim$CO$_2$Me	(PhS)$_2$, AIBN, C$_6$H$_6$, hv, 80°	(cyclopentane with CO$_2$Me, vinyl, Ph, CO$_2$Me) (58)	327

TABLE I. MONOCYCLIC RINGS CONTAINING ONLY CARBON (*Continued*)

Reactant	Conditions	Product(s) and Yield(s) (%)	Refs.
+ alkene with R^1, R^2 R^1 | R^2 H | OBu-i H | CH$_2$OH Me | OTMS H | Bu-n	(BuS)$_2$, AIBN	 *cis:trans* (81) 56:44 (57) 83:17 (57) 61:39 (42) 77:23	339
	Ph$_3$SnH, AIBN	(35)	340
+ alkene R^2, R^3 R^1 | R^2 | R^3 Et | H | C$_6$H$_{13}$ Et | Me | (CH$_2$)$_2$OC$_6$H$_{13}$ Me | Me | CH$_2$Ts	Mn(OAc)$_3$, Cu(OAc)$_2$.	 (80) (60) (55)	210

417

TABLE I. MONOCYCLIC RINGS CONTAINING ONLY CARBON (*Continued*)

Reactant	Conditions	Product(s) and Yield(s) (%)	Refs.
C19			
	1. Bu₃SnH, AIBN, C₆H₆, 80° 2. AcOH, MeOH	(56)	341
	Bu₃SnH, AIBN, C₆H₆, 80°	(76) *cis:trans* = 55:45	342
	Bu₃SnH, AIBN, PhMe, 110°	(85)	343
	Me₃SnCl, AIBN, Na(CN)BH₃, t-BuOH, 80°	(95) *cis:trans* < 4:1	330
C20			
	(Bu₃Sn)₂, hv	(82)	85

418

TABLE I. MONOCYCLIC RINGS CONTAINING ONLY CARBON (*Continued*)

Reactant	Conditions	Product(s) and Yield(s) (%)	Refs.

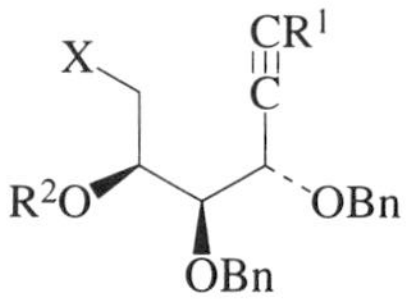

Reactant structures and data:

R^1, R^2, X table:

R^1	R^2	X
H	H	I
Me	H	I
TMS	H	I
CO_2Me	Ac	Br

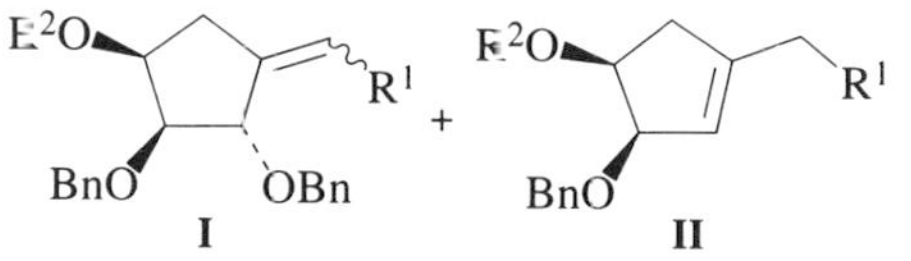

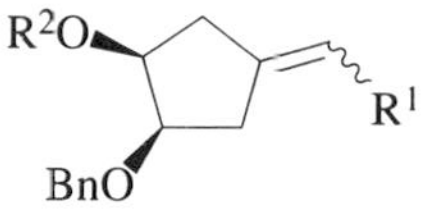

Bu$_3$SnH, AIBN, C$_6$H$_6$, 80°

Products **I**, **II**, **III**

R^1	**II + III**	**II:III**
(8)	(80)	78:22
(16)	(80)	62:38
(3)	(74)	86:14
(—)	(44)	30:70

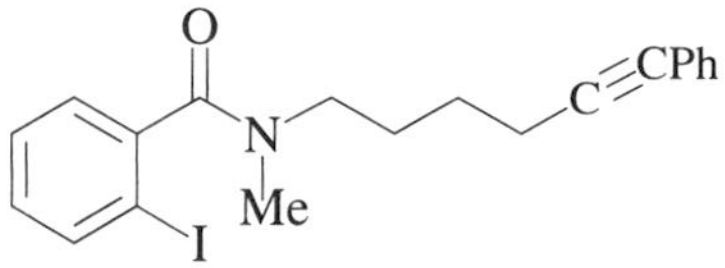

344

Bu$_3$SnH, AIBN, C$_6$H$_6$, 80°

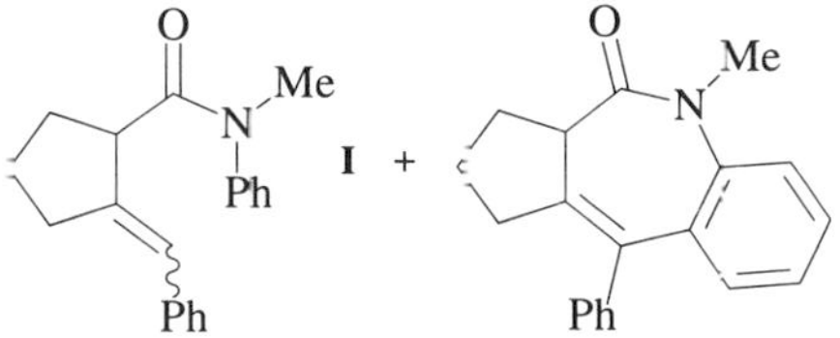

I + II (52)   **I:II** 1:1.3

199

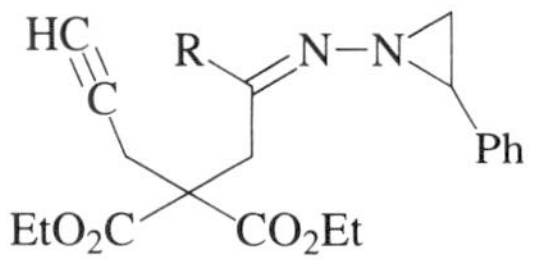

Bu$_3$SnH, AIBN, PhMe, 110°

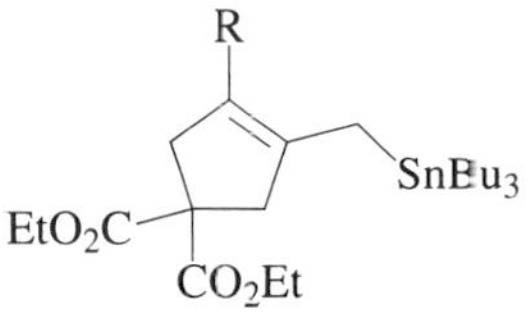

R = H (84)
R = Me (96)

343

419

TABLE I. MONOCYCLIC RINGS CONTAINING ONLY CARBON (*Continued*)

Reactant	Conditions	Product(s) and Yield(s) (%)	Refs.
	Bu₃SnH, AIBN, C₆H₆, 80°	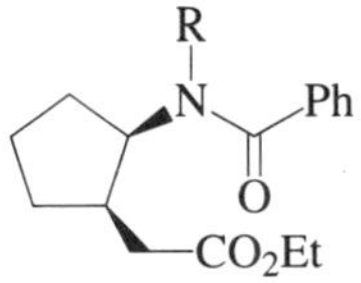	345

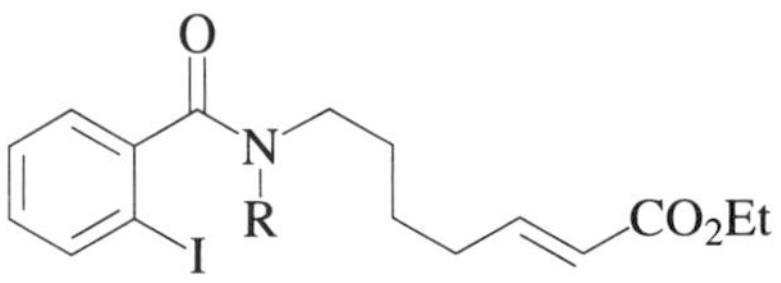

$$\frac{R}{\begin{array}{l}C_6H_{11}\\ Bu\\ Bu\\ (CH_2)_4CH=CHCO_2Et\end{array}}$$

(27)
(43)
(38)
(82)

| | Bu₃SnH, AIBN, PhMe, 110° | | 343 |

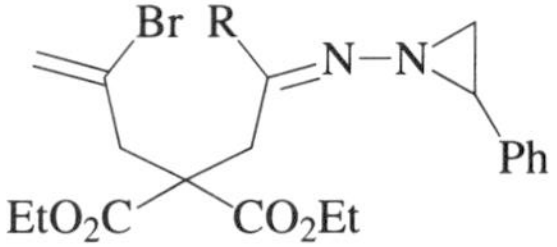

$$\frac{R}{\begin{array}{l}H\\ Me\end{array}}$$

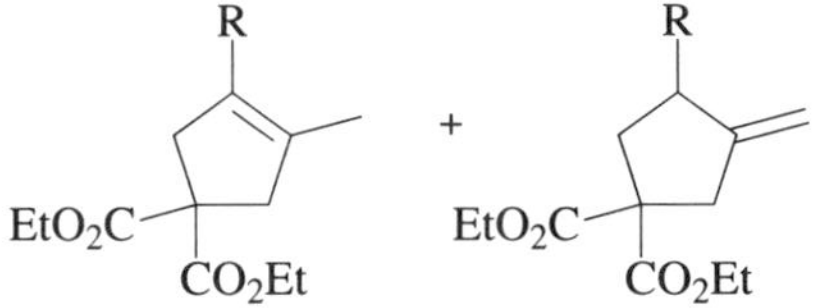

	I + II	I:II
	(67)	87:13
	(92)	94:6

C₂₁			
	Cp₂TiCl, THF, rt	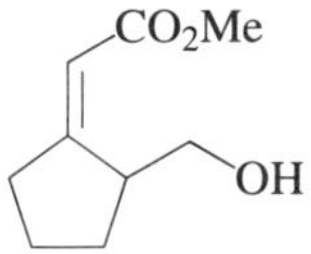 (83)	346
	Cp₂TiCl, THF, rt	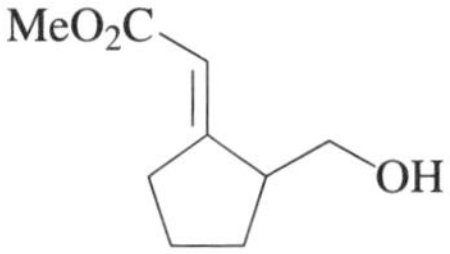 (83)	346

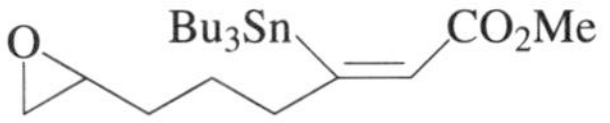

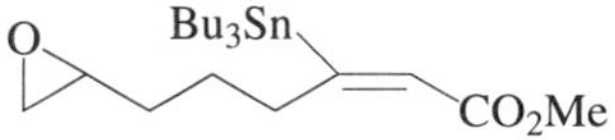

420

TABLE I. MONOCYCLIC RINGS CONTAINING ONLY CARBON (*Continued*)

Reactant	Conditions	Product(s) and Yield(s) (%)	Refs.

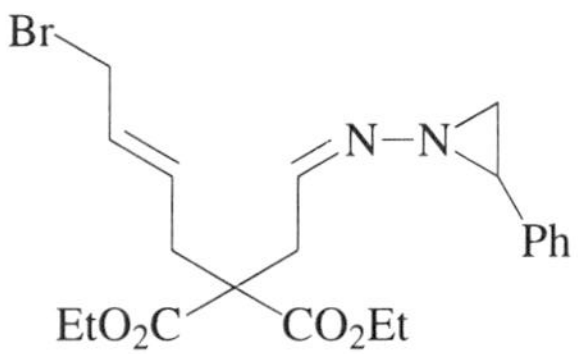

Row 1:
- Conditions: Bu$_3$SnH, AIBN, PhMe, 110°
- Product: (89)
- Refs.: 343

Row 2:
- Conditions:
  1. I C≡CH
  C$_6$H$_6$, *hv*, 80°
  2. Bu$_3$SnH
- Product: I + II
  I + II (48) I: 99:1 mixture of diastereomers
  I:II = 71:29   (major isomer is shown)
  II: one diastereomer
- Refs.: 347

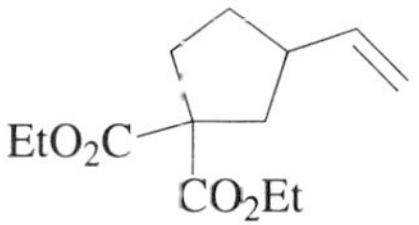

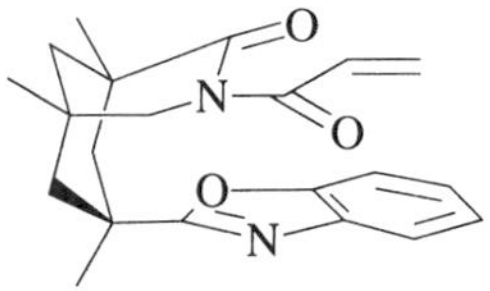

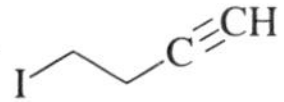

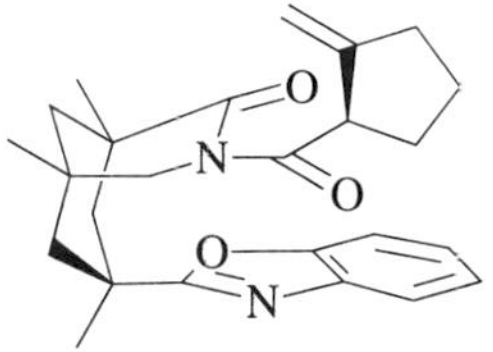

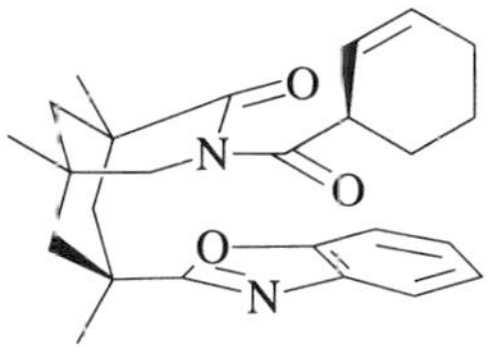

Row 3 (C$_22$):
- Conditions: Ce(NH$_4$)$_2$(NO$_2$)$_6$
- Product: (42)
- Refs.: 348

Reactant structures include: EtO$_2$C, CO$_2$Et, Br-, N–N, Ph; OSi(thexyl)Me$_2$, Bu-*t*

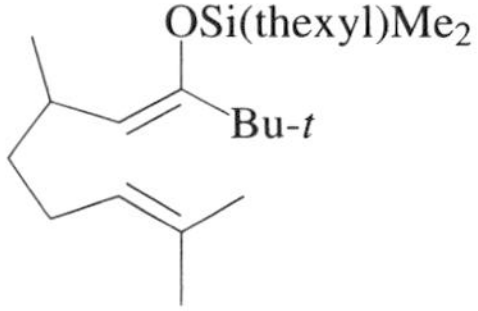

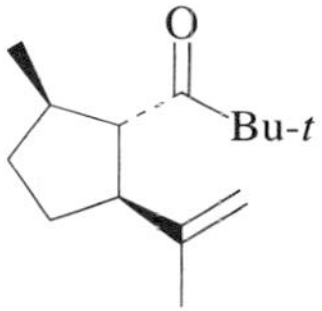

421

TABLE I. MONOCYCLIC RINGS CONTAINING ONLY CARBON (*Continued*)

Reactant	Conditions	Product(s) and Yield(s) (%)	Refs.
C23			
(structure: S–Im thioimidazolide ester, C≡CC5H11, Ph chain)	Bu3SnH, AIBN, C6H6, 80°	(cyclopentane with =CHCH3... C5H11 and CH2CH2Ph substituents) (81)	38
(structure: R1, EtO2C–C(CO2Et), allyl and isopropenyl groups) + (R2, CH2Ts, isopropenyl)	A. TsCl, BPO, PhMe B. (Bu3Sn)2, C6H6, hv	(cyclopentane product with R2, R1, CH2Ts, EtO2C, CO2Et)	349

R1 / R2 table:

R1	R2		
H	CO2Et	A	(64)
H	CO2Et	B	(51)
CO2Et	CO2Et	A	(53) 1.1:1
H	Me	B	(65)
H	Cl	B	(86)
H	H	B	(49)

C24			
(structure: GePh3, cyclohexenone-type with terminal alkene)	THF, *hv*, 25°	(cyclopentanone with CH2CH2GePh3) (92)	350
(structure: MeO2C, CO2Me diene) + (O=C(p-MeOC6H4)CH2SePh)	Bu3SnH, AIBN, C6H6, 80°	(cyclopentane with CO2Me, C(O)CH2–p-MeOC6H4, CH2CO2Me) (61) *cis:trans* = 2.4:1	209

TABLE I. MONOCYCLIC RINGS CONTAINING ONLY CARBON (*Continued*)

Reactant	Conditions	Product(s) and Yield(s) (%)	Refs.

Row 1:

Conditions: Bu₃SnH, AIBN, C₆H₆, 80°

Product: **I** + **II**

R = Bn **I:II** = 62:38   **I + II** (93)
R = Me **I:II** = 50:31   **I + II** (82)

Refs.: 37

Row 2:

Conditions: PhMe, 110°

Products:
(64)
(60) 9.6:1
(53) 1.1:1

R¹ / R²:
H / CO₂Et
H / COMe
CO₂Et / CO₂Et

Refs.: 351, 210, 349

Row 3:

Conditions: Bu₃SnH, AIBN, 80°

Product: **I** + **II**

R¹	R²	R³	X		**I + II**	**I:II**
C₆H₁₁	H	H	PhOC(=S)O	C₆H₆	(84)	52:48
C₆H₁₁	H	Me	PhOC(=S)O	PhMe	(74)	69:31
C₆H₁₁	Me	H	Br	C₆H₆	(63)	78:22
p-MeOC₆H₄	H	H	PhOC(=S)O	C₆H₆	(42)	>98:2
BnOMe	H	H	PhOC(=S)O	PhMe	(59)	50:50

Refs.: 37

423

TABLE I. MONOCYCLIC RINGS CONTAINING ONLY CARBON (*Continued*)

Reactant	Conditions	Product(s) and Yield(s) (%)	Refs.
C$_{25}$	Ph$_3$SnH, AIBN	(64)	352
C$_{26}$ R^1 R^2 Bn H Bn Me *n*-C$_{10}$H$_{21}$ H	PhMe, heat	 (50) (52) (63)	11
	Ph$_3$SnH, AIBN, C$_6$H$_6$, 80°	(79)	353, 354, 354a

TABLE I. MONOCYCLIC RINGS CONTAINING ONLY CARBON (*Continued*)

Reactant	Conditions	Product(s) and Yield(s) (%)	Refs.
C$_{27}$ (structure with Ph, N–N–aziridine-Ph, EtO$_2$C, CO$_2$Et)	Bu$_3$SnH, AIBN, PhMe, 110°	(cyclopentane with Bu$_3$Sn, Ph, EtO$_2$C, CO$_2$Et) (92)	343
(structure with R, N–N–aziridine-Ph, EtO$_2$C, CO$_2$Et)	Bu$_3$SnH, AIBN, PhMe, 110°	(cyclopentene with R, EtO$_2$C, CO$_2$Et) R = H (62), R = Me (82)	343
C$_{29}$ (menthyl ester structure with SPh, Ph, *t*-Bu)	Bu$_3$SnH, AIBN, C$_6$H$_6$, 80°, 6 h	R*O$_2$C $\overset{S}{\cdots}$ (I) + R*O$_2$C $\overset{R}{\cdots}$ (II) I + II (90), I:II = 4:1 R* = 8-phenylmenthyl I: *cis:trans* = 64:36 II: *cis:trans* = 35:65	84
(Co(salophen)py structure with O, CO$_2$Et)	Heat, sunlamp	(cyclopentanone with CO$_2$Et) (25) + (cyclopentanone with CO$_2$Et) (28)	355

425

TABLE I. MONOCYCLIC RINGS CONTAINING ONLY CARBON (*Continued*)

Reactant	Conditions	Product(s) and Yield(s) (%)	Refs.

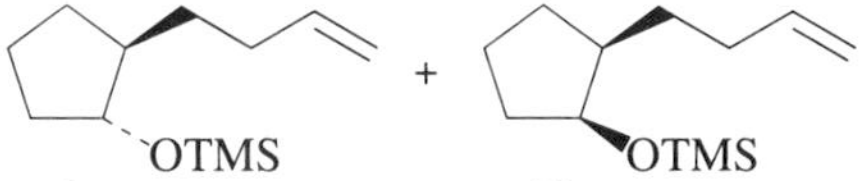

C$_{30}$

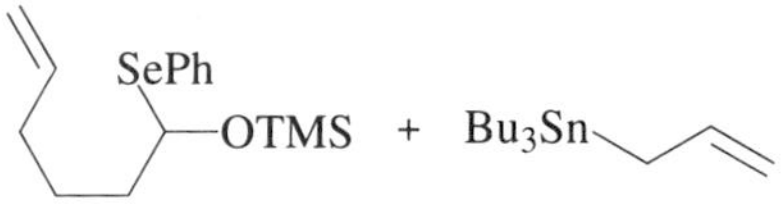

(first row)
AIBN, C$_6$H$_6$, 80°

I + **II** (67) **I:II** = 7:1

80

(second row)
Me$_3$SnCl, Na(CN)BH$_3$, AIBN, *t*-BuOH, 80°

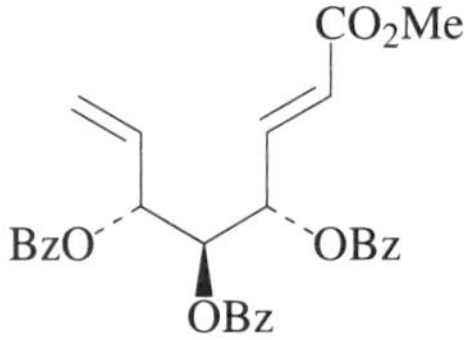

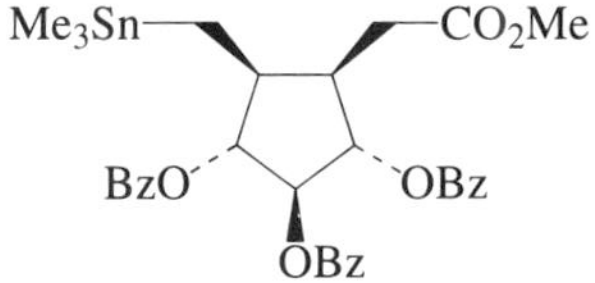

(52)

330

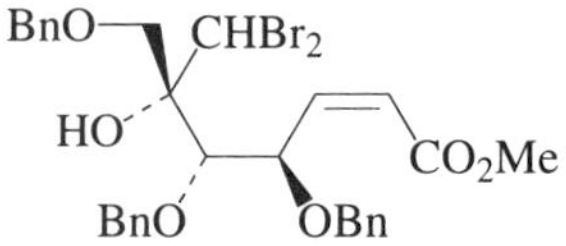

(third row)
Bu$_3$SnH, AIBN, C$_6$H$_6$, 80°

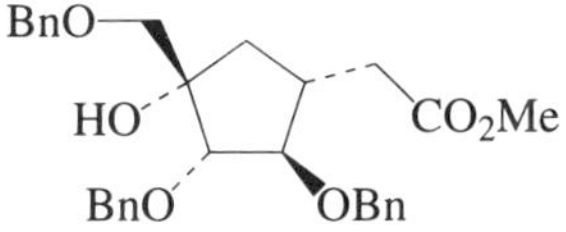

(80)

356, 357

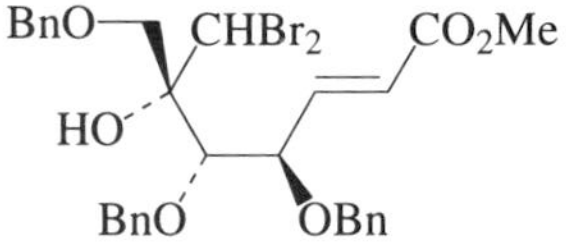

(fourth row)
Bu$_3$SnH, AIBN, C$_6$H$_6$, 80°

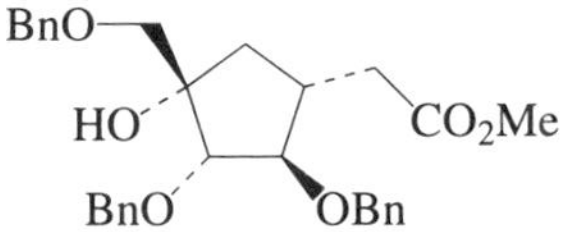

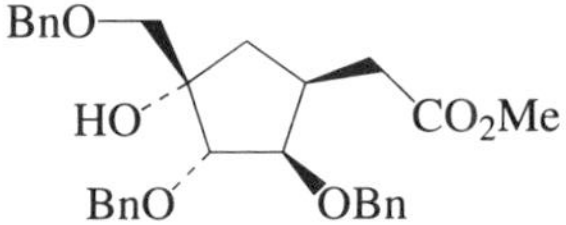

I + **II** (66), **I:II** = 1:4.5

356, 357

TABLE I. MONOCYCLIC RINGS CONTAINING ONLY CARBON (*Continued*)

Reactant	Conditions	Product(s) and Yield(s) (%)	Refs.
C₃₁	Bu₃SnH, AIBN, C₆H₆, 80°		358
C₃₂	Heat, sunlamp	(42)	355
C₃₃	Bu₃SnH, AIBN, C₆H₆, 80°	I + II (77) I:II = 8:1	359

427

TABLE I. MONOCYCLIC RINGS CONTAINING ONLY CARBON (*Continued*)

Reactant	Conditions	Product(s) and Yield(s) (%)	Refs.

C$_{39}$

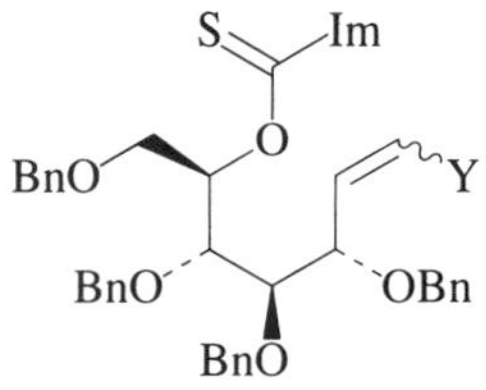

Bu$_3$SnH, AIBN, PhMe, 110°

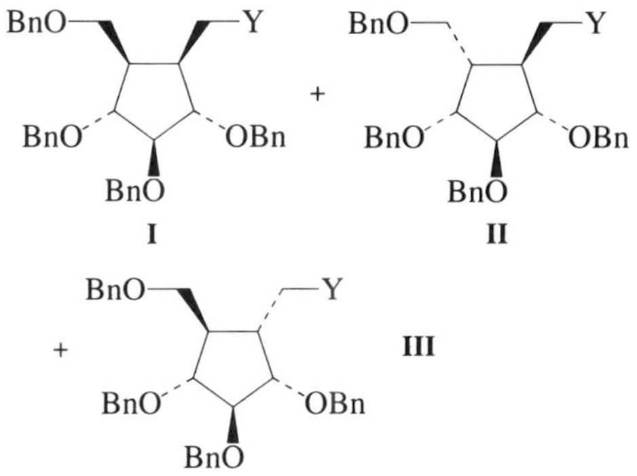

I + II + III (61) I:II:III = 74:14:12
I + II + III (63) I:II:III = 78:20:2

82, 360

Y = H
Y = OMe

D. 6-Membered Rings

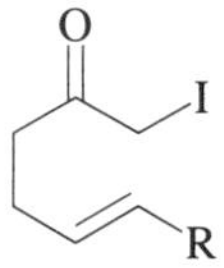

(Me$_3$Sn)$_2$, C$_6$H$_6$, *hv*

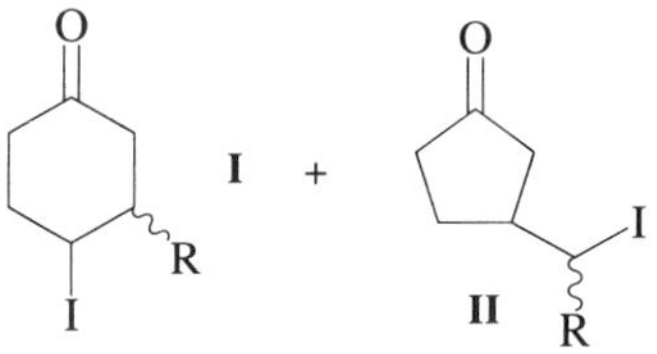

18

	I + II	I:II
	(56)	>97:3
	(73)	75:25

C$_6$

R
H
Me

428

TABLE I. MONOCYCLIC RINGS CONTAINING ONLY CARBON (*Continued*)

Reactant	Conditions	Product(s) and Yield(s) (%)	Refs.
C₇			
	PhSH, AIBN, TMTHF, reflux	FhS (70)	361
C₈			
HC≡CCO₂Et + R $\frac{R}{\begin{array}{c}CO_2Me\\CN\end{array}}$	Bu₃SnH, AIBN, C₆H₆, 80°	CO₂Et (1:1 mixture) (43) (48)	362
	Carbon rod cathode, MeOH, dioxane, Et₄NOTs	HO (75) **I** + HO **II** I + II (91) I:II = 36:1	246 178
$\frac{R}{\begin{array}{c}OMe\\Me\end{array}}$	Mn(OAc)₃, Cu(OAc)₂, AcOH	OH O R (94) (96)	363, 321

TABLE I. MONOCYCLIC RINGS CONTAINING ONLY CARBON (*Continued*)

Reactant	Conditions	Product(s) and Yield(s) (%)	Refs.
	A. (PhS)₂, AIBN, *hv* B. Bu₃SnH, AIBN		314
	$\dfrac{R}{H}$ A (82) H B (81) Me A (85) 46:37 Me B (87)		
	Bu₃SnH, AIBN, C₆H₆, 80°	(82)	200
C₉	(Bu₃Sn)₂, C₆H₆, *hv*	 **I** + **II** + **III** (91) **I**:(**II** + **III**) = 55:45	364
	Mn(OAc)₃, Cu(OAc)₂, AcOH	(71)	363

430

TABLE I. MONOCYCLIC RINGS CONTAINING ONLY CARBON (*Continued*)

Reactant	Conditions	Product(s) and Yield(s) (%)	Refs.
	Mn(OAc)$_3$, Cu(OAc)$_2$, AcOH	(38)	363
	Mn(OAc)$_3$, Cu(OAc)$_2$, AcOH	(78)	363
	Mn(OAc)$_3$, Cu(OAc)$_2$, AcOH	(70)	363
	SmI$_2$, THF, HMPA	(86) diastereomeric ratio = 4:1	178
	SmI$_2$, THF, HMPA	(89) diastereomeric ratio = 6:1	178

TABLE I. MONOCYCLIC RINGS CONTAINING ONLY CARBON (*Continued*)

Reactant	Conditions	Product(s) and Yield(s) (%)	Refs.
C$_{10}$ OHC⌒⌒⌒⌒CO$_2$Me	Hg cathode, Et$_4$NOTs, H$_2$O, CH$_2$(CO$_2$Et)$_2$	(72) *trans:cis* 1.8:1	365
I—C(CN)(CN)—ring—⌇⌒	AIBN, C$_6$H$_6$, 80°, 10 h	(95) 52:48 *ratio*	251
HC≡CCO$_2$Et + ⌇CO$_2$Me	Bu$_3$SnH, AIBN, C$_6$H$_6$, 80°	(46)	362
CO$_2$Et—Br⌒⌒⌒	Bu$_3$SnH, AIBN, C$_6$H$_6$, *hv*	(100)	366
O=⌇⌒CO$_2$Me	Mn(OAc)$_3$, Cu(OAc)$_2$, AcOH	(91)	363

TABLE I. MONOCYCLIC RINGS CONTAINING ONLY CARBON (*Continued*)

Reactant	Conditions	Product(s) and Yield(s) (%)	Refs.
(structure with CO$_2$Me ketone)	Mn(OAc)$_3$, Cu(OAc)$_2$, AcOH	(phenol with CO$_2$Me) (40)	363
(acyl-X structure) $\begin{array}{c} \underline{X} \\ \text{SePh} \\ \text{SPh} \\ \text{Cl} \end{array}$	Bu$_3$SnH, AIBN, C$_6$H$_6$, 80°, 2.5 h	(cyclohexanone) (84) / no reaction / (59)	35, 367
(methylenecyclopropane structure) $\begin{array}{c|c} R^1 & R^2 \\ \hline Bn & H \\ H & TMS \end{array}$	Bu$_3$SnH, AIBN, PhMe, 110°	(methylenecyclohexane, R^1, R^2) (71) / (4)	368
(bromide structure)	(Bu$_3$Sn)$_2$–resin, Me$_2$CO, *i*-PrOH, *hv* (300 nm), 40°	(89)	369

433

TABLE I. MONOCYCLIC RINGS CONTAINING ONLY CARBON (*Continued*)

Reactant	Conditions	Product(s) and Yield(s) (%)	Refs.
R^{1}O, Br, CO$_2$Me, R^2, R^3	Bu$_3$SnH, AIBN, PhMe, 110°	I (R^{1}O, R^2, R^3, CO$_2$Me) + II (R^{1}O, R^2, R^3, CO$_2$Me)	370

R^1	R^2	R^3	I + II	I : II
Ac	OBn	OMe	(100)	34 : 68
TBDMS	OBn	OMe	(90)	68 : 34
Ac	H	OBn	(80)	37 : 63
TBDMS	H	OBn	(91)	55 : 45
Ac	OBn	H	(72)	50 : 50
TBDMS	OBn	H	(70)	74 : 26
Ms	H	H	(81)	35 : 65

C$_{11}$

Reactant	Conditions	Product(s) and Yield(s) (%)	Refs.
I, CO$_2$Me, CO$_2$Me (methyl)	(Me$_3$Sn)$_2$, Bu$_3$SnH, hv	CO$_2$Me, CO$_2$Me cyclohexane (84)	18
O, CO$_2$Me, HO (chain)	Mn(OAc)$_3$, Cu(OAc)$_2$, AcOH, 60°, 1 h	CO$_2$Me, HO α-OH (41) β-OH (8)	271

434

TABLE I. MONOCYCLIC RINGS CONTAINING ONLY CARBON (*Continued*)

Reactant	Conditions	Product(s) and Yield(s) (%)	Refs.
EtO₂C—CN (structure)	BPO, c-C₆H₁₂, 85°	(structure) CN, CO₂Et (75)	22
(structure with I, HO)	Bu₃SnH, C₆H₆, hv, heat	(structure, HO) (72)	24
(structure) O—OMe + O=CH (acetaldehyde)	BPO, Ph₂CO, hv	**I** (OMe) + **II** (OMe)	371
		I + II **I:II**	
	−25°	(45) 1:1.4	
	4°	(57) 1:1.3	
	25°	(41) 1:1.2	
	78°	(43) 1:1.2	
O—CO₂Me (structure)	Mn(OAc)₃, Cu(OAc)₂, AcOH, 60°	(structure) CO₂Me (41)	271, 192

435

TABLE I. MONOCYCLIC RINGS CONTAINING ONLY CARBON (*Continued*)

Reactant	Conditions	Product(s) and Yield(s) (%)	Refs.
(ketone, CH_2CO_2Me, dimethylalkenyl side chain)	1. Mn(OAc)$_3$, Cu(OAc)$_2$, 2. AcOH, LiCl, 100°	(phenol, OH, CO_2Me, trimethyl aromatic) (50)	363
(ketone, CH_2CO_2Me, Z/E alkene)	Mn(OAc)$_3$, Cu(OAc)$_2$, AcOH, 50°	I (cyclohexanone, CO_2Me, propenyl) + II (cyclohexene, OH, CO_2Me, propenyl) **I + II** (71) **I + II** (64)	192
Z, E			
(MeSe ketone, isopropenyl chain)	Bu$_3$SnH, AIBN, C$_6$H$_6$, 80°	**I, II** (isopropyl cyclohexanone) **I + II** (85) **I:II = 2:1**	34
(silyl enol ether R^1, Me, Si, O, Me, R^2, hexenyl)	DCA, MeCN, hv, 65°	(cyclohexyl C(O)R^1)	372

R^1	R^2	
Ph	Me	(38)
Ph	t-Bu	(31)
C$_6$H$_{11}$	Me	(27)
Me	Me	(4)

436

TABLE I. MONOCYCLIC RINGS CONTAINING ONLY CARBON (*Continued*)

Reactant	Conditions	Product(s) and Yield(s) (%)	Refs.

C₁₂

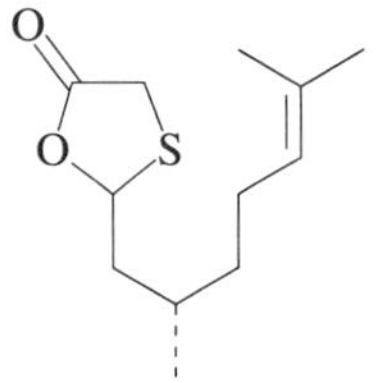

Bu₃SnH, AIBN, C₆H₆, 80°

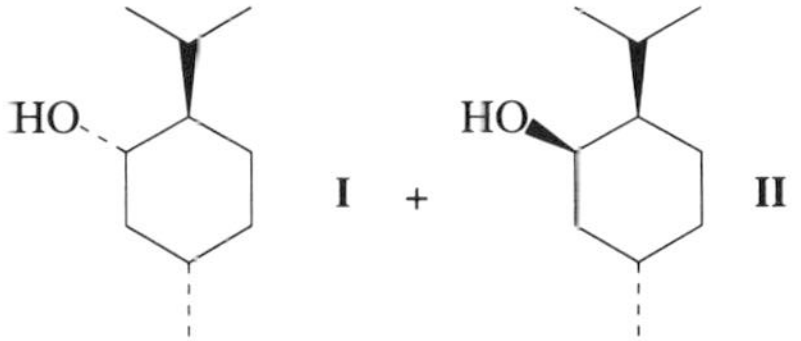

I + **II** (80) **I:II** = 1:1

117

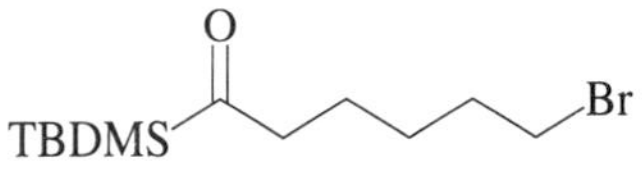

Bu₃SnH, AIBN, C₆H₆, 80°

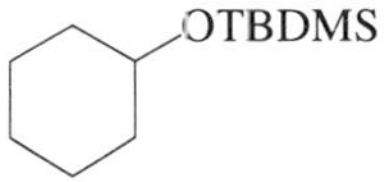

OTBDMS

(64)

337

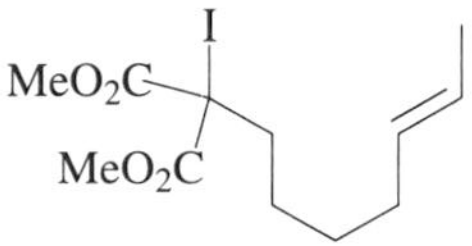

1. (Me₃Sn)₂, C₆H₆, *hv*
2. Bu₃SnH, AIBN, C₆H₆, 80°

(62)

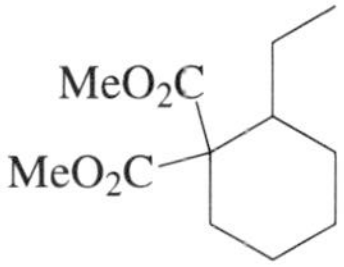

18

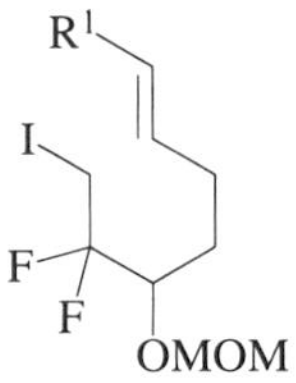

R¹
Ph
CO₂Et
BnOCH₂

Bu₃SnH, AIBN, C₆H₆, 80°

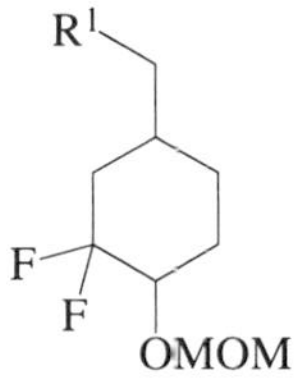

(53)
(79)
(64)

317

437

TABLE I. MONOCYCLIC RINGS CONTAINING ONLY CARBON (*Continued*)

Reactant	Conditions	Product(s) and Yield(s) (%)	Refs.
C_{13} [structure with CO$_2$Me]	Hg cathode, Et$_4$NOTs	[structure with OH, CO$_2$Me] (70) *trans:cis* 1:2.9	365
[structure with SePh]	Bu$_3$SnH, AIBN, C$_6$H$_6$, 80°	[cyclohexanone structure] (76)	19
Z—C=C=CH$_2$ [with I] Z: S(O)Ph, SO$_2$Ph, POPh$_2$	Bu$_3$SnH, AIBN, C$_6$H$_6$, 80°	[cyclohexene Z structure] (72) (81) (85)	287
[structure with CO$_2$Et] Z / E	Mn(OAc)$_3$, Cu(OAc)$_2$, AcOH, 50°	[structure with CO$_2$Et] (73) (62)	192

438

TABLE I. MONOCYCLIC RINGS CONTAINING ONLY CARBON (*Continued*)

Reactant	Conditions	Product(s) and Yield(s) (%)	Refs.
[structure: cyclohexanone with CO₂Me and pentenyl chain]	Mn(OAc)₃, Cu(OAc)₂, AcOH, 60°, 1 h	[structure] CO₂Me (75)	271
[structure: Br, CO₂Me, t-Bu chain]	Bu₃SnH, AIBN, C₆H₆, 80°	[structure with t-Bu] CO₂Me (85)	17
[structure: Me–Si(Me)(O–t-Bu), C≡C, Br]	1. Ph₃SnH, AIBN, C₆H₆ 2. H₂O₂	[structure] HO (95) + [structure] HO, OH (5)	373
[structure: C≡CPh, I chain]	CrCl₂, DMF, 25°	[structure] Ph (85)	274
C₁₄ [structure: CF₃, OBz, I chain]	Bu₃SnH, AIBN, C₆H₆, 80°	[structure] CF₃, OBz **I** + [structure] CF₃, OBz **II** **I + II** (94) **I:II** = 11:1	313

TABLE I. MONOCYCLIC RINGS CONTAINING ONLY CARBON (*Continued*)

Reactant	Conditions	Product(s) and Yield(s) (%)	Refs.
	Bu₃SnH, AIBN, C₆H₆, 80°	(75)	24
	hv	(60) + (20)	88
	Bu₃SnH, AIBN, C₆H₆, 80°	(85)	41, 40
	Bu₃SnH, AIBN, C₆H₆, 80°	(60)	374
	Bu₃SnH, AIBN, C₆H₆, 80°	(52)	375

440

TABLE I. MONOCYCLIC RINGS CONTAINING ONLY CARBON (*Continued*)

Reactant	Conditions	Product(s) and Yield(s) (%)	Refs.

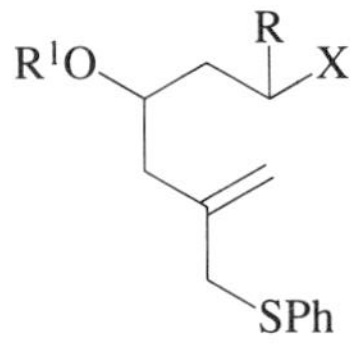

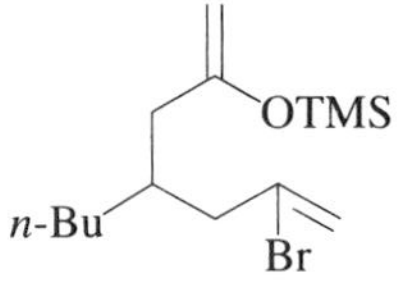

Reactant: structure with R, X, R¹O, SPh

R	R¹	X
H	Me	Br
H	(CH₂)₂OH	Br
Ph	Me	Cl
n-C₅H₁₁	Me	Br
Me	Me	Br
Me	Bz	Br
Me	Bz	OC(S)Im

Conditions:

A. (Bu₃Sn)₂, C₆H₆, hν, 10°
B. (Me₃Sn)₂, Ph₂CO, C₆H₆, hν, 10°
C. (Me₃SnOCPh₂)₂, C₆H₆, 80°

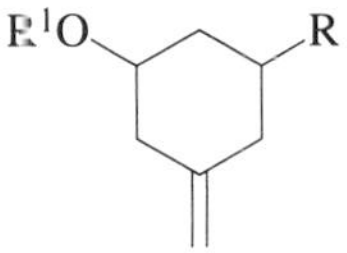

		cis:trans
A	(35)	—
A	(75)	—
A	(60)	1.2:1
A	(44)	1.3:1
A	(55)	1.1:1
B	(61)	1.3:1
C	(61)	1.2:1

Product R¹O— / R cyclohexane with methylene — **376**

C₁₅

Reactant: structure with OTMS, Br, n-Bu

Conditions: Bu₃SnH, AIBN, C₆H₆, 80°

Product: (72) — OH, n-Bu cyclohexane with methylene — **30**

Reactant: X—...—C₆H₄Me-p structure

Conditions: Bu₃SnCl, AIBN, Na(CN)BH₄

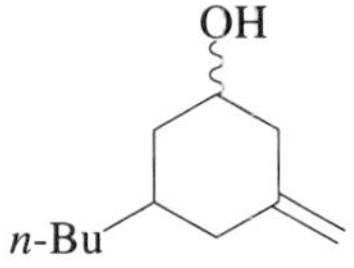

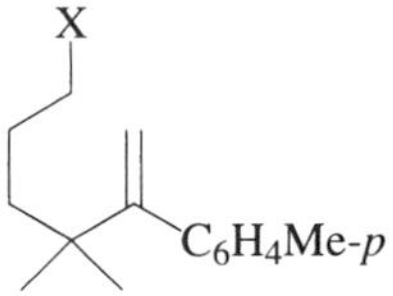

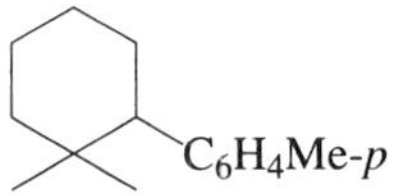

X = Br (70) — **333**

Conditions: Bu₃SnH, AIBN, C₆H₆, 80°

X = I (90) — C₆H₄Me-p cyclohexane product — **377**

441

TABLE I. MONOCYCLIC RINGS CONTAINING ONLY CARBON (*Continued*)

Reactant	Conditions	Product(s) and Yield(s) (%)	Refs.
	Bu₃SnH, AIBN, PhMe, 110°	(48)	233
	Bu₃SnH, AIBN, C₆H₆, 80°	**I** + **II** **I + II** (100) **I:II** = 1:1 **I + II** (98) **I:II** = 94:6	378
	  hv, rt, 15 min	 (77) (53) (70) (74)	379
	Bu₃SnH, AIBN, C₆H₆, heat	(93)	313

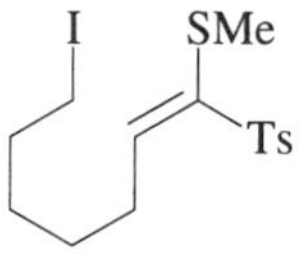

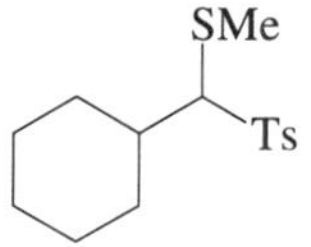

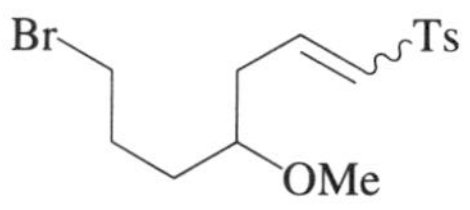

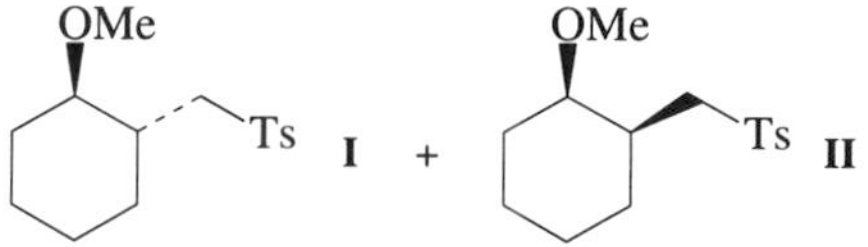

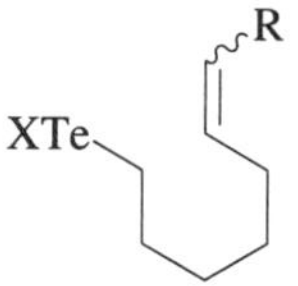

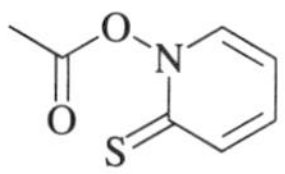

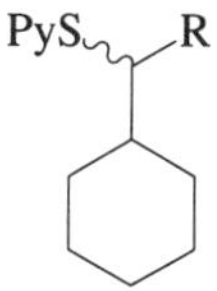

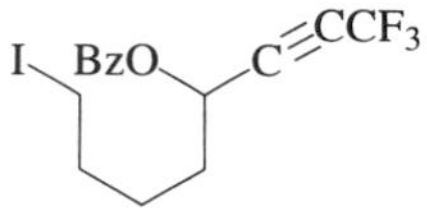

For reactant 3: *E* / *Z*

For reactant 6:

X	R
4-MeOC₆H₄	SO₂Ph
4-MeOC₆H₄	P(O)(OEt)₂
Ph	CO₂Me
4-MeOC₆H₄	CO₂Me

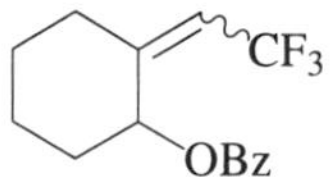

442

TABLE I. MONOCYCLIC RINGS CONTAINING ONLY CARBON (*Continued*)

Reactant	Conditions	Product(s) and Yield(s) (%)	Refs.

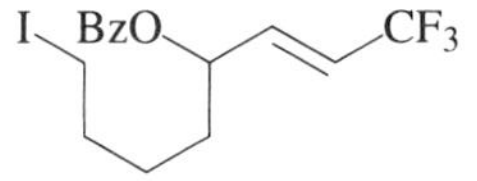

Bu₃SnH, AIBN, C₆H₆, heat

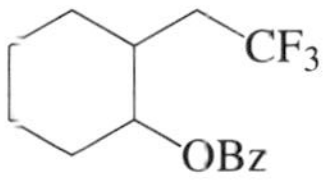

I (90)

313

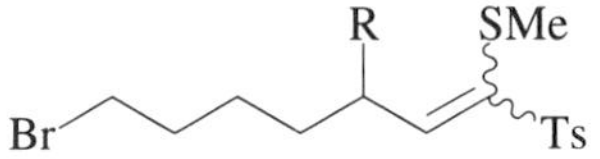

Bu₃SnH, AIBN, C₆H₆, 80°

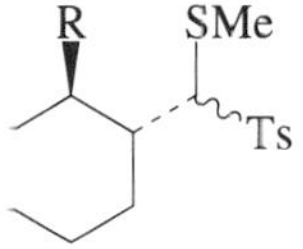

I +

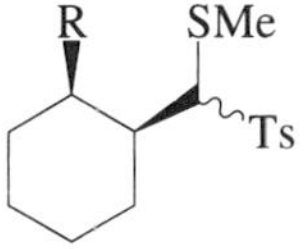

II

378

R		**I** diastereomeric ratio		**II** diastereomeric ratio
OH	$E:Z = 46:54$	(68) (54:46)	(7)	(1:1)
OAc	$E:Z = 67:33$	(85) (78:22)	(6)	(100:0)
OMe	E	(81) (73:27)	(—)	( )
OMe	Z	(82) (74:26)	( )	( )
OPr-i	Z	(78) (71:29)	( )	( )
Me	$E:Z = 60:40$	(70) (61:39)	(—)	(—)

C₁₆

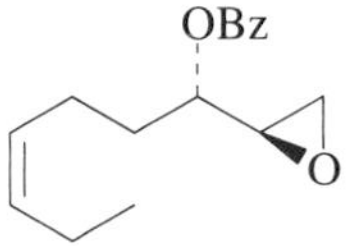

1. MgI₂
2. Bu₃SnH, AIBN, C₆H₆, 80°

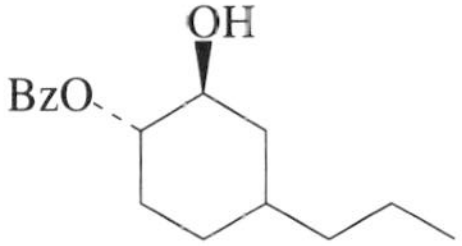

(50)

238

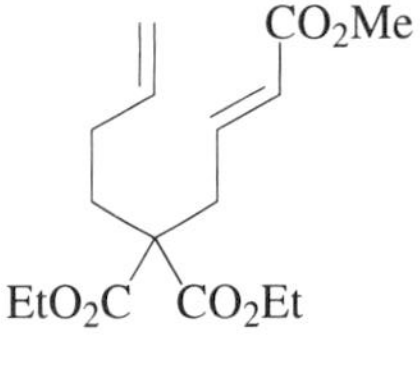

Me₂SnCl, Na(CN)BH₃, AIBN, t-BuOH

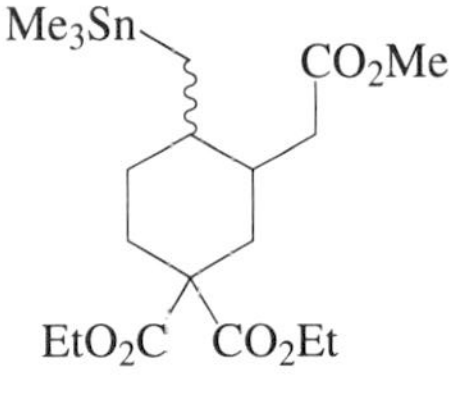

I, II **I + II** (97)
I:II = 1:1

330

TABLE I. MONOCYCLIC RINGS CONTAINING ONLY CARBON (*Continued*)

Reactant	Conditions	Product(s) and Yield(s) (%)	Refs.
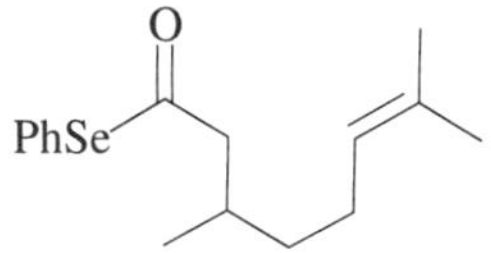	Bu_3SnH, AIBN, C_6H_6, 80°	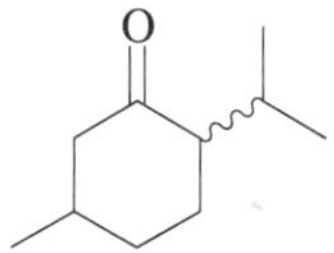 (84)	34
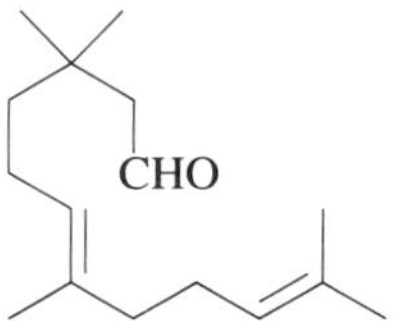	$(t\text{-}BuO)_2$, $c\text{-}C_6H_{12}$, 85°	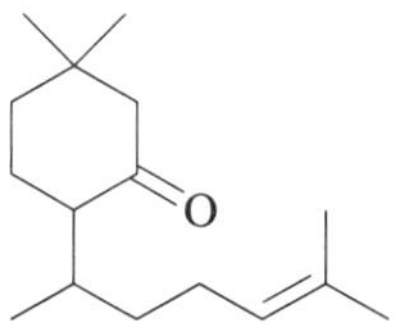 (40)	380
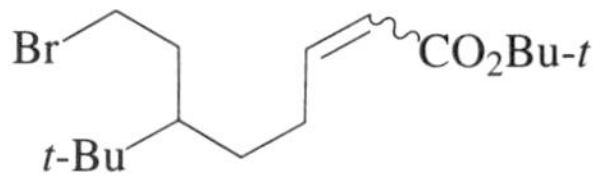	Bu_3SnH, AIBN, C_6H_6, 80°	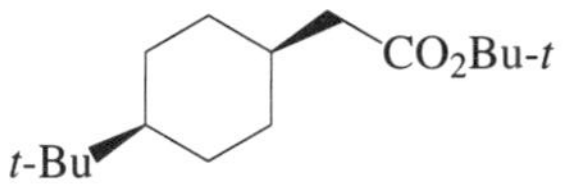 **I** + 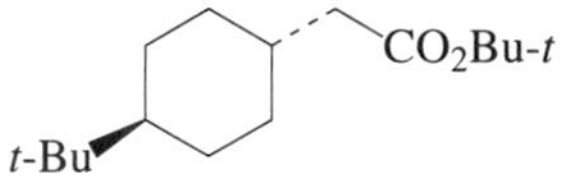 **II** Z: **I** + **II** (75) **I**:**II** = 1:9 E: **I** + **II** (85) **I**:**II** = 1:3	17
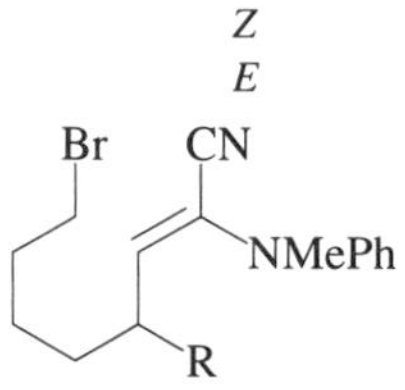 R: Ph *E*; Ph *Z*; Me *E*	Bu_3SnH, AIBN, C_6H_6, 80°	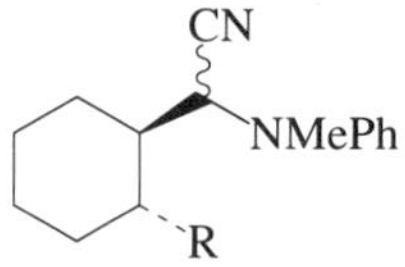 **I**, **II** **I** / **II**: (87) / (5); (83) / (4); (70) / (15)	325

TABLE I. MONOCYCLIC RINGS CONTAINING ONLY CARBON (*Continued*)

Reactant	Conditions	Product(s) and Yield(s) (%)	Refs.

C_{17}

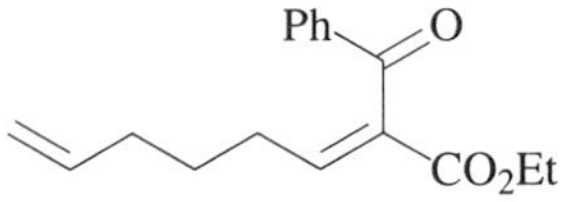

1. $Hg(OAc)_2$, AcOH
2. $NaBH(OMe)_3$

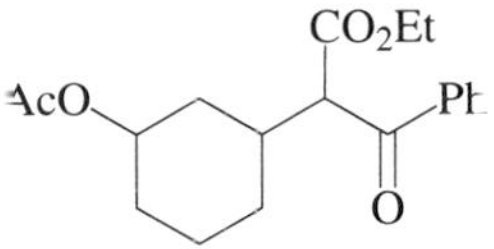

(77)

299

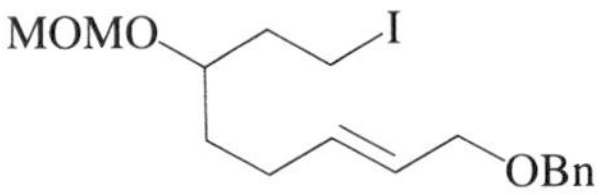

Bu_3SnH, AIBN, C_6H_6, 80°

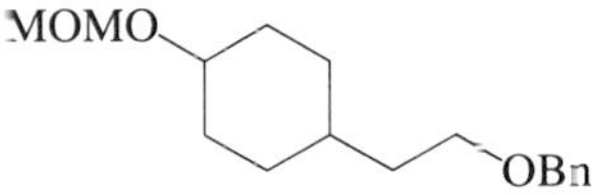

(62)

317

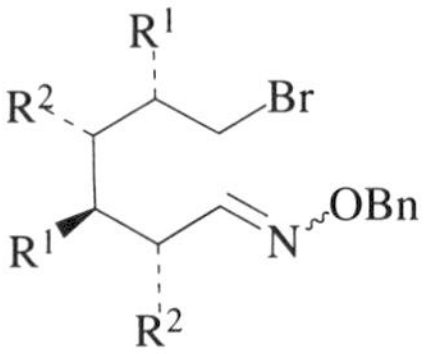

Bu_3SnH, AIBN, PhMe, 110°

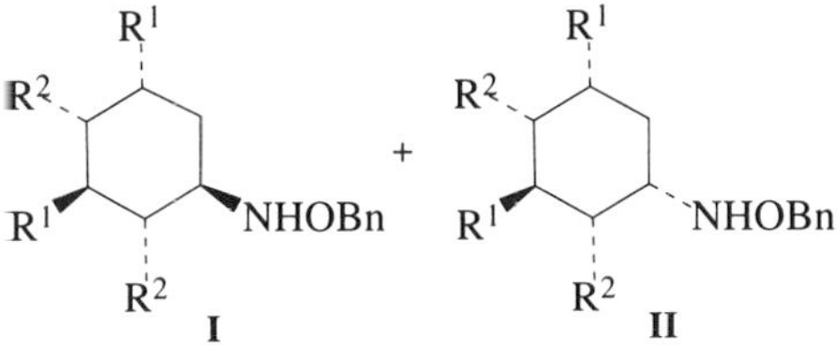

R¹	R²	I + II	I:II
OAc	OAc	(52)	33:17
OBz	OBz	(55)	75:25
OBn	OH	(40)	30:20
OBn	OAc	(42)	78:22

217, 381

C_{19}

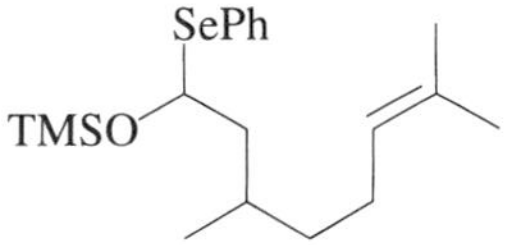

Bu_3SnH, AIBN, C_6H_6, 80°

(61)

80

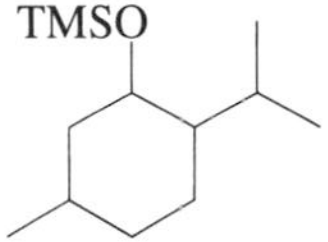

445

TABLE I. MONOCYCLIC RINGS CONTAINING ONLY CARBON (Continued)

Reactant	Conditions	Product(s) and Yield(s) (%)	Refs.
C$_{20}$ (structure with SiPh$_2$Me and Br)	allyl-SnBu$_3$, AIBN, C$_6$H$_6$, 87°	MePh$_2$SiO (allyl cyclohexane) (60)	338
(structure: OBn, =N, X, R^1, R^2)	Bu$_3$SnH, AIBN, reflux	**I** + **II**	37

R^1	R^2	X
C$_6$H$_{11}$	H	PhOC(S)O
C$_6$H$_{11}$	Me	Br
p-MeOC$_6$H$_4$	H	PhOC(S)O

(I: OR^4, HN, R^1, R^2, R^3 cyclohexane)   (II: OR^4, HN, R^2, R^1, R^3 cyclohexane)

	I + II	I:II
PhMe	(71)	33:67
C$_6$H$_6$	(68)	33:67
C$_6$H$_6$	(18)	<2:98

| (aziridine structure: Ph, N–N, EtO$_2$C, CO$_2$Et, Br) | Bu$_3$SnH, AIBN, PhMe, 110° | (EtO$_2$C, CO$_2$Et cyclohexane) (85) | 343 |
| C$_{21}$ (acyl selenide SePh, Ph) + CO$_2$Me | Bu$_3$SnH, AIBN, C$_6$H$_6$, 80° | **I** + **II** (71)   **I:II** = 1.1:1 | 209 |

I (cyclohexanone, CO$_2$Me, CH$_2$Ph) + II (cyclohexanone, CO$_2$Me, CH$_2$Ph)

446

TABLE I. MONOCYCLIC RINGS CONTAINING ONLY CARBON (*Continued*)

Reactant	Conditions	Product(s) and Yield(s) (%)	Refs.
py(dmgH)$_2$Co, HO, HO	C$_6$H$_6$, hv, 25°	(51)	168
C$_{22}$ MeO$_2$C, SnBu$_3$, I, Z, E	Bu$_3$SnH, AIBN, C$_6$H$_6$, 80°	(35) Z:E = 14:1 (48) Z:E = 1:77	382
MeO$_2$C, Br, TBDMSO, OBn	Bu$_3$SnH, AIBN, C$_6$H$_6$, 80°	I + II, I + II (91), I:II = 55:45	383
C$_{23}$ py(dmgH)$_2$Co, HO, HO	C$_6$H$_6$, hv, 25°	(94)	168

TABLE I. MONOCYCLIC RINGS CONTAINING ONLY CARBON (*Continued*)

Reactant	Conditions	Product(s) and Yield(s) (%)	Refs.

Reactant:

EtO₂C— (chain with Br, vinyl, TBDMSO, OTBDMS substituents)

Conditions: Bu₃SnH, AIBN, C₆H₆, 80°

Product: (structure with EtO₂C, methylene, TBDMSO, OTBDMS) (94)

Refs. 384

Reactant: (N–N hydrazone, Ph aziridine) + $\equiv$Y ; EtO₂C, CO₂Et

Y = CN
Y = CO₂Me

Conditions: Bu₃SnH, AIBN, PhMe, 110°

Product: (cyclohexane with Y and CO₂Et, EtO₂C) (86) (87)

Refs. 343

C₂₄

Reactant: (thionoester S–Im, Ph, Ph, methylenecyclopropane)

Conditions: Bu₃SnH, AIBN, xylene, 140°

Product: (cyclohexene with CH₂Ph, Ph) + (methylenecyclohexane with CH₂Ph, Ph) (30–40) (6)

Refs. 385

Reactant: (CH₂Ts, CO₂Me, MeO₂C, methylene) + $\equiv$R¹

R¹
C₆H₁₃
CH₂Ts

Conditions: Mn(OAc)₃, Cu(OAc)₂,

Product: (cyclohexane with R¹, MeO₂C, CO₂Me, methylene) (35) (28)

Refs. 210

448

TABLE I. MONOCYCLIC RINGS CONTAINING ONLY CARBON (*Continued*)

Reactant	Conditions	Product(s) and Yield(s) (%)	Refs.
C_{25}	THF, *hv*, 25°	GePh₃ (86)	350
C_{28} (N−N aziridine, Ph, EtO₂C, CO₂Et)	Bu₃SnH, AIBN, PhMe, 110°	(65) EtO₂C, CO₂Et	343
C_{28} py(dmgH)₂Co, HO, HO	C_6H_6, *hv*, 25°	(87) HO, HO	168
C_{29} PhSe, TBDPSO	Bu₃SnH, AIBN, C_6H_6, 80°	**I, II** + **III** TBDPSO; TBDPSO I, II (41) III (11) I:II = 1:1	386, 387
C_{31} RO, RO, RO, SePh, PhS, R = TBDMS	Bu₃SnH, AIBN, C_6H_6, 80°	SPh, OTBDMS, TBDMSO I, II (90) I:III = 1:1	388, 389

TABLE I. MONOCYCLIC RINGS CONTAINING ONLY CARBON (*Continued*)

Reactant	Conditions	Product(s) and Yield(s) (%)	Refs.
C35 R = TBDMS	Bu3SnH, AIBN, C6H6, 80°	SC6H4Cl-*p* (67)	384, 390

E. 7-Membered Rings

Reactant	Conditions	Product(s) and Yield(s) (%)	Refs.
C8	Bu3SnH, AIBN, C6H6, 80°	**I** + **II** **I + II** (90) I:II = 45:51	391
C9	SmI2, THF, HMPA	(52)	178
	Bu3SnH, AIBN, C6H6, 80°	(73)	74, 75, 200
C10	Mn(OAc)3, Cu(OAc)2, AcOH	**I** + **II** **I + II** (68) I:II = 2.8:1	193

TABLE I. MONOCYCLIC RINGS CONTAINING ONLY CARBON (*Continued*)

Reactant	Conditions	Product(s) and Yield(s) (%)	Refs.

C_{11}

Mn(OAc)$_3$, Cu(OAc)$_2$, AcOH   (35)   194

Mn(OAc)$_3$, Cu(OAc)$_2$, AcOH   **I** + **II**   194

C_{15}

Bu$_3$SnH, AIBN, C$_6$H$_6$, 80°   **I** + **II**   **I** + **II** (86)   **I:II** = 93:7   313

Bu$_3$SnH, AIBN, C$_6$H$_6$, 80°   **I** + **II**   392, 387

		I	**II**
		(32)	(55)
		(72)	(12)
		(17)	(54)
		(24)	(—)
		(27)	(1)

R^1	R^2	R^3	R^4
H	H	O(CH$_2$)$_2$O	
	O(CH$_2$)$_2$O	H	H
OTBDPS	H	H	H
	O(CH$_2$)$_2$O	OTBDPS	H
H	H	H	OEt

451

TABLE I. MONOCYCLIC RINGS CONTAINING ONLY CARBON (*Continued*)

Reactant	Conditions	Product(s) and Yield(s) (%)	Refs.

F. 8-Membered Rings

C_{10}

| 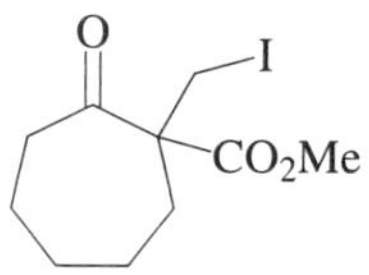 | Bu_3SnH, AIBN, C_6H_6, 80° | 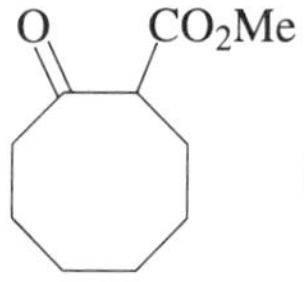 (90) | 200 |
| 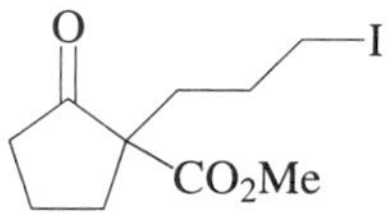 | Bu_3SnH, AIBN, C_6H_6, 80° | 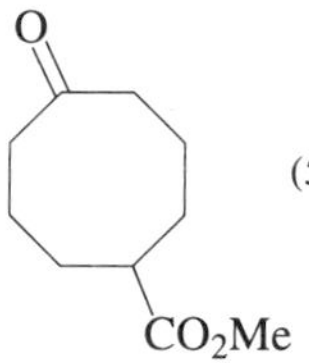 (55) | 200 |

C_{11}

| 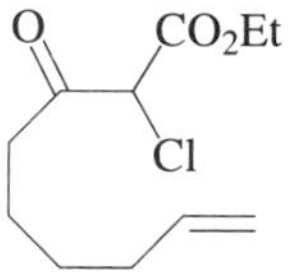 | $Mn(OAc)_3$, $Cu(OAc)_2$, AcOH | 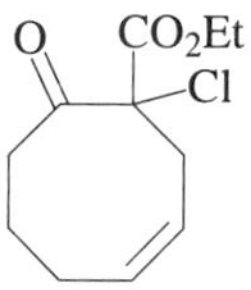 (47) | 193 |
| 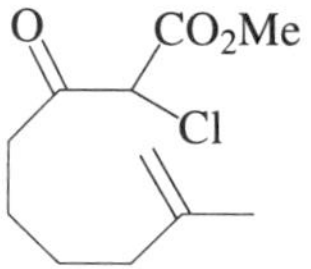 | $Mn(OAc)_3$, $Cu(OAc)_2$, AcOH | 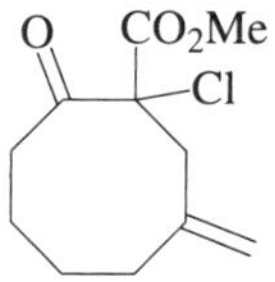 **I** 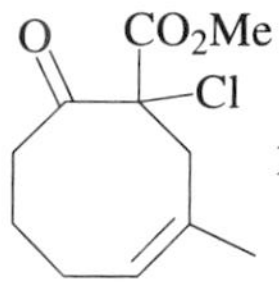 + **II** I + II (69) I:II = 2.5:1 | 194 |

TABLE I. MONOCYCLIC RINGS CONTAINING ONLY CARBON (*Continued*)

Reactant	Conditions	Product(s) and Yield(s) (%)	Refs.
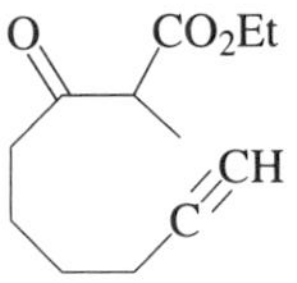 C₁₂ 	Mn(OAc)₃, Cu(OAc)₂, AcOH	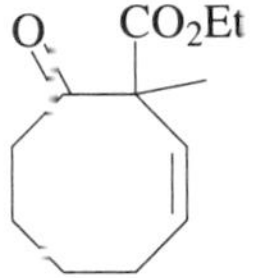 (35)	194
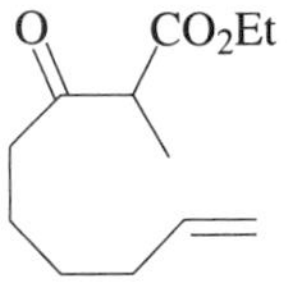	Mn(OAc)₃, Cu(OAc)₂, AcOH	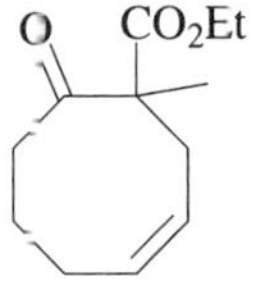 (38)	193

G. Macrocycles

C₁₀ 	Bu₃SnH, AIBN, C₆H₆, 80°	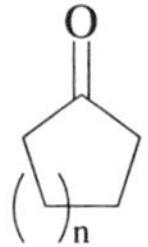 n = 7 (15) n = 11 (55-65) n = 15 (55-65)	78, 393
C₁₄ 	Bu₃SnH, AIBN, C₆H₆, 80°	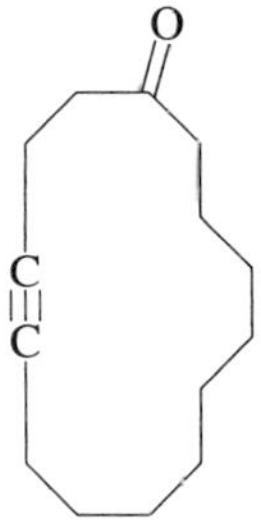 (70-80)	78, 393

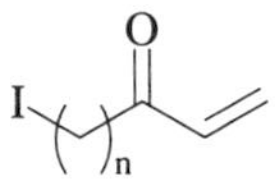

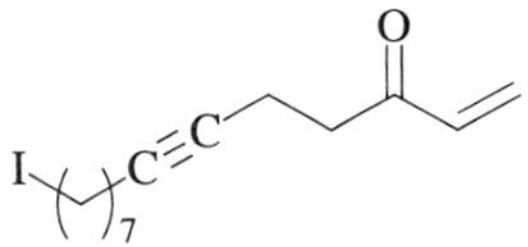

453

TABLE I. MONOCYCLIC RINGS CONTAINING ONLY CARBON (*Continued*)

Reactant	Conditions	Product(s) and Yield(s) (%)	Refs.
	Bu$_3$SnH, AIBN, C$_6$H$_6$, 80°	(70-80)	78, 393
	Bu$_3$SnH, AIBN, C$_6$H$_6$, 80°	(75)	78, 393
C$_{17}$	Bu$_3$SnH, AIBN, C$_6$H$_6$, 80°	I + II (52) I:II = 3:1	394, 395

454

TABLE I. MONOCYCLIC RINGS CONTAINING ONLY CARBON (*Continued*)

Reactant	Conditions	Product(s) and Yield(s) (%)	Refs.
C₂₀			

Reactant	Conditions	Product(s) and Yield(s) (%)	Refs.
	Bu₃SnH, AIBN, C₆H₆, 80°	(54)	208
	Bu₃SnH, AIBN, C₆H₆, 80°	I + I + II (40) I:II = 4:1 II	394, 395

TABLE I. MONOCYCLIC RINGS CONTAINING ONLY CARBON (*Continued*)

Reactant	Conditions	Product(s) and Yield(s) (%)	Refs.
C_{23}	Bu$_3$SnH, AIBN, C$_6$H$_6$, 80°	(50)	78
C_{26}	Bu$_3$SnH, AIBN, C$_6$H$_6$, 80°	(40)	396
C_{29}	Bu$_3$SnH, AIBN, C$_6$H$_6$, 80°	(89)	76

456

TABLE II. MONOCYCLIC RINGS CONTAINING ONE OR MORE HETEROATOMS

Reactant	Conditions	Product(s) and Yield(s) (%)	Refs.

A. 3-Membered Rings

C_9

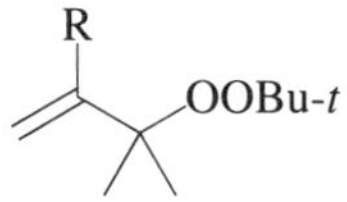

$(t\text{-BuO})_2$, ZH,

C_6H_6, $h\nu$, 30°

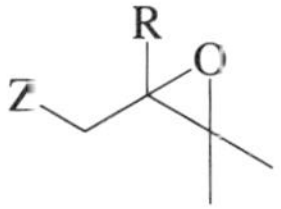

Z	R = H	R = Me
$HC(CO_2Me)_2$	(75)	(80)
$C(CO_2Et)_3$	(50)	(58)
$HC(CN)CO_2Et$	(50)	(56)
$CH_2(CO_2Me)$	(54)	(60)

397

C_{10}

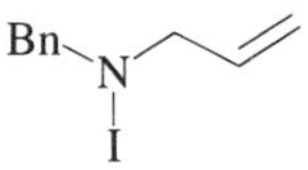

Bu_3SnH, $h\nu$

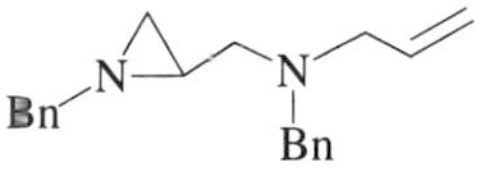

(45)

398

B. 4-Membered Rings

C_{19}

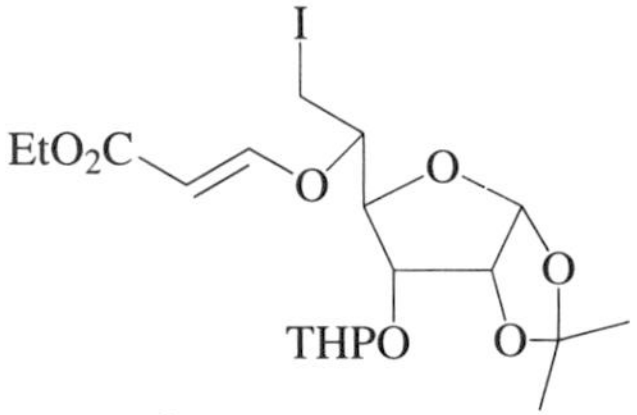

Bu_3SnH, AIBN, C_6H_6, 80°

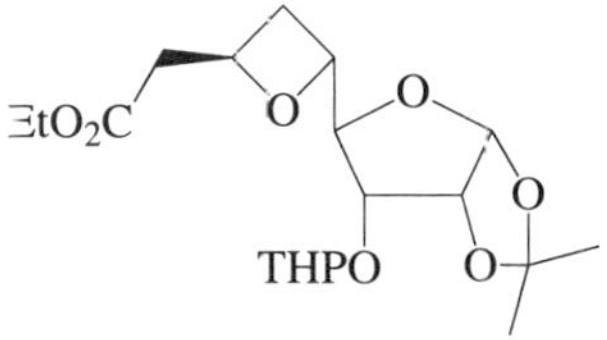

(71)

399

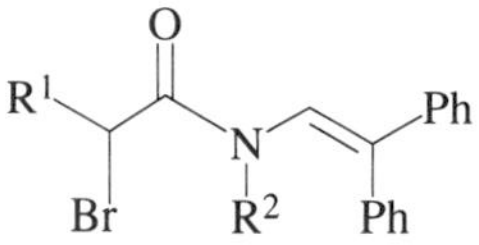

Bu_3SnH, AIBN, C_6H_6, 80°

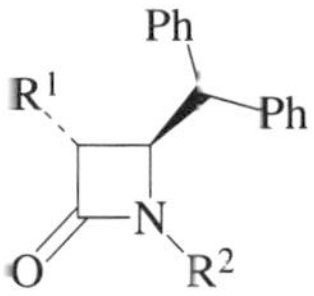

R^1	R^2	
Et	Me	(40)
Et	Bu-t	(58)
Et	C_6H_{11}	(70)
Me	C_6H_{11}	(69)

400

457

TABLE II. MONOCYCLIC RINGS CONTAINING ONE OR MORE HETEROATOMS (*Continued*)

Reactant	Conditions	Product(s) and Yield(s) (%)	Refs.
C$_{20}$			
	Bu$_3$SnH, AIBN, C$_6$H$_6$, 80°		401
R^1 R^2 / Et H / H SPh / Et SPh		(R^1 = Et, R^2 = H) (45) / (R^1 = H, R^2 = SPh) (45) + (R^1 = H, R^2 = H) (24) / (R^1 = Et, R^2 = SPh) (45) + (R^1 = Et, R^2 = H) (3)	
C$_{22}$			
	Bu$_3$SnH, AIBN, C$_6$H$_6$, 80°	I + II (69) I:II = 1:1 / I + II (64) I:II = 2:1	401
R^1 / H / SPh		**I** + **II**	
	1. Bu$_3$SnH, AIBN, C$_6$H$_6$, 80° / 2. NaOH, Py	(50)	401
	CH$_2$Cl$_2$, *hv*	(45)	402

TABLE II. MONOCYCLIC RINGS CONTAINING ONE OR MORE HETEROATOMS (*Continued*)

Reactant	Conditions	Product(s) and Yield(s) (%)	Refs.
C_4 O=C(SBu-*t*)... Co(dmgH)$_2$py, mesityl	*hv* solvent C$_6$H$_6$ (92) MeCN (69) MeOH (84) CHCl$_3$ (16)	β-thiolactone, mesityl	403

C. 5-Membered Rings

BrCH$_2$CH$_2$C(=O)N=C=O	Et$_3$GeH, MeOH	succinimide (80)	404
pentenol (OH)	HgO, I$_2$, *hv*	ICH$_2$-tetrahydrofuran (50)	61
pentenyl nitrite (ONO)	C$_6$H$_6$, *hv*	HON=CH-tetrahydrofuran (63)	405

TABLE II. MONOCYCLIC RINGS CONTAINING ONE OR MORE HETEROATOMS (Continued)

Reactant	Conditions	Product(s) and Yield(s) (%)	Refs.
C₆			
HS⟋R² / N(R¹)(allyl) R¹ | R² H | Me Et | Me Bz | H	C₆H₁₂, $h\nu$, 80°	I + II + III I + II + III | I:(II + III) (46) | 66:34 (70) | 30:70 (87) | 60:40	406, 407
HC≡C–C(=O)–O–CH₂CH=CH–R R | H, Me, Ph	Ph₃SnH, AIBN, C₆H₆, 80°	I + II I + II (40) (42) (62)	408
allyl ether structure	TsSePh, AIBN, PhMe, 110°	(97) cis:trans = 3.2:1	281
	TsBr, AIBN, MeCN, $h\nu$, 25°	(58) cis:trans = 4.2:1	409

TABLE II. MONOCYCLIC RINGS CONTAINING ONE OR MORE HETEROATOMS (*Continued*)

Reactant	Conditions	Product(s) and Yield(s) (%)	Refs.

Reactant 1 (Y-linked diallyl)

Conditions: γ-Irradiation, *i*-PrOH

Products: **I** + **II** 410

Y	I + II	I:II
O	(97)	4:1
CH$_2$	(93)	3.2:1
C(CO$_2$H)$_2$	(83)	5.7:1
NH	(80)	4.6:1
NMe	(44)	4.9:1
NCH$_2$CH=CH$_2$	(49)	4.3:1
NMe$_2$$^+Cl^-$	(80)	4:1

Reactant 2

Conditions:
A. (PPh$_3$)$_2$ReCl$_3$, CCl$_4$, reflux
B. (PPh$_3$)$_2$RuCl$_2$, CCl$_4$, reflux

Products: **I** + **II** 282

Y		I + II	I:II
O	A	(87)	4.3:1
NC(O)Me	A	(64)	6:1
	B	(77)	6:1

Reactant 3 (diallyl sulfone)

Conditions: TsBr, CH$_2$Cl$_2$, *hv*

Products: (40) *cis:trans* = 1.2:1 411

TABLE II. MONOCYCLIC RINGS CONTAINING ONE OR MORE HETEROATOMS (*Continued*)

Reactant	Conditions	Product(s) and Yield(s) (%)	Refs.
	$(t\text{-BuO})_2$, 135°	(72) + (13)	412
	MeOH, hv, H$^+$	**I** + **II** (83) **I:II** = 4:1	413
	CuCl, MeCN, 110°, 3 h, sealed tube	**I** + **II**	90
		R **I + II** **I:II**	
		Me (60) 89:11	
		Et (50) 92:8	
		i-Pr (80) 98:2	
	$(Bu_3Sn)_2$, C_6H_6, hv, 80°	(87)	51

TABLE II. MONOCYCLIC RINGS CONTAINING ONE OR MORE HETEROATOMS (*Continued*)

Reactant	Conditions	Product(s) and Yield(s) (%)	Refs.

C₇

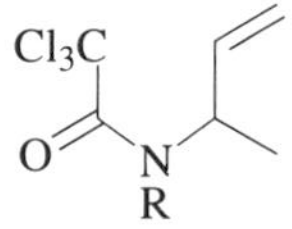

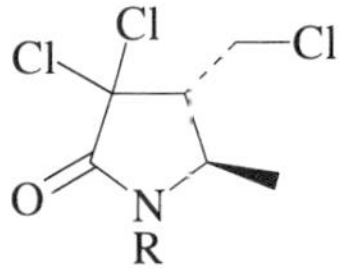

Reactant (structure): Cl₃C–C(=O)–N(R)–CH(CH₃)–CH=CH₂

R	Conditions	I + II	I:II
Bn	CuCl, MeCN, 80°	(72)	86:14
Bn	CuCl, CH₂Cl₂, bipyridine, -15°	(93)	95:5
Me		(98)	9C:10
Ts		(90)	15:85
Ms		(98)	20:80
Cbz		(76)	20:80
Boc		(80)	14:86

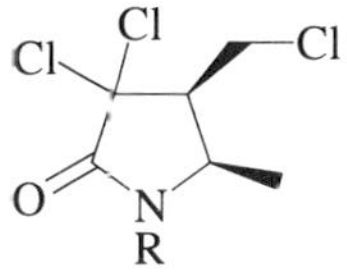

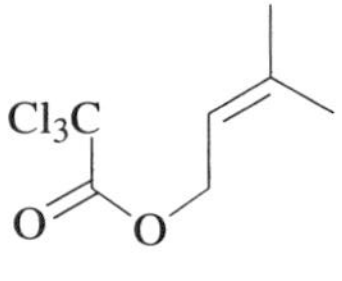

Products: I + II — Refs. 91, 414

Reactant: Cl₃C–C(=O)–O–CH₂–CH=C(CH₃)₂

CuCl, MeCN, 140° → (54) — Ref. 415

Reactant: (allyl-substituted) –OOH

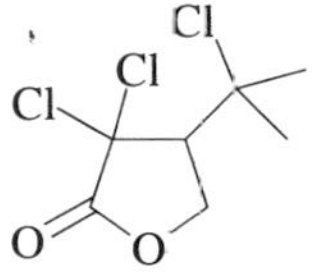

NIS → (54) — Ref. 166

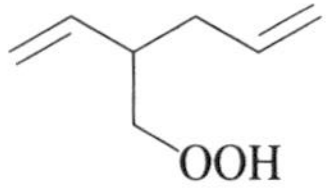

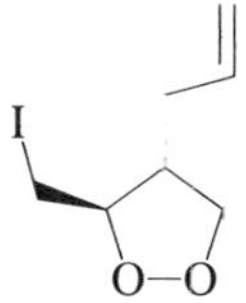

463

Reactant	Conditions	Product(s) and Yield(s) (%)	Refs.
	Bu$_3$SnH, AIBN, C$_6$H$_6$, 80°	*trans:cis* 	R			
---	---	---
Me	(68)	71:29
Bn	(80)	72:28
Ph	(90)	83:17		89
  	R^1	R^2
---	---
Me	H
Me	H
H	Me
H	Me		CuCl, CuCl$_2$, HOAc, H$_2$O FeSO$_4$, HOAc, H$_2$O CuCl, CuCl$_2$, HOAc, H$_2$O FeSO$_4$, HOAc, H$_2$O	**I** + **II** |	I+II	I:II
---	---	---
	(79)	9:91
	(93)	0:100
	(81)	100:0
	(74)	100:0		416
  	R	
---	---
H	
Et *E*	
Et *Z*			Cu(bpy)Cl, reflux CH$_2$Cl$_2$ THF MeOAc	**I** + **II** |	I+II	I:II
---	---	---
	(84)	71:29
	(85)	60:40
	(76)	67:33		417

TABLE II. MONOCYCLIC RINGS CONTAINING ONE OR MORE HETEROATCMS (*Continued*)

Reactant	Conditions	Product(s) and Yield(s) (%)	Refs.
MeO–C(=S)–S–CH₂CH₂–O–CH₂CH=CH₂	(*t*-BuO)₂, PhCl, 135°	THF–CH₂–S–C(=S)–OMe (57)	255
H–Ge(Me)(Me)–(chain)–CH=CH₂	AIBN, heat	Ge(Me)(Me) ring (70)	418
R–CH(O–Si(Me)(Me)–CH₂Br)–CH=CH₂	Bu₃SnH, AIBN, C₆H₆, 80°, 2 h	**I** + **II** (see below)	56
R–CH(I)–C(=O)–O–CH(CH=CH₂)₂	Ph₃SnH, AIBN, C₆H₆	**I** + **II** (lactones, see below)	419

For the silyl product (Ref. 56):

R	I + II	I:II
Me	(>84)	4.6 1
i-Pr	(>82)	100 0
t-Bu	(>71)	100 0
CH₂=CH	(>72)	5.5 1
Ph	(>59)	11 1

For the lactone product (Ref. 419):

R	I + II	E:II
H	(74)	81:19
Me	(86)	73:22

465

TABLE II. MONOCYCLIC RINGS CONTAINING ONE OR MORE HETEROATOMS (*Continued*)

Reactant	Conditions	Product(s) and Yield(s) (%)	Refs.
	(Bu$_3$Sn)$_2$, C$_6$H$_6$, *hv*, heat	(68)	420
	Bu$_3$SnH, AIBN, C$_6$H$_6$, 80°	$\begin{array}{ccc|c} R^1 & R^2 & R^3 & \\ \hline H & Me & Me & \\ H & H & H & (88\text{-}92) \\ Me & H & H & \\ Me & H & Me & \end{array}$	421
	ClCo(dmgH)$_2$py, NaBH$_4$, MeOH, NaOH, 50°	$\begin{array}{ccc|c} R^1 & R^2 & R^3 & \\ \hline H & H & Ph & (78) \\ H & Me & Ph & (73) \\ H & Ph & Ph & (85) \\ \multicolumn{2}{c}{-(CH_2)_3-} & H & (48) \\ \multicolumn{2}{c}{-(CH_2)_4-} & H & (64) \end{array}$	223
C$_8$	Bu$_3$SnH, AIBN, C$_6$H$_6$, 65-70°	$\begin{array}{c|c} R & \\ \hline Ph & (82) \\ p\text{-ClC}_6\text{H}_4 & (85) \\ Et & (76) \\ Ph(CH_2)_2 & (88) \end{array}$	241

Reactant	Conditions	Product(s) and Yield(s) (%)	Refs.
	Bu₃SnH, AIBN, C₆H₆, 80°	(74)	89
	1. Hg(OAc)₂ 2. NaBr 3. NaOH 4. NaBH₄	(82)	422
	1. PhSh, O₂ 2. Ph₃P	(59) (49)	423
	Bu₃SnH, AIBN, C₆H₆, 80°	H (95) Me (95)	424
	Bu₃SnH, AIBN, PhMe, 110°	(95) (—) (—)	425

$$Bu_3SnH, AIBN, C_6H_6, 80°$$

$$Bu_3SnH, AIBN, PhMe, 110°$$

R¹	R²
H	Me
Me	H

	R
	H (95)
	Me (95)

Y	R¹	R²
NH	Me	Me
O	Me	Me
O	Me	Ph

467

TABLE II. MONOCYCLIC RINGS CONTAINING ONE OR MORE HETEROATOMS (*Continued*)

Reactant	Conditions	Product(s) and Yield(s) (%)	Refs.

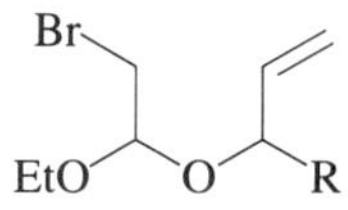

Bu₃SnH, AIBN, C₆H₆, 80°

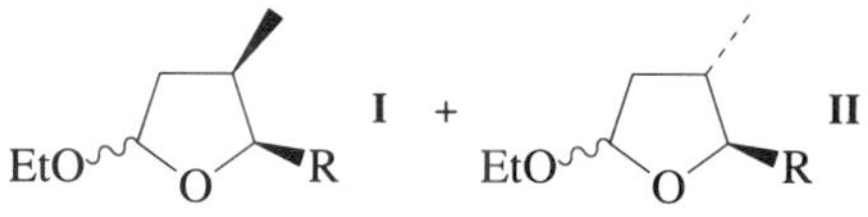

R	I + II	I:II
Me	(52)	96:4
Et	(71)	96:4
i-Pr	(66)	98:2

426

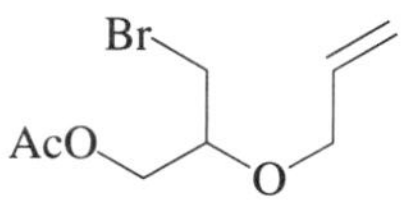

Bu₃SnCl, AIBN, Na(CN)BH₃, t-BuOH, 80°

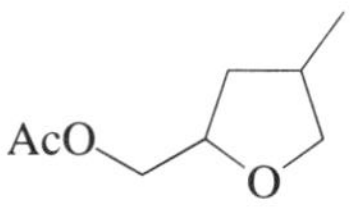

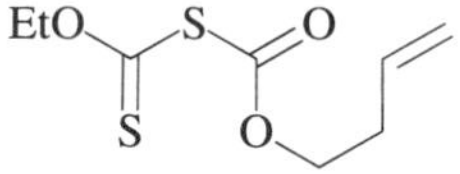

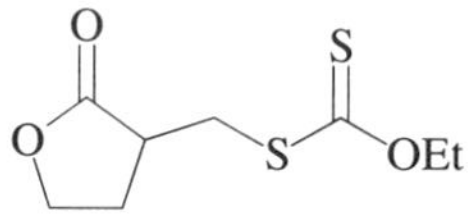

(70)

427

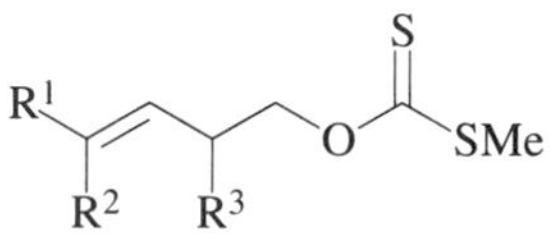

Bu₃SnH, AIBN, PhMe, 80°

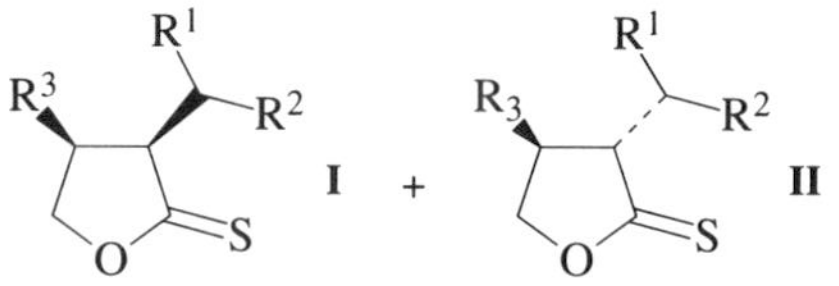

R¹	R²	R³		I + II	I:II
H	Et	H		(50)	—
Et	H	H		(72)	—
Me	H	Et		(80)	4:96
Et	H	Et		(77)	4:96
m-MeOC₄H₆	H	m-MeOC₆H₄		(75)	10:90
H	-(CH₂)₃-			(71)	99:1

428

C₇H₁₆, hv, reflux

(84)

264

TABLE II. MONOCYCLIC RINGS CONTAINING ONE OR MORE HETEROATOMS (*Continued*)

Reactant	Conditions	Product(s) and Yield(s) (%)	Refs.
	Cu(bipy)Cl, AcOMe, reflux	**I** + **II** $\quad$ **I + II** (89) $\quad$ **I:II** = 5.8:1	417
R^1, R^2, R^3 table: R^1 H, R^2 n-C_5H_{11}, R^3 H R^1 n-C_8H_{17}, R^2 H, R^3 H R^1 H, R^2 Me, R^3 Me R^1 n-C_4H_9, R^2 n-C_4H_9, R^3 H	Ph_3SnH, AIBN, C_6H_6, 80°	(>77) (>62) (>69) (>65)	429
	AcOH, H_2SO_4, Fe^{2+}	(66)	430
R^1, R^2, R^3 table: R^1 Ph, R^2 H, R^3 H R^1 Ph, R^2 H, R^3 Me R^1 Me_2, R^2 CH_2OH, R^3 H	Bu_3Sn AIBN, C_6H_6	(35) (25) (37)	431

469

TABLE II. MONOCYCLIC RINGS CONTAINING ONE OR MORE HETEROATOMS (*Continued*)

Reactant	Conditions	Product(s) and Yield(s) (%)	Refs.

C$_9$

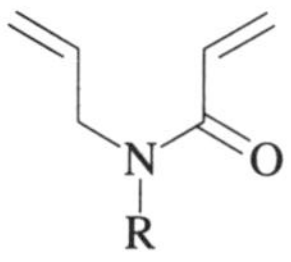

TsX

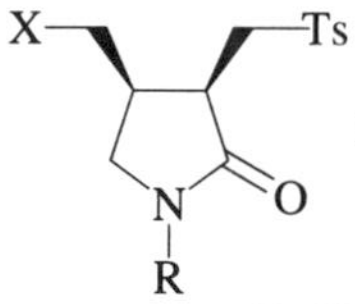

I + II

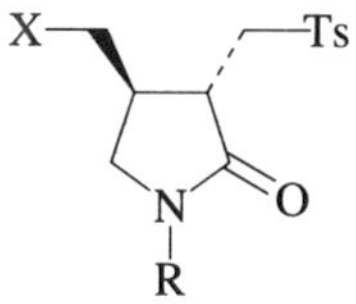

X		**I+II**	**I:II**
Br	MeCN, hv, 18°	(65)	24:76
Cl	MeCN, hv, 18°	(60)	10:90
Cl	PhMe, 110°	(65)	26:74
Br	MeCN, hv, 18°	(66)	20:80
Br	MeCN, hv, 18°	(53)	5:95

R
Bn
Bn
Bn
CH$_2$CH=CH$_2$
t-Bu

219

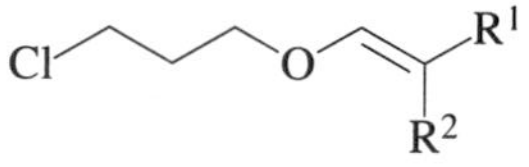

Bu$_3$SnH, AIBN

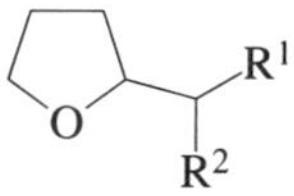

R^1	R^2	
CN	CO$_2$Et	(74)
COMe	CO$_2$Me	(75)
COMe	COPh	(54)
CO$_2$Et	CO$_2$Et	(81)

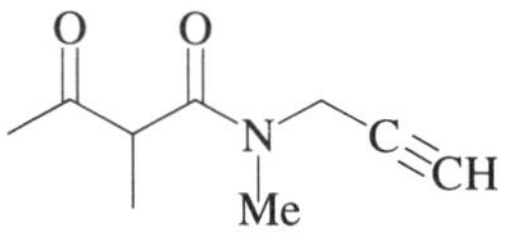

432

Mn(OAc)$_3$

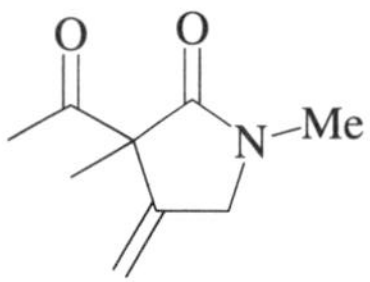

(40)

433

Reactant	Conditions	Product(s) and Yield(s) (%)	Refs.
(R = CO$_2$Bu-t, CH=CHCO$_2$Me, Ph)	(Ph$_2$Se)$_2$, O$_2$, AIBN, MeCN, hv, 0°	I + II $\begin{array}{ccc} & \textbf{I+II} & \textbf{I:II} \\ & (41) & 1:1.8 \\ & (88) & >95:5 \\ & (88) & >95:5 \end{array}$	434, 435
(Br·CH·CH$_3$–CO–O–CH$_2$–C≡CTMS)	Bu$_3$SnH, AIBN, PhMe, 110°	(80)	425
(thiohydroxamate allyl carbonate)	THF, hv, 25° THF, (PhSe)$_2$, hv, 25° THF, t-BuSH, hv, 25°	$\begin{array}{cc} \textbf{Y} & \\ \text{S-(2-pyridyl)} & (85) \\ \text{SePh} & (70) \\ \text{H} & (100) \end{array}$	436
(propargyl ether of methylbutenol)	Ph$_3$SnH, AIBN, C$_6$H$_6$, 80°	(96)	27

471

Reactant	Conditions	Product(s) and Yield(s) (%)	Refs.
	Bu₃SnH, Et₃B, THF, –78°	(66) (53)	437
	Bu₃SnH, AIBN, C₆H₆, 80°	I + II (85) I:II = 70:30	111, 438
	Bu₃SnH, AIBN, C₆H₆, 80°	I + II (90) I:II = 65:35	111, 438
	1. MeOH, *hv* 2. MeC≡CMe	(41) + (39)	439

472

TABLE II. MONOCYCLIC RINGS CONTAINING ONE OR MORE HETEROATOMS (Continued)

Reactant	Conditions	Product(s) and Yield(s) (%)	Refs.
	ClCo(dmgH)$_2$py, NaBH$_4$, MeOH, NaOH, 0°	(60)	439a
	AcOH, H$_2$SO$_4$, Fe^{2+}	(55)	430
Substrate table: Substrate / R^1 / R^2 / R^3 A H H H B Me H H C H Me H D H Me Me	(PhS)$_2$, hv		440
C$_{10}$	Bu$_3$SnH, AIBN, C$_6$H$_6$, heat	(48)	441

Product table (for row 3):

Product	R^1	R^2	R^3	R^4	R^5
I	H	H	H	SPh	H
II	H	H	H	H	SPh
III	Me	H	H	SPh	H
IV	Me	H	H	H	SPh
V	H	Me	H	H	SPh
VI	H	Me	Me	H	SPh

Substrate Product (%) cis:trans

A I (57) 0:100 II (22) 1:1
B III (89) 1:1 IV (2)
C V (37) 2:1
D VI (63) 2:1

473

TABLE II. MONOCYCLIC RINGS CONTAINING ONE OR MORE HETEROATOMS (*Continued*)

Reactant	Conditions	Product(s) and Yield(s) (%)	Refs.

Reactant:

$$EtO_2C\text{—}CH(NC)\text{—}CH_2\text{—}CH=CH\text{—}R^1 \quad + \quad R^2SH$$

R^1	R^2
H	Et
H	MeO$_2$C(CH$_2$)$_2$
Ph	MeO$_2$C(CH$_2$)$_2$

Conditions: AIBN, hv

Product: ring with R^1, CO$_2$Et, R^{2}S, N

Conditions	Yield
C$_6$H$_6$, 40°	(98)
PhMe, 110°	(93)
C$_6$H$_6$, 40°	(86)

Refs. 442

Reactant: H, HO(CH$_2$)$_2$

Conditions: 1. C$_6$H$_6$, 40°   2. H$_2$O

Product: lactam CO$_2$Et (98)

Refs. 111, 438

Reactant (Et, NO$_2$, CH≡C, AcO, O)

Conditions: Bu$_3$SnH, AIBN, C$_6$H$_6$, 80°

Product: (Et, AcO, O) (78)

Refs. 52, 53

Reactant (Br, OEt, O, isopropyl, vinyl)

Conditions: Bu$_3$SnH, AIBN, C$_6$H$_6$, 80°

Product: (OEt, O, isopropyl) (81)

Refs. 443

Reactant (Br, OBu-n, O, vinyl, R)

R
H
Me

Conditions: Bu$_3$SnH, AIBN, C$_6$H$_6$, 80°

Product: (OBu-n, O, R)

Yield
(62)
(65)

TABLE II. MONOCYCLIC RINGS CONTAINING ONE OR MORE HETEROATOMS (*Continued*)

Reactant	Conditions	Product(s) and Yield(s) (%)	Refs.

Row 1

Reactant:

CO_2^- structure with N-allyl, R^1, R^2CO, plus $R^3CO_2^-$

R^1	R^2
H	Me
H	Me
H	H
H	H
Me	Me
Me	H
Me	H

Conditions: Pt electrode, MeOH, 40-45°

Product:

Pyrrolidine with R^1 on N, R^2CO, R^3

R^3	
Me	(58)
C_5H_{11}	(46)
Me	(58)
C_5H_{11}	(45)
Me	(56)
Me	(67)
C_5H_{11}	(63)

Refs.: 167

Row 2

Reactant: structure with R^1, R^2, R^3, Br, EtO-O

R^1	R^2	R^3
H	H	H
H	H	Me
H	Me	H
Me	Me	H

Conditions: Bu$_3$SnH, AIBN, PhMe, 110° *or* Bu$_3$SnCl, AIBN, Na(CN)BH$_3$, t-BuOH, 80°

Product: tetrahydrofuran with R^1, R^2, R^3, EtO, O

(60)
(55)
(54)
(50)

Refs.: 444

Row 3

Reactant: epoxide structure with Br

Conditions: Bu$_3$SnH, 85°

Product: **I, II**

I + II (70-80) **I:II = 1:1**

Refs.: 445

Reactant	Conditions	Product(s) and Yield(s) (%)	Refs.
	Bu₃SnCl, AIBN, Na(CN)BH₃, t-BuOH	 R n-C₆H₁₃ (89) n-Bu (67) CO₂Et (65) (67) (75)	446
	(Ph₂Se)₂, O₂, AIBN, MeCN, hv, 0°	I + II | R	R¹	I + II	I:II | CH=CHCO₂Me H (55) 6.4:1 CH=CHCO₂Me Me (73) >95:5 Ph Me (63) >95:5	435
	Bu₃SnH, (slow addition), AIBN, C₆H₆, 80°, 6 h	I + II R TBDMS MOM I+II I:II (89) 1:1.6 (95) 1:0.7	424

TABLE II. MONOCYCLIC RINGS CONTAINING ONE OR MORE HETEROATOMS (*Continued*)

Reactant	Conditions	Product(s) and Yield(s) (%)	Refs.

R		I + II	I:II
Ph		(63)	> 10:1
CO_2Bu-t		(66)	1:1.3
CO_2Ph		(63)	1:3
$CO_2C_6F_6$		(60)	1:5.4
$CO_2CH(CF_3)_2$		(70)	1:6
$CO_2CH(CF_3)_2$		(65)	> 1:10

$(Ph_2Se)_2$, O_2, AIBN, MeCN, hv, 0°

447

$Cr(OAc)_2$, THF, 25°

448, 449

R^1	R^2		I + II	I:II
H	H		(79)	1:3.9
Me	H		(83)	1:0.9
H	*E/Z*-Me		(83)	1:17
n-C_5H_{11}	H		(73)	1:1
i-C_3H_7	H		(57)	1:1.1
H	*E/Z*-Ph		(76)	1:0.6
Ph	H		(99)	1:0.9
Me	Me		(85)	1:0.3

TABLE II. MONOCYCLIC RINGS CONTAINING ONE OR MORE HETEROATOMS (*Continued*)

Reactant	Conditions	Product(s) and Yield(s) (%)	Refs.

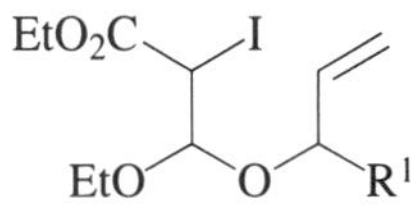

Cr(OAc)$_2$, THF, 25°

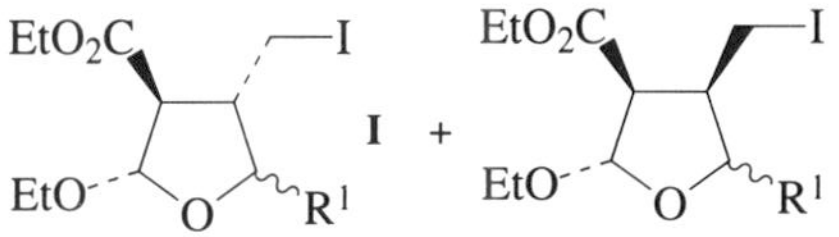

I + **II**

	I + II	I:II
R = H	(93)	4:1
Me	(93)	1:1

262

Cr(OAc)$_2$, THF, 25°

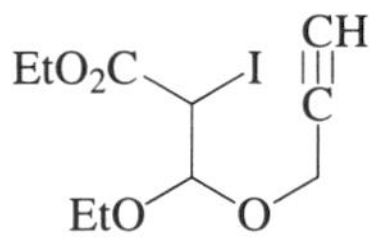

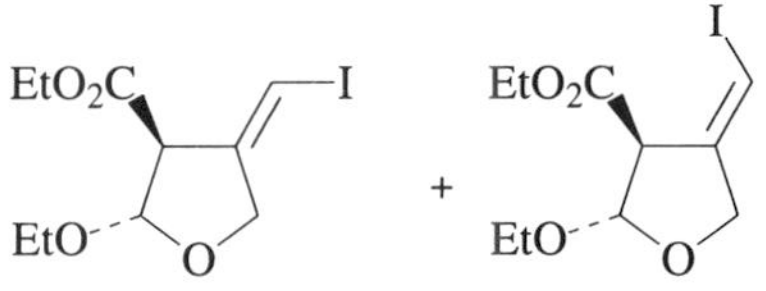

I + **II** + **III**

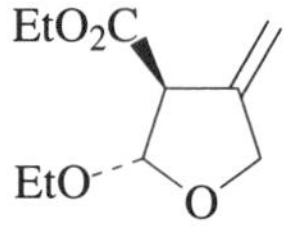

I + II + III (92)

I:II:III = 2:1:1.1

262

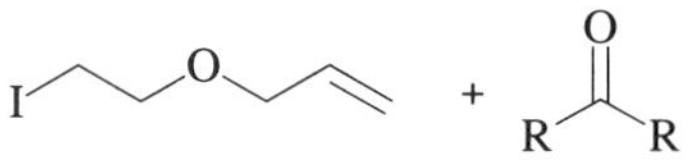

SmI$_2$, HMPA, THF

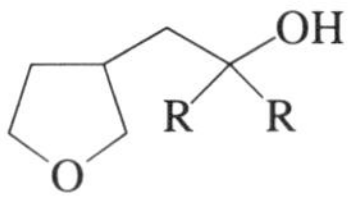

III

R		Diastereomeric ratio
Et	(57)	
Pr	(38)	
–(CH$_2$)$_4$–	(53)	
–(CH$_2$)$_5$–	(52)	
–CH(Me)(CH$_2$)$_4$–	(74)	1.3:1
–(CH$_2$)$_2$CH(Bu-t)(CH$_2$)$_2$–	(55)	5:1

176

TABLE II. MONOCYCLIC RINGS CONTAINING ONE OR MORE HETEROATOMS (Continued)

Reactant	Conditions	Product(s) and Yield(s) (%)	Refs.
C₁₁ (structure)	BPO, C₆H₆, 80°	(75)	160
(structure) R¹ / R² Cl H Cl Cl I H SPh SPh	Bu₃SnH, AIBN, C₆H₆, 80°	(structure) (17) (—) (47) (75)	450
(structure) X Y Z Br H H Cl Cl Cl Cl Cl H	Bu₃SnH, AIBN, C₆H₆, 80°	I + II I+II I:II (83) 34:66 (92) 90:10 (78) 5:95	451

TABLE II. MONOCYCLIC RINGS CONTAINING ONE OR MORE HETEROATOMS (*Continued*)

Reactant	Conditions	Product(s) and Yield(s) (%)	Refs.
  R / H / Ph	Bu₃SnH, O₂, PhMe, 0°, 5-24 h	 *trans:cis* (84) 100:0 (74) 87:13	260
	Bu₃SnH, AIBN, C₆H₆, heat	 R H (87) Me (85)	452
  Y / NAc / O	CH₂Cl₂, *hv*	 (77) (—)	453
	Mn(OAc)₃, AcOH	 (40)	433

TABLE II. MONOCYCLIC RINGS CONTAINING ONE OR MORE HETEROATOMS (*Continued*)

Reactant	Conditions	Product(s) and Yield(s) (%)	Refs.
$+ \ ArN_2{}^+Cl^-$	$TiCl_3$, MeOH, 0°	Ar—NH $\dfrac{Ar}{p\text{-}ClC_6H_4 \quad (40)}$ $p\text{-}MeOC_6H_4 \quad (30)$ $Ph \quad (45)$	454
Ph—S—O	Bu_3SnH, AIBN, t-BuPh, 80°	(73)	455
	$Cl(CH_2)_2Cl$, 85° Me_2CO, reflux	**I** + **II** **I + II** (78) **I:II** = 80:20 **I + II** (83) **I:II** = 11:89	456, 417
	t-BuO$_2$H, AIBN, O$_2$, 2,2,4-trimethyl-pentane, 60°	(73) + (19)	457

TABLE II. MONOCYCLIC RINGS CONTAINING ONE OR MORE HETEROATOMS (Continued)

Reactant	Conditions	Product(s) and Yield(s) (%)	Refs.
	(PhSe)$_2$, AIBN, O$_2$, MeCN, $h\nu$, 0°	**I** + **II**	447, 435

R	
Me	
t-Bu	

	I + II	**I:II**
	(70)	1:1.7
	(84)	1:1

Ph$_3$SnH, AIBN, C$_6$H$_6$, 80°

458, 215, 459, 460

R^1	R^2	R^3	**I + II**	*E:Z*
Me	Me	CH$_2$CH=CH$_2$	(67)	70:30
Me	Me	(CH$_2$)$_2$CH=CH$_2$	(75)	75:25
Me	Me	(CH$_2$)$_2$CH=CH$_2$	(70)	95:5
Me	Me	(CH$_2$)$_3$OTHP	(60)	0:100
H	n-Bu	n-Bu	(65)	100:0
Me	Me	Ph	(84)	25:75
Me	Me	TMS	(85)	35:65

TABLE II. MONOCYCLIC RINGS CONTAINING ONE OR MORE HETEROATOMS (*Continued*)

Reactant	Conditions	Product(s) and Yield(s) (%)	Refs.
R: H, Me	NaCo(dmgH)₂py, MeOH, NaOH, 0-25°		461
	Bu₂SnH₂, THF, ultrasound 6° -55°	(60) (80) (60-70) *trans-cis* 87:13 94:6	86
C₁₂	Bu₃SnH, AIBN, C₆H₆, 70°	R: H (67), Me (56), n-Pr (63)	462
	Bu₃SnH, AIBN, C₆H₆, 80°	(49)	463
	(Ph₃Sn)₂, t-BuNC, C₆H₆, hv, 50°	(61)	206

483

TABLE II. MONOCYCLIC RINGS CONTAINING ONE OR MORE HETEROATOMS (Continued)

Reactant	Conditions	Product(s) and Yield(s) (%)	Refs.
	Zn, NH$_4$Cl B$_{12}$ (2 mol%), DMF, 20°	(77)	464
	BPO, C$_6$H$_6$, 80°	(63)	160
	Me$_3$SnCl, Na(CN)BH$_3$ AIBN, t-BuOH	I, II I + II (84) I:II = 2:1	218
	MeC(O)SH, AIBN	(67)	452
	Bu$_3$SnH, AIBN, C$_6$H$_6$, 80°	(99)	36, 465

484

Reactant	Conditions	Product(s) and Yield(s) (%)	Refs.

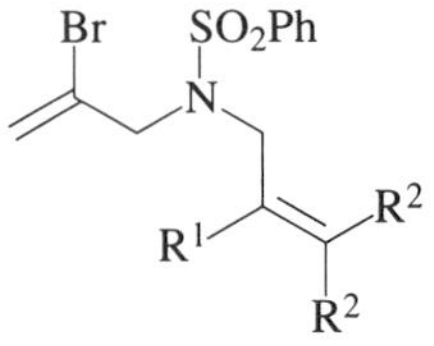

MeC(O)SH, AIBN

I + II

	I	**II**	**I:II** = 3:1
	(—)	(—)	
	(57)	(—)	
	(—)	(80)	

452

R^1	R^2
H	H
H	Me
Me	H

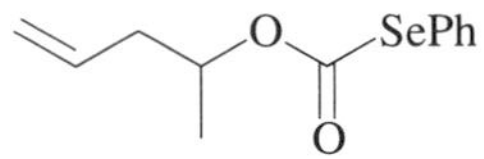

Bu₃SnH, AIBN, C₆H₆, 80°

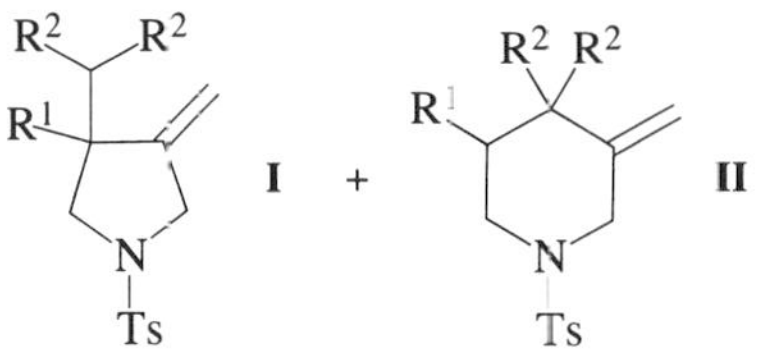

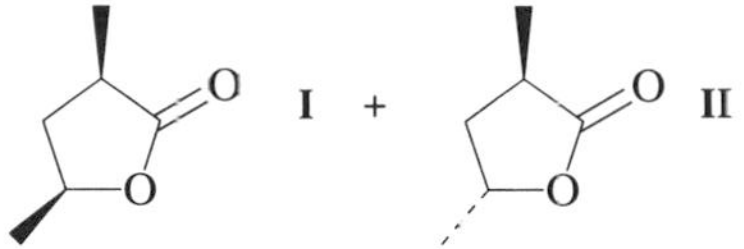

I + II (92) **I:II** = 2.5:1

36, 465

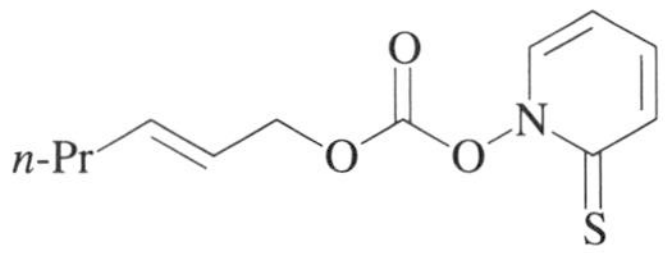

C₆H₆, *hν*
t-BuSH, reflux

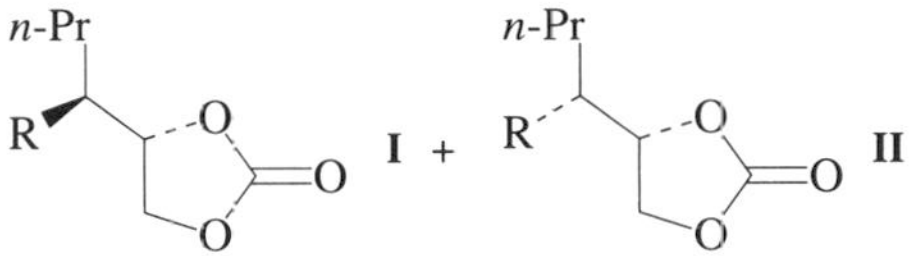

R = (*S*)-2-pyridyl; **I + II** (70), **I:II** = 3:1
R = H; **I + II** (65)

466

485

TABLE II. MONOCYCLIC RINGS CONTAINING ONE OR MORE HETEROATOMS (*Continued*)

Reactant	Conditions	Product(s) and Yield(s) (%)	Refs.
	Bu$_3$SnCl, AIBN, NaBH$_4$, C$_6$H$_6$, EtOH, hv	**I + II** (60) **I:II** = 86:14	426
	Bu$_3$SnH, AIBN, C$_6$H$_6$, 80°	(77)	467, 463
	MeC(O)SH, AIBN, C$_6$H$_6$, 80°	$\dfrac{R}{\text{H}}$ (100) *cis:trans* = 2:1 Me (64)	452
$\dfrac{R}{\text{Ph}}$ p-MeC$_6$H$_4$ p-MeOC$_6$H$_4$	BuLi	(52) (48) (46)	468, 469
	(TMS)$_3$SiH, PhMe, AIBN, 88-90°, 2 h	**I, II** **I + II** (82) **I:II** = 2.5:1	470

Reactant	Conditions	Product(s) and Yield(s) (%)	Refs.
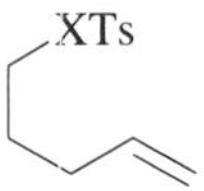 $X = O, S$	BPO, 80–90°	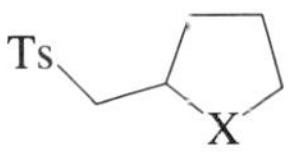 (100)	471
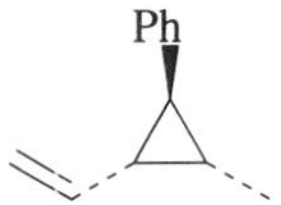	(Ph$_2$Se)$_2$, AIBN, MeOH, $h\nu$, 50°	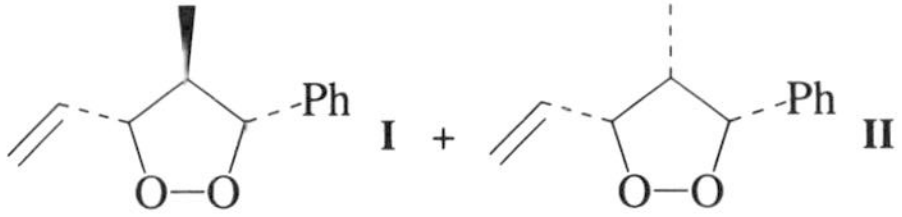 **I** + **II** (73) **I:II** = 2.8:1	434, 435
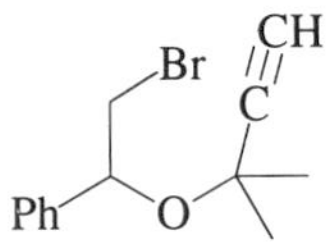	Bu$_3$SnCl, Na(CN)BH$_3$, AIBN, t-BuOH, 80°	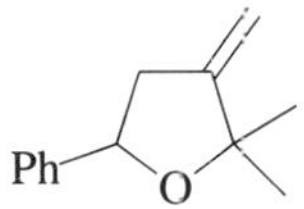 (90)	472
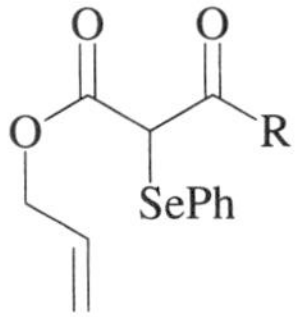	C$_6$H$_6$, $h\nu$	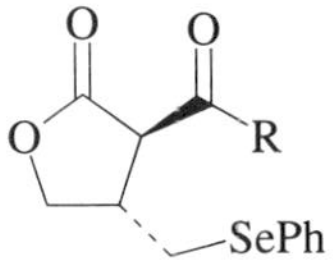 $\dfrac{\text{R}}{\text{OEt} \ (41)}$ Me (47)	88
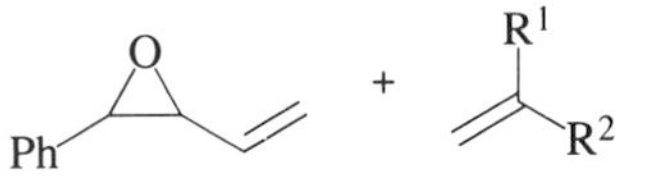	PhSH, C$_6$H$_6$, 80°	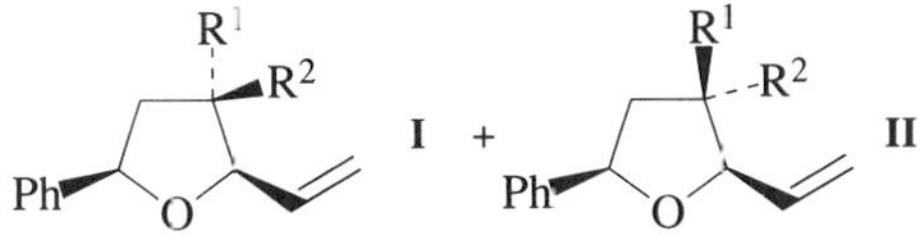 **I** + **II**	473

C$_{13}$

R^1	R^2		**I+II**	**I:II**
CN	H		(53)	2.1:1
CO$_2$Me	H		(56)	1.8:1
Ph	H		(57)	4:1
CO$_2$Me	Me		(55)	1:1

487

Reactant	Conditions	Product(s) and Yield(s) (%)	Refs.
	TsSePh, AIBN, CHCl$_3$, 65°	(96) *cis:trans* = 60:40	281
	TsBr, AIBN, MeCN, 25°	(74) *cis:trans* = 77:23	409
	Bu$_3$SnH, AIBN	(84)	452
diastereomeric ratio = 5:1 diastereomeric ratio = 1:2	Bu$_3$SnH, AIBN, C$_6$H$_6$, 80°	**I** (62) (21) **II** (12) (42)	313
	Bu$_3$SnH, AIBN, C$_6$H$_6$, 80°	(40)	474

488

TABLE II. MONOCYCLIC RINGS CONTAINING ONE OR MORE HETEROATOMS (Continued)

Reactant	Conditions	Product(s) and Yield(s) (%)	Refs.

Bu₃SnH, AIBN, C₆H₆, 80°

Y	R^1	R^2	R^3		cis:trans
SO	H	H	H	(70)	9:1
S	H	H	H	(55)	9:1
SO₂	H	H	H	(51)	9:1
SO	H	H	Ph	(80)	4:1
S	H	H	Ph	(59)	4:1
SO₂	H	H	Ph	(66)	4:1
S	H	Me	Me	(67)	3:1
SO	H	H	Cl	(84)	3:1
SO	H	H	Cl	(62)	3:1
SO	Cl	H	H	(25)	1:1

475

Bu₃SnH, AIBN, C₆H₆, 2 h

I + II

	I + II	I : II
	<(85)	84 : 16
	<(94)	100 : 0

R^1	R^2
H	Ph
Ph	H

56

TABLE II. MONOCYCLIC RINGS CONTAINING ONE OR MORE HETEROATOMS (*Continued*)

Reactant	Conditions	Product(s) and Yield(s) (%)	Refs.
C$_{14}$			
	Ph$_3$SnH, AIBN, C$_6$H$_6$, 80°	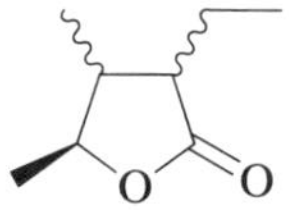 (66)	49
	Bu$_3$SnH, AIBN, C$_6$H$_6$, 80°	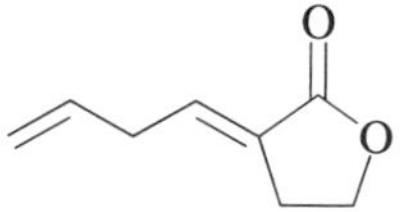 (70) 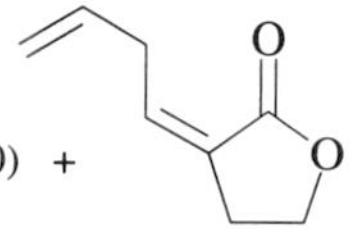 + (11)	465
	Bu$_3$SnH, AIBN, C$_6$H$_6$, 80°	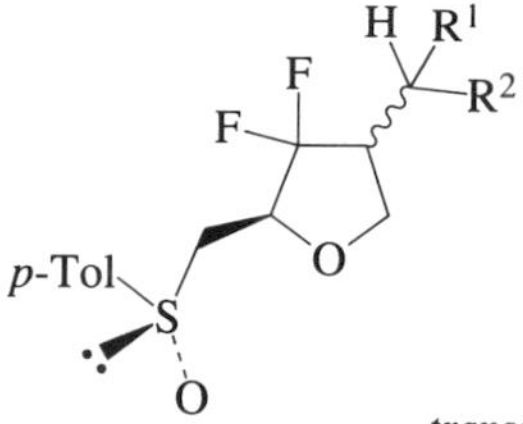 *trans:cis*	476

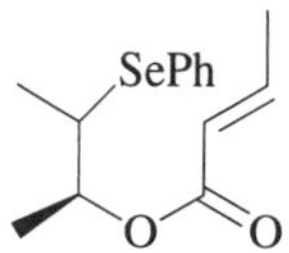

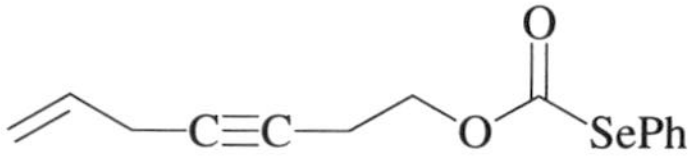

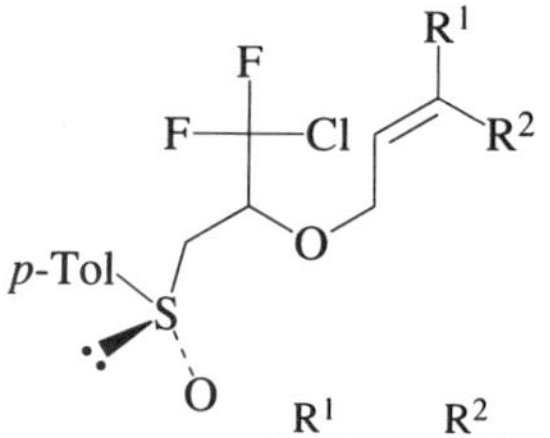

R^1 R^2
H   H   (87)   9:1
H   H   (70)   9:1
CO$_2$Et   H   (92)   4:1
Me   Me   (85)   2:1

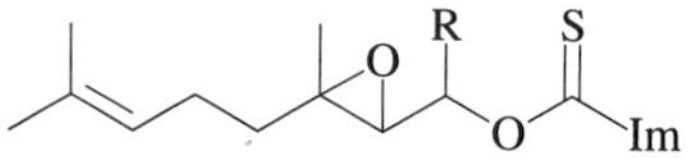

| | Bu$_3$SnH, AIBN, C$_6$H$_6$, 80° | 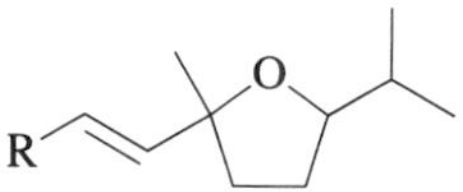 | 198 |

R
H   (14)
Me   (53)
n-Bu   (63)

490

TABLE II. MONOCYCLIC RINGS CONTAINING ONE OR MORE HETEROATOMS (*Continued*)

Reactant	Conditions	Product(s) and Yield(s) (%)	Refs.
	(PhSe)$_2$, AIBN, O$_2$, MeCN, 0°	**I** + **II** $\quad$ **I** + **II** (48) $\quad$ **I**:**II** = 1:1.4	434, 435
	Bu$_3$SnH, AIBN, C$_6$H$_6$, 80°	(70)	463
	Bu$_3$SnH, AIBN, C$_6$H$_6$, heat	(82)	477
	Bu$_3$SnH, AIBN, C$_6$H$_6$, heat	R $\quad$ Me (69) $\quad$ H (71)	478
	Bu$_3$SnH, AIBN, C$_6$H$_6$, heat	(53) $\quad$ *cis:trans* = 57:43	479

Reactant	Conditions	Product(s) and Yield(s) (%)	Refs.
	Bu₃SnH, AIBN, C₆H₆, 80°	(90) *cis:trans* = 3:1	465
	t-BuSH, CH₂(CO₂H)₂, MeCN, hv, 25°	(70)	68
	C₆H₆, hv C₆H₆, t-BuSH, hv C₆H₆, (PhSe)₂, hv		480
	CH₂Cl₂, hv	(68)	453
	ClCo(dmgH)₂py, MeOH, NaOH, NaBH₄, 0°	(60)	439a

492

TABLE II. MONOCYCLIC RINGS CONTAINING ONE OR MORE HETEROATOMS (*Continued*)

Reactant	Conditions	Product(s) and Yield(s) (%)	Refs.
C₁₅			

C₁₅

Reactant 1 (TBDMSO, SMe, Br, SMe structure)
Conditions: Bu₃SnH, AIBN, C₆H₆, 80°
Product: (TBDMSO, MeS, SMe lactone) (68)
Refs.: 478

Reactant 2 (Cl, CO₂Me, Et, dioxolane structure)
Conditions: Ph₃SnH, AIBN, C₆H₆, heat
Product: I (lactone + CO₂Me, Et, dioxolane) + II (lactone + CO₂Me)
I + II (74) I:II = 8:92
Refs.: 419

Reactant 3 (Cl, CO₂Me, Et, dioxolane structure)
Conditions: Ph₃SnH, AIBN, C₆H₆, heat
Product: I + II
I + II (77) I:II = 96:4
Refs.: 419

Reactant 4 (CO₂Bu-t + SH structure)
Conditions: AIBN, CCl₄, t-BuOH, hν
Product: (thiolane, CO₂Bu-t)
(80)
(60)
Refs.: 481

493

TABLE II. MONOCYCLIC RINGS CONTAINING ONE OR MORE HETEROATOMS (*Continued*)

Reactant	Conditions	Product(s) and Yield(s) (%)	Refs.
(bromo ether, C≡CC$_8$H$_{17}$-n)	Bu$_3$SnH, AIBN, C$_6$H$_6$, heat	(tetrahydrofuran, $=$C$_8$H$_{17}$-n) (70)	195
(fluoro diether, C$_6$H$_{13}$-n)	Bu$_3$SnH, AIBN, C$_6$H$_6$, 80°	(F-tetrahydrofuran, BuO, C$_6$H$_{13}$-n) (78)	482
(C≡CMe, SPh, CO$_2$Et ether)	Bu$_3$SnH, AIBN, C$_6$H$_6$, 80°	(CO$_2$Et furanone) (88) E:Z = 50:50	479
(pyridine-2-thione N-Bu-n amide)	t-BuSH, CH$_2$(CO$_2$H)$_2$, MeCN, hv, 25°		68, 483
	THF, BF$_3$•OEt$_2$, 22°, hv		
	THF, BF$_3$•OEt$_2$, (Ph$_2$Se)$_2$, −78°, hv		

Y	
Me	(92)
(S)-2-pyridyl	(98)
SePh	(80)

(pyrrolidine, n-Bu–N, CH$_2$Y)

TABLE II. MONOCYCLIC RINGS CONTAINING ONE OR MORE HETEROATOMS (*Continued*)

Reactant	Conditions	Product(s) and Yield(s) (%)	Refs.
	Bu₃SnH, AIBN, C₆H₆, 65°	(77)	484
	Me₃SnCl, Na(CN)BH₃, AIBN, *t*-BuOH, 80°	(92) *cis:trans* = 2:1	218
	CH₂Cl₂, *hv*, 25°	(—)	485
	Bu₃SnH, AIBN, C₆H₆	(46) *syn:anti* = 1:1	478

495

TABLE II. MONOCYCLIC RINGS CONTAINING ONE OR MORE HETEROATOMS (*Continued*)

Reactant	Conditions	Product(s) and Yield(s) (%)	Refs.
C$_{16}$			
	Pt electrode, MeCN, Et$_4$NClO$_4$	(81)	486
	THF, BF$_3$•Et$_2$O, *hv*, –78°	(63-74)	483
	1. Bu$_3$SnCl, Na(CN)BH$_3$, AIBN 2. TsH		487

R^1	R^2	
n-C$_6$H$_{13}$	Ph	(50)
n-C$_6$H$_{13}$	*p*-MeC$_6$H$_4$	(60)
n-C$_6$H$_{13}$	*i*-Pr	(55)
n-Bu	Ph	(52)
n-Bu	*p*-MeC$_6$H$_4$	(53)

TABLE II. MONOCYCLIC RINGS CONTAINING ONE OR MORE HETEROATCMS (*Continued*)

Reactant	Conditions	Product(s) and Yield(s) (%)	Refs.
(pyridine-2-thione N–O–C(=O)–N(R)–CH2CH=CH2) + CH2=CH–OEt; R: n-Pr, n-C7H15, n-Bu	H+, hv; t-BuSH; t-BuSH	**I, II** (pyrrolidine, 3-OEt, 4-CH2X, N-R) 	488

Product table for row 1:

R	I+II	I:II	X
n-Pr	(38)	—	Spy
n-C7H15	(56)	1:1.4	H
n-Bu	(48)	2:1	H

Reactant	Conditions	Product(s) and Yield(s) (%)	Refs.
MeC≡C–...–SPh, N(CO2Me)(CH(CO_2Me))	Bu3SnH, AIBN, PhMe, heat	(pyrrolidine, =CH–CO_2Me, 2-CO_2Me, N) (75) E:Z = 59:41	489, 113
Et–CH=CH–...–SPh, O, CO_2Et (E, Z)	Bu3SnH, AIBN, C_6H_6, heat	(tetrahydrofuran, Pr-n, CO_2Et) (77) cis:trans = 35:65 (65) cis:trans = 54:46	479
EtO_2C–CHBr–CH(Ar)–O–$CH_2C\equiv CH$; Ar = 3,4-(MeO)2C6H3	Bu3SnH, AIBN, C_6H_6, 80°	(tetrahydrofuran, =CH2, EtO_2C, 3,4-(MeO)2C6H3) (80–82)	490

497

Reactant	Conditions	Product(s) and Yield(s) (%)	Refs.
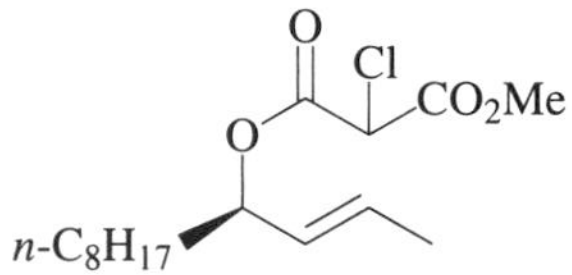	$Mn(OAc)_3$, $Cu(OAc)_2$, AcOH, 75°	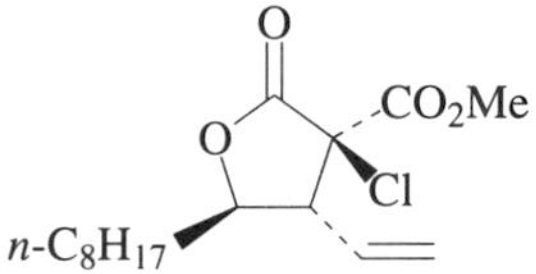 **I** + 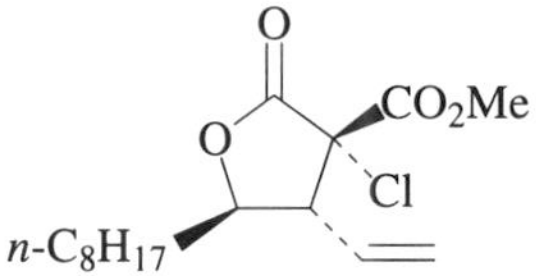 **II**; **I** + **II** (82), **I**:**II** = 3.1:1	491
C_{17}			
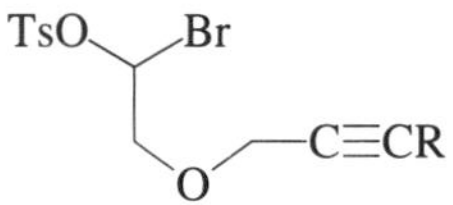 R: Ph; n-C_5H_{11}	Bu_3SnH, AIBN, C_6H_6, 80°	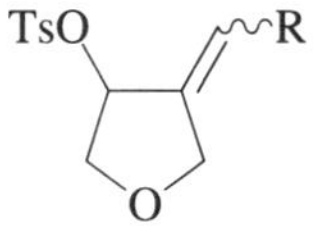 R = Ph (85), *cis* : *trans* 54 : 46; R = n-C_5H_{11} (93), *cis* : *trans* 0 : 100	462
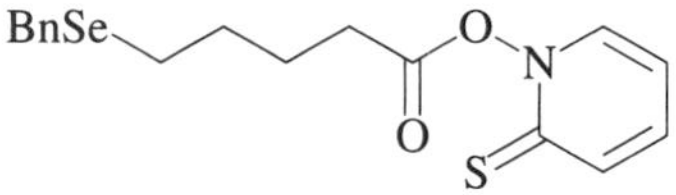	C_6H_6, $h\nu$, 80°	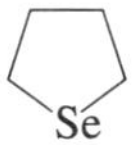 (79)	492, 493
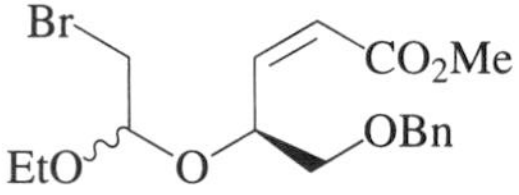	Bu_3SnH, AIBN, C_6H_6, 80°	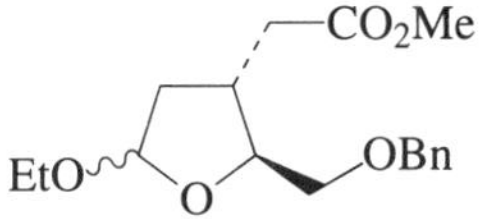 (92)	494

Reactant	Conditions	Product(s) and Yield(s) (%)	Refs.
	ClCo(dmgH)₂py, MeOH, NaOH, Py, NaBH₄, 0°	**I** + **II** **I + II** (50) **I:II** = 9:1	495
	Bu₃SnH, AIBN, C₆H₆, 80°	(87)	465, 496
	Bu₃SnH, AIBN, C₆H₆, 80°	**I** + **II** $\dfrac{\text{I+II}\quad\text{I:II}}{(95)\quad 2.6:1}$ (83) 2.2:1	497, 48
	Bu₃SnH, AIBN, C₆H₆, heat	(36)	452

499

TABLE II. MONOCYCLIC RINGS CONTAINING ONE OR MORE HETEROATOMS (*Continued*)

Reactant	Conditions	Product(s) and Yield(s) (%)	Refs.
	Bu₃SnH, AIBN, C₆H₆, 80°	(80)	36, 465
	Bu₃SnH, AIBN, C₆H₆, 80°	(47) + (9) + (12)	463
	Bu₃SnH, AIBN, PhMe, 80°	$\dfrac{cis{:}trans}{}$ (93) 35:65 (91) 30:70	489, 113, 456
	Bu₃SnH, AIBN, C₆H₆, heat	(72) *cis:trans* = 44:56	479

$$\begin{array}{c|c} R^1 & R^2 \\ \hline Et & H \\ H & Et \end{array}$$

Reactant	Conditions	Product(s) and Yield(s) (%)	Refs.

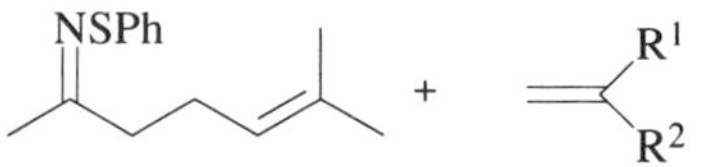

Bu$_3$SnH, AIBN, C$_6$H$_6$, heat

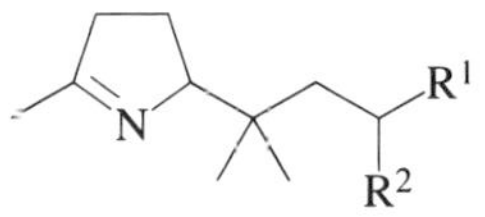

R^1	R^2	
CO$_2$Me	H	(83)
CO$_2$Bu-*t*	CO$_2$Bu-*t*	(68)
CN	H	(60)

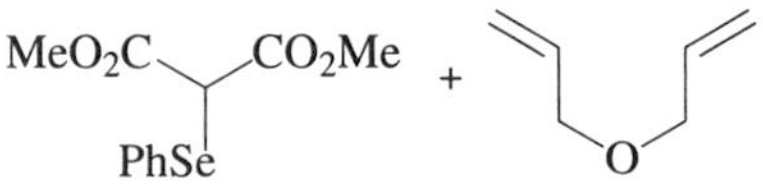

69

C$_6$H$_6$, *hv*, 12 h

(86)

cis:trans = 1.8:1

498

Ph$_3$SnH, AIBN, C$_6$H$_6$, 80°, 8 h

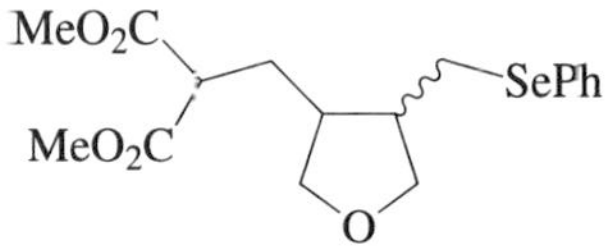

R^1	R^2	*cis:trans*	
H	H	62:38	(90)
H	Me	60:40	(85)
H	Pr	60:40	(75)
H	Ph	55:45	(85)
Me	H	—	(83)

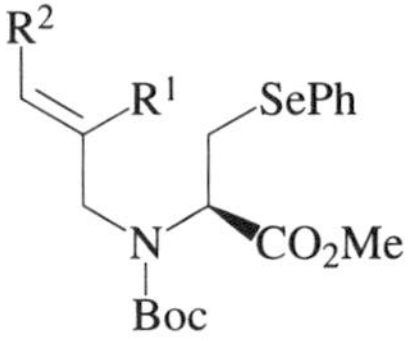

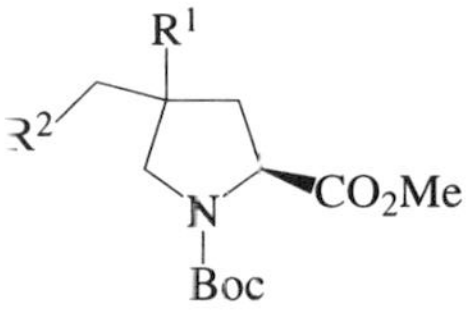

87

501

TABLE II. MONOCYCLIC RINGS CONTAINING ONE OR MORE HETEROATOMS (*Continued*)

Reactant	Conditions	Product(s) and Yield(s) (%)	Refs.
PhSe— (CMe≡C, O, Ph)	Bu₃SnH, AIBN, C₆H₆, 80°	(83) 1.5:1 isomer mixture	497, 48
R¹—(Co(dmgH)₂py, R²)	Cl₃CSO₂Cl, CH₂Cl₂, *hv*, 15°	**I** + **II**	499

R¹ + R² structures (I and II):

R¹	R²	**I**	**II**
H	H	(82)	(—)
Me	H	(22)	(—)
Me	Me	(48)	(19)
H	Me	(60)	(17)

| PhSe—(Ph) | Bu₃SnH, AIBN, C₆H₆, 80° | **I** + **II** I + II (>98) I:II = 3.2:1 | 497, 48 |
| TsO—(R, X) | Bu₃SnH, AIBN, NaI, DME, 80°, 1-2 h | | 500 |

R	X	
H	NTs	(80)
Ph	NTs	(82)
Ph	O	(76)

502

TABLE II. MONOCYCLIC RINGS CONTAINING ONE OR MORE HETEROATOMS (*Continued*)

Reactant	Conditions	Product(s) and Yield(s) (%)	Refs.
C$_{19}$			
	DCA, MeOH, MeCN, $h\nu$	$\begin{array}{ll} \underline{Y} & \\ \text{Me} & (83) \\ \text{OMe} & (88) \end{array}$	188
	Bu$_3$SnH, AIBN, PhMe, heat	(58)	501
C$_{20}$			
	Bu$_3$SnH, AIBN, THF, heat	(45)	204
	(Ph$_2$Se)$_2$, AIBN, O$_2$, MeCN, 0°, $h\nu$		434, 435

TABLE II. MONOCYCLIC RINGS CONTAINING ONE OR MORE HETEROATOMS (Continued)

Reactant	Conditions	Product(s) and Yield(s) (%)	Refs.
C22			
[structure: PhS, SPh, O, N–Me, pyridinyl]	Bu$_3$SnH, AIBN, PhMe, 110°	[structure: lactam O, N–Me, pyridinyl] (77)	450
[structure: TBDMSO, I, N–CO$_2$Ph, isopropenyl]	ClCo(dmgH)$_2$py, NaBH$_4$, MeOH, NaOH, 0°	[structure: TBDMSO, N–CO$_2$Ph pyrrolidine] (60) *cis:trans* = 58:42	502
C23			
[structure: alkene + $=$SO$_2$Ph, O, N–Bu, thiopyridone]	H$^+$, $h\nu$	[structure: N–Bu-n, PhO$_2$S, S-pyridyl] (56)	503
[structure: $=$ SO$_2$Ph, CO$_2$Bu-t + Br, O, n-BuO, allyl]	I$_2$CoL, Zn, DMF, $h\nu$, 20°	[structure: CO$_2$Bu-t, n-BuO, O] **I** + [structure: CO$_2$Bu-t, n-BuO, O] **II** **I** + **II** (66) **I:II** = 81:19	184

504

TABLE II. MONOCYCLIC RINGS CONTAINING ONE OR MORE HETEROATOMS (*Continued*)

Reactant	Conditions	Product(s) and Yield(s) (%)	Refs.
C₂₄	(Ph₂Se)₂, AIBN, O₂, MeCN, *hv*, 0°	(38)	434, 435
	C₇H₁₆, *hv*, 25°	(75)	350
	C₇H₁₆, *hv*, 25°	(82)	350
	Bu₃SnH, AIBN, C₆H₆, 80°		419

R	X
H	Cl, I
Me	Cl
i-Pr	Br

I + II	I:II
(81)	88:12
(81)	80:20
(91)	91:9

Reactant	Conditions	Product(s) and Yield(s) (%)	Refs.

C₂₆

Reactant (with + $TsCH_2C(=CH_2)C(O)R^3$):

R^1	X	R^2	R^3
H	O	Bn	OEt
H	O	Bn	NHBn
Me	O	Bn	OEt
H	H_2	C(O)Bu-t	NHBn
H	H_2	C(O)Ph	OEt

Conditions: TsCl, BPO, PhMe, 110°

Product:

	ratio
(51)	3.6:1
(46)	3.5:1
(51)	1.5:1
(53)	1.8:1
(58)	1.6:1

Refs.: 351

C₂₈

R	X	R^1
Bn	O	CO_2Et
Bn	O	CONHBn
Ts	H_2	CO_2Et
Bn	O	CO_2Et

Conditions:
A. TsCl, BPO, PhMe
B. $(Bu_3Sn)_3$, C_6H_6, $h\nu$, 4 h

Product: **I, II**

I + II	I:II
(51)	3.6:1
(46)	3.5:1
(68)	1.8:1
(54)	2:1

Refs.: 504

Conditions: $ClCo(dmgH)_2py$, $NaBH_4$, MeOH, NaOH, 0°

Product: (80) *cis:trans* = 62:38

Refs.: 502

TABLE II. MONOCYCLIC RINGS CONTAINING ONE OR MORE HETEROATOMS (*Continued*)

Reactant	Conditions	Product(s) and Yield(s) (%)	Refs.

C_{29}

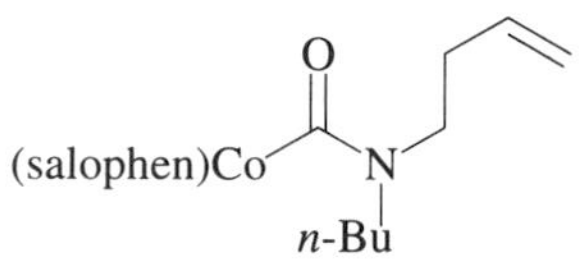

(salophen)Co, N, *n*-Bu

PhMe, 110°

I + **II**

I + II (62)

402

C_{32}

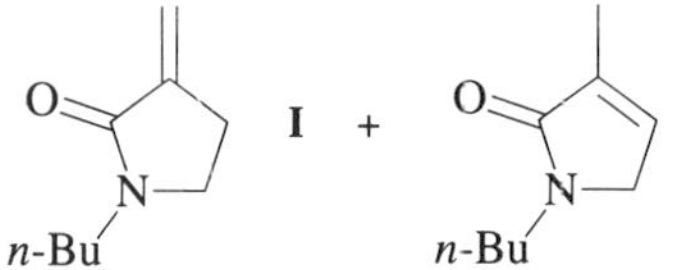

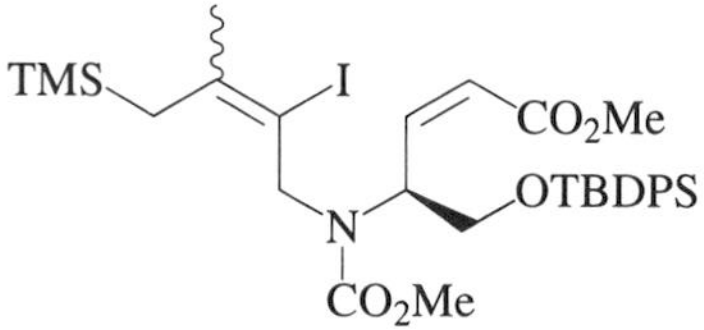

TMS, I, CO$_2$Me, OTBDPS, N, CO$_2$Me

Bu$_3$SnH, AIBN, C$_6$H$_6$, 80°

TMS, CO$_2$Me, OTBDPS, N, CO$_2$Me

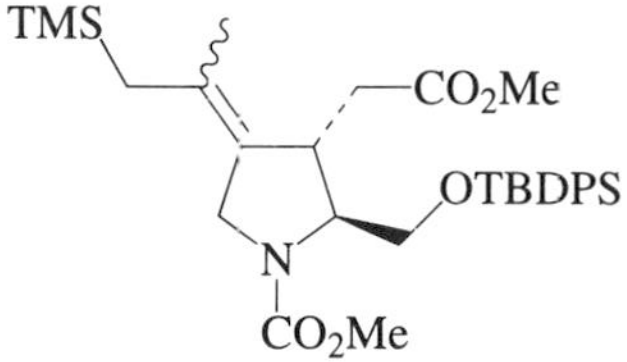

(86)

505

D. 6-Membered Rings

C_5

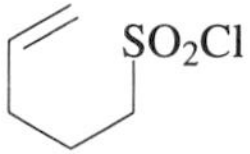

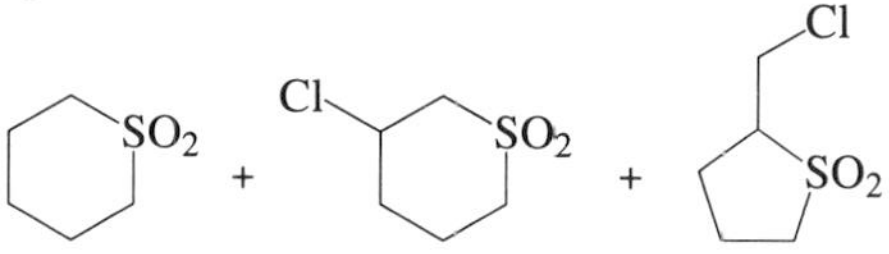

SO$_2$Cl

CuCl$_2$, AIBN, 170°
Bu$_3$SnH, AIBN, 80°
Bu$_3$SnH, ultrasound
Ph$_3$SiH, *hv*

I + **II** + **III**

	I+II+III	I:II:III
	(17)	0:100:0
	(—)	7:93:0
	(—)	100:0:0
	(—)	14:78:8

506

HC≡C, SH

C$_5$H$_{12}$, *hv*, 36°

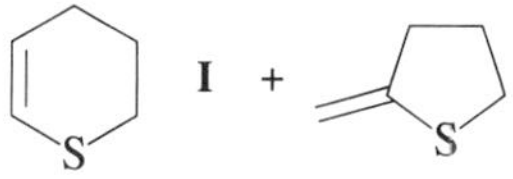

I + **II**

I + II (61)

I:II = 70:30

507

507

TABLE II. MONOCYCLIC RINGS CONTAINING ONE OR MORE HETEROATOMS (Continued)

	Reactant	Conditions	Product(s) and Yield(s) (%)	Refs.
C_6	HOO—	1. BPO, O_2, C_6H_6, rt 2. PPh$_3$	OH (30)	508
	SH—S (dithiol)	Cyclohexane, hv, 80° C_6H_{14}, hv, −65°	**I** + **II** **I + II** (46) **I:II** = 97:3 **I + II** (61) **I:II** = 23:77	509
C_9	Br— —CO$_2$Et, R	Bu$_3$SnH, AIBN, C_6H_6, 80°	—CO$_2$Et R = H (96) R = Me (96)	424
C_{10}	X, Y, SO_2—O—C≡CH	Bu$_3$SnH, AIBN,	X—SO$_2$, Y	510

For the C_{10} products:

X	Y	
CO$_2$Me	H	(61)
H	Me	(35)
F	F	(39)

TABLE II. MONOCYCLIC RINGS CONTAINING ONE OR MORE HETEROATOMS (*Continued*)

Reactant	Conditions	Product(s) and Yield(s) (%)	Refs.
	Bu₃SnH, AIBN, C₆H₆, heat	I + II (85) I:II = 4.7:1	511
	Bu₃SnH, AIBN, Na(CN)BH₃, C₆H₆, benzo-15-crown-5, 80°		512

R	I + II	I:II
MeO	(40)	75:25
EtO	(53)	80:20
	(53)	75:25
	(35)	85:15
	(33)	90:10

509

TABLE II. MONOCYCLIC RINGS CONTAINING ONE OR MORE HETEROATOMS (*Continued*)

Reactant	Conditions	Product(s) and Yield(s) (%)	Refs.

Reactant (C$_{11}$): thioether with SH group, cyclopropane bearing R^1, R^2

Conditions: AIBN, C$_6$H$_6$, *hv*, 80°

Product:

R^1	R^2		*cis:trans*
Cl	Cl	(54)	1:1
Br	Br	(50)	1.1:1
Ph	H	(51)	1:1
OBn	H	(65)	2:1

513

Conditions: Bu$_3$SnH, AIBN, C$_6$H$_6$, 80°

Product: **I** + **II** (29) **I:II** = 20:80

514

Conditions:
1. Bu$_3$SnH, AIBN, C$_6$H$_6$, 80°
2. TsOH, 80°

Product: **I** + **II**

R^1	R^2		I + II	I:II
H	H		(80)	3:1
Me	H		(74)	1:3
H	Me		(74)	1:3
Me	Me (*anti*)		(64)	1:2
Me	Me (*syn*)		(80)	1:3

515

TABLE II. MONOCYCLIC RINGS CONTAINING ONE OR MORE HETEROATOMS (Continued)

Reactant	Conditions	Product(s) and Yield(s) (%)	Refs.
(structure with OOH)	$(t\text{-BuON=})_2$, O₂, C₆H₆, 30°	(peroxide structure with OH) (70)	516
C₁₂ (structure with O–C(=O)–SePh)	Bu₃SnH, AIBN, C₆H₆, 80°	(lactone structure) (55)	465
PhS (structure) NCS	Bu₃SnH, AIBN, PhMe 75°, sunlight 30° 10°, Hg lamp	(SPh thiolactam structure) (17) (33) (64)	517
TMS–N(R²) (enone with C(=O)R¹)	DCA, $h\nu$	(piperidine I, R¹C(=O)) + (pyrrolidine II, R¹C(=O), N–R²)	518, 188

R¹	R²
Me	Me
Me	Bn
OMe	Me
OMe	Bn

I	II
(90)	—
(90)	—
(55)	(48)
(40)	(41)

Reactant	Conditions	Product(s) and Yield(s) (%)	Refs.
C₁₃			
	Bu₃SnH, AIBN, C₆H₆, heat	R / Me (88) / Et (97)	519, 93
	DCN, *i*-PrOH, *hv*	(70–78)	520
	Bu₃SnH, AIBN, C₆H₆, 80°	X / N (57) / CH (53)	510
	TiCl₃, AcOH, H₂O, 0°	(92)	521
	Bu₃SnH, AIBN, C₆H₆, 80°	(74)	465

TABLE II. MONOCYCLIC RINGS CONTAINING ONE OR MORE HETEROATOMS (Continued)

Reactant	Conditions	Product(s) and Yield(s) (%)	Refs.
 	R^1	R^2	E:Z
SPh	CO$_2$Me	1.8:1
CO$_2$Bu-t	CO$_2$Me	1.3:1
SPh	Ph	12:1		Bu$_3$SnH, AIBN, PhMe, 75°	 (90) (93) (69)	517
(C$_{14}$)	Bu$_3$SnH, AIBN, C$_6$H$_6$, 80°	 I + II	522, 93
	(PhS)$_2$, hv	 (87) cis:trans = 6:1	440
	Bu$_3$SnCl, Na(CN)BH$_3$, AIBN, t-BuOH, 80°	 | R	
H	(90)
Me	(89)		72

513

Reactant	Conditions	Product(s) and Yield(s) (%)	Refs.
C$_{15}$			
[structure with SePh]	Bu$_3$SnH, AIBN, C$_6$H$_6$, 80°	(86)	465
[structure with Ts, SMe] $\dfrac{R}{\text{Me } E}$ $\text{Ph } Z$	Bu$_3$SnH, AIBN, C$_6$H$_6$, 80°	**I, II** $\dfrac{\textbf{I + II}}{(97)\quad(82)}$ diastereomeric ratio (60:40) (74:26)	378
C$_{17}$			
[structure with MeO$_2$C, CO$_2$Me, OTBDMS]	Bu$_3$SnH, AIBN, C$_6$H$_6$, heat	**I** + **II** **I + II** (85) **I:II** = 80:20	511

TABLE II. MONOCYCLIC RINGS CONTAINING ONE OR MORE HETEROATOMS (*Continued*)

Reactant	Conditions	Product(s) and Yield(s) (%)	Refs.
	Bu₃SnH, AIBN, C₆H₆, 80°, 3.5–27 h	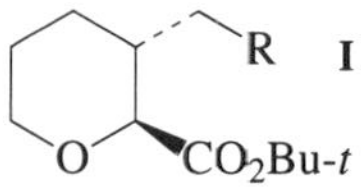 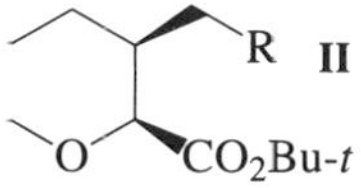	94

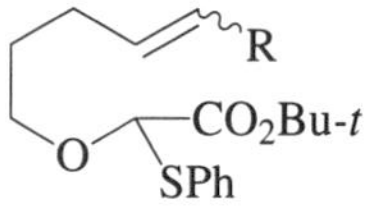

R	I + II	*trans cis* I:II
H	(85)	1.6:1
cis-TMS	(91)	5:1
trans-TBDPS	(66)	1.3:1
cis-TBDPS	(82)	1:0

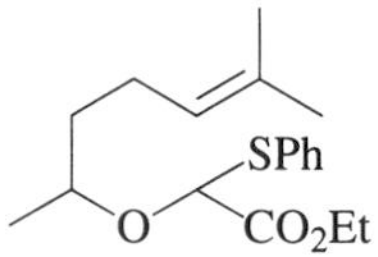

| | Bu₃SnH, AIBN, C₆H₆, heat | 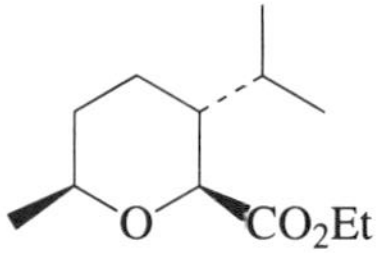 (62) | 479 |

| | Bu₃SnH, AIBN, C₆H₆, 80° | I, II + III
I, II + III (97) I, II:III = 14.3:1 | 48, 497 |

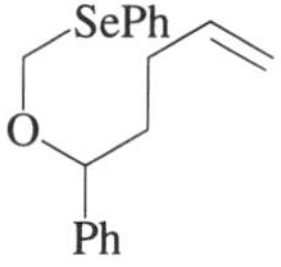

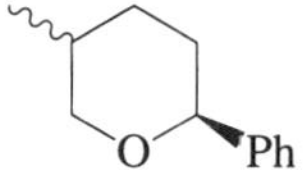

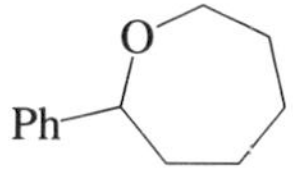

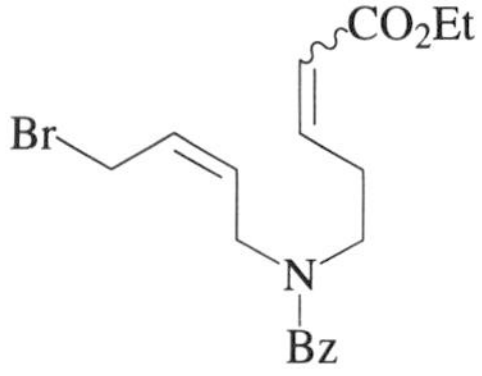

| | Bu₃SnH, AIBN, C₆H₆, 80° | I +
I + II (90), I:II = 1:1
I + II (90), I:II = 1:1 | 523 |

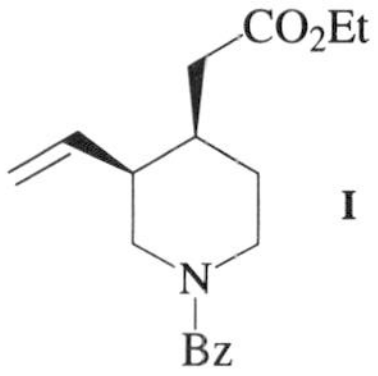

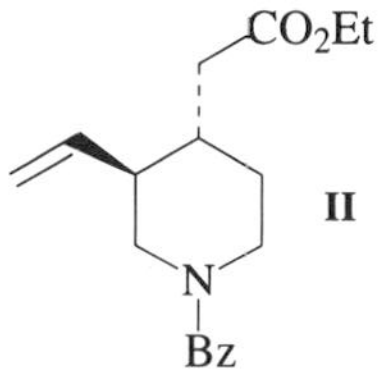

C₁₈

Z
E

TABLE II. MONOCYCLIC RINGS CONTAINING ONE OR MORE HETEROATOMS (Continued)

Reactant	Conditions	Product(s) and Yield(s) (%)	Refs.

C_{19}

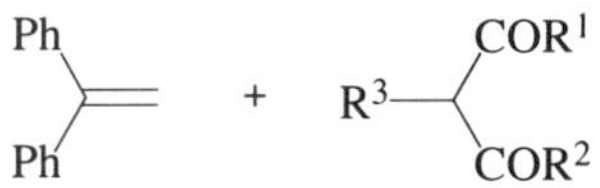

Mn(OAc)$_3$, Cu(OAc)$_2$, AcOH, O$_2$, 23°

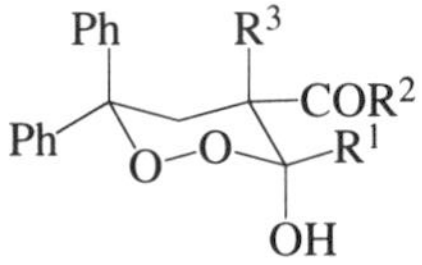

R^1	R^2	R^3	
Me	OMe	H	(90)
Me	OPr	H	(83)
Me	OBu	H	(81)
Me	OBu-t	H	(65)
Ph	OEt	H	(68)
Me	OCH$_2$CH$_2$O		(85)

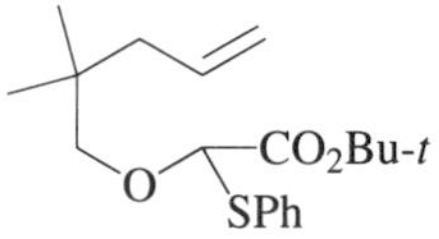

524

Bu$_3$SnH, AIBN, C$_6$H$_6$, 80°, 3.5-27 h

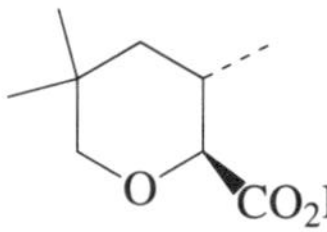

I + **II** (74) **I:II** = 1:1

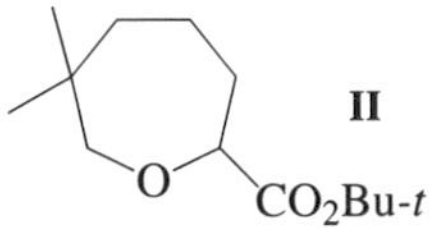

94

DCA, MeOH, MeCN, hv

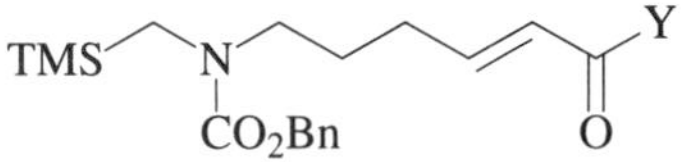

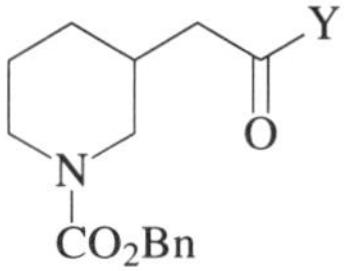

Y	
Me	(75)
OMe	(78)

188

516

Reactant	Conditions	Product(s) and Yield(s) (%)	Refs.

C_{21}

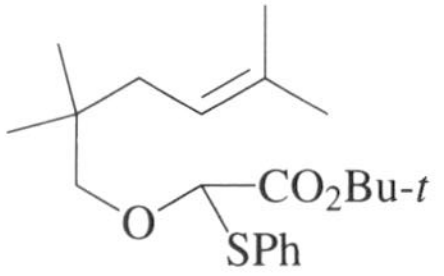

Bu$_3$SnH, AIBN, C$_6$H$_6$, 80°, 3.5-27 h

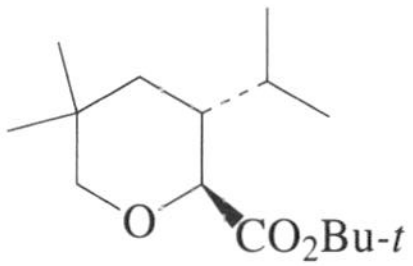

(74)

94

C_{25}

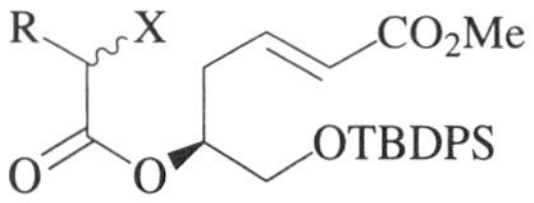

Ph$_3$SnH, AIBN

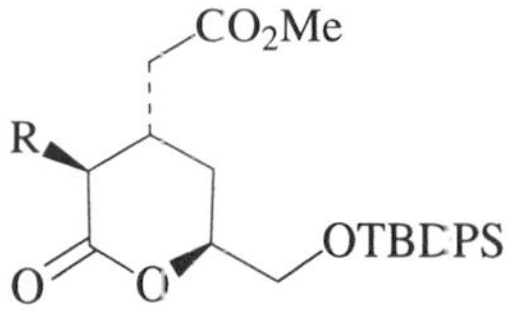

	I:II
I + II	23:77
(51)	
(63)	78:22

I + II

419

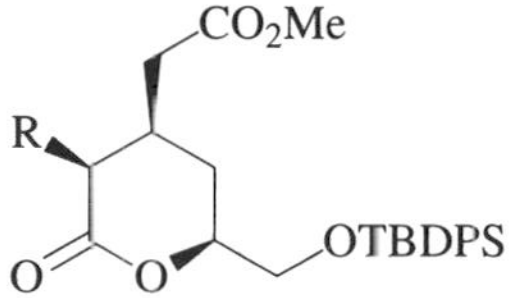

R	X
H	I
Me	Cl

C_{29}

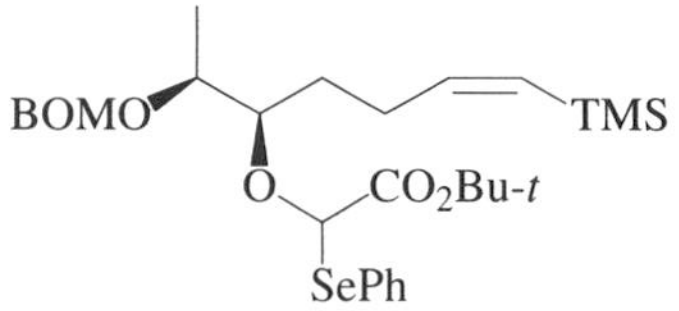

Bu$_3$SnH, AIBN, C$_6$H$_6$, 80°, 3-27 h

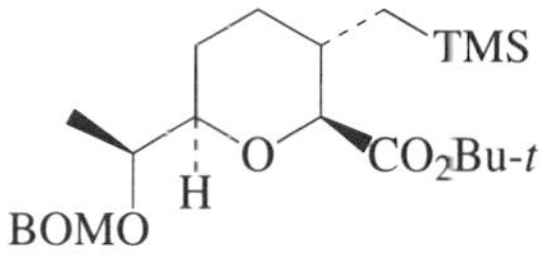

(61)

94

517

Reactant	Conditions	Product(s) and Yield(s) (%)	Refs.

C₃₀

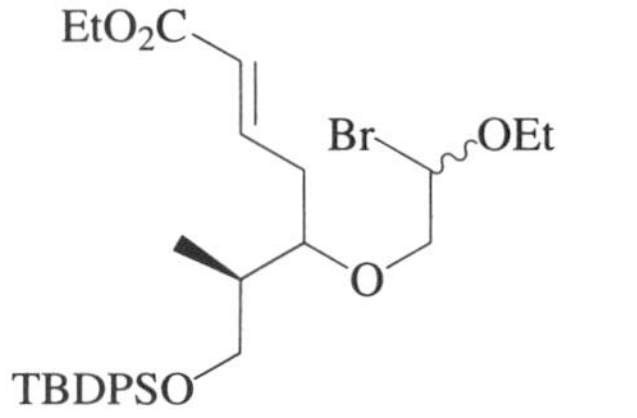

Bu₃SnCl, Na(CN)BH₃, AIBN, *t*-BuOH, 1 h

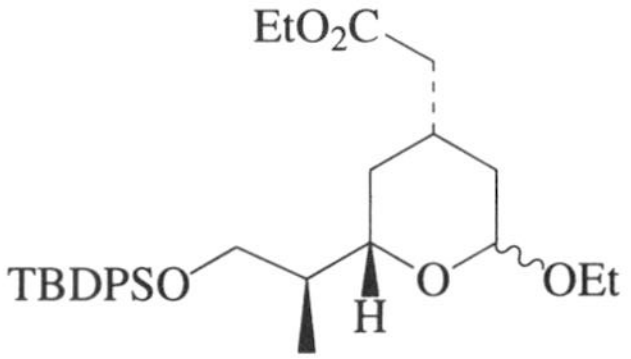

(78)

92

C₃₁

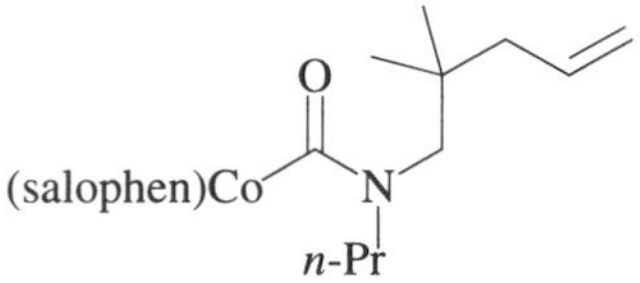

PhMe, 110°

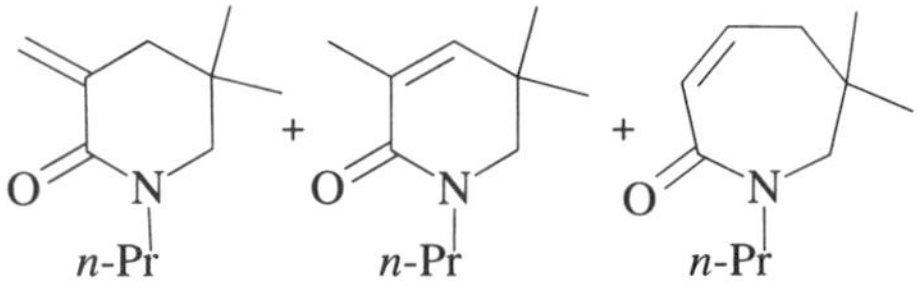

I + II + III (65), **I:II:III** = 78:11:11

402

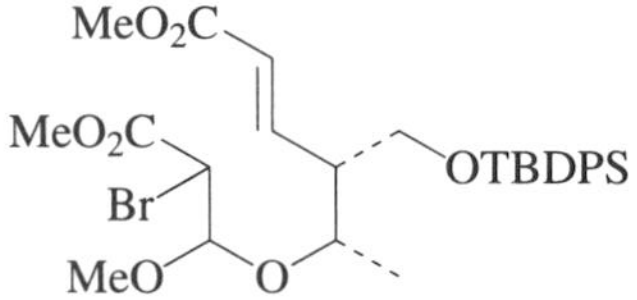

1. Bu₃SnH, AIBN, C₆H₆, 80°
2. TsOH, 80°

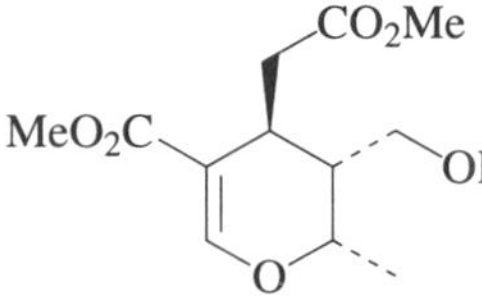

I + II (58)
I:II = 4:1

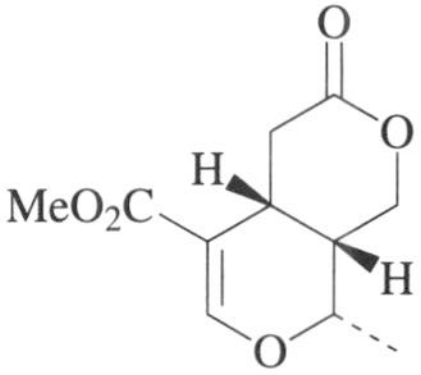

515

518

TABLE II. MONOCYCLIC RINGS CONTAINING ONE OR MORE HETEROATOMS (Continued)

Reactant	Conditions	Product(s) and Yield(s) (%)	Refs.

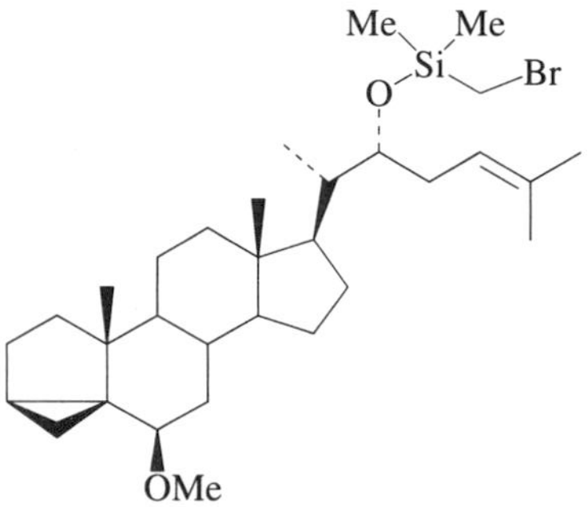

Bu₃SnCl, Na(CN)BH₃, AIBN, t-BuOH, 80°

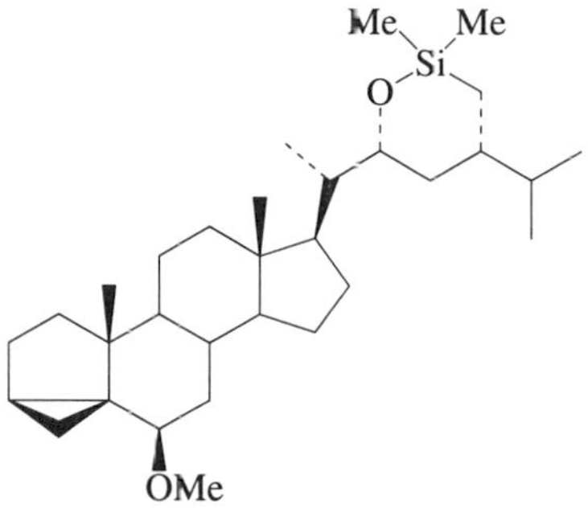

(95)

72

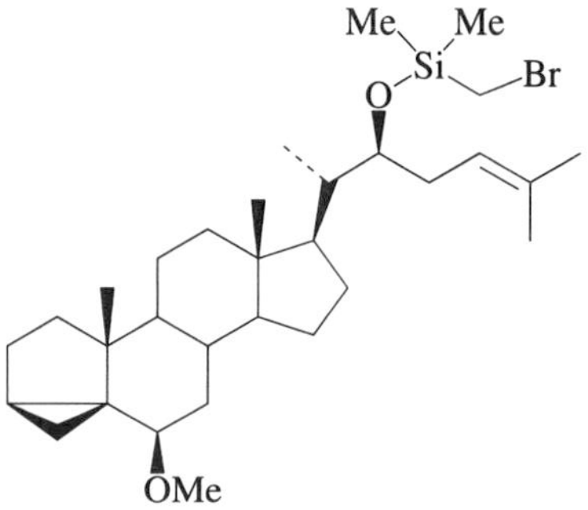

Bu₃SnCl, Na(CN)BH₃, AIBN, t-BuOH, 80°

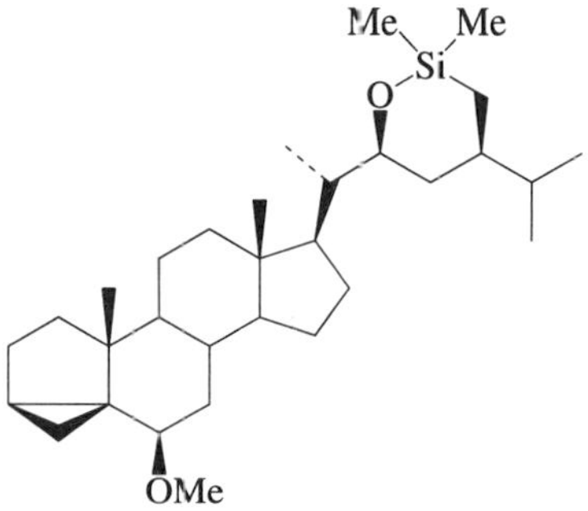

(98)

72

E. 7-Membered Rings

C₅

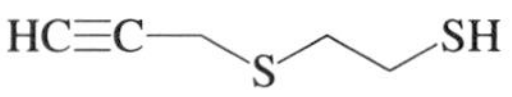

HC≡C—S—SH

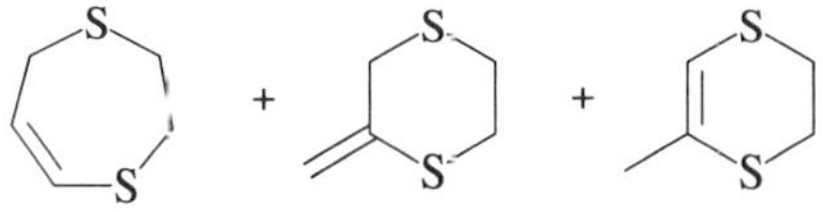

C₅H₁₂, hv, 36°

I + II + III (51)   I:II:III = 77:3:2

507

519

TABLE II. MONOCYCLIC RINGS CONTAINING ONE OR MORE HETEROATOMS (*Continued*)

Reactant	Conditions	Product(s) and Yield(s) (%)	Refs.
C₇	(*t*-BuO)₂, PhCl, 132°	$\dfrac{R^1}{\text{H} \quad (60)}$ $\text{Me} \quad (58)$	525
C₁₂	Bu₃SnH, AIBN, C₆H₆, 80°	(25) + (12)	392
C₁₃	Bu₃SnCl, Na(CN)BH₃, AIBN, *t*-BuOH, 80°	(92)	72
C₁₄	Bu₃SnH, AIBN, C₆H₆, 80°	**I** + **II** (51) **I:II** = 93:7	526

TABLE II. MONOCYCLIC RINGS CONTAINING ONE OR MORE HETEROATOMS (Continued)

Reactant	Conditions	Product(s) and Yield(s) (%)	Refs.
C_{16}	Bu_3SnH, AIBN, C_6H_6, 80°	**I** + **II** $\quad$ **I:II** = 91:9 $\quad$ **I + II** (63)	526
C_{19}	Bu_3SnH, AIBN, PhMe, 60°	Ph $\quad$ (60-70)	527
C_{28}	Bu_3SnCl, Na(CN)BH_3, AIBN, t-BuOH, 80°	(95)	72

521

Reactant	Conditions	Product(s) and Yield(s) (%)	Refs.
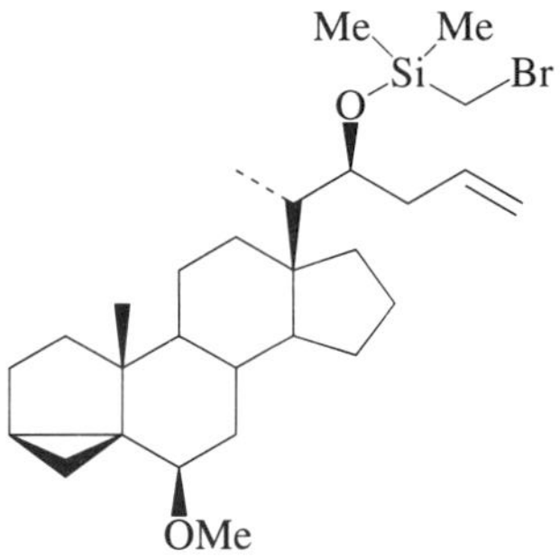	Bu$_3$SnCl, Na(CN)BH$_3$, AIBN, *t*-BuOH, 80°	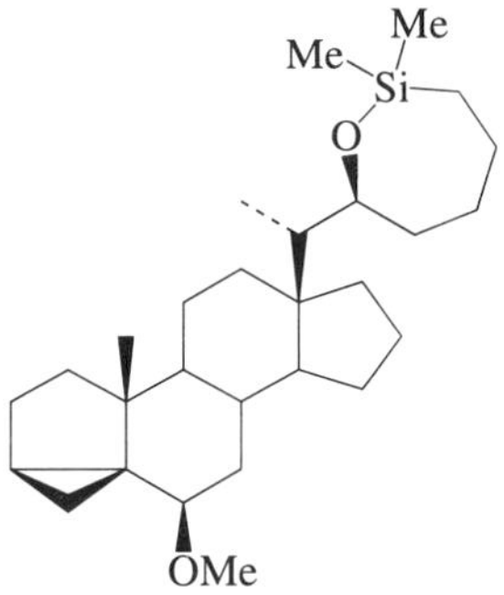 (95)	72

F. 8-Membered Rings

C$_{14}$ 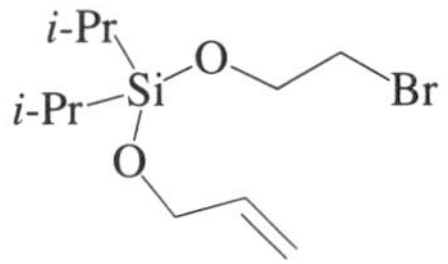	Bu$_3$SnH, AIBN, C$_6$H$_6$, 80°	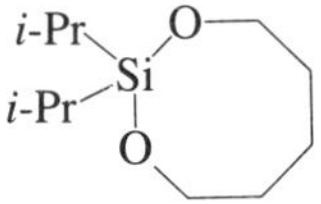 (59)	526
	C$_6$H$_6$, *hv*	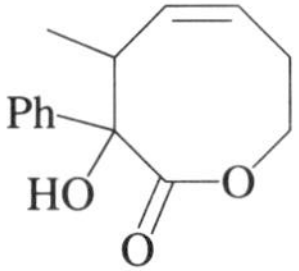 (49)	528
C$_{15}$ 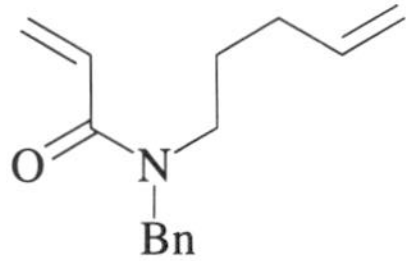	(PhS)$_2$, *hv*	(15)	440

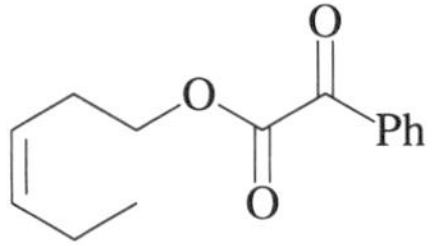

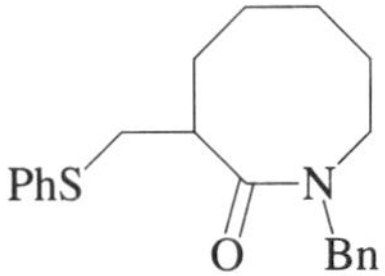

522

Reactant	Conditions	Product(s) and Yield(s) (%)	Refs.

G. Macrocycles

C$_6$	Bu$_3$Sn$_2$, C$_6$H$_6$, hv	(66)	420
C$_9$ $+$ $\begin{array}{c} CR \\ \| \\ CH \end{array}$ R = Me, Bu, CH$_2$OAc	Pr$_3$B, O$_2$	(30)	529
C$_{18}$	Bu$_3$SnH, AIBN, C$_6$H$_6$, 80°	**I** + **II**	95

		n	I + II	I:II
X	Z			
CH$_2$	CO$_2$Et	10	(57)	>98:<2
CH$_2$	CONEt$_2$	9	(40-45)	10:1
CH$_2$		9	(40-45)	14:1:1:1
CH$_2$	"	10	(40-45)	13:1:1:1
O	CO$_2$Et	11	(40-45)	2.5:1

TABLE II. MONOCYCLIC RINGS CONTAINING ONE OR MORE HETEROATOMS (*Continued*)

Reactant	Conditions	Product(s) and Yield(s) (%)	Refs.
C_{19}	Bu$_3$SnH, AIBN, C$_6$H$_6$, 80°	$\dfrac{n}{16}$ (63-67) 20 (68-76)	77
C_{21}	Bu$_3$SnH, AIBN, C$_6$H$_6$, 80°	(68)	530
	Bu$_3$SnH, AIBN, C$_6$H$_6$, 80°	$\dfrac{R}{H}$ (30) Me (74)	530
	Bu$_3$SnH, AIBN, C$_6$H$_6$, 80°	(74)	530

524

TABLE II. MONOCYCLIC RINGS CONTAINING ONE OR MORE HETEROATOMS (Continued)

Reactant	Conditions	Product(s) and Yield(s) (%)	Refs.
C_{28} n = 5, 6, 7, 8, 9, 10	Bu_3SnH, AIBN, C_6H_6, 80°	Ring size 10 (54) 11 (<6) 12 (61) 13 (50) 14 (80) 15 (72)	531
C_{32}	Bu_3SnH, AIBN, C_6H_6, 80°	(54)	532

Reactant	Conditions	Product(s) and Yield(s) (%)	Refs.

C$_{42}$

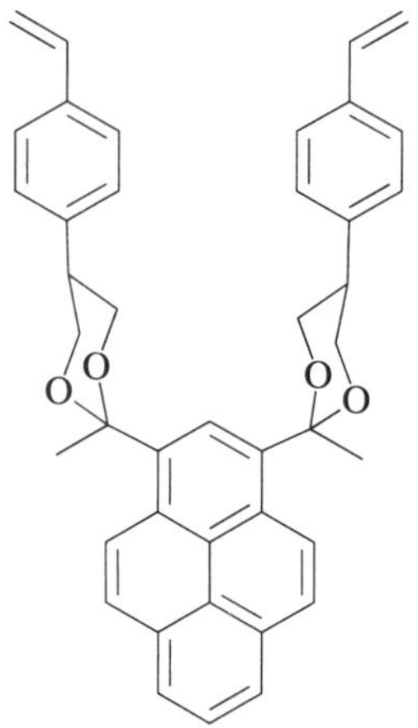

Bu$_3$SnH, AIBN,
C$_6$H$_6$, 80°

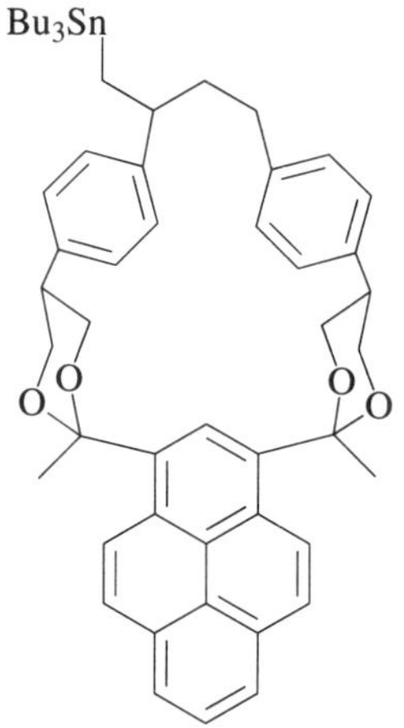

(74)

532

TABLE III. BICYCLIC RINGS CONTAINING ONLY CARBON

Reactant	Conditions	Product(s) and Yield(s) (%)	Refs.

A. (3+n)-Membered Rings

Reactant	Conditions	Product(s) and Yield(s) (%)	Refs.
C7	Bu3SnH, AIBN, C6H6, 80°	I + II (63) I:II = 5:1	533
C19 Co(dmgH)2py	TsI, CH2Cl2, 40°	(80) ratio = 1:1	139
C20 Co(dmgH)2py	TsI, CH2Cl2, 40° Me2NSO2Cl, CH2Cl2, 50-60°	R = Ts (76) R = Me2NSO2 (72)	139
C21 Co(dmgH)2py	TsI, CH2Cl2, 40°	(77) ratio = 1:1	139
Co(dmgH)2py	TsI, CH2Cl2, 40° Me2NSO2Cl, CH2Cl2, 50-60°	R = Ts (78) R = Me2NSO2 (68)	139

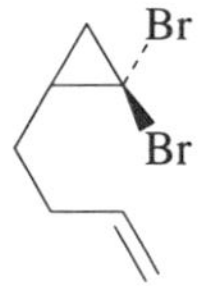

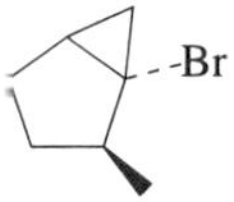

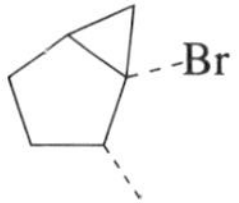

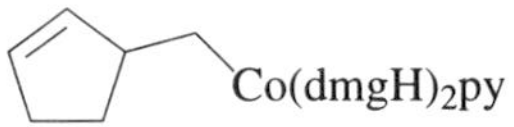

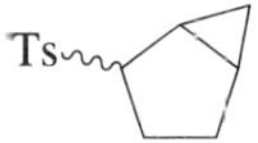

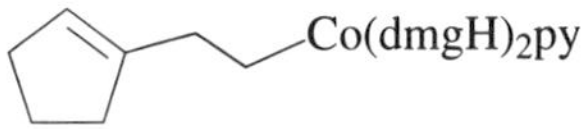

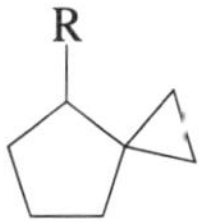

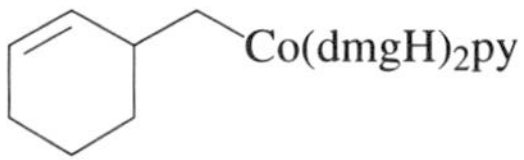

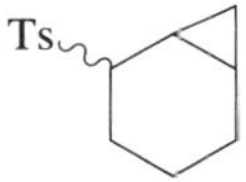

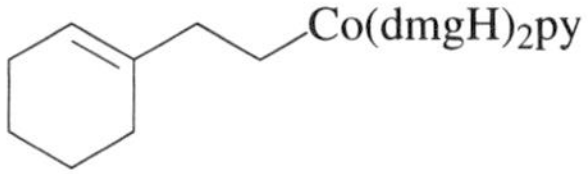

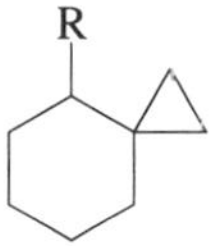

527

TABLE III. BICYCLIC RINGS CONTAINING ONLY CARBON (*Continued*)

Reactant	Conditions	Product(s) and Yield(s) (%)	Refs.
	B. (4 + 5)-Membered Rings		
C₇	*p*-Xylene, *hv*, heat		534
		(100) (43)	
C₈	Bu₃SnH, AIBN, C₆H₆, 80°	(85)	535
	C. (5 + 5)-Membered Rings		
C₈	Sn cathode, *i*-PrOH, Et₄NOTs, 25°	(68)	536
C₈	A. Carbon cathode, MeOH, dioxane, Et₄NOTs B. HMPA, *hv*	**A.** (69) **B.** (67)	246 180

TABLE III. BICYCLIC RINGS CONTAINING ONLY CARBON (*Continued*)

Reactant	Conditions	Product(s) and Yield(s) (%)	Refs.
C₉			
	Hg cathode, DMF, rt	(41)	225
	NaC₁₀H₈	(53)	225
	Bu₃SnH, AIBN, C₆H₆, 80°	(60)	537
	Carbon cathode, MeOH dioxane, Et₄NOTs	(87)	246
	HMPA, *hv*	(81)	180

529

TABLE III. BICYCLIC RINGS CONTAINING ONLY CARBON (*Continued*)

Reactant	Conditions	Product(s) and Yield(s) (%)	Refs.
	SmI$_2$, THF, HMPA	$\begin{array}{cc} & \text{ratio} \\ (90) & >150:1 \\ (92) & 93:5:2 \\ (89) & >150:1 \\ (85) & 2:1:1 \end{array}$	178
n m 1 1 2 1 1 2 2 2			
	Bu$_3$SnH, AIBN, C$_5$H$_{12}$, reflux	**I** + **II** **I + II** (50) **I:II** = 8.1:1	538
	(Bu$_3$Sn)$_2$ (cat.), C$_6$H$_6$, sunlamp	(40) 1.5:1 ratio of isomers	213
C$_{10}$			
	Bu$_3$SnH, AIBN, C$_6$H$_6$, 80°	$\begin{array}{cc} \dfrac{R}{} & \\ H & (47) \\ TMS & (78) \end{array}$	537

TABLE III. BICYCLIC RINGS CONTAINING ONLY CARBON (*Continued*)

Reactant	Conditions	Product(s) and Yield(s) (%)	Refs.
	1. Bu₃SnH, AIBN, C₆H₆, 80° 2. NaOH, EtOH	 **I + II** (95) **I:II** = 7:3	117, 116
	Hg cathode, DMF, rt NaC₁₀H₈	 (37) (23)	225
  R / H H MOM	Bu₃SnH, AIBN, C₆H₆, 80°	 $\dfrac{\text{I + II}}{(45)}$ (64)	539
	1. Cp₂TiCl 2. I₂	 (63)	172

TABLE III. BICYCLIC RINGS CONTAINING ONLY CARBON (*Continued*)

Reactant	Conditions	Product(s) and Yield(s) (%)	Refs.
C_{11}			
	Zn, TMSCl, 2,6-lutidine, THF, reflux	(82)	540
R = H, R = Me	Bu$_3$SnH, AIBN, C$_6$H$_6$, 80°	**I** + **II** **I** + **II** (61) **I:II** = 81:19 **I** + **II** (64) **I:II** = 20:1	208
	Bu$_3$SnH, AIBN, C$_6$H$_6$, 80°	R = H (83) R = Me (63)	291
	Zn, TMSCl, THF, 2,6-lutidine, reflux HMPA, *hv* Et$_3$N, MeCN, *hv*	(77) (80) (86)	540 180 180

532

TABLE III. BICYCLIC RINGS CONTAINING ONLY CARBON (*Continued*)

Reactant	Conditions	Product(s) and Yield(s) (%)	Refs.
	$(PhS)_2$, AIBN, C_6H_6, $h\nu$, $80°$	(70) 3:2	541
	Zn, TMSCl, THF, 2,6-lutidine, reflux	**I** + **II** (82) **I:II** = 83:17	540
	HMPA, $h\nu$	**I** + **II** (90) **I:II** = 97:3	180
	Hg cathode, Et_4NOTs, MeCN, H_2O, $25°$	**I** + **II** (79) **I:II** = 92:8	365
	Bu_3SnH, AIBN, C_6H_6, $80°$	**I** + **III** (69) **I:III** = 75:24	115

533

TABLE III. BICYCLIC RINGS CONTAINING ONLY CARBON (*Continued*)

Reactant	Conditions	Product(s) and Yield(s) (%)	Refs.
	Bu₃SnH, AIBN, C₆H₆, 80°	$\dfrac{R}{(CH_2)_2SH \quad (66)}$ H (after ester cleavage) (93)	117
C₁₂	Mn(OAc)₃, EtOH	$\dfrac{R^1 \quad R^2}{Me \quad H \quad (35)}$ H Me (32)	296
	SmI₂, Me₂CO, THF, -30° to rt	(66)	179
	Bu₃SnH, AIBN, C₆H₆, 80°	(90)	17

534

TABLE III. BICYCLIC RINGS CONTAINING ONLY CARBON (Continued)

Reactant	Conditions	Product(s) and Yield(s) (%)	Refs.
(TMS, Br, alkene ketone structure)	1. Bu$_3$SnH, AIBN, C$_6$H$_6$, 80° 2. TBAF, THF	(81)	342
(Br, alkyne enone structure)	Bu$_3$SnH, AIBN, C$_6$H$_6$, 80°	**I** + **II** (44) **I:II** = 3:1	542
(epoxide thiocarbonyl imidazolide structure)	Bu$_3$SnH, AIBN, C$_6$H$_6$, 80°	(52) ratio = 1.3:1	96
C$_{13}$ (methylenecyclopentane + CO$_2$Me, NC structure)	Mn(OAc)$_3$, EtOH, H$^+$	CO$_2$Me (54)	543

TABLE III. BICYCLIC RINGS CONTAINING ONLY CARBON (*Continued*)

Reactant	Conditions	Product(s) and Yield(s) (%)	Refs.
(structure with CO_2Me) *trans/cis* 6:1	Bu_3SnH, AIBN, C_6H_6, 80°	**I** + **II** I + II (56)	544
(iodide with C≡CTMS)	Bu_3SnH, AIBN, C_6H_6, 80°	(75) Z:E = 4:1	537
(R, Br, C≡CTMS structure)	Bu_3SnH, AIBN, C_6H_6, 80°	R — H (71), Me (73)	545
(cyclopentanone, MeO_2C, CO_2Me)	Zn, TMSCl, THF, 2,6-lutidine, reflux	(76)	540
C_{14} (SePh structure)	Ph_3SnH, AIBN, C_6H_6, 80°	(64)	19

TABLE III. BICYCLIC RINGS CONTAINING ONLY CARBON (*Continued*)

Reactant	Conditions	Product(s) and Yield(s) (%)	Refs.
	Mn(OAc)$_3$, Cu(OAc)$_2$, AcOH	R $\quad$ OMe (75) $\quad$ OEt (100)	211
(SnMe$_3$, Br, CO$_2$Me)	Bu$_3$SnH, AIBN, C$_6$H$_6$, 80°	CO$_2$Me (90)	546
CO$_2$Me + CO$_2$Me	(Bu$_3$Sn)$_2$, C$_6$H$_6$, *hv*, 25°	MeO$_2$C CO$_2$Me (81)	323
	PhMe, 110°	N S (52)	547
C$_{15}$ TBDMSO, CH	Bu$_3$SnH, Et$_3$B, air, C$_6$H$_{14}$, rt, 3 h	H H TBDMSO (76)	548

537

TABLE III. BICYCLIC RINGS CONTAINING ONLY CARBON (*Continued*)

Reactant	Conditions	Product(s) and Yield(s) (%)	Refs.

C. (5 + 5)-Membered Rings

C_{15}

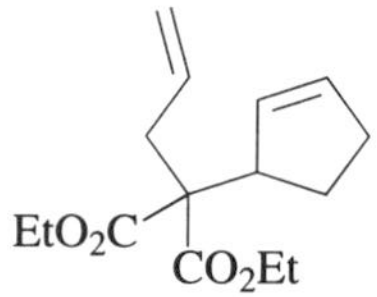

Me$_3$SnCl, Na(CN)BH$_3$, AIBN, *t*-BuOH

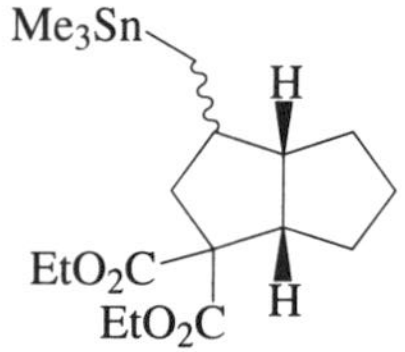

(54) 6:1 *ratio*

330

C_{16}

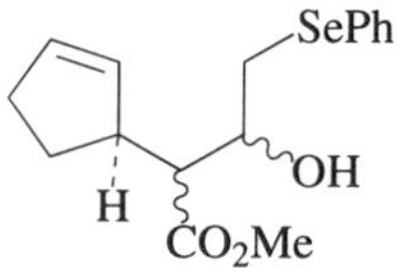

Ph$_3$SnH, AIBN, C$_6$H$_6$, 80°

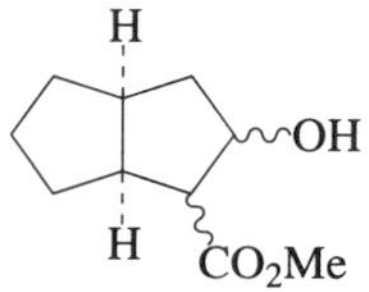

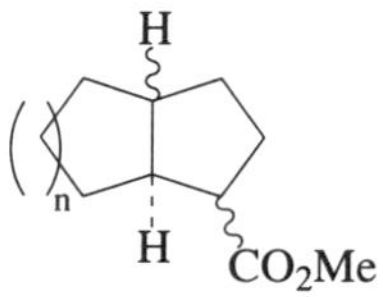

(89)

549

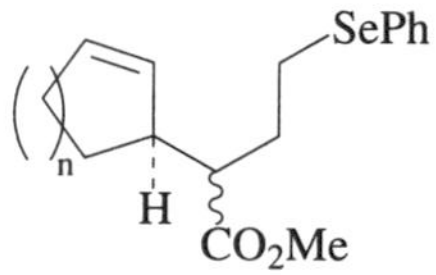

n
1
2
3

Ph$_3$SnH, AIBN, C$_6$H$_6$, 80°

Ring fusion	Isomer ratio	
cis	80:20	(93)
cis	90:10	(83)
cis and *trans*	72:14:11:2	(71)

102

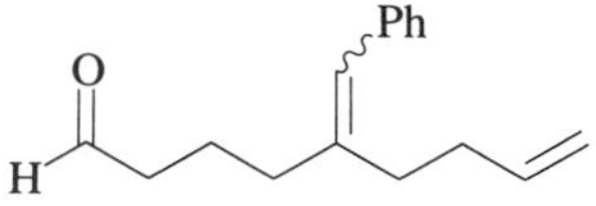

Bu$_3$SnH, AIBN, C$_6$H$_6$, 80°

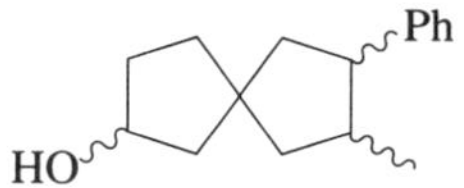

(75)

four isomers 5.3:3:2.5:1

544

TABLE III. BICYCLIC RINGS CONTAINING ONLY CARBON (*Continued*)

Reactant	Conditions	Product(s) and Yield(s) (%)	Refs.
	Bu_3SnH, AIBN, C_6H_6, 80°		340

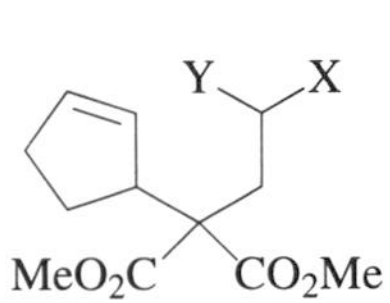

$$\begin{array}{c|c} X & Y \\ \hline SO_2Ph & Br \\ SO_2Bu\text{-}t & Br \\ CO_2Me & PhSe \end{array}$$

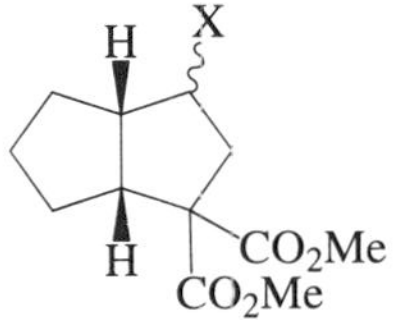

(61)
(74)
(80)

| 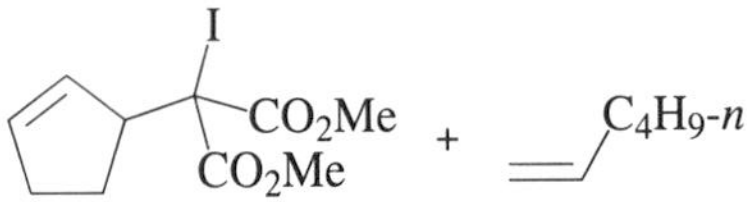 | 1. $(Bu_3Sn)_2$, C_6H_6, hv 2. Bu_3SnH, AIBN | 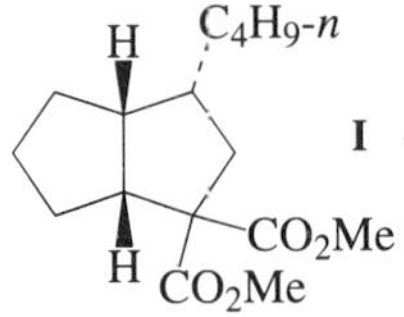 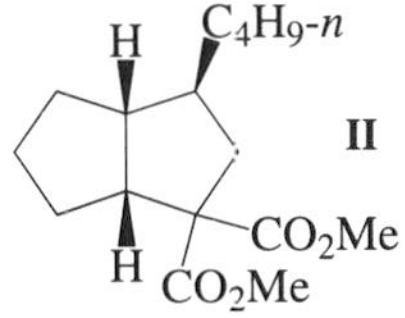 I + II (73) **I:II** = 89:11 | 256 |

I + II   **I:II** = 89:11

II

| 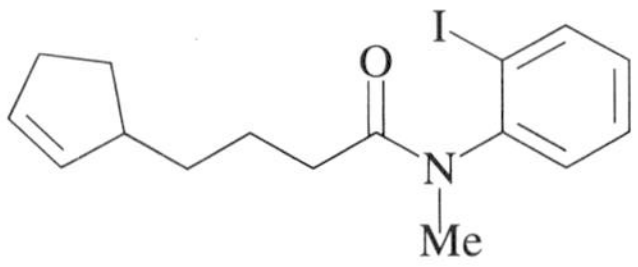 | Bu_3SnH, AIBN, C_6H_6, 80° | 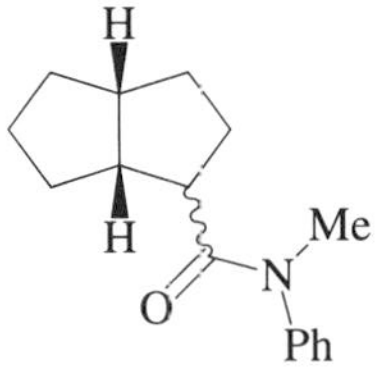 (60) *cis:trans* = 1:1.2 | 549a |

| | Bu_3SnH, AIBN, C_6H_6, 80° | I + II + III (75) **I:II:III** = 45:20:35 | 539 |

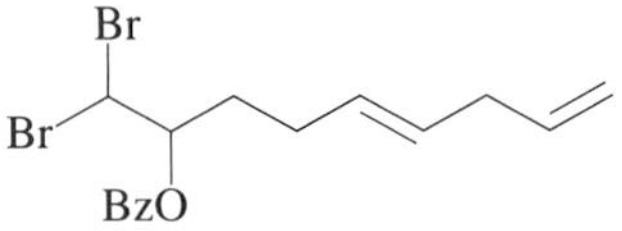

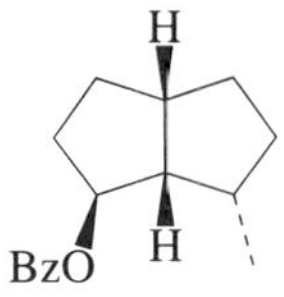

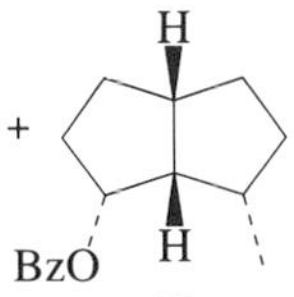

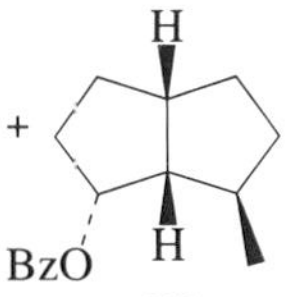

I + II + III (75) **I:II:III** = 45:20:35

539

TABLE III. BICYCLIC RINGS CONTAINING ONLY CARBON (*Continued*)

Reactant	Conditions	Product(s) and Yield(s) (%)	Refs.

C₁₇ section:

| | Bu₃SnH, AIBN, C₆H₆, 80° | (49) | 80 |
| | Mn(OAc)₃, Cu(OAc)₂, AcOH | I + II (49) I:II = 8:1 | 211 |

C₁₈ section:

	1. Ph₃SnH, AIBN, C₆H₆, 80° 2. H₂O₂	(51)	215 550
	Bu₃SnH, AIBN	(85)	551
	1. Bu₃SnH, AIBN, C₆H₆, 80° 2. AcOH, MeOH	(90)	341

540

TABLE III. BICYCLIC RINGS CONTAINING ONLY CARBON (*Continued*)

Reactant	Conditions	Product(s) and Yield(s) (%)	Refs.
	Bu$_3$SnH, AIBN, C$_6$H$_6$, 80°	(80)	552
C$_{19}$	Bu$_3$SnH, AIBN, C$_6$H$_6$, 80°	(95) 1:1 *ratio*	553
	Bu$_3$SnH, AIBN, C$_6$H$_6$, 80°	(80)	553
C$_{20}$	$h\nu$	(74)	554

TABLE III. BICYCLIC RINGS CONTAINING ONLY CARBON (*Continued*)

	Reactant	Conditions	Product(s) and Yield(s) (%)	Refs.
C$_{21}$		TsCl, BPO, PhMe	(49)	351
C$_{22}$		Ph$_3$SnH, AIBN, C$_6$H$_6$, 80°		353, 354, 354a
			R: CN (89), SO$_2$Ph (30)	
		Ph$_3$SnH, AIBN, C$_6$H$_6$, 80°	(68)	352
		Bu$_3$SnH, AIBN, C$_6$H$_6$, 80°	R: H (62), Me (71)	199

542

TABLE III. BICYCLIC RINGS CONTAINING ONLY CARBON (*Continued*)

Reactant	Conditions	Product(s) and Yield(s) (%)	Refs.
C$_{26}$ (TBDMSO, C≡CH, I, THPO, OTHP structure)	Bu$_3$SnH, AIBN, C$_6$H$_6$, 80°	(82) (OTBDMS, THPO, THPO structure)	555
C$_{29}$ (TBDMSO, C$_5$H$_{11}$-n, C≡CH, TBDMSO, CHO structure)	VCl$_3$(THF)$_3$, Zn, CH$_2$Cl$_2$, 0°	(71) α:β = 4.2:1 (C$_5$H$_{11}$-n, TBDMSO, OH structure)	252
C$_{44}$ (PhMe$_2$Si, OC(S)SMe, C≡C, CO$_2$Me, TBDMSO, OTBDMS structure)	Bu$_3$SnH, (*t*-BuO)$_2$, C$_6$H$_6$, 65°, 4 h	(86) Z:E = 1:1 (PhMe$_2$Si, CO$_2$Me, OTBDMS, TBDMSO structure)	556

TABLE III. BICYCLIC RINGS CONTAINING ONLY CARBON (*Continued*)

Reactant	Conditions	Product(s) and Yield(s) (%)	Refs.
		D. (5 + 6)-Membered Rings	
C$_8$	Bu$_3$SnH, AIBN, C$_6$H$_6$, 80°	(80)	557, 558
C$_9$	Mn(OAc)$_3$, Cu(OAc)$_2$	(38)	192
	SmI$_2$, THF, HMPA	(88) diastereomeric ratio = 17:1	178
	Zn, TMSCl, THF, 2,6-lutidine, reflux	(78)	540
	Sn cathode, *i*-PrOH, Et$_4$NOTs, 25°	(63)	536

TABLE III. BICYCLIC RINGS CONTAINING ONLY CARBON (Continued)

Reactant	Conditions	Product(s) and Yield(s) (%)	Refs.
	Sn cathode, i-PrOH, Et₄NOTs, 25°	(60)	536
	HMPA, hv	(76)	180
	Carbon cathode, MeOH, dioxane, Et₄NOTs	(65) one isomer; stereochemistry not determined	246
	1. Hg(OAc)₂, THF, H₂O, CaO, rt, 1 h 2. NaBH₄, MeOH, 0°	R¹ R² Me OH (35) OH Me (35)	559

C₁₀

TABLE III. BICYCLIC RINGS CONTAINING ONLY CARBON (*Continued*)

Reactant	Conditions	Product(s) and Yield(s) (%)	Refs.
	Bu₃SnH, AIBN, C₆H₆, 80°	(69)	560, 561
	Bu₃SnH, AIBN, C₆H₆, 80°	(76)	118
	1. Bu₃SnH, AIBN, C₆H₆, 80° 2. Ester cleavage	**I** + **II** **I + II** (81) **I:II** = 2:1	117
	Mn(OAc)₃, Cu(OAc)₂	(48)	192

546

TABLE III. BICYCLIC RINGS CONTAINING ONLY CARBON (Continued)

Reactant	Conditions	Product(s) and Yield(s) (%)	Refs.

	1. MgI$_2$ 2. Bu$_3$SnH, AIBN, C$_6$H$_6$, 80°	(40) + (40)	238
	Bu$_3$SnCl, Na(CN)BH$_3$, AIBN, t-BuOH, 110°, 3 d	(63)	332
	Bu$_3$SnH, AIBN, C$_6$H$_6$, 80°	R = H (83) R = Me (85)	562
	Bu$_3$SnH, AIBN, C$_6$H$_6$, 80°	(>88)	563
	Bu$_3$SnH, AIBN, C$_6$H$_6$, 80°	(56) + (16)	97

TABLE III. BICYCLIC RINGS CONTAINING ONLY CARBON (*Continued*)

Reactant	Conditions	Product(s) and Yield(s) (%)	Refs.

C11

Reactant: (epoxide with R side chain)

R	
Me	
Ph	

Conditions: $(Bu_3Sn)_2$, C_6H_6, $h\nu$

Product:

(66)	2.5:1
(75)	6.2:1

Refs.: 213

Reactant: (cyclohexanone with ketene side chain)

Conditions: Hg cathode, DMF, rt

Product: (43)

Refs.: 225

Reactant: (methylenecyclohexane with CO_2Et and Br)

Conditions: Bu_3SnH, AIBN, PhMe, 80°

Product: CO_2Et (76)

Refs.: 564

Reactant:

R^1	R^2
CO_2Me	H
Me	H
H	CO_2Me
H	Me

Conditions: Bu_3SnH, AIBN, C_6H_6, 80°

Product (indane with R^1, R^2): (90–98)

cis:trans

1.4:1
2.1:1
0.14:1
0.27:1

Refs.: 565

TABLE III. BICYCLIC RINGS CONTAINING ONLY CARBON (*Continued*)

Reactant	Conditions	Product(s) and Yield(s) (%)	Refs.
	Bu$_3$SnH, AIBN, C$_6$H$_6$, 70°	(82) 1 : 1 m xture of isomers	566
	Bu$_3$SnH, AIBN, C$_6$H$_6$, 65°	(50)	567
	1. DTBP, Cl$_3$SiH, C$_6$H$_6$, 140°, sealed tube 2. H$_3$O$^+$	(68)	568
	Mn(OAc)$_3$, Cu(OAc)$_2$, AcOH, 25°		569

R	
Cl	(—)
OPO(OEt)$_2$	(72)
Me	(77)

TABLE III. BICYCLIC RINGS CONTAINING ONLY CARBON (*Continued*)

Reactant	Conditions	Product(s) and Yield(s) (%)	Refs.

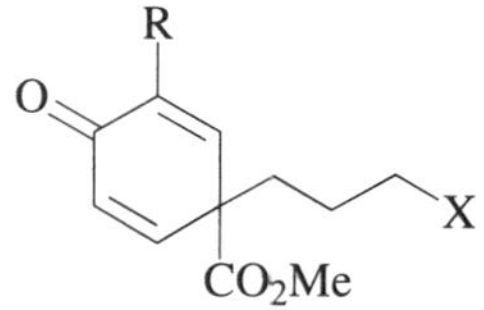

Row 1:

Reactant: (cyclohexadienone with CO2Me and R, side chain with X)

R X
H I
Me Br

Conditions: Bu₃SnH, AIBN, C₆H₆, 80°

Products: I + II

$$\frac{I + II}{(81) \quad (65)} \quad \frac{I:II}{100:0 \quad 42:58}$$

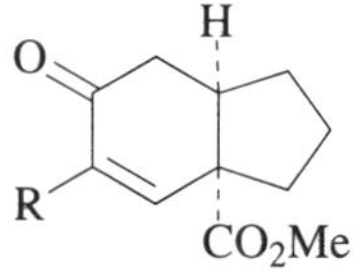

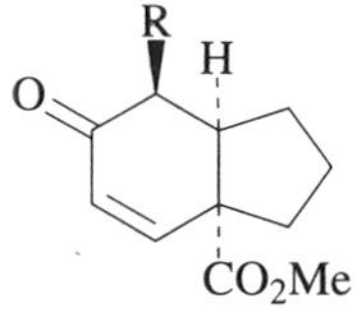

Refs: 570

Row 2:

Conditions: Bu₃SnH, AIBN, C₆H₆, 80°

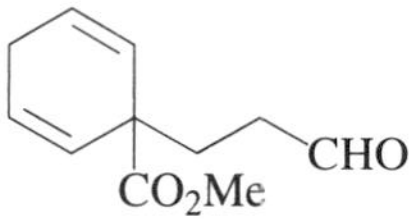

Product: (90) α:β = 45:55

Refs: 570

Row 3:

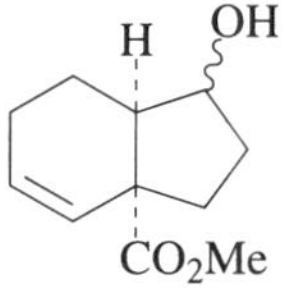

Conditions: Bu₃SnH, AIBN, C₆H₆, 80°

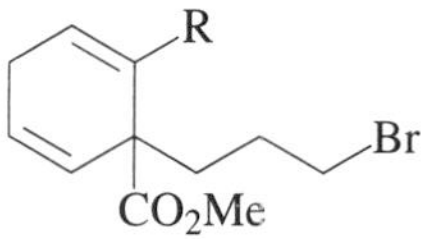

Product:

R	
H	(96)
Me	(85)
OMe	(92)

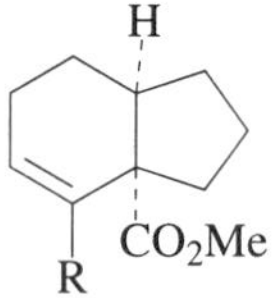

Refs: 570, 567

Row 4:

Conditions: Mn(OAc)₃, EtOH, H⁺

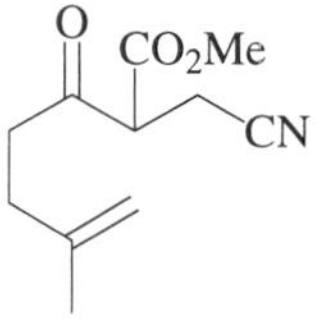

Product: (51)

Refs: 543

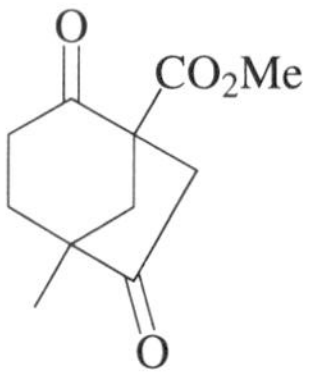

550

TABLE III. BICYCLIC RINGS CONTAINING ONLY CARBON (*Continued*)

Reactant	Conditions	Product(s) and Yield(s) (%)	Refs.
X = Br, X = I (with CO₂Me, alkene group)	Bu₃SnH, AIBN, C₆H₆, 80°	MeO₂C structure (91) (93)	562
R-substituted alkene with Br	Bu₃SnH, AIBN, C₆H₆, 80°	bicyclic R,H product (70) + R=H (93), R=Me (59)	293
epoxide bicyclic	Bu₃SnH, AIBN, C₆H₆, 80°	(10) + OH product	571
R,R alkene with epoxide; R = H, Me	Bu₃SnH, AIBN, C₆H₆, 80°	I + II	572

For reactant row 2:

$$\frac{R}{\begin{array}{l}H \quad (93)\\ Me \quad (59)\end{array}}$$

For reactant row 4:

	I	II
	(31)	(9)
	(26)	(15)

TABLE III. BICYCLIC RINGS CONTAINING ONLY CARBON (*Continued*)

Reactant	Conditions	Product(s) and Yield(s) (%)	Refs.
	Bu$_3$SnH, AIBN, C$_6$H$_6$, 80°	**I** + **II** **I + II** (71) **I:II** = 2:1	24
	HMPA, *hv* Et$_3$N, MeCN, *hv*	**I** + **I + II** (65) **I:II** = 90:10 **I + II** (74) **I:II** = 77:23	573
	DTBP, Et$_3$SiH, C$_6$H$_6$, 140°	(71)	568
	(Bu$_3$Sn)$_2$, C$_6$H$_6$, *hv*	(40)	213

552

TABLE III. BICYCLIC RINGS CONTAINING ONLY CARBON (*Continued*)

Reactant	Conditions	Product(s) and Yield(s) (%)	Refs.

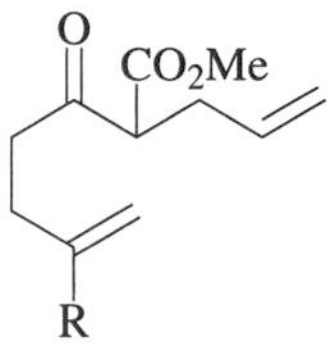

Conditions column (first entry):
Mn(OAc)$_3$, Cu(OAc)$_2$, AcOH

R	I	II
Me	(86)	(—)
OPO(OEt)$_2$	(77)	(—)
OCH$_2$O(CH$_2$)$_2$OMe	(52)	(—)
H	(48)	(18)

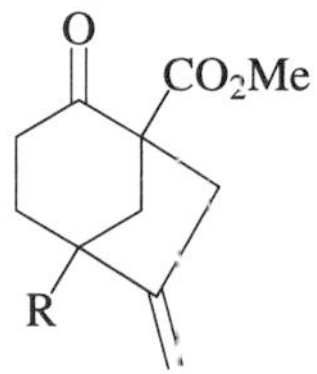

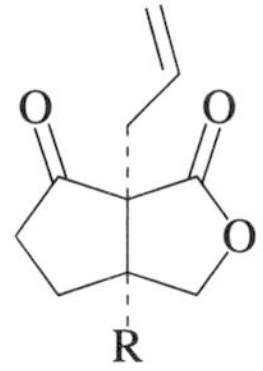

Refs.: 574, 296, 196

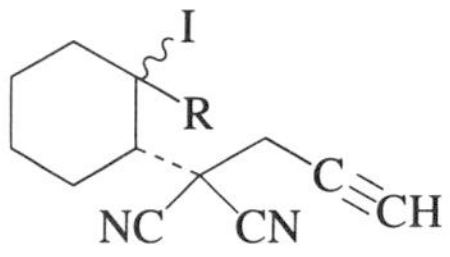

R	
H	(79)
Me	(45)

Refs.: 291, 208

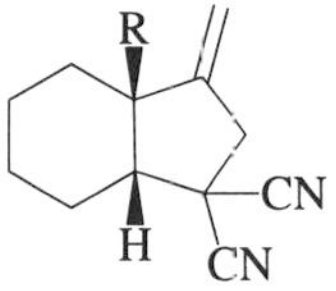

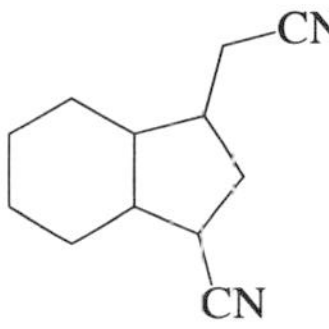

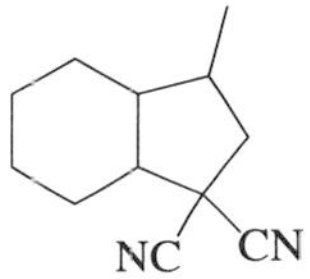

4 isomers 14:2:2:1 (55)

C$_{12}$

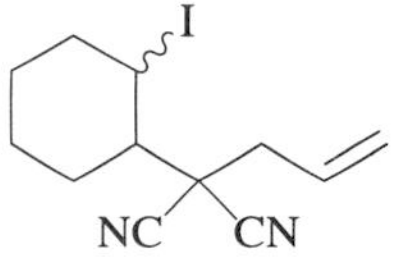

cis:trans = 1.5:1

Bu$_3$SnH, AIBN, C$_6$H$_6$, 80°

Bu$_3$SnH, AIBN, C$_6$H$_6$, 80°

TABLE III. BICYCLIC RINGS CONTAINING ONLY CARBON (*Continued*)

Reactant	Conditions	Product(s) and Yield(s) (%)	Refs.
	SmI_2, THF, DMPU	 $\dfrac{R}{CO_2Me}$ (77) H (36)	203
  $\dfrac{R}{Me}$ OMe	Bu$_3$SnH, AIBN, C$_6$H$_6$, 80°	 **I** + **II** **I + II I : II α-R : II β-R** (80) 32 : 14 : 54 (89) 48 : 13 : 39	570, 567
	Bu$_3$SnH, AIBN, C$_6$H$_6$, 80°, *hv*	 (70)	24
	Bu$_3$SnH, AIBN, C$_6$H$_6$, 80°	 **I** + **II** **I + II** (80) **I:II** = 1:1	575

TABLE III. BICYCLIC RINGS CONTAINING ONLY CARBON (*Continued*)

Reactant	Conditions	Product(s) and Yield(s) (%)	Refs.
	Et$_3$SiH, C$_6$H$_6$, 140°, sealed tube	(58)	568
	(Me$_3$Sn)$_2$, AIBN, C$_6$H$_6$, 80°	(65)	18
	Bu$_3$SnH, AIBN, C$_6$H$_6$, 80°	I + II (88) I:II 58:42	115
	Bu$_3$SnH, AIBN, C$_6$H$_6$, 80°	(70)	118

TABLE III. BICYCLIC RINGS CONTAINING ONLY CARBON (*Continued*)

Reactant	Conditions	Product(s) and Yield(s) (%)	Refs.

TABLE III. BICYCLIC RINGS CONTAINING ONLY CARBON (*Continued*)

Reactant	Conditions	Product(s) and Yield(s) (%)	Refs.
	Hg cathode, Et₄NOTs	$I + II$ (74) $I:II = 2.5:1$	365
R Me CH₂=CH Ph	(Ph₂S)₂, AIBN, C₆H₆, 80°, *hv*	$\dfrac{I+II}{(73)} \quad \dfrac{I:II}{3:2}$ $(82) \quad 13:1$ $(60) \quad 4:1$	541
C₁₃ *E:Z* = 1:1 R COPh TBDMS Ms	Bu₃SnH, AIBN, PhMe, 110°	$\dfrac{I+II}{(70)} \quad \dfrac{I:II}{90:10}$ $(83) \quad 90:10$ $(64) \quad 100:0$	370

TABLE III. BICYCLIC RINGS CONTAINING ONLY CARBON (*Continued*)

Reactant	Conditions	Product(s) and Yield(s) (%)	Refs.
	Bu$_3$SnH, AIBN, C$_6$H$_6$, 80°	(86) (92) (58)	293
	Bu$_3$SnH, AIBN, C$_6$H$_6$, 80°	(67)	38
	SmI$_2$, THF, DMPU	TMS (79) H (57)	203
	BPO, *c*-C$_6$H$_{12}$, heat	(37)	576

R^1	R^2	R^3
H	H	CO$_2$Me
Me	H	CO$_2$Me
H	OAc	H

TABLE III. BICYCLIC RINGS CONTAINING ONLY CARBON (*Continued*)

Reactant	Conditions	Product(s) and Yield(s) (%)	Refs.
	Mn(OAc)$_3$, EtOH, H$^+$	(57)	543
	Bu$_3$SnH, AIBN, C$_6$H$_6$, 80°	(>76)	18
	Bu$_3$SnH, AIBN, C$_6$H$_6$, 80°	(60) α:β 26:74	570
	Mn(OAc)$_3$, Cu(OAc)$_2$, AcOH	(65) + (5) 2:1 mixture	574

559

TABLE III. BICYCLIC RINGS CONTAINING ONLY CARBON (*Continued*)

Reactant	Conditions	Product(s) and Yield(s) (%)	Refs.
	1. TsNHNH$_2$, THF, 2. ZnCl$_2$, NaCNBH$_3$	 **I** + **II** **I + II** (66) **I:II** = 1:3.5	577
	SmI$_2$, THF, DMPU	(36)	203
	Mn(OAc)$_3$, Cu(OAc)$_2$ 1. (Bu$_3$Sn)$_2$, C$_6$H$_6$, *hv* 2. DBU, 120°	(67) 11:1 mixture (40) single isomer	364
	Bu$_3$SnH, AIBN, C$_6$H$_6$, 80°	X = Br (82) X = I (86)	562

560

TABLE III. BICYCLIC RINGS CONTAINING ONLY CARBON (*Continued*)

Reactant	Conditions	Product(s) and Yield(s) (%)	Refs.
	Mn(OAc)$_3$, Cu(OAc)$_2$	**I + II** (67) **I:II** = 25:1 **I + II** (46) **I:II** = 2:1	574, 569, 196
	Bu$_3$SnH, AIBN, C$_6$H$_6$, 80°	(43)	537
	Bu$_3$SnH, AIBN, C$_6$H$_6$, 80°	(73) $Z:E$ = 6:1	537
	Bu$_3$SnH, AIBN, C$_6$H$_6$, 80°	(90)	578

TABLE III. BICYCLIC RINGS CONTAINING ONLY CARBON (*Continued*)

Reactant	Conditions	Product(s) and Yield(s) (%)	Refs.
C_{14} 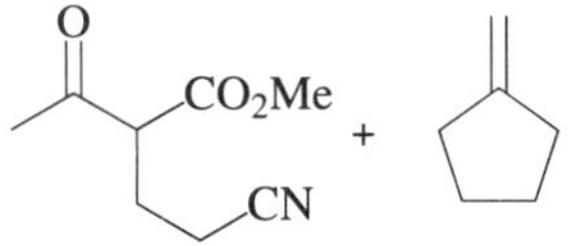	$Mn(OAc)_3$, EtOH, AcOH	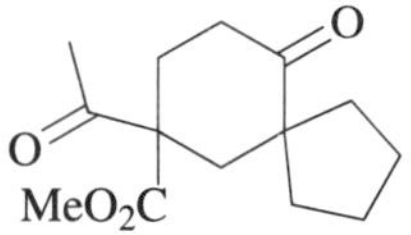 (41)	543
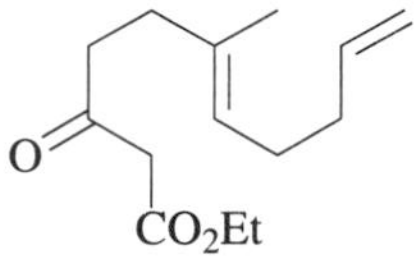	$Mn(OAc)_3$, $Cu(OAc)_2$	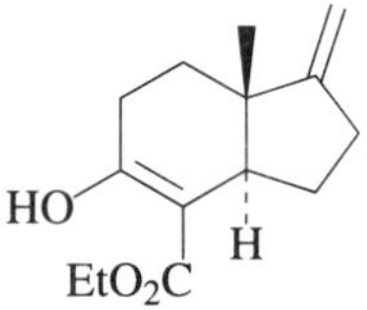 (41)	579
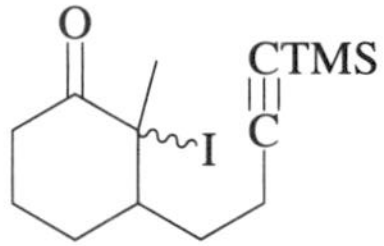	Bu_3SnH, AIBN, C_6H_6, 80°	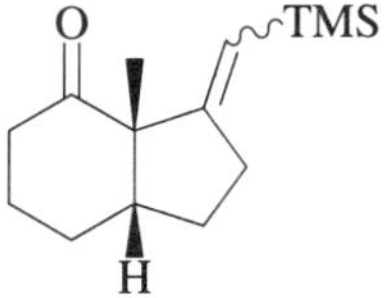 (65) $Z{:}E = 5{:}1$	537
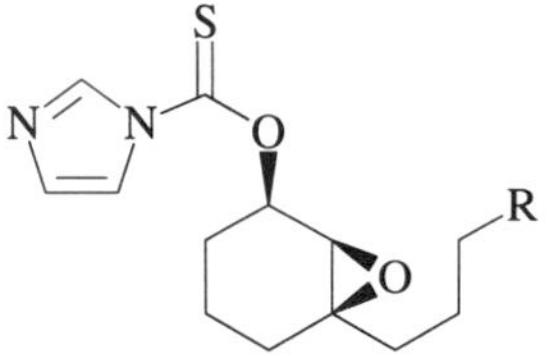	Bu_3SnH, AIBN, C_6H_6, 80°	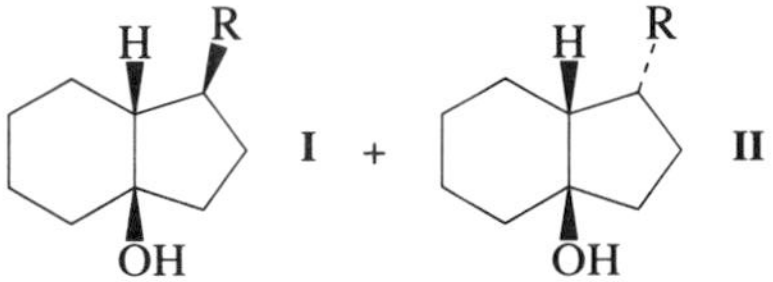	96

R	I + II	I:II
Me	(68)	2.7:1
CO_2Me	(69)	2.7:1
Ph	(47)	2.7:1

TABLE III. BICYCLIC RINGS CONTAINING ONLY CARBON (*Continued*)

Reactant	Conditions	Product(s) and Yield(s) (%)	Refs.
	(Ph₃P)₃RuCl₂, CCl₄, reflux	(48)	282
	Bu₃SnH, AIBN, C₆H₆, 80°	(58)	33
	Bu₃SnH, AIBN, C₆H₆, 80°	(69)	33
	Bu₃SnH, AIBN, C₆H₆, 80°		580

| | Conditions | | |

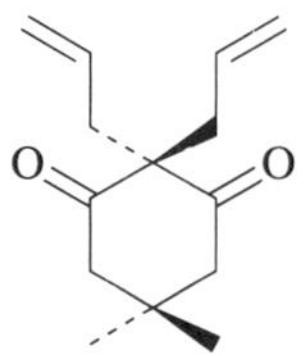

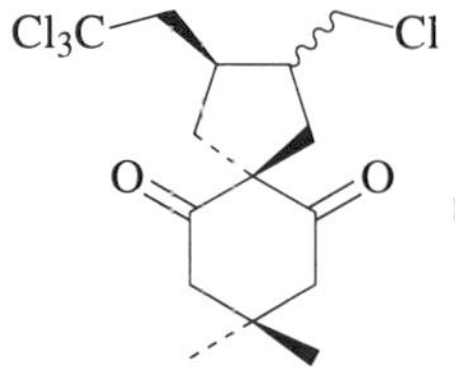

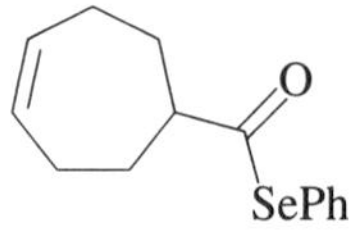

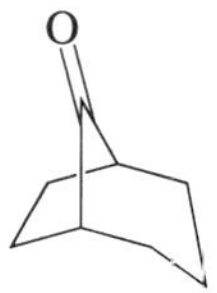

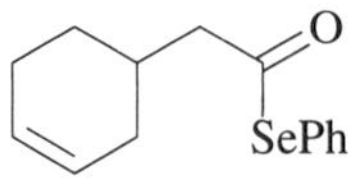

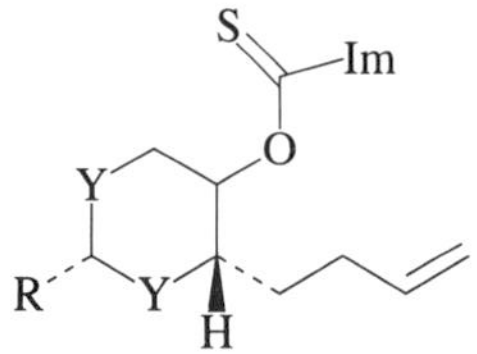

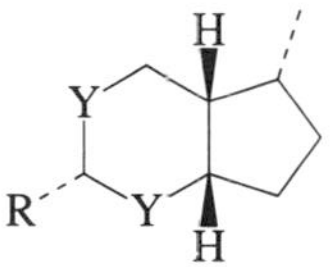

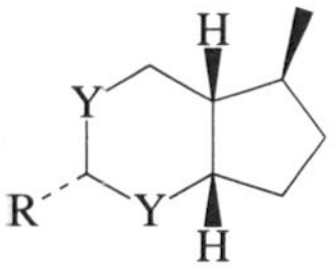

I + II

R	Y
Ph	CH₂
t-Bu	CH₂
H	CH₂
Ph	O

I + II	*cis:trans*
(29)	87:13
(37)	86:14
(—)	78:22
(55)	70:30

563

TABLE III. BICYCLIC RINGS CONTAINING ONLY CARBON (*Continued*)

Reactant	Conditions	Product(s) and Yield(s) (%)	Refs.

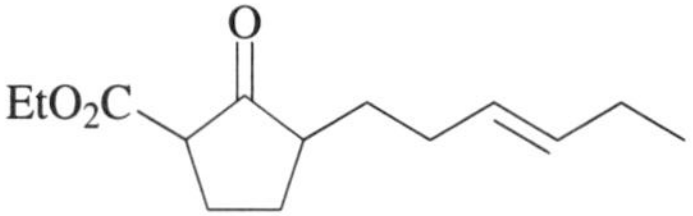

C$_{15}$

Mn(OAc)$_3$, Cu(OAc)$_2$, AcOH

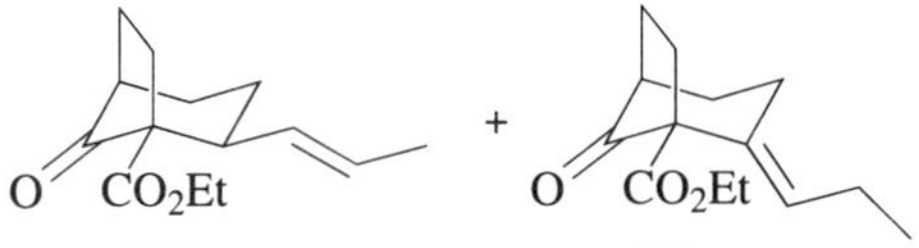

(78)

+

(4)

192

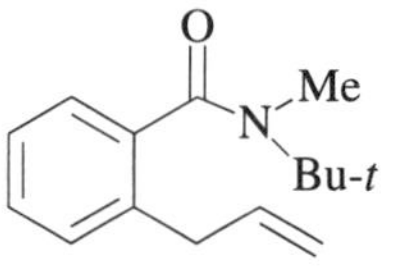

Li, THF, ultrasound

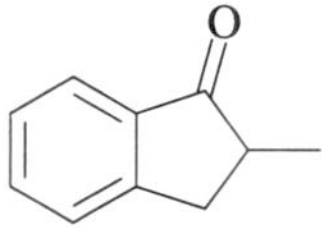

(84)

581

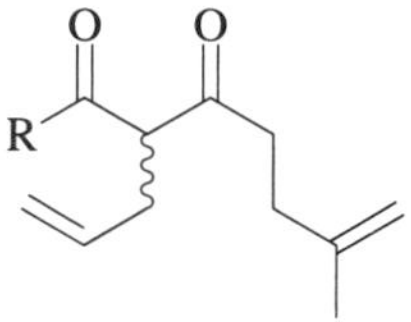

Mn(OAc)$_3$, Cu(OAc)$_2$, AcOH

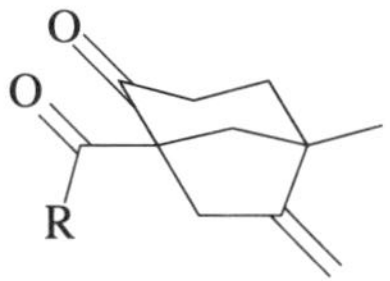

R = NEt$_2$ (45)

R = (90) 86% de

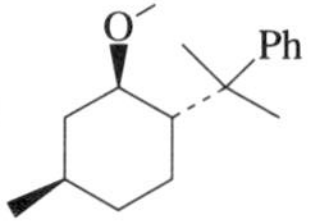

R = (89) 60% de

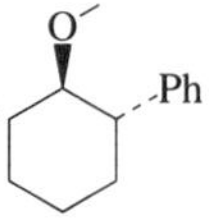

582

TABLE III. BICYCLIC RINGS CONTAINING ONLY CARBON (*Continued*)

Reactant	Conditions	Product(s) and Yield(s) (%)	Refs.

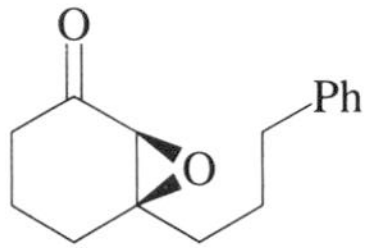

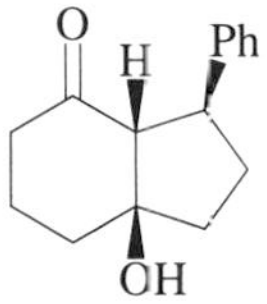

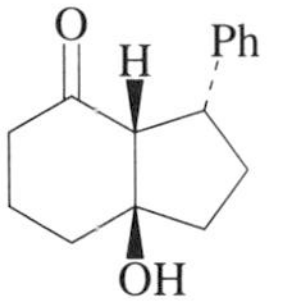

(Me₃Sn)₂, AIBN
C₆H₆, hv, 6 h
Bu₃SnH, AIBN,
C₆H₆, 80°

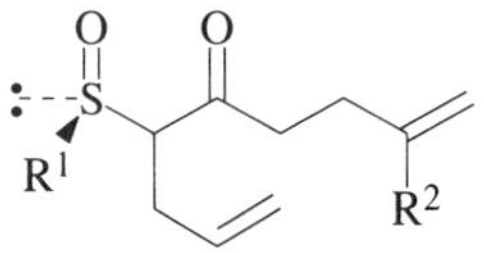

I (56)
I + II (72) I:II = 22:1

571

	R¹	R²
	Ph	H
	Ph	Me
	p-MeC₆H₄	Me

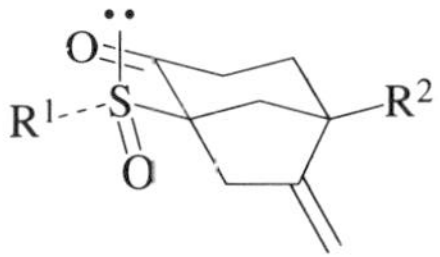

Mn(OAc)₃,
Cu(OAc)₂,
AcOH, 25°

(40)
(44)
(44)

583,
584

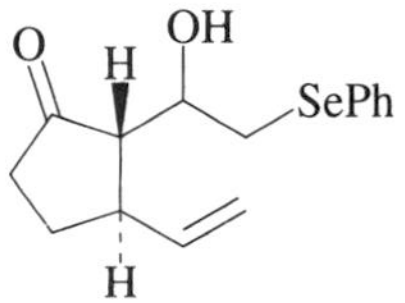

Ph₃SnH, AIBN,
C₆H₆, 80°

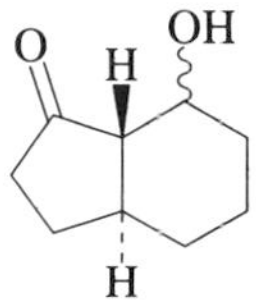

(34)

549

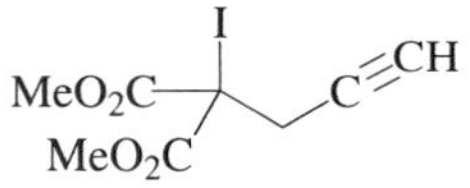

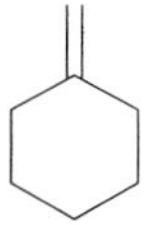

(Bu₃Sn)₂, C₆H₆, hv

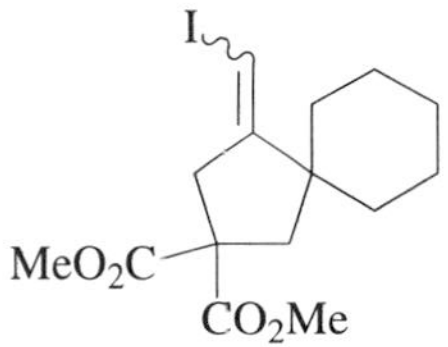

(30) E:Z = 3:1

256

565

TABLE III. BICYCLIC RINGS CONTAINING ONLY CARBON (*Continued*)

Reactant	Conditions	Product(s) and Yield(s) (%)	Refs.
	Ph₃SnH, AIBN, C₆H₆, 80° Bu₃SnH, AIBN, C₆H₆, 80°	(64) (72)	585, 19
	1,4-CHD, C₆H₄Cl₂, 170–230°	(54)	586
	Bu₃SnH, AIBN, C₆H₆, 80°	(86) (68) (49)	33
	Bu₃SnH, AIBN, C₆H₆, 80°	(86)	35

TABLE III. BICYCLIC RINGS CONTAINING ONLY CARBON (*Continued*)

Reactant	Conditions	Product(s) and Yield(s) (%)	Refs.
	(Bu₃Sn)₂, C₆H₆, *hv*	(72)	256
	Bu₃SnH, AIBN, C₆H₆, 80°	I + II (70) I:II = 2:1	117
	Bu₃SnH, AIBN, C₆H₆, 80°	I + II (82) I:II = 1:1	288
	Bu₃SnH, AIBN, C₆H₆, 80°	(58)	33, 35
C₁₆	Ph₃SnH, AIBN, C₆H₆, 80°		549

567

TABLE III. BICYCLIC RINGS CONTAINING ONLY CARBON (*Continued*)

Reactant	Conditions	Product(s) and Yield(s) (%)	Refs.

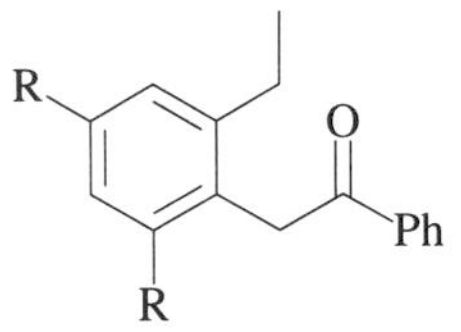

First row reactant (aryl ketone with R substituents, $CH_2C(=O)Ph$), hv:

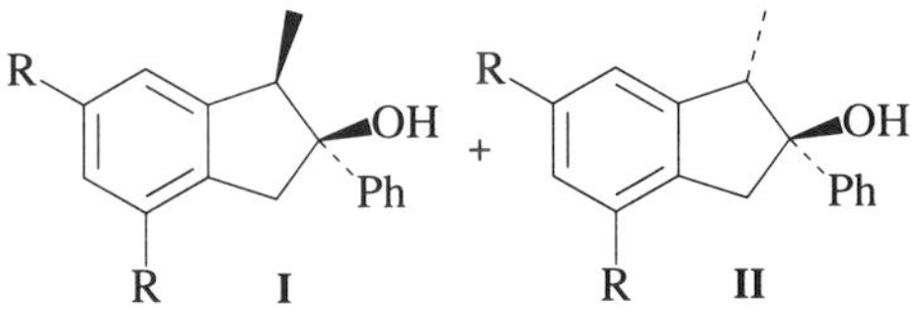

Products I and II (indanol with Ph, OH, and R substituents):

	$I + II$	$I:II$
C_6H_6-d_6	(100)	20:1
MeOH-d_6	(100)	2:1
C_6H_6-d_6	(100)	30:1
MeOH-d_6	(100)	4.3:1

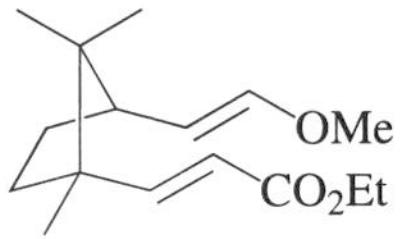

R: H, H, Et, Et

587

1. Hg(OAc)$_2$, MeOH
2. NaBH$_4$

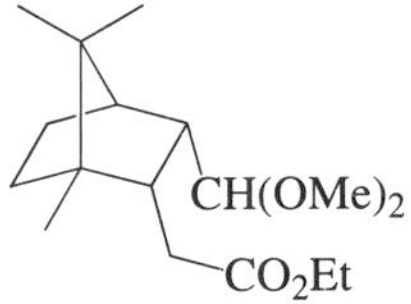

$CH(OMe)_2$, CO_2Et product (70)

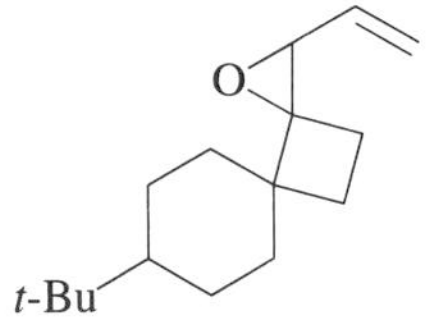

131

(PhS)$_2$, AIBN, hv

t-Bu product (60) diastereomeric ratio = 5:1

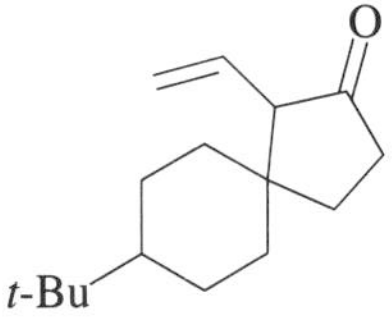

314

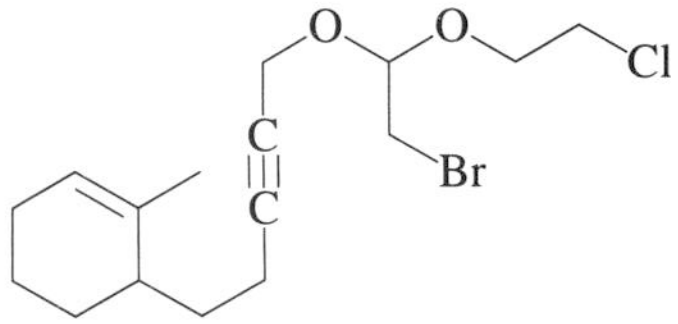

Bu$_3$SnH, AIBN, C$_6$H$_6$, 80°

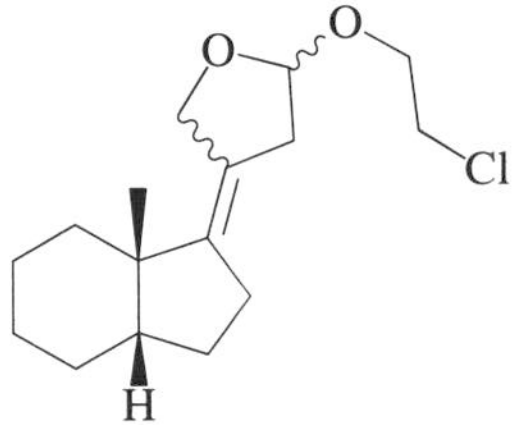

(75)

195

TABLE III. BICYCLIC RINGS CONTAINING ONLY CARBON (*Continued*)

Reactant	Conditions	Product(s) and Yield(s) (%)	Refs.

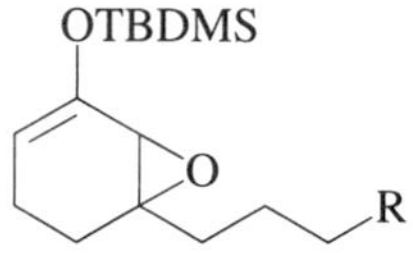

Bu₃SnH, AIBN, C₆H₆, 80°

R	I + II	I : II
Ph	(68)	00 : 0
OBn	(80)	88 : 12
CH=CH₂	(61)	95 : 5
Me	(67)	—

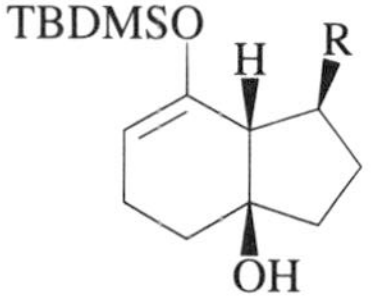

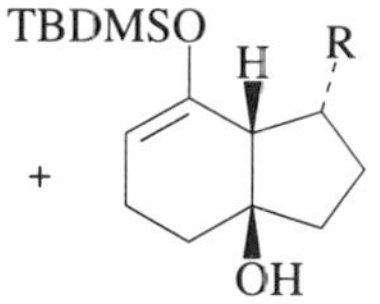

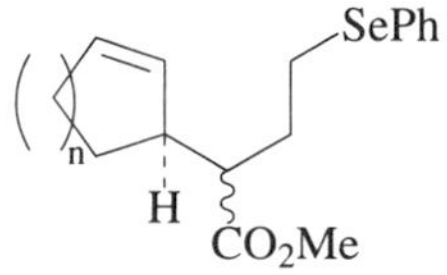

588

Ph₃SnH, AIBN, C₆H₆, 80°

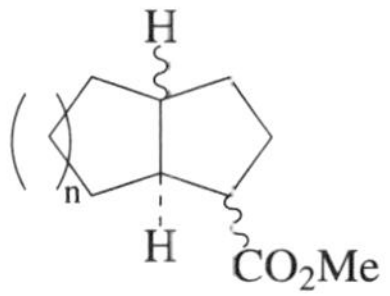

n	ring fusion	isomer ratio
1	*cis*	80:20
2	*cis*	90:10
3	*cis* and *trans*	72:14:11:2

(93) *cis*
(83) *cis*
(71) *cis* and *trans*

102

C₁₇

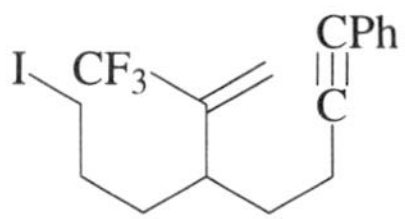

Bu₃SnH, AIBN, C₆H₆, 80°

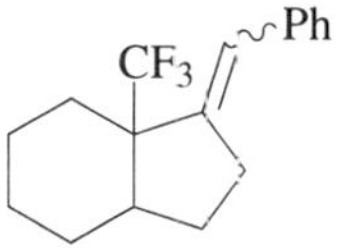

(71)

313

569

TABLE III. BICYCLIC RINGS CONTAINING ONLY CARBON (*Continued*)

Reactant	Conditions	Product(s) and Yield(s) (%)	Refs.
	Bu₃SnH, AIBN, C₆H₆, 80°	(86) *cis:trans* > 97:3	589
	Ph₃SnH, AIBN, C₆H₆, 80°	(64)	354
	Ph₃SnH, AIBN, C₆H₆, 80°	(53)	340
	Bu₃SnH, AIBN, PhMe, 110°	(47) + (25)	590

TABLE III. BICYCLIC RINGS CONTAINING ONLY CARBON (*Continued*)

Reactant	Conditions	Product(s) and Yield(s) (%)	Refs.
	Ph₃SnH, AIBN, C₆H₆, 80°	I + II I+II / I:II I+II (69) / 1.4:1 (81) / 1:1.6 β-OH (69) 1.4:1 α-OH (81) 1:1.6	549
	Ph₃SnH, AIBN, C₆H₆, 80°	(51)	549
	Bu₃SnH, AIBN, C₆H₆, 80°	I + II	80
	DCA, MeCN, *hv*	(59)	372

TABLE III. BICYCLIC RINGS CONTAINING ONLY CARBON (*Continued*)

Reactant	Conditions	Product(s) and Yield(s) (%)	Refs.
	1. Bu₃SnH, PhCl, -40° 2. O₂, -40 to 0° 3. Ph₃P, *i*-PrOH, 0° 4. PyTs, MeOH, 23°	**I + II** (60-90) **I:II** = 2:1	591, 592
	1. (Bu₃Sn)₂, C₆H₆, *hv* 2. Bu₃SnH, AIBN, 80°	**I** + **II**	256
	Ph₃SnH, Et₃B, air	(92)	593
C₁₈	1. Ph₃SnH, Et₃B, air 2. PCC	(58)	594

572

TABLE III. BICYCLIC RINGS CONTAINING ONLY CARBON (*Continued*)

Reactant	Conditions	Product(s) and Yield(s) (%)	Refs.
	Bu$_3$SnH, Et$_3$B, $-30°$	(97)	595
	Bu$_3$SnH, AIBN		580
	Bu$_3$SnH, AIBN, C$_6$H$_6$, 80°		209

For entry 580:

	I+II	I:II
R = Ph	(21)	31:63
R = t-Bu	(—)	22:60

For entry 209:

R	Y	
H	CO$_2$Me	(63)
H	CN	(68)
H	Ph	(52)
Me	CO$_2$Me	(60)

TABLE III. BICYCLIC RINGS CONTAINING ONLY CARBON (*Continued*)

Reactant	Conditions	Product(s) and Yield(s) (%)	Refs.
TBDMSO, I, CO_2Me (structure)	Bu_3SnH, AIBN, C_6H_6, 80°	CO_2Me, OTBDMS (structure) (61–78)	332
Br, CO_2Me, TBDMSO (structure)	Bu_3SnH, Et_3B, –30°	H, CO_2Me, TBDMSO (structure) (87) *trans:cis* 95:5	595
C_{19} EtO_2C, EtO_2C + (methylenecyclohexane)	$(BuS)_2$, *hv*, 25°	(structure) (77)	328
Ts, Br (structure)	Bu_3SnH, AIBN, C_6H_6, 80°	Ts, EtO_2C EtO_2C (structure) (61)	112
NOBn, C≡CH, CO_2Me (structure)	1. Bu_3SnH, AIBN, C_6H_6, 80° 2. AcOH, MeOH	BnO—NH, CO_2Me (structure) (82)	341

TABLE III. BICYCLIC RINGS CONTAINING ONLY CARBON (Continued)

Reactant	Conditions	Product(s) and Yield(s) (%)	Refs.
OTMS, SePh (with allyl-substituted benzene)	Bu$_3$SnH, AIBN, C$_6$H$_6$, 80°	OH (51) + OH (16) trans:cis = 2:1	80
TBDMSO, I, (enone)	Bu$_3$SnH, AIBN, C$_6$H$_6$, heat	TBDMSO (spiro ketone) (45)	332
(methylenecyclohexane) + (isopropylidene cyclopropane, PhSO$_2$)	(BuS)$_2$, $h\nu$	(spiro, PhSO$_2$) (59)	34
CO$_2$Me, CO$_2$Me, PhSe	Ph$_3$SnH, AIBN, C$_6$H$_6$, 80°	CO$_2$Me, CO$_2$Me, H, H (80)	112

C$_{20}$

TABLE III. BICYCLIC RINGS CONTAINING ONLY CARBON (*Continued*)

Reactant	Conditions	Product(s) and Yield(s) (%)	Refs.
C_{21}	Bu$_3$SnH, AIBN, C$_6$H$_6$, 80°	TMS (71)	202, 596
	Bu$_3$SnH, AIBN, C$_6$H$_6$, 80°	TMS (81)	202, 596
	Ph$_3$SnH, AIBN, C$_6$H$_6$, 80°	(93) (98)	340
	Ph$_3$SnH, AIBN, C$_6$H$_6$, 80°	R Ph (96) Pr (82)	597

TABLE III. BICYCLIC RINGS CONTAINING ONLY CARBON (*Continued*)

Reactant	Conditions	Product(s) and Yield(s) (%)	Refs.

C22

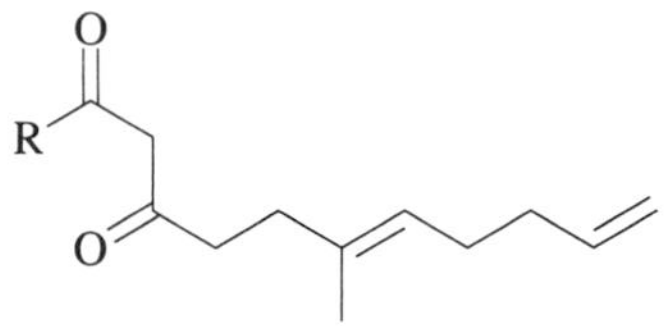

R = D-Camphorsultam

Mn(OAc)₃, Cu(OAc)₂, AcOH, 25°

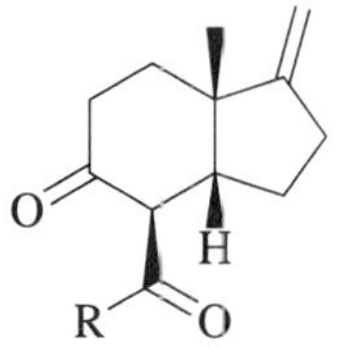

(49) + (17)   598

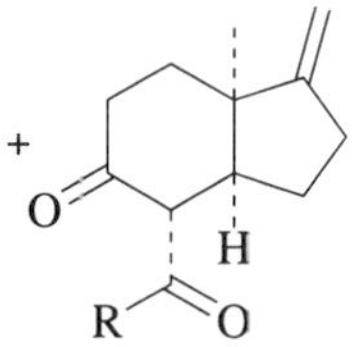

C23

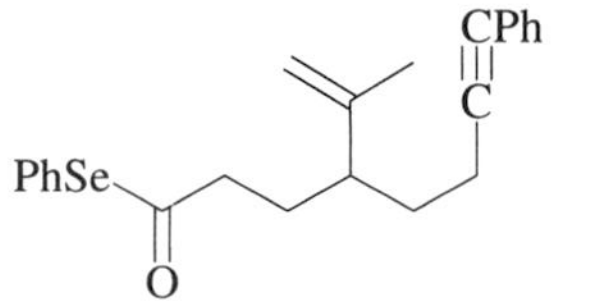

Bu₃SnH, AIBN, C₆H₆, 80°

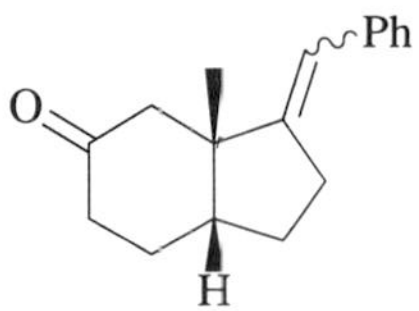

(77)   209

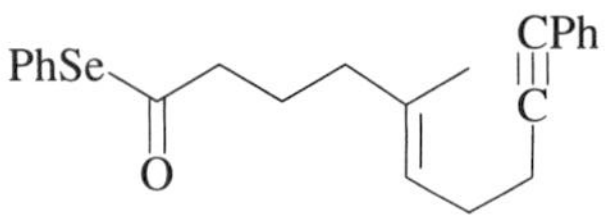

Bu₃SnH, AIBN, C₆H₆, 80°

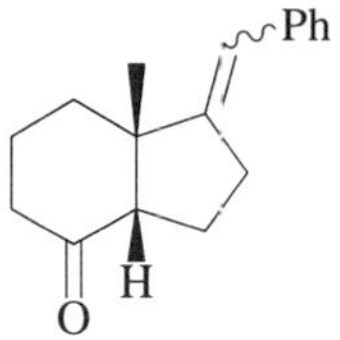

(72)   209

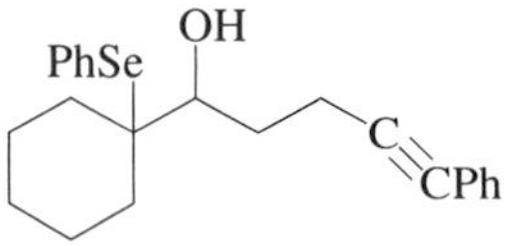

Ph₃SnH, AIBN, C₆H₆, 80°

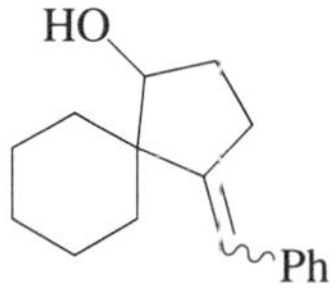

(75)   585, 352

577

TABLE III. BICYCLIC RINGS CONTAINING ONLY CARBON (*Continued*)

Reactant	Conditions	Product(s) and Yield(s) (%)	Refs.
	$Mn(OAc)_3$, $Cu(OAc)_2$,	(90) 86% de R = (89) 60% de R = (87) 23% de R =	584
	Ph_3SnH, AIBN, C_6H_6, 80°	(94)	354, 354a

TABLE III. BICYCLIC RINGS CONTAIN NG ONLY CARBON (*Continued*)

Reactant	Conditions	Product(s) and Yield(s) (%)	Refs.
C_{24}	**A.** TsCl, BPO, PhMe **B.** (Bu$_3$Sn)$_2$, C$_6$H$_6$, hv, 4 h **A** **B**	 Yield Ratio (63) 6.7:1 (81) 7.5:1	504, 351
C_{27}	Ph$_3$SnH, AIBN, C$_6$H$_6$, 80°	(86)	352
C_{28}	C$_7$H$_{16}$, hv, 25°	(73)	350
	Ph$_3$SnH, AIBN, C$_6$H$_6$, 80°	(55)	353

TABLE III. BICYCLIC RINGS CONTAINING ONLY CARBON (*Continued*)

Reactant	Conditions	Product(s) and Yield(s) (%)	Refs.
	C_7H_{16}, hv, 25°	(75)	350
	Hg cathode, MeCN, n-Bu$_4$NBr, CH$_2$(CO$_2$Me)$_2$	(90)	310
	Ph$_3$SnH, AIBN, C$_6$H$_6$, 80°	(78)	597
C$_{32}$	Ph$_3$SnH, AIBN, C$_6$H$_6$, 80°	(65)	353

TABLE III. BICYCLIC RINGS CONTAINING ONLY CARBON (Continued)

Reactant	Conditions	Product(s) and Yield(s) (%)	Refs.
E. (5 + 7)-Membered Rings			
C$_9$	Bu$_3$SnH, AIBN, C$_6$H$_6$, 80°	(10) + (70)	557, 558
C$_{10}$	SmI$_2$, MeOH Bu$_3$SnH, AIBN	**I** I (68) I (60)	599
	SmI$_2$, t-BuOH, HMPA Bu$_3$SnH, AIBN	I (40) I (56)	599
	Sn cathode, i-PrOH, Et$_4$NOTs, 25°	(76) 2:1 mixture of isomers	536
	Bu$_3$SnH, AIBN, C$_6$H$_6$, 80°	**I** + **II** + **III** I + II + III (84) I:II:III = 16:4:1	600, 601

TABLE III. BICYCLIC RINGS CONTAINING ONLY CARBON (*Continued*)

Reactant	Conditions	Product(s) and Yield(s) (%)	Refs.
	Bu₃SnH, AIBN, C₆H₆, 80°	 **I + II + III** (—) **I:II:III** = 6.2:1:2	600, 601
	Hg(OAc)₂, CaO, NaBH₄, MeOH	 (70) 1:1 ratio	602
	Mn(OAc)₃, Cu(OAc)₂	 (33) Δ3:Δ4 = 1:1	192
C₁₁	IBDA, I₂, *hv*	 R Me (78) H (40)	603

TABLE III. BICYCLIC RINGS CONTAINING ONLY CARBON (*Continued*)

Reactant	Conditions	Product(s) and Yield(s) (%)	Refs.
	Mn(pic)$_3$,^{a} Bu$_3$SnH DMF, 0°	(75)	103
	Mn(OAc)$_3$, Cu(OAc)$_2$	(68)	196, 194
	Bu$_3$SnH, AIBN, C$_6$H$_6$, 80°	(65)	38
	Mn(pic)$_3$,^{a} Bu$_3$SnH DMF, 0°	(66)	103

583

TABLE III. BICYCLIC RINGS CONTAINING ONLY CARBON (*Continued*)

Reactant	Conditions	Product(s) and Yield(s) (%)	Refs.
C$_{15}$	Na, MeOH HMPA, *hv* TiCl$_4$, Mg/Hg	**I** + **I** + **II** **II** **I + II** (71) **I:II** = 1:1 **I + II** (—) **I:II** = 100:0 **I + II** (—) **I:II** = 0:100	604
C$_{16}$	Ph$_3$SnH, AIBN	(72)	585, 19
	Bu$_3$SnH, AIBN, C$_6$H$_6$, 80°	(52)	605
C$_{18}$	Bu$_3$SnH, AIBN, C$_6$H$_6$, 80°	(80) ring fusion *cis:trans* 44:56	589

584

TABLE III. BICYCLIC RINGS CONTAINING ONLY CARBON (*Continued*)

Reactant	Conditions	Product(s) and Yield(s) (%)	Refs.
	Bu$_3$SnH, AIBN, C$_6$H$_6$, 80°	(58) ratio = 1:1	96
C$_{22}$	SnCl$_4$, THF, −70°, 2 h	I + II	606
	Na, C$_{10}$H$_8$, THF, 25°, 3 min		
	SmI$_2$, MeOH, THF, 25°, 5 min		
	Ph$_3$SnH, AIBN, C$_6$H$_6$, 80°	(79)	597

	I	II
	(57)	(6)
	(56)	(0)
	(51)	(13)

TABLE III. BICYCLIC RINGS CONTAINING ONLY CARBON (*Continued*)

Reactant	Conditions	Product(s) and Yield(s) (%)	Refs.
C_{24}	Ph_3SnH, AIBN, C_6H_6, 80°	(91)	354, 354a, 353
C_{25}	Mn(pic)_3[a], DMF, 0°	(81)	103

F. (5 + (n+7))-Membered Rings

C_{11}	Bu_3SnH, AIBN, C_6H_6, 80°	(54) + (19)	600, 601
	Bu_3SnH, AIBN, C_6H_6, 80°	(18) (22) (51)	607

R	
H	
CO_2Et	
OSiEt_3	

TABLE III. BICYCLIC RINGS CONTAINING ONLY CARBON (*Continued*)

Reactant	Conditions	Product(s) and Yield(s) (%)	Refs.

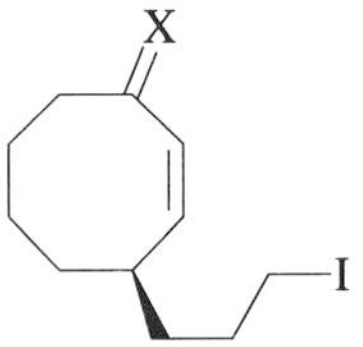

Bu₃SnH, AIBN, C₆H₆, 80°

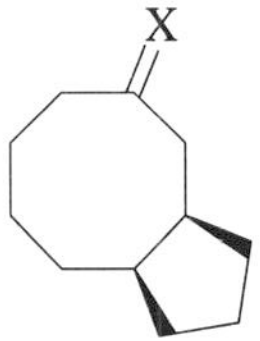

$$\frac{X}{\textit{cis}\text{-OH} \quad (70)}{O \quad (79)}$$

105

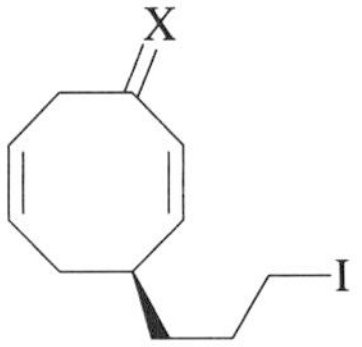

Bu₃SnH, AIBN, C₆H₆, 80°

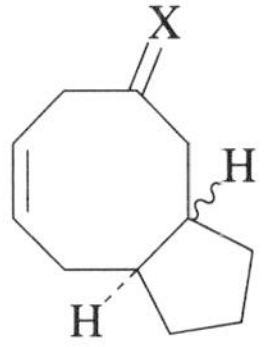

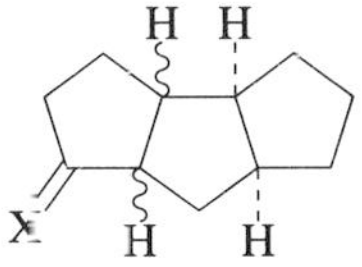

I +

II

	I	**II**
	trans	*cis:trans:cis*
	73:11:11:5	

	I + II	
X		
H₂	(—)	
cis-OH	(65)	>95:<1:0:0
trans-OH	(70)	50:10:20:20
O	(65)	95:5:0:0

105

Bu₃SnH, AIBN, C₆H₆, 80°

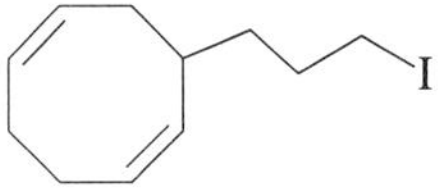

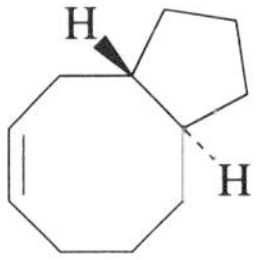

I +

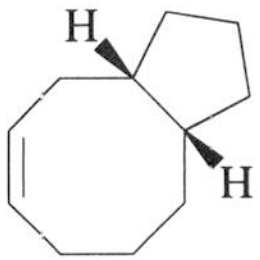

II

I + II (51)   **I:II** = 73:27

104

587

TABLE III. BICYCLIC RINGS CONTAINING ONLY CARBON (Continued)

Reactant	Conditions	Product(s) and Yield(s) (%)	Refs.
C$_{12}$	Bu$_3$SnH, AIBN, C$_6$H$_6$, 80°	R: Me (20), H (50)	603
C$_{14}$	Hg cathode, DMF, Bu$_4$N$^+$BF$_4^-$, DMP$^+$BF$_4^-$	(29) + (20)	608
A + B	Bu$_3$SnH, AIBN, C$_6$H$_6$, 80°	I + II; I + II cis:trans (55) 6.4:1, (81) 0.63:1	78
C$_{17}$	Bu$_3$SnH, AIBN, C$_6$H$_6$, 80°	(79)	33

588

TABLE III. BICYCLIC RINGS CONTAIN NG ONLY CARBON (Continued)

Reactant	Conditions	Product(s) and Yield(s) (%)	Refs.
C$_{19}$ 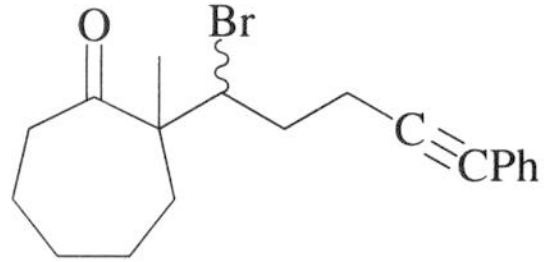	Bu$_3$SnH, AIBN, C$_6$H$_6$, 80°	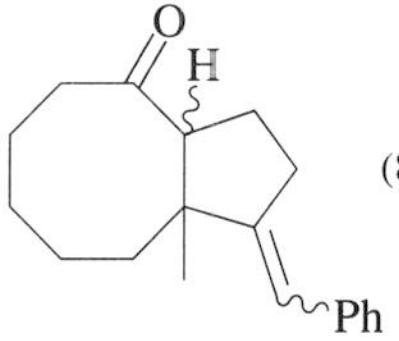 (85)	589
C$_{22}$ 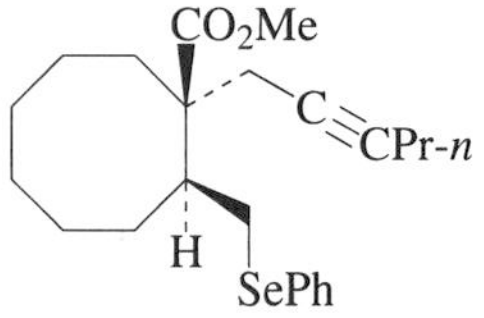	Ph$_3$SnH, AIBN, C$_6$H$_6$, 80°	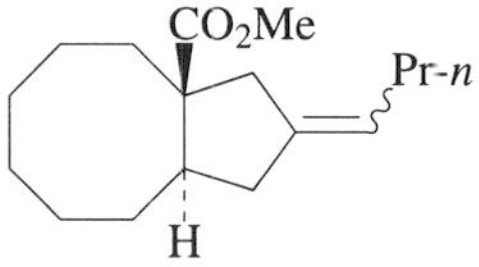 (77)	597
C$_{26}$ 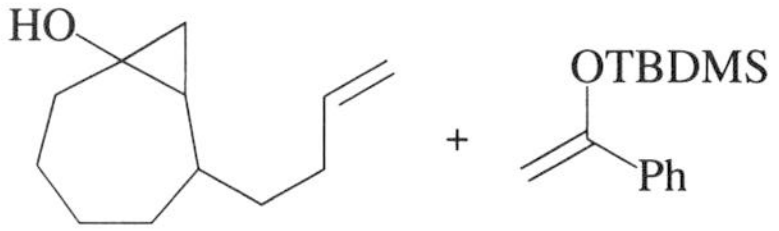	Mn(pic)$_3$[a], DMF, 0°	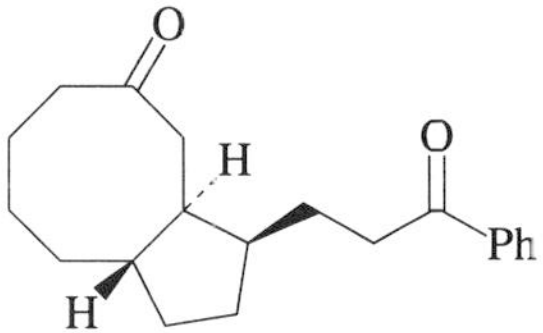 (63)	103

G. (6 + 6)-Membered Rings

Reactant	Conditions	Product(s) and Yield(s) (%)	Refs.
C$_{10}$ 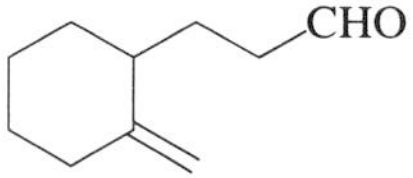	Mg, TMSCl, THF, I$_2$, rt, 60 h	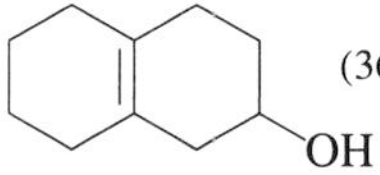 (36) 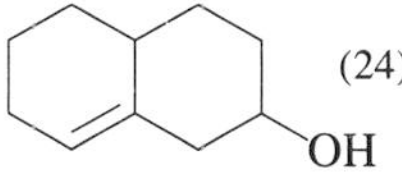 + (24)	609
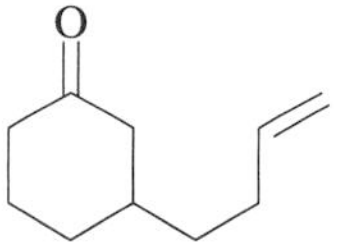	SmI$_2$, THF, HMPA	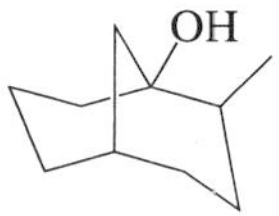 (66) diastereomeric ratio 17:1	178

589

TABLE III. BICYCLIC RINGS CONTAINING ONLY CARBON (*Continued*)

Reactant	Conditions	Product(s) and Yield(s) (%)	Refs.
C_{11}			
	Sn cathode, *i*-PrOH, Et$_4$NOTs, 25°	**I** + **II** **I** + **II** (60) **I:II** = 2:1	536
	SmI$_2$, THF, HMPA	(85) diastereomeric ratio 2:1:1	178
	A. Hg cathode, DMF, Bu$_4$N$^+$BF$_4^-$, DMP$^+$BF$_4^-$ **B.** Sn cathode, *i*-PrOH, Et$_4$NOTs, rt	**A** R = Me (80) **B** R = Me (70) **B** R = Et (55)	244 610 610
	Sn cathode, *i*-PrOH, Et$_4$NOTs, rt	R 7-Me (56) 6-Me (19) 5-Me (29) 7-OMe (73) 8-OMe (17) 7-Cl (63)	610

590

TABLE III. BICYCLIC RINGS CONTAINING ONLY CARBON (*Continued*)

Reactant	Conditions	Product(s) and Yield(s) (%)	Refs.
C$_{12}$			
  $\dfrac{R}{CN}$ $CO_2Pr\text{-}i$	Sn cathode, *i*-PrOH, Et$_4$NOTs, rt	 $\mathbf{I}$ + $\mathbf{II}$ $\begin{array}{cc} \mathbf{I} & \mathbf{II} \\ \hline (32) & (24) \\ (15) & (34) \end{array}$	610
  $\begin{array}{ccc} R^1 & R^2 & R^3 \\ \hline Me & H & H \\ H & Me & H \\ H & H & Me \end{array}$	Sn cathode, *i*-PrOH, Et$_4$NOTs, rt	 (60) (68) (60)	610
	Bu$_3$SnH, AIBN, C$_6$H$_6$, 80°	 (76)	570
	Bu$_3$SnH, AIBN, C$_6$H$_6$, 80°	 $\mathbf{I}$ + $\mathbf{II}$ $\mathbf{I} + \mathbf{II}$ (79) $\mathbf{I}$:$\mathbf{II}$ = 85:15	570

591

TABLE III. BICYCLIC RINGS CONTAINING ONLY CARBON (Continued)

Reactant	Conditions	Product(s) and Yield(s) (%)	Refs.

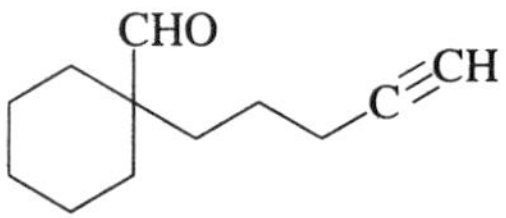

Reactant 1 (CHO, C≡CH cyclohexane)

HMPA, hv
Et₃N, MeCN, hv

Product (HO, methylene spiro): (55) (68)

573

CO₂Me vinyl cyclohexanone reactant

Mn(OAc)₃, Cu(OAc)₂

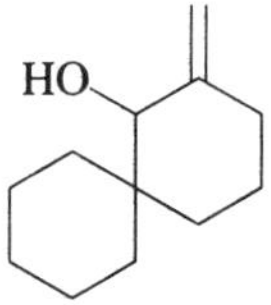

Product: OH, CO₂Me tetralin (46)

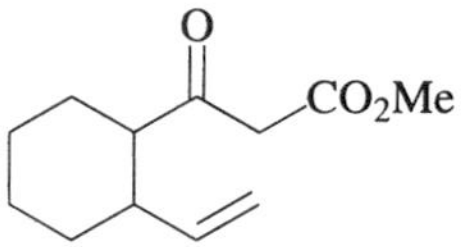

363

CO₂Me ketone cyclohexene reactant

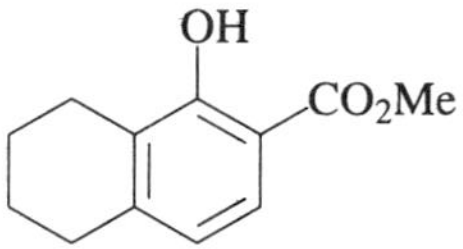

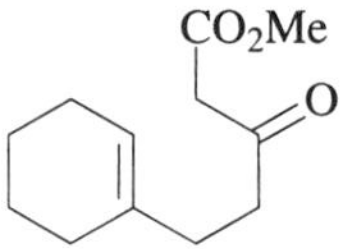

Mn(OAc)₃, Cu(OAc)₂, AcOH, LiCl

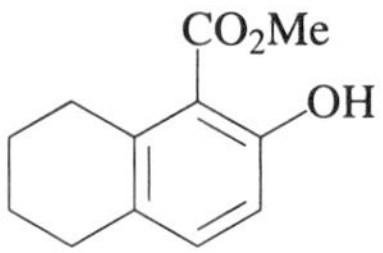

Product: CO₂Me, OH tetralin (37)

363

C₁₃

Br, OTMS, R aryl reactant

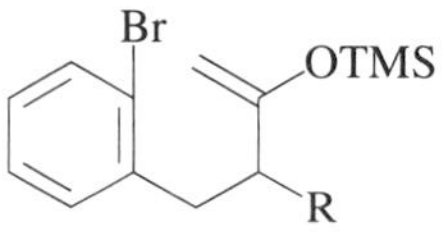

R
H
n-Bu

1. Bu₃SnH, AIBN, C₆H₆, 80°
2. HCl, MeOH

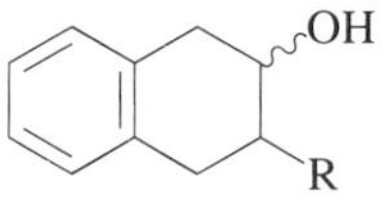

Product: OH, R tetralin
(72)
(74) 1:1 mixture

30

TABLE III. BICYCLIC RINGS CONTAINING ONLY CARBON (*Continued*)

Reactant	Conditions	Product(s) and Yield(s) (%)	Refs.
(structure with I, F₃C, OAc, O)	Bu₃SnH, AIBN, C₆H₆, 80°	(93) H, F_3C, OAc, O	578
(structure with Br) + (CN, acrylonitrile)	Bu₃SnH, AIBN, C₆H₆, 80°	**I** + NC— **II** (structure, NC)	611
EtO_2C CO_2Et (structure)	Pb(OAc)₄, MeCN	CO_2Et (80)	612
Br, CO_2Me (structure)	Bu₃SnH, AIBN, C₆H₆, 80°	CO_2Me CO_2Me (84)	613
$C{\equiv}CTMS$, I, O (structure)	Bu₃SnH, AIBN, C₆H₆, 80°	TMS, O, H, H (40) $Z{:}E = 9{:}1$	537

C₁₄

TABLE III. BICYCLIC RINGS CONTAINING ONLY CARBON (*Continued*)

Reactant	Conditions	Product(s) and Yield(s) (%)	Refs.

C$_{15}$

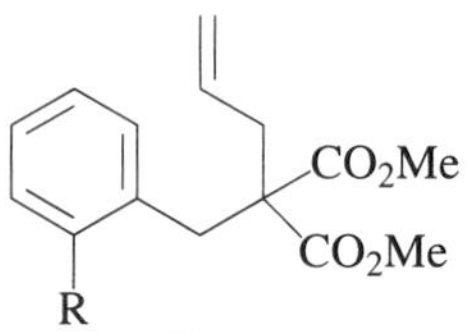

R	
H	
CN	

TsNa, Cu(OAc)$_2$, AcOH, 90°

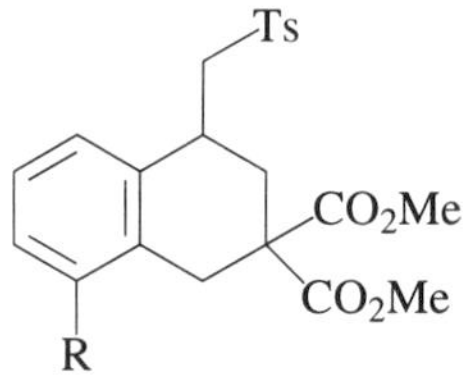

(72)
(82)

280

Mn(OAc)$_3$, Cu(OAc)$_2$, AcOH

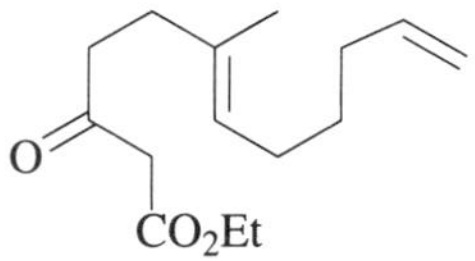

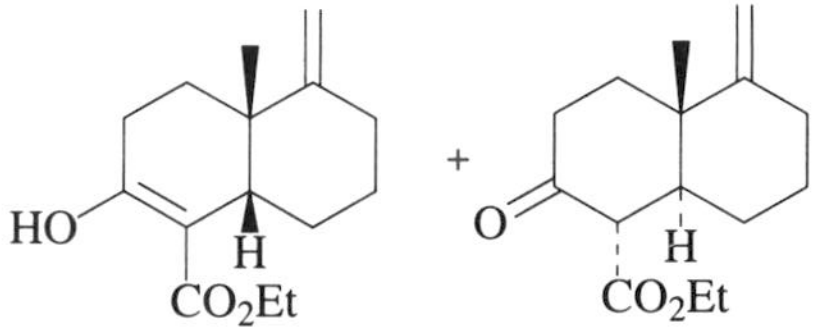

(43)

+

(12)

614

C$_{16}$

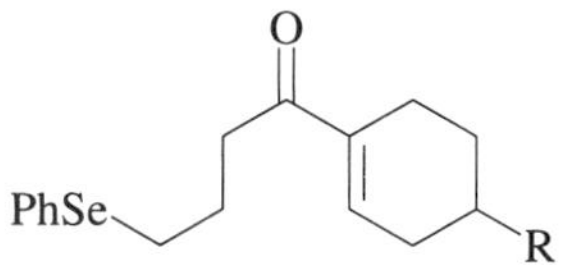

Bu$_3$SnH, AIBN, C$_6$H$_6$, 80°

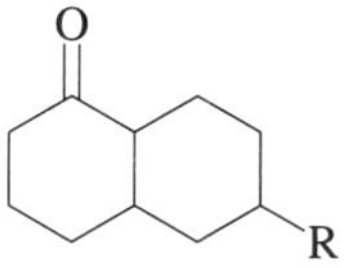

R	
H	(82)
t-Bu	(83)

605

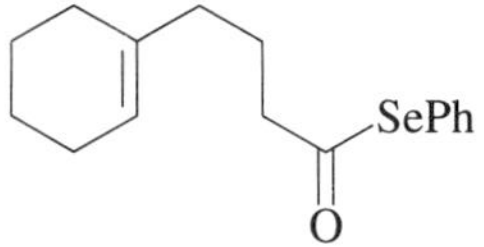

Bu$_3$SnH, AIBN, C$_6$H$_6$, 80°

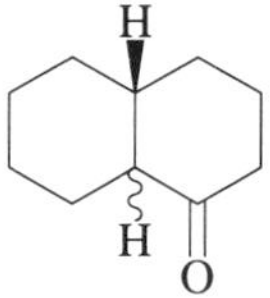

(82) *cis:trans* = 38:62

35, 33

594

TABLE III. BICYCLIC RINGS CONTAINING ONLY CARBON (*Continued*)

Reactant	Conditions	Product(s) and Yield(s) (%)	Refs.
C$_{17}$			
	Bu$_3$SnH, AIBN, C$_6$H$_6$, 80°	(48)	202
	Bu$_3$SnH, AIBN, C$_6$H$_6$, 80°	$\dfrac{R}{\text{H} \quad (76)}$ $\text{CO}_2\text{Me} \quad (84)$	35, 33
	Bu$_3$SnH, AIBN, PhMe	I + II I + II (80) I:II = 1:1.5	590
C$_{18}$			
	1. Ph$_3$SnH, Et$_3$B, air 2. PCC	(73)	594

595

TABLE III. BICYCLIC RINGS CONTAINING ONLY CARBON (*Continued*)

Reactant	Conditions	Product(s) and Yield(s) (%)	Refs.
C_{19}	Bu_3SnH, AIBN, C_6H_6, 80°	 $$\begin{array}{c|cc} & R^1 & R^2 \\ \hline \mathbf{I} & H & Me \\ \mathbf{II} & Me & H \\ \mathbf{III} & Me & H \\ \mathbf{IV} & H & Me \end{array}$$ **I, II** + **III, IV** **I + II + III + IV** (71) **I:II:III:IV** = 57:11:11:21	209
	Bu_3SnH, AIBN, C_6H_6, 80°	(83)	35, 33
	Bu_3SnH, AIBN, C_6H_6, 80°	(37) + (42)	35
C_{20}	Bu_3SnH, AIBN	(85)	615

TABLE III. BICYCLIC RINGS CONTAINING ONLY CARBON (*Continued*)

Reactant	Conditions	Product(s) and Yield(s) (%)	Refs.
C$_{21}$	Ph$_3$SnH, AIBN, C$_6$H$_6$, 80°	(79)	38
C$_{22}$	Bu$_3$SnH, AIBN, C$_6$H$_6$, 80°	(62)	33
C$_{23}$	C$_6$H$_6$, *hv*	$\dfrac{R}{}$ Et (81) *t*-Bu (86)	403
C$_{24}$	Bu$_3$SnH, AIBN, C$_6$H$_6$, 80°	(82)	209
C$_{28}$	Bu$_3$SnH, AIBN, PhMe, heat, 2 h	(86) *cis:trans* = 1:7	613

TABLE III. BICYCLIC RINGS CONTAINING ONLY CARBON (*Continued*)

Reactant	Conditions	Product(s) and Yield(s) (%)	Refs.

H. ((n+6) + (n+6))-Membered Rings

C₁₁

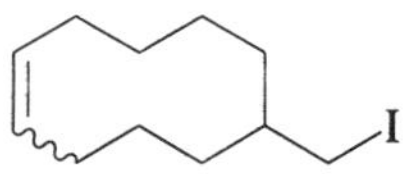

AIBN

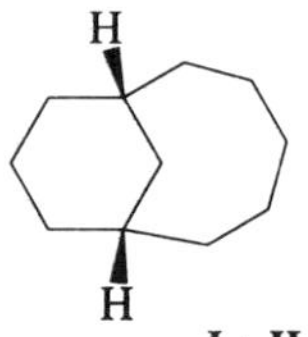

I + II

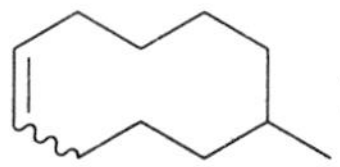

	I + II	I:II
cis	Bu₃SnH, 80°	(99)	2.4:1
cis	(TMS)₃SiH, 80°	(100)	4.0:1
cis	Bu₃SnH, 145°	(100)	4.1:1
trans	Bu₃SnH, 80°	(82)	0.2:1
trans	Bu₃SnH, 145°	(94)	0.4:1

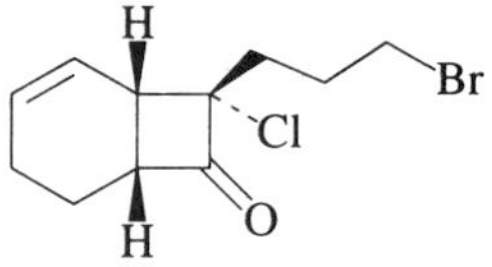

616

Bu₃SnH, AIBN, C₆H₆, 80°

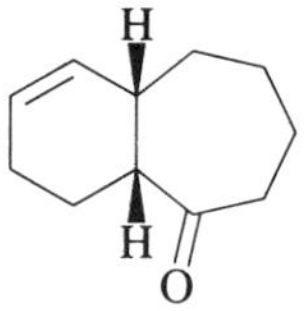

(39)

600, 601

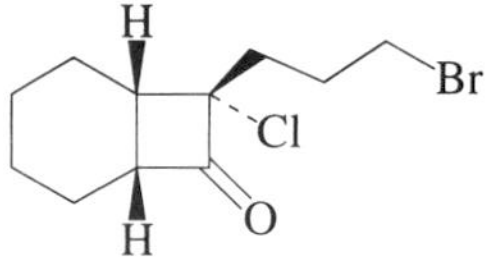

Bu₃SnH, AIBN, C₆H₆, 80°

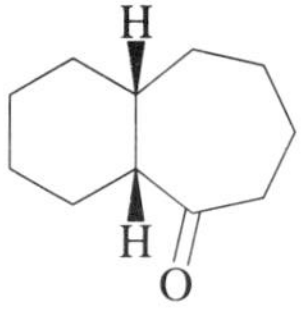

(63)

600, 601

C₁₂

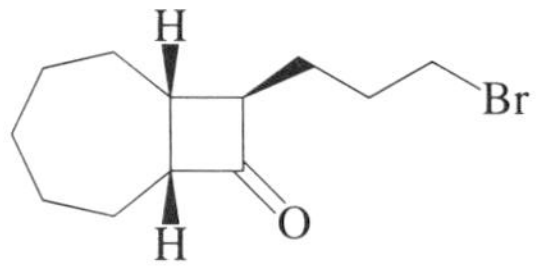

Bu₃SnH, AIBN, C₆H₆, 80°

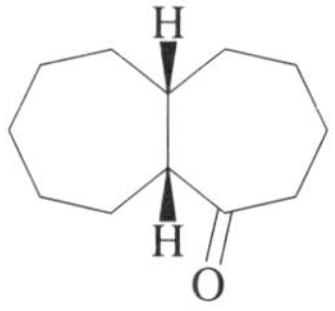

(64)

600, 601

598

TABLE III. BICYCLIC RINGS CONTAINING ONLY CARBON (*Continued*)

Reactant	Conditions	Product(s) and Yield(s) (%)	Refs.
C₁₃	Mn(OAc)₃, Cu(OAc)₂	I + II (35) I:II = 1:1.1	192
	NaC₁₀H₈	(28)	225
	Bu₃SnH, AIBN, C₆H₆, 80°	(43)	600, 601
C₁₅	Mn(OAc)₃, Cu(OAc)₂, AcOH, 25°	(61)	617, 618
C₁₇	Bu₃SnH, AIBN, C₆H₆, 80°	(76)	605

TABLE III. BICYCLIC RINGS CONTAINING ONLY CARBON (*Continued*)

Reactant	Conditions	Product(s) and Yield(s) (%)	Refs.
C$_{18}$	Bu$_3$SnH, AIBN, C$_6$H$_6$, 80°	(82)	605
C$_{19}$	Bu$_3$SnH, AIBN, C$_6$H$_6$, 80°	R / H (74) / CO$_2$Me (92)	35, 33
	1. Ph$_3$SnH, Et$_3$B, air 2. PCC	n / 1 (80) / 2 (77)	617, 618
C$_{20}$	Bu$_3$SnH, AIBN, C$_6$H$_6$, 80°	(71)	35, 33
C$_{21}$	Bu$_3$SnH, AIBN, C$_6$H$_6$, 80°	(80)	615

^{a} Mn(pic)$_3$ = Manganese(III) tris(2-pyridinecarboxylate).

600

TABLE IV. BICYCLIC RINGS CONTAINING ONE OR MORE HETEROATOMS

Reactant	Conditions	Product(s) and Yield(s) (%)	Refs.

A. (3+n)-Membered Rings

C8

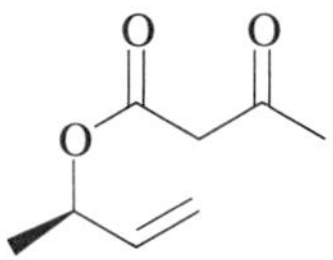

Mn(OAc)₃,
Cu(OAc)₂,
AcOH, 25°

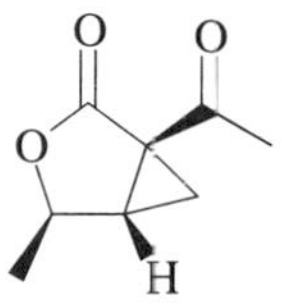

I + II (62)   I:II = 9.3:1

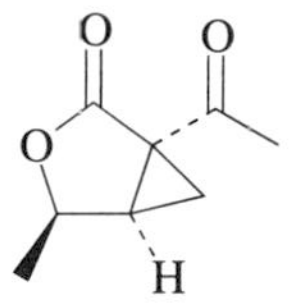

491

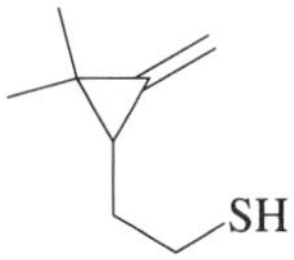

C₅H₁₂, hv

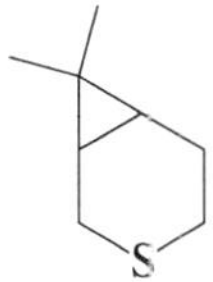

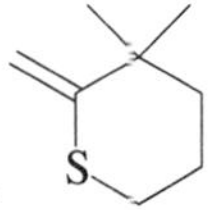

-70°
80°

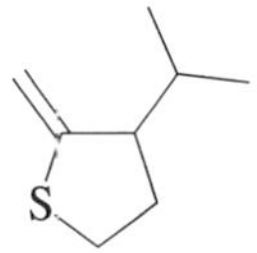

I + II + III (65)   I:II:III = 98:2:0
I + II + III (85)   I:II:III = 36:3:61

619

C9

TiCl₃, AcOH, H₂O, -5°
MeOH, hv

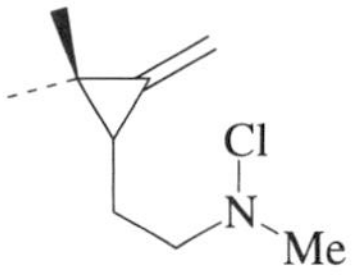

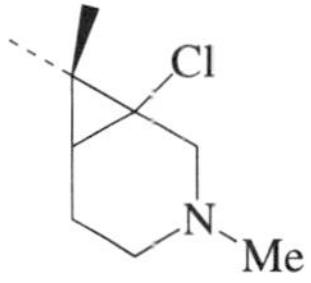

(55)
(18)

620

C14

hv

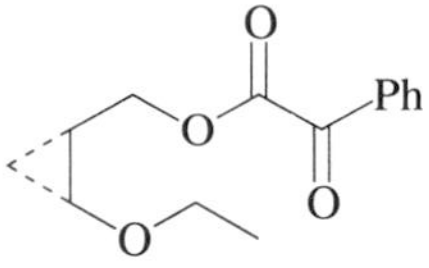

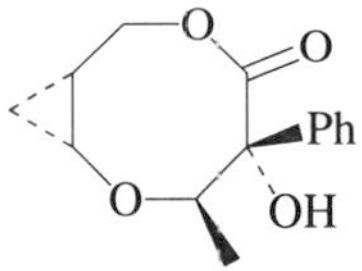

(51)

528

TABLE IV. BICYCLIC RINGS CONTAINING ONE OR MORE HETEROATOMS (*Continued*)

Reactant	Conditions	Product(s) and Yield(s) (%)	Refs.

C$_{17}$

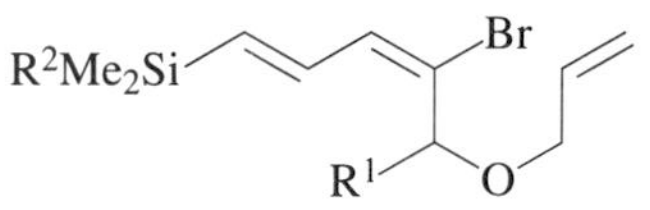

R^1	R^2	
Me	Ph	(58)
Ph	Me	(54)
C$_6$H$_{13}$	Me	(48)

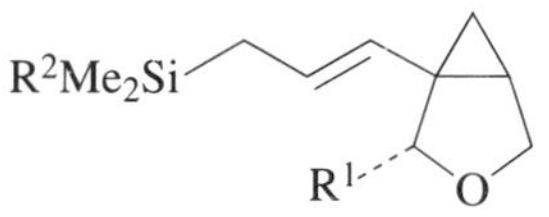

Bu$_3$SnH, AIBN, C$_6$H$_6$, 80° — 621

B. (4+5)-Membered Rings

C$_8$

Bu$_3$SnH, AIBN

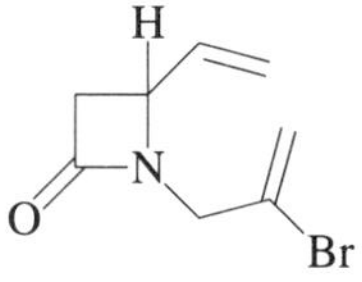

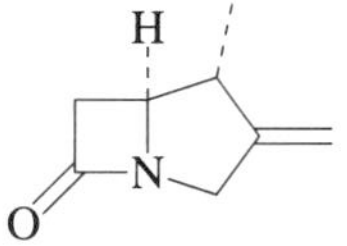

C$_6$H$_6$, 80°   **II** (50)
C$_6$H$_6$, *hv*, 25°   **I** (58)

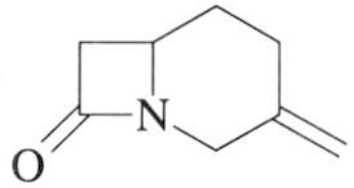

I + **II**   622, 623

C$_9$

Bu$_3$SnH, AIBN

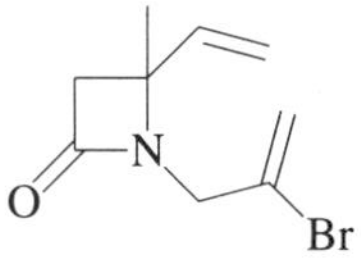

C$_6$H$_6$, 80°
C$_6$H$_6$, *hv*, 25°

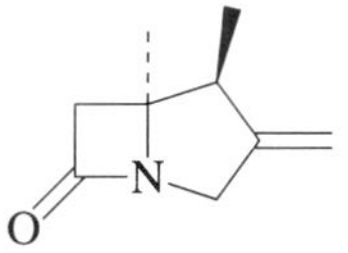

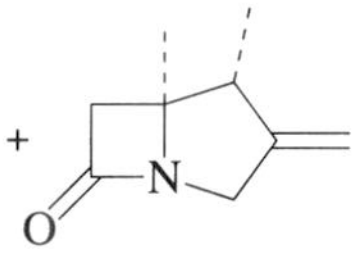

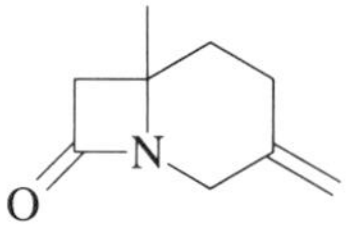

I + **II** + **III**   622, 623

I	**II**	**III**
(59)	(3)	(30)
(58)	(10)	(10)

TABLE IV. BICYCLIC RINGS CONTAINING ONE OR MORE HETEROATOMS (*Continued*)

Reactant	Conditions	Product(s) and Yield(s) (%)	Refs.
C_{18}	Bu$_3$SnH, AIBN, C$_6$H$_6$, 80°	$\mathbf{I + II}$ (66) $\mathbf{I:II}$ = 2:1	624
	Bu$_3$SnH, AIBN, C$_6$H$_6$, 80°	$\mathbf{I + II}$ (62) $\mathbf{I:II}$ = 1.1:1	625
C_{19}	Bu$_3$SnH, AIBN, C$_6$H$_6$, 80°	$\mathbf{I + II}$ (80) $\mathbf{I:II}$ = 1.8:1	625
C_{27}	Bu$_3$SnH, AIBN, PhMe, 90°	(68)	626

TABLE IV. BICYCLIC RINGS CONTAINING ONE OR MORE HETEROATOMS (*Continued*)

Reactant	Conditions	Product(s) and Yield(s) (%)	Refs.

C_33

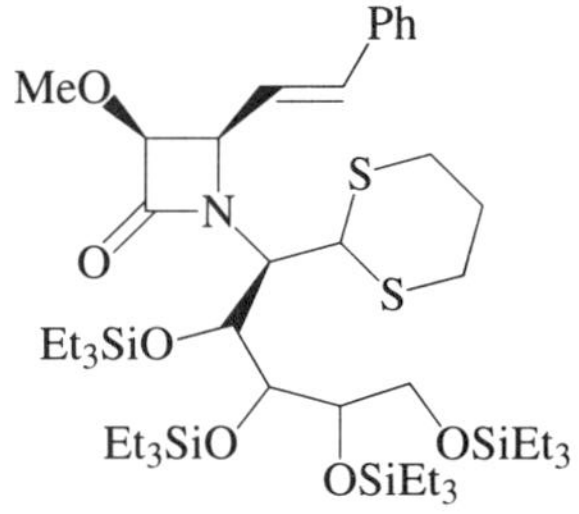

Bu_3SnH, AIBN, PhMe, 90°

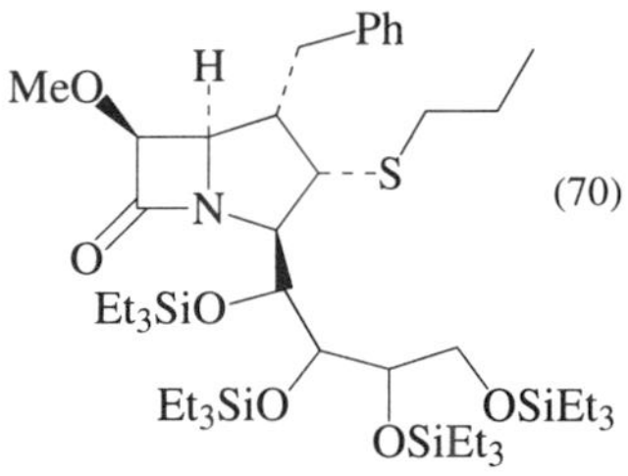

(70)

626

C. (4+6)-Membered Rings

C_8

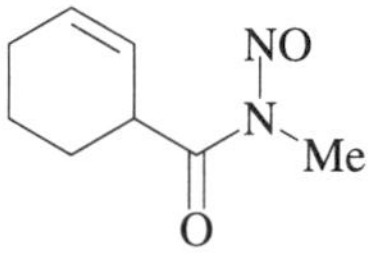

C_6H_6, *hv*

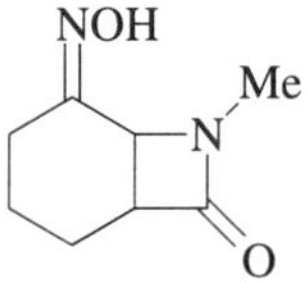

(44)

164

C_12

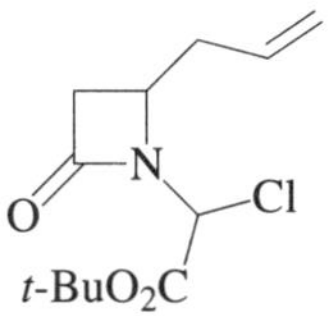

Bu_3SnH, AIBN, C_6H_6, 80°

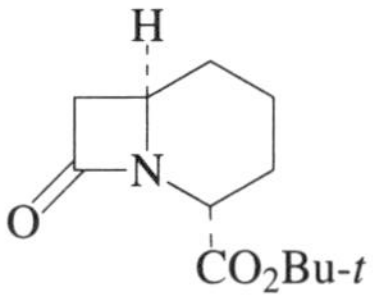

(50)

625

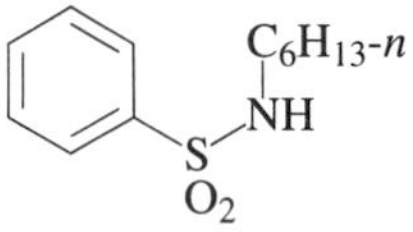

Na_2S_2O_8, CuCl_2•2 H_2O, H_2O, 90°, 5 h

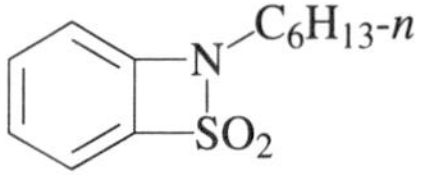

(60)

627

604

TABLE IV. BICYCLIC RINGS CONTAINING ONE OR MORE HETEROATOMS (Continued)

Reactant	Conditions	Product(s) and Yield(s) (%)	Refs.
C_{15}	Bu₃SnH, AIBN, C_6H_6, 80°	(43), (59)	628, 629
R = SPh, SePh			
C_{17}	Bu₃SnH, AIBN, C_6H_6, 80°	(55)	630
C_{17}	Bu₃SnH, AIBN, C_6H_6, 80°	(50-60) ca. 2:1	629, 631
R¹ = Et, CH₂CH(OTBDMS)Me, CH₂CH(OTBDMS)Me; R² = Me, Me, Bn			
C_{18}	Bu₃SnH, AIBN, C_6H_6, 80°	I + II (64) I:II = 1:1.3	632

TABLE IV. BICYCLIC RINGS CONTAINING ONE OR MORE HETEROATOMS (Continued)

Reactant	Conditions	Product(s) and Yield(s) (%)	Refs.
R = COCH$_2$COPh	Fe(ClO$_4$)$_3$, Ac$_2$O, MeCN, 0-5°	**I** + **II** I + II (42) **I:II** = 80:20	633
	Fe(NO$_3$)$_3$•9H$_2$O, Ac$_2$O, MeCN, 0°, 3 h	**I + II** (49)	634
C$_{25}$	Bu$_3$SnH, AIBN, PhMe, 90°	**I** + **II** **I, II** (—) α:β = 4:1	635

TABLE IV. BICYCLIC RINGS CONTAINING ONE OR MORE HETEROATOMS (*Continued*)

D. (4+7)-Membered Rings

Reactant	Conditions	Product(s) and Yield(s) (%)	Refs.
C$_7$ 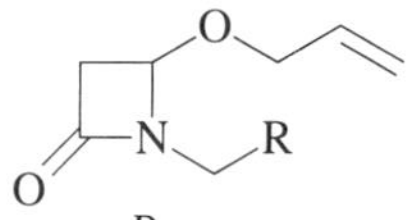 $\dfrac{R}{\text{Cl}}$ SePh SPh	Bu$_3$SnH, AIBN, C$_6$H$_6$, 80°	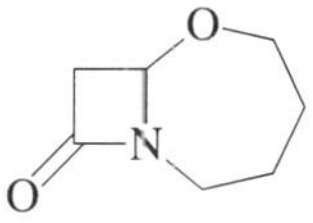 (57) (38) (56)	636, 637
C$_9$ 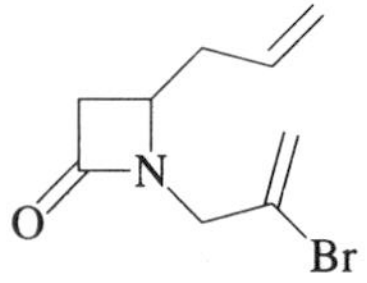	Bu$_3$SnH, AIBN, C$_6$H$_6$, 80°	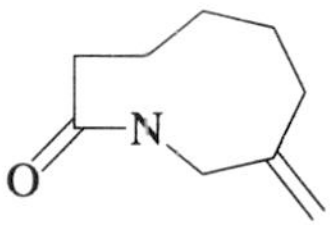 (77)	623
C$_{12}$ 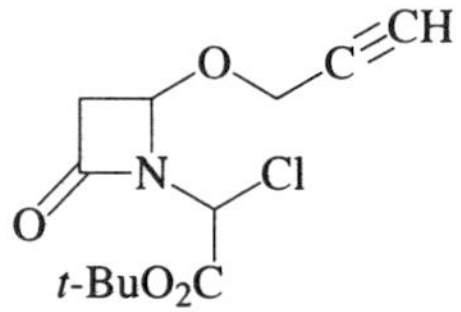	Bu$_3$SnH, AIBN, C$_6$H$_6$, 80°	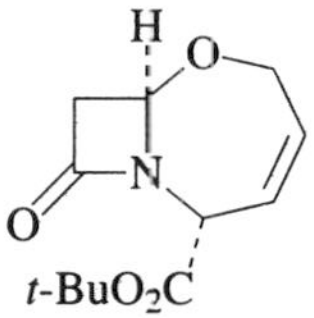 (49)	636
C$_{12}$ $\dfrac{R}{\text{H}}$ CO$_2$Me Ph	Bu$_3$SnH, AIBN, C$_6$H$_6$, 80°	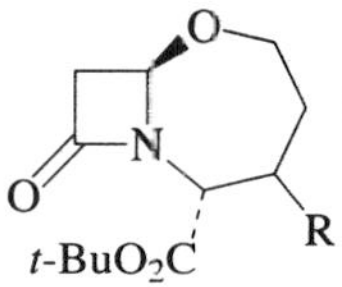 **I** 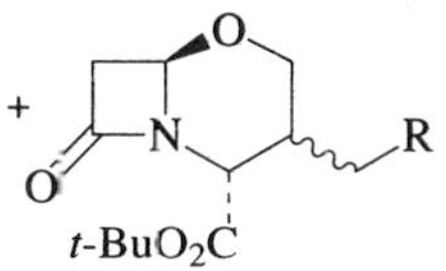 + **II** 	I	II	 (47) (—) (4) (68) (—) (68)	638

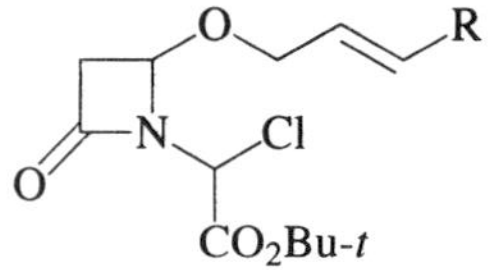

607

TABLE IV. BICYCLIC RINGS CONTAINING ONE OR MORE HETEROATOMS (Continued)

Reactant	Conditions	Product(s) and Yield(s) (%)	Refs.

E. (5+5)-Membered Rings

C6

Reactant: allyl–X–but-3-enyl system

Conditions:
- **A.** (TMS)$_3$SiH, Et$_3$B, 25°
- **B.** (TMS)$_3$SiH, AIBN, PhMe, 80°

Products: **I** (bicyclic with TMS–Si–TMS) + **II** (Si(TMS)$_3$) + **III** (TMS)

X	Conditions	I (%) cis:trans	II (%) cis:trans	III (%) cis:trans	Refs.
C(CO$_2$Me)$_2$	A	(11)	(87)	(—)	639
C(CO$_2$Me)$_2$	B	(71) 15:1	(17) 1:2	(8) 5:1	639
CH$_2$	B	(62) 6:1	(27) 1:5	(4) 3:1	639
O	B	(53) 100:0	(26) 1:8	(3) 3:1	639
O	B	(55)	(15)	(15)	142
CH$_2$	B	(35)	(19)	(21)	142

C7

Bu$_3$SnH, AIBN, C$_6$H$_6$, 80°	(>90)

640

Bu$_3$SnH, AIBN, C$_6$H$_6$, 80°	(>90)

640

608

TABLE IV. BICYCLIC RINGS CONTAINING ONE OR MORE HETEROATOMS (*Continued*)

Reactant	Conditions	Product(s) and Yield(s) (%)	Refs.
	CuCl, MeCN, 110–140° RuCl$_2$(PPh$_3$)$_3$, C$_6$H$_6$, 110–140°	 $\dfrac{R}{\text{H}}$ (71) Bn (89) H (71) Bn (88)	641
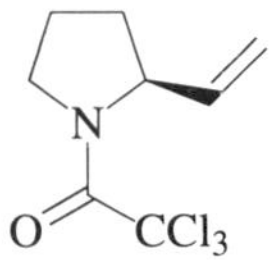	CuCl, MeCN, 150°	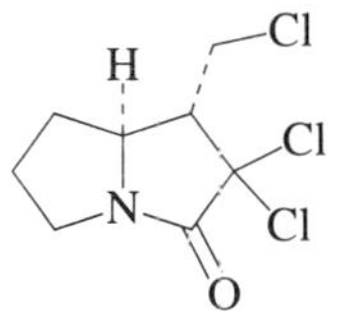 (93)	222
	Bu$_3$SnH, AIBN, C$_6$H$_6$, 80°	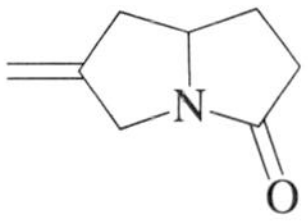 (84)	642
	Bu$_3$SnH, AIBN, C$_6$H$_6$, 80°	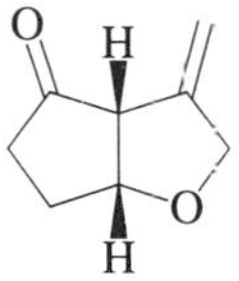 (88–92)	421

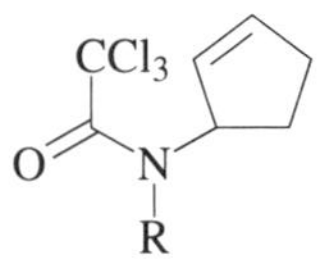

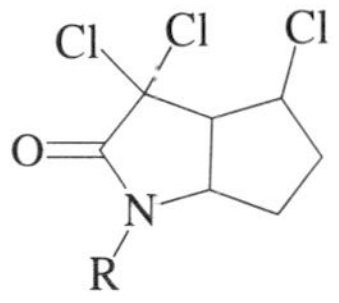

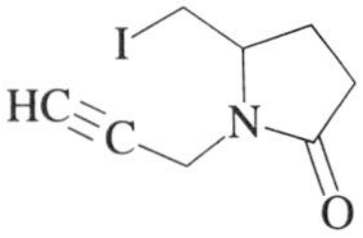

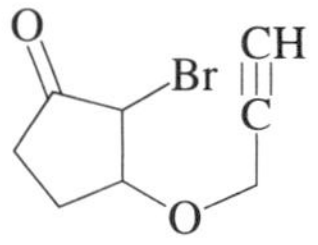

C$_8$

609

TABLE IV. BICYCLIC RINGS CONTAINING ONE OR MORE HETEROATOMS (*Continued*)

Reactant	Conditions	Product(s) and Yield(s) (%)	Refs.

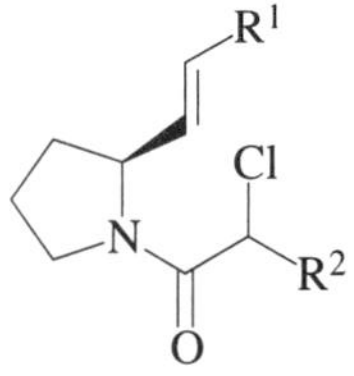

Bu$_3$SnH, AIBN, C$_6$H$_6$, 80°

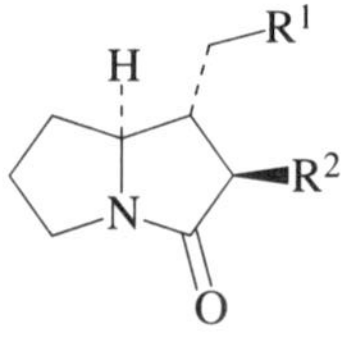

R^1	R^2	
H	H	(56)
H	SPh	(49)
H	SMe	(60)
CO$_2$Et	SPh	(77)

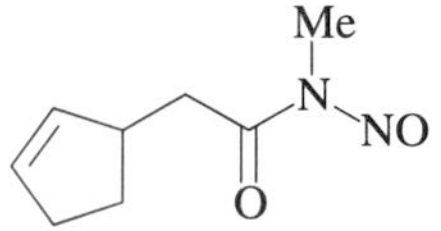

643

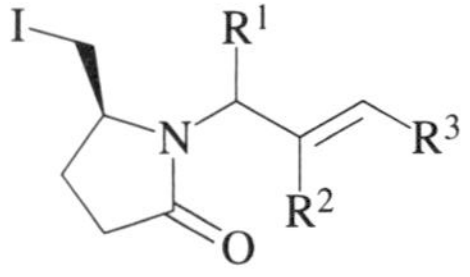

BrCCl$_3$, *hv*

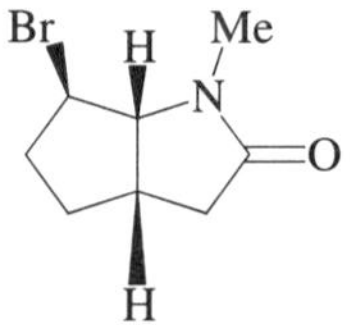

(89)

164

Bu$_3$SnH, AIBN, C$_6$H$_6$, 80°

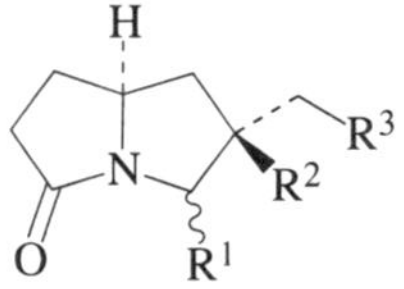

R^1	R^2	R^3	
H	H	H	(70)
H	H	Ph	(72)
Me	H	H	(54) 65:35
(CH$_2$)$_3$		H	(58)
H	H	Me	(72)
H	Me	H	(52)

644, 645

610

TABLE IV. BICYCLIC RINGS CONTAINING ONE OR MORE HETEROATOMS (*Continued*)

Reactant	Conditions	Product(s) and Yield(s) (%)	Refs.
	(Bu₃Sn)₂, EtI, *hv*	I + II (58) I:II = 30:1	646, 51
	AcOH, H₂O, TiCl₃ (15%), –10°	(66)	647
	HgO, I₂, *hv*	(68)	61
C₉	2,4,6-Collidine, Ph₂CO (cat.), MeOH, *i*-PrOH, *hv*, 350 nm	(88)	648
	Bu₃SnH, AIBN, C₆H₆, 80°	(88)	300, 298

TABLE IV. BICYCLIC RINGS CONTAINING ONE OR MORE HETEROATOMS (*Continued*)

Reactant	Conditions	Product(s) and Yield(s) (%)	Refs.
	ClCo(dmgH)₂py, Pt cathode, Et₄NOTs, NaOH, MeOH	$\dfrac{R}{\text{Et} \quad (87)}$ $C_5H_{11} \quad (84)$	649
	BPO, C₆H₆, 80°	$\dfrac{R}{\text{Me} \quad (88)}$ $\text{TMS} \quad (85)$	160
	Bu₃SnH, AIBN, C₆H₆, 80°	(70)	228
	Bu₃SnH, AIBN, C₆H₆, 80°	SMe (60)	650
	(Ph₃Sn)₂, C₆H₆, *t*-BuNC, *hv*, 50°	$\dfrac{R}{\text{H} \quad (62)}$ $\text{Me} \quad (65) \ 6{:}1$	206

612

TABLE IV. BICYCLIC RINGS CONTAINING ONE OR MORE HETEROATOMS (*Continued*)

Reactant	Conditions	Product(s) and Yield(s) (%)	Refs.
C_{10}			
	$Mn_2(OAc)_7$, AcOH, 64°	(64)	191
	Pt anode, AcOH, MeOH, 45°	I + II (60) I:II = 2:1	167
	$Mn(OAc)_3$, EtOH, rt, 1 h	R / H (60) / Me (55)	433
	Bu_3SnH, AIBN, C_6H_6, 80°	R / H (30) / Et (70)	651
	SmI_2, t-BuOH, THF, −78 to 0°	(87)	286

TABLE IV. BICYCLIC RINGS CONTAINING ONE OR MORE HETEROATOMS (*Continued*)

Reactant	Conditions	Product(s) and Yield(s) (%)	Refs.

SmI$_2$, Me$_2$CO, THF, −30° to rt

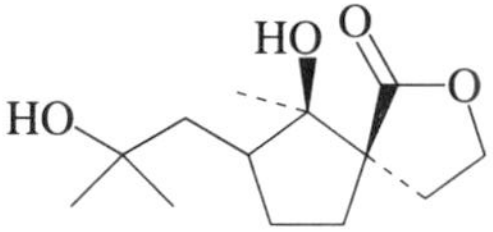

(85) + diastereomer 11:1

286

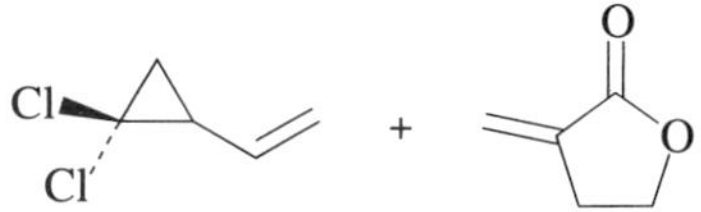

(PhS)$_2$, AIBN, rt, 4 h

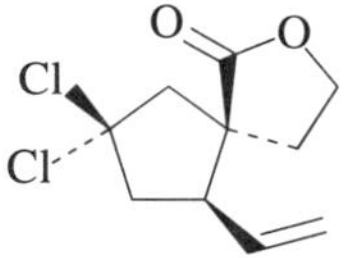

I + **II** (61) **I:II** = 15:1

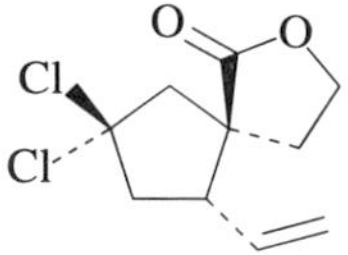

227

Bu$_3$SnH, AIBN, C$_6$H$_6$, 80°

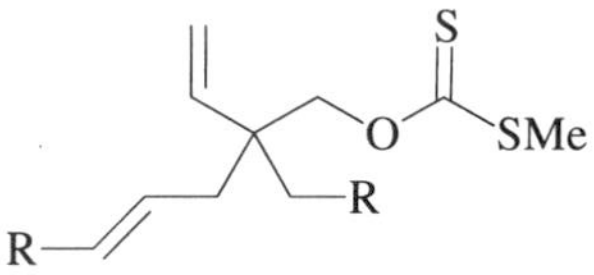

R		α:β
H	(76)	76:24
Me	(67)	86:14
Ph	(53)	58:42

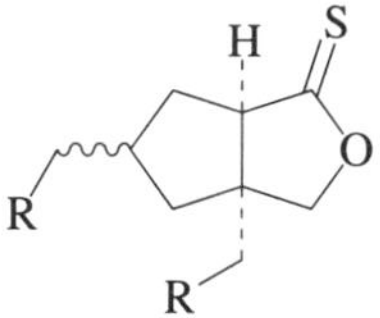

652

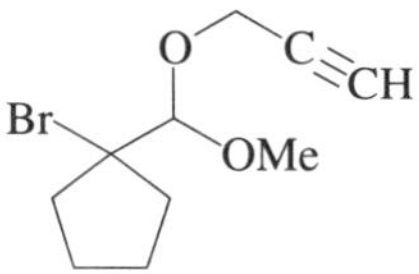

1. Bu$_3$SnCl, AIBN, Na(CN)BH$_3$, t-BuOH, sealed tube, 110°
2. Jones reagent

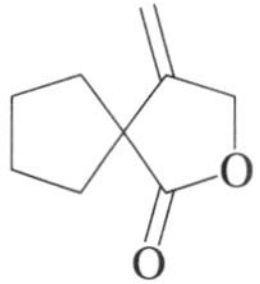

(79)

653

R
H
Me
Ph

614

TABLE IV. BICYCLIC RINGS CONTAINING ONE OR MORE HETEROATOMS (*Continued*)

Reactant	Conditions	Product(s) and Yield(s) (%)	Refs.

C_{11}

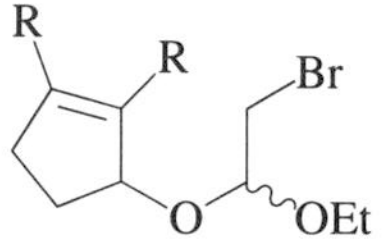

Bu$_3$SnH, AIBN, C$_6$H$_6$, 80°

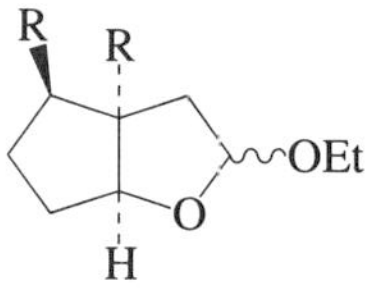

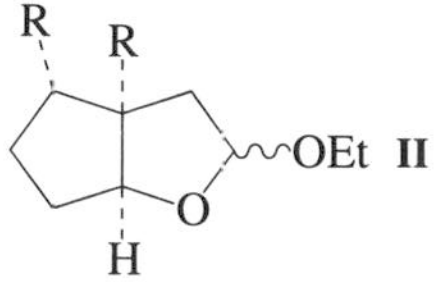

	I + II	**I:II**
	(80)	100:0
	(—)	≤0:50

298

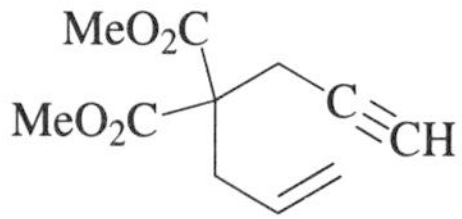

(TMS)$_3$SiH, AIBN, C$_6$H$_6$, reflux

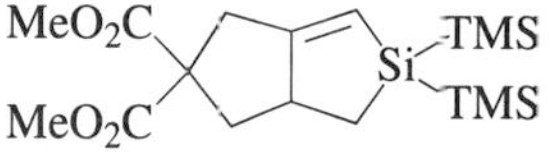

(48) +

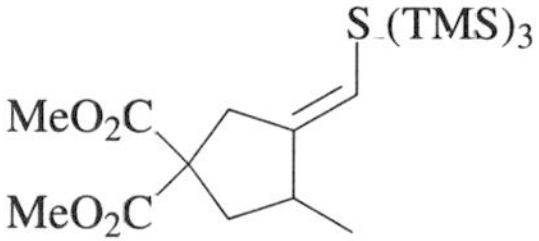

(6) +

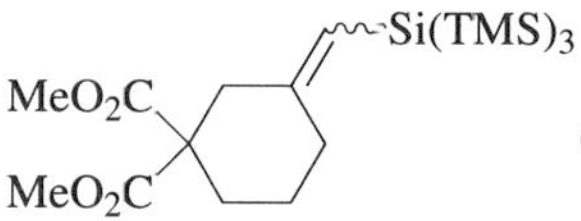

(39)

639

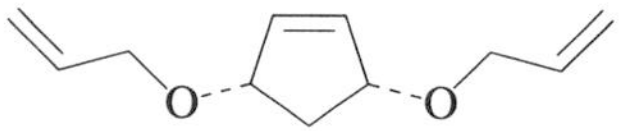

Me$_3$SnCl, Na(CN)BH$_3$, AIBN, *t*-BuOH

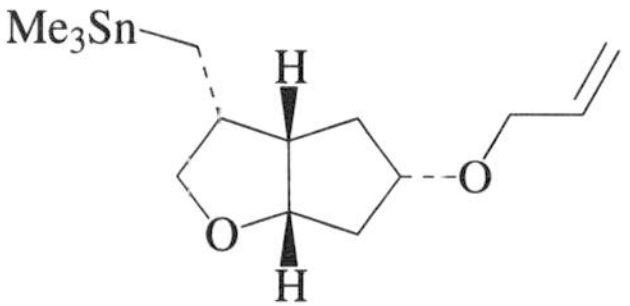

(42)

330

TABLE IV. BICYCLIC RINGS CONTAINING ONE OR MORE HETEROATOMS (*Continued*)

Reactant	Conditions	Product(s) and Yield(s) (%)	Refs.
	Ph$_2$CO, C$_6$H$_6$, $h\nu$	X Y S S (78) S O (15)	654
	1,4-Dicyanonaphthalene, *i*-PrOH, $h\nu$	**I** + **II** **I + II** (90) **I:II** = 97:3	655
C$_{12}$			
	Bu$_3$SnH, AIBN, C$_6$H$_6$, 80° or Bu$_3$SnCl, Na(CN)BH$_3$, AIBN, *t*-BuOH, 80°	(>80)	656
	MeCOSH, AIBN	(55)	452
	Bu$_3$SnH, AIBN, C$_6$H$_6$, 80°	(44)	652

TABLE IV. BICYCLIC RINGS CONTAINING ONE OR MORE HETEROATOMS (*Continued*)

Reactant	Conditions	Product(s) and Yield(s) (%)	Refs.
	Bu$_3$SnCl, Na(CN)BH$_3$, AIBN, *t*-BuOH, reflux	(38)	197
+ RCO$_2$H $\quad$ R: Me, (CH$_2$)$_2$CO$_2$Me	Pt anode, MeOH, 40-45°	(35) (33)	187
	Ph$_2$CO, C$_6$H$_6$, *hv*	(58)	654
R: H, Me	Mn(OAc)$_3$, EtOH, rt, 1 h	I + II $\quad$ $\dfrac{\text{I} \quad \text{II}}{(47) \; (3)}$ $\quad$ (47) (3)	433

TABLE IV. BICYCLIC RINGS CONTAINING ONE OR MORE HETEROATOMS (*Continued*)

Reactant	Conditions	Product(s) and Yield(s) (%)	Refs.

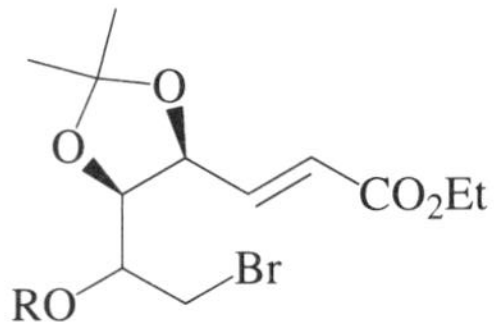

Bu₃SnH, AIBN, C₆H₆, 80°

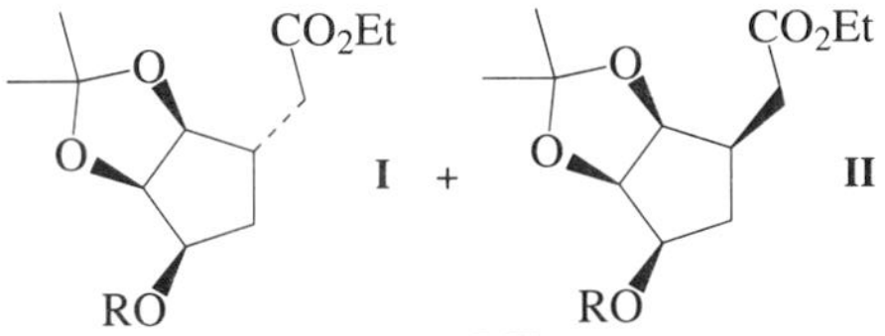

I + II

R		I + II	I:II
H	Z	(80)	6:1
H	E	(80)	2:1
COMe	Z	(80)	5:1
COMe	E	(82)	1:1
COPh	Z	(89)	10:1
COPh	E	(87)	1:1.2
COBu-*t*	Z	(87)	11:1

657, 658

C₁₃

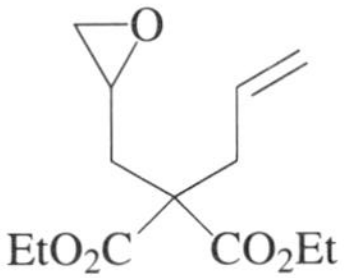

1. Cp₂TiCl
2. I₂

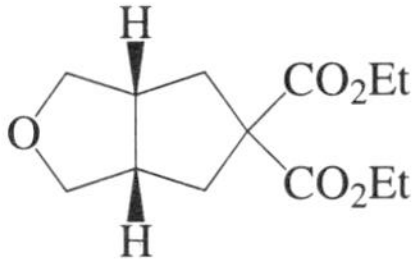

(52)

172

Bu₃SnH, AIBN, C₆H₆, 80°

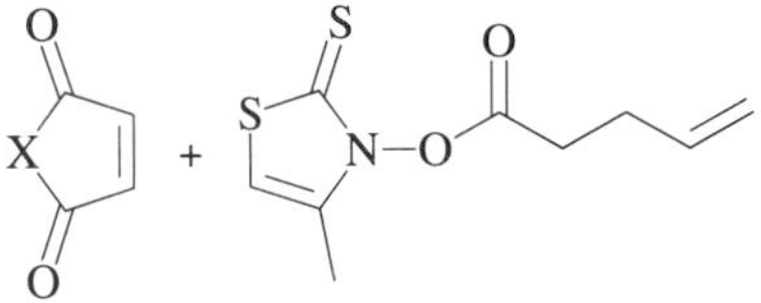

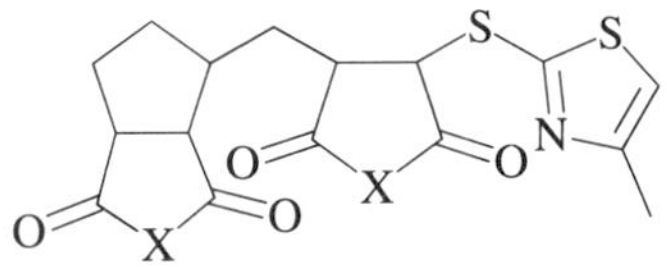

X	
NPh	(60)
NBu-*t*	(63)
O	(61)

243

618

TABLE IV. BICYCLIC RINGS CONTAINING ONE OR MORE HETEROATOMS (*Continued*)

Reactant	Conditions	Product(s) and Yield(s) (%)	Refs.
	AIBN, c-C$_6$H$_{12}$, heat 24 h	(47)	659
	Bu$_3$SnH, AIBN, C$_6$H$_6$, 80°	**I** + **II** (89) **I:II** = 4.8:1	660
	(Bu$_3$Sn)$_2$, C$_6$H$_6$, hv, 25°	(80)	323
	Bu$_3$SnH, AIBN	**I** + **II**	661

	I + II	**I:II**
R		
CN	(71)	4.8:1
CO$_2$Et	(92)	10.5:1
COPh	(57)	4.2:1
n-C$_5$H$_{11}$	(76)	2.5:1

619

TABLE IV. BICYCLIC RINGS CONTAINING ONE OR MORE HETEROATOMS (*Continued*)

Reactant	Conditions	Product(s) and Yield(s) (%)	Refs.
C$_{14}$	MeCN, H$^+$, 25°	(82)	152
	CuCl, bipyridine, 25°	(61)	91
	AIBN, C$_6$H$_{14}$, 2 h	(92)	659
	Bu$_3$SnH, AIBN, C$_6$H$_6$, 80°	(55) (60) (80)	662, 663

R	R^1
H	H
H	H
Me	Me

620

TABLE IV. BICYCLIC RINGS CONTAINING ONE OR MORE HETEROATOMS (*Continued*)

Reactant	Conditions	Product(s) and Yield(s) (%)	Refs.

SmI_2, CSA, MeCN — I + II — 664

Bu_3SnH, AIBN, C_6H_6, 80° — (52) + (14) — 665

Bu_3SnH, AIBN, C_6H_6, 80° — I + II — 665

	I	II
	(45)	(24)
	(42)	(22)
	(45)	(4)
	(60)	(0)

R^1	R^2	R^3
H	H	H
Me	H	H
H	Me	H
Me	Me	H

Bu_3SnH, AIBN, C_6H_6, 80° — I + II — 666

I + II (84) I:II = 10:1

TABLE IV. BICYCLIC RINGS CONTAINING ONE OR MORE HETEROATOMS (*Continued*)

Reactant	Conditions	Product(s) and Yield(s) (%)	Refs.
	Bu₃SnH, AIBN, C₆H₆, 65°	**I** + **II** I + II (58) I:II = 88:12	484
	Bu₃SnH, AIBN, C₆H₆, 65°	(95)	484
	AIBN, c-C₆H₁₂, 2 h	(29)	659
	Bu₃SnH, AIBN, C₆H₆, 80°	(67)	197
	MeCN, t-BuSH, H⁺, 25° MeCN, H⁺, 25° CH₂Cl₂, BF₃•OEt₂, hv, –78°	R = H (56); ξ–S–(pyridin-2-yl) (90); " (70)	483, 152, 67

TABLE IV. BICYCLIC RINGS CONTAINING ONE OR MORE HETEROATOMS (Continued)

Reactant	Conditions	Product(s) and Yield(s) (%)	Refs.
	MeCN, t-BuSH, H⁺, 25° MeCN, H⁺, 25°	 R: H (68), S-pyr (96)	152
	Bu₃SnH, AIBN, C₆H₆, 80°	(75)	667
C₁₅	Bu₃SnH, AIBN, C₆H₆, 80°	**I** + **II** + **III** + reduced product	668

R	I	II	III
Me	(—)	(27)	(61)
TMS	(48)	(22)	(—)
t-Bu	(49)	(—)	(—)
C(Me)₂OMe	(20)	(4)	(65)
i-Pr	(49)	(24)	(—)
n-Pr	(53)	(4)	(—)

TABLE IV. BICYCLIC RINGS CONTAINING ONE OR MORE HETEROATOMS (*Continued*)

Reactant	Conditions	Product(s) and Yield(s) (%)	Refs.

Row 1

Reactant: cyclic lactam with PhS, N, O; substituents R^1, R^2

R^1	R^2
Me	H
H	Me

Conditions: Bu$_3$SnH, AIBN, C$_6$H$_6$, 80°

Products: **I** + **II** + **III**

I + II + III	I:II:III
(69)	7:61:33
(56)	12:81:7

Refs.: 666

Row 2

Reactant: cyclic lactam with PhS, N, O; substituent R

R
CO$_2$Bu-t
CN

Conditions: Bu$_3$SnH, AIBN, C$_6$H$_6$, 80°

Products: **I** + **II**

I + II	I:II
(72)	9:1
(85)	9:1

Refs.: 669

Row 3

Reactant: cyclohexane bearing Ph and O$_2$C substituents

Products: (85) 9:1 (no significant induction)

Row 4

Reactant: structure with S, SMe, O

Conditions: Bu$_3$SnH, AIBN, C$_6$H$_6$, 80°

Products: (78)

Refs.: 652

624

TABLE IV. BICYCLIC RINGS CONTAINING ONE OR MORE HETEROATOMS (*Continued*)

Reactant	Conditions	Product(s) and Yield(s) (%)	Refs.
	Bu$_3$SnH, AIBN, C$_6$H$_6$, 80°	(94)	670
	Bu$_3$SnH, AIBN, C$_6$H$_6$, 80°	(74)	671
C$_{16}$	ClCo(dmgH)$_2$py, NaBH$_4$, NaOH, 1 MeOH, 0°	(67)	672
	Bu$_3$SnH, AIBN, C$_6$H$_6$, heat	(79)	673

TABLE IV. BICYCLIC RINGS CONTAINING ONE OR MORE HETEROATOMS (*Continued*)

Reactant	Conditions	Product(s) and Yield(s) (%)	Refs.
	Ph₃SnH, AIBN	(73)	352
	Co(OAc)₂, rt Ag(OAc)₂, rt	 **I + II** (60) **I:II** = 95:5 **I + II** (51) **I:II** = 94:6	674
C₁₇	Bu₃SnH, AIBN. C₆H₆, 80°	(88)	675
	Bu₃SnH, AIBN, C₆H₆, 80°		329

626

TABLE IV. BICYCLIC RINGS CONTAINING ONE OR MORE HETEROATOMS (*Continued*)

Reactant	Conditions	Product(s) and Yield(s) (%)	Refs.
$\dfrac{R}{H}$ $\dfrac{\ }{Me}$	Bu$_3$SnH, AIBN, C$_6$H$_6$, 80°	**I** + **II** $\quad$ $\dfrac{\textbf{I}}{(65)}\ \dfrac{\textbf{II}}{(9)}$ $\quad$ (49) (—)	676
$\dfrac{R}{H}$ $\dfrac{\ }{Me}$	Bu$_3$SnH, AIBN, C$_6$H$_6$, 80°	**I** + **II** $\quad$ $\dfrac{\textbf{I}}{(41)}\ \dfrac{\textbf{II}}{(9)}$ $\quad$ (50) (—)	676
	Bu$_3$SnH, AIBN, C$_6$H$_6$, 80°	$\dfrac{R}{H}\ (64)$ $\quad$ Me (86) $\quad$ mixture of isomers	669

TABLE IV. BICYCLIC RINGS CONTAINING ONE OR MORE HETEROATOMS (*Continued*)

Reactant	Conditions	Product(s) and Yield(s) (%)	Refs.
(cyclopentene-SPh-N(CH(CO$_2$Me)) with CO$_2$Me)	Bu$_3$SnH, AIBN, C$_6$H$_6$, 80°	(89) α:β = 13:87	113
(pyrrolidinone with C≡CTMS, Cl, SO$_2$Ph)	Bu$_3$SnH, AIBN, C$_6$H$_6$, 80°	(72)	677
(oxazolidinone with SPh, R^2, R^3, C≡CTMS) R^1, R^2, R^3	Bu$_3$SnH, AIBN, C$_6$H$_6$, 80°	(bicyclic product with R^1, R^2 H, R^3, TMS)	678

R^1 R^2 R^3

R^1	R^2	R^3	
Me	H	H	(75)
Et	H	H	(70)
Ph	H	H	(32)
Me	Me	H	(74)
H	H	Me	(72)

TABLE IV. BICYCLIC RINGS CONTAINING ONE OR MORE HETEROATOMS (Continued)

Reactant	Conditions	Product(s) and Yield(s) (%)	Refs.
	MeCN or MeOH, hv	 X H (47) F (47) OMe (46–65)	679
	Bu$_3$SnH, AIBN, PhMe, 110°, 5 h	 **I + II** (76) **I:II** = 88:12	680
	Bu$_3$SnH, AIBN, C$_6$H$_6$, 80°	(70)	681
C$_{18}$	ClCo(dmgH)$_2$py, NaBH$_4$, NaOH, MeOH, 0°	(69) $cis:trans$ = 1:1	672

629

TABLE IV. BICYCLIC RINGS CONTAINING ONE OR MORE HETEROATOMS (*Continued*)

Reactant	Conditions	Product(s) and Yield(s) (%)	Refs.
$\overset{O}{\underset{}{\parallel}}$ ...SiPh$_2$Me ...I	1. Bu$_3$SnH, AIBN, $\overset{}{\underset{}{}}CO_2$Me C$_6H_6$, heat 2. F$^-$	(64)	682
Im–S, CO$_2$Me (furanose)	Bu$_3$SnH, AIBN, PhMe, 110°	CO$_2$Me, 7, H, H (80) 4 isomers in ratio 11.3:2.3:2:1 main isomer 7-β	680
C$_{19}$ C$_{11}$H$_{23}$ CO$_2$Me I	(Bu$_3$Sn)$_2$, C$_6$H$_6$, *hv*, 3 d	C$_{11}$H$_{23}$ I + C$_{11}$H$_{23}$ II + MeO$_2$C, C$_{11}$H$_{23}$, I III I:II:III 33:28:39 I+II+III (50)	683
	AgOAc, SnCl$_2$, MeOH, *hv*, rt	35:31:34 (68)	

TABLE IV. BICYCLIC RINGS CONTAINING ONE OR MORE HETEROATOMS (*Continued*)

Reactant	Conditions	Product(s) and Yield(s) (%)	Refs.
	Bu₃SnH, AIBN, C_6H_6, 80°	(76)	684
	Bu₃SnH, AIBN, C_6H_6, 80°	(61)	652
	1. Mn(OAc)₃, EtOH 2. AcOH, H₂O	**I** + **II** (72) **I:II** = 1:1	685
	Bu₃SnH, AIBN, C_6H_6, 80°	(60-71)	678, 668

631

TABLE IV. BICYCLIC RINGS CONTAINING ONE OR MORE HETEROATOMS (Continued)

Reactant	Conditions	Product(s) and Yield(s) (%)	Refs.
(structure: thiocarbonate with SMe, Ph, R groups; R = H, Me)	Bu$_3$SnH, AIBN, C$_6$H$_6$, 80°	(structure) $\dfrac{\alpha:\beta}{> 99:1 \quad 90:10}$ (71) (58)	652
C$_{20}$ (structure with Ph, N–SPh)	Bu$_3$SnH, AIBN, C$_6$H$_6$, 80°	(49) + (8)	686
(structure with pyrrole, CO$_2$Et, CO$_2$Et, Br, Ph, C=O)	Et$_3$B, O$_2$, C$_6$H$_6$ Mn(OAc)$_3$, AcOH, NaOAc, 80°	(structure) (75) (95)	687
(structure with N, C$_7$H$_{15}$, C=S, O) + (dihydrofuran)	hv, H$^+$	(structure) (52)	488

632

TABLE IV. BICYCLIC RINGS CONTAINING ONE OR MORE HETEROATOMS (*Continued*)

Reactant	Conditions	Product(s) and Yield(s) (%)	Refs.

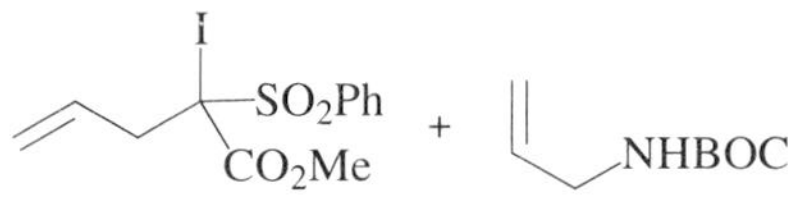

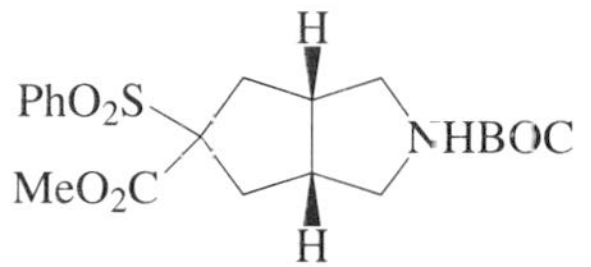

C$_{21}$

1. (Bu$_3$Sn)$_2$, C$_6$H$_6$, $h\nu$
2. Et$_3$N

(50)

688

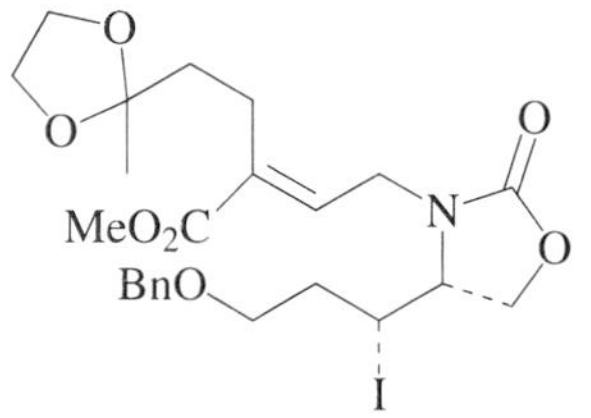

ClCo(dmgH)$_2$py, NaBH$_4$, NaOH, MeOH

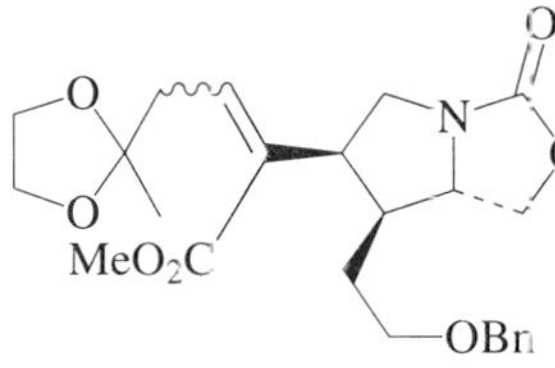

(64)

689

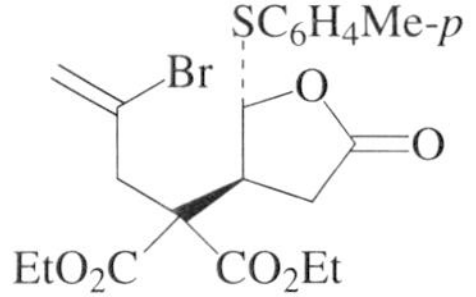

Bu$_3$SnH, AIBN, PhMe, 110°

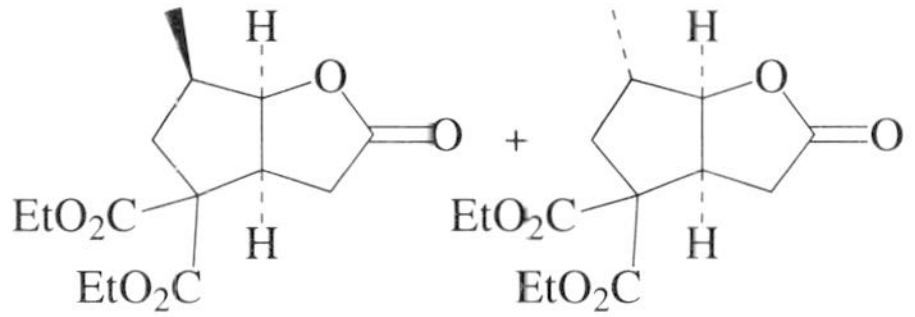

I + II (87) I:II = 2:1

690

C$_{22}$

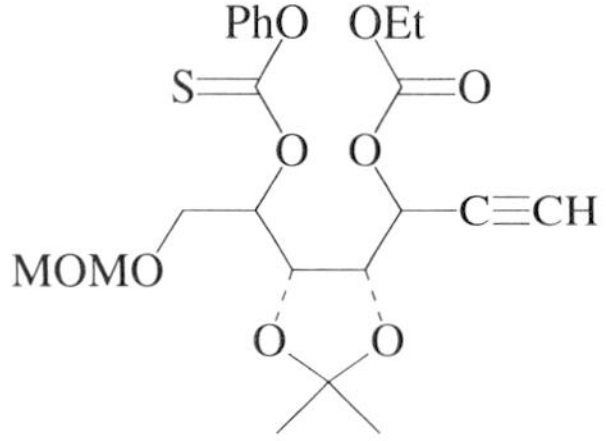

Bu$_3$SnH, AIBN, C$_6$H$_6$, 80°

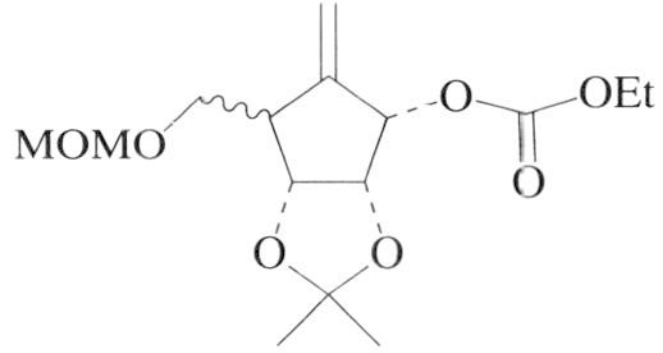

(63)

endo:exo = 6.4:1

691

TABLE IV. BICYCLIC RINGS CONTAINING ONE OR MORE HETEROATOMS (*Continued*)

Reactant	Conditions	Product(s) and Yield(s) (%)	Refs.

C_{23}

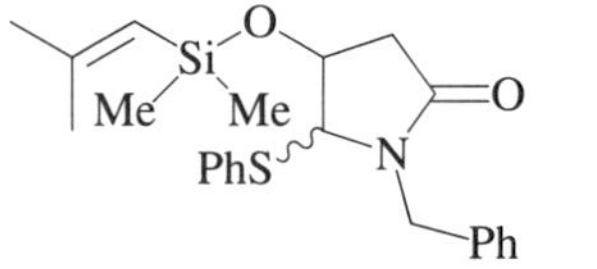

Bu$_3$SnH, AIBN, C$_6$H$_6$, 80°

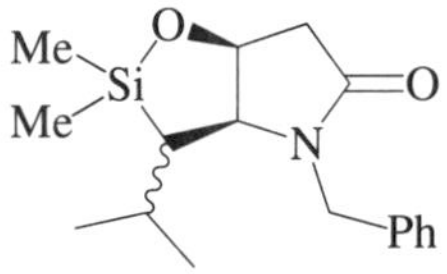

(72)

692

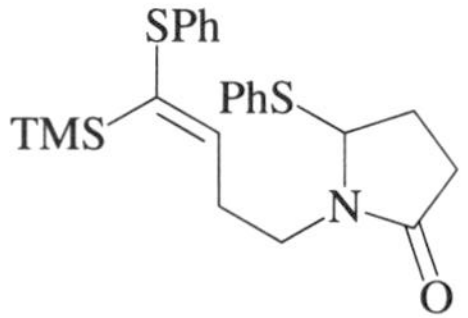

Bu$_3$SnH, AIBN, C$_6$H$_6$, 80°

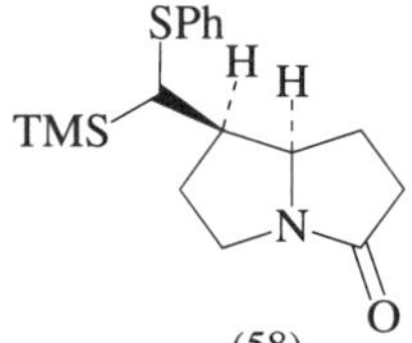

(58)   +   (13)

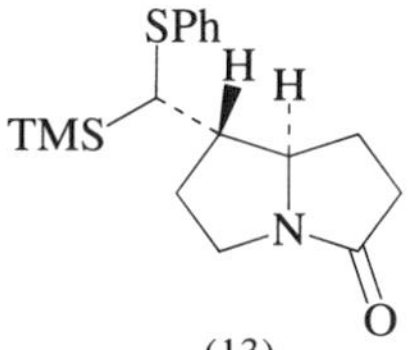

669

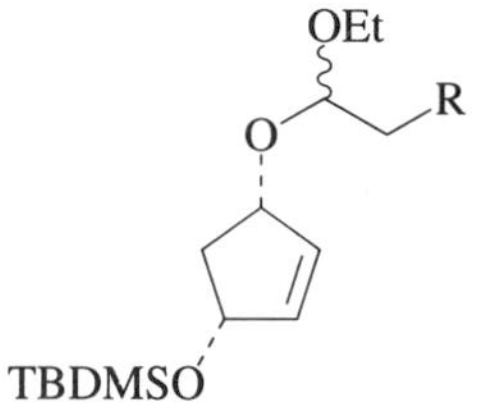

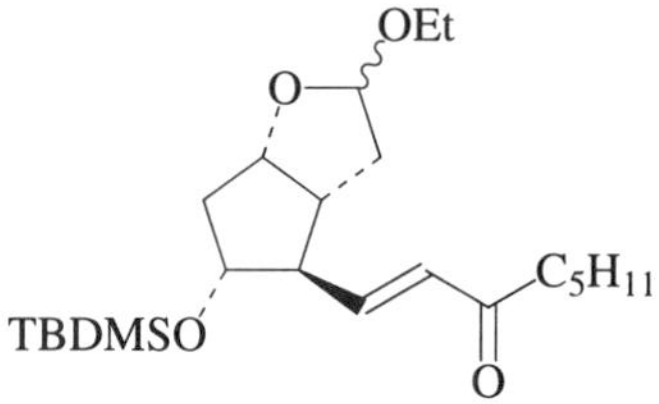

$R = I$

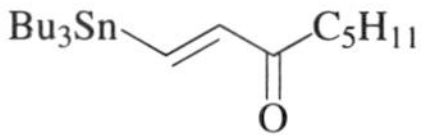

Bu$_3$SnH, ACN, PhMe, 110°

(72)

205

$R = Br$

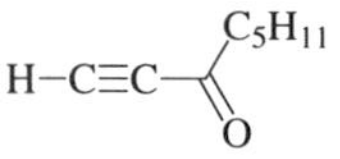

Vitamin B$_{12}$, C-felt cathode, LiClO$_4$, DMF, *hv*

(>55)

224

634

TABLE IV. BICYCLIC RINGS CONTAINING ONE OR MORE HETEROATOMS (Continued)

Reactant	Conditions	Product(s) and Yield(s) (%)	Refs.

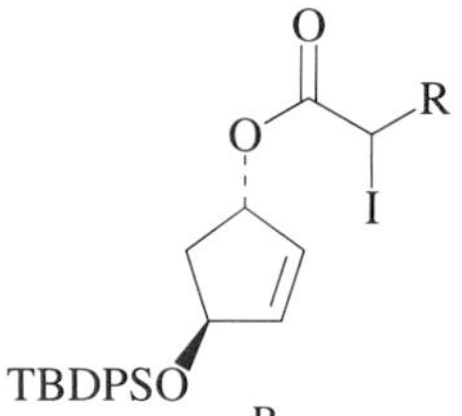

Ph₃SnH, AIBN, C₆H₆, 80°

$$ \text{Ph}_3\text{SnH, AIBN,} \quad \text{C}_6\text{H}_6, 80° $$

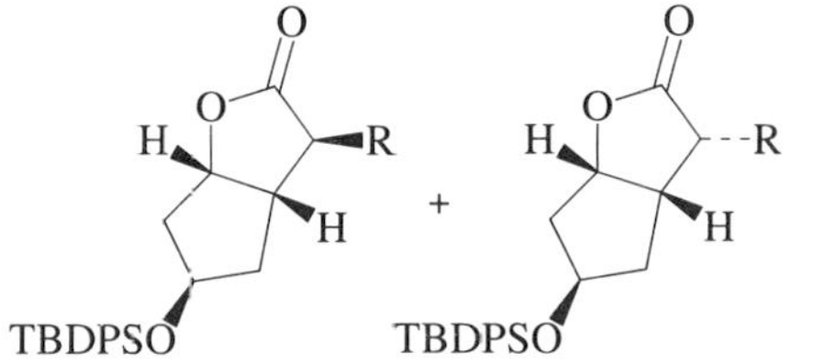

(54)
(37)

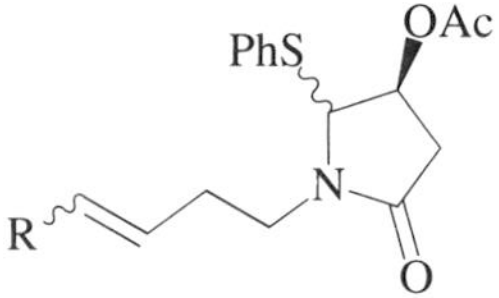

419

Bu₃SnH, AIBN, C₆H₆, 80°

693

	I	II	III
	(63)	(10)	(18)
	(64)	(11)	(12)

$\dfrac{R}{H}$
Me

R = PhMe₂Si

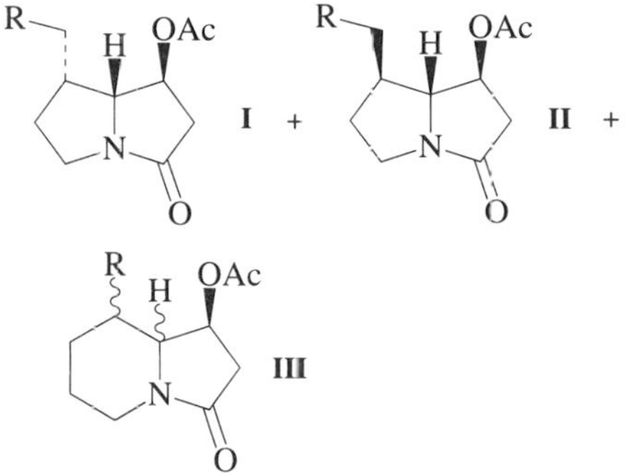

C₂₄

cis:trans = 18:1
cis:trans = 1.3:1

635

TABLE IV. BICYCLIC RINGS CONTAINING ONE OR MORE HETEROATOMS (*Continued*)

Reactant	Conditions	Product(s) and Yield(s) (%)	Refs.
	Bu₃SnH, AIBN, C₆H₆, 80°	 **I + II** (74) **I:II** = 6.4:1	497
	Bu₃SnH, AIBN, PhMe, 110°	(85) *E:Z* = 1.3:1	690
C₂₅	Bu₃SnH, AIBN, C₆H₆, 80°	(50)	694
C₂₆	Bu₃SnCl, THF, Na(CN)BH₃, *hv*	(58)	100

636

TABLE IV. BICYCLIC RINGS CONTAINING ONE OR MORE HETEROATOMS (*Continued*)

Reactant	Conditions	Product(s) and Yield(s) (%)	Refs.
C27	hv	**I + II** (45) **I:II** = 11.3:1	132, 695
C29	Bu3SnH, ACN, PhMe, 110°	(74)	205
C33	Bu3SnH, AIBN, C6H6, 80°	(<95)	696
C34	Bu3SnH, AIBN, C6H6, 80°	(<75)	696

TABLE IV. BICYCLIC RINGS CONTAINING ONE OR MORE HETEROATOMS (*Continued*)

Reactant	Conditions	Product(s) and Yield(s) (%)	Refs.

F. (5+6)-Membered Rings

C7

	PhCl, *t*-(BuO)$_2$, heat		697
	or		
	c-C$_6$H$_{12}$, *hv*, heat		

R^1	R^2		I + II	I:II
H	H		(82)	100:0
H	Me		(80)	100:0
H	Et		(72)	100:0
H	Bn		(70)	100:0
Me	Me		(35)	100:0
Me	Me	MeOH, *hv*	(45)	60:40
Me	Me	MeOH, H$^+$, *hv*	(43)	55:45

C8

| | RuCl$_2$(PPh$_3$)$_3$ (1-5%), xylene, 140° | (71) | 698 |
| | CuCl, CH$_2$Cl$_2$, bipyridine, 25° | (98) | 91 |

| | Bu$_3$SnH, AIBN, C$_6$H$_6$, 80° | (63) | 642 |

638

TABLE IV. BICYCLIC RINGS CONTAINING ONE OR MORE HETEROATOMS (*Continued*)

Reactant	Conditions	Product(s) and Yield(s) (%)	Refs.
	Bu$_3$SnH, AIBN, C$_6$H$_6$, 80°	I$^-$	699, 700
		R = H (65)	
		R = Me (65)	
	(Bu$_3$Sn)$_2$, C$_6$H$_6$, hv, 80°	(82)	156, 51
	ClCo(dmgH)$_2$py, Et$_4$NOTs, NaOH, MeOH, Pt cathode	R^1 = H, R^2 = H (35)	649
		R^1 = C$_5$H$_{11}$, R^2 = H (70)	
		R^1 = C$_5$H$_{11}$, R^2 = Me (82)	
	Et$_3$B, C$_6$H$_{14}$, 25°	R = H (75)	275
		R = TMS (90)	

TABLE IV. BICYCLIC RINGS CONTAINING ONE OR MORE HETEROATOMS (Continued)

Reactant	Conditions	Product(s) and Yield(s) (%)	Refs.
	ClCo(dmgH)₂py, Et₄NOTs, NaOH, MeOH, Pt cathode	(44)	649
	Polymer-SnCl, AIBN, NaBH₄, C₆H₆, EtOH, *hv*	" (73)	426
	Bu₃SnH, AIBN, C₆H₆, 80°	" (50)	426
C₉	PhSiH₂, BPO, C₈H₁₈, 110°		272
RC≡CPh	(PhS)₂, 100°		701

TABLE IV. BICYCLIC RINGS CONTAINING ONE OR MORE HETEROATOMS (Continued)

Reactant	Conditions	Product(s) and Yield(s) (%)	Refs.
(structure: 2-iodophenyl allyl, X = CH₂, O, NMe) $\dfrac{X}{\begin{array}{l}CH_2\\O\\NMe\end{array}}$	Bu₃SnH, AIBN, C₆H₆	(structure) 3-methyl-2,3-dihydrobenzofuran analog (50) (99) (78)	29
(structure: 2-iodophenyl allyl ether)	CoI(salen) or CoI(salophen)py, hv	(structure with CoL) $\dfrac{L}{\begin{array}{l}(\text{salen}) \quad (65)\\(\text{salophen})py \quad (55)\end{array}}$	186
	CoI(salophen)py, NaHg, dark	(structure with Co(salophen)py) (70)	169
	Ni(II)complex, DMF, NH₄ClO₄, Et₄ClO₄, Hg cathode	(structure) (75)	702
(structure: X = Br, I, Cl, F allyl ether)	10-Me-9,10-dihydro-acridine (cat.), DMF, NaBH₄, hv	(structure) X = Br (65) X = I (82) X = Cl (48) X = F (21)	703

TABLE IV. BICYCLIC RINGS CONTAINING ONE OR MORE HETEROATOMS (*Continued*)

Reactant	Conditions	Product(s) and Yield(s) (%)	Refs.
	SmI$_2$, E$^+$		174

E$^+$	E	
H$_2$O	H	(—)
D$_2$O	D	(80)
I$_2$	I	(70)
(PhS)$_2$	SPh	(65)
(PhSe)$_2$	SePh	(72)
Bu$_3$SnI	Bu$_3$Sn	(82)

Reactant	Conditions	Product(s) and Yield(s) (%)	Refs.
	N-Oxide **A** or **B**, Me$_2$CO, 60°		704

N-Oxide	X	R^1	R^2	
A	O	H	DBA	(76)
B	O	H	TMP	(84)
A	O	Me	DBA	(68)
B	O	Me	TMP	(82)
A	NAc	H	DBA	(75)
B	NAc	H	TMP	(80)
A	NAc	Me	DBA	(35)
B	NAc	Me	TMP	(51)

TABLE IV. BICYCLIC RINGS CONTAINING ONE OR MORE HETEROATOMS (*Continued*)

Reactant	Conditions	Product(s) and Yield(s) (%)	Refs.

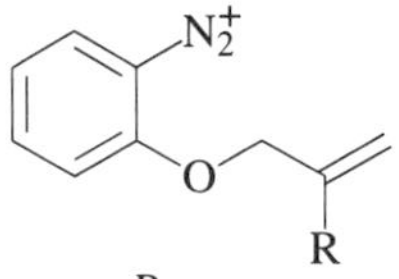

X	
Br	(82)
Br	(89)
Cl	(63)
CN	(40)
SPh	(60)
SPh	(53)
SBu-*n*	(64)
SC(S)OEt	(75)

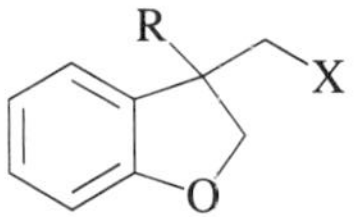

R	
H	CuBr$_2$
Me	CuBr$_2$
Me	CuCl$_2$
Me	Cu(CN)$_2$
Me	NaSPh
H	NaSPh
Me	Cu, NaSBu-*n*
Me	NaSC(S)OEt

705, 706

NaI, Me$_2$CO, 25°

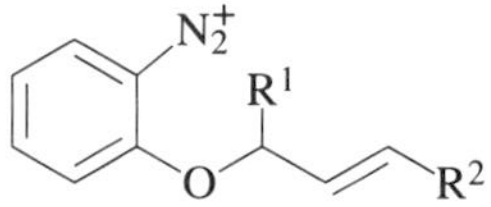

R^1	R^2	
H	H	(86)
Me	H	(89)
H	Me	(73)

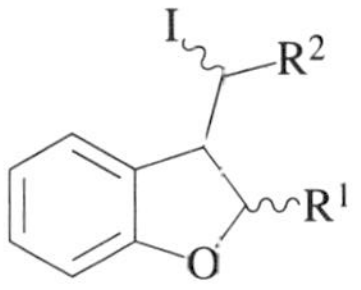

707

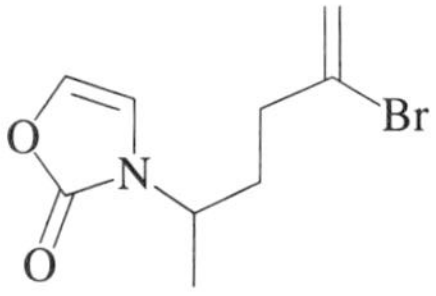

Bu$_3$SnH, AIBN, C$_6$H$_6$, 80°

(72)

708

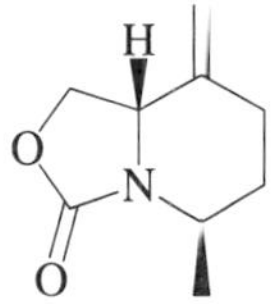

643

TABLE IV. BICYCLIC RINGS CONTAINING ONE OR MORE HETEROATOMS (*Continued*)

Reactant	Conditions	Product(s) and Yield(s) (%)	Refs.

Reactant 1 (brominated aniline allyl structure with R¹, R² substituents):

X	R¹	R²	
H	H	H	(70)
H	H	Ph	(59)
Ac	H	H	(88)
Ac	Me	Me	(99)

Conditions: SmI$_2$, HMPA, THF, MeCN

Product: indoline with R¹, R², N–X

Refs. 175

Reactant 2 (cyclohexenyl propargyl, X linker):

Conditions: Ph$_2$PH, AIBN, C$_6$H$_6$, 80°

Product (PPh$_2$ vinylidene bicyclic):
X = O (66) 1.7:1
X = NCO$_2$Me (66) 6.1:1

Refs. 301

Reactant 3 (cyclohexenyl propargyl, O linker):

Conditions: Ph$_3$SnH, Et$_3$B

Product (SnPh$_3$ vinylidene bicyclic O): (60)

Refs. 709, 27

Reactant 4 (iodo sugar with allyl ether):

Conditions: Bu$_3$SnH, AIBN, C$_6$H$_6$, 80°

Product (bicyclic sugar): (94)

Refs. 710

644

TABLE IV. BICYCLIC RINGS CONTAINING ONE OR MORE HETEROATOMS (*Continued*)

Reactant	Conditions	Product(s) and Yield(s) (%)	Refs.
	Bu₃SnH, AIBN, PhMe, *hv*		607

$$\begin{array}{cc} X & \\ \hline O & \\ O & \\ NTs & \end{array}$$

$$\begin{array}{cc} R & \\ \hline SnBu_3 & (72) \\ H & (28) \\ H & (22) \end{array}$$

	−20°		
	−20°		
	17°		
	Bu₃SnH, AIBN, PhMe, heat	$\dfrac{R}{\text{H (92)}}$ Me (47)	711, 712
	Bu₃SnCl, Na(CN)BH₃, AIBN, *t*-BuOH	(96)	713
	Bu₃SnH, AIBN, C₆H₆, 80°		714, 715

$$\begin{array}{cc} R & \\ \hline H & (63) \\ Me & (73) \\ Ph & (75) \end{array}$$

645

TABLE IV. BICYCLIC RINGS CONTAINING ONE OR MORE HETEROATOMS (*Continued*)

Reactant	Conditions	Product(s) and Yield(s) (%)	Refs.
	Bu₃SnH, AIBN, C₆H₆, 80°	(71)	228
	Bu₃SnH, AIBN, C₆H₆, 80°	**I** + **II** + **III** (69) **I:II:III** = 19:39:42	115
	TsSePh, AIBN, C₆H₆, 80°	(53) 1.5:1	302
	Bu₃SnH, AIBN, C₆H₆, 80°	(88–92)	421

646

TABLE IV. BICYCLIC RINGS CONTAINING ONE OR MORE HETEROATOMS (*Continued*)

Reactant	Conditions	Product(s) and Yield(s) (%)	Refs.
	Bu$_3$SnH, AIBN, C$_6$H$_6$, 80°	**I** + reduced product **II** **I + II** (80) **I:II** = 1:2	646
	(Bu$_3$Sn)$_2$, EtI, C$_6$H$_6$, *hv*	**Ia, Ib** + reduced **II** product **I + II** (88) $\begin{array}{ccc} & R^1 & R^2 \\ \textbf{Ia} & I & H & (57) \\ \textbf{Ib} & H & I & (11) \\ \textbf{II} & H & H & (20) \end{array}$	646
C$_{10}$	(Bu$_3$Sn)$_2$, C$_6$H$_6$, *hv*, 80° Me$_2$CO,	**I** (60–70) R^1 = H, R^2 = Me (**72**) R^1 = R^2 = Me (quant., NMR)	51 716
	Bu$_3$SnH, AIBN, C$_6$H$_6$, 80°	(50)	52

647

TABLE IV. BICYCLIC RINGS CONTAINING ONE OR MORE H ETEROATOMS (*Continued*)

Reactant	Conditions	Product(s) and Yield(s) (%)	Refs.
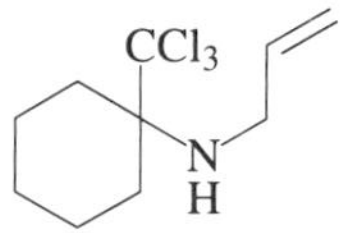	Bu_3SnH, AIBN, C_6H_6, 80°	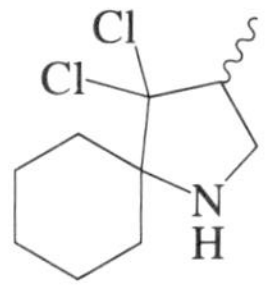 (75)	717
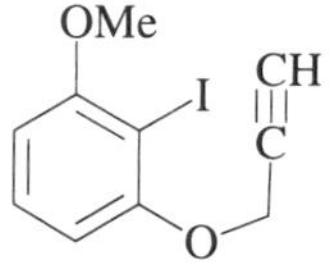	Bu_3SnH, AIBN, C_6H_6, 80°	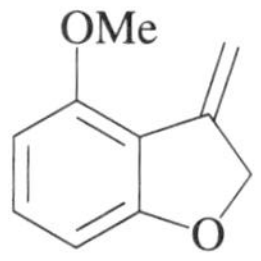 (52)	718
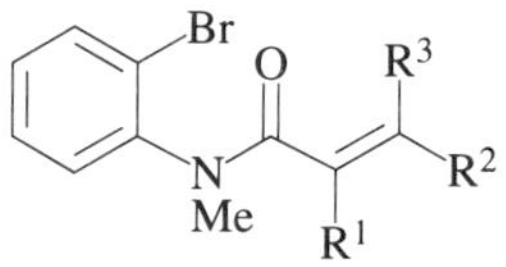 R¹ / R² / R³: H / H / H H / Me / H H / Me / Me Me / H / H $(CH_2)_4$ / H	Bu_3SnH, AIBN, C_6H_6, 80°	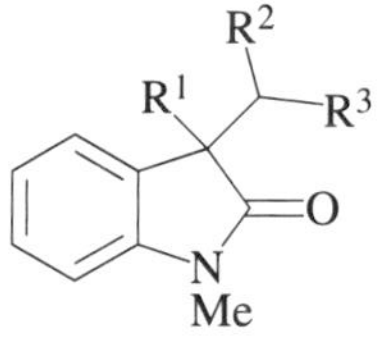 **I** + 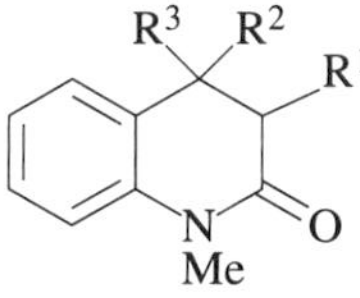 **II** **I** / **II**: (79) / (—) (72) / (—) (80) / (—) (72) / (18) (69) / (22)	719
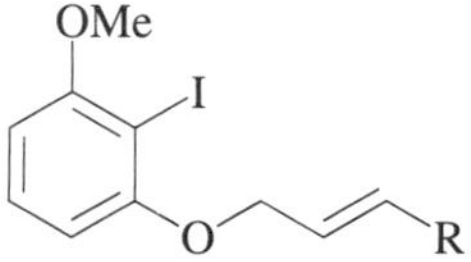	Bu_3SnH, AIBN, C_6H_6, 80°	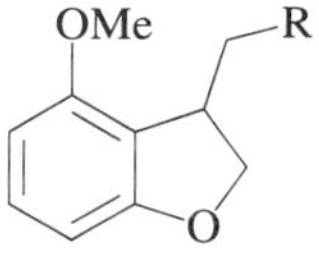 R: H (80) Ph (88) CO_2Et (75)	718

TABLE IV. BICYCLIC RINGS CONTAINING ONE OR MORE HETEROATOMS (*Continued*)

Reactant	Conditions	Product(s) and Yield(s) (%)	Refs.
	Bu$_3$SnH, AIBN, C$_6$H$_6$, 80°	R^1 R^2 Me Me (88) H CO$_2$Et (75)	718
	1. HgO, I$_2$, C$_6$H$_6$ 2. *hv*	R^1 R^2 H H (41) H Cl (55) H OMe (54) OMe OMe (67)	720
	Bu$_3$SnH, AIBN, C$_6$H$_6$, 80°	R C$_6$H$_{11}$ (92) *t*-Bu (80) *n*-Bu (70) *i*-Pr (49) Me (39) Ph (<5)	721
	2,4,6-Collidine, Ph$_2$CO (cat.) MeOH, *i*-PrOH, *hv*	(54)	648
	Bu$_3$SnH, AIBN, C$_6$H$_6$, 80°	(85)	722

TABLE IV. BICYCLIC RINGS CONTAINING ONE OR MORE HETEROATOMS (*Continued*)

Reactant	Conditions	Product(s) and Yield(s) (%)	Refs.
(structure)	Et$_3$N, MeCN, *hv*	X: O (50), CH$_2$ (40)	181
(structure)	Bu$_3$SnH, AIBN, C$_6$H$_6$, 80°	(50) + (16)	723
(structure)	Bu$_3$SnH, AIBN, C$_6$H$_6$, 80°	(43) + (5)	723
(structure)	Bu$_3$SnH, AIBN, C$_6$H$_6$, 80°	(72)	723
(structure)	BPO, C$_6$H$_6$, 80°	R: Me (59), TMS (93)	160

650

TABLE IV. BICYCLIC RINGS CONTAINING ONE OR MORE HETEROATOMS (Continued)

Reactant	Conditions	Product(s) and Yield(s) (%)	Refs.
	1. Bu$_3$SnCl, Na(CN)BH$_3$, AIBN, t-BuOH 2. TsOH, C$_6$H$_6$	 R^1 R^2 R^3 H H Me (38) H H Et (42) H Me Me (45) Me H Me (45)	724
	Bu$_3$SnH, AIBN, C$_6$H$_6$, 80°	(63)	725
	Cu(bipy)Cl, MeOAc, reflux	 I + II (63) I:II = 2.5:1	417
X = Br	Bu$_3$SnH, AIBN, C$_6$H$_6$, 80°	I R = H (50)	52
X = Br	Bu$_3$SnCl, Na(CN)BH$_3$, AIBN, t-BuOH, t-BuNC	I R = CN (50)	122
X = Br	(Ph$_3$Sn)$_2$, t-BuNC, C$_6$H$_6$, hv, 50°	I R = CN (58)	122

TABLE IV. BICYCLIC RINGS CONTAINING ONE OR MORE HETEROATOMS (*Continued*)

Reactant	Conditions	Product(s) and Yield(s) (%)	Refs.
X = I + 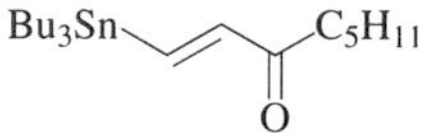	Bu_3SnH, ACN, PhMe, 110°	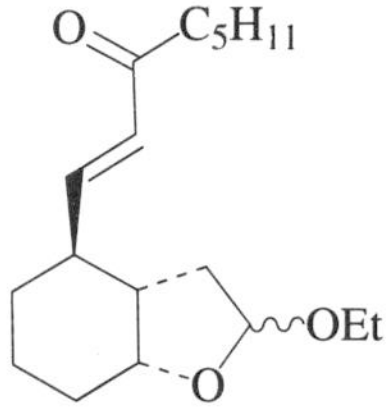 (60)	205
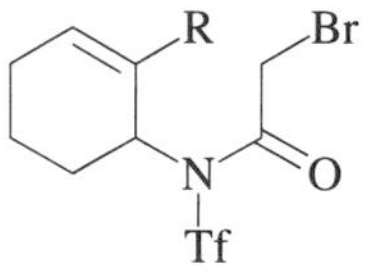	1. Ph_3GeH, AIBN 2. H^+	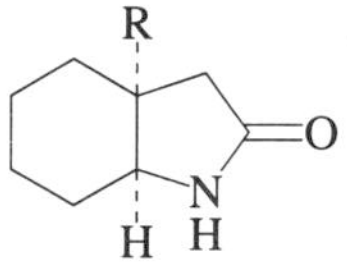 R: Me (82); Ph (60)	54
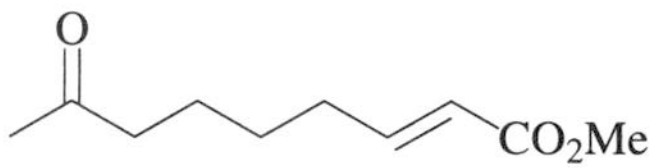	SmI_2, THF, reflux	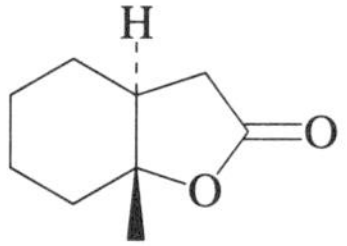 **I** + 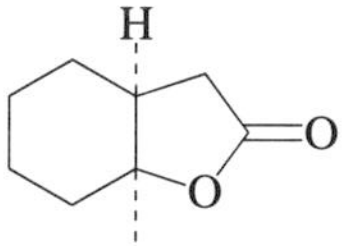 **II** **I** + **II** (56) **I**:**II** = 3:7	726
C_{11} 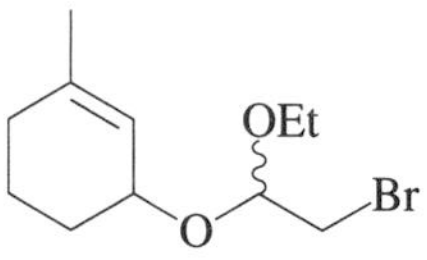	Vitamin B_{12}, Zn, NH_4Cl, MeOH, H_2O	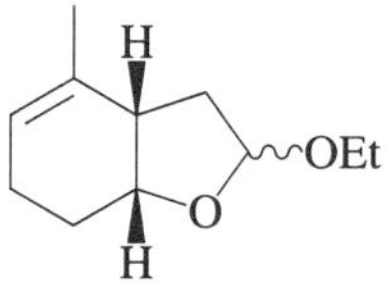 (80)	186
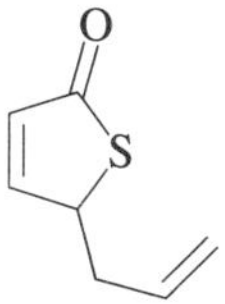 + MeC≡CMe	MeOH, *hν*	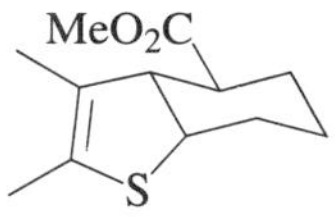 (45) + 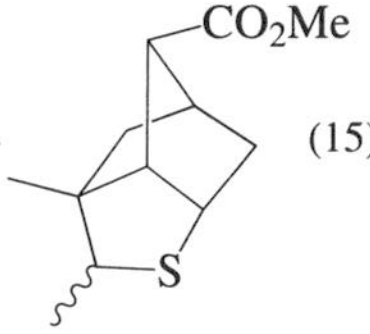 (15)	727

TABLE IV. BICYCLIC RINGS CONTAINING ONE OR MORE HETEROATOMS (*Continued*)

Reactant	Conditions	Product(s) and Yield(s) (%)	Refs.
	Bu₃SnH, AIBN, PhMe, 110°	(63)	728
	Bu₃SnH, AIBN	**I** + **II** **I + II** (99)	729
	SmI₂, HMPA, THF	(47)	175
	Bu₃SnH, AIBN		729

R^1	R^2	R^3	
H	H	H	(91)
H	Me	H	(93)
Me	H	H	(93)
H	Ph	H	(92)
H	Me	Me	(93)
OMe	H	CO₂Et	(31)

TABLE IV. BICYCLIC RINGS CONTAINING ONE OR MORE HETEROATOMS (*Continued*)

Reactant	Conditions	Product(s) and Yield(s) (%)	Refs.
	NaI, Me₂CO, 25°	(84)	707
	Bu₃SnH, AIBN, C₆H₆, 80°	X–Y: CHMe (93), CH=CH₂ (39), CHMe (35)	730
	ClCo(dmgH)₂py, NaBH₄, EtOH, NaOH (0.1 M), 40°	I (18) II (47)	731
	ClCo(dmgH)₂py, NaBH₄, EtOH, NaOH (0.01 M), 40°	II (29)	731

654

TABLE IV. BICYCLIC RINGS CONTAINING ONE OR MORE HETEROATOMS (*Continued*)

Reactant	Conditions	Product(s) and Yield(s) (%)	Refs.
	Bu₃SnH, AIBN, C₆H₆, 80°	(53)	478
	(Me₃Sn)₂, hv	**I** + **II** (66) **I:II** = 60:40	18
	TsSePh, AIBN, C₆H₆, heat	(73) 6:1 ratio	301
	TsSePh, AIBN, C₆H₆, heat	(94) 1.3:1 ratio	301

655

TABLE IV. BICYCLIC RINGS CONTAINING ONE OR MORE HETEROATOMS (*Continued*)

Reactant	Conditions	Product(s) and Yield(s) (%)	Refs.
	1. Bu$_3$SnH, AIBN, C$_6$H$_6$, 80° 2. TsOH (cat.), C$_6$H$_6$	$\dfrac{R}{H}$ (45) Me (46)	732
	Bu$_3$SnH, AIBN, C$_6$H$_6$, 80°	$\dfrac{R}{H}$ (88–92) Me	421
	1. Bu$_3$SnCl, Na(CN)BH$_3$, AIBN, *t*-BuOH 2. Jones reagent	(96)	653
	Bu$_3$SnH, AIBN, C$_6$H$_6$, 80°	(13) + (25)	733
	Bu$_3$SnH, AIBN, PhMe, 110°	(85)	425

656

TABLE IV. BICYCLIC RINGS CONTAINING ONE OR MORE HETEROATOMS (*Continued*)

	Reactant	Conditions	Product(s) and Yield(s) (%)	Refs.
C$_{12}$				

Reactant 1 (with Br, allyl ether, benzo ring) + CH$_2$=CH–R

Bu$_3$SnH, AIBN, PhMe, 110°

Product: dihydrobenzofuran with propyl–R chain

R	
CO$_2$Et	(60)
CN	(57)
Ph	(58)
SO$_2$Ph	(60)

Refs. 207, 734

Reactant 2 (HO–sugar, O–Si(Me)–Me–Br) + CH$_2$=CH–CN

Bu$_3$SnCl, Na(CN)BH$_3$, AIBN, *t*-BuOH, heat, 18 h

Product: bicyclic with CN chain, Si(Me)Me; HO

(—)

Ref. 735

Reactant 3 (cyclohexenone with N(R^1)(R^2) side chain)

MeCN or MeOH, *hv*

Product: spirocyclic ketone with pyrrolidine, R^1, R^2

R^1	R^2		R^1	R^2	
Ph	CH=CH$_2$		CH=CH$_2$	Ph	(66)
Ph	CO$_2$Me		Ph	CO$_2$Me	(44)
Ph	TMS		Ph	TMS	(65)
Ph	TMS		H	Ph	(71)
H	TMS		TMS	H	(76)
H	CH=CH$_2$		CH=CH$_2$	H	(79)

Ref. 736

TABLE IV. BICYCLIC RINGS CONTAINING ONE OR MORE HETEROATOMS (*Continued*)

Reactant	Conditions	Product(s) and Yield(s) (%)	Refs.
	Bu$_3$SnH, AIBN, C$_6$H$_6$, 80°	 **I + II** (90) **I:II** = 85:15	52
	RuCl$_2$(PPh$_3$)$_3$, xylene, 140°, 8 h	 (64)	154
	1,4-Dicyano-naphthalene, *i*-PrOH, *hv*	 **I + II** (85) **I:II** = 2:98	655
	1,4-Dicyano-naphthalene, *i*-PrOH, *hv*	 **I + II** (87) **I:II** = 95:5	655

TABLE IV. BICYCLIC RINGS CONTAINING ONE OR MORE HETEROATOMS (*Continued*)

Reactant	Conditions	Product(s) and Yield(s) (%)	Refs.
	SmI$_2$, HMPA, THF, MeCN	(61)	175
	SmI$_2$, HMPA, THF,	(55)	175
	MeCN, Et$_4$NClO$_4$, Pt electrode	(60)	486
	Bu$_3$SnH, AIBN, C$_6$H$_6$, 80°	**I** + **II**	31

R	I	II
CHO	(—)	(50)
CH$_2$OH	(47)	(—)
CN	(60)	(40)

TABLE IV. BICYCLIC RINGS CONTAINING ONE OR MORE HETEROATOMS (*Continued*)

Reactant	Conditions	Product(s) and Yield(s) (%)	Refs.

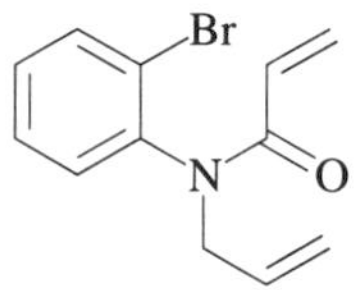

Bu₃SnH, AIBN, PhMe, heat — (77) — 737

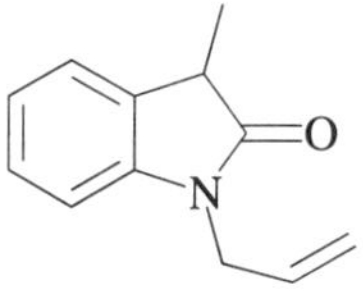

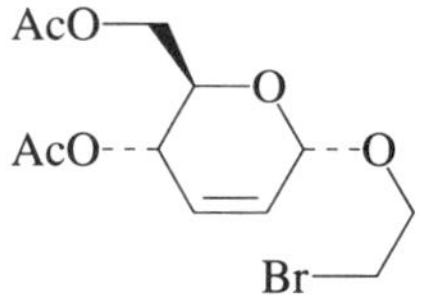

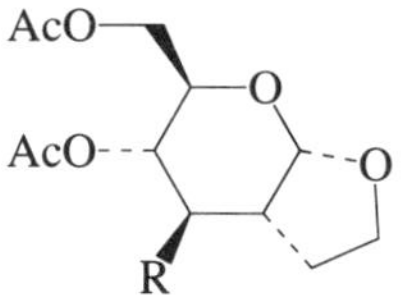

Bu₃SnH, AIBN, C₆H₆, 80° — **I** R = H (76) — 738

Bu₃SnCl, NaCNBH₃, AIBN, t-BuOH — **I** R = H (88) — 738

AIBN, C₆H₆, 80° + ⌇SnBu₃ — **I** R = CH₂CH=CH₂ (56) — 739

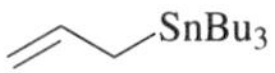

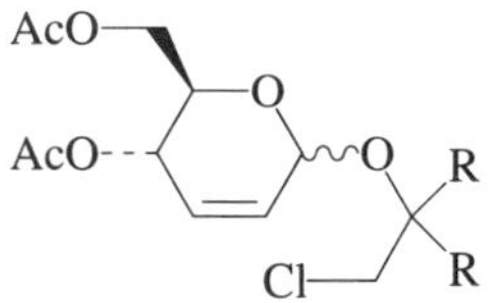

Bu₃SnH, AIBN, C₆H₆, 80° — 740

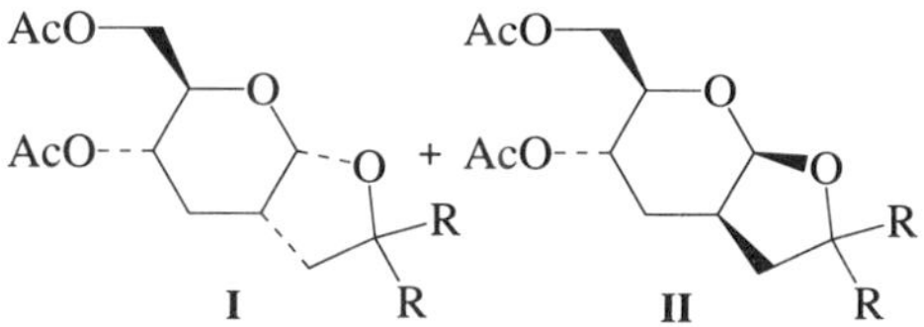

	I + II	I:II
R	(96)	86:14
H	(96)	86:14
Me		

α:β = 86:14

660

TABLE IV. BICYCLIC RINGS CONTAINING ONE OR MORE HETEROATOMS (*Continued*)

Reactant	Conditions	Product(s) and Yield(s) (%)	Refs.
(X/N-Me acrylamide structure)	A. Co(I)(salen) B. Bu₃SnH	**I** + **II** + **III**	741, 742
$\dfrac{X}{\text{Br}}$ Br		**I + II + III** **I:II:III** A (66) 72:2:26 A (40) 72:2:26 B (65) 0:72:28	
(OMOM / OMe / Br cyclohexene ether structure)	Bu₃SnH, AIBN	MOMO····OMe furan bicyclic (50)	743
(N-Me allyl cyclohexenone structure)	MeOH, *hv*	N–Me spirocyclic (79) 1:1 ratio	226
(Br / O / N-R cyclohexene amide structure)	Bu₃SnH, AIBN, C₆H₆, 80°	bicyclic N-R lactam $\dfrac{R}{}$ *t*-Bu (77) PhCMe₂ (73) Ts (80)	54

where for the first reactant X = I, Br, Br.

TABLE IV. BICYCLIC RINGS CONTAINING ONE OR MORE HETEROATOMS (Continued)

Reactant	Conditions	Product(s) and Yield(s) (%)	Refs.
(dithiane with R and CO_2Et side chain)	Ph_2CO, C_6H_6, hv	(spiro dithiane with CO_2Et, R) R : H (60) ; Me (55)	654
(cyclohexene with OR^1, Br, R^4, OR^3, R^2) R^1 : i-Bu, i-Bu, Me, MEM, TBDMS R^2 : H, Me, Me, Me, t-Bu R^3 : Et, Et, Et, Et, Me R^4 : H, H, H, H, Me	Bu_3SnH, AIBN, C_6H_6, 80°	(bicyclic furan with R^4, OR^3, R^1O, R^2) (91) (78) (56) (90) (60)	743
(cyclohexane with AcO, NO_2, O, allyl)	Bu_3SnH, AIBN, C_6H_6, 80°	I + II **I + II** (74) **I:II** = 85:15	111 438
(ester with I, R, O, dienyl chain)	Ph_3SnH, AIBN, C_6H_6, 80°	(bicyclic lactone with R, isopropyl) R : H (53) ; Me (52)	419

TABLE IV. BICYCLIC RINGS CONTAINING ONE OR MORE HETEROATOMS (*Continued*)

Reactant	Conditions	Product(s) and Yield(s) (%)	Refs.
	Bu₃SnCl, Na(CN)BH₃, AIBN, *t*-BuOH, 110°, 8 h	(90)	744
	AIBN, C₆H₆, 80°	I + II $\dfrac{\text{I:II}}{90:10}$ 66:34 44:56	745
	Bu₃SnH 0.5 M Bu₃SnH 0.05 M Bu₃SnH 0.02 M		
	Pb(OAc)₄		746
	AcOH, 20° CH₂Cl₂, 20° CH₂Cl₂, 20°	(86) (60) (34)	
	Bu₃SnH, Et₃B, THF, -78°	(71)	437

663

TABLE IV. BICYCLIC RINGS CONTAINING ONE OR MORE HETEROATOMS (*Continued*)

Reactant	Conditions	Product(s) and Yield(s) (%)	Refs.
R^1, R^2 alkene $+$ 2-methyl-1,3-cyclopentanedione $+ O_2$	A. AIBN, MeCN, 50–60° B. Carbon anode, Et$_4$NOTs, MeCN		747
R^1 = Ph, R^2 = H	A	(66)	
	B	(79)	
R^1 = Ph, R^2 = Me	A	(90)	
	B	(48)	
R^1 = CH$_2$TMS, R^2 = H	A	(73)	
	B	(72)	
C$_{13}$			
	Bu$_3$SnCl, Na(CN)BH$_3$, AIBN, t-BuOH	(70)	748
	Bu$_3$SnH, AIBN or Bu$_3$SnCl, Na(CN)BH$_3$, AIBN, t-BuOH		749, 738

R^1	R^2	X	
Ac	OMe	Br	(96)
Ac	OEt	I	(83)
Ac	OMe	HgOAc	(90)
Ac	H	Br	(55)
Bz	H	Br	(75)

TABLE IV. BICYCLIC RINGS CONTAINING ONE OR MORE HETEROATOMS (*Continued*)

Reactant	Conditions	Product(s) and Yield(s) (%)	Refs.

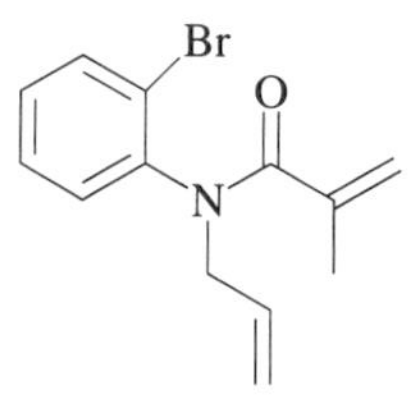

Ph₃SnH, AIBN, C₆H₆, 80°

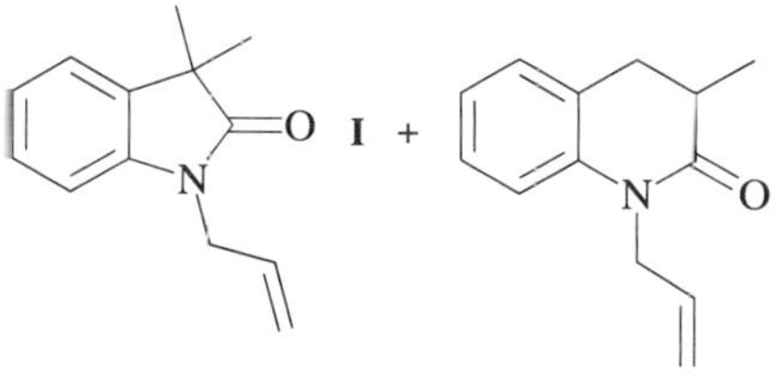

I + **II** (98) **I:II** = 3:1

737

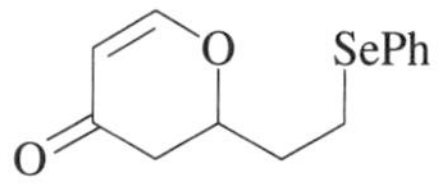

Ph₃SnH, AIBN, C₆H₆, heat

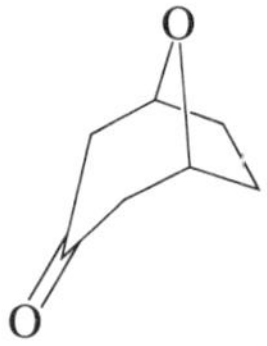

(88)

112

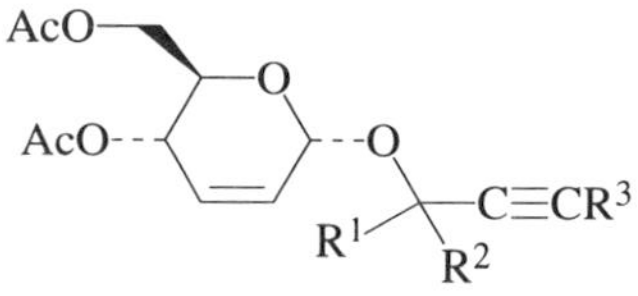

Bu₃SnH, AIBN, C₆H₆, 80°

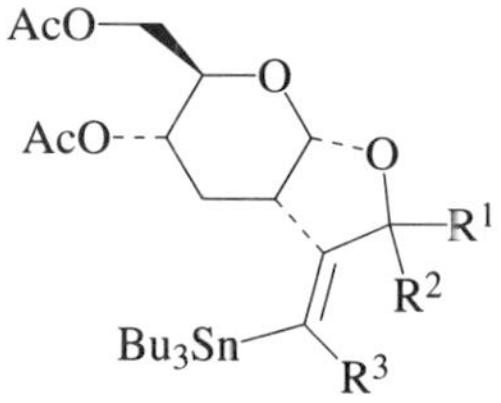

R¹	R²	R³		Z:E
H	H	H	(75)	100:0
H	H	CH₂OAc	(77)	5:1
Me	H	H	(65)	5:1
H	Me	H	(63)	4:1

750, 740, 738

TABLE IV. BICYCLIC RINGS CONTAINING ONE OR MORE HETEROATOMS (*Continued*)

Reactant	Conditions	Product(s) and Yield(s) (%)	Refs.
	Bu₃SnH, AIBN, C₆H₆, 80°	(89)	740
	TsBr, *hv*	(70) *endo:exo* = 7:1	751
	Mn(OAc)₃, EtOH, rt, 1 h	(30) + (10)	433
	Et₃N, MeCN, *hv*	(50)	752

TABLE IV. BICYCLIC RINGS CONTAINING ONE OR MORE HETEROATOMS (*Continued*)

Reactant	Conditions	Product(s) and Yield(s) (%)	Refs.
	Bu₃SnH, AIBN, C₆H₆, 80°	(52)	708
	MeOH, *hv*	CO₂Me (70) + CO₂Me (25)	727
	Bu₃SnH, AIBN, C₆H₆, 80°	(33)	228
	(TMS)₃SiH, AIBN, PhMe, 88–90°	(79) 2.4:1	470

667

TABLE IV. BICYCLIC RINGS CONTAINING ONE OR MORE HETEROATOMS (*Continued*)

Reactant	Conditions	Product(s) and Yield(s) (%)	Refs.
	MeOH, *hv*	(72)	226
	MeCN, *hv*	(76) 4:3	226
C$_{14}$	Bu$_3$SnH, AIBN, C$_6$H$_6$, 80°	$$\frac{R}{\begin{array}{l} H \quad\quad (80) \\ n\text{-}C_6H_{13} \quad (50) \end{array}}$$	753
$$\frac{R}{\begin{array}{l}(CH_2)_3NEt_2 \\ (CH_2)_3Cl \\ (CH_2)_3CH=CH_2 \end{array}}$$	SmI$_2$, HMPA, THF, 25°	(76) (10) (38)	176

668

TABLE IV. BICYCLIC RINGS CONTAINING ONE OR MORE HETEROATOMS (*Continued*)

Reactant	Conditions	Product(s) and Yield(s) (%)	Refs.
R: $(CH_2)_3CO_2Et$		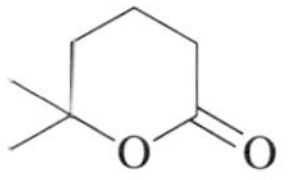 (40)	176
		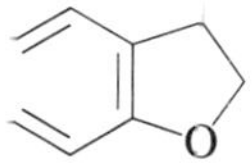	
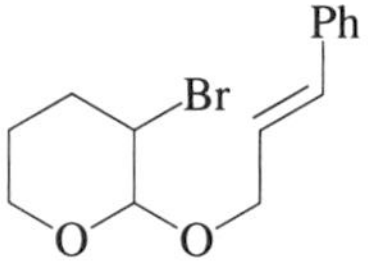	$CoI(dmgH)_2py$, Et_4NOTs, NaOH, MeOH, Pt cathode	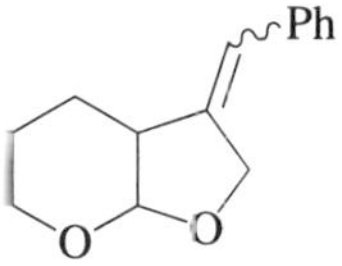 (81)	649
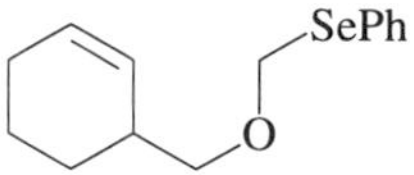	Bu_3SnH, AIBN, C_6H_6, 80°	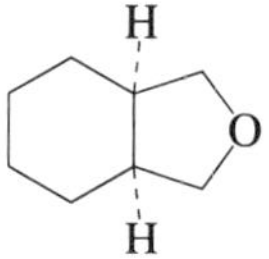 (80)	48, 497
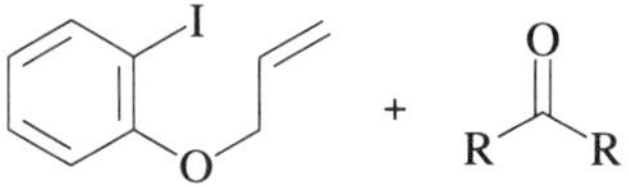	SmI_2, HMPA, THF, 25°	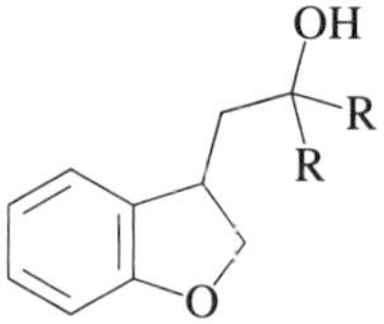	176
R			
Et		(69)	
n-C_3H_7		(81)	
$(CH_2)_4$		(68)	
$(CH_2)_5$		(65)	
$CH(Me)(CH_2)_4$		(70)	
$(CH_2)_2CH(Bu\text{-}t)(CH_2)_2$		(67)	

TABLE IV. BICYCLIC RINGS CONTAINING ONE OR MORE HETEROATOMS (*Continued*)

Reactant	Conditions	Product(s) and Yield(s) (%)	Refs.
	Bu₃SnH, AIBN, C₆H₆, 80°	(70) (68) (58)	754
R Cl CO₂Me OAc			
	1. (Bu₃Sn)₂, *hv* 2. ⟍⟍SnBu₃	(72)	593
	Ph₃SnH, Et₃B, air	(92)	593

TABLE IV. BICYCLIC RINGS CONTAINING ONE OR MORE HETEROATOMS (*Continued*)

Reactant	Conditions	Product(s) and Yield(s) (%)	Refs.
(2-iodophenyl)-N-Me cyclohexenecarboxamide	Co(I)(salen)	**I** + **II** (70) **I:II** = 74:26	741
(2-bromophenyl)-N-Me cyclohexenecarboxamide	Bu$_3$SnH, AIBN, C$_6$H$_6$, 80°	**I** + **II** (91) **I:II** = 74:26	719
CO$_2$Me, OTMS cyclohexene with propargyl chain (C≡CH)	Bu$_3$SnH, AIBN, PhMe, 110°	(85) β:α = 70:30	755, 756

671

TABLE IV. BICYCLIC RINGS CONTAINING ONE OR MORE HETEROATOMS (*Continued*)

Reactant	Conditions	Product(s) and Yield(s) (%)	Refs.
	Bu₃SnH, AIBN		661
	Bu₃SnH, AIBN, C₆H₆, 80°	(39) (24) (80)	757
	Ph₃SnH, AIBN	(90)	419
	Polymer-SnCl, AIBN, NaBH₄, C₆H₆, EtOH, *hv*	(73)	426

672

TABLE IV. BICYCLIC RINGS CONTAINING ONE OR MORE HETEROATOMS (*Continued*)

Reactant	Conditions	Product(s) and Yield(s) (%)	Refs.
C$_{15}$	Bu$_3$SnH, AIBN, C$_6$H$_6$, 80°	(91)	185
	ClCo(dmgH)$_2$py, EtOH, NaBH$_4$, NaOH, 50-60°	" (71)	185
	Bu$_3$SnH, AIBN, C$_6$H$_6$, 80°	(64)	443
	Bu$_3$SnH, AIBN, C$_6$H$_6$, 80°	(72)	758
	SmI$_2$, HMPA, THF, MeCN	(89)	175

TABLE IV. BICYCLIC RINGS CONTAINING ONE OR MORE HETEROATOMS (*Continued*)

Reactant	Conditions	Product(s) and Yield(s) (%)	Refs.
(2-iodophenyl)ethyl SeBn	(TMS)$_3$SiH, AIBN, Et$_3$N, C$_6$H$_6$, 80° Bu$_3$SnH, AIBN, C$_6$H$_6$, 80°	2,3-dihydrobenzo[b]selenophene (63) (77)	759
R^1, OH, SeBn, R^2 substituted (2-iodophenyl)	(TMS)$_3$SiH, AIBN, C$_6$H$_6$, 80°	benzoselenophene with R^1, R^2 R^1 R^2 H H (80) H Ph (86) Me H (82) Ph H (83)	141, 759
CH$_2$=C=C– with N–SePh pyrrolidinone	Bu$_3$SnH, AIBN, C$_6$H$_6$, 80°	bicyclic lactam (60) Δ7:Δ8 = 7:1	665
PhSe– ketone with dioxolane and butenyl	Bu$_3$SnH, AIBN, C$_6$H$_6$, 80°	spiro dioxolane cycloheptanone (12) + spiro dioxolane cyclohexanone (72)	386, 387

674

TABLE IV. BICYCLIC RINGS CONTAINING ONE OR MORE HETEROATOMS (*Continued*)

Reactant	Conditions	Product(s) and Yield(s) (%)	Refs.
	Bu₃SnH, AIBN, C₆H₆, 80°	**I + II** (70) **I:II** = 3.7:1	666
	SmI₂, CSA, MeCN	**I + II** (60) **I:II** = 1:1	664
	CuCl, CuCl₂, THF, AcOH, H₂O, -45°	**I + II** (66) **I:II** = 4.8:1	162
	Bu₃SnH, AIBN, C₆H₆, 80°	(64) (85) two diastereomers 3.7:1	760

675

TABLE IV. BICYCLIC RINGS CONTAINING ONE OR MORE HETEROATOMS (*Continued*)

Reactant	Conditions	Product(s) and Yield(s) (%)	Refs.
	Bu$_3$SnH, (Bu$_3$Sn)$_2$	 $\dfrac{\text{R}}{\begin{array}{ll}\text{H} & (70)\\ \text{I} & (80)\end{array}}$	710
	Bu$_3$SnH, AIBN, C$_6$H$_6$, 80°	(99)	710
	Bu$_3$SnH, AIBN, C$_6$H$_6$, 80°	 $\dfrac{\begin{array}{ll}\text{R}^1 & \text{R}^2\end{array}}{\begin{array}{ll}\text{Ac} & \text{OAc}\ (60)\\ \text{TBDMS} & \text{H}\ (80)\end{array}}$	710

TABLE IV. BICYCLIC RINGS CONTAINING ONE OR MORE HETEROATOMS (*Continued*)

Reactant	Conditions	Product(s) and Yield(s) (%)	Refs.

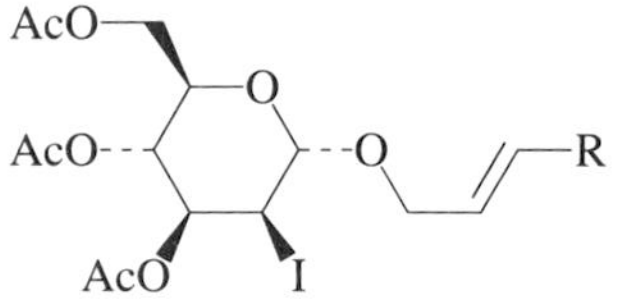

Bu₃SnH, AIBN, C₆H₆, 80°

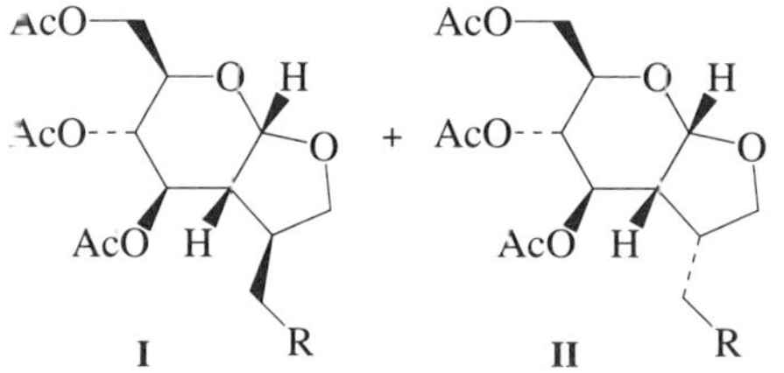

$$\frac{R}{H} \quad \frac{}{Ph}$$

	I + II	I:II
	(85)	53:47
	(80)	75:25

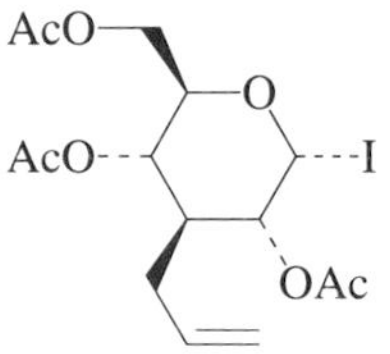

761

Bu₃SnH, AIBN, C₆H₆, 80°

I + II (53)   I:II = 9:1

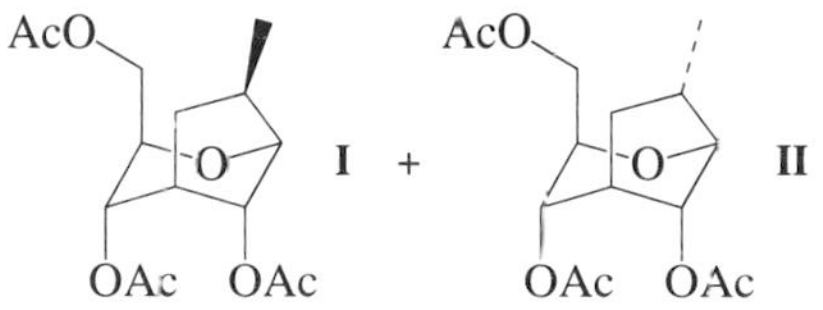

762

CuCl, MeCN, 110-140°
RuCl₂(PPh₃)₃,
C₆H₆, 110-140°

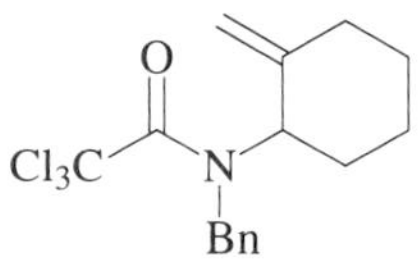

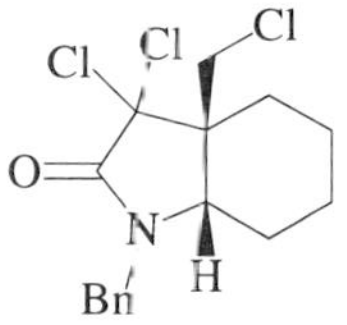

(85)
(88)

641

C₁₆

677

TABLE IV. BICYCLIC RINGS CONTAINING ONE OR MORE HETEROATOMS (*Continued*)

Reactant	Conditions	Product(s) and Yield(s) (%)	Refs.
	CuCl, MeCN, 110-140° RuCl$_2$(PPh$_3$)$_3$, C$_6$H$_6$, 110-140°	 (81) (89)	641
	Bu$_3$SnH, AIBN, C$_6$H$_6$, 80°	 $\begin{array}{cc|c} R & X & \\ \hline H & NH & (96) \\ Me & NH & (56) \\ H & O & (75) \end{array}$	763
	Bu$_3$SnH, Et$_3$B, THF, -78°	(95)	437
	Bu$_3$SnH, AIBN, C$_6$H$_6$, heat	(81) + (5)	313

TABLE IV. BICYCLIC RINGS CONTAINING ONE OR MORE HETEROATOMS (*Continued*)

Reactant	Conditions	Product(s) and Yield(s) (%)	Refs.
	Bu₃SnH, AIBN, C₆H₆, heat	(70) + (5)	313
	Bu₃SnH, AIBN, C₆H₆, 80°	(32)	331
	Bu₃SnH, AIBN, C₆H₆, 80°	I + II (72) I:II = 6:1	764
	Bu₃SnH, AIBN, C₆H₆, heat	(67)	482
	Bu₃SnH, AIBN, C₆H₆, heat	(10) + (19)	765, 766

679

TABLE IV. BICYCLIC RINGS CONTAINING ONE OR MORE HETEROATOMS (*Continued*)

Reactant	Conditions	Product(s) and Yield(s) (%)	Refs.
	Bu$_3$SnH, AIBN, C$_6$H$_6$, 80°	**I** (72) *cis:trans* = 1:4 + **II + III** (12)	489, 113
	Bu$_3$SnH, AIBN, C$_6$H$_6$, 80°	(65)	760
	Bu$_3$SnH, AIBN, C$_6$H$_6$, 80°	(25) + (32)	386, 392
	SmI$_2$, HMPA, THF, MeCN	(87)	175

680

TABLE IV. BICYCLIC RINGS CONTAINING ONE OR MORE HETEROATOMS (*Continued*)

Reactant	Conditions	Product(s) and Yield(s) (%)	Refs.
(AcO-substituted pyran with O–CH₂CH₂CH₂Br + CH₂=CHR¹) $$\begin{array}{ll} R^1 \\ \hline CO_2Me \\ " \\ CH_2SnBu_3 \\ " \end{array}$$	Bu₃SnH, AIBN, C₆H₆, 80° " AIBN, C₆H₆, 80° "	$$\begin{array}{ll} R^2 \\ \hline CH_2CH_2CO_2Me & (53) \\ \\ CH_2CH(CO_2Me)(CH_2)_2CO_2Me & (7) \\ CH_2CH=CH_2 & (56) \\ H & (27) \end{array}$$	749
	BPO, c-C₆H₁₂, 80°, 2 h	(55)	659
	Bu₃SnH, AIBN, C₆H₆, 80°	(66)	767
	Bu₃SnH, AIBN, C₆H₆, 80°	$$\begin{array}{ll} R \\ \hline H & (97) \\ Me & (95) \end{array}$$	768

TABLE IV. BICYCLIC RINGS CONTAINING ONE OR MORE HETEROATOMS (*Continued*)

Reactant	Conditions	Product(s) and Yield(s) (%)	Refs.

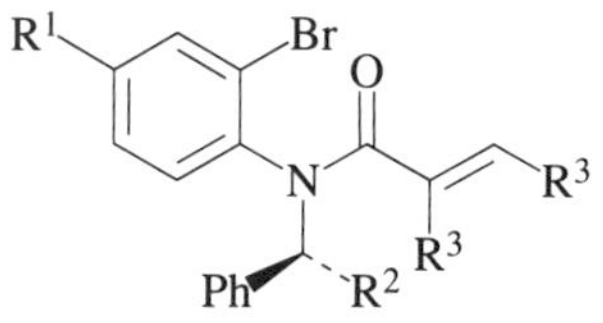

Bu$_3$SnH, AIBN, C$_6$H$_6$, 80°

R^1	R^2	R^3		% de
OMe	Me	H	(64)	2
OMe	Me	Me	(59)	39
H	OMe	H	(83)	7
H	OMe	Me	(79)	14

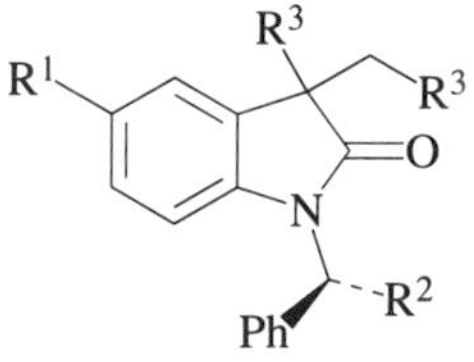

769

C$_17$

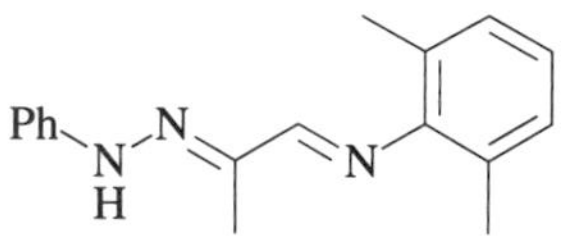

600°, 10^{-2} Torr

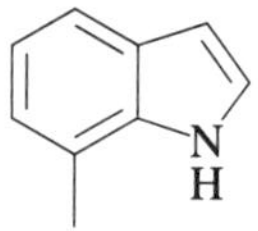

(62) + (11)

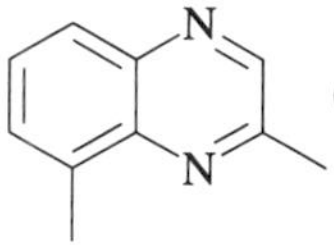

770, 771

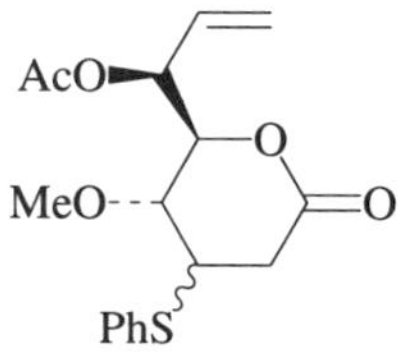

Bu$_3$SnH, AIBN, C$_6$H$_6$

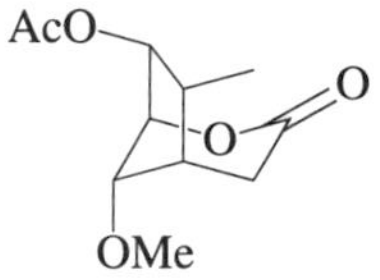

I + II

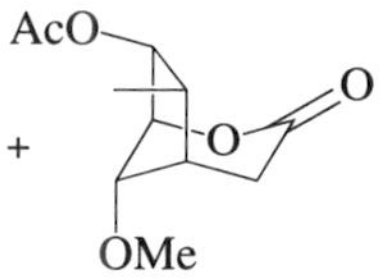

I + II (92) I:II = 4:1

764

TABLE IV. BICYCLIC RINGS CONTAINING ONE OR MORE HETEROATOMS (*Continued*)

Reactant	Conditions	Product(s) and Yield(s) (%)	Refs.
	Bu₃SnH, AIBN, C₆H₆, 80°	**I + II + III** (78) **I:II:III** = 2.5:1.2:1	764
	SmI₂, CSA, MeCN	(78)	664
	TsNa, AcOH, H₂O, 100°	**I + II** (78) **I:II** = 5:2	772, 136
	TsNa, AcOH, H₂O, 100°	**I + II** (83) **I:II** = 10:1	772, 136

TABLE IV. BICYCLIC RINGS CONTAINING ONE OR MORE HETEROATOMS (Continued)

Reactant	Conditions	Product(s) and Yield(s) (%)	Refs.
	Bu$_3$SnH, AIBN, C$_6$H$_6$, 80°	(38)	673
	Bu$_3$SnH, AIBN, C$_6$H$_6$, 80°	(30)	673
	Ph$_3$SnH, AIBN, C$_6$H$_6$, 80°	R — H (69), t-Bu (86)	352
	Bu$_3$SnH, AIBN, C$_6$H$_6$, 80°	(91) + (3)	766
	Bu$_3$SnH, AIBN, C$_6$H$_6$, 80°	(66)	768

684

TABLE IV. BICYCLIC RINGS CONTAINING ONE OR MORE HETEROATOMS (*Continued*)

Reactant	Conditions	Product(s) and Yield(s) (%)	Refs.
	Bu₃SnH, AIBN, C₆H₆, 80°	(68) *cis trans* = 53:47	479
	Bu₃SnH, AIBN, C₆H₆, 80°	I + II (95) I:II = 4.3:1	48, 497
	Bu₃SnH, AIBN, C₆H₆, 80°	(75)	773
	ClCo(dmgH)₂py, Et₄NOTs, NaOH, MeOH, Pt-cathode	(75)	649
	Bu₃SnH, AIBN, C₆H₆, 80°	(92)	774

685

TABLE IV. BICYCLIC RINGS CONTAINING ONE OR MORE HETEROATOMS (*Continued*)

Reactant	Conditions	Product(s) and Yield(s) (%)	Refs.
[structure: CO_2Me, OTBDMS, OHC, acetonide]	SmI_2, THF, MeOH, $-78°$	[structure with CO_2Me, OH, OTBDMS] (69)	775
R = TBDMS; [structure: CO_2Me, *E*, *Z*, OHC, RO, acetonide]	SmI_2, THF, MeOH, $-78°$	[structure CO_2Me, OH, RO — **I**] + [structure CO_2Me, OH, RO — **II**] **I + II** (64) **I:II** = 4:1 **I + II** (73) **I:II** = 1:100	775
[structure: SPh, $N=\!\!<R$]	DPDC, C_6H_6	[benzothiazole structure with R: S, N, R]	776

R

R	
Ph	(64)
4-ClC_6H_4	(64)
4-$MeOC_6H_4$	(55)
3-$O_2NC_6H_4$	(50)
2-thienyl	(30)
2-naphthyl	(50)

TABLE IV. BICYCLIC RINGS CONTAINING ONE OR MORE HETEROATOMS (*Continued*)

Reactant	Conditions	Product(s) and Yield(s) (%)	Refs.
C$_{18}$			
	Bu$_3$SnH, AIBN, C$_6$H$_6$, 80°	(95)	777
	SmI$_2$, HMPA, THF, 25°	 $\dfrac{R}{\text{Bn} \quad (27)}$ $\text{CH}_2\text{CH=CH}_2 \quad (38)$	176
	MeCN, Et$_4$NClO$_4$, Pt electrode	ClO$_4^-$ (72)	486
	Bu$_3$SnH, AIBN, C$_6$H$_6$, 80°	(65) 1:1 ratio	621
	Bu$_3$SnH, AIBN, C$_6$H$_6$, 80°	(65)	31

687

TABLE IV. BICYCLIC RINGS CONTAINING ONE OR MORE HETEROATOMS (*Continued*)

Reactant	Conditions	Product(s) and Yield(s) (%)	Refs.
α:β = 82:18	Bu₃SnH, AIBN, C₆H₆, 80°	**I + II** (66) **I:II** = 97:3	740
α:β = 86:14	Bu₃SnH, AIBN, C₆H₆, 80°	**I + II** (83) **I:II** = 88:12	740
R* =	Bu₃SnH, AIBN, PhMe, 110°	R % d.e. H — (0) Me 15 (65)	769
	Bu₃SnH, AIBN, C₆H₆, 80°	(35) + (22)	479

688

TABLE IV. BICYCLIC RINGS CONTAINING ONE OR MORE HETEROATOMS (Continued)

Reactant	Conditions	Product(s) and Yield(s) (%)	Refs.
	Bu₃SnH, AIBN, C₆H₆, 80°	**I + II** (95) **I:II** = 3:1	778, 779
	Vitamin B₁₂ (cat.), C felt cathode, LiClO₄, MeOH	**I – II** (>75) **I:II** = >95:<5	183, 779
	Me₃SnO ⌒ OSnMe₃, Ph Ph, Ph Ph, C₆H₆, 80°	**I + II** (85) **I:II** = >95:<5	779
R = TMS, R = Ph	Bu₃SnH, AIBN, C₆H₆, 80°	**I + II** (80) **I:II** = 1:12, **I + II** (90) **I:II** = 5:1	496

689

TABLE IV. BICYCLIC RINGS CONTAINING ONE OR MORE HETEROATOMS (*Continued*)

Reactant	Conditions	Product(s) and Yield(s) (%)	Refs.
	Bu₃SnH, AIBN, C₆H₆, 80°	(93)	780
	SmI₂, THF, MeOH, −78°	(65)	781
	Bu₃SnH, AIBN	(39) (23)	661
	Bu₃SnH, AIBN, C₆H₆, 80°	(>76)	782

TABLE IV. BICYCLIC RINGS CONTAINING ONE OR MORE HETEROATOMS (*Continued*)

Reactant	Conditions	Product(s) and Yield(s) (%)	Refs.

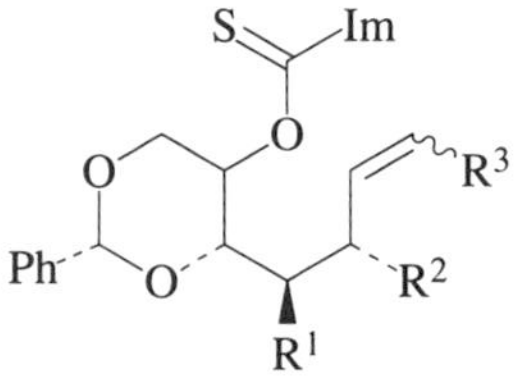

Bu₃SnH, AIBN

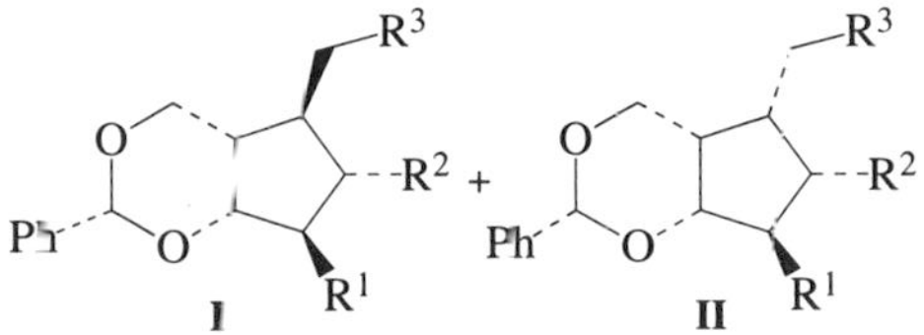

R¹	R²	R³
OBn	OBn	OMe
H	OBn	H
H	H	H
OBn	H	H
OBn	H	OMe

I + II	I:II
(62)	100:0
(67)	92:8
(55)	30:70
(32)	23:77
(51)	14:86

82

C₁₉

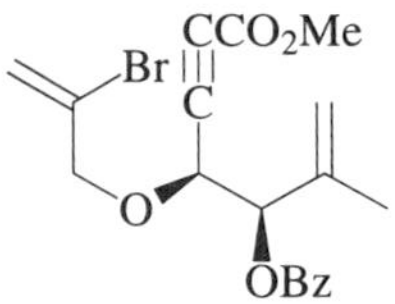

Bu₃SnH, AIBN, C₆H₆, 80°

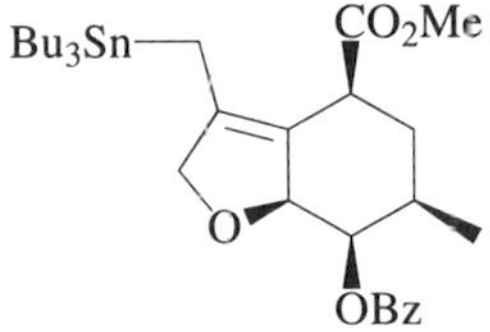

(43)

783

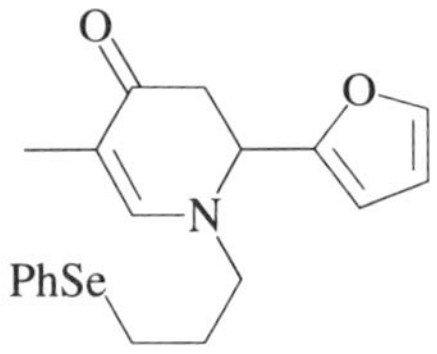

Ph₃SnH, AIBN, C₆H₆, heat

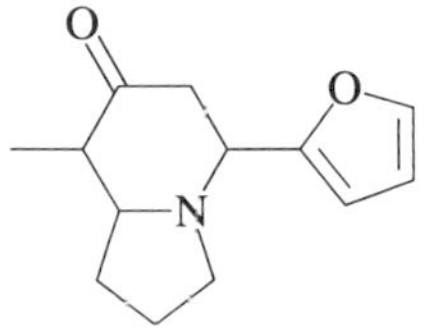

(83)

784

691

TABLE IV. BICYCLIC RINGS CONTAINING ONE OR MORE HETEROATOMS (*Continued*)

Reactant	Conditions	Product(s) and Yield(s) (%)	Refs.
(structure, *E* / *Z*)	Bu₃SnH, AIBN, C₆H₆, 80°	I + II (structures with Bu-*t*) **I + II** (95) **I:II** = 72:28 **I + II** (85) **I:II** = 90:10	785
(structure) R: Ph, *n*-Bu, *t*-Bu	Bu₃SnH, AIBN, C₆H₆, 80°	I + II (structures with R) **I + II** (85) **I:II** = 74:26 **I + II** (90) **I:II** = 71:29 **I + II** (91) **I:II** = 36:64	761
(structure)	AIBN, C₆H₆, 80° Bu₃SnH, ⟍⟋SnBu₃	(structure) R: H (82), CH₂CH=CH₂ (84)	739

692

TABLE IV. BICYCLIC RINGS CONTAINING ONE OR MORE HETEROATOMS (*Continued*)

Reactant	Conditions	Product(s) and Yield(s) (%)	Refs.
	Bu₃SnH, AIBN, C₆H₆, 80°	(70) 5:1 mixture	42
	Bu₃SnH, AIBN, C₆H₆, 80°	I + II (84) I:II = 3:4	786
	MeOH, *hv*	(73) 1:1 ratio	226
	MeOH, *hv*	(71)	226, 518, 736

TABLE IV. BICYCLIC RINGS CONTAINING ONE OR MORE HETEROATOMS (*Continued*)

Reactant	Conditions	Product(s) and Yield(s) (%)	Refs.
	CF$_3$CO$_2$H, t-BuSH	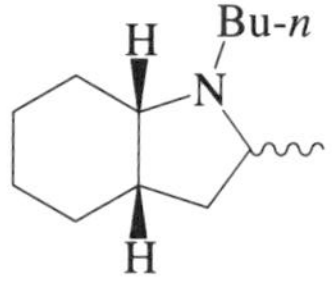 (60)	67
	A. Co(I)(salen), 20° B. Bu$_3$SnH, 20°	I (40) 15% de II + III (79), II 15% de	741

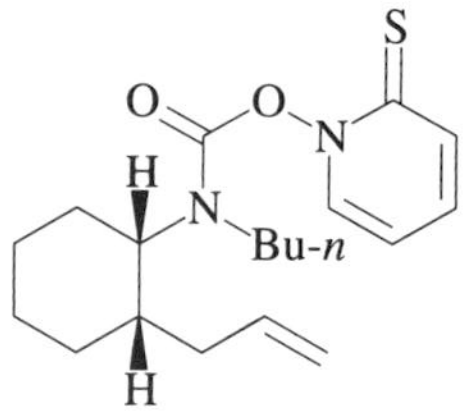

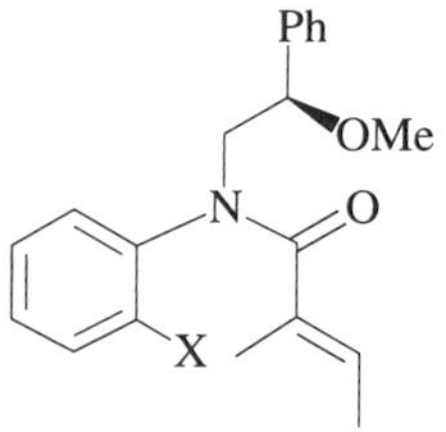

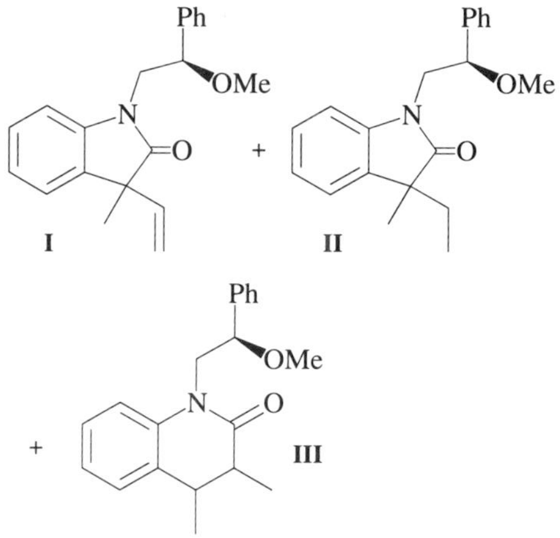

C$_{20}$

X = I   A

X = Br   B

694

TABLE IV. BICYCLIC RINGS CONTAINING ONE OR MORE HETEROATOMS (*Continued*)

Reactant	Conditions	Product(s) and Yield(s) (%)	Refs.

Row 1:

Reactant: (MEMO-substituted cyclohexene bromide) + $CH_2=CH-R$

R
CO_2Me
CN

Conditions: Bu_3SnCl, $Na(CN)BH_3$, AIBN, *t*-BuOH, 80°

Products **I** and **II**:

	I + II	**I:II**
	(56)	1:4
	(69)	1:2

Refs.: 787

Row 2:

Reactant: (oxazolidinone with SPh, Ph, R substituents)

Conditions: Bu_3SnH, AIBN, C_6H_6, 80°

Products:

R	
Me	(57)
$CH_2-CH=CH_2$	(73)

Refs.: 788

Row 3:

Reactant: (cyclohexene carboxamide, N-allyl, N-CH₂Ph, NH)

Conditions: 1. $Mn(OAc)_3$ 2. H_3O^+

Products **I** and **II**: **I** – **II**
I + II (70) **I:II** = 67:33

Refs.: 685

Row 4:

Reactant: (2-iodobenzamide, N-Bu-*t*, CO_2Et chain)

Conditions: Bu_3SnH, AIBN, C_6H_6, 80°

Product: (isoindolinone with CO_2Et, N-Bu-*t*) (38)

Refs.: 345

TABLE IV. BICYCLIC RINGS CONTAINING ONE OR MORE HETEROATOMS (*Continued*)

Reactant	Conditions	Product(s) and Yield(s) (%)	Refs.
	Bu$_3$SnH, AIBN, C$_6$H$_6$, 80°	(50)	217, 381
	Bu$_3$SnH, AIBN, C$_6$H$_6$, 80°	**I** + **II** (52) **I:II** = 97:3	217
C$_{21}$			
	Bu$_3$SnH, AIBN, C$_6$H$_6$, 80°	(60)	714
	Bu$_3$SnH, AIBN, C$_6$H$_6$, 80°	(75) + (15)	465

TABLE IV. BICYCLIC RINGS CONTAINING ONE OR MORE HETEROATOMS (*Continued*)

Reactant	Conditions	Product(s) and Yield(s) (%)	Refs.
	Bu₃SnH, AIBN, C₆H₆, 80°	(72)	551
	C₆H₆, heat, *hv*		789
	Heat, *hv*	(64)	212
	Bu₃SnH, AIBN, C₆H₆, 80°	(72)	386

For the second entry:

R	
H	(54)
Me	(45)
Et	(60)

697

TABLE IV. BICYCLIC RINGS CONTAINING ONE OR MORE HETEROATOMS (*Continued*)

Reactant	Conditions	Product(s) and Yield(s) (%)	Refs.
	Bu$_3$SnH, AIBN, C$_6$H$_6$, 80°	(81)	69
	Bu$_3$SnH, AIBN, C$_6$H$_6$, 80°	I + II (85) I:II = 75:25	785
	Bu$_3$SnH, AIBN, C$_6$H$_6$, 80°	I + II (86) I:II = 46:54	785
	THF, *hv*, rt	(49–64) (69)	150

698

TABLE IV. BICYCLIC RINGS CONTAINING ONE OR MORE HETEROATOMS (*Continued*)

Reactant	Conditions	Product(s) and Yield(s) (%)	Refs.

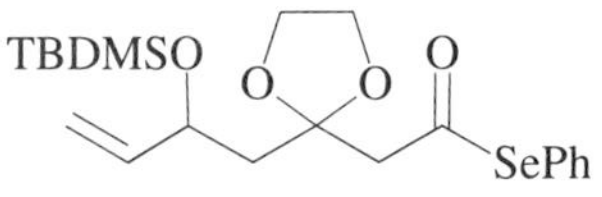

Bu₃SnH, AIBN, C₆H₆, 80°

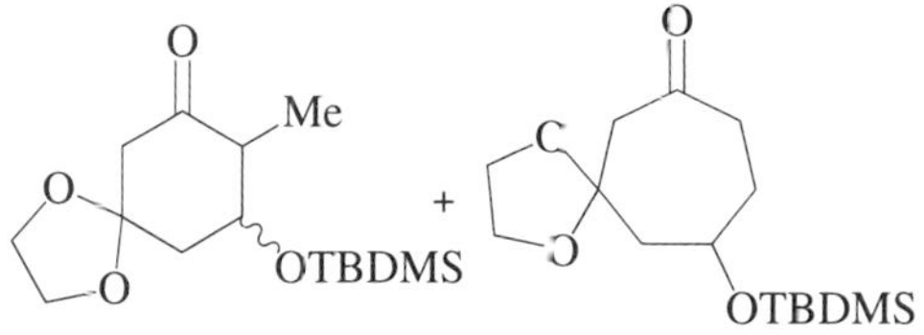

(72)   β:α = 2:1   (24)

387

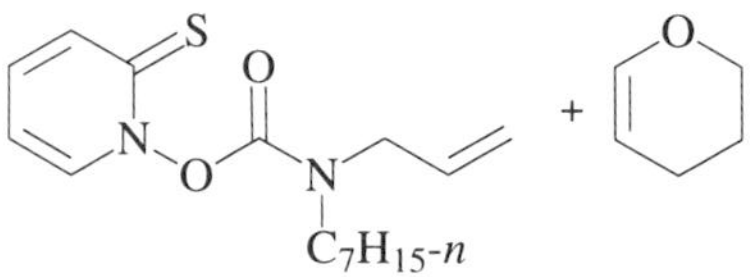

t-BuSH, H⁺, *hv*

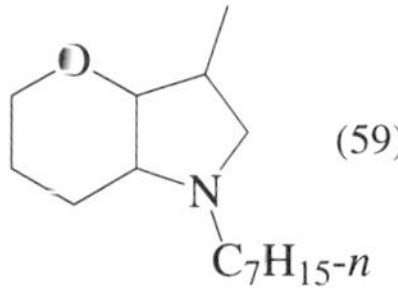

(59)

488

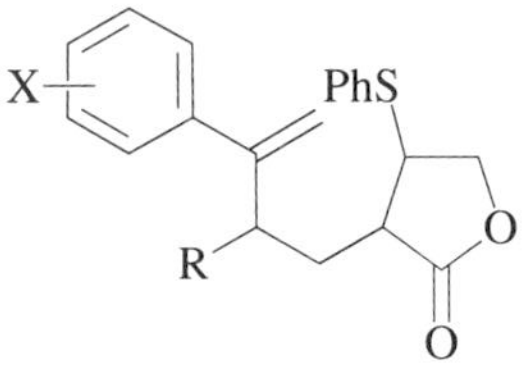

Bu₃SnH, AIBN, C₆H₆, 80°

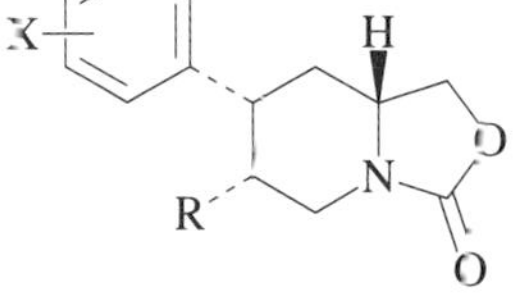

790

X	R	
p-OMe	Me	(57)
p-OMe	Et	(60)
p-OMe	Allyl	(60)
p-OMe	Bn	(62)
m-OMe	Me	(58)
m-OMe	Et	(63)

699

TABLE IV. BICYCLIC RINGS CONTAINING ONE OR MORE HETEROATOMS (*Continued*)

Reactant	Conditions	Product(s) and Yield(s) (%)	Refs.
(structure: 2-iodo-*N*-Bn cinnamamide, Ph, I, N—Bn, O)	Bu$_3$SnH, AIBN, C$_6$H$_6$, 80°	(structure: 3-benzyl-*N*-Bn oxindole, Ph—CH$_2$, O, N—Bn) (88)	791
(structure: SPh, Br, N—SO$_2$Ph)	Bu$_3$SnH, AIBN, C$_6$H$_6$, 80°	(structure: 3-vinyl indoline, N—SO$_2$Ph) (82)	758
(structure: SePh, cyclohexane ring, N, R, C≡CPh, CN)	Bu$_3$SnH, AIBN, C$_6$H$_6$, 80°	(structure: Ph, H, R, CN) $$\begin{array}{cc} \text{R} & \\ \hline \text{H} & (87) \\ \text{Me} & (85) \end{array}$$	768
(structure: glycoside, AcO, AcO, O, OAc, I, O—CH$_2$CH=CH—C$_6$H$_4$CN-*p*)	Bu$_3$SnH, AIBN, C$_6$H$_6$, 80°	(structures I and II) C$_6$H$_4$CN-*p*; AcO, AcO, AcO, O, H, H; **I** + **II** **I + II** (89) **I:II = 43:57**	785

TABLE IV. BICYCLIC RINGS CONTAINING ONE OR MORE HETEROATOMS (*Continued*)

Reactant	Conditions	Product(s) and Yield(s) (%)	Refs.

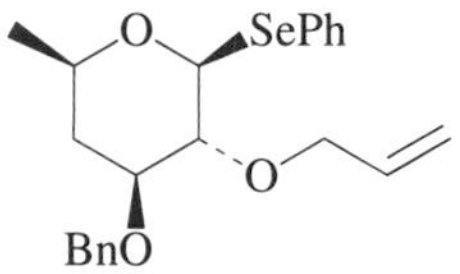

Bu₃SnH, AIBN

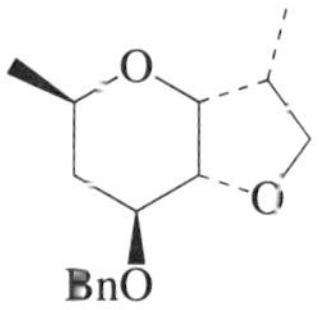

I + II

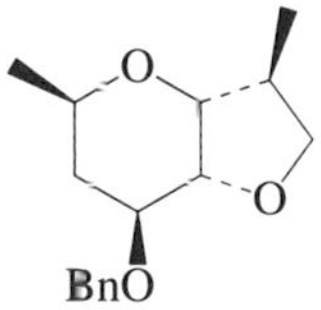

I + II (83) **I:II** = 2.3:1
I + II (88) **I:II** = 9:1

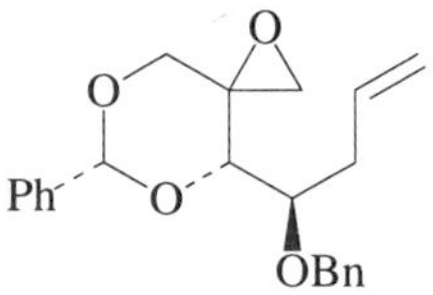

792

C₆H₆, 80°
PhMe, *hv*, −78°

1. Cp₂TiCl
2. H⁺

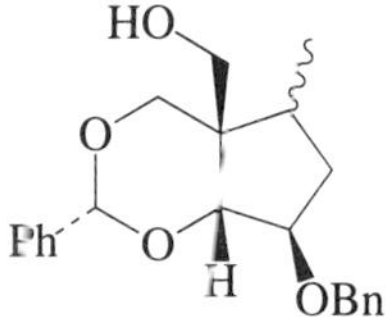

(70) *endo:exo* = 83:17

172

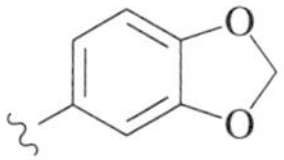

RuCl₂(PPh₃)₃,
C₆H₆, 150°,
sealed tube

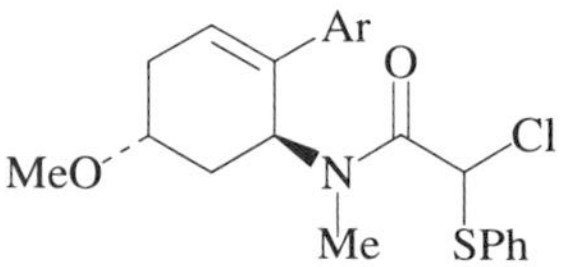

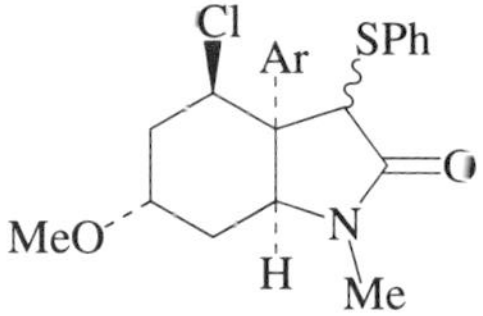

(57)

793

C₂₃

Ar =

TABLE IV. BICYCLIC RINGS CONTAINING ONE OR MORE HETEROATOMS (*Continued*)

Reactant	Conditions	Product(s) and Yield(s) (%)	Refs.

C$_{24}$

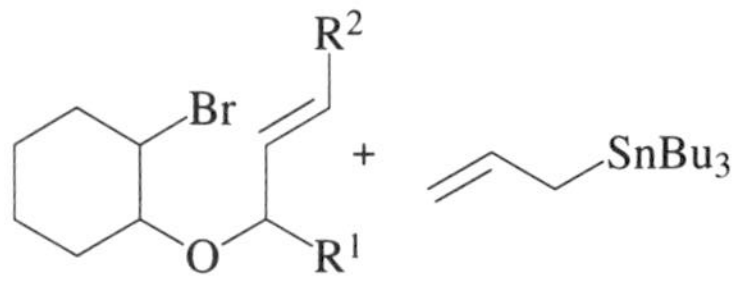

Bu$_3$SnH, AIBN, C$_6$H$_6$, 80°

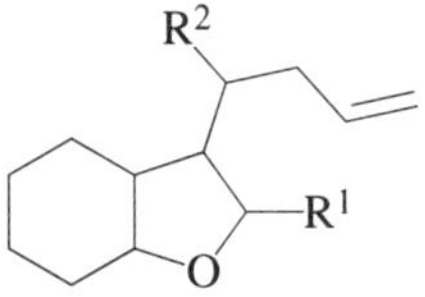

R^1	R^2	
H	H	(49)
Me	H	(48)
H	Me	(43)

431

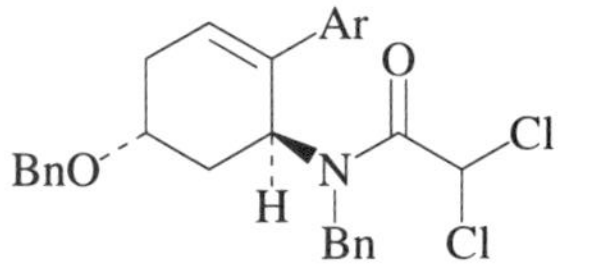

Ar = 3,4-(MeO)$_2$C$_6$H$_3$

Bu$_3$SnH, AIBN, C$_6$H$_6$, 80°

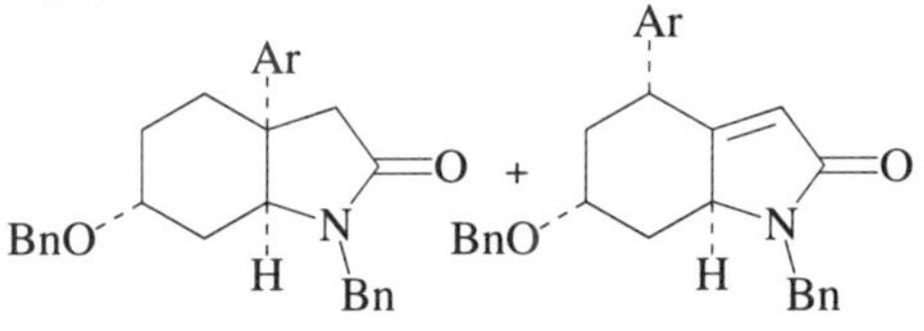

(51) + (30)

712

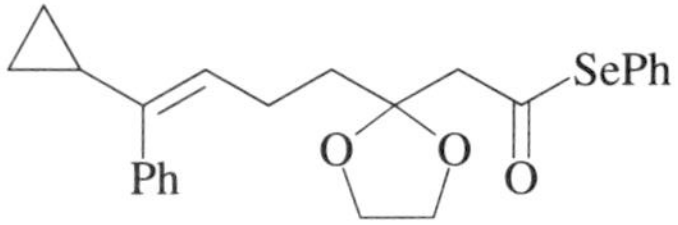

Bu$_3$SnH, AIBN, C$_6$H$_6$, reflux

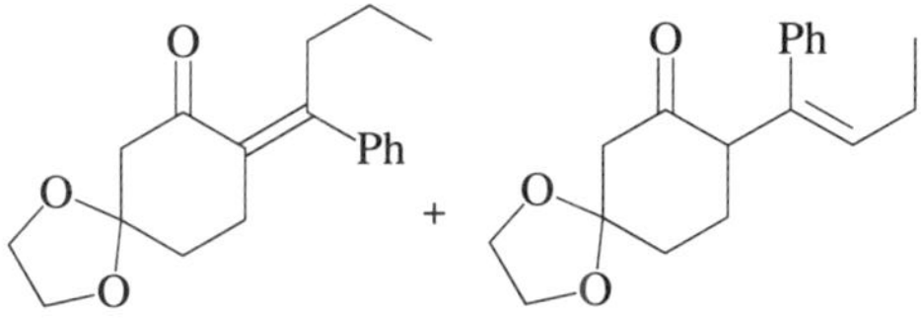

(56) + (30)

777

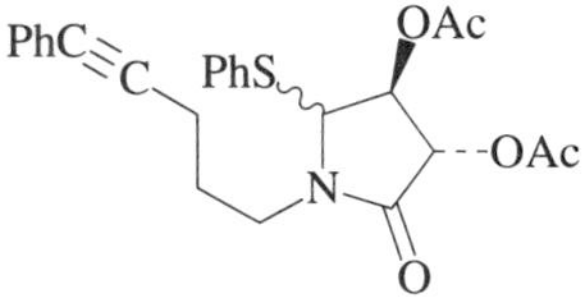

Bu$_3$SnH, AIBN, C$_6$H$_6$, 80°

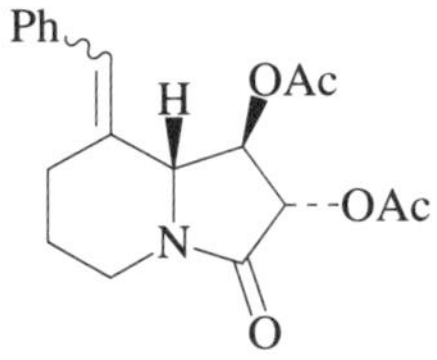

(81) 011*E:Z* = 1.2:1

794

TABLE IV. BICYCLIC RINGS CONTAINING ONE OR MORE HETEROATOMS (*Continued*)

Reactant	Conditions	Product(s) and Yield(s) (%)	Refs.

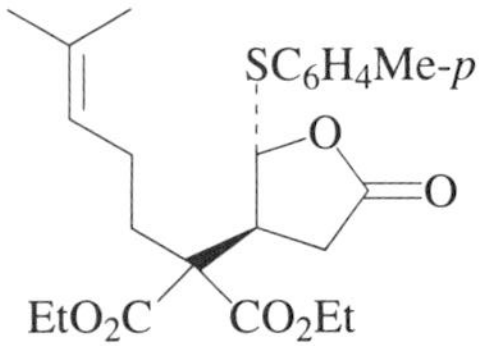

C$_{25}$

R^1	R^2	
Bz	Et	*E*
Bn	*t*-Bu	*E*
Bn	*t*-Bu	*Z*

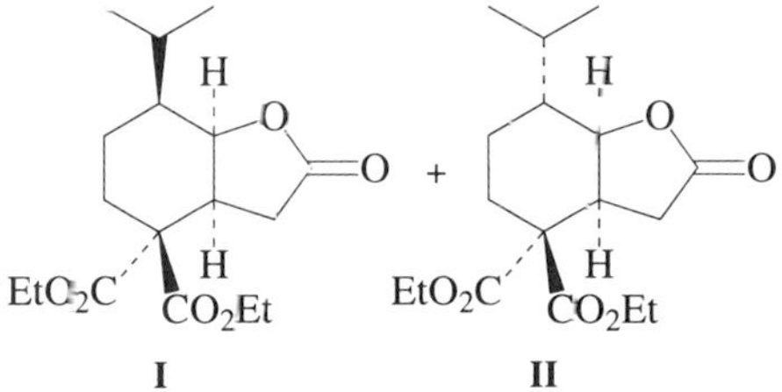

Bu$_3$SnH, AIBN, PhMe, 110°

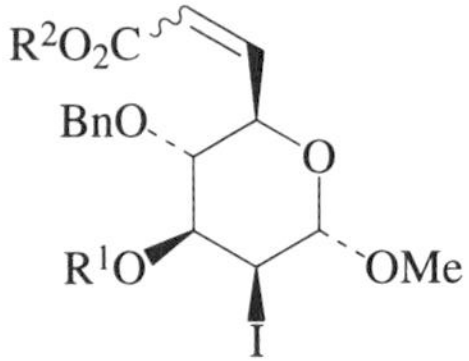

I   +   **II**

I + II	**I:II**
(91)	1.8:1
(80)	4.5:1
(80)	5.5:1

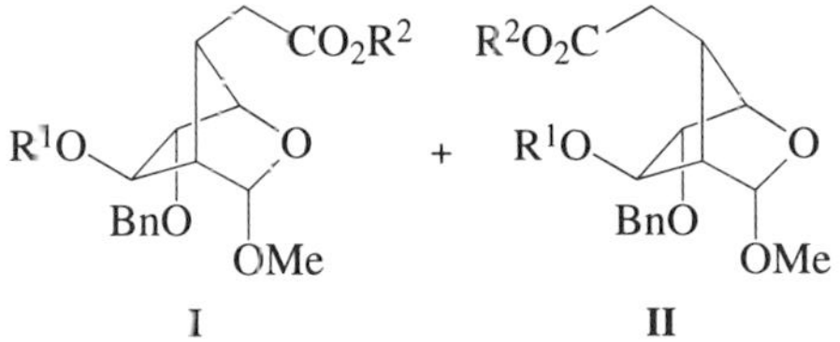

795, 796

Bu$_3$SnH, AIBN, PhMe, 110°

I   +   **II**

I + II (55)   **I:II** = 1:4.5

690

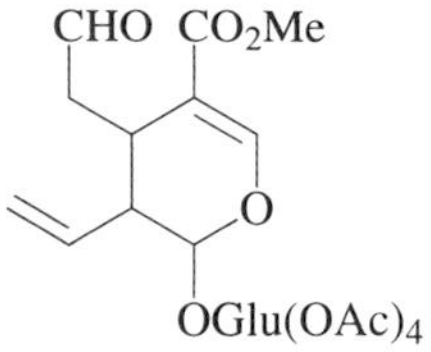

Mg, TMSCl, I$_2$ (cat.), THF, rt, 60 h

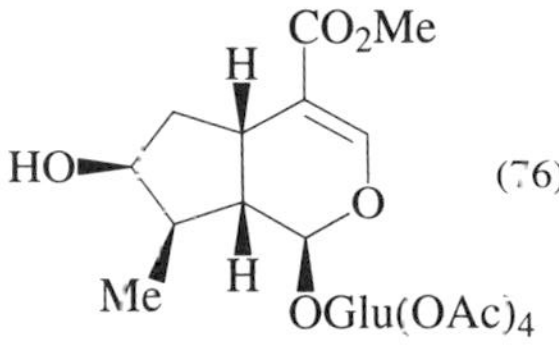

(76) + C-6 and C-7 epimers

609

703

TABLE IV. BICYCLIC RINGS CONTAINING ONE OR MORE HETEROATOMS (*Continued*)

Reactant	Conditions	Product(s) and Yield(s) (%)	Refs.
	Bu₃SnH, AIBN, PhMe, 110°	(100)	696
	Bu₃SnH, AIBN, C₆H₆, 80°	(43)	797
C₂₆	Bu₃SnH, AIBN, C₆H₆, 80°	(62-73)	798
	Bu₃SnH, AIBN, PhMe, heat	(58)	799

TABLE IV. BICYCLIC RINGS CONTAINING ONE OR MORE HETEROATOMS (*Continued*)

Reactant	Conditions	Product(s) and Yield(s) (%)	Refs.
	Bu₃SnH, AIBN, C₆H₆, 80°	(96) 79 : 21 ratio	785
	C₆H₆, *hv*	(85)	461
C₂₇ +	Bu₃SnH, AIBN, C₆H₆, 80°	R: CH₂CH(CO₂Bu-*t*)₂ (65); H (30)	69

705

TABLE IV. BICYCLIC RINGS CONTAINING ONE OR MORE HETEROATOMS (*Continued*)

Reactant	Conditions	Product(s) and Yield(s) (%)	Refs.
C$_{28}$	Bu$_3$SnH, AIBN, C$_6$H$_6$, 80°	**I** + **II** (20) **I:II** = 1:1	764
C$_{28}$	Bu$_3$SnH, AIBN, C$_6$H$_6$, 80°	(92) (91)	800, 801
C$_{29}$	C$_6$H$_6$, *hv*	(72)	597
C$_{30}$	Bu$_3$SnH, AIBN, C$_6$H$_6$, 80°	(90) *Z:E* = 6.5:1	750

TABLE IV. BICYCLIC RINGS CONTAINING ONE OR MORE HETEROATOMS (*Continued*)

Reactant	Conditions	Product(s) and Yield(s) (%)	Refs.
	Bu₃SnH, AIBN, C₆H₆, 80°	**I** + **II** (80) **I:II** = 1:1	785
	Bu₃SnH, AIBN, C₆H₆, 80°	(>71)	782
C₃₁	**A**. Bu₃SnH, AIBN, C₆H₆, reflux **B**. Bu₃SnCl, Na(CN)BH₃, AIBN, *t*-BuOH	**A** (85) **B** (78)	738

707

TABLE IV. BICYCLIC RINGS CONTAINING ONE OR MORE HETEROATOMS (*Continued*)

Reactant	Conditions	Product(s) and Yield(s) (%)	Refs.
C32	Bu₃SnH, AIBN, PhMe, heat	(50)	360, 82
C33	Bu₃SnH, AIBN, C₆H₆, 80°	OEt (60) *Z:E* = 16:1	750
	Bu₃SnH, AIBN, C₆H₆, 80°	OEt (84)	738
C35	AIBN, C₆H₆, 80°	(84)	749

708

TABLE IV. BICYCLIC RINGS CONTAINING ONE OR MORE HETEROATOMS (*Continued*)

Reactant	Conditions	Product(s) and Yield(s) (%)	Refs.

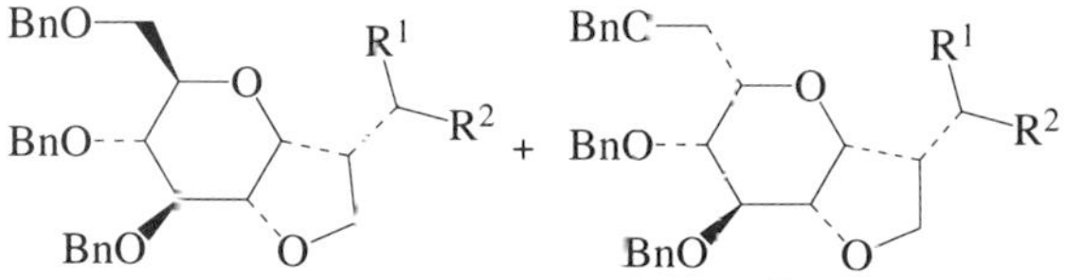

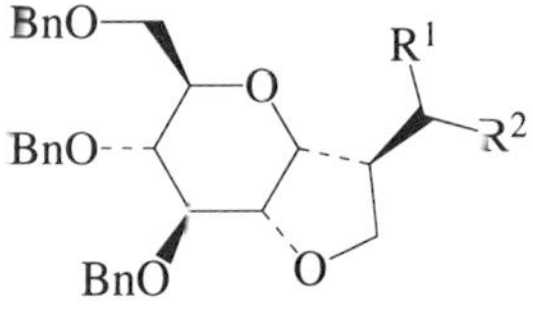

C_{36}

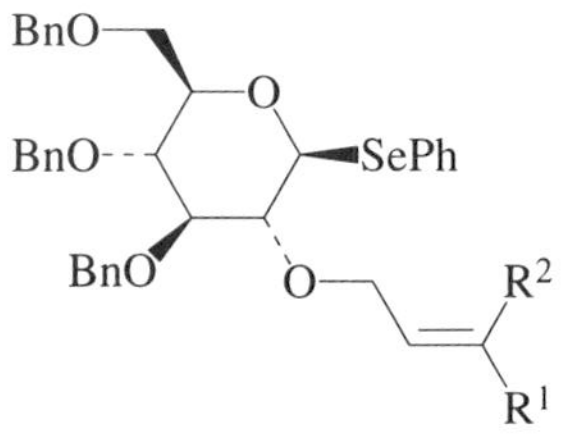

R^1	R^2
H	H
H	H
H	Me
Me	H
Ph	H

Bu₃SnH, AIBN

	I + II + III	I:II:III
C₆H₆, 80°, 0.01 M	(90)	29:15:56
C₆H₆, 80°, 0.3 M	(95)	48:–:52
PhMe, *hv*, −78°	(88)	55:–:45
C₆H₆, 80°	(90)	46:–:54
C₆H₆, 80°	(89)	29:–:71

Refs: 802,802a / 792 / 792 / 802,802a / 802,802a

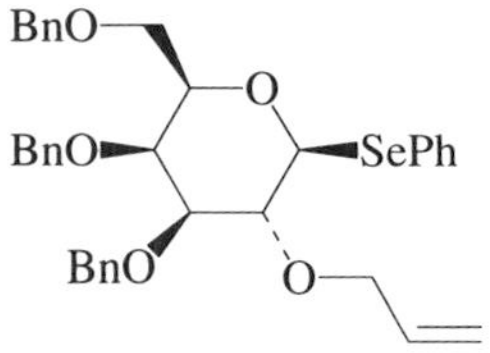

Bu₃SnH, AIBN

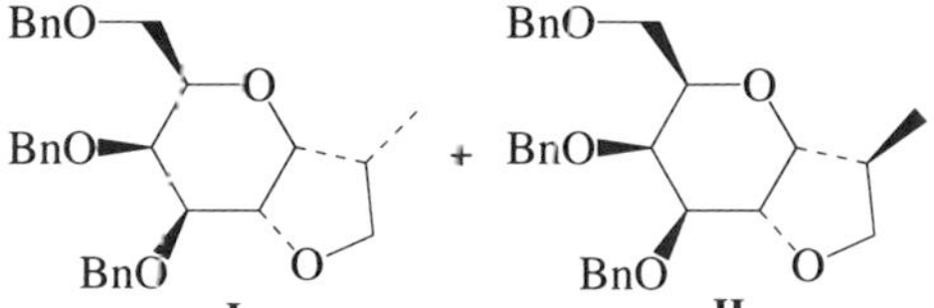

I + III

| C₆H₆, 80° | I + II (86) | I:II = 6.1:1 |
| PhMe, *hv*, −78° | I + II (95) | I:II = 99:1 |

792

TABLE IV. BICYCLIC RINGS CONTAINING ONE OR MORE HETEROATOMS (Continued)

Reactant	Conditions	Product(s) and Yield(s) (%)	Refs.

C_{37}

First reactant (structure with SePh, BnO groups, R substituents):

R
H
H
Me

Conditions / Products (I + II):

Bu$_3$SnH, AIBN

Conditions	Products	Refs.
C$_6$H$_6$, 80°	**I + II** (92) **I:II** = 10.1:1	792
PhMe, $h\nu$, -30°	**I + II** (92) **I:II** = 32.3:1	792
C$_6$H$_6$, 80°	**I + II** (90) **I:II** = 62:38	802,802a

Second reactant (SePh, BnO, OMe allyl structure):

Bu$_3$SnH, AIBN, C$_6$H$_6$, 80°

I, II + **III**

Refs: 802, 802a

	I (*anti*) : **II** (*syn*) : **III**
0.01 mol/L	30 : 48 : 22
0.1 mol/L	45 : 50 : 5
0.3 mol/L	50 : 50 : —

TABLE IV. BICYCLIC RINGS CONTAINING ONE OR MORE HETEROATOMS (Continued)

Reactant	Conditions	Product(s) and Yield(s) (%)	Refs.
	Bu$_3$SnH, AIBN, C$_6$H$_6$, 80°	(77)	803
	HgO, I$_2$, hv	(53)	61
C$_{39}$	Bu$_3$SnH, AIBN, C$_6$H$_6$, 80°	I + II (95) I:II = 82:18	802, 802a

TABLE IV. BICYCLIC RINGS CONTAINING ONE OR MORE HETEROATOMS (*Continued*)

Reactant	Conditions	Product(s) and Yield(s) (%)	Refs.

C$_{43}$

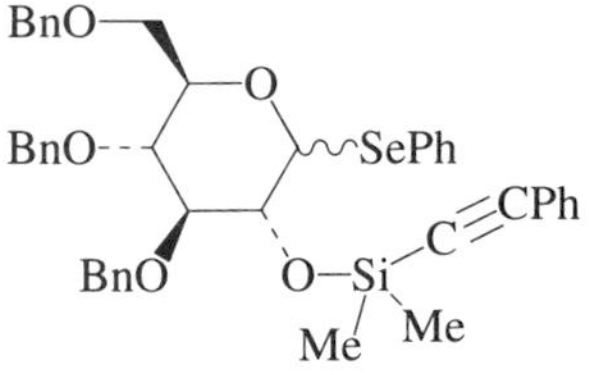

Bu$_3$SnH, AIBN,
C$_6$H$_6$, 80°

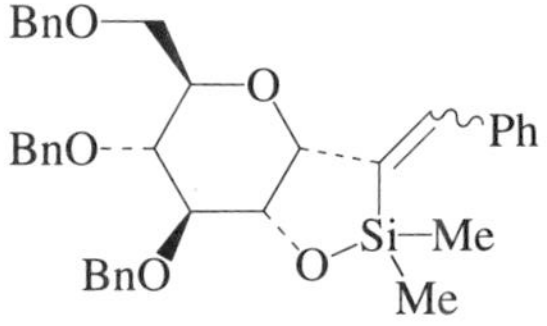

(>85)

696

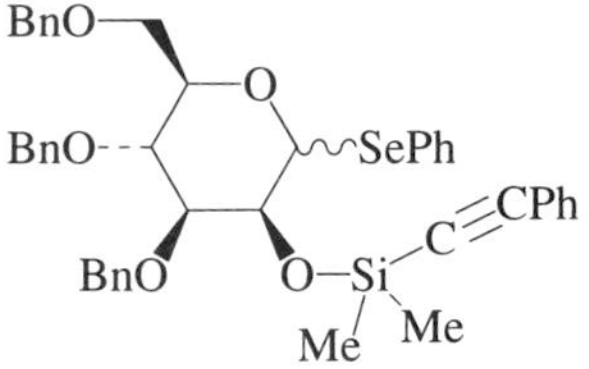

Bu$_3$SnH, AIBN,
C$_6$H$_6$, 80°

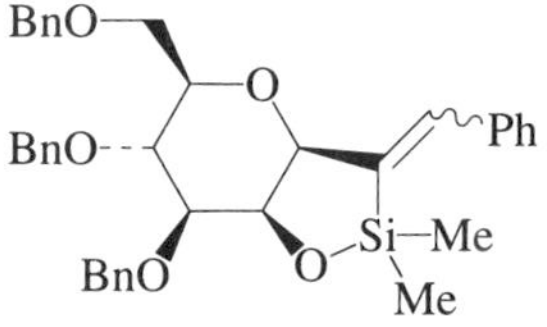

(>69)

696

G. (5+(n+5))-Membered Rings

C$_8$

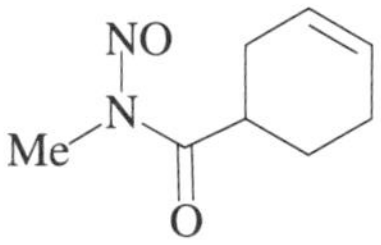

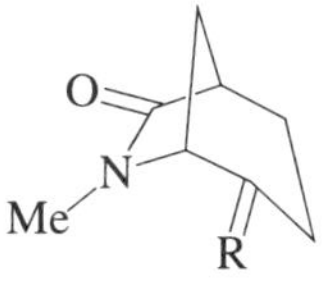

C$_6$H$_6$, *hv*
C$_6$H$_6$, O$_2$, *hv*

R	
NOH	
H, ONO$_2$	

164

TABLE IV. BICYCLIC RINGS CONTAINING ONE OR MORE HETEROATOMS (*Continued*)

Reactant	Conditions	Product(s) and Yield(s) (%)	Refs.
	C_6H_6, $h\nu$	(72)	164
	MeOH, HCl, $h\nu$	**I, II, III** |	R	
---	---	---
I	O	(11)
II	(*E*)-NOH	(37)
III	(*Z*)-NOH	(13)		413
	$(t\text{-BuO})_2$, O_2	(29)	804
C_9	Bu$_3$SnH, AIBN, C_6H_6, 80°	(31) + (53)	805

TABLE IV. BICYCLIC RINGS CONTAINING ONE OR MORE HETEROATOMS (*Continued*)

Reactant	Conditions	Product(s) and Yield(s) (%)	Refs.
C$_{10}$	C$_6$H$_6$, *hv*	(49)	164
C$_{11}$	Bu$_3$SnH, AIBN, C$_6$H$_6$, 80°	(88–92)	421
	Bu$_3$SnH, AIBN, C$_6$H$_6$, 80°	(25) + (25) + (23)	805
C$_{12}$	(Me$_3$Sn)$_2$, *hv*	(71)	18

714

TABLE IV. BICYCLIC RINGS CONTAINING ONE OR MORE HETEROATOMS (*Continued*)

Reactant	Conditions	Product(s) and Yield(s) (%)	Refs.

C_{14}

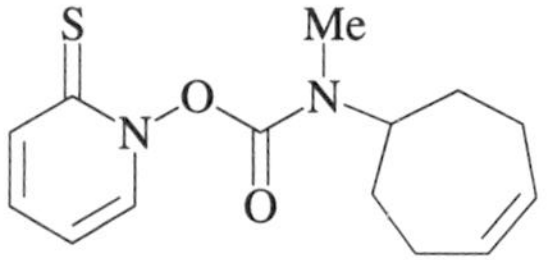

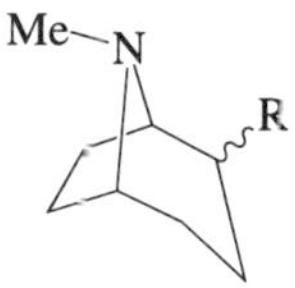

AcOH, MeCN, 25°

R		
$-SC_5H_4N$	(94)	
$-SC_5H_4N$	(92)	
H	(67)	
Br	(60)	
SePh	(82)	

AcOH, MeCN, 25°

AcOH, C_6H_6, 4°

AcOH, C_6H_6, *t*-BuSH, 4°

AcOH, C_6H_6, CBr_4, 4°

AcOH, C_6H_6, $(PhSe)_2$, 4°

152
153
153
503
503

C_{15}

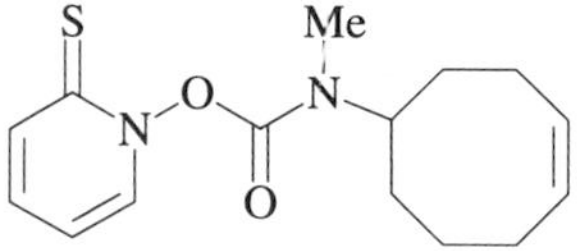

AcOH, MeCN, 25°

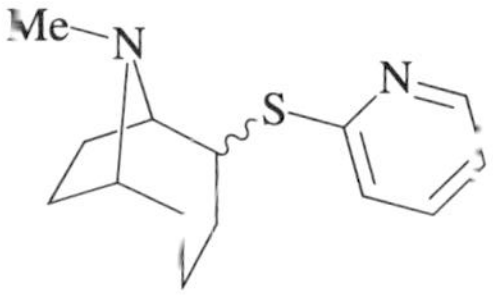

(50)

152

TABLE IV. BICYCLIC RINGS CONTAINING ONE OR MORE HETEROATOMS (*Continued*)

Reactant	Conditions	Product(s) and Yield(s) (%)	Refs.

C_{16}

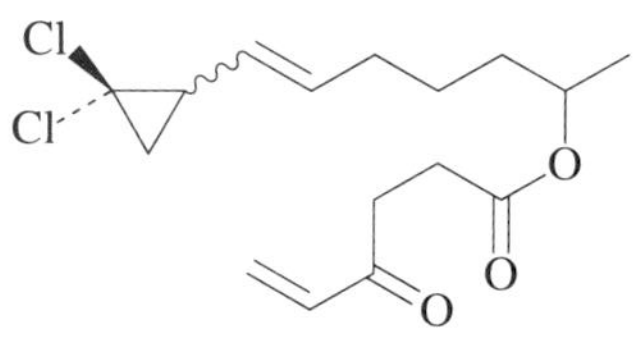

(PhS)$_2$, AIBN, heat

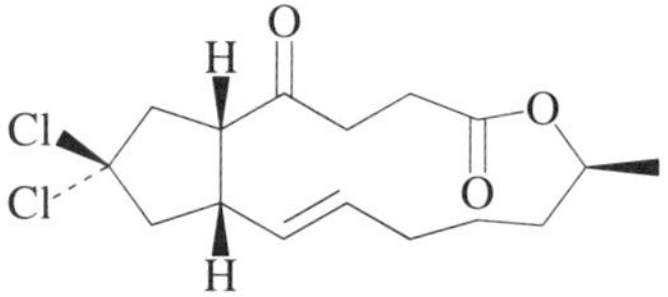

I +

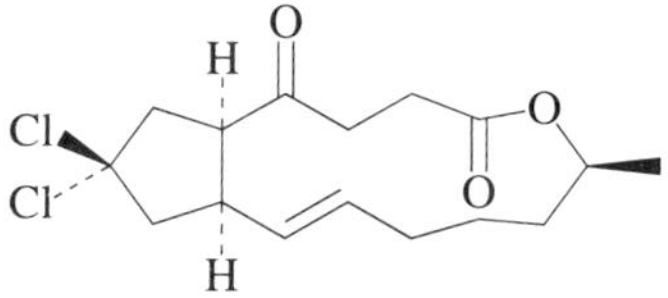

II

	I + II	**I:II**
	(61)	2:1
	(56)	1:2

C_6H_6
c-C_6H_{12}

806

C_{19}

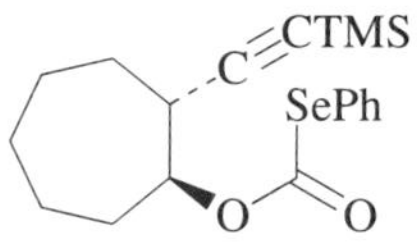

Bu$_3$SnH, AIBN,
C_6H_6, 80°

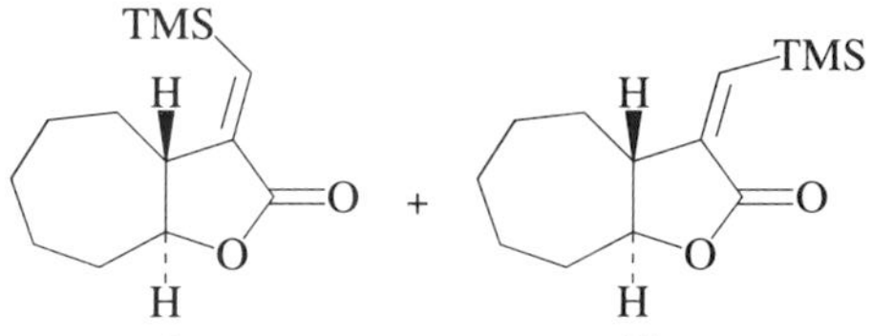

I +   **II**

I + II (92) **I:II** = 1:10

496

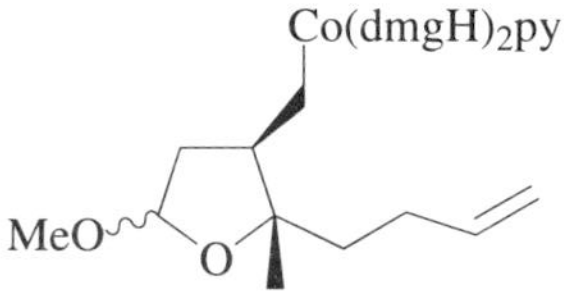

Co(dmgH)$_2$py

C_6H_6, hv

(58)

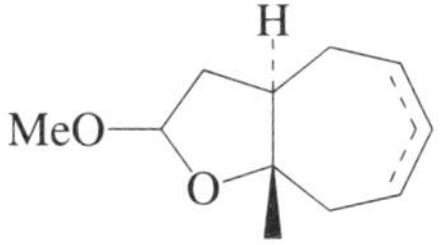

461

716

TABLE IV. BICYCLIC RINGS CONTAINING ONE OR MORE HETEROATOMS (*Continued*)

Reactant	Conditions	Product(s) and Yield(s) (%)	Refs.

H. (6+(n+6))-Membered Rings

C₉

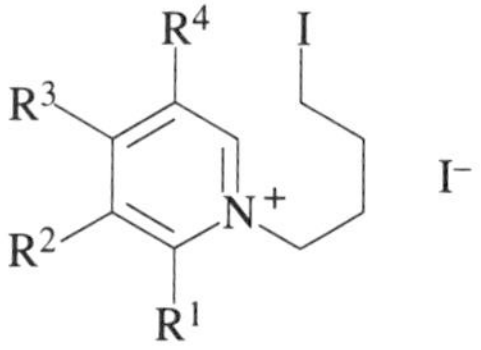

Bu₃SnH, AIBN, heat

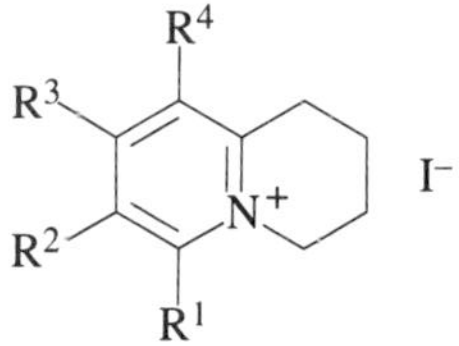

R^1	R^2	R^3	R^4	
H	H	H	H	(60)
Me	H	H	H	(58)
H	H	Me	H	(58)
H	H	H	Me	(67)

699

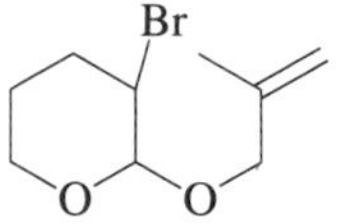

Ph₂SnH₂, BPO, C₈H₁₈, 110°

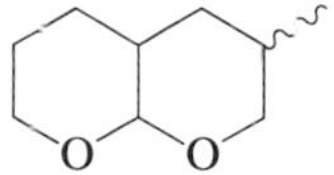

(62)

272

C₁₀

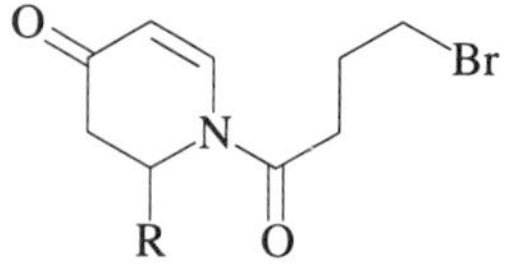

Bu₃SnH, AIBN, C₆H₆, 80°

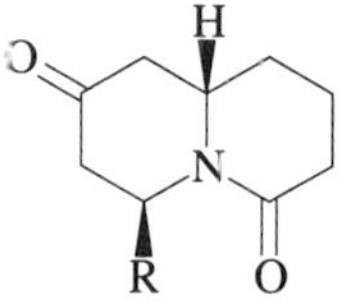

R	
Me	(91)
Ph	(89)
3,4-(MeO)₂C₆H₃	(91)

757

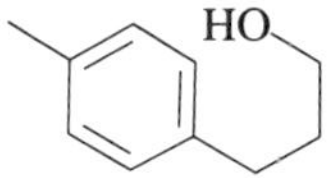

FeS₂O₈

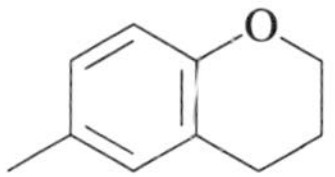

I + **II**

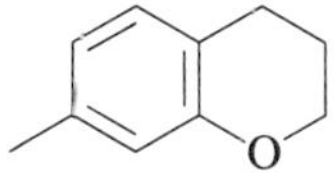

pH	**I**	**II**
6.0	(5)	(42)
5.4	(12)	(46)
5.1	(17)	(45)
3.5	(12)	(22)

807

TABLE IV. BICYCLIC RINGS CONTAINING ONE OR MORE HETEROATOMS (*Continued*)

Reactant	Conditions	Product(s) and Yield(s) (%)	Refs.
	NaI, Me$_2$CO, 25°	(65)	707
	1. Ph$_3$GeH, AIBN 2. H$^+$	(85)	54
	Et$_3$N, MeCN, *hv*	(53)	181
	Bu$_3$SnH, AIBN, PhMe, heat		808

C$_{11}$

R^1	R^2	
Me	H	(49)
Me	Me	(79)
H	Me	(100)
H	CO$_2$Me	(100)

TABLE IV. BICYCLIC RINGS CONTAINING ONE OR MORE HETEROATOMS (*Continued*)

Reactant	Conditions	Product(s) and Yield(s) (%)	Refs.
	Bu₃SnCl, Na(CN)BH₃, AIBN, t-BuOH, 80°		809

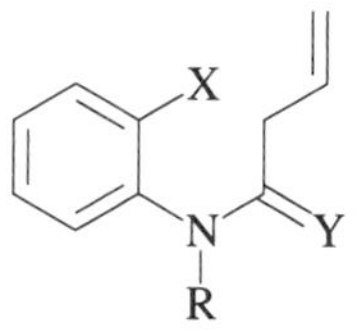

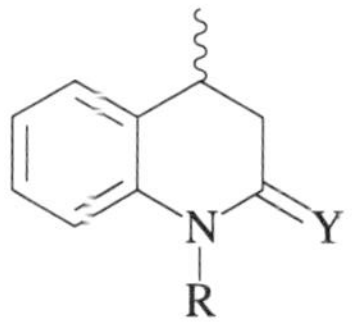

$$\begin{array}{ccc} R & X & Y \\ \hline Me & I & O \\ Bn & Br & O \\ Ac & Br & H_2 \end{array}$$

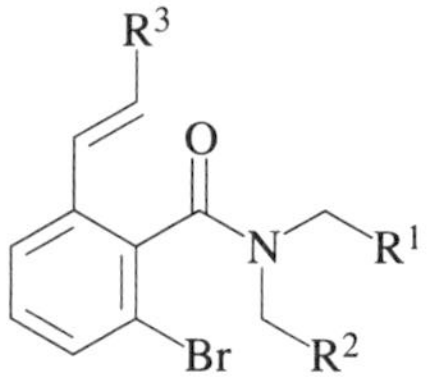

(64)
(58)
(70)

| | Bu₃SnH, AIBN, C₆H₆, 80° | | 345 |

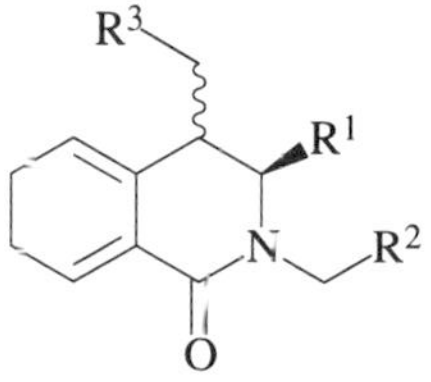

$$\begin{array}{ccc} R^1 & R^2 & R^3 \\ \hline Me & Me & H \\ Me & Me & Ph \\ Me & Me & CO_2Et \\ -(CH_2)_2- & & H \\ -(CH_2)_3- & & H \\ H & H & H \end{array}$$

(67)
(60)
(60)
(45)
(45)
(36)

719

TABLE IV. BICYCLIC RINGS CONTAINING ONE OR MORE HETEROATOMS (*Continued*)

Reactant	Conditions	Product(s) and Yield(s) (%)	Refs.
	TsNa, Cu(OAc)$_2$, AcOH, 90°	(73) (68)	280
	Bu$_3$SnH, AIBN, C$_6$H$_6$, 80°	(43)	723
	Bu$_3$SnH, AIBN, C$_6$H$_6$, 80°	(60) (70) (78)	810
	NaI, Me$_2$CO, 25°	(57)	707

TABLE IV. BICYCLIC RINGS CONTAINING ONE OR MORE HETEROATOMS (*Continued*)

Reactant	Conditions	Product(s) and Yield(s) (%)	Refs.
C₁₃	FeCl₃, PhC≡CH		811

R¹	R²	
H	Ph	(46)
MeO	Ph	(55)
NO₂	Ph	(61)
H	Ph	(76)
MeO	Ph	(70)
NO₂	Ph	(86)
H	p-O₂NC₆H₄	(79)

Reactant	Conditions	Product(s) and Yield(s) (%)	Refs.
	1,4-Dicyano-naphthalene, *i*-PrOH, *hv*	(88)	655
	Bu₃SnCl, Na(CN)BH₃, AIBN, *t*-BuOH, sealed tube, 11 0°, 8 h	(80)	744
	MeOH, *hv*	(91) 1.2:1 mixture	226, 518

TABLE IV. BICYCLIC RINGS CONTAINING ONE OR MORE HETEROATOMS (*Continued*)

Reactant	Conditions	Product(s) and Yield(s) (%)	Refs.
C$_{14}$			
	Ph$_3$SnH, AIBN, C$_6$H$_6$, heat	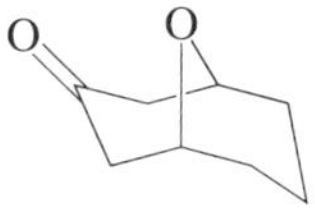 (99)	112
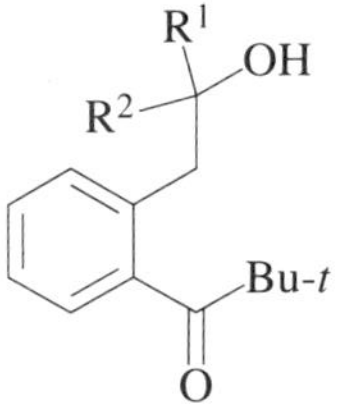		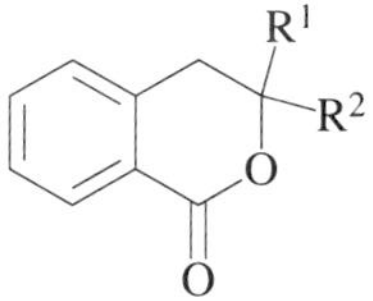	812
		R^1 / R^2	
	CHCl$_3$, hv	Me / H (62)	
	CHCl$_3$, hv	p-MeC$_6$H$_4$ / H (46)	
	CHCl$_3$, hv	Me / Me (64)	
	CHCl$_3$, hv	—(CH$_2$)$_5$— (42)	
	HgO, I$_2$, C$_6$H$_6$, hv	Ph / Ph (64)	
	MeCN, H$_2$O, hv		813

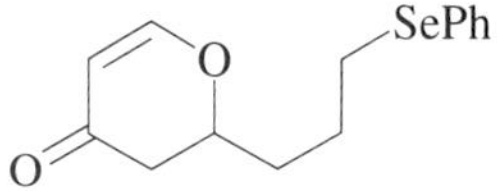

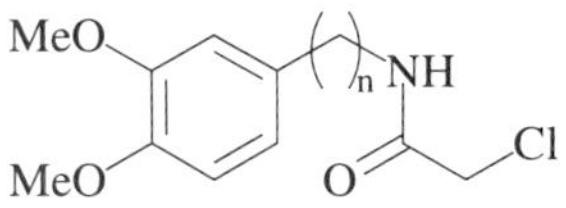

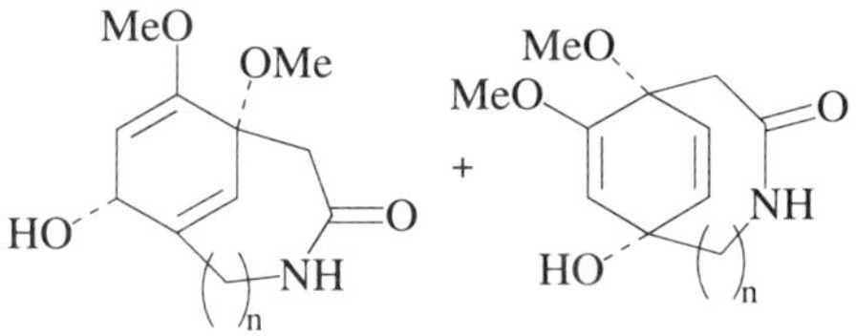

I / II
(15) (6)
(23) (20)

n
4
5

TABLE IV. BICYCLIC RINGS CONTAINING ONE OR MORE HETEROATOMS (*Continued*)

Reactant	Conditions	Product(s) and Yield(s) (%)	Refs.
C$_{15}$	Bu$_3$SnH, AIBN, C$_6$H$_6$, 80°, 8 h	(>84)	59
C$_{16}$	Bu$_3$SnH, AIBN, C$_6$H$_6$, 80°, 8 h	X / O (52) / S (66)	814
	1. hv, 8° 2. (PhSe)$_2$, hv, 8°	(61)	815
	Bu$_3$SnH, AIBN, C$_6$H$_6$, 80°, 8 h	R^1 R^2 / H H (90) / H Me (87) / Me H (79) / H Ph (91)	367

TABLE IV. BICYCLIC RINGS CONTAINING ONE OR MORE HETEROATOMS (*Continued*)

Reactant	Conditions	Product(s) and Yield(s) (%)	Refs.

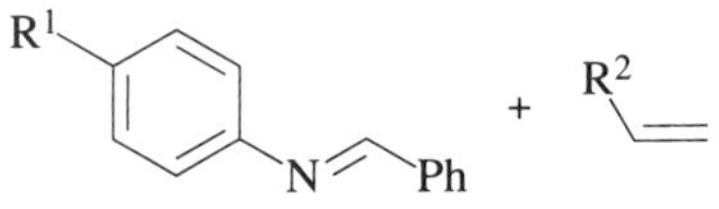

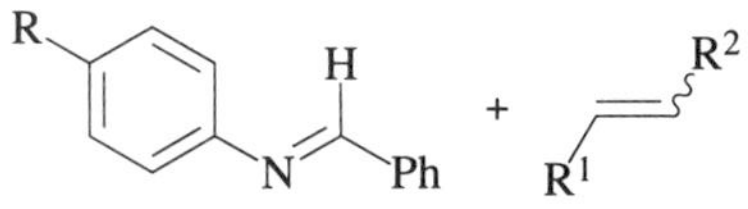

DPDC, C_6H_6, 60°

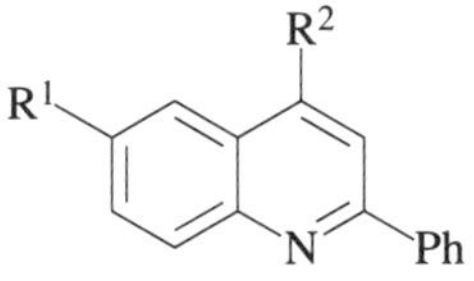

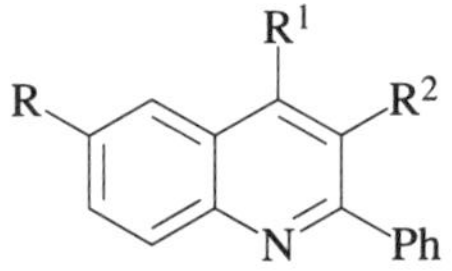

R^1	R^2	
H	CN	(63)
H	Ph	(42)
H	CO_2Me	(41)
Cl	CN	(34)
OMe	CN	(57)
OMe	Ph	(42)
OMe	CO_2Me	(44)
Me	CN	(55)

816

C_{17}

DPDC, C_6H_6, 60°

R	R^1	R^2	
H	CO_2Me	CO_2Me	(37)
H	CN	CN	(38)
OMe	Ph	Ph	(20)
OMe	CN	CN	(42)
H	Ph	Ph	(40)

816

TABLE IV. BICYCLIC RINGS CONTAINING ONE OR MORE HETEROATOMS (*Continued*)

Reactant	Conditions	Product(s) and Yield(s) (%)	Refs.
(imine) + $R^1C{\equiv}CH$	DPDC, C_6H_6, 60°	(quinoline with R^1, R)	817

R	R^1	
Ph	Ph	(75)
m-$O_2NC_6H_4$	Ph	(65)
p-ClC_6H_4	Ph	(65)
p-$MeOC_6H_4$	Ph	(75)
Ph	COMe	(70)
p-ClC_6H_4	COMe	(65)
p-$MeOC_6H_4$	COMe	(85)
Ph	CO_2Et	(75)
p-ClC_6H_4	CO_2Et	(75)
p-$MeOC_6H_4$	CO_2Et	(80)

Reactant	Conditions	Product(s) and Yield(s) (%)	Refs.
(oxime ester of cinnamaldehyde)	H_2O, heat	(quinoline with R)	818

R	
Ph	(91)
c-MeC_6H_4	(75)

Reactant	Conditions	Product(s) and Yield(s) (%)	Refs.
(allyl bromide ester substrate)	Bu_3SnH, AIBN, C_6H_6, 80°, 8 h	(macrocyclic lactone) (61)	127

725

TABLE IV. BICYCLIC RINGS CONTAINING ONE OR MORE HETEROATOMS (*Continued*)

Reactant	Conditions	Product(s) and Yield(s) (%)	Refs.
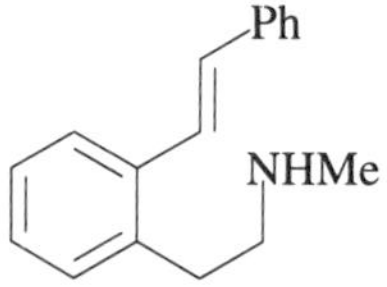	MeCN, $h\nu$	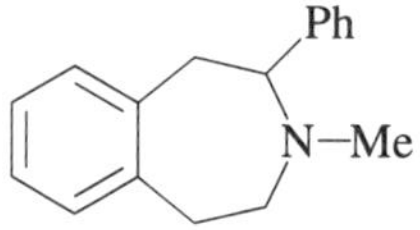 (65)	819
C_{18} 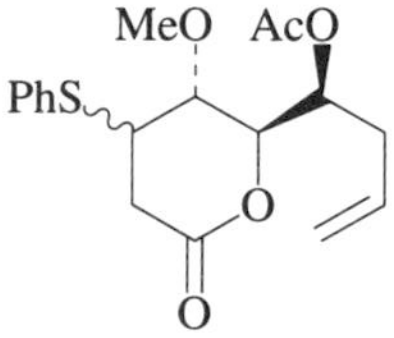	Bu_3SnH, AIBN, C_6H_6, 80°, 8 h	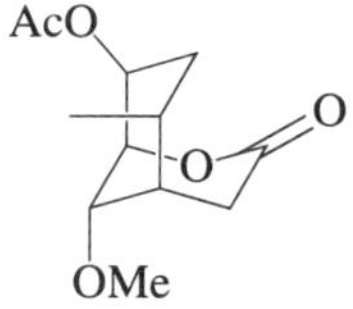 **I** + 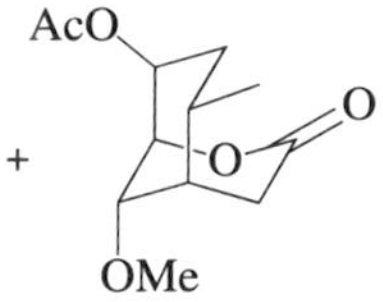 **II** **I** + **II** (83) **I**:**II** = 1:12.5	764
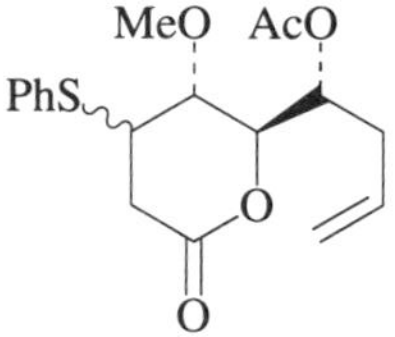	Bu_3SnH, AIBN, C_6H_6, 80°, 8 h	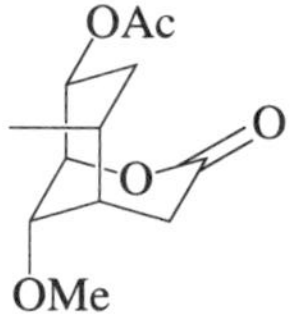 **I** + 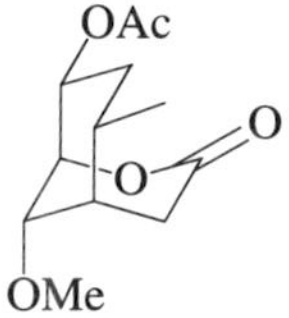 **II** **I** + **II** (71) **I**:**II** = 11:3	764
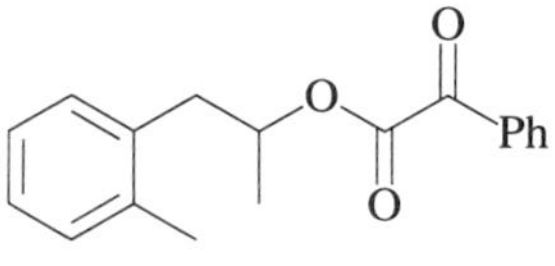	$h\nu$	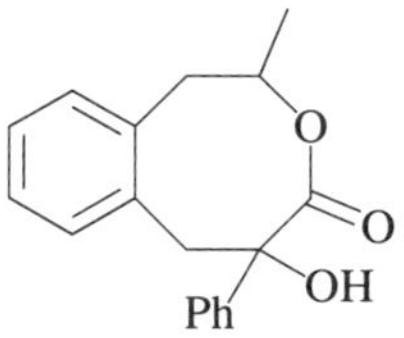 (74)	528

TABLE IV. BICYCLIC RINGS CONTAINING ONE OR MORE HETEROATOMS (*Continued*)

Reactant	Conditions	Product(s) and Yield(s) (%)	Refs.
	DPDC, C_6H_6, 60°		820

X	R^1	R^2	
CH	Ph	H	(45)
CH	Ph	OMe	(72)
CH	p-MeOC$_6$H$_4$	H	(47)
CH	p-MeOC$_6$H$_4$	OMe	(73)
CH	t-Bu	OMe	(57)
N	Ph	OMe	(47)
N	Ph	OMe	(47)

	Bu$_3$SnH, AIBN, PhMe, 110°		821

R^1	R^2	**I**	**II**
\-OCH$_2$O\-		(56)	(10)
OMe	OMe	(51)	(9)

TABLE IV. BICYCLIC RINGS CONTAINING ONE OR MORE HETEROATOMS (*Continued*)

Reactant	Conditions	Product(s) and Yield(s) (%)	Refs.
C₁₉			
	HgO, EtOH, 78°	R — H (51), OMe (89)	822
	Bu₃SnH, AIBN, C₆H₆, 80°	(69)	33
	Bu₃SnH, AIBN, C₆H₆, 80°	(61) *cis:trans* = 7:1 + (29)	489

728

TABLE IV. BICYCLIC RINGS CONTAINING ONE OR MORE HETEROATOMS (Continued)

Reactant	Conditions	Product(s) and Yield(s) (%)	Refs.
C_20			
	DCA, MeCN, hv	(89)	226
	(TMS)$_3$SiH, AIBN, PhMe, 85°, 10 h	(55)	127, 823
	Bu$_3$SnH, AIBN, C$_6$H$_6$, 80°	 R R^1 Ac CH$_2$OMe (67) Boc H (55) Ac CO$_2$Et (25)	824
C_21			
	Bu$_3$SnH, AIBN, C$_6$H$_6$, 80°	(73; 3:1)	551

729

TABLE IV. BICYCLIC RINGS CONTAINING ONE OR MORE HETEROATOMS (*Continued*)

Reactant	Conditions	Product(s) and Yield(s) (%)	Refs.
C$_{25}$	Bu$_3$SnH, AIBN, C$_6$H$_6$, 80°	(>73)	696
C$_{34}$	Bu$_3$SnH, AIBN, C$_6$H$_6$, 80°	**I** 7:1 mixture **II** 5:1 mixture **I + II** (100)	795, 796
C$_{39}$	Bu$_3$SnH, AIBN, C$_6$H$_6$, 80°	**I** 5:1 mixture **II** 3:1 mixture $\dfrac{\textbf{I + II}}{(82)}$ (83) *S* *R*	795, 796

730

TABLE V. OLIGOCYCLIC RINGS CONTAINING ONLY CARBON

Reactant	Conditions	Product(s) and Yield(s) (%)	Refs.
	A. Tricyclic Systems		
C$_9$	Bu$_3$SnH, AIBN, C$_6$H$_6$, hv	(60)	366
	Bu$_3$SnH, AIBN, C$_6$H$_6$, 65°	(98)	825
C$_{10}$	Bu$_3$SnH, AIBN, C$_6$H$_6$, 80°	(85)	535
	Bu$_3$SnH, AIBN, C$_6$H$_6$, 80°	(74)	601
C$_{11}$	Bu$_3$SnH, AIBN, C$_6$H$_6$, 80°	(45)	826

TABLE V. OLIGOCYCLIC RINGS CONTAINING ONLY CARBON (*Continued*)

Reactant	Conditions	Product(s) and Yield(s) (%)	Refs.
	Bu₃SnH, AIBN, C₆H₆, 80°	(75)	827
	Bu₃SnH, AIBN, C₆H₆, 80°	(40)	826
	Bu₃SnH, AIBN, C₆H₆, 80°	(61)	535
	Bu₃SnH, AIBN, C₆H₆, 80°	(88)	546

C₁₂

TABLE V. OLIGOCYCLIC RINGS CONTAINING ONLY CARBON (Continued)

Reactant	Conditions	Product(s) and Yield(s) (%)	Refs.
	Bu₃SnH, AIBN, C₆H₆, 80°	(51) ratio 87:13	535
	Bu₃SnH, AIBN, C₆H₆, 80°	(77) + (10)	107
	Bu₃SnH, AIBN, C₆H₆, 80°	**I** + **II** **I + II** (63) **I:II = 4:1**	828
C₁₃	Bu₃SnH, AIBN, C₆H₆, 80°	(84) 4 isomers: 6.3:5:3.3:1	294

733

TABLE V. OLIGOCYCLIC RINGS CONTAINING ONLY CARBON (*Continued*)

Reactant	Conditions	Product(s) and Yield(s) (%)	Refs.
	Bu$_3$SnH, AIBN, C$_6$H$_6$, 80°	(85)	546
	K$_3$Fe(CN)$_6$, KOH, CH$_2$Cl$_2$	(70)	829
	Bu$_3$SnCl, Na(CN)BH$_3$, AIBN, *t*-BuOH, 85°	(77)	830
C$_{14}$	Et$_4$NOTs, *i*-PrOH, Sn cathode, rt	(74)	610

734

TABLE V. OLIGOCYCLIC RINGS CONTAINING ONLY CARBON (*Continued*)

Reactant	Conditions	Product(s) and Yield(s) (%)	Refs.
	Bu₃SnH, C₆H₆, PhCO₃Bu-*t*, *hν*, 36°	 **I + II** (62) **I:II** = 3:2	831
	Mn(OAc)₃•H₂O, AcOH, 35°, 15 h	(79)	196
	Bu₃SnH, AIBN, C₆H₆, 80°	 **I + II** (95) **I:II** = 9:1	832
	Et₄NOTs, *i*-PrOH, Sn cathode, rt	(45)	610

735

TABLE V. OLIGOCYCLIC RINGS CONTAINING ONLY CARBON (*Continued*)

Reactant	Conditions	Product(s) and Yield(s) (%)	Refs.
	Bu₃SnH, AIBN, C₆H₆, 80°	(70)	27
	Bu₃SnH, AIBN, C₆H₆, 80°	(80)	833
	K₃Fe(CN)₆, KOH, CH₂Cl₂	(66)	829
	Bu₃SnH, AIBN, C₆H₆, 80°	(53)	834

736

TABLE V. OLIGOCYCLIC RINGS CONTAINING CNLY CARBON (*Continued*)

Reactant	Conditions	Product(s) and Yield(s) (%)	Refs.
	MeCN, Et$_3$N, *hv*	(58)	835, 836
	Bu$_3$SnH, AIBN, C$_6$H$_6$, 80°	(85) *anti:syn* = 4:1	537
	Bu$_3$SnH, AIBN, C$_6$H$_6$, 80°	**I** + **II** **I + II** (81) **I:II** = 1.5:1	291

TABLE V. OLIGOCYCLIC RINGS CONTAINING ONLY CARBON (*Continued*)

Reactant	Conditions	Product(s) and Yield(s) (%)	Refs.
	PhI(OAc)$_2$, I$_2$, *hv*	(58)	837
	Bu$_3$SnH, AIBN, C$_6$H$_6$, 80°	(52)	838
	Mn(OAc)$_3$•H$_2$O, AcOH, 35°, 17 h	(81)	196
C$_{15}$ + TsCN	AIBN	(72) + (6)	839
	Bu$_3$SnH, AIBN, C$_6$H$_6$, 65°	(85)	567

TABLE V. OLIGOCYCLIC RINGS CONTAINING ONLY CARBON (*Continued*)

Reactant	Conditions	Product(s) and Yield(s) (%)	Refs.
	Bu₃SnH, AIBN, C₆H₆, 80°	(55-65)	840
		R¹ R²	
		H H	
		Me Me	
		H OMe	
		Me OMe	
	Bu₃SnH, AIBN, C₆H₆, 80°	(86)	19
	Bu₃SnH, AIBN, C₆H₆, 80°	(97)	570
	Bu₃SnH, AIBN, C₆H₆, 80°	(92)	570

TABLE V. OLIGOCYCLIC RINGS CONTAINING ONLY CARBON (*Continued*)

Reactant	Conditions	Product(s) and Yield(s) (%)	Refs.
	(PhS)$_2$, AIBN, C$_6$H$_6$, *hv*, 80°	(60)	541
	Bu$_3$SnH, AIBN, C$_6$H$_6$, 80°	(61)	841
	Bu$_3$SnH, AIBN, C$_6$H$_6$, 80°	(64)	107
	Bu$_3$SnH, AIBN, C$_6$H$_6$, 80°	(73)	559
	(TMS)$_3$SiH, AIBN, PhMe, 90°	(72)	559
	(Me$_3$Sn)$_2$, C$_6$H$_6$, 80°	(74) *E:Z* = 6:1	159

TABLE V. OLIGOCYCLIC RINGS CONTAINING ONLY CARBON (*Continued*)

Reactant	Conditions	Product(s) and Yield(s) (%)	Refs.
	SmI_2, HMPA, THF, 25-30°, 1 h	(69)	559
	$NaC_{10}H_8$	(40)	842
	$(t\text{-}BuO)_2$, $c\text{-}C_6H_{12}$, 150°, sealed tube	**I** + **II** (60) **I:II** = 3:1	104
	Bu_3SnH, AIBN, C_6H_6, 80°	**I** + **II** (66) **I:II** = 3:1	158, 828

TABLE V. OLIGOCYCLIC RINGS CONTAINING ONLY CARBON (*Continued*)

Reactant	Conditions	Product(s) and Yield(s) (%)	Refs.
	$Mn(OAc)_3$, $Cu(OAc)_2$	CO_2Me (73)	574
	$Mn(OAc)_3$, $Cu(OAc)_2$, AcOH, 25°	CO_2Me (61)	569
	(*t*-BuO)$_2$, MeCHO, 130°, sealed tube	**I** + **II** + **III** + **IV** (54) **I:II:III:IV** = 5:1:1:5	843

742

TABLE V. OLIGOCYCLIC RINGS CONTAINING ONLY CARBON (*Continued*)

Reactant	Conditions	Product(s) and Yield(s) (%)	Refs.
	MeSH, (MeS)$_2$, C$_6$H$_6$, $h\nu$	(10!) X = H and SMe *ratio* 1:1	279
	1,4-CHD, PhCl, 191-210°.	R^1 R^2 CO$_2$Me Me (93) Me CO$_2$Me (53) CH$_2$OH H (>99) OAc H (83) OMe H (95)	844
	Bu$_3$SnH, AIBN, C$_6$H$_6$, 80°	(5J) CO$_2$Et	845
C$_{16}$	Bu$_3$SnH, AIBN, C$_6$H$_6$, 80°	R Ph (85) *p*-MeC$_6$H$_4$ (93) *p*-MeOC$_6$H$_4$ (85) *o*-MeC$_6$H$_4$ (60) PhC≡C (82)	846

743

TABLE V. OLIGOCYCLIC RINGS CONTAINING ONLY CARBON (*Continued*)

Reactant	Conditions	Product(s) and Yield(s) (%)	Refs.

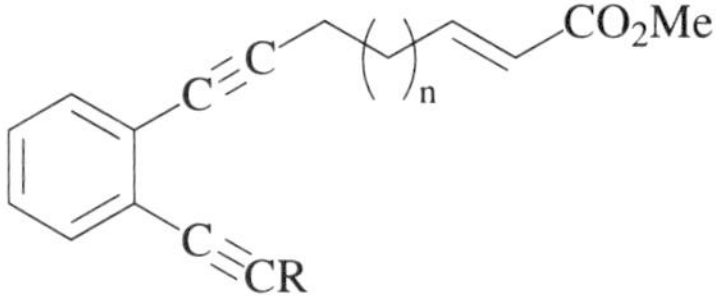

1,4-CHD, PhCl, 210°, 19-24 h

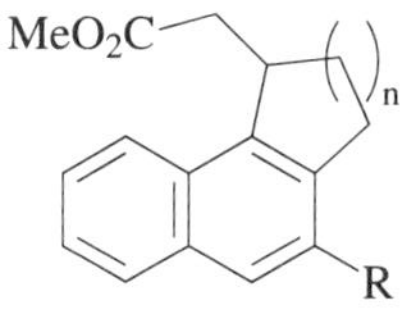

R	n		
H	1	(72)	
CH₂OTBDMS	1	(58)	
H	2	(53)	

847

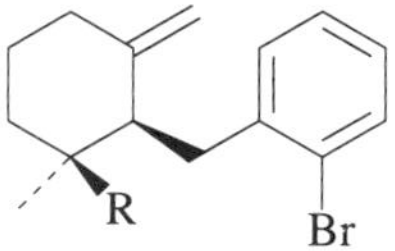

Bu₃SnH, AIBN, C₆H₆, 80°

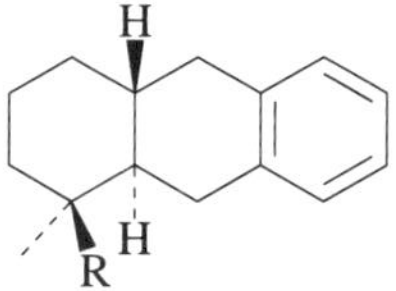

R	
Me	(95)
CO₂Me	(85)

832

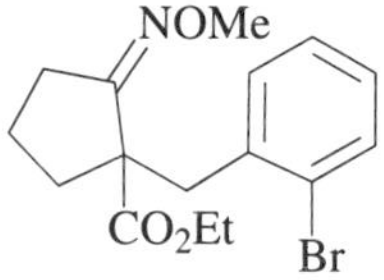

Bu₃SnH, AIBN, C₆H₆, 80°

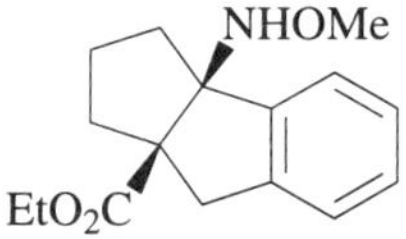

(75)

560, 561

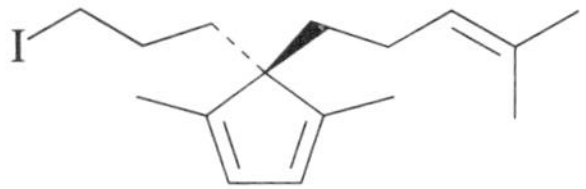

Bu₃SnH, AIBN, C₆H₆, 80°

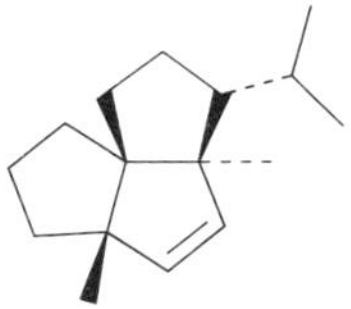

I + II (65) **I:II** = 1:5

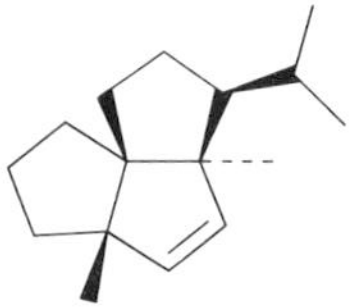

34

744

TABLE V. OLIGOCYCLIC RINGS CONTAINING ONLY CARBON (*Continued*)

Reactant	Conditions	Product(s) and Yield(s) (%)	Refs.
C$_{17}$			
(structure: Br, methylenecyclohexane)	Bu$_3$SnH, AIBN, C$_6$H$_6$, 80°, 10 h	(57)	840
(structure: MeO$_2$C, OH, HC≡C)	Bu$_3$SnH, AIBN, C$_6$H$_6$, 80°	(75) MeO$_2$C, OH	848
(structure: CH≡C, O O, CHO)	1. SmI$_2$, solvent, 0° 2. TsOH, Me$_2$CO	**I** R^1 = OH, R^2 = H **II** R^1 = H, R^2 = OH	849

	I + II	I:II
THF, HMPA (20:1)	(63)	100:0
THF, DMPU (20:1)	(69)	91:9

Reactant	Conditions	Product(s) and Yield(s) (%)	Refs.
(structure: I, C≡CTMS)	Bu$_3$SnH, AIBN, C$_6$H$_6$, 80°	(72) TMS	546

TABLE V. OLIGOCYCLIC RINGS CONTAINING ONLY CARBON (*Continued*)

Reactant	Conditions	Product(s) and Yield(s) (%)	Refs.

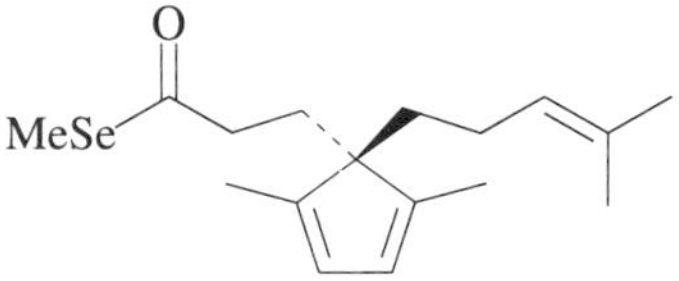

Bu$_3$SnH, AIBN, C$_6$H$_6$, 80°

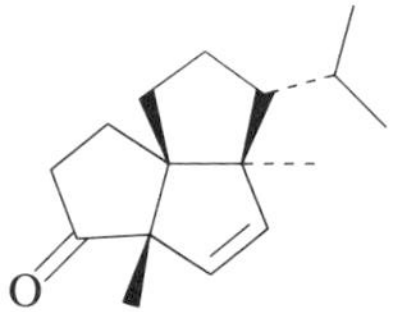

I + II 34

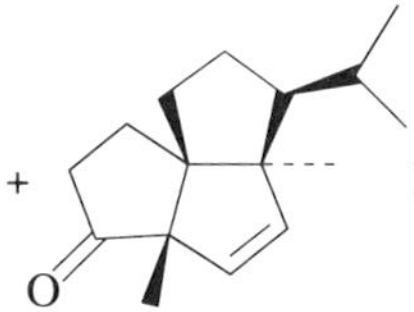

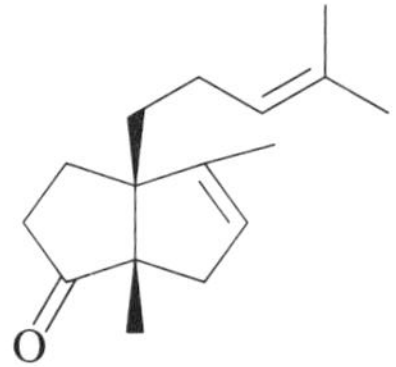

I + II + III (62)
I:II:III = 1:5:2

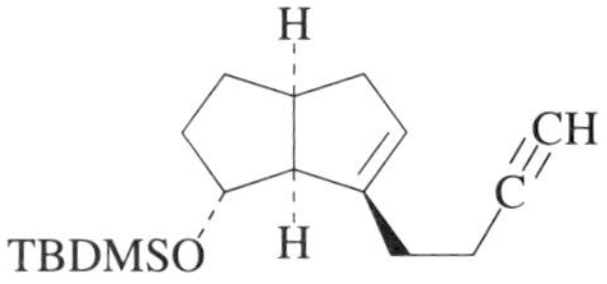

1. Ph$_3$SnH, AIBN, PhMe, 110°
2. Silica gel

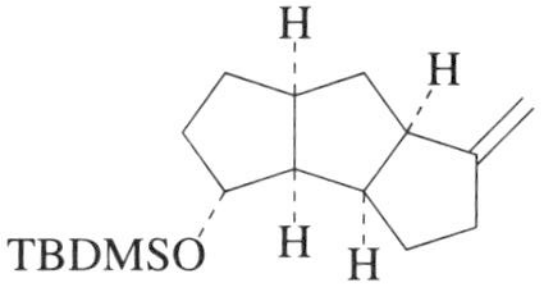

(79)   548

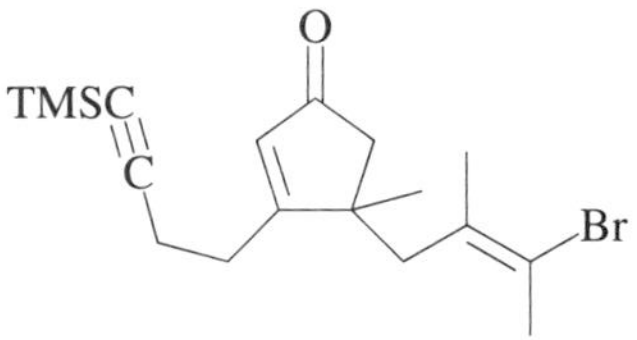

Bu$_3$SnH, AIBN, C$_6$H$_6$, 80°

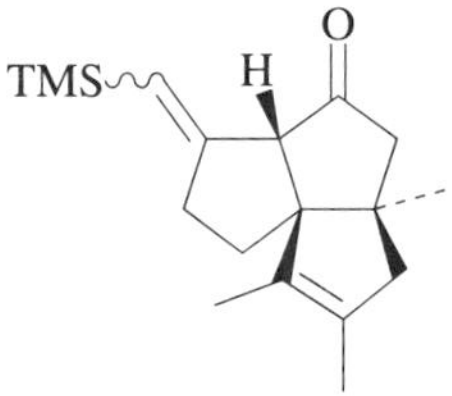

(78)   828

C$_{18}$

TABLE V. OLIGOCYCLIC RINGS CONTAINING ONLY CARBON (*Continued*)

Reactant	Conditions	Product(s) and Yield(s) (%)	Refs.
	BPO, c-C$_6$H$_{12}$, 80°	(89)	850
C$_{19}$	1,4-CHD, PhCl, 245°, 3 h	(98) 1:1 ratio	586
	Bu$_3$SnH, AIBN, PhMe, 110°	I + II (46) I:II = 1:1	851
C$_{20}$	Bu$_3$SnH, AIBN, C$_6$H$_6$, 80°	(86)	33

747

TABLE V. OLIGOCYCLIC RINGS CONTAINING ONLY CARBON (*Continued*)

Reactant	Conditions	Product(s) and Yield(s) (%)	Refs.
	C$_6$H$_6$, 80°	(69) (63)	852
$\dfrac{R}{H}$ Me			
	NaC$_{10}$H$_8$, THF, 25°	(41)	853, 854
	Bu$_3$SnH, AIBN, C$_6$H$_6$, 80°, 8 h	(62)	855
C$_{21}$	Ph$_3$SnH, AIBN, C$_6$H$_6$, 80°	(80)	340

748

TABLE V. OLIGOCYCLIC RINGS CONTAINING ONLY CARBON (Continued)

Reactant	Conditions	Product(s) and Yield(s) (%)	Refs.

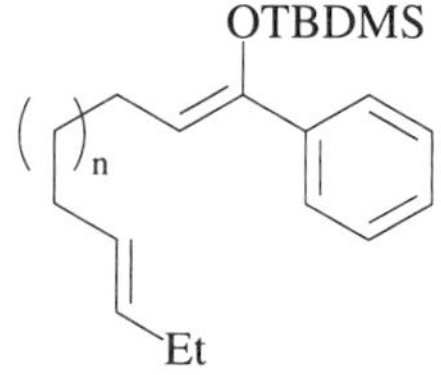

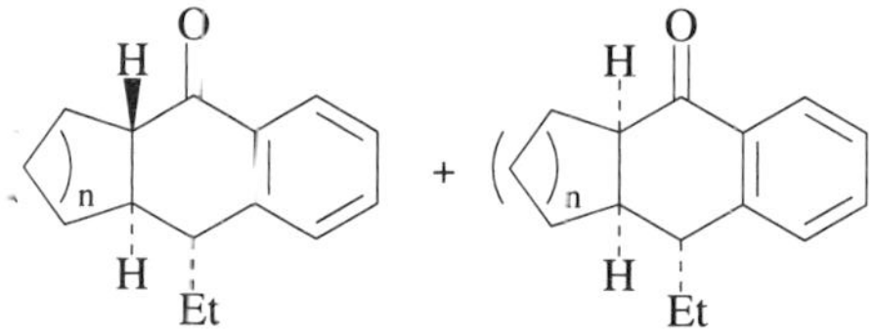

C_{22}

C_{23}

Reactant (top): OTBDMS

n
1
2

Conditions: Cu_2O, $Cu(OTf)_2$

Products I + II:

	I + II	I:II
	(90)	20:1
	(87)	20:1

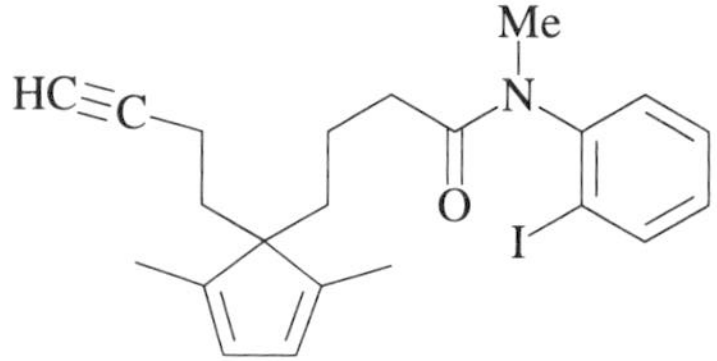

Refs.: 348

Conditions: Bu_3SnH, AIBN, C_6H_6, 80°

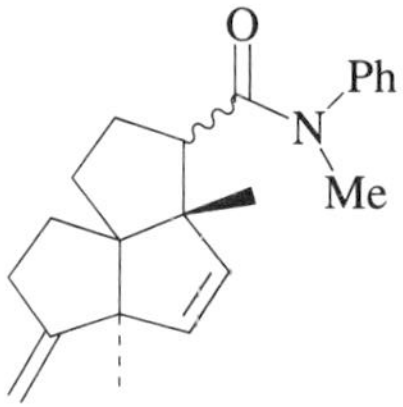

Product: (75) 1:1 mixture

Refs.: 199

Conditions: Bu_3SnH, AIBN

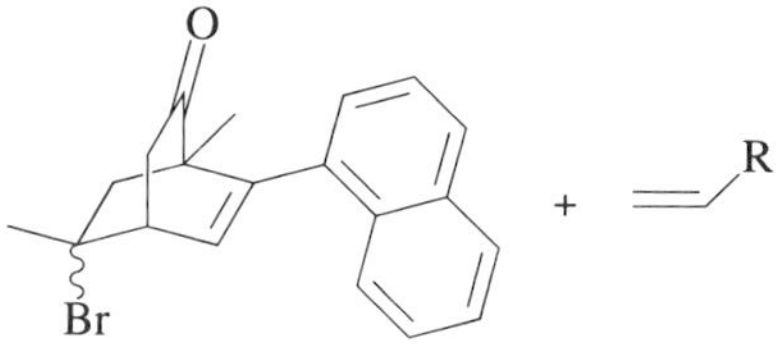

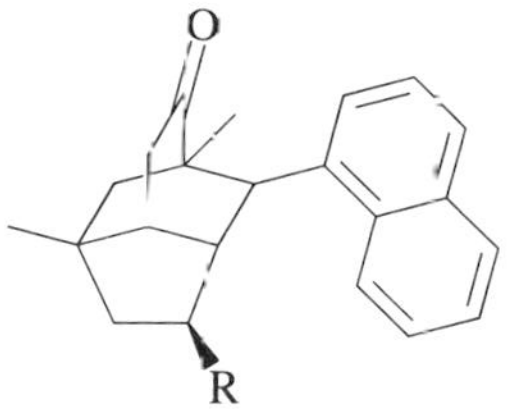

R	
CO_2Me	(50)
CN	(58)
COMe	(30)
Ph	(60)

Refs.: 856, 857

TABLE V. OLIGOCYCLIC RINGS CONTAINING ONLY CARBON (*Continued*)

Reactant	Conditions	Product(s) and Yield(s) (%)	Refs.
C_{24}	Bu$_3$SnH, AIBN, C$_6$H$_6$, 80°	(75)	858, 859
	Mn(OAc)$_3$, AcOH, rt	(70)	860
	Bu$_3$SnH, AIBN, C$_6$H$_6$, 80°	(65)	34
	TsI, CH$_2$Cl$_2$, 40°	(80)	139

TABLE V. OLIGOCYCLIC RINGS CONTAINING ONLY CARBON (*Continued*)

Reactant	Conditions	Product(s) and Yield(s) (%)	Refs.
C_{25}			
	Bu_3SnH, AIBN, C_6H_6, 80°	(35)	199
	Bu_3SnH, AIBN, C_6H_6, 80°	(42)	861
C_{26}			
	NaC_{10}H_8, THF, MeOH, 25°	(50)	862
	Ph_3SnH, AIBN, C_6H_6, 80°	(91)	352, 585

TABLE V. OLIGOCYCLIC RINGS CONTAINING ONLY CARBON (*Continued*)

Reactant	Conditions	Product(s) and Yield(s) (%)	Refs.
C_{28} 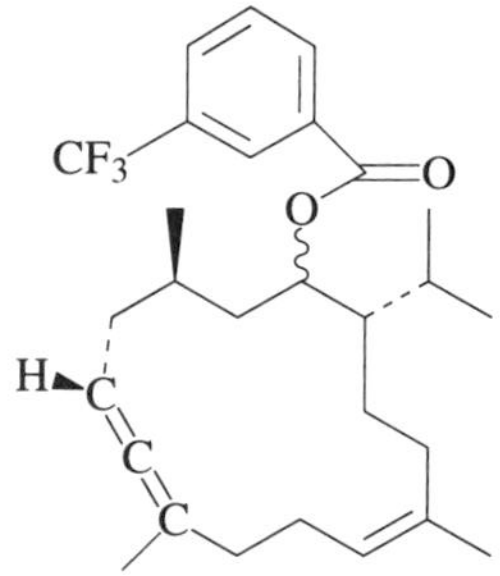	1. *N*-Methyl-carbazole, 1,4-CHD, THF, H_2O, $h\nu$, 55°, 5 h 2. PhSH, C_7H_{16}, AIBN, 50°, 30 min	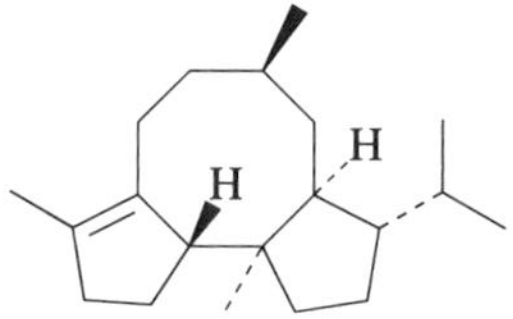 (46)	863

B. Tetracyclic Systems

C_{10} 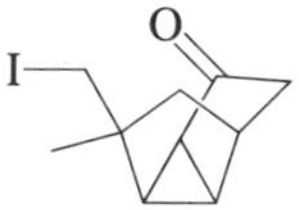	$(Me_3Sn)_2$, C_6H_6, $h\nu$	 OH (59)	864
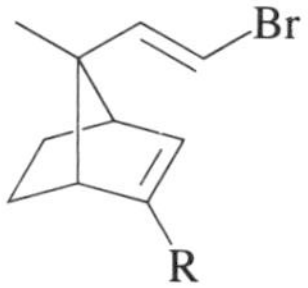	Bu_3SnH, AIBN, C_6H_6, $h\nu$	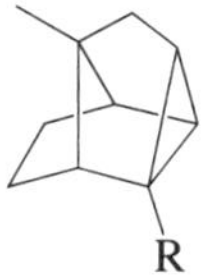 R = H (100) R = OTBDMS (100)	366
C_{11} 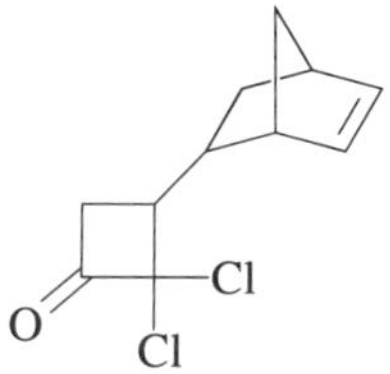	Bu_3SnH, AIBN, C_6H_6, 80°	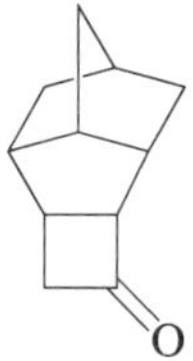 (58)	601

752

TABLE V. OLIGOCYCLIC RINGS CONTAINING ONLY CARBON (*Continued*)

Reactant	Conditions	Product(s) and Yield(s) (%)	Refs.
C_{14}	Bu$_3$SnH, AIBN, C$_6$H$_6$, 80°	(85)	827
C_{15}	*t*-BuOH, C$_6$H$_6$, *hv*	(90)	865
C_{16}	Bu$_3$SnH, AIBN, C$_6$H$_6$, 80°	(86)	827
	DMF, TBAL, e$^-$	$\dfrac{\textbf{I} \quad \textbf{II}}{(57) \quad (0)}$ $(0) \quad (55)$	866

TABLE V. OLIGOCYCLIC RINGS CONTAINING ONLY CARBON (*Continued*)

Reactant	Conditions	Product(s) and Yield(s) (%)	Refs.
C₁₈	K₃Fe(CN)₆, Na₂CO₃	(46)	867
	K₃Fe(CN)₆, KOH	(88)	868
C₁₉	Bu₃SnH, AIBN, C₆H₆, 80°	(67)	869
	(PPh₃)₂ReCl₃, CCl₄, reflux	(79) *cis:trans* = 4.2:1	282

TABLE V. OLIGOCYCLIC RINGS CONTAINING ONLY CARBON (*Continued*)

Reactant	Conditions	Product(s) and Yield(s) (%)	Refs.
	$KC_{10}H_8$, THF	(69)	870
C$_{22}$	Bu_3SnCl, $NaBH_4$, EtOH, hv	(67)	871
C$_{25}$	$Mn(OAc)_3 \cdot 2\ H_2O$, $Cu(OAc)_2 \cdot H_2O$, AcOH, rt, 20 h	(31)	872

755

TABLE V. OLIGOCYCLIC RINGS CONTAINING ONLY CARBON (*Continued*)

Reactant	Conditions	Product(s) and Yield(s) (%)	Refs.
C$_{28}$	NaC$_{10}$H$_8$, THF	(89)	870
C$_{29}$	Mn(OAc)$_3$, Cu(OAc)$_2$, AcOH, rt	(79)	873
C$_{33}$	Bu$_3$SnH, AIBN, C$_6$H$_6$, 80°	(79)	585, 874

TABLE VI. OLIGOCYCLIC RINGS CONTAINING ONE OR MORE HETEROATOMS

Reactant	Conditions	Product(s) and Yield(s) (%)	Refs.
		A. Tricyclic Systems	
C$_8$	RuCl$_2$(PPh$_3$)$_3$, C$_6$H$_6$, 160°, sealed tube	(88)	221
C$_9$	Mn$_2$(OAc)$_7$, AcOH, 23°	(63)	191
C$_{10}$	Mn$_2$(OAc)$_7$, AcOH, 23°	(61)	191
	Bu$_3$SnH, AIBN, C$_6$H$_6$, 80°	$\begin{array}{c|c} R & \\ \hline H & (86) \\ Me & (58) \end{array}$	875
	Bu$_3$SnH, AIBN, C$_6$H$_6$, 80°	(70)	642

Reactant	Conditions	Product(s) and Yield(s) (%)	Refs.

C$_{11}$

	Et$_3$B, O$_2$, PhMe, 100°	(38) + (5)	731
	Mn$_2$(OAc)$_7$, AcOH, 23°	(63)	191
	NaI, Me$_2$CO, rt	(45)	876
	ClCo(dmgH)$_2$py, NaBH$_4$, MeOH, NaOH, 50°	(90) (95)	877

Reactant	Conditions	Product(s) and Yield(s) (%)	Refs.
	Bu$_3$SnH, AIBN, C$_6$H$_6$, 80°	(30)	878
	Bu$_3$SnH, AIBN, C$_6$H$_6$, 80°	(62)	642
R ⎯ Me / OMe / H	Bu$_3$SnH, AIBN	I + II	879

I + II as a function of R:

R	I + II	I:II
Me	(66)	2:1
OMe	(68)	1.7:1
H	(88)	2.1:1

| | Me$_3$SnCl, Na(CN)BH$_3$, AIBN, *t*-BuOH | I + II (86) I:II = 2:1 | 218 |

759

TABLE VI. OLIGOCYCLIC RINGS CONTAINING ONE OR MORE HETEROATOMS (*Continued*)

Reactant	Conditions	Product(s) and Yield(s) (%)	Refs.
	ClCo(dmgH)₂py, NaBH₄, MeOH, NaOH, 50°	I + II (48)	731
	Et₃B, O₂, PhMe, 100°	I + II (50)	731
	Bu₃SnH, AIBN, C₆H₆, 80°	I + II + III (90), I:II = 76:24, (I, II):III = 85:15	671

760

TABLE VI. OLIGOCYCLIC RINGS CONTAINING ONE OR MORE HETEROATOMS (*Continued*)

Reactant	Conditions	Product(s) and Yield(s) (%)	Refs.
	Bu$_3$SnH, AIBN, C$_6$H$_6$, 80°	(55)	880
	Bu$_3$SnH, AIBN, C$_6$H$_6$, 80°	I + II I + II (64) I:II = 1:3	671
C$_{12}$	Bu$_3$SnH, AIBN, C$_6$H$_6$, 80°		881

Y	X	R	
I	O	Me	(63)
Br	CH$_2$	Me	(40)
I	NMe	Me	(39)
I	O	OMe	(24)
I	O	F	(50)
I	NMe	F	(44)

TABLE VI. OLIGOCYCLIC RINGS CONTAINING ONE OR MORE HETEROATOMS (*Continued*)

Reactant	Conditions	Product(s) and Yield(s) (%)	Refs.
	BPO, CHCl$_3$, *hv*	 R = H (40) R = Me (40)	882
	ClCo(dmgH)$_2$py, NaBH$_4$, MeOH, NaOH, 50°	 **I** (95)	877
	Ni(II)complex, Hg cathode, NH$_4$ClO$_4$, DMF	**I** (86)	702
	Bu$_3$SnH, AIBN, C$_6$H$_6$, 80°		883

R	R^1	R^2	R^3	
OMe	H	H	H	(85)
Me	H	H	H	(82)
Cl	H	H	H	(82)
OMe	Me	H	H	(80)
OMe	H	Me	H	(78)
OMe	H	H	Me	(75)

TABLE VI. OLIGOCYCLIC RINGS CONTAINING ONE OR MORE HETEROATOMS (*Continued*)

Reactant	Conditions	Product(s) and Yield(s) (%)	Refs.
	Bu$_3$SnH, AIBN, C$_6$H$_6$, 80°	 (6) (81) (73) (52)	884
$\dfrac{R^1 \quad R^2}{\begin{array}{cc} H & H \\ H & CN \\ H & CO_2Me \\ OMe & H \end{array}}$			
	Bu$_3$SnH, AIBN, C$_6$H$_6$, 80°	 (79) (74)	885
$\dfrac{R \quad X}{\begin{array}{cc} H & I \\ OMe & Br \end{array}}$			
	Bu$_3$SnH, AIBN, C$_6$H$_6$, 80°	 **I + II** (80) **I:II** = 5:1	878

TABLE VI. OLIGOCYCLIC RINGS CONTAINING ONE OR MORE HETEROATOMS (*Continued*)

Reactant	Conditions	Product(s) and Yield(s) (%)	Refs.
	Bu₃SnH, AIBN, C₆H₆, 80°	(73)	886
	Bu₃SnH, AIBN, C₆H₆, 80°	(63)	887
	Bu₃SnH, AIBN, C₆H₆, 80°	I + II + III (83) I:II:III = 6:1:2	888

Reactant	Conditions	Product(s) and Yield(s) (%)	Refs.
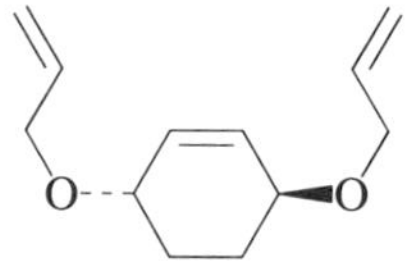	Me_3SnCl, $Na(CN)BH_3$, AIBN, *t*-BuOH	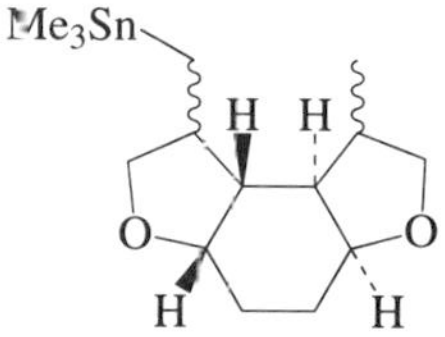 (85)	330
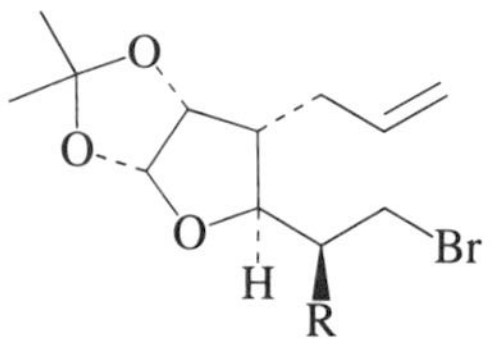 R: OH; OBz	Bu_3SnH, AIBN, C_6H_6, 80°	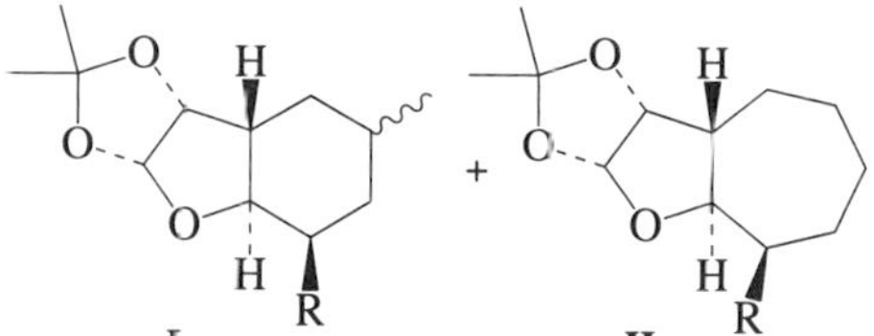 I + II R = OH: I (33), II (44), I, α:β ratio 70:30 R = OBz: I (34), II (46), I, α:β ratio —	889
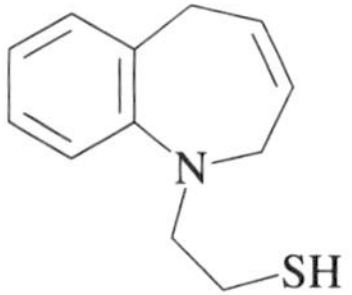	$(t\text{-BuO})_2$, PhCl, 132°	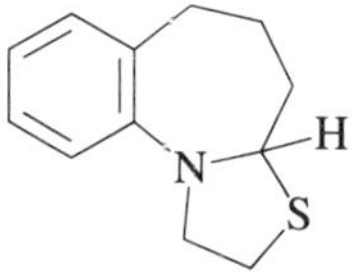 (50)	890
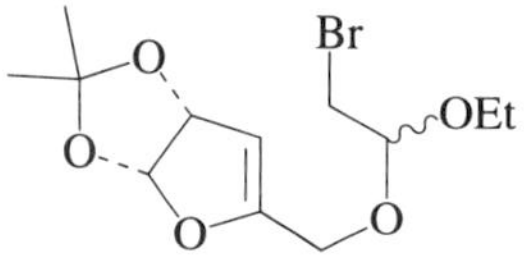	Bu_3SnH, AIBN, C_6H_6, 80°	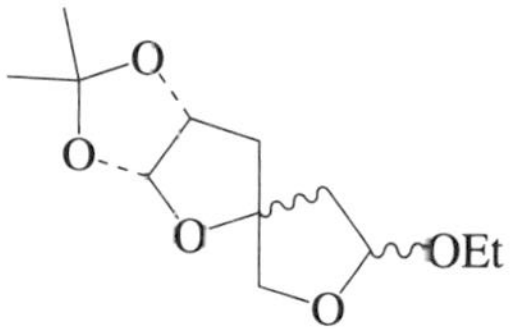 (66)	891

TABLE VI. OLIGOCYCLIC RINGS CONTAINING ONE OR MORE HETEROATOMS (*Continued*)

Reactant	Conditions	Product(s) and Yield(s) (%)	Refs.
C₁₃			
	Bu₃SnH, AIBN, C₆H₆, 80°		757
	MeOH, *hv*		892
	MeOH, *hv*		892
	1. Hg(OAc)₂ 2. NaBH₄		893

TABLE VI. OLIGOCYCLIC RINGS CONTAINING ONE OR MORE HETEROATOMS (*Continued*)

Reactant	Conditions	Product(s) and Yield(s) (%)	Refs.
	$K_3Fe(CN)_6$, KOH	(84)	868
	SmI_2, HMPA, THF	(31)	175
	SmI_2, HMPA, THF, MeCN	(89)	175
	$(t\text{-BuO})_2$, PhCl, 110°		894

R^1	R^2	R^3	R^4	
H	H	H	H	(83)
Cl	H	H	H	(85)
MeO	H	H	H	(90)
H	H	NO_2	H	(55)
H	Me	H	Me	(95)

TABLE VI. OLIGOCYCLIC RINGS CONTAINING ONE OR MORE HETEROATOMS (*Continued*)

Reactant	Conditions	Product(s) and Yield(s) (%)	Refs.

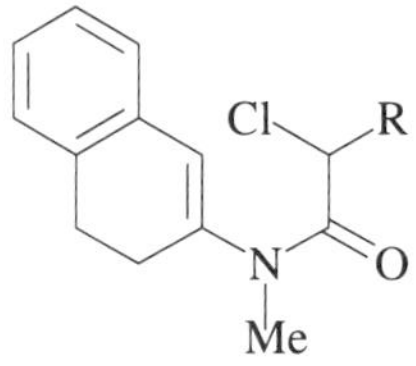

	I	**II**
Me	(42)	(35)
Ph	(0)	(76)
H	(50)	(0)

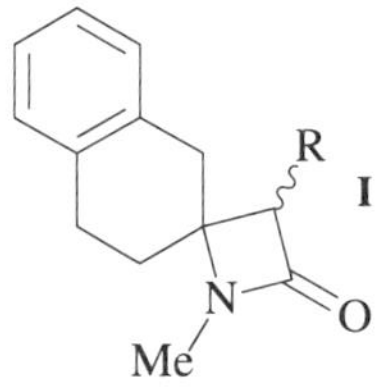

Bu₃SnH, AIBN, C₆H₆, 80°

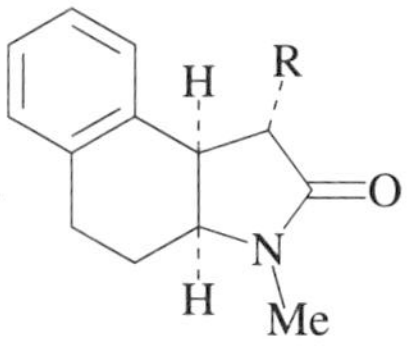

R: Me, Ph, H

714, 715

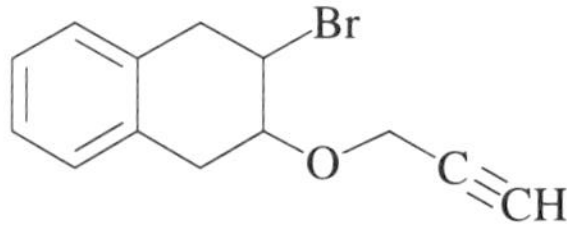

ClCo(dmgH)₂py, NaBH₄, MeOH, NaOH, 50° — (80) — 877

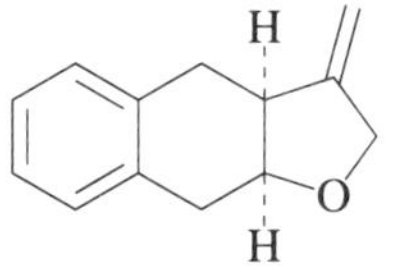

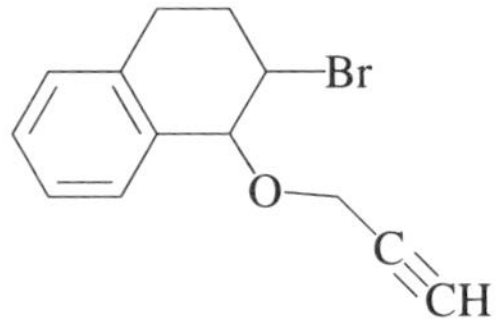

ClCo(dmgH)₂py, NaBH₄, MeOH, NaOH, 50° — (62) — 877

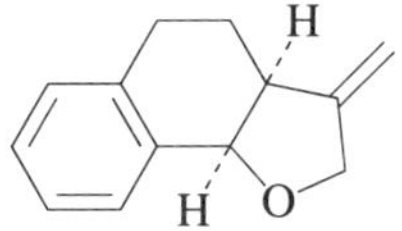

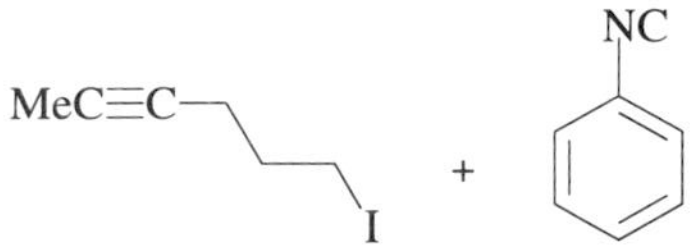

(Bu₃Sn)₂, *hv*, 150° — (70) — 895

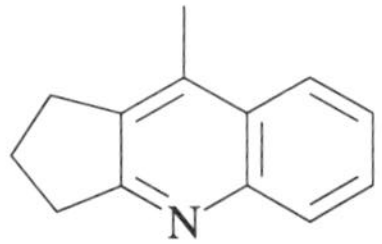

Reactant	Conditions	Product(s) and Yield(s) (%)	Refs.
	Bu₃SnH, AIBN, C₆H₆, 80°	(46)	718
	Bu₃SnH, AIBN, C₆H₆, 80°	(7)	896
	Bu₃SnH, AIBN, C₆H₆, 80°	(34)	897
	Bu₃SnH, AIBN, C₆H₆, 80°	(36)	878

769

TABLE VI. OLIGOCYCLIC RINGS CONTAINING ONE OR MORE HETEROATOMS (*Continued*)

Reactant	Conditions	Product(s) and Yield(s) (%)	Refs.
R = OMe R = SPr-*i*	Bu₃SnH, DIPHOS, AIBN, C₆H₆, 80°	**I + II** (85) **I:II** = 5:1 **I + II** (80) **I:II** = 8:1	878
	Bu₃SnH, AIBN, C₆H₆, 80°	(68)	898
	(Bu₃Sn)₂, C₆H₆, *hν*	n = 1 (75) n = 2 (66)	256
	Bu₃SnH, AIBN, C₆H₆, heat	(25)	889

770

TABLE VI. OLIGOCYCLIC RINGS CONTAINING ONE OR MORE HETEROATOMS (*Continued*)

Reactant	Conditions	Product(s) and Yield(s) (%)	Refs.
	Bu$_3$SnH, AIBN, C$_6$H$_6$, 80°		399

R^1	R^2
H	CO$_2$Et
CO$_2$Me	CO$_2$Me
CO$_2$Me	CO$_2$Me
Me	CO$_2$Me

R^3	R^4	
CH$_2$CO$_2$Et	H	(69)
CO$_2$Me	CH$_2$CO$_2$Me	(67)
CH$_2$CO$_2$Me	CO$_2$Me	(15)
Me, CH$_2$CO$_2$Me	Me, CH$_2$CO$_2$Me	(89)
	diastereomeric mixture, ca. 4:1	

Reactant	Conditions	Product(s) and Yield(s) (%)	Refs.
	Bu$_3$SnH, AIBN, C$_6$H$_6$, 80°		889

R
OH
OBz

	α:β
(82)	85:15
(80)	80:20

Reactant	Conditions	Product(s) and Yield(s) (%)	Refs.
	Bu$_3$SnH, AIBN, C$_6$H$_6$, 80°	(31) *endo:exo* = 82:18	671

Reactant	Conditions	Product(s) and Yield(s) (%)	Refs.
$\dfrac{R}{H}$ $\dfrac{}{Me}$	Bu$_3$SnH, AIBN, PhMe, 110°	**I** + **II** (81) **I:II** = 86:14 **I** + **II** (79) **I:II** = 47:53	419, 899
	Bu$_3$SnH, AIBN, PhMe, 110°	(75)	900
	Na, NH$_3$, (NH$_4$)$_2$SO$_4$	(90)	901
	Bu$_3$SnH, AIBN, xylenes, 155°	(77)	902

TABLE VI. OLIGOCYCLIC RINGS CONTAINING ONE OR MORE HETEROATOMS (*Continued*)

Reactant	Conditions	Product(s) and Yield(s) (%)	Refs.
	Bu₃SnH, AIBN, C₆H₆, 80°	(60)	722
	Ph₃SnH, AIBN, PhMe, 100°	I + II (85) I:II = 1:3.25	903
	Bu₃SnH, AIBN, C₆H₆, 80°	(60)	897
	Bu₃SnH, AIBN, C₆H₆, 80°	(100)	904

773

Reactant	Conditions	Product(s) and Yield(s) (%)	Refs.
	Bu₃SnH, AIBN, C₆H₆, 80° Bu₃SnH, =CN, AIBN, C₆H₆, 80° Bu₃Sn, AIBN, C₆H₆, 80°	 R = H (85) R = (CH₂)₂CN (80) R = CH₂CH=CH₂ (70)	905
	Bu₃SnH, AIBN, C₆H₆, 80°	(100)	904
C₁₄ 	Bu₃SnH, AIBN, C₆H₆, 80°	(90)	757

774

TABLE VI. OLIGOCYCLIC RINGS CONTAINING ONE OR MORE HETEROATOMS (*Continued*)

Reactant	Conditions	Product(s) and Yield(s) (%)	Refs.
MeO-substituted 2-(2-methoxyphenyl)benzoic acid (MeO, OH, O)	t-BuOI, t-BuOH, hv Pb(OAc)$_4$	MeO dibenzo lactone (O, O) (76) (68)	906
Cyclopentane-fused indole with CN and Br-alkyl chain; N–Me	Bu$_3$SnH, AIBN, C$_6$H$_6$, 80°	Spiro-cyclopentane indoline CN, N–Me (81)	907
NC–C$_6$H$_4$–OMe + MeC≡C–(CH$_2$)$_n$–I	(Bu$_3$Sn)$_2$, AIBN, hv, 150°	**I** (OMe, Me, N tricyclic) + **II** (OMe, Me, N tricyclic) I + II (62) I:II = 79:21	895

Reactant	Conditions	Product(s) and Yield(s) (%)	Refs.
	MeCN, H$_2$O, *hv*	Y = H, X = H (34) Y = OH, X = H (58) YX = O (—)	908
	MeCN, *hv* H$_2$O MeOH NH$_2$OH	Y = OH (58) Y = OMe (63) Y = NHOH (51)	908
	Bu$_3$SnH, AIBN	(45)	909

Reactant	Conditions	Product(s) and Yield(s) (%)	Refs.

Row 1

Reactant:

R—N with cyclohexenyl and o-bromobenzyl groups

R
n-Bu
Et
Me

Conditions: Bu₃SnH, AIBN

Products: **I** + **II** + **III**

	I	**II**	**III**	**I:III**
	(30)	(11)	(9)	2.3:1
	(31)	(13)	(28)	2.9:1
	(12)	(—)	(28)	>10:1

Refs.: 910

Row 2

Reactant: MeO₂C-substituted N-(o-bromobenzyl) tetrahydropyridine

Conditions: Bu₃SnH, AIBN, C₆H₆, 80°

Products: **I** + **II** (63) **I:II** = 1:3.2

Refs.: 723

Row 3

Reactant: bicyclic oxa compound with TBDMSO and O-acryloyl groups

Conditions: Bu₃SnH, AIBN, C₆H₆, 25°, *hv*

Products: tricyclic product with OTBDMS

R	
SO₂Ph	(84)
CO₂Me	(62)
CN	(76)

ClCH₂SO₂Ph
PhSCH₂CO₂Me
PhSCH₂CN

Refs.: 911

TABLE VI. OLIGOCYCLIC RINGS CONTAINING ONE OR MORE HETEROATOMS (*Continued*)

Reactant	Conditions	Product(s) and Yield(s) (%)	Refs.
	Bu₃SnH, AIBN, C₆H₆, 80°	I + II (89) I:II = 8:1 I + II (97) I:II = 100:0	795
	Bu₃SnH, AIBN, C₆H₆	(41)	714
	Bu₃SnH, AIBN, C₆H₆, 80°	(75) 5:1 mixture of isomers (77) 9:1 mixture of isomers	399

R¹	R²
H	CO₂Et
Me	CO₂Me

778

TABLE VI. OLIGOCYCLIC RINGS CONTAINING ONE OR MORE HETEROATOMS (*Continued*)

Reactant	Conditions	Product(s) and Yield(s) (%)	Refs.
	Bu₃SnH, AIBN, C₆H₆, 80°	(65)	57
$E:Z = 1:1$ $\frac{R}{\text{OMe}}$ CO_2Me	Bu₃SnH, AIBN, C₆H₆, 80°	I + II I:II (84) 92:8 (86) 93:7	217
	Bu₃SnH, AIBN, C₆H₆, 80°	(80)	399

779

TABLE VI. OLIGOCYCLIC RINGS CONTAINING ONE OR MORE HETEROATOMS (Continued)

Reactant	Conditions	Product(s) and Yield(s) (%)	Refs.

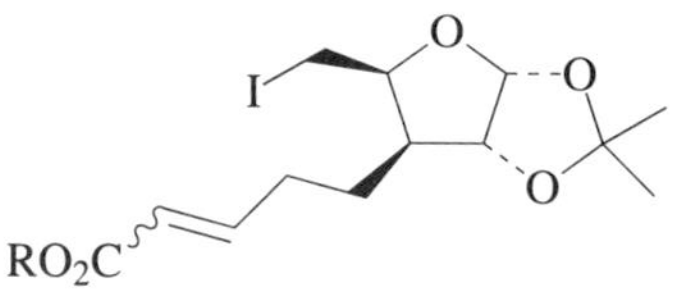

C_{15}

Z R = Me
E R = Et

Bu$_3$SnH, AIBN, C$_6$H$_6$, 80°

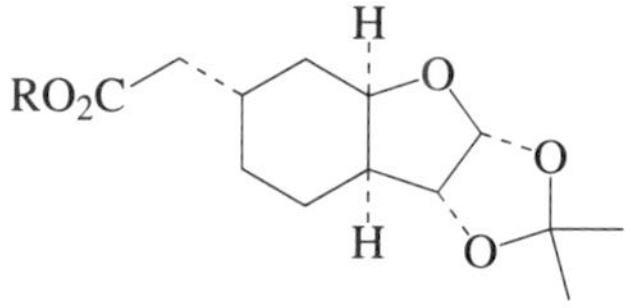

I +

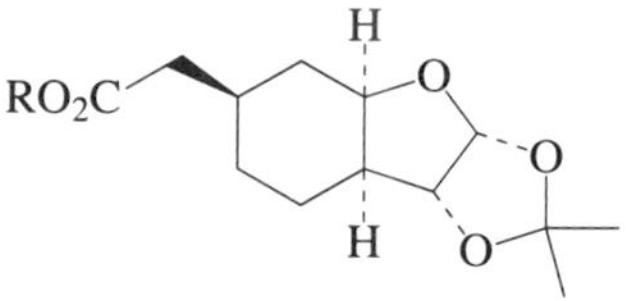

II

I + II (85)   I:II = 9:1
I + II (74)   I:II = 2:1

880

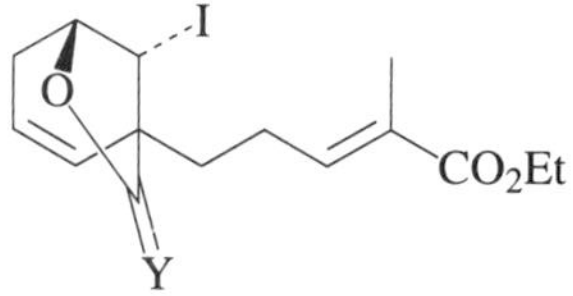

Bu$_3$SnH, AIBN, C$_6$H$_6$, 60°

I, II +

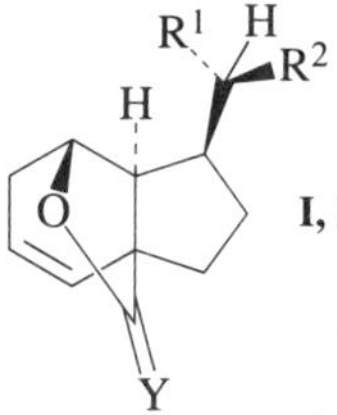

III, IV

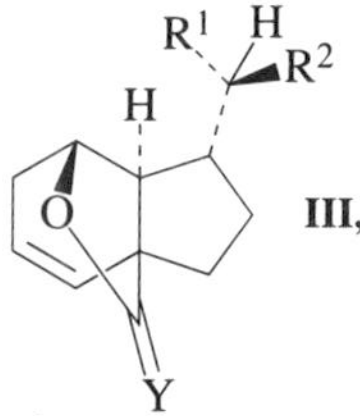

	R^1	R^2
I, III	Me	CO$_2$Et
II, IV	CO$_2$Et	Me

I-IV (92)   I:II:III:IV = 83:1:11:4
I-IV (79)   I:II:III:IV = 61:7:24:8

Y	
O	
CH$_2$	

912

TABLE VI. OLIGOCYCLIC RINGS CONTAINING ONE OR MORE HETEROATOMS. *Continued*

Reactant	Conditions	Product(s) and Yield(s) (%)	Refs.
	Bu₃SnH, AIBN, C₆H₆, 80°	I + II (84) I:II = 7.4:1	913
	Bu₃SnH, AIBN, C₆H₆, 80°	(97)	914
	A: Pt foil anode, CH₂(CO₂R)₂, n-Bu₄NBr, MeCN A A + LiClO₄ A + Mg(ClO₄)₂	I + II (90) I:II = 1:1 I + II (77) I:II = 3:1 I + II (62) I:II = 11.4:1	915

781

Reactant	Conditions	Product(s) and Yield(s) (%)	Refs.
	$S_2O_8^{2-}$, Fe^{2+}	 R H (71) Me (60) Ph (66)	916
	Bu$_3$SnCl, Na(CN)BH$_3$, AIBN, t-BuOH, 80°	(85)	713
	Bu$_3$SnH, AIBN, C$_6$H$_6$, 80°	(73)	497, 48
	Bu$_3$SnH, AIBN, C$_6$H$_6$, 80°	(95)	917

TABLE VI. OLIGOCYCLIC RINGS CONTAINING ONE OR MORE HETEROATOMS (Continued)

Reactant	Conditions	Product(s) and Yield(s) (%)	Refs.
$R^1 = CO_2Et$, $R^2 = Me$ $R^1 = Me$, $R^2 = CO_2Et$ (structures)	Mn(OAc)$_3$, AcOH, NaOAc, 110°	**I** + **II** I + II : I:II (38) : 1.8:1 (77) : 2.7:1	918
(structure)	Bu$_3$SnH, AIBN, C$_6$H$_6$, 80°	**I** + **II** R^3 | R^4 | I | II Me | H | (4) | (4) H | Me | (4) | (81) Me | H | (12) | (1) H | Me | (15) | (37)	98, 912
(structure with OAc, NO$_2$, O)	Bu$_3$SnH, AIBN, C$_6$H$_6$, 80°	**I** + **II** (67); **I:II** = 65:35	111, 438

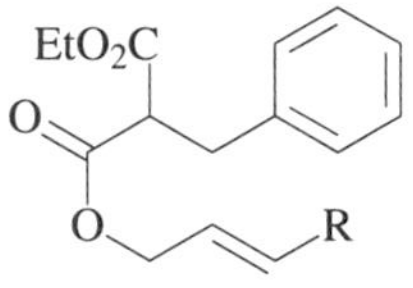

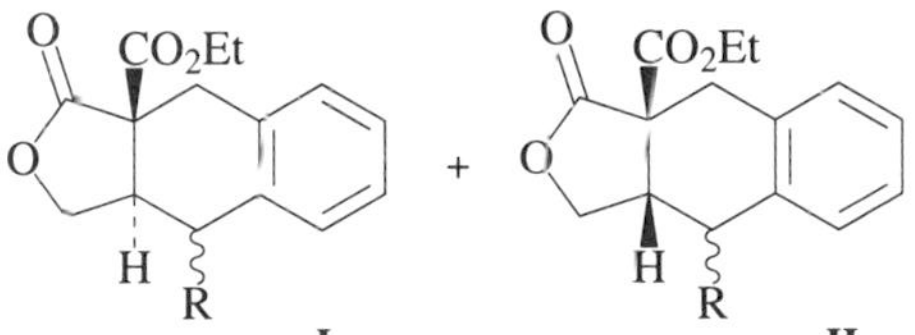

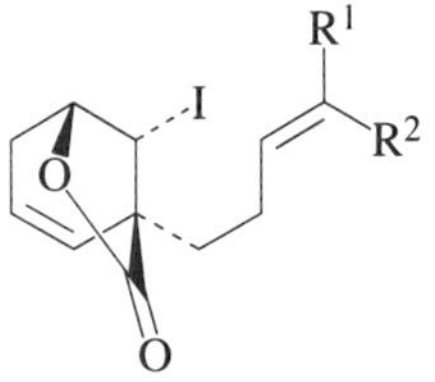

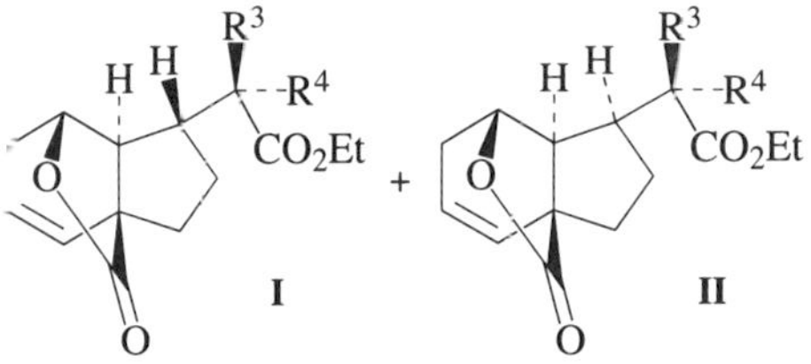

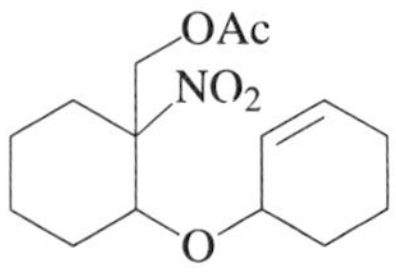

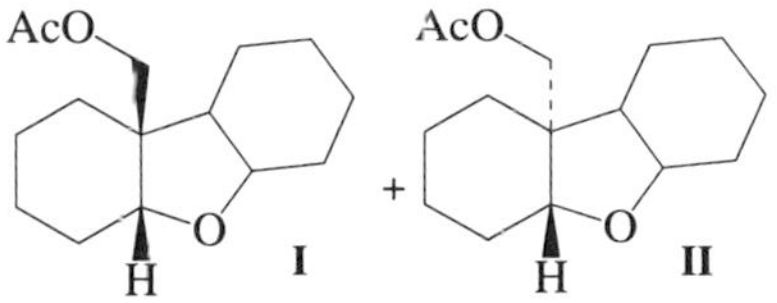

Reactant column detail (row 1):

EtO$_2$C — (structure)

R
H
Me

Reactant	Conditions	Product(s) and Yield(s) (%)	Refs.

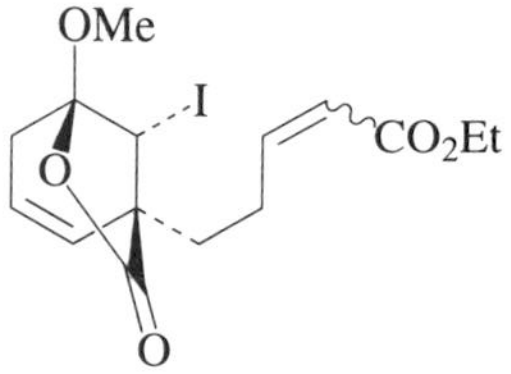

Bu₃SnH, AIBN, C₆H₆, 80°

	I + II	I:II
	(61)	10:1
	(56)	4.5:1

919

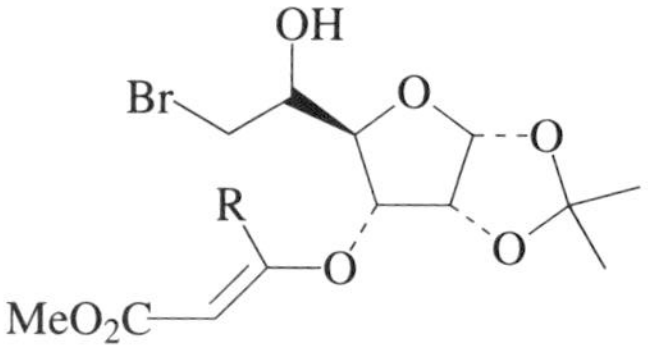

E
E/Z mixture 5:1

Bu₃SnH, AIBN, C₆H₆, 80°

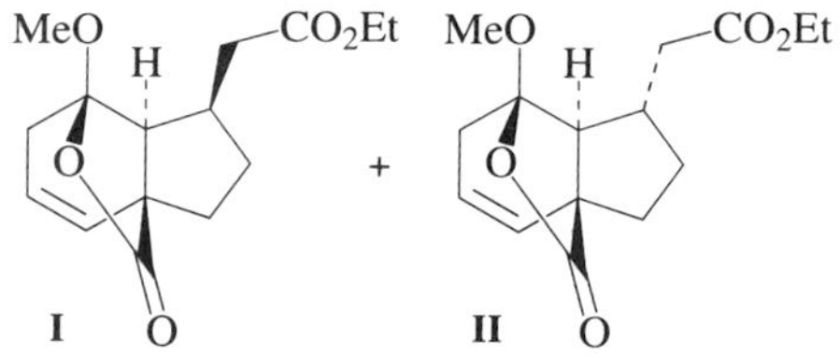

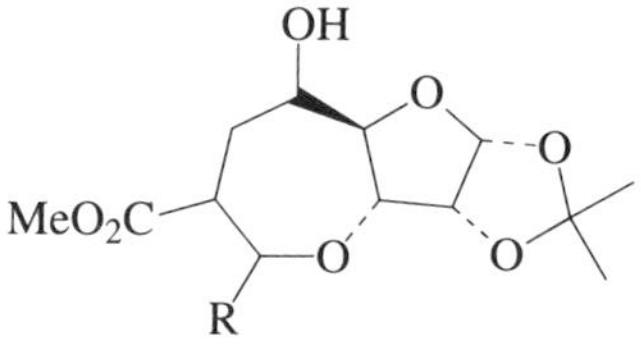

R = CO₂Me, one diastereomer, (61)
R = Me, one diastereomer, (61)

399

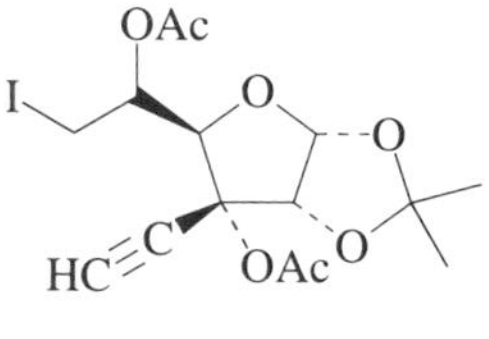

Bu₃SnH, AIBN, C₆H₆, 80°

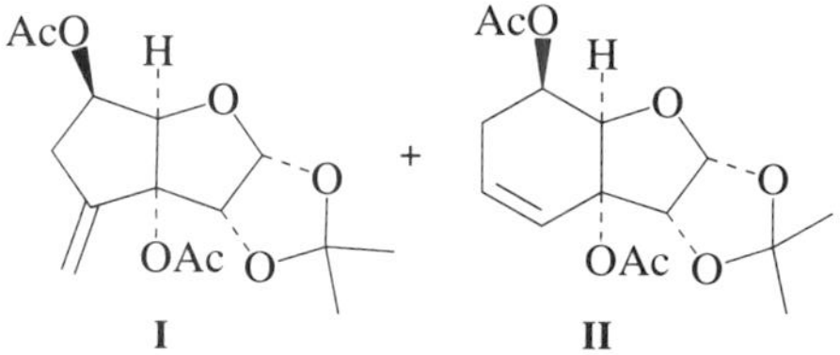

I + II (—)   I:II = 7:3

671

Reactant	Conditions	Product(s) and Yield(s) (%)	Refs.
  R OC(S)SMe OC(S)Im	Bu$_3$SnH, AIBN, C$_6$H$_6$, 80°	 **I** + **II** 	I : II	 88:12 88:12 	I + II	 (73) (88)		671
	(Me$_3$Sn)$_2$, *hv*	(42)	18
	Mn(OAc)$_3$, AcOH, O$_2$, rt	(30)	618

785

TABLE VI. OLIGOCYCLIC RINGS CONTAINING ONE OR MORE HETEROATOMS (*Continued*)

Reactant	Conditions	Product(s) and Yield(s) (%)	Refs.
	Bu₃SnH, AIBN, c-C₆H₁₂, heat	$\dfrac{R^1 \quad R^2}{Me \quad H} \quad \dfrac{I}{I} \quad \dfrac{I+II (67)}{II} \quad I:II = 3:7$	23
	Bu₃SnH, Et₃B	(67)	920
	Bu₃SnH, AIBN, C₆H₆, 80°	(52)	723
	Bu₃SnH, AIBN	(72)	921

C₁₆

TABLE VI. OLIGOCYCLIC RINGS CONTAINING ONE OR MORE HETEROATOMS (*Continued*)

Reactant	Conditions	Product(s) and Yield(s) (%)	Refs.
R = α-Me	Bu₃SnH, AIBN	(95)	914, 922
R = β-Me		(95)	
	Bu₃SnH, AIBN, C₆H₆, 80°	(50)	897
C₁₇	Bu₃SnH, AIBN, PhMe, 110°, 2 h	(81)	923

TABLE VI. OLIGOCYCLIC RINGS CONTAINING ONE OR MORE HETEROATOMS (*Continued*)

Reactant	Conditions	Product(s) and Yield(s) (%)	Refs.
	Bu₃SnH, AIBN, C₆H₆, 80°	(68)	924
	Bu₃SnH, AIBN, C₆H₆, 80°	$\dfrac{R}{\begin{array}{l}H \quad (45)\\ TMS \quad (60)\end{array}}$	925
	Bu₃SnH, AIBN, xylene, 140°	(60)	463
	Bu₃SnH, AIBN, C₆H₆, 80°	(80)	905

TABLE VI. OLIGOCYCLIC RINGS CONTAINING ONE OR MORE HETEROATOMS (*Continued*)

Reactant	Conditions	Product(s) and Yield(s) (%)	Refs.
	Bu₃SnH, AIBN, C₆H₆, 80°	(73)	888
	Bu₃SnH, AIBN, C₆H₆, 80°	I + II (89) I:II = 4:1	99
	Bu₃SnH, AIBN, C₆H₆, 80°	I + II (75) I:II = 91:9	217
	UV, (450-W Hanovia), THF, rt, 45 min	(33)	926

TABLE VI. OLIGOCYCLIC RINGS CONTAINING ONE OR MORE HETEROATOMS (*Continued*)

Reactant	Conditions	Product(s) and Yield(s) (%)	Refs.
	Bu₃SnH, AIBN, C₆H₆, 80°		671
	Bu₃SnH, AIBN, C₆H₆, 80°, 2 h	(93) >10:1	542
	Bu₃SnH, AIBN, C₆H₆, 80°	(>90)	122

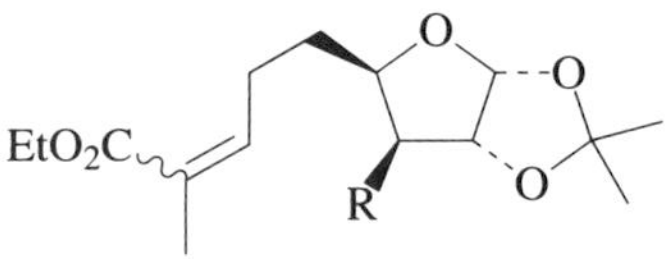

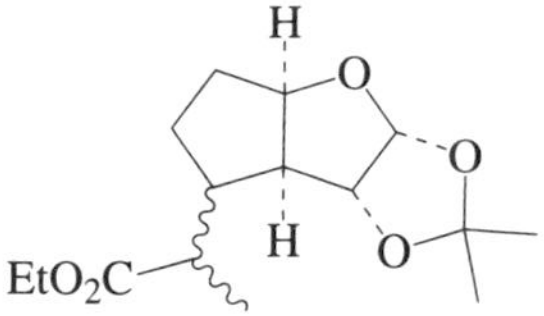

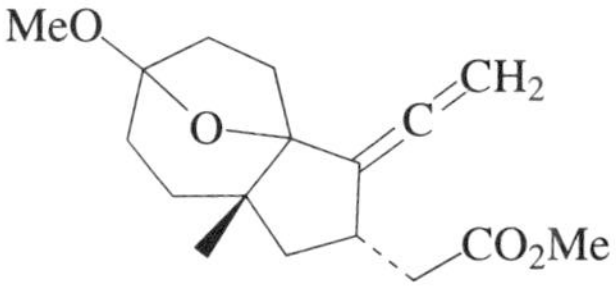

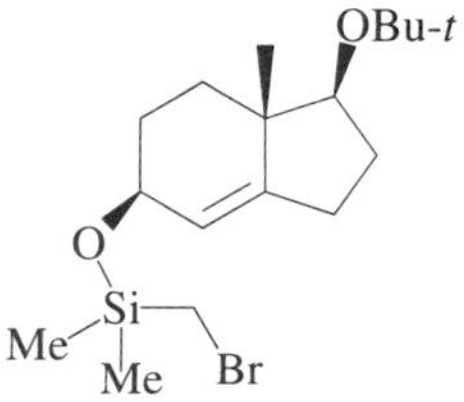

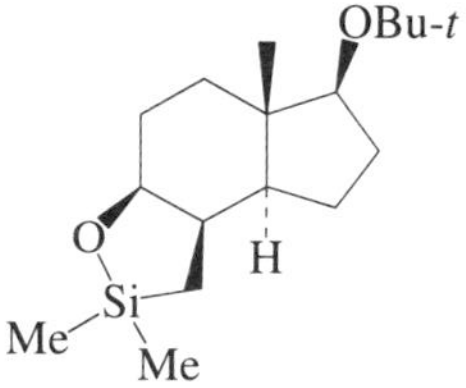

$$\frac{R}{\text{OC(S)SMe}}$$
OC(S)Im

4 isomers in ratio 68:14:12:6
(83)
(76)

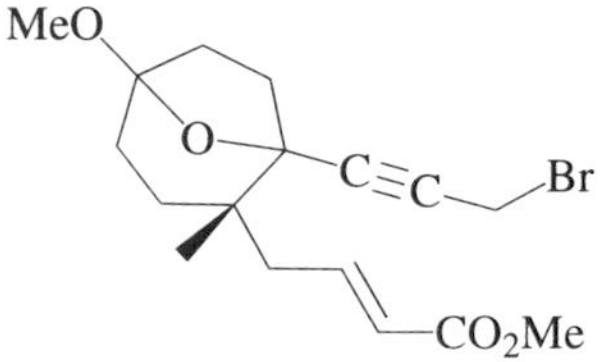

790

TABLE VI. OLIGOCYCLIC RINGS CONTAINING ONE OR MORE HETEROATOMS (*Continued*)

Reactant	Conditions	Product(s) and Yield(s) (%)	Refs.
	1. KOH, MeOH, 2. Mn₃(OAc)₇, AcOH	(63)	927
	Pt-foil anode, CH₂(CO₂R)₂, *n*-Bu₄NBr, MeCN	 **I +** **I + II** (90) **I:II** = 1:1 **II**	915
C₁₈ 	RuCl₂(PPh₃)₃, C₆H₆, 80°, 72 h	(93)	928

TABLE VI. OLIGOCYCLIC RINGS CONTAINING ONE OR MORE HETEROATOMS (*Continued*)

Reactant	Conditions	Product(s) and Yield(s) (%)	Refs.
	Bu₃SnH, AIBN	**I** + **II**	99
	NaI, Me₂CO, -20°	R = 4-MeC₆H₄ (84) / R = 4-ClC₆H₄ (90)	929, 930
	(Bu₃Sn)₂, *hv*, 150°	**I** + **II** (57) **I:II** = 93:7	895
	Bu₃SnH, AIBN, PhMe, 110°	(71)	931

TABLE VI. OLIGOCYCLIC RINGS CONTAINING ONE OR MORE HETEROATOMS (*Continued*)

Reactant	Conditions	Product(s) and Yield(s) (%)	Refs.
	Bu₃SnH, AIBN, C₆H₆, 80°	(47)	560, 561
	Bu₃SnH, AIBN, C₆H₆, DIPHOS	(60) 1 : 1 mixture of isomers	878
(1:1)	Bu₃SnH, AIBN, C₆H₆, 80°	(30) α:β = 90:10	889
	Bu₃SnH, AIBN, C₆H₆, 80°	(>69)	753

Reactant	Conditions	Product(s) and Yield(s) (%)	Refs.

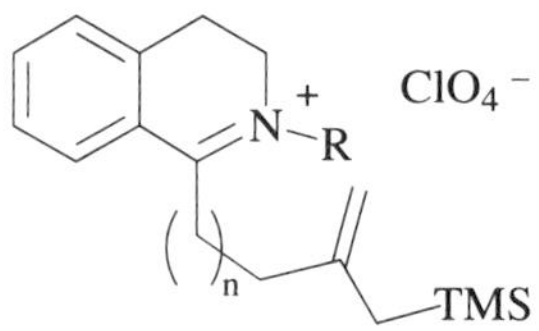

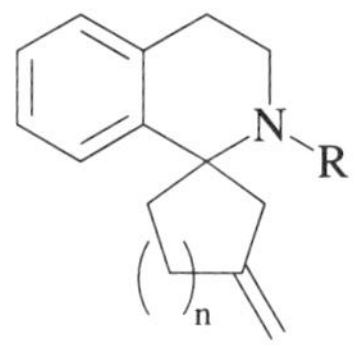

The reactant (isoquinolinium perchlorate bearing a TMS-allyl chain, ClO_4^-, N–R) gives product (tetrahydroisoquinoline spiro with exocyclic methylene, N–R):

n	R		
1	Me	MeOH, $h\nu$	(87)
1	CH$_2$CO$_2$Et	MeCN, $h\nu$	(70)
2	CH$_2$CO$_2$Et	MeOH, $h\nu$	(71)
3	Me	MeOH, $h\nu$	(66)

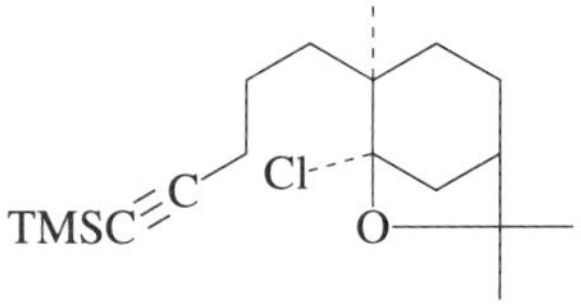

932

C$_{19}$

Bu$_3$SnH, AIBN — (72) — 23

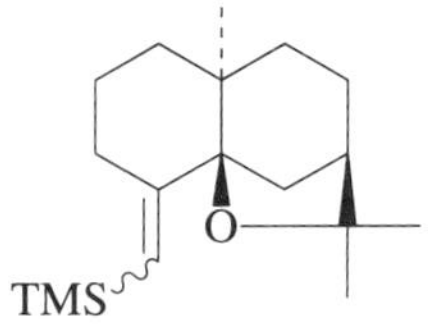

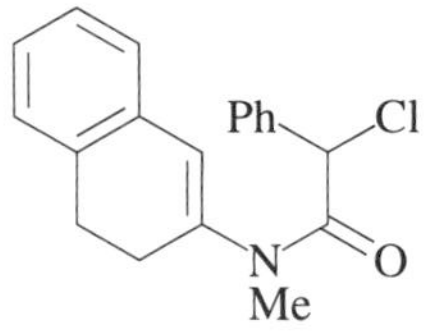

Bu$_3$SnH, AIBN
PhMe, 110° — (76) — 715

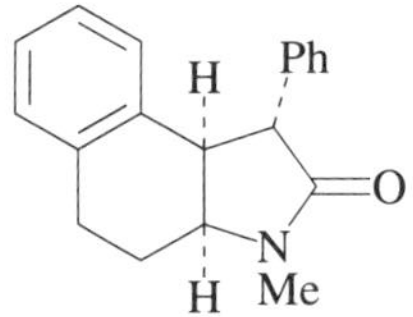

794

TABLE VI. OLIGOCYCLIC RINGS CONTAINING ONE OR MORE HETEROATOMS (*Continued*)

Reactant	Conditions	Product(s) and Yield(s) (%)	Refs.
	DPDC, C_6H_6, 60°		933
		R R^1	
		H Ph (85)	
		Cl Ph (90)	
		OMe 4-ClC$_6$H$_4$ (85)	
		H 4-O$_2$NC$_6$H$_4$ (86)	
		H 4-MeOC$_6$H$_4$ (86)	
	$K_2S_2O_8$, H_2O, 100°	(97)	934
	CuCl$_2$, CuCl, THF, AcOH, H$_2$O	I + II (84) I:II = 2.2:1	163

795

Reactant	Conditions	Product(s) and Yield(s) (%)	Refs.
	Bu₃SnH, AIBN, C₆H₆, 80°	 **I** + **II** **I + II** (60) **I:II** = 3.3:1	935
	1. NaI 2. Bu₃SnH, AIBN	 (91)	555
	Bu₃SnH, AIBN, C₆H₆, 80°	 (71)	931

TABLE VI. OLIGOCYCLIC RINGS CONTAINING ONE OR MORE HETEROATOMS (*Continued*)

Reactant	Conditions	Product(s) and Yield(s) (%)	Refs.
	Bu₃SnH, AIBN, C₆H₆, 80°	I + II (50) I:II = 85:15	936
	Bu₃GeH, AIBN, C₆H₆, 80°	I + II (73) I:II = 1:2.4	50
	Bu₃SnH, AIBN, C₆H₆, 80°	(85)	42

Reactant	Conditions	Product(s) and Yield(s) (%)	Refs.
C$_{20}$			
	Bu$_3$SnH, C$_6$H$_6$, *hv*	R = Ph or SPh (60-85)	937
	Bu$_3$SnH, AIBN, C$_6$H$_6$, 80°	(85) 2:1 ratio	42
C$_{21}$			
	Bu$_3$GeH, AIBN, C$_6$H$_6$, 80°	(83)	50
	MeCN, *hv*	(47) *trans:cis* = 3:1	908

Reactant	Conditions	Product(s) and Yield(s) (%)	Refs.
	Bu₃SnH, AIBN	**I** + **II** $\dfrac{R \quad I \quad II}{Me \quad (30) \quad (50)}$ Ph (75) (13)	938
	Bu₃SnH, AIBN, C₆H₆, 80°	**I** + **II** **I + II** (45) **I:II** = 83:17	671

799

Reactant	Conditions	Product(s) and Yield(s) (%)	Refs.
	Bu₃SnH, AIBN, C₆H₆, 80°	(85)	79
	MeOH, *hv*	CO₂Me (60)	932, 939
	t-BuSH, *hv*	(57)	940
	Ph₃SnH, AIBN, C₆H₆, heat	(77)	112

C₂₂

Reactant	Conditions	Product(s) and Yield(s) (%)	Refs.
	A: (Ph₂S)₂, AIBN, C₆H₆, 40°, *hv* B: (Ph₂COSnMe₃)₂, C₆H₆, 80° A B	 (94) (85)	941
	Bu₃SnH, AIBN, C₆H₆, 80°	 I + II (92) I:II = 10:1	942
	Bu₃SnH, AIBN, C₆H₆, 80°	 (98)	943
	Bu₃SnH, AIBN, C₆H₆, 80°	 (36)	345

TABLE VI. OLIGOCYCLIC RINGS CONTAINING ONE OR MORE HETEROATOMS (Continued)

Reactant	Conditions	Product(s) and Yield(s) (%)	Refs.
C_{23}	Bu$_3$SnH, AIBN, C$_6$H$_6$, 80°	 R: H (57); 3-OMe (63); 4-OMe (65)	944
	Bu$_3$SnH, AIBN, C$_6$H$_6$, 80°	(>80)	696
	Bu$_3$SnH, Et$_3$B, C$_6$H$_{14}$, rt	**I, II** **I + II** (75) **I:II** = 1:1	945

802

TABLE VI. OLIGOCYCLIC RINGS CONTAINING ONE OR MORE HETEROATOMS (*Continued*)

Reactant	Conditions	Product(s) and Yield(s) (%)	Refs.
	1. Ph₃SnH, AIBN, C₆H₆, 80° 2. [O]	**I** (52)	945
C₂₅	1. Bu₃SnH, AIBN, C₆H₆, 80° 2. CrO₃, py, CH₂Cl₂	**I** (58) *E:Z* = 5.5:1	134
	Bu₃SnH, AIBN, C₆H₆, 80°	(>62)	946
SEM = TMS(CH₂)₂OCH₂	Bu₃SnH, AIBN, PhMe, 110°	(70)	947

803

Reactant	Conditions	Product(s) and Yield(s) (%)	Refs.

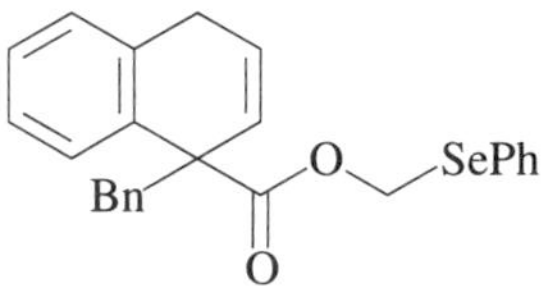

C$_{26}$

Bu$_3$SnH, AIBN, C$_6$H$_6$, 80°

(66)

50

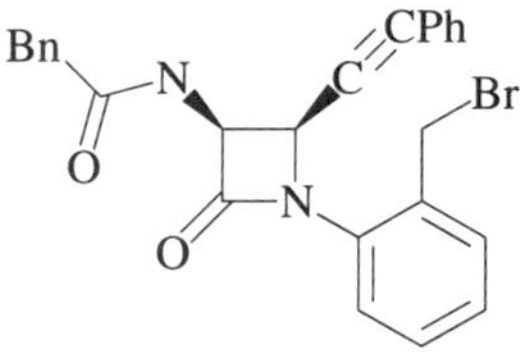

Bu$_3$SnH, AIBN, C$_6$H$_6$, 80°

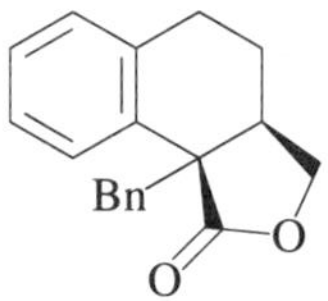

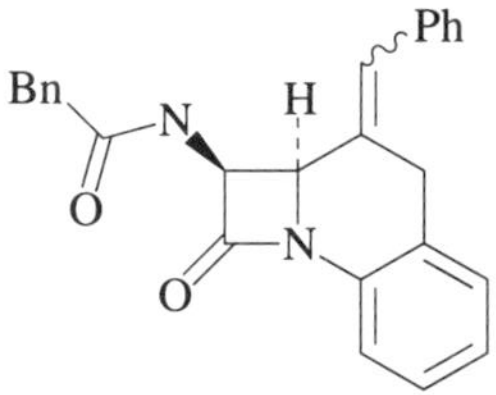

(75) *E:Z* = 1:1

948

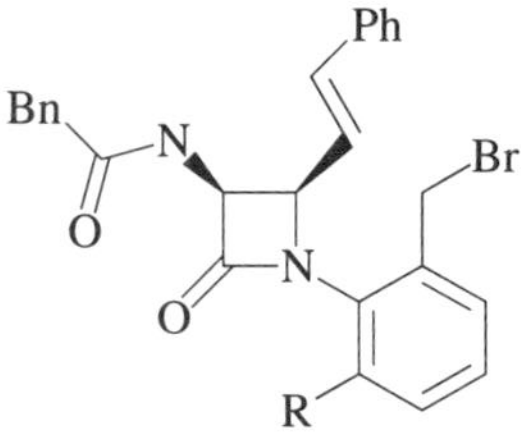

Bu$_3$SnH, AIBN, C$_6$H$_6$, 80°

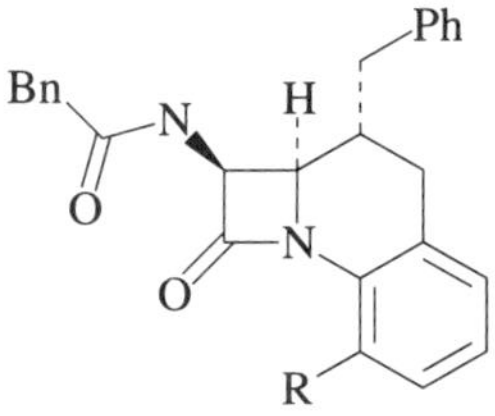

R	
H	(65)
OBn	(70)

948

Reactant	Conditions	Product(s) and Yield(s) (%)	Refs.

C₂₉

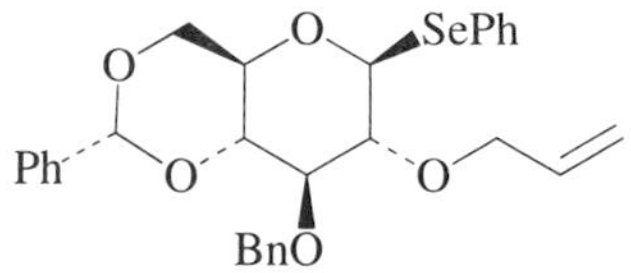

Bu₃SnH, AIBN,

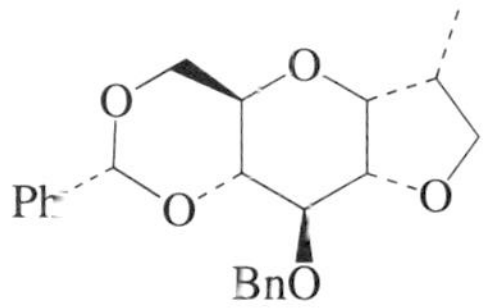

C₆H₆, 80°
PhMe, −78°, *hv*

I +

II

I + II (95) **I:II** = 1.6:?
I + II (95) **I:II** = 3.8:1

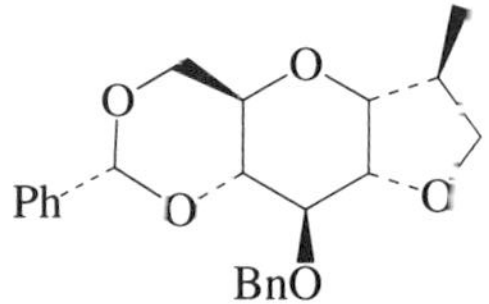

792

Bu₃SnH, AIBN,

C₆H₆, 80°
PhMe, −78°, *hv*

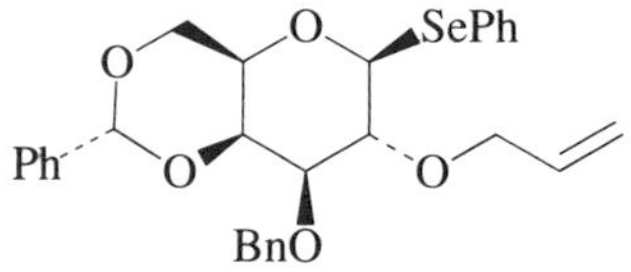

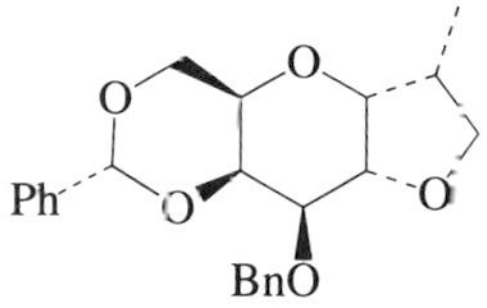

I +

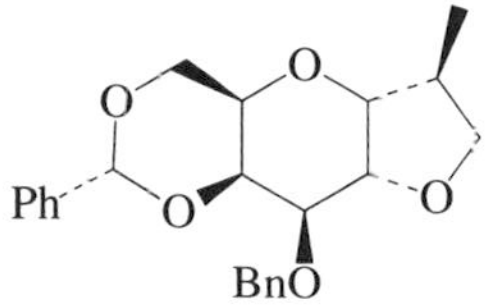

II

I + II (96) **I:II** = 8.1:1
I + II (96) **I:II** = >150:1

792

Reactant	Conditions	Product(s) and Yield(s) (%)	Refs.

C$_{37}$

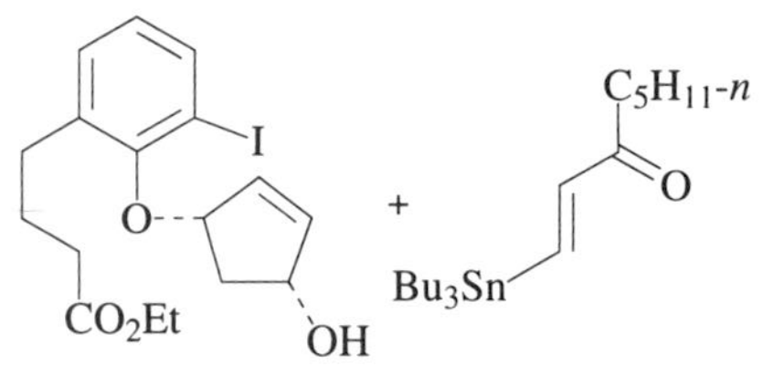

Bu$_3$SnH, AIBN, PhMe, 90°

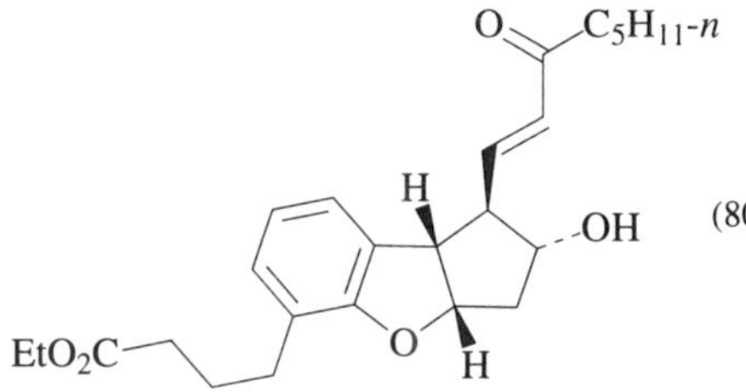

(80)   949

C$_{41}$

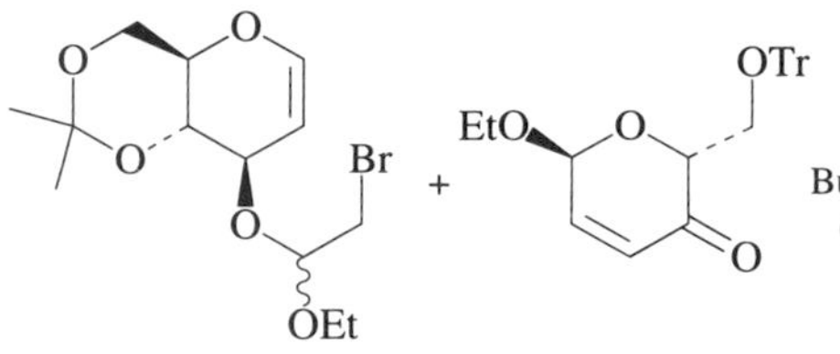

Bu$_3$SnH, AIBN, C$_6$H$_6$, 80°

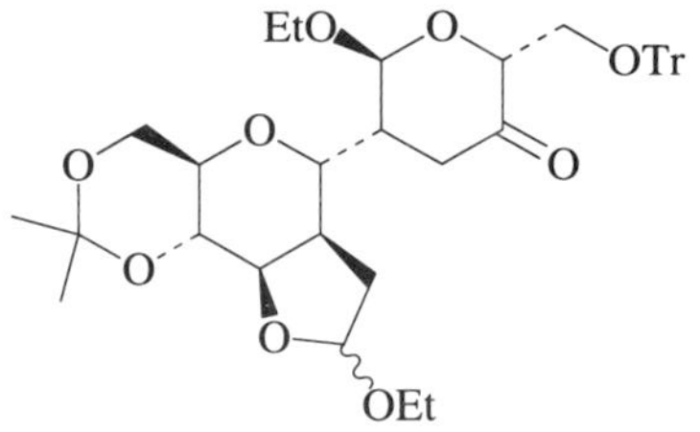

(70)   905

C$_{44}$

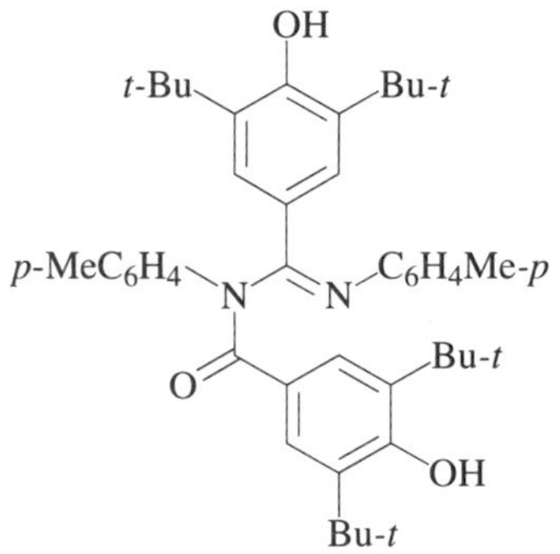

K$_2$Fe(CN)$_6$, OH$^-$

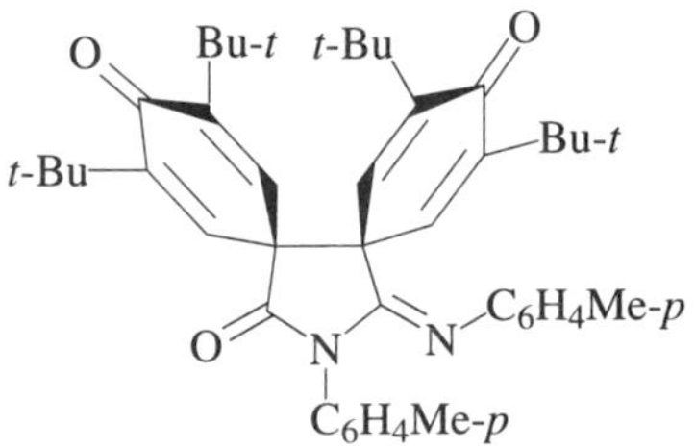

(95)   950

TABLE VI. OLIGOCYCLIC RINGS CONTAINING ONE OR MORE HETEROATOMS (*Continued*)

Reactant	Conditions	Product(s) and Yield(s) (%)	Refs.

C_{48}

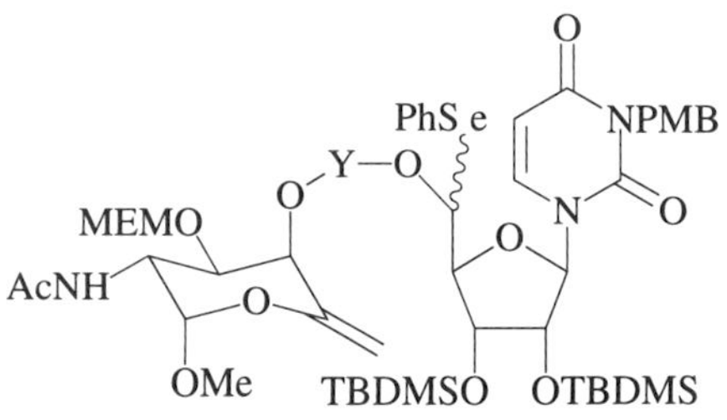

Bu$_3$SnH, Et$_3$B

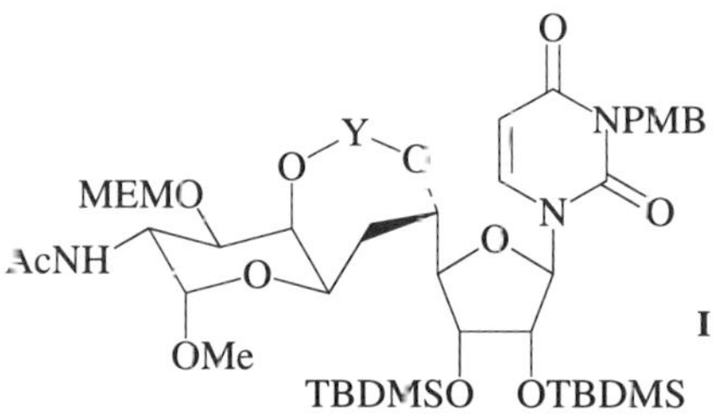

I + 527

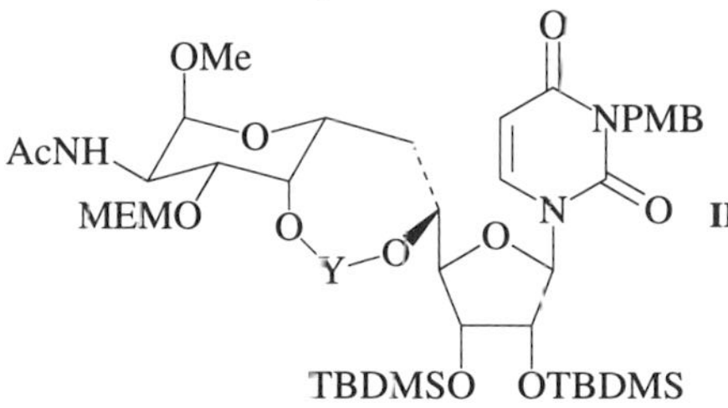

Y		Conditions		**I + II**	**I:II**
SiMe$_2$		MeCN, −10 to 23°		(80)	1:3.4
SiMe$_2$		PhMe, 0°		(74)	6:1
Si(OMe)$_2$		PhMe, 0°		(—)	3:1
Me$_2$SiOSiMe$_2$		PhMe, 0°		(—)	>95:5

TABLE VI. OLIGOCYCLIC RINGS CONTAINING ONE OR MORE HETEROATOMS (*Continued*)

Reactant	Conditions	Product(s) and Yield(s) (%)	Refs.
C$_{56}$	Bu$_3$SnH, AIBN, C$_6$H$_6$, 80°	(>40)	951
C$_{57}$	Bu$_3$SnH, AIBN, PhMe, 110°	(>35)	952

TABLE VI. OLIGOCYCLIC RINGS CONTAINING ONE OR MORE HETEROATOMS (*Continued*)

Reactant	Conditions	Product(s) and Yield(s) (%)	Refs.

C_{60}

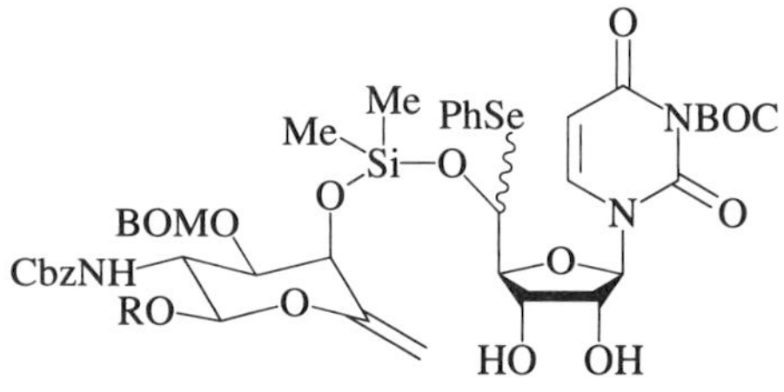

Bu₃SnH, Et₃B, 0°

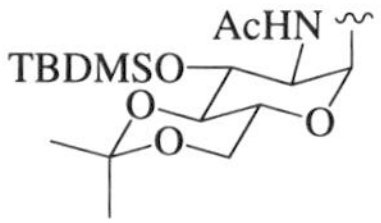

I + 953

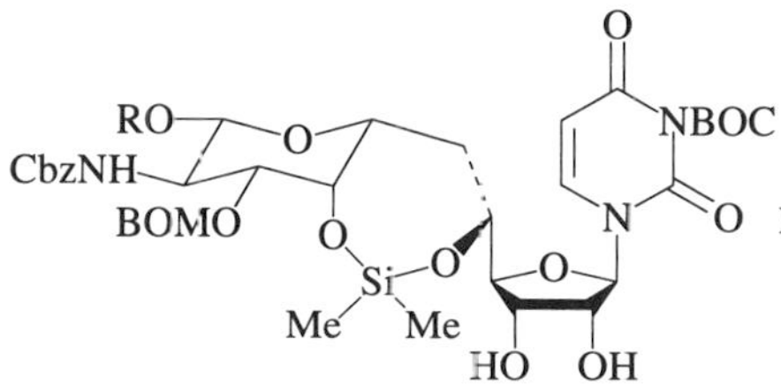

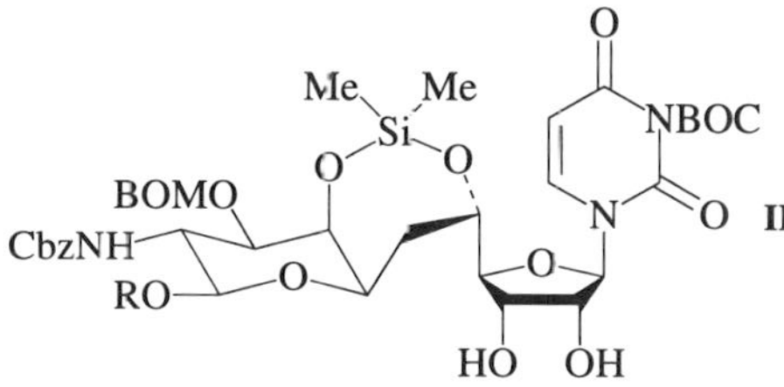

II

	I:II
I + II	
PhMe	(>68)	7.5:1
MeOH	(—)	1.6:1

809

Reactant	Conditions	Product(s) and Yield(s) (%)	Refs.

B. Tetracyclic Systems

C₁₃

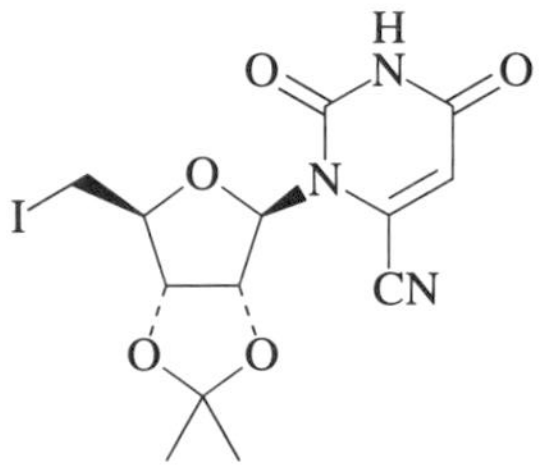

1. Bu₃SnH, AIBN, C₆H₆, 80°
2. AcOH, H₂O

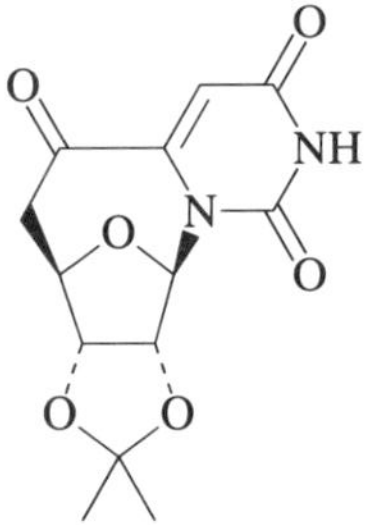

(35)

954

C₁₄

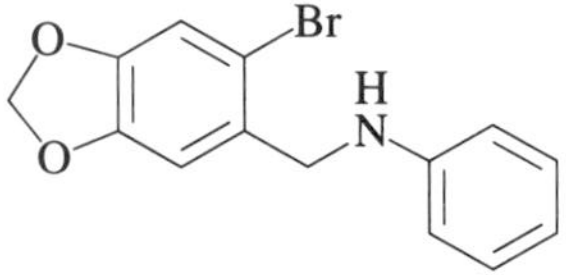

Bu₃SnH, AIBN, C₆H₆, 80°

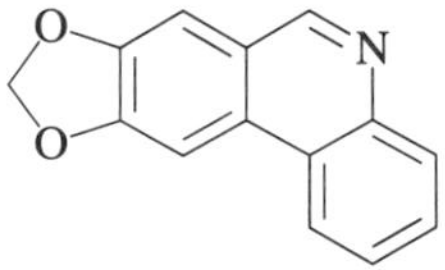

(66)

955

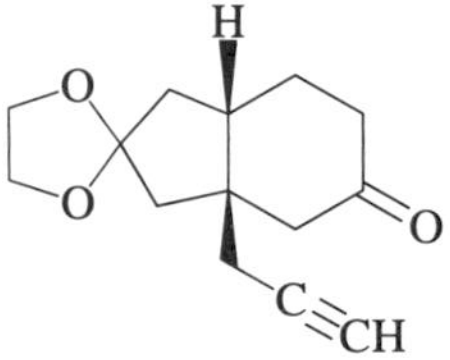

NH₃, K, THF, (NH₄)₂SO₄

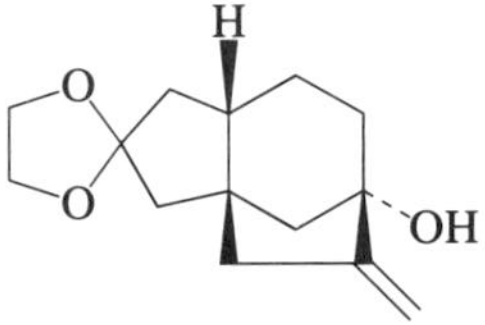

(75)

956

C₁₅

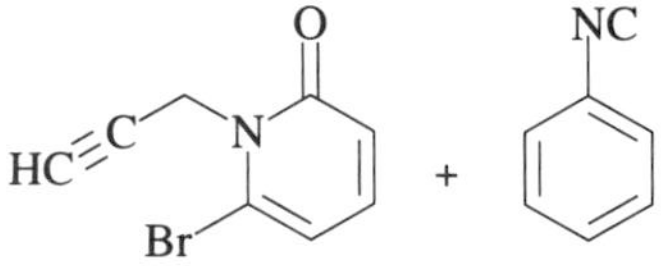

(Me₃Sn)₂, C₆H₆, hv, 80°

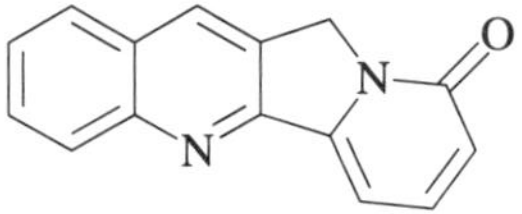

(40)

957

TABLE VI. OLIGOCYCLIC RINGS CONTAINING ONE CR MORE HETEROATOMS (*Continued*)

Reactant	Conditions	Product(s) and Yield(s) (%)	Refs.
	Bu₃SnH, AIBN, C₆H₆, 80°	(82) (90) (88)	958
	Bu₃SnH, AIBN, C₆H₆, heat	(51)	887
	Bu₃SnH, AIBN, C₆H₆, 80°	(68)	887

R	R¹	R²
H	H	H
OMe	–CH=CHCH=CH–	
Cl	–CH=CHCH=CH–	

Reactant	Conditions	Product(s) and Yield(s) (%)	Refs.
C$_{16}$			
	Bu$_3$SnH, AIBN, C$_6$H$_6$, 80°, *hv*	(53)	959
	Bu$_3$SnH, AIBN, C$_6$H$_6$, 80°, *hv*	(35)	959
	Bu$_3$SnH, AIBN, C$_6$H$_6$, 80°	(55)	887

Reactant	Conditions	Product(s) and Yield(s) (%)	Refs.
	Bu₃SnH, AIBN, C₆H₆, 80°	(81)	960
	Bu₃SnH, AIBN, C₆H₆, 80°	(31) + (40)	961
C₁₇	Bu₃SnH, AIBN, C₆H₆, 80°	(67)	962
	K₃Fe(CN)₆, KOH	(57)	868

813

Reactant	Conditions	Product(s) and Yield(s) (%)	Refs.
	1. Bu₃SnH, AIBN 2. [O]	(82) 3.8:1 mixture	963
	Carbon felt anode	Y OH (70) OH (70) OAc (73)	964
R Ac CO₂Bu-*t* Me	Bu₃SnH, AIBN	I + II (65) I:II = 2.5:1	158, 828, 965

TABLE VI. OLIGOCYCLIC RINGS CONTAINING ONE OR MORE HETEROATOMS (*Continued*)

Reactant	Conditions	Product(s) and Yield(s) (%)	Refs.
	Bu$_3$SnH, AIBN, C$_6$H$_6$, 80°	(90)	686
	Bu$_3$SnH, AIBN	(62)	966
	1. Cycloaddition 2. Bu$_3$SnH, AIBN	(70)	962
C$_{18}$	Bu$_3$SnH, AIBN, C$_6$H$_6$, 80°	I + II	345

Y	Z
H	H
TMS	H
H	CO$_2$Et

I	II
(21) *cis* Me, H	(35) *trans* Me, H
(—)	(54)
(36)	(36)

Reactant	Conditions	Product(s) and Yield(s) (%)	Refs.

C$_{19}$

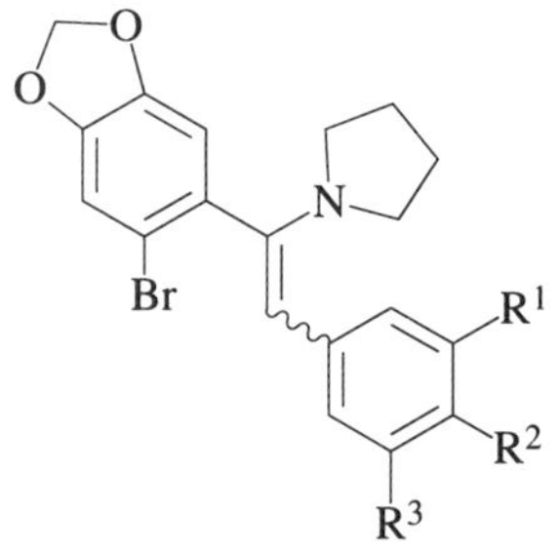

Bu$_3$SnH, AIBN, C$_6$H$_6$, 80°

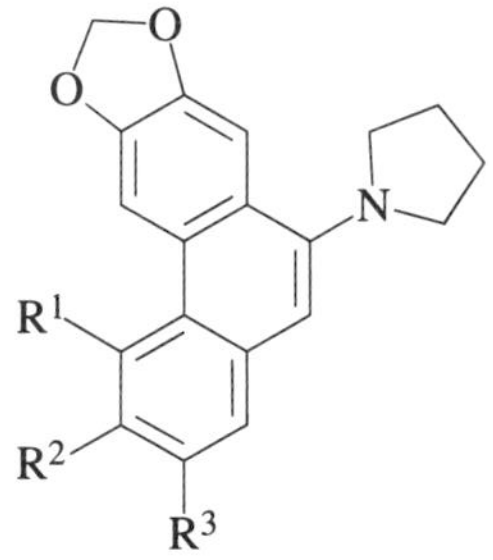

R^1	R^2	R^3
OMe	OMe	OMe
H	H	H
H	—OCH$_2$O—	
—OCH$_2$O—		

R^1	R^2	R^3	
OMe	OMe	OMe	(60)
H	H	H	(60–80)
H	—OCH$_2$O—		(60–80) 6:1
—OCH$_2$O—		H	

967

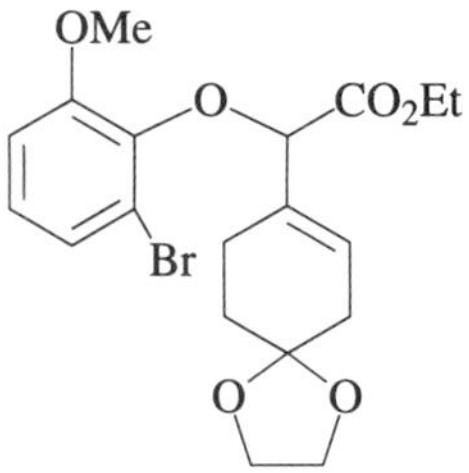

Bu$_3$SnH, AIBN, *o*-xylene, 140°

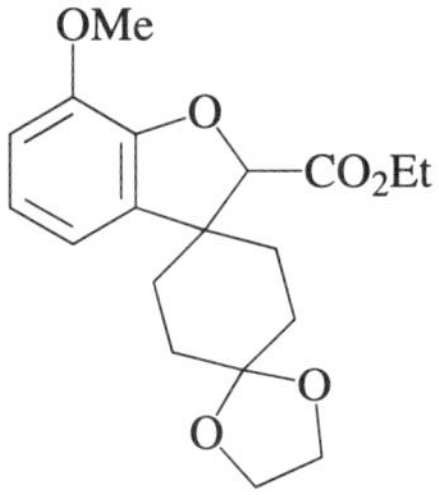

(91)

923

816

Reactant	Conditions	Product(s) and Yield(s) (%)	Refs.
	Bu₃SnH, AIBN PhMe, 110°	R = H (61) R = OMe (62)	968
	Bu₃SnH, AIBN C₆H₆, 80°	(53)	887
	Bu₃SnH, AIBN PhMe, 110°, 20 h	(74)	969

817

TABLE VI. OLIGOCYCLIC RINGS CONTAINING ONE OR MORE HETEROATOMS (*Continued*)

Reactant	Conditions	Product(s) and Yield(s) (%)	Refs.
	1. Bu₃SnH, AIBN 2. H⁺	(80)	943
	Bu₃SnH, AIBN, C₆H₆, 80°	**I** **II** **I + II** (91) **I:II** = 1:4	970
	(Me₃Sn)₂, *hv*, 80°	(45)	957
	Bu₃SnH, AIBN, C₆H₆, 80°	X O (49) NH (60)	971, 972

C₂₀

TABLE VI. OLIGOCYCLIC RINGS CONTAINING ONE OR MORE HETEROATOMS (Continued)

Reactant	Conditions	Product(s) and Yield(s) (%)	Refs.

C_21

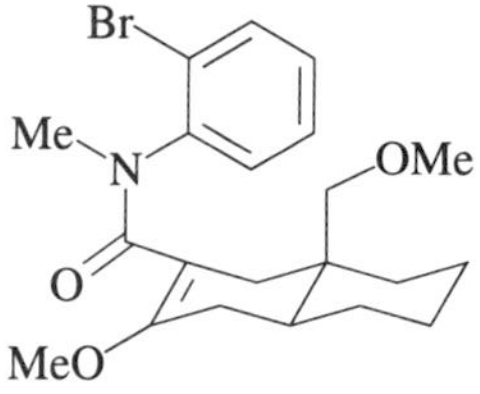

Bu_3SnH, AIBN, C_6H_6, 80°

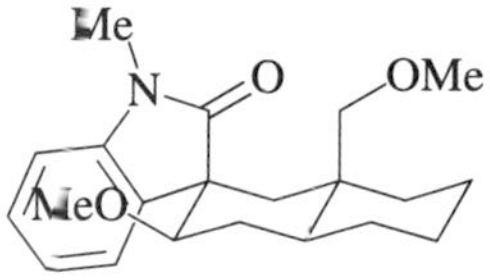

I +

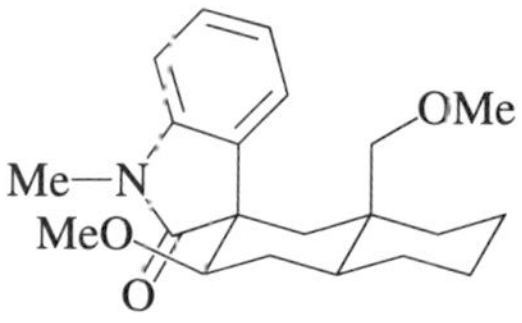

II

I + II (84)
I:II = 1:1

917

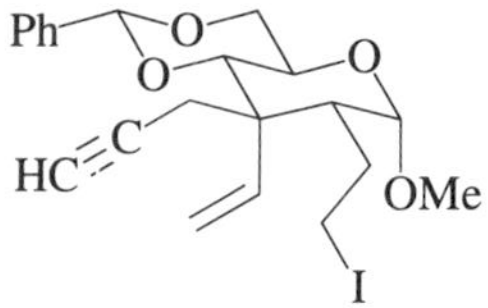

Bu_3SnH, AIBN, C_6H_6, 80°

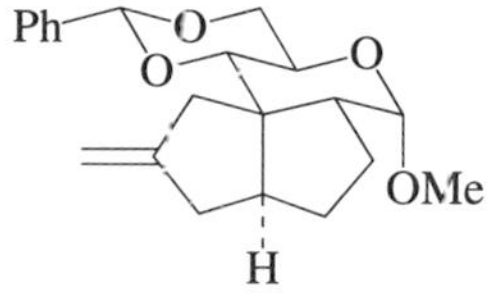

(91)

970

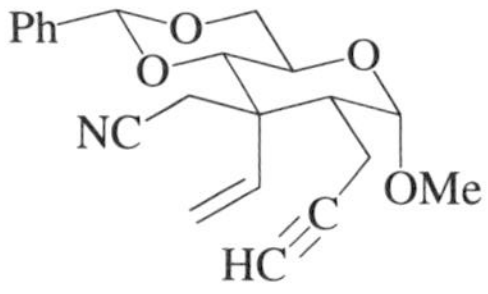

1. Bu_3SnH, AIBN, C_6H_6, 80°
2. CH_2Cl_2, silica gel

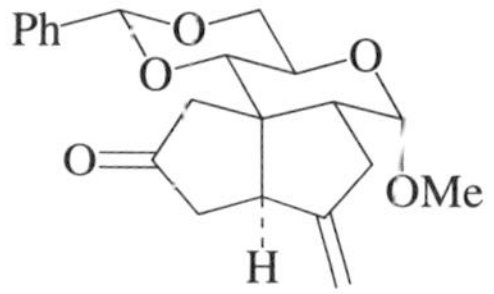

(85)

973

Reactant	Conditions	Product(s) and Yield(s) (%)	Refs.
	Bu₃SnH, AIBN, PhMe, 110°	R = –CH₂O– (51) R = Me (32)	968
	MeOH, *hv*	(80) (84) (83) (61) (52)	974, 975
	Bu₃SnH, AIBN, C₆H₆, 80°	(94)	943

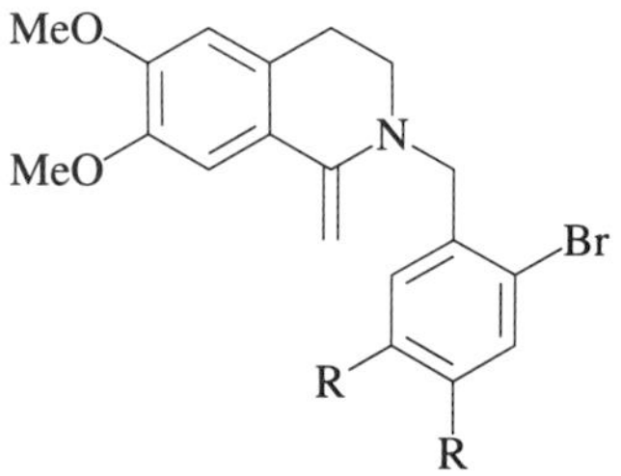

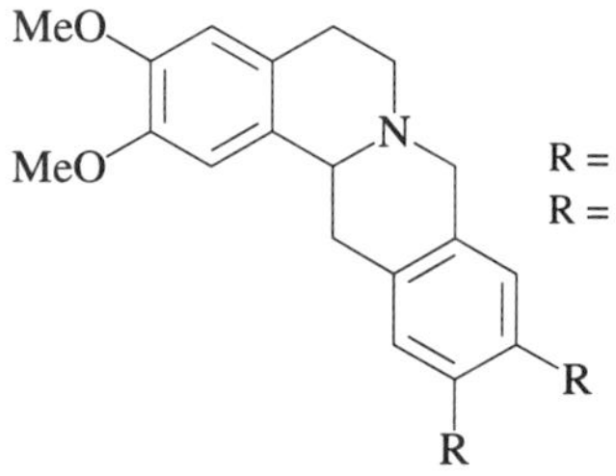

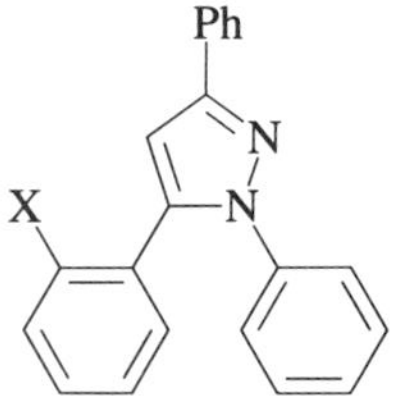

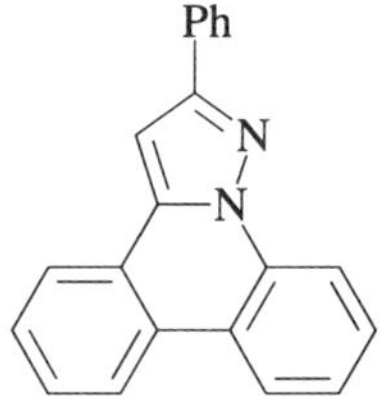

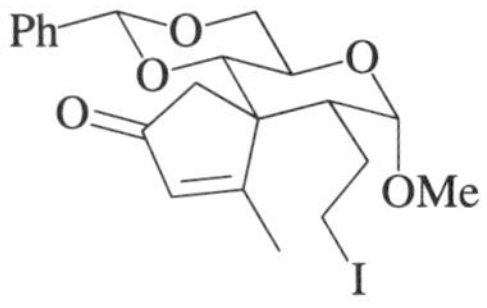

X / Cl / Br / I / OAc / OMe

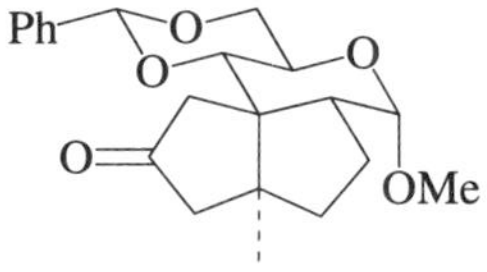

820

TABLE VI. OLIGOCYCLIC RINGS CONTAINING ONE OR MORE HETEROATOMS (*Continued*)

Reactant	Conditions	Product(s) and Yield(s) (%)	Refs.
	MeOH, MeCN (1:1) *hv*	(55) + (16)	976
	MeOH, MeCN (1:1) *hv*	(57) + (34)	976

TABLE VI. OLIGOCYCLIC RINGS CONTAINING ONE OR MORE HETEROATOMS (*Continued*)

Reactant	Conditions	Product(s) and Yield(s) (%)	Refs.
C$_{22}$			
	1. Bu$_3$SnH, AIBN, C$_6$H$_6$, 80° 2. CH$_2$Cl$_2$, silica gel	(92)	973
	1. Bu$_3$SnH, AIBN, C$_6$H$_6$, 80° 2. CH$_2$Cl$_2$, silica gel	(91)	977
	Bu$_3$SnH, AIBN, C$_6$H$_6$, 80°	(82)	978
	Bu$_3$SnH, AIBN, C$_6$H$_6$, 80°	(>63)	979

Reactant	Conditions	Product(s) and Yield(s) (%)	Refs.
C₂₃			
	Bu₃SnH, AIBN, C₆H₆, 80°	(94)	980
	1. Bu₃SnH, AIBN, PhMe, 110° 2. DDQ	R = Ph (84) R = H (86)	981
	1. Bu₃SnCl, NaH 2. Bu₃SnH, AIBN, PhMe, 110°	(73)	981

TABLE VI. OLIGOCYCLIC RINGS CONTAINING ONE OR MORE HETEROATOMS (*Continued*)

Reactant	Conditions	Product(s) and Yield(s) (%)	Refs.
C₂₄	Bu₃SnH, AIBN	(51) + (28)	982
	Bu₃SnH, AIBN, C₆H₆, 80°	(65)	71
C₂₅	Bu₃SnH, AIBN	(48)	982

824

TABLE VI. OLIGOCYCLIC RINGS CONTAINING ONE OR MORE HETEROATOMS (*Continued*)

Reactant	Conditions	Product(s) and Yield(s) (%)	Refs.
	Bu$_3$SnH, AIBN, C$_6$H$_6$, 80°	(50)	983
C$_{26}$			
	Bu$_3$SnH, AIBN, C$_6$H$_6$, 80°	(70)	984
	SmI$_2$, THF, 25°	(92)	985

Reactant	Conditions	Product(s) and Yield(s) (%)	Refs.
C₃₀	Bu₃SnH, AIBN, C₆H₆, 80°, 18 h	(52)	969
C₃₁	Bu₃SnH, AIBN, C₆H₆, heat, 35 h	(35)	986
	Bu₃SnH, AIBN, C₆H₆, 80°	(70)	987, 988

TABLE VI. OLIGOCYCLIC RINGS CONTAINING ONE OR MORE HETEROATOMS (*Continued*)

Reactant	Conditions	Product(s) and Yield(s) (%)	Refs.
C_{33}	Bu$_3$SnH, AIBN, C$_6$H$_6$, 80°	R = H (57) R = Me (82)	987, 988
C_{38}	Bu$_3$SnH, AIBN, C$_6$H$_6$, 80°	(65)	133

C. Polycyclic Systems

C_{16}	Bu$_3$SnH, AIBN, C$_6$H$_6$, 80°	(59)	887
C_{17}	Bu$_3$SnH, AIBN, C$_6$H$_6$, 80°	X = O (76) X = NMe (54) X = SO (67)	887, 962

Reactant	Conditions	Product(s) and Yield(s) (%)	Refs.
	Bu$_3$SnH, AIBN, C$_6$H$_6$, 80°	(67)	962
	Ph$_3$SnH, AIBN, C$_6$H$_6$, 80°	(55) + (37)	989
C$_{20}$			
	Bu$_3$SnH, AIBN, C$_6$H$_6$, 80°	(56)	886

828

TABLE VI. OLIGOCYCLIC RINGS CONTAINING ONE OR MORE HETEROATOMS (*Continued*)

Reactant	Conditions	Product(s) and Yield(s) (%)	Refs.

C_{22}

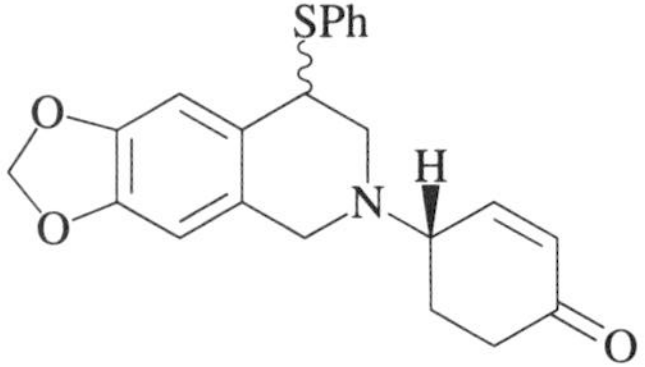

Bu₃SnH, AIBN, xylene

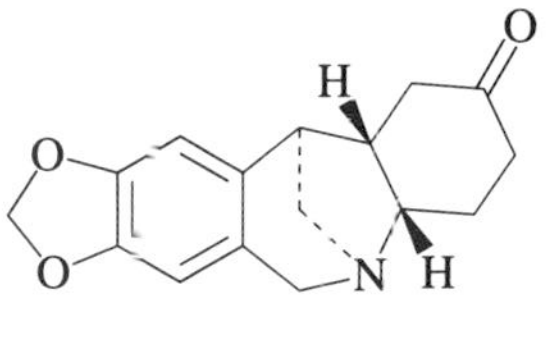

(80)

990

C_{25}

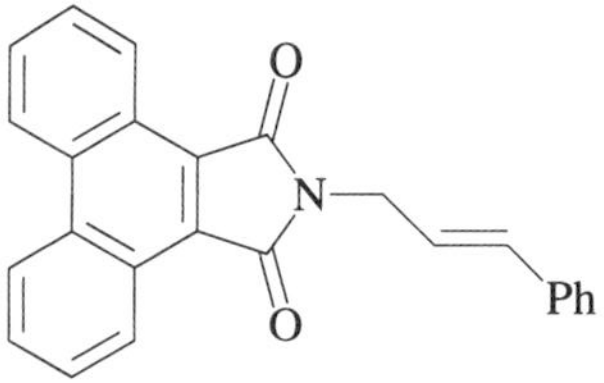

MeOH, MeCN (1:1), *hv*

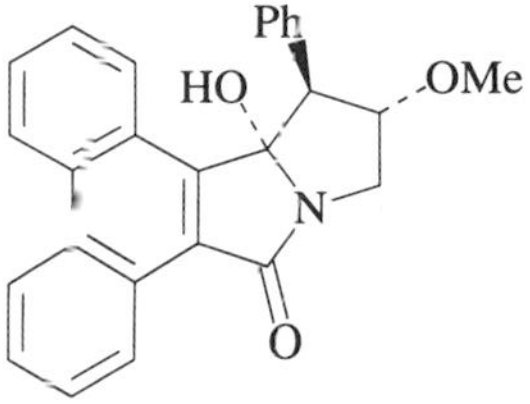

(49)

976

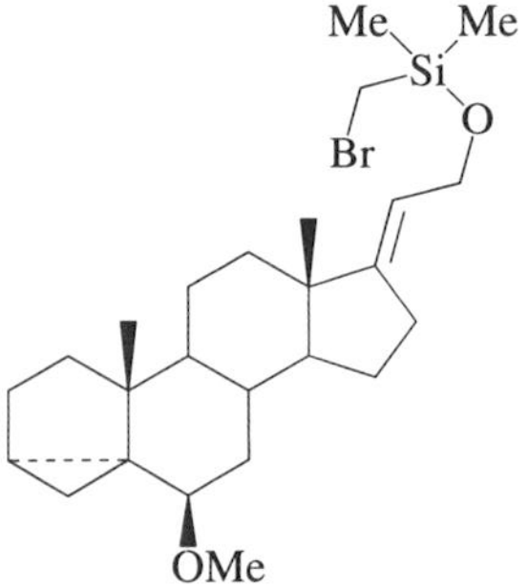

Bu₃SnH, AIBN, C_6H_6, 80°

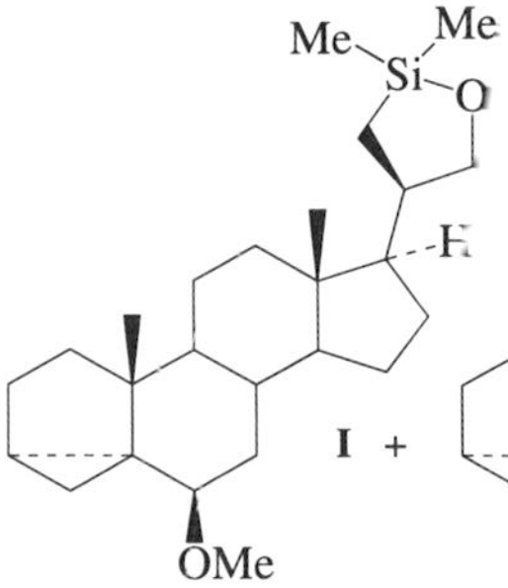

I + II

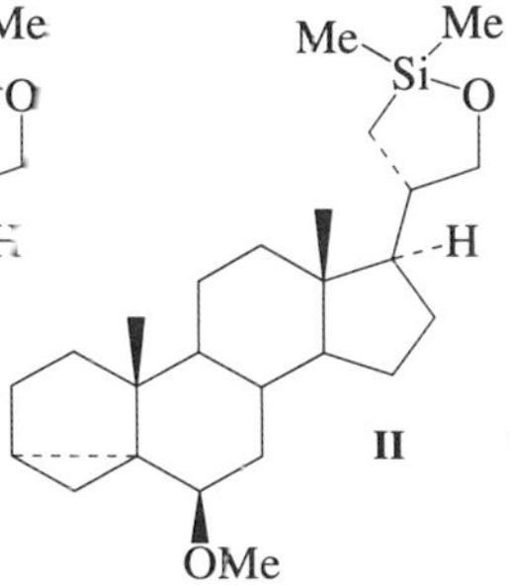

I + II (>55) I:II = 49:51
I + II (>48) I:II = 43:57

991

E
Z

TABLE VI. OLIGOCYCLIC RINGS CONTAINING ONE OR MORE HETEROATOMS (*Continued*)

Reactant	Conditions	Product(s) and Yield(s) (%)	Refs.
C_{27}	Bu₃SnH, AIBN, PhMe, 110°	R = H (68) R = OAc (70)	992
	Bu₃SnH, AIBN, C₆H₆, 80°	(52) + (16)	993
C_{29}	t-BuSH, THF, hv, 80°	(60)	994

TABLE VI. OLIGOCYCLIC RINGS CONTAINING ONE OR MORE HETEROATOMS (*Continued*)

Reactant	Conditions	Product(s) and Yield(s) (%)	Refs.
	Bu₃SnH, AIBN, C₆H₆, 80°	(70)	58
	Bu₃SnH, AIBN, C₆H₆, 80°	(65)	58
	Bu₃SnH, AIBN, C₆H₆, 80°	(>75)	101

TABLE VI. OLIGOCYCLIC RINGS CONTAINING ONE OR MORE HETEROATOMS (*Continued*)

Reactant	Conditions	Product(s) and Yield(s) (%)	Refs.
C$_{30}$	A: Li/NH$_3$, THF, -78° B: Bu$_3$SnH, AIBN, PhMe, 110°		995

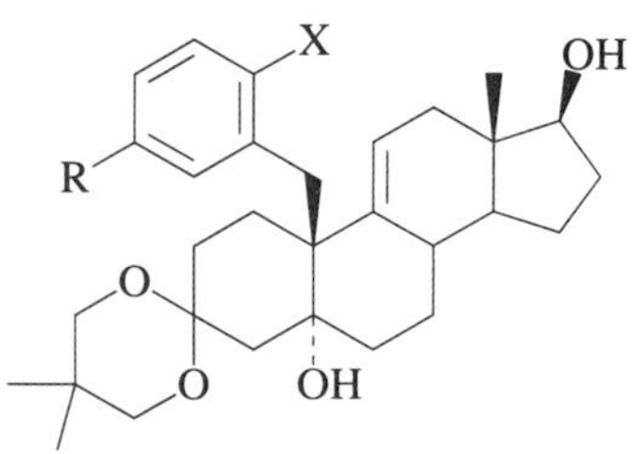

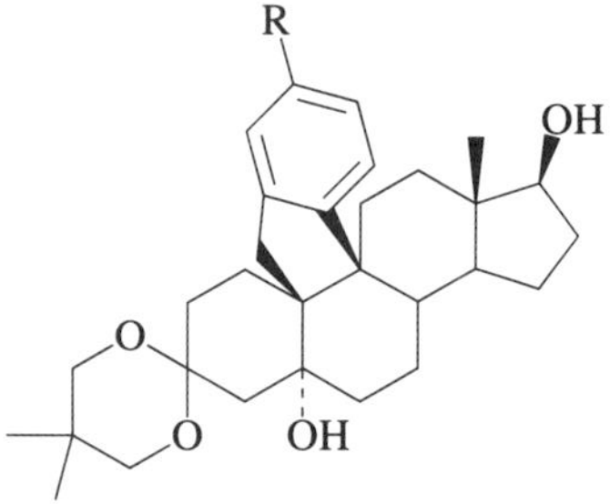

R	X	
H	F	A (82)
H	Cl	A (81)
H	Br	A (85)
H	Br	B (89)
OMe	Br	A (89)
OMe	Br	B (92)

| C$_{31}$ | Bu$_3$SnH, AIBN,
C$_6$H$_6$, 80° | (73) | 996 |

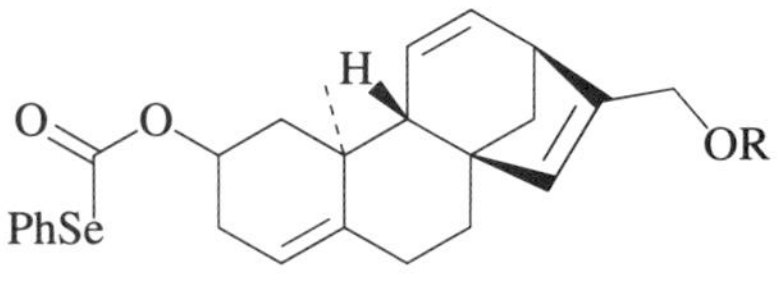

R = TBDMS

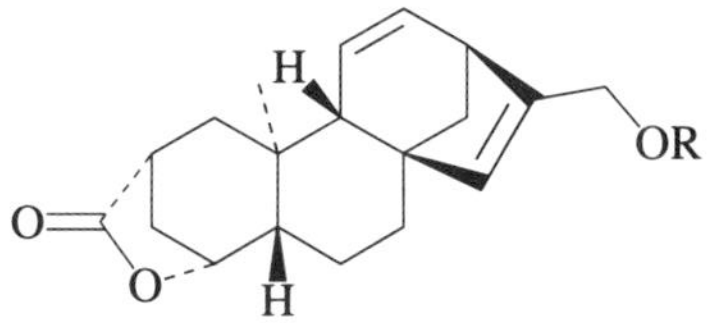

832

TABLE VI. OLIGOCYCLIC RINGS CONTAINING ONE OR MORE HETEROATOMS (Continued)

Reactant	Conditions	Product(s) and Yield(s) (%)	Refs.
C_{30}			
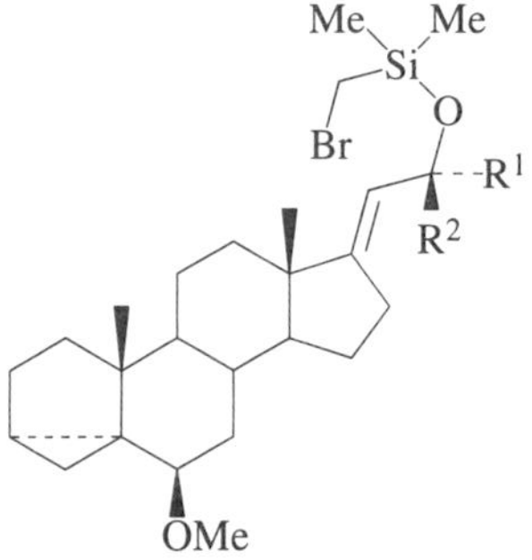 R^1 / R^2 H / $(CH_2)_2Pr\text{-}i$ $(CH_2)_2Pr\text{-}i$ / H	Bu_3SnH, AIBN, C_6H_6, 80°	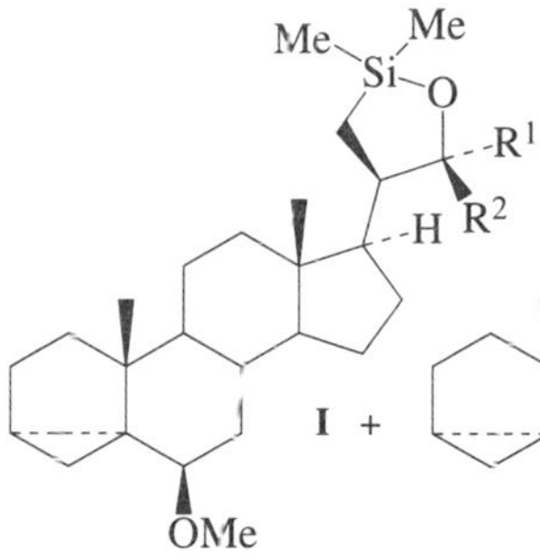 **I** + 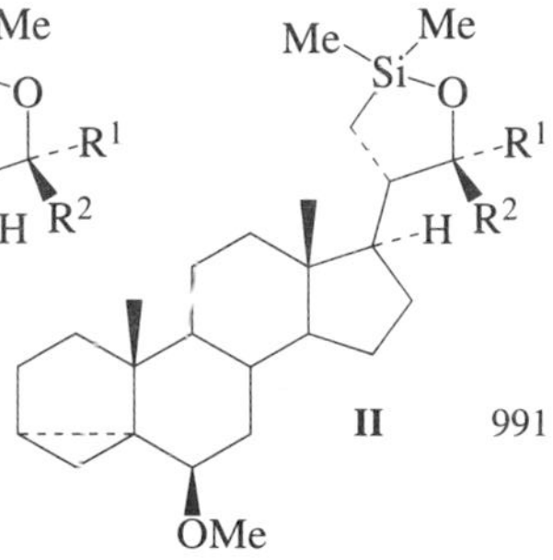 **II** **I + II** (>62) **I:II** = 100:0 **I + II** (>56) **I:II** = 0:100	991
C_{33}			
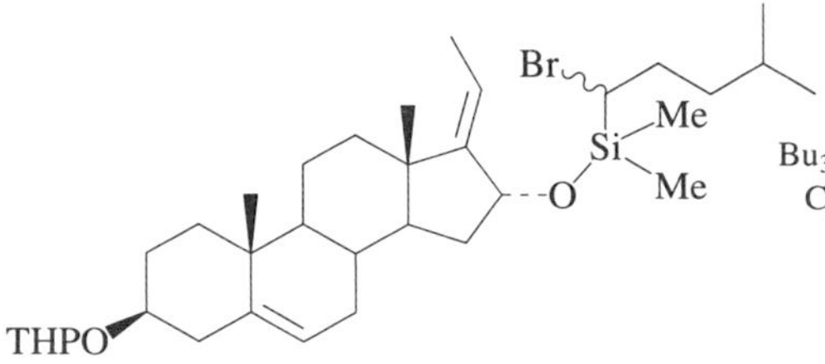	Bu_3SnH, AIBN, C_6H_6, 80°	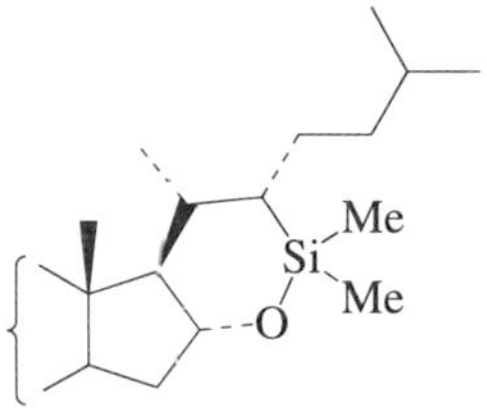 **I** + 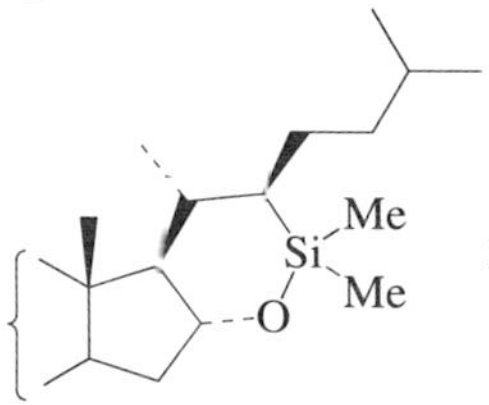 **II** **I + II** (65) **I:II** = 4:1	997

TABLE VI. OLIGOCYCLIC RINGS CONTAINING ONE OR MORE HETEROATOMS (*Continued*)

Reactant	Conditions	Product(s) and Yield(s) (%)	Refs.
C_{34}	Bu_3SnH, AIBN, PhMe, 110°	(50)	991
C_{36}	Bu_3SnH, AIBN, C_6H_6, 80°	(46)	917

REFERENCES

[1] J.-M. Surzur in *Reactive Intermediates*, R. A. Abramovitch, Ed., Vol. 2, Plenum Press, New York, 1982, p. 121.

[2] D. P. Curran, *Synthesis*, **417**, 489 (1988).

[3] M. Ramaiah, *Tetrahedron*, **43**, 3541 (1987).

[4] D. P. Curran in *Comprehensive Organic Synthesis*, B. M. Trost, and I. Fleming, Eds.; Vol. IV, Pergamon Press, London, 1991, p. 779.

[5] C. P. Jasperse, D. P. Curran, and T. L. Fevig, *Chem. Rev.*, **91**, 1237 (1991).

[6] B. Giese in Radicals in *Organic Synthesis, Formation of Carbon–Carbon Bonds*, Pergamon Press, Oxford, 1986.

[7] M. Regitz and B. Giese in *Houben–Weyl, Methoden der Organischen Chemie*, Vol. E19a, G. Thieme Verlag, Stuttgart, 1989.

[8] M. B. Motherwell and D. Crich in *Best Synthetic Methods, Free Radical Chain Reactions in Organic Synthesis*, Academic Press, London, 1991.

[9] A. Effio, D. Griller, K. U. Ingold, A. L. J. Beckwith, and A. K. Serelis, *J. Am. Chem. Soc.* **102**, 1734 (1980).

[10] A. L. J. Beckwith and K. U. Ingold in *Rearrangements in Ground and Excited States*, P. de Mayo, Ed., Academic Press, New York, 1980.

[11] Z. Cekovic and R. Saicic, *Tetrahedron Lett.*, **31**, 6085 (1990).

[12] M. E. Jung, J. D. Trifunovich, and N. Lensen, *Tetrahedron Lett.*, **33**, 6719 (1992).

[13] A. L. J. Beckwith and C. H. Schiesser, *Tetrahedron*, **41**, 3925 (1985).

[14] A. L. J. Beckwith, *Tetrahedron*, **37**, 3073 (1981).

[15] D. C. Spellmeyer and K. N. Houk, *J. Org. Chem.*, **52**, 959 (1987).

[16] A. L. J. Beckwith, J. C. Christopher, T. Lawrence, and A. K. Serelis, *Aust. J. Chem.*, **36**, 545 (1983).

[17] S. Hanessian, D. S. Dhanoa, and P. L. Beaulieu, *Can. J. Chem.*, **65**, 1859 (1987).

[18] D. P. Curran and C.-T. Chang, *J. Org. Chem.*, **54**, 3140 (1989).

[19] D. L. J. Clive and D. R. Cheshire, *J. Chem. Soc., Chem. Commun.*, **1987**, 1520.

[20] M. Julia, *Acc. Chem. Res.*, **4**, 386 (1971).

[21] M. Julia, *Pure Appl. Chem.*, **40**, 553 (1974).

[22] M. Julia and M. Maumy, *Org. Synth.*, **55**, 57 (1976); *Org. Synth., Coll. Vol.*, **6**, 586 (1988).

[23] G. Büchi and H. Wüest, *J. Org. Chem.*, **44**, 546 (1979).

[24] G. Stork and N. H. Baine, *J. Am. Chem. Soc.*, **104**, 2321 (1982).

[25] G. Stork and R. Mook, Jr., *Tetrahedron Lett.*, **27**, 4529 (1986).

[26] A. L. J. Beckwith and D. M. O'Shea, *Tetrahedron Lett.*, **27**, 4525 (1986).

[27] K. Nozaki, K. Oshima, and K. Utimoto, *J. Am. Chem. Soc.*, **109**, 2547 (1987).

[28] G. Stork and N. H. Baine, *Tetrahedron Lett.*, **26**, 5927 (1985).

[29] A. L. J. Beckwith and W. B. Gara, *J. Chem. Soc., Perkin. Trans. 2*, **1975**, 795.

[30] H. Urabe and I. Kuwajima, *Tetrahedron Lett.*, **27**, 1355 (1986).

[31] K. A. Parker, D. M. Spero, and K. C. Inman, *Tetrahedron Lett.*, **27**, 2833 (1986).

[32] A. N. Abeywickrema, A. L. J. Beckwith, and S. Gerba, *J. Org. Chem.*, **52**, 4072 (1987).

[33] D. L. Boger and R. J. Mathvink, *J. Org. Chem.*, **57**, 1429 (1992).

[34] C. E. Schwartz and D. P. Curran, *J. Am. Chem. Soc.*, **112**, 9272 (1990).

[35] D. L. Boger and R. J. Mathvink, *J. Org. Chem.*, **53**, 3377 (1988).

[36] M. D. Bachi and E. Bosch, *Heterocycles*, **28**, 579 (1989).

[37] P. A. Bartlett, K. L. McLaren, and P. C. Ting, *J. Am. Chem. Soc.*, **110**, 1633 (1988).

[38] D. L. J. Clive, P. L. Beaulieu, and L. Set, *J. Org. Chem.*, **49**, 1313 (1984).

[39] A. L. J. Beckwith and B. P. Hay, *J. Am. Chem. Soc.*, **111**, 230 (1989).

[40] R. Tsang, J. K. Dickson, Jr., H. Pak, R. Walton, and B. Fraser–Reid, *J. Am. Chem. Soc.*, **104**, 3484 (1987).

[41] R. A. Walton and B. Fraser–Reid, *J. Am. Chem. Soc.*, **113**, 5791 (1991).

[42] R. Tsang and B. Fraser–Reid, *J. Am. Chem. Soc.*, **108**, 8102 (1986).

[43] A. Padera, H. Nimmesgern, and G. S. K. Wong, *Tetrahedron Lett.*, **26**, 957 (1985).

[44] T. W. Smith and G. B. Butler, *J. Org. Chem.*, **43**, 6 (1978).

[45] A. L. J. Beckwith, I. A. Blair, and G. Phillipou, *Tetrahedron Lett.*, **96**, 1613 (1974).

[46] A. L. J. Beckwith and S. A. Glover, *Aust. J. Chem.*, **40**, 157 (1987).

[47] A. Padwa, W. Dent, H. Nimmesgern, M. K. Venkatramanan, and G. S. K. Wong, *Chem. Ber.*, **119**, 813 (1986).

[48] V. H. Rawal, S. P. Singh, C. Dufour, and C. Michoud, *J. Org. Chem.*, **56**, 5245 (1991).

[49] D. L. J. Clive and P. L. Beaulieu, *J. Chem. Soc., Chem. Commun.*, **1983**, 307.

[50] A. L. J. Beckwith and P. E. Pigou, *J. Chem. Soc., Chem. Commun.*, **1986**, 85.

[51] D. P. Curran and J. Tamine, *J. Org. Chem.*, **56**, 2746 (1991).

[52] G. Stork, R. Mook, Jr., S. A. Biller, and S. D. Rychnovsky, *J. Am. Chem. Soc.*, **105**, 3741 (1983).

[53] Y. Ueno, K. Chinom, M. Watanabe, O. Moriya, and M. Okawara, *J. Am. Chem. Soc.*, **104**, 5565 (1982).

[54] G. Stork and R. Mah, *Heterocycles*, **28**, 723 (1989).

[55] J. W. Wilt, *Tetrahedron*, **41**, 3979 (1985).

[56] H. Nishiyama, T. Kitajima, M. Matsumoto, and K. Itoh, *J. Org. Chem.*, **49**, 2298 (1984).

[57] G. Stork and M. Kahn, *J. Am. Chem. Soc.*, **107**, 500 (1985).

[58] M. Koreeda and I. A. George, *J. Am. Chem. Soc.*, **108**, 8098 (1986).

[59] M. T. Crimmins and R. O'Mahony, *J. Org. Chem.*, **54**, 1157 (1989).

[60] J. M. Surzur, M. P. Bertrand, and R. Nouguier, *Tetrahedron Lett.*, **1969**, 4197.

[61] G. A. Kraus and J. Thurston, *Tetrahedron Lett.*, **28**, 4011 (1987).

[62] A. L. J. Beckwith and B. P. Hay, *J. Am. Chem Soc.*, **110**, 4415 (1988).

[63] M. Newcomb, M. T. Burchill, and T. M. Deeb, *J. Am. Chem. Soc.*, **110**, 3163 (1988).

[64] R. Sutcliffe and K. U. Ingold, *J. Am. Chem. Soc.*, **104**, 6071 (1982).

[65] F. Minisci, *Synthesis*, **1973**, 1.

[66] F. Stella, Angew. Chem., 95, 368 (1983); *Angew. Chem. Int. Ed. Engl.*, **22**, 337 (1983).

[67] M. Newcomb and T. M. Deeb, *J. Am. Chem. Soc.*, **109**, 3163 (1987).

[68] M. Newcomb and K. A. Weber, *J. Org. Chem.*, **56**, 1309 (1991).

[69] J. Boisin, E. Fouquet, and S. Z. Zard, *Tetrahedron Lett.*, **31**, 3545 (1990).

[70] B. B. Snider and J. E. Merritt, *Tetrahedron*, **47**, 8663 (1991).

[71] J. H. Rigby and M. N. Qabar, *J. Org. Chem.*, **58**, 4473 (1993).

[72] M. Koreeda and L. G. Hamann, *J. Am. Chem. Soc.*, **112**, 8175 (1990).

[73] P. Dowd and W. Zhang, *Chem. Rev.*, **93**, 2091 (1993).

[74] P. Dowd and S.-C. Choi, *J. Am. Chem. Soc.*, **109**, 3493 (1987).

[75] P. Dowd and S.-C. Choi, *Tetrahedron*, **45**, 77 (1989).

[76] J. Baldwin, R. M. Adlington, and J. Robertson, *Tetrahedron*, **45**, 909 (1989).

[77] N. A. Porter and V. H. T. Chang, *J. Am. Chem. Soc.*, **109**, 4976 (1987).

[78] N. A. Porter, V. H.-T. Chang, D. R. Magnin, and B. T. Wright, *J. Am. Chem. Soc.*, **110**, 3554 (1988).

[79] F. E. Ziegler and Z. L. Zheng, *Tetrahedron Lett.*, **28**, 5973 (1987).

[80] G. E. Keck and A. M. Tafesh, *Synlett*, **1990**, 257.

[81] P. Renaud, *Tetrahedron Lett.*, **31**, 4601 (1990).

[82] T. V. RajanBabu, T. Fukunaga, and G. S. Reddy, *J. Am. Chem. Soc.*, **111**, 1759 (1989).

[83] T. V. RajanBabu, *Acc. Chem. Res.*, **24**, 139 (1991).

[84] M.-Y. Chen, J.-M. Fang, Y.-M. Tsai, and R.-L. Yeh, *J. Chem. Soc., Chem. Commun.*, **1991**, 1603.

[85] D. P. Curran, W. Shen, J. Zhang, and T. A. Heffner, *J. Am. Chem. Soc.*, **112**, 6738 (1990).

[86] E. Nakamura, D. Machii, and T. Inubushi, *J. Am. Chem. Soc.*, **111**, 6849 (1989).

[87] F. Soucy, D. Wernic, and P. Beaulieu, *J. Chem. Soc., Perkin Trans. 1*, **1991**, 2885.

[88] J. H. Byers, T. G. Gleason, and K. S. Knight, *J. Chem. Soc., Chem. Commun.*, **1991**, 354.

[89] T. Sato, Y. Wada, M. Nishimoto, H. Ishibashi, and M. Ikeda, *J. Chem. Soc., Perkin Trans. 1*, **1989**, 879.

[90] H. Nagashima, K. Seki, N. Ozaki, H. Wakamatsu, K. Itoh, Y. Tomo, and J. Tsuji, *J. Org. Chem.*, **55**, 985 (1990).

[91] H. Nagashima, N. Ozaki, M. Ishii, K. Seki, M. Washiyama, and K. Itoh, *J. Org. Chem.*, **58**, 464 (1993).

[92] A. V. R. Rao, G. V. M. Sharma, and M. N. Bhanu, *Tetrahedron Lett.*, **33**, 3907 (1992).

[93] M. Ikara, K. Yasai, N. Tanigachi, and K. Fukumoto, *J. Chem. Soc., Perkin Trans. 1*, **1990**, 1469.

[94] S. D. Burke and J. Rancourt, *J. Am. Chem. Soc.*, **113**, 2335 (1991).

[95] N. A. Porter, B. Lacher, V. H.-T. Chang, and D. R. Magnin, *J. Am. Chem. Soc.*, **111**, 8309 (1989).

[96] V. H. Rawal, R. C. Newton, and V. Krishnamurthy, *J. Org. Chem.*, **55**, 5181 (1990).

[97] A. L. J. Beckwith, G. Phillipou, and A. K. Serelis, *Tetrahedron Lett.*, **29**, 2811 (1981).

[98] D. J. Hart and H. C. Huang, *Tetrahedron Lett.*, **26**, 3749 (1985).

[99] C. P. Chuang, J. C. Galluci, and D. J. Hart, *J. Org. Chem.*, **53**, 3210 (1988).

[100] G. Stork, P. M. Sher, and H. L. Chen, *J. Am. Chem. Soc.*, **108**, 6384 (1986).

[101] G. Stork and M. J. Sofia, *J. Am. Chem. Soc.*, **108**, 6826 (1986).

[102] A. Y. Mohammed and D. L. J. Clive, *J. Chem. Soc., Chem. Commun.*, **1986**, 588.

[103] N. Iwasawa, M. Funahashi, S. Hayakawa, and K. Narasaka, *Chem. Lett.*, **1993**, 545.

[104] J. D. Winkler and V. Sridar, *J. Am. Chem. Soc.*, **108**, 1708 (1986).

[105] J. D. Winkler and V. Sridar, *Tetrahedron Lett.*, **29**, 6219 (1988).

[106] D. P. Curran and D. Kim, *Tetrahedron Lett.*, **27**, 5821 (1986).

[107] D. P. Curran and D. M. Rakiewicz, *Tetrahedron*, **41**, 3943 (1985).

[108] W. P. Neumann, *Synthesis*, **1987**, 665.

[109] C. Walling, *Tetrahedron*, **41**, 3890 (1985).

[110] K. Miura, Y. Ichinose, K. Nozaki, K. Fugami, K. Oshima, and K. Utimoto, *Bull. Chem. Soc. Jpn.*, **62**, 143 (1989).

[111] N. Ono, H. Miyake, and A. Kaji, *Chem. Lett.*, **1985**, 635.

[112] D. L. J. Clive and R. J. Bergstra, *J. Org. Chem.*, **55**, 1786 (1990).

[113] P. M. Esch, H. Hiemstra, R. F. De Boer, and W. N. Speckamp, *Tetrahedron*, **48**, 4659 (1992).

[114] D. Crich and L. Quintero, *Chem. Rev.*, **89**, 1413 (1988).

[115] E. J. Enholm and G. Prasad, *Tetrahedron Lett.*, **30**, 4939 (1989).

[116] V. Yadav and A. G. Fallis, *Tetrahedron Lett.*, **30**, 3283 (1989).

[117] V. Yadav and A. G. Fallis, *Can. J. Chem.*, **69**, 779 (1991).

[118] G. Stork and R. Mook Jr., *J. Am. Chem. Soc.*, **109**, 2829 (1987).

[119] R. Mook and P. M. Sher, *Org. Synth.*, **66**, 75 (1987).

[120] J. Ardisson, J. P. Férézou, M. Julia, L. Lenglet, and A. Pancrazi, *Tetrahedron Lett.*, **28**, 1997 (1987).

[121] E. J. Corey and J. W. Suggs, *J. Org. Chem.*, **40**, 2554 (1975).

[122] G. Stork and P. M. Sher, *J. Am. Chem. Soc.*, **108**, 303 (1986).

[123] J. M. Berge and S. M. Roberts, *Synthesis*, **1979**, 471.

[124] J. Lusztyk, B. Maillard, S. Deycard, D. A. Lindsay, and K. U. Ingold, *J. Org. Chem.*, **52**, 3509 (1987).

[125] C. Chatgilialoglu, J. Dickhaut, and B. Giese, *J. Org. Chem.*, **56**, 6399 (1991).

[126] C. Chatgilialoglu, *Acc. Chem. Res.*, **25**, 188 (1992).

[127] S. A. Hitchcock and G. Pattenden, *J. Chem. Soc., Perkin Trans. 1*, **1992**, 1323.

[128] G. A. Russel and D. Guo, *Tetrahedron Lett.*, **25**, 5239 (1984).

[129] B. Giese and G. Kretzschmar, *Chem. Ber.*, **117**, 3160 (1984).

[130] J. Barluenga and M. Yus, *Chem. Rev.*, **88**, 487 (1988).

[131] K. Weinges and W. Sipos, *Chem. Ber.*, **121**, 363 (1988).

[132] G. E. Keck and E. J. Enholm, *Tetrahedron Lett.*, **26**, 3311 (1985).

[133] S. J. Danishefsky and J. S. Panek, *J. Am. Chem. Soc.*, **109**, 917 (1987).

[134] D. L. J. Clive and A. C. Joussef, *J. Org. Chem.*, **55**, 1096 (1990).

[135] J. W. Harvey and D. H. Whitman, *J. Chem. Soc., Perkin Trans. 1*, **1993**, 185.

[136] T. A. K. Smith and G. H. Whitham, *J. Chem. Soc., Chem. Commun.*, **1985**, 897.

[137] G. N. Schrauzer, *Angew. Chem., Int. Ed. Engl.*, **15**, 417 (1976).

[138] M. R. Ashcraft, A. Buny, C. J. Cooksey. A. G. Davies, B. D. Gupta, M. D. Johnson, and H. Morris, *J. Organomet. Chem.*, **195**, 89 (1980).

[139] M. R. Ashcroft, P. Bougeard, A. Bury, C. J. Cooksey, and M. D. Johnson, *J. Organomet. Chem.*, **289**, 403 (1985).

[140] P. Bougeard, C. J. Cooksey, M. D. Johnson, M. J. Lewin, S. Mitchell, and P. A. Owens, *J. Organomet. Chem.*, **288**, 389 (1985).

[141] C. H. Schiesser and K. Sutej, *Tetrahedron Lett.*, **33**, 5137 (1992).

[142] K. J. Kulicke, C. Chatgilialoglu, B. Kopping, and B. Giese, *Helv. Chim. Acta*, **75**, 935 (1992).

[143] D. Crich, *Aldrichim. Acta*, **20**, 35 (1987).

[144] D. H. R. Barton and S. Z. Zard, *Pure Appl. Chem.*, **58**, 675 (1986).

[145] D. H. R. Barton and M. Samadi, *Tetrahedron*, **48**, 7083 (1992).

[146] D. H. R. Barton, Y. Herv, P. Potier, and J. Thierry, *Tetrahedron*, **47**, 6127 (1987).

[147] D. H. R. Barton, N. Ozbalik, and B. Vacher, *Tetrahedron*, **44**, 3501 (1988).

[148] D. H. R. Barton, D. Crich, and P. Potier, *Tetrahedron Lett.*, **26**, 5943 (1985).

[149] D. H. R. Barton, D. Bridon, E. Fernandez–Picot, and S. Z. Zard, *Tetrahedron*, **43**, 2733 (1987).

[150] D. H. R. Barton, J. Guilhem, Y. Herv, P. Potier, and J. Thierry, *Tetrahedron Lett.*, **28**, 1413 (1987).

[151] M. Newcomb, S.–U. Park, J. Kaplan, and D. J. Marquardt, *Tetrahedron Lett.*, **26**, 5651 (1985).

[152] M. Newcomb, D. J. Marquardt, and T. M. Deeb, *Tetrahedron*, **46**, 2329 (1990).

[153] M. Newcomb and D. J. Marquardt, *Heterocycles*, **28**, 129 (1989).

[154] J. Iqbal, B. Bhatia, and N. K. Nayyar, *Chem. Rev.*, **94**, 519 (1994).

[155] G. M. Lee, M. Parvez, and S. M. Weinreb, *Tetrahedron*, **44**, 4671 (1988).

[156] D. P. Curran and C. T. Chang, *Tetrahedron Lett.*, **28**, 2477 (1987).

[157] D. P. Curran, E. Bosch, J. Kaplan, and M. Newcomb, *J. Org. Chem.*, **54**, 1826 (1989).

[158] D. P. Curran and S. C. Kuo, *J. Am. Chem. Soc.*, **108**, 1106 (1986).

[159] D. P. Curran, M. H. Chen, and D. Kim, *J. Am. Chem. Soc.*, **108**, 2489 (1986).

[160] G. Haaima and R. T. Weavers, *Tetrahedron Lett.*, **29**, 1085 (1988).

[161] L. Stella, *Angew. Chem.*, **95**, 368 (1982); *Angew. Chem. Int. Ed. Engl.*, **22**, 337 (1982).

[162] C. A. Broka and K. K. Eng, *J. Org. Chem.*, **51**, 5043 (1986).

[163] C. A. Broka and J. F. Gerlits, *J. Org. Chem.*, **53**, 2144 (1988).

[164] Y. L. Chow and R. A. Perry, *Can. J. Chem.*, **63**, 2203 (1985).

[165] D. H. R. Barton and M. Akthar, *J. Am. Chem. Soc.*, **86**, 1528 (1964).

[166] A. J. Bloodworth, R. J. Curtis, and N. Mistry, *J. Chem. Soc., Chem. Commun.*, **1989**, 954.

[167] L. Becking and H. J. Schäfer, *Tetrahedron Lett.*, **29**, 2797 (1988).

[168] D. C. Harrowven and G. Pattenden, *Tetrahedron Lett.*, **32**, 243 (1991).

[169] V. F. Patel and G. Pattenden, *Tetrahedron Lett.*, **28**, 1451 (1987).

[170] G. Pattenden, *Chem. Soc. Rev.*, **17**, 361 (1988).

[171] H. Fischer, *J. Am. Chem. Soc.*, **108**, 3925 (1986).

[172] W. A. Nugent and T. V. RajanBabu, *J. Am. Chem. Soc.*, **110**, 8561 (1988).

[173] T. V. RajanBabu and W. A. Nugent, *J. Am. Chem. Soc.*, **111**, 4525 (1989).

[174] D. P. Curran, T. L. Fevig, and M. J. Totleben, *Synlett*, **1990**, 773.

[175] J. Inanaga, O. Ujikawa, and M. Yamaguchi, *Tetrahedron Lett.*, **32**, 1737 (1991).

[176] G. A. Molander and L. S. Harring, *J. Org. Chem.*, **55**, 6171 (1990).

[177] H. B. Kagan and J. L. Namy, *Tetrahedron*, **42**, 6573 (1986).

[178] G. A. Molander and J. A. McKie, *J. Org. Chem.*, **57**, 3132 (1992).

[179] G. A. Molander and C. Kenny, *J. Org. Chem.*, **56**, 1439 (1991).

[180] D. Belotti, J. Cossy, J. P. Pete, and C. Portella, *J. Org. Chem.*, **51**, 4966 (1986).

[181] J. Mattay, A. Banning, E. W. Bischof, A. Heidbreder, and J. Runsink, *Chem. Ber.*, **125**, 2119 (1992).

[182] R. Scheffold, S. Abrecht, R. Orlinski, H.–R. Ruf, P. Stamouli, O. Tinembart, L. Walder, and C. Weymoth, *Pure Appl. Chem.*, **59**, 363 (1987).

[183] J. H. Hutchinson, G. Pattenden, and P. L. Myers, *Tetrahedron Lett.*, **28**, 1313 (1987).

[184] B. Giese, P. Erdmann, T. Gĸbel, and R. Springer, *Tetrahedron Lett.*, **33**, 4545 (1992).

[185] M. Ladlow and G. Pattenden, *Tetrahedron Lett.*, **25**, 4317 (1984).

[186] H. Bhandal, V. F. Patel, G. Pattenden, and J. J. Russel, *J. Chem. Soc., Perkin Trans. 1*, **1990**, 2691.

[187] L. Becking and H. J. Schfer, *Tetrahedron Lett.*, **29**, 2801 (1988).

[188] Y. T. Jeon, C. P. Lee, and P. S. Mariano, *J. Am. Chem. Soc.*, **113**, 8847 (1991).

[189] B. B. Snider, R. Mohan, and S. A. Kates, *Tetrahedron Lett.*, **28**, 841 (1987).

[190] J.-M. Surzur and M. P. Bertrand, *Pure Appl. Chem.*, **60**, 1659 (1988).

[191] E. J. Corey and M. Kang, *J. Am. Chem. Soc.*, **106**, 5384 (1984).

[192] S. A. Kates, M. A. Dombroski, and B. B. Snider, *J. Org. Chem.*, **55**, 2427 (1990).

[193] J. E. Merritt, M. Sasson, S. A. Kates, and B. B. Snider, *Tetrahedron Lett.*, **29**, 5209 (1988).

[194] B. B. Snider and J. E. Meritt, *Tetrahedron*, **47**, 8663 (1991).

[195] G. Stork and R. Mook, Jr., *J. Am. Chem. Soc.*, **105**, 3720 (1983).

[196] B. B. Snider, Q. Zhang, and A. M. Dombroski, *J. Org. Chem.*, **57**, 4195 (1992).

[197] D. P. Curran, D. Kim, H. T. Liu, and W. Shen, *J. Am. Chem. Soc.*, **110**, 5900 (1988).

[198] A. Johns and J. A. Murphy, *Tetrahedron Lett.*, **29**, 837 (1988).

[199] D. P. Curran, A. C. Abraham, and H. Lin, *J. Org. Chem.*, **56**, 4335 (1991).

[200] A. L. J. Beckwith, D. M. O'Shea, and S. W. Westwood, *J. Am. Chem. Soc.*, **110**, 2565 (1988).

[201] W. B. Motherwell and J. D. Harling, *J. Chem. Soc., Chem. Commun.*, **1988**, 1380.

[202] R. A. Batey, J. D. Harling, and W. B. Motherwell, *Tetrahedron*, **48**, 8031 (1992).

[203] R. A. Batey and W. B. Motherwell, *Tetrahedron Lett.*, **32**, 6649 (1991).

[204] A. Johns, J. A. Murphy, and M. S. Sherburn, *Tetrahedron*, **45**, 7835 (1989).

[205] G. E. Keck and D. A. Burnett, *J. Org. Chem.*, **52**, 2958 (1987).

[206] G. Stork and P. M. Sher, *J. Am. Chem. Soc.*, **105**, 6765 (1983).

[207] H. Togo and O. Kikuchi, *Tetrahedron Lett.*, **29**, 4133 (1988).

[208] D. P. Curran and C. M. Jeong, *J. Am. Chem. Soc.*, **112**, 9401 (1990).

[209] D. L. Boger and R. J. Mathvink, *J. Am. Chem. Soc.*, **112**, 4003 (1990).

[210] C.-P. Chuang, *Synlett*, **1991**, 859.

[211] B. B. Snider and B. D. Buckman, *Tetrahedron*, **45**, 6969 (1989).

[212] J. Boivin, E. Crepon, and S. Z. Zard, *Tetrahedron Lett.*, **32**, 199 (1991).

[213] V. H. Rawal and S. Iwasa, *Tetrahedron Lett.*, **33**, 4687 (1992).

[214] D. A. Singleton and K. M. Church, *J. Org. Chem.*, **55**, 4780 (1990).

[215] M. Journet and M. Malacria, *J. Org. Chem.*, **57**, 3085 (1992).

[216] D. A. Singleton, K. M. Church, and M. J. Lucero, *Tetrahedron Lett.*, **31**, 5551 (1990).

[216a] C. Thebtaranonth and Y. Thebtaranonth, *Tetrahedron*, **46**, 1385 (1990) and references cited therein.

[216b] J. K. Sutherland in *Comprehensive Organic Synthesis*, B. M. Trost, I. Fleming, Eds., Vol. 3, Pergamon Press, New York, 1991, p. 341.

[216c] W. F. Bailey, J. J. Patricia, V. C. DelGobbo, R. M. Jarret, and P. J. Okarma, *J. Org. Chem.*, **50**, 1999 (1985).

[216d] W. F. Bailey, T. T. Nurmi, J. J. Patricia, and W. Wang, *J. Am. Chem. Soc.*, **109**, 2442 (1987).

[216e] R. D. Little, M. R. Masjedizadeh, O. Wallquist, and J. I. McLaughlin, *Org. React.*, **47**, 315 (1995) and references cited therein.

[216f] B. M. Trost, *Angew. Chem., Int. Ed. Engl.*, **29**, 1173 (1989).

[217] J. Marco-Contelles, C. Pozuelo, M. L. Jimeno, L. Martinez, and A. Martinez-Grau, *J. Org. Chem.*, **57**, 2625 (1992).

[218] S. Hanessian and R. J. Leger, *Synlett*, **1992**, 402.

[219] I. De Riggi, S. Gastaldi, J. M. Surzur, M. P. Bertrand, and A. Virgili, *J. Org. Chem.*, **57**, 6118 (1992).

[220] D. H. R. Barton, D. Crich, and G. Kretzschmar, *J. Chem. Soc., Perkin Trans. 1*, **1986**, 39.

[221] T. K. Hayes, R. Villani, and S. M. Weinreb, *J. Am. Chem. Soc.*, **110**, 5533 (1988).

[222] J. A. Seijas, M. P. Vazquez-Tato, L. Castedo, R. J. Estevez, M. G. Onega, and M. Ruiz, *Tetrahedron*, **48**, 1637 (1992).

[223] M. Okabe, M. Abe, and M. Tada, *J. Org. Chem.*, **47**, 1775 (1982).

[224] S. Busato, O. Tinembert, Z. Zhang, and R. Scheffold, *Tetrahedron*, **46**, 3155 (1990).

[225] G. Pattenden and G. M. Robertson, *Tetrahedron*, **41**, 4001 (1985).

[226] W. Xu, X. M. Zhang, and P. S. Mariano, *J. Am. Chem. Soc.*, **113**, 8863 (1991).

[227] K. S. Feldmann, H. M. Berven, and P. H. Weinreb, *J. Am. Chem. Soc.*, **115**, 11364 (1993).

[228a] S. Iwasa, M. Yamamoto, S. Kohmoto, and K. Yamada, *J. Org. Chem.*, **56**, 2849 (1991).

[228b] S. Iwasa, M. Yamamoto, S. Kohmoto, and K. Yamada, *J. Chem. Soc., Perkin Trans. 1*, **1991**, 1173.

[229] A. Bury and M. D. Johnson, *J. Chem. Soc., Chem. Commun.*, **1980**, 498.

[230] A. Bury, M. D. Johnson, and M. J. Stewart, *J. Chem. Soc., Chem. Commun.*, **1980**, 622.

[231] A. Bury, S. T. Corker, and M. D. Johnson, *J. Chem. Soc., Perkin Trans. 1*, **1982**, 645.

[232] S. U. Park, T. R. Varick, and M. Newcomb, *Tetrahedron Lett.*, **31**, 2975 (1990).

[233] K. Ogura, N. Sumitani, A. Kuyano, H. Iguchi, and M. Fujita, *Chem. Lett.*, **1992**, 1487.

[234] E. Hasegawa and D. P. Curran, *Tetrahedron Lett.*, **34**, 1717 (1993).

[235] E. J. Walsh, Jr., J. M. Messinger, II, D. A. Grusdoski, and C. A. Allchin, *Tetrahedron Lett.*, **27**, 4409 (1980).

[236] M. Ballestri, C. Chatgilialoglu, N. Cardi, and A. Sommazzi, *Tetrahedron Lett.*, **33**, 1787 (1992).

[237] D. P. Curran, M.-H. Chen, and D. Kim, *J. Am. Chem. Soc.*, **108**, 2489 (1986); *J. Am. Chem. Soc.*, **111**, 6265 (1989).

[238] C. Bonini, R. Di Fabio, S. Merozzi, and G. Righi, *Tetrahedron Lett.*, **31**, 5369 (1990).

[239] C. Walling and A. Cioffari, *J. Am. Chem. Soc.*, **94**, 6059 (1972).

[240] M. Julia and E. Colomer, *C. R. Acad. Sci., Ser. C*, **270**, 1305 (1970).

[241] Y. Watanabe, T. Yokozawa, T. Takata, and T. Endo, *J. Fluorine Chem.*, **39**, 431 (1988).

[242] J. M. Aurrecoechea and A. Fernandez–Acebes, *Tetrahedron Lett.*, **34**, 549 (1993).

[243] R. N. Saicic and Z. Cekovic, *Tetrahedron*, **46**, 3627 (1990).

[244] J. E. Swartz, T. J. Mahachi, and E. Kariv–Miller, *J. Am. Chem. Soc.*, **110**, 3622 (1988).

[245] E. Kariv–Miller and T. J. Mahachi, *J. Org. Chem.*, **51**, 1041 (1986).

[246] T. Shono, I. Nishiguchi, H. Ohmizu, and M. Mitani, *J. Am. Chem. Soc.*, **100**, 545 (1978).

[247] W. R. Bowman and S. W. Jackson, *Tetrahedron Lett.*, **30**, 1857 (1989).

[248] D. P. Curran, M.-H. Chen, and D. Kim, *J. Am. Chem. Soc.*, **111**, 6265 (1989).

[249] I. Ryu, K. Kusano, M. Hasegawa, N. Kambe, and N. Sonoda, *J. Chem. Soc., Chem. Commun.*, **1991**, 1018.

[249a] A. L. J. Beckwith, I. A. Blair, and G. Phillipou, *Tetrahedron Lett.*, **1974**, 2251.

[250] N. O. Brace, *J. Org. Chem.*, **32**, 2711 (1967).

[251] D. P. Curran and M. Shu, *Bull. Soc. Chim. Fr.*, **130**, 314 (1993).

[252] T. Inokuchi, H. Kawafuchi, and S. Torii, *J. Org. Chem.*, **56**, 4983 (1991).

[253] G. Pandey and B. B. V. Setzhar, *J. Chem. Soc., Chem. Commun.*, **1993**, 780.

[254] J. Hatem, C. Henriet–Bernard, J. Grimaldi, and R. Maurin, *Tetrahedron Lett.*, **33**, 1057 (1992).

[255] J. E. Forbes, C. Tailhan, and S. Z. Zard, *Tetrahedron Lett.*, **31**, 2565 (1990).

[256] D. P. Curran, M.-H. Chen, E. Spletzer, C. M. Seong, and C.-T. Chang, *J. Am. Chem. Soc.*, **111**, 8872 (1989).

[257] E. Lee, C. H. Hur, and J. H. Park, *Tetrahedron Lett.*, **30**, 7219 (1989).

[258] A. L. J. Beckwith and G. Moad, *J. Chem. Soc., Perkin Trans. 2*, **1975**, 1726.

[259] I. De Riggi, R. Mouguier, J. M. Surzur, C. Lesueur, M. Bertrand, C. Jaime, and A. Virgili, *Bull. Soc. Chim. Fr.*, **130**, 229 (1993).

[260] E. Nakamura, T. Inubushi, S. Aoki, and D. Machii, *J. Am. Chem. Soc.*, **113**, 8980 (1991).

[261] J. K. Crandall and M. Mualla, *Tetrahedron Lett.*, **27**, 2243 (1986).

[262] T. Lübbers and H. J. Schäfer, *Synlett*, **1990**, 861.

[263] G. A. Molander and C. Kenny, *J. Am. Chem. Soc.*, **111**, 8236 (1989).

[264] J. E. Forbes and S. Z. Zard, *J. Am. Chem. Soc.*, **112**, 2034 (1990).

[265] Y. Ichinose, K. Oshima, and K. Utimoto, *Chem. Lett.*, **1988**, 1437.

[266] C. B. Ziegler, Jr., *J. Org. Chem.*, **55**, 2983 (1990).

[267] M. Julia and M. Barreau, *C. R. Hebd. Seances Acad. Sci., Ser. C*, **280**, 957 (1975).

[268] M. Yamamoto, A, Furusawa, S. Iwasa, S. Kohmoto, and K. Yamada, *Bull. Chem. Soc. Jpn.*, **65**, 1550 (1992).

[269] E. J. Enholm and K. S. Kinter, *J. Am. Chem. Soc.*, **113**, 7784 (1991).

[270] K. S. Feldman, A. L. Romanelli, R. E. Ruckle, Jr., and G. Jean, *J. Org. Chem.*, **57**, 100 (1992).

[271] B. B. Snider, R. Mohan, and S. A. Kates, *J. Org. Chem.*, **50**, 3661 (1985).

[272] C.-P. Chuang and V.-J. Jiang, *J. Chin. Chem. Soc.*, **36**, 177 (1989).

[273] S. A. Dodson and R. D. Stipanovic, *J. Chem. Soc., Perkin Trans. 1*, **1975**, 410.

[274] J. K. Crandall and W. I. Michaely, *J. Org. Chem.*, **49**, 4244 (1984).

[275] Y. Ichinose, S. Matsanuga, K. Fugami, K. Oshima, and K. Utimoto, *Tetrahedron Lett.*, **30**, 3155 (1989).

[276] J. K. Crandall and D. J. Keyton, *Tetrahedron Lett.*, **21**, 1653 (1969).

[277] N. O. Brace and J. E. Van Elswyk, *J. Org. Chem.*, **41**, 766 (1976).

[278] I. De Riggi, J.-M. Surzur, M. P. Bertrand, A. Archavlis, and R. Faure, *Tetrahedron*, **46**, 5285 (1990).

[279] M. E. Kuehne and R. E. Damon, *J. Org. Chem.*, **42**, 1825 (1977).

[280] C.-P. Chuang, *Tetrahedron Lett.*, **33**, 6311 (1992).

[281] C.-P. Chuang, *Synth. Commun.*, **22**, 3151 (1992).

[282] R. Grigg, J. Devlin, A. Ramasubbu, R. M. Scott, and P. Stevenson, *J. Chem. Soc., Perkin Trans. 1*, **1987**, 1515.

[283] K. Weinges and W. Sipos, *Angew. Chem.*, **99**, 1177 (1987); *Angew. Chem., Int. Ed. Engl.*, **27**, 1152 (1987).

[284] Z. Cekovic and R. Saicic, *Tetrahedron Lett.*, **27**, 5981 (1986).

[285] V. K. Yadav and A. G. Fallis, *Tetrahedron Lett.*, **29**, 897 (1988).

[286] G. A. Molander and C. Kenny, *Tetrahedron Lett.*, **28**, 4367 (1987).

[287] J. K. Crandall and T. A. Ayers, *Tetrahedron Lett.*, **32**, 3659 (1991).

[288] F. Barth and C. O-Yang, *Tetrahedron Lett.*, **32**, 5873 (1991).

[289] K. Miura, K. Fugami, K. Oshima, and K. Utimoto, *Tetrahedron Lett.*, **29**, 5135 (1988).

[290] C.-P. Chuang and T. H. J. Ngoi, *J. Chem. Res. (S)*, **1991**, 1.

[291] D. P. Curran and C. M. Seong, *Tetrahedron*, **48**, 2157 (1992).

[292] M. Julia, B. Mansour, and D. Mansuny, *Tetrahedron Lett.*, **1976**, 3443.

[293] G. Stork and M. E. Reynolds, *J. Am. Chem. Soc.*, **110**, 6911 (1988).

[294] D. P. Curran and C. M. Seong, *Tetrahedron*, **48**, 2175 (1992).

[295] D. P. Curran and W. Shen, *J. Am. Chem. Soc.*, **115**, 6051 (1993).

[296] B. B. Snider, J. E. Merritt, M. A. Dombroski, and B. O. Buckman, *J. Org. Chem.*, **56**, 5544 (1991).

[297] K. S. Feldman, R. E. Ruckle, Jr., and A. L. Romanelli, *Tetrahedron Lett.*, **30**, 5845 (1989).

[298] T. Harrison, P. L. Myers, and G. Pattenden, *Tetrahedron*, **45**, 5247 (1989).

[299] S. Danishefsky, S. Chackalamannil, and B. J. Uang, *J. Org. Chem.*, **47**, 2231 (1982).

[300] T. Harrison, G. Pattenden, and P. L. Myers, *Tetrahedron Lett.*, **29**, 3869 (1988).

[301] J. E. Brumwell, N. S. Simpkins, and N. K. Terrett, *Tetrahedron Lett.*, **34**, 1215 (1993).

[302] J. E. Brumwell, N. S. Simpkins, and N. K. Terrett, *Tetrahedron Lett.*, **34**, 1219 (1993).

[303] G. A. Kraus and S. Liras, *Tetrahedron Lett.*, **31**, 5265 (1990).

[304] A. Ogawa, H. Yokoyama, K. Yokoyama, T. Masawaki, N. Kambe, and N. Sonoda, *J. Org. Chem.*, **56**, 5721 (1991).

[305] A. Arnone, P. Bravo, G. Cavicchio, M. Frigerio, and F. Viani, *Tetrahedron*, **48**, 8523 (1992).

[306] A. Arnone, P. Bravo, G. Cavicchio, M. Frigerio, and F. Viani, *Tetrahedron*, **49**, 6873 (1993).

[307] T. Kataoka, M. Yoshimatsu, Y. Noda, T. Sato, H. Shimizu, and M. Hori, *J. Chem. Soc., Perkin Trans. 1*, **1993**, 121.

[308] L. Moens, M. M. Baizer, and R. D. Little, *J. Org. Chem.*, **51**, 4497 (1986).

[309] R. Mook, Jr. and P. M. Sher, *Org. Synth.*, **66**, 75 (1988).

[310] C. G. Sowell, R. L. Wolin, and R. D. Little, *Tetrahedron Lett.*, **31**, 485 (1990).

[311] B. B. Snider, L. Armanetti, and R. Baggio, *Tetrahedron Lett.*, **34**, 1701 (1993).

[312] C. Chatgilialoglu, B. Giese, and B. Kopping, *Tetrahedron Lett.*, **31**, 6013 (1990).

[313] T. Morikawa, T. Nishiwaki, and Y. Kobayashi, *Tetrahedron Lett.*, **30**, 2407 (1989).

[314] S. Kim and S. Lee, *Tetrahedron Lett.*, **32**, 6575 (1991).

[315] A. Srikrishna and G. Sundarababu, *Tetrahedron*, **46**, 3601 (1990).

[316] T. A. K. Smith and G. H. Whitham, *J. Chem. Soc., Perkin Trans. 1*, **1989**, 319.

[316a] T. A. K. Smith and G. H. Whitham, *J. Chem. Soc., Perkin Trans. 1*, **1989**, 313.

[317] T. Morikawa, Y. Kodama, J. Uchida, M. Takano, Y. Washio, and T. Taguchi, *Tetrahedron*, **48**, 8915 (1992).

[318] M. J. Begley, N. Housden, A. Johns, and J. A. Murphy, *Tetrahedron*, **47**, 8417 (1991).

[319] F. E. Ziegler, C. A. Metcalf, III, and G. Schulte, *Tetrahedron Lett.*, **33**, 3117 (1992).

[320] S. Kim and J. R. Cho, *Synlett*, **1992**, 629.

[321] J. R. Peterson, R. S. Egler, D. B. Horsley, and T. J. Winter, *Tetrahedron Lett.*, **28**, 6109 (1987).

[322] C.-P. Chuang, S.-S. Hou, and T. H. J. Ngoi, *J. Chin. Chem. Soc.*, **37**, 85 (1990).

[323] C.-P. Chuang, S.-S. Hou, and T. H. J. Ngoi, *J. Chem. Res. (S)*, **1991**, 216.

[324] Y.-M. Tsai, B.-W. Ke, and C.-H. Lin, *Tetrahedron Lett.*, **31**, 6074 (1990).

[324a] D. A. Singleton and K. M. Church, *J. Org. Chem.*, **55**, 4780 (1990).

[325] J. M. Fang, H. T. Chang, and C. C. Lin, *J. Chem. Soc., Chem. Commun.*, **1988**, 1385.

[326] Y.-M. Tsai, F.-C. Chang, J. Huang, and C.-L. Shiu, *Tetrahedron Lett.*, **30**, 2121 (1989).

[327] K. S. Feldman, A. L. Romanelli, R. E. Ruckle, Jr., and R. F. Miller, *J. Am. Chem. Soc.*, **110**, 3300 (1988).

[328] D. A. Singleton, C. C. Huval, K. M. Church, and E. S. Priestley, *Tetrahedron Lett.*, **32**, 5765 (1991).

[329] D. P. Curran and P. A. van Elburg, *Tetrahedron Lett.*, **30**, 2501 (1989).

[330] S. Hanessian and R. J. Leger, *J. Am. Chem. Soc.*, **114**, 3115 (1992).

[331] D. Crich and S. M. Fortt, *Tetrahedron Lett.*, **28**, 2895 (1987).

[332] A. D. Borthwick, S. Caddick, and P. J. Parsons, *Tetrahedron Lett.*, **31**, 6911 (1990).

[333] A. Srikrishna and G. Sundarababu, *Tetrahedron*, **47**, 481 (1991).

[334] A. Srikrishna and G. Sunderbabu, *Tetrahedron Lett.*, **30**, 3561 (1989).

[335] A. L. J. Beckwith, D. M. Cliff, and C. H. Schiesser, *Tetrahedron*, **48**, 4641 (1992).

[336] T. Morikawa, M. Uejima, and Y. Kobayashi, *Chem. Lett.*, **1989**, 623.

[337] Y.-M. Tsai and C.-D. Cherng, *Tetrahedron Lett.*, **32**, 3515 (1991).

[338] D. P. Curran, W.-T. Jiaang, M. Palovich, and Y.-M. Tsai, *Synlett*, **1993**, 403.

[339] D. A. Singleton, K. M. Church, and M. I. Lucero, *Tetrahedron Lett.*, **31**, 5551 (1990).

[340] D. L. J. Clive and T. L. B. Boivin, *J. Org. Chem.*, **54**, 1997 (1989).

[341] E. J. Enholm, J. A. Burroff, and L. M. Jaramillo, *Tetrahedron Lett.*, **31**, 3727 (1990).

[342] Y. M. Tsai, K. H. Tang, and W. T. Jiaang, *Tetrahedron Lett.*, **34**, 1303 (1993).

[343] S. Kim, I. S. Kee, and S. Lee, *J. Am. Chem. Soc.*, **113**, 9882 (1991).

[344] I. Rochigneux, M. L. Fontanel, J. C. Malanda, and A. Doutheau, *Tetrahedron Lett.*, **32**, 2017 (1991).

[345] V. Snieckus, J. C. Cuevas, C. P. Sloan, H. Lin, and D. P. Curran, *J. Am. Chem. Soc.*, **112**, 896 (1990).

[346] T. B. Lowinger and L. Weiler, *Can. J. Chem.*, **68**, 1636 (1990).

[347] J. G. Stack, D. P. Curran, S. V. Geib, J. Rebek, Jr., and P. Ballester, *J. Am. Chem. Soc.*, **114**, 7007 (1992).

[348] B. B. Snider and T. Kwon, *J. Org. Chem.*, **57**, 2399 (1992).

[349] C.-P. Chuang, *Tetrahedron*, **47**, 5425 (1991).

[350] S. Kiyooka, Y. Kaneko, H. Matsue, and R. Fujiyama, *J. Org. Chem.*, **55**, 5562 (1990).

[351] C.-P. Chuang, *Synlett*, **1990**, 527.

[352] L. Set, D. R. Cheshire, and D. L. J. Clive, *J. Chem. Soc., Chem. Commun.*, **1985**, 1205.

[353] D. L. J. Clive, T. L. B. Boivin, and A. G. Angoh, *J. Org. Chem.*, **52**, 4943 (1987).

[354] A. G. Angoh and D. L. J. Clive, *J. Chem. Soc., Chem. Commun.*, **1985**, 941.

[354a] A. G. Angoh and D. L. J. Clive, *J. Chem. Soc., Chem. Commun.*, **1985**, 980.

[355] D. J. Coveney, V. F. Patel, and G. Pattenden, *Tetrahedron Lett.*, **28**, 5949 (1987).

[356] J. J. Gaudino and C. S. Wilcox, *Carbohydr. Res.*, **206**, 233 (1990).

[357] C. S. Wilcox and J. J. Gaudino, *J. Am. Chem. Soc.*, **108**, 3102 (1986).

[358] N. S. Simpkins, S. Stokes, and A. J. Whittle, *Tetrahedron Lett.*, **33**, 793 (1992).

[359] S. M. Roberts and K. A. Shoberu, *J. Chem. Soc., Perkin Trans. 1*, **1992**, 2625.

[360] T. V. RajanBabu, *J. Am. Chem. Soc.*, **109**, 609 (1987).

[361] C. A. Broka and D. E. C. Reichert, *Tetrahedron Lett.*, **28**, 1503 (1987).

[362] E. Lee and C. U. Hur, *Tetrahedron Lett.*, **32**, 5101 (1991).

[363] B. B. Snider and J. J. Patricia, *J. Org. Chem.*, **54**, 38 (1989).

[364] D. P. Curran, T. M. Morgan, C. E. Schwartz, B. B. Snider, and M. A. Dombroski, *J. Am. Chem. Soc.*, **113**, 6607 (1991).

[365] R. D. Little, D. P. Fox, L. Van Hijfte, R. Dannecker, G. Sowell, R. L. Wolin, L. Moens, and M. M. Baizer, *J. Org. Chem.*, **53**, 2287 (1988).

[366] R. C. Denis, J. Rancourt, E. Ghiro, F. Boutonnet, and D. Gravel, *Tetrahedron Lett.*, **34**, 2091 (1993).

[367] M. D. Bachi and D. Denenmark, *Heterocycles*, **28**, 583 (1989).

[368] C. Destabel, J. D. Kilburn, and J. Knight, *Tetrahedron Lett.*, **34**, 3151 (1993).

[369] M. Harendza, K. Lexmann, and W. P. Neumann, *Synlett*, **1993**, 283.

[370] J. Marco–Contelles and B. Snchez, *J. Org. Chem.*, **58**, 4293 (1993).

[371] P. Gottschalk and D. C. Neckers, *J. Org. Chem.*, **50**, 3498 (1985).

[372] A. Heidbrecher and J. Mattay, *Tetrahedron Lett.*, **33**, 1973 (1992).

[373] M. Journet, E. Magnuol, G. Aguel, and M Malacria, *Tetrahedron Lett.*, **31**, 4445 (1990).

[374] A. Arnone, P. Bravo, G. Cavicchio, M. Frigerio, and F. Viani, *Tetrahedron· Asymmetry*, **3**, 9 (1992).

[375] A. Arnone, P. Bravo, and F. Viani, *Tetrahedron: Asymmetry*, **2**, 399 (1992).

[376] J. E. Ward and B. F. Kaller, *Tetrahedron Lett.*, **32**, 843 (1991).

[377] W. F. Bailey and A. D. Khanolkar, *Tetrahedron*, **47**, 7727 (1991).

[378] K. Ogura, A. Kayano, T. Fujino, N. Sumitani, and M. Fujita, *Tetrahedron Lett.*, **34**, 8313 (1993).

[379] D. H. R. Barton, P. J. Dalko, and S. D. Gero, *Tetrahedron Lett.*, **32**, 4713 (1991).

[380] M. Chatzopoulos and J. P. Montheard, *Rev Roum. Chim.*, **26**, 275 (1981).

[381] J. Marco–Contelles, L. Martinez, A. Martinez–Grau, C. Pozuelo, and M. J. Jimeno, *Tetrahedron Lett.*, **32**, 6437 (1991).

[382] F. L. Harris and L. Weiler, *Tetrahedron Lett.*, **28**, 2941 (1987).

[383] J. Marco–Contelles, B. Snchez, and P. Ruiz, *Natural Product Letters 1*, **1992**, 167.

[384] C. Chen and D. Crich, *Tetrahedron Lett.*, **33**, 1945 (1993).

[385] C. Destabel and J. D. Kilburn, *J. Chem. Soc., Chem. Commun.*, **1992**, 596.

[386] D. Crich, K. A. Eustache, S. M. Fortt, and T. J. Ritchie, *Tetrahedron*, **46**, 2135 (1990).

[387] D. Crich and S. M. Fortt, *Tetrahedron*, **45**, 6581 (1989).

[388] D. Batty, D. Crich, and S. M. Fortt, *J. Chem. Soc., Chem. Commun.* **1989**, 1366.

[389] D. Batty, D. Crich, and S. M. Fortt, *J. Chem. Soc., Perkin Trans 1*, **1990**, 2875.

[390] D. Batty and D. Crich, *J. Chem. Soc., Perkin Trans. 1*, **1991**, 2894.

[391] M. Apparu and J. K. Crandall, *J. Org. Chem.*, **49**, 2125 (1984).

[392] D. Crich and S. M. Fortt, *Tetrahedron Lett.*, **29**, 2585 (1988).

[393] N. A. Porter, D. R. Magnin, and B. T. Wright, *J. Am. Chem. Soc.*, **108**, 2787 (1986).

[394] N. J. G. Cox, G. Pattenden, and S. D. Mills, *Tetrahedron Lett.*, **30**, 621 (1989).

[395] N. J. G. Cox, S. D. Mills, and G. Pattenden, *J. Chem. Soc., Perkin Trans. 1*, **1992**, 1313.

[396] M. P. Astley and G. Pattenden, *Synlett*, **1991**, 335.

[397] H. S. Dang and B. P. Roberts, *J. Chem. Soc., Perkin Trans. 1*, **8**, 891 (1993).

[398] D. L. Flynn and D. L. Zabrowski, *J. Org. Chem.*, **55**, 3673 (1990).

[399] Y. Araki, T. Endo, Y. Arai, M. Tanji, and Y. Ishido, *Tetrahedron Lett.*, **30**, 2892 (1989).

[400] S. L. Fremont, J. L. Belletiere, and D. M. Ho, *Tetrahedron Lett.*, **32**, 2335 (1991).

[401] H. Ishibashi, C. Kameoka, A. Yoshikawa, R. Ueda, K. Kodama, T. Sato, and M. Ikeda, *Synlett*, **1993**, 649.

[402] G. B. Gill, G. Pattenden, and S. J. Reynolds, *Tetrahedron Lett.*, **30**, 3229 (1989).

[403] M. Tada, T. Nakamura, and M. Matsumoto, *Chem. Lett.*, **1987**, 409.

[404] P. Kanshal and B. P. Roberts, *J. Chem. Soc., Perkin Trans. 2*, **1989**, 1559.

[405] R. D. Rieke and N. A. Moore, *Tetrahedron Lett.*, **25**, 2035 (1969).

[406] M. Kaafarani, M. P. Crozet, and J. M. Surzur, *Bull. Soc. Chim. Fr.*, **1981**, 449.

[407] M. P. Crozet, M. Kaafarani, and J. M. Surzur, *Bull. Soc. Chim. Fr.*, **1984**, 390.

[408] E. Lee, S. B. Ko, K. W. Jung, and M. H. Chang, *Tetrahedron Lett.*, **30**, 827 (1989).

[409] I. De Riggi, J. M. Surzur, and M. P. Bertrand, *Tetrahedron*, **44**, 7119 (1988).

[410] A. Naim, G. Mills, and P. B. Shevlin, *Tetrahedron Lett.*, **33**, 6779 (1992).

[411] A. C. Serra, C. M. M. da Silva Corrаa, M. A. M. S. A. Viera, and M. A. Gomes, *Tetrahedron*, **46**, 3061 (1990).

[412] T. J. Barton and A. Revis, *J. Am. Chem. Soc.*, **106**, 3802 (1984).

[413] R. A. Perry, S. C. Chen, B. C. Menon, K. Hanaya, and Y. L. Chow, *Can. J. Chem.*, **54**, 2385 (1976).

[414] H. Nagashima, N. Ozaki, K. Seki, M. Ishii, and K. Itoh, *J. Org. Chem.*, **54**, 4497 (1989).

[415] S. Takano, S. Nishizawa, M. Akiyama, and K. Ogasawara, *Synthesis*, **1984**, 949.

[416] J. L. Bougeois, L. Stella, and J. M. Surzur, *Tetrahedron Lett.*, **22**, 61 (1981).

[417] J. H. Udding, H. Hiemstra, M. N. A. van Zanden, and W. N. Speckamp, *Tetrahedron Lett.*, **32**, 3123 (1991).

[418] K. Mochida and K. Asami, *J. Organomet. Chem.*, **232**, 13 (1982).

[419] S. Hanessian, R. Di Fabio, J.-F. Marcoux, and M. Prud'homme, *J. Org. Chem.*, **55**, 3436 (1990).

[420] F. Barth and C. O-Yang, *Tetrahedron Lett.*, **31**, 1121 (1990).

[421] J. P. Dulcere, J. Rodriguez, M. Santelli, and J. P. Zahra, *Tetrahedron Lett.*, **28**, 2009 (1987).

[422] J. L. Courtneidge, M. Bush, and L. S. Loh, *Tetrahedron*, **48**, 3835 (1992).

[423] A. L. J. Beckwith and R. D. Wagner, *J. Chem. Soc., Chem. Commun.*, **1980**, 485.

[424] E. Lee, J. S. Tae, C. Lee, and C. M. Park, *Tetrahedron Lett.*, **34**, 4831 (1993).

[425] J. M. Clough, G. Pattenden, and P. G. Wight, *Tetrahedron Lett.*, **30**, 7469 (1989).

[426] Y. Ueno, O. Moriya, K. Chino, M. Watanabe, and M. Okawara, *J. Chem. Soc., Perkin Trans. 1*, **1986**, 1351.

[427] A. Srikrishna and K. Krishnan, *J. Org. Chem.*, **54**, 3981 (1984).

[428] M. Yamamoto, T. Uruma, S. Iwasa, S. Kohmoto, and K. Yamada, *J. Chem. Soc., Chem. Commun.*, **1989**, 1265.

[429] E. Magnol and M. Malacria, *Tetrahedron Lett.*, **27**, 2255 (1986).

[430] J. M. Surzur, L. Stella, and P. Tordo, *Bull. Soc. Chim. Fr.*, **1975**, 1429.

[431] O. Moriya, M. Kakihana, Y. Urata, T. Sugizaki, T. Kageyama, Y. Ueno, and T. Endo, *J. Chem. Soc., Chem. Commun.*, **1985**, 1401.

[432] O. Moriya, Y. Urata, Y. Ikeda, Y. Ueno, and T. Endo, *J. Org. Chem.*, **51**, 4708 (1986).

[433] J. Cossy and C. Leblanc, *Tetrahedron Lett.*, **30**, 4531 (1989).

[434] K. S. Feldman, R. E. Simpson, and M. Parvez, *J. Am. Chem. Soc.*, **108**, 9328 (1986).

[435] K. S. Feldman and R. E. Simpson, *J. Am. Chem. Soc.*, **111**, 4878 (1989).

[436] M. Newcomb, M. U. Kumar, J. Boivin, E. Crpon, and S. Z. Zard, *Tetrahedron Lett.*, **32**, 45 (1991).

[437] K. Nozaki, K. Oshima, and K. Utimoto, *Bull. Chem. Soc. Jpn.*, **63**, 2578 (1990).

[438] N. Ono, H. Miyake, A. Kamimura, I. Itamamoto, R. Tamura, and A. Kaji, *Tetrahedron*, **41**, 4013 (1985).

[439] R. Kiesewettter and P. Margaretha, *Helv. Chim. Acta*, **70**, 125 (1987).

[439a] J. E. Baldwin, M. G. Molonay, and A. F. Passons, *Tetrahedron*, **47**, 155 (1991).

[440] T. Naito, Y. Houda, O. Miyata, and I. Ninomiya, *Heterocycles*, **32**, 2319 (1991).

[441] B. Maillard, C. Gardrat, and M. J. Bourgeois, *J. Organomet. Chem.*, **236**, 61 (1982).

[442] M. D. Bachi and D. G. Lasanow, *Synlett*, **1990**, 551.

[443] O. Moriya, M. Okawara, and Y. Ueno, *Chem. Lett.*, **1984**, 1437.

[444] J. S. Yadav and V. R. Gadgil, *J. Chem. Soc., Chem. Commun.*, **1989**, 1824.

[445] R. C. Gash, F. MacCorquodale, and J. C. Walton, *Tetrahedron*, **45**, 5531 (1989).

[446] A. Sririshina and G. Sunderababu, *Chem. Lett.*, **1988**, 371.

[447] K. S. Feldman and C. M. Kraebel, *J. Org. Chem.*, **57**, 4574 (1992).

[448] T. Lрbbers and H. J. Schfer, *Synlett*, **1990**, 44.

[449] C. Hackmann and H. J. Schfer, *Tetrahedron*, **49**, 4559 (1993).

[450] T. Sato, N. Machigashira, H. Ishibashi, and M. Ikeda, *Heterocycles*, **33**, 139 (1992).

[451] Y. Watanabe and T. Endo, *Tetrahedron Lett.*, **29**, 321 (1988).

[452] A. Padwa, H. Nimmesgern, and G. S. K. Wong, *J. Org. Chem.*, **50**, 5620 (1985).

[453] E. Castagnino, S. Corsano, and D. H. R. Barton, *Tetrahedron Lett.*, **30**, 2983 (1989).

[454] A. Citterio, M. Ramperti, and E. Vismara, *J. Heterocycl. Chem.*, **18**, 763 (1981).

[455] A. L. J. Beckwith, B. P. Hay, and G. M. Williams, *J. Chem. Soc., Chem. Commun.*, **1989**, 1202.

[456] J. H. Udding, J. M. Tuijp, H. Hiemstra, and W. N. Speckamp, *J. Chem. Soc., Perkin Trans. 2*, **1992**, 857.

[457] J. L. Courtneidge, *J. Chem. Soc., Chem. Commun.*, **1992**, 1270.

[458] M. Journet and M. Malacria, *Tetrahedron Lett.*, **33**, 1893 (1992).

[459] M. Journet, E. Magual, W. Smadja, and M. Malacria, *Synlett*, **1991**, 58.

[460] G. Agnel and M. Malacria, *Synthesis*, **1989**, 687.

[461] A. Ali, D. C. Harrowven, and G. Pattenden, *Tetrahedron Lett.*, **33**, 2851 (1992).

[462] Y. Ueno, R. K. Khare, and M. Okawara, *J. Chem Soc., Perkin Trans. 1*, **1983**, 2637.

[463] M. D. Bachi, E. Bosch, D. Denenmark, and D. Girsh, *J. Org. Chem.*, **57**, 6803 (1992).

[464] K. Yamamoto, S. Abrecht, and R. Scheffold, *Chimia*, **45**, 86 (1991).

[465] M. D. Bachi and E. Bosch, *J. Org. Chem.*, **57**, 4696 (1992).

[466] A. L. J. Beckwith and G. E. Davison, *Tetrahedron Lett.*, **32**, 49 (1991).

[467] M. D. Bachi and E. Bosch, *J. Org. Chem.*, **54**, 1234 (1989).

[468] M. Tokuda, Y. Yamada, T. Takagi, H. Suginome, and A. Furusaki, *Tetrahedron*, **43**, 281 (1987).

[469] M. Tokuda, Y. Yamada, T. Takagi, H. Suginome, and A. Furusaki, *Tetrahedron Lett.*, **26**, 6085 (1985).

[470] P. Arya and D. D. M. Wayner, *Tetrahedron Lett.*, **32**, 6265 (1991).

[471] A. Serra and C. M. M. da Silva Correa, *Tetrahedron Lett.*, **32**, 6653 (1991).

[472] A. Srikrishna, *Ind. J. Chem.*, **29B**, 479 (1990).

[473] K. S. Feldman and T. E. Fisher, *Tetrahedron*, **45**, 2969 (1989).

[474] J. L. Belletire and N. O. Mahmoodi, *J. Nat. Prod.*, **55**, 194 (1992).

[475] G. Cavicchio, V. Marchetti, A. Arnone, P. Bravo, and V. Fiorenza, *Tetrahedron*, **47**, 9439 (1991).

[476] G. Cavicchio, V. Marchetti, A. Arnone, P. Bravo, and F. Viani, *Gazz. Chim. Ital.*, **121**, 423 (1991).

[477] S. Adhikari and S. Roy, *Tetrahedron Lett.*, **33**, 6025 (1992).

[478] S. Iwasa, M. Yamamoto, A. Furusawa, S. Kohmoto, and K. Yamada, *Chem. Lett.*, **1991**, 1457.

[479] L. D. M. Lolkema, H. Hiemstra, A. A. A. Ghouch, and W. N. Speckamp, *Tetrahedron Lett.*, **32**, 1491 (1991).

[480] M. Newcomb and J. L. Esker, *Tetrahedron Lett.*, **32**, 1035 (1991).

[481] R. Kiesewetter and P. Margaretha, *Helv. Chim. Acta*, **72**, 83 (1989).

[482] T. Morikawa, T. Nishiwaki, Y. Iitaka, and Y. Kobayashi, *Tetrahedron Lett.*, **28**, 671 (1987).

[483] M. Newcomb and C. Ha, *Tetrahedron Lett.*, **32**, 6493 (1991).

[484] J. L. Esker and M. Newcomb, *Tetrahedron Lett.*, **34**, 6877 (1993).

[485] J. Boivin, E. Fouquet, and S. Z. Zard, *Tetrahedron Lett.*, **32**, 4299 (1991).

[486] J. Gunic, I. Tabakovic, and Z. Samicanin, *Electrochim. Acta*, **35**, 225 (1990).

[487] A. Srikrishna and G. Sundarababu, *Tetrahedron*, **46**, 7901 (1990).

[488] M. Newcomb and M. U. Kumar, *Tetrahedron Lett.*, **31**, 1675 (1990).

[489] P. M. Esch, H. Hiemstra, and W. N. Speckamp, *Tetrahedron Lett.*, **31**, 759 (1990).

[490] S. C. Roy and S. Adhikari, *Tetrahedron Lett.*, **49**, 8415 (1993).

[491] B. B. Snider and B. A. McCarthy, *Tetrahedron*, **49**, 9447 (1993).

[492] C. H. Schiesser and K. Sutej, *J. Chem. Soc., Chem. Commun.*, **1992**, 57.

[493] L. Bejamin, C. H. Schiesser, and K. Sutej, *Tetrahedron*, **49**, 2557 (1993).

[494] J. F. Lavallee and G. Just, *Tetrahedron Lett.*, **32**, 3469 (1991).

[495] J. E. Baldwin, M. G. Moloney, and A. F. Parsons, *Tetrahedron*, **46**, 7263 (1990).

[496] M. D. Bachi and E. Bosch, *Tetrahedron Lett.*, **27**, 641 (1986).

[497] V. H. Rawal, S. P. Singh, C. Dufour, and C. Michoud, *J. Org. Chem.*, **58**, 7718 (1993).

[498] J. H. Byers and G. C. Lane, *J. Org. Chem.*, **58**, 3355 (1993).

[499] M. R. Ashcroft, P. Bougeard, A. Bury, C. J. Cooksey, M. D. Johnson, J. M. Hungerford, and G. M. Lampman, *J. Org. Chem.*, **49**, 1751 (1984).

[500] Y. Ueno, C. Tanaka, and M. Okawara, *Chem. Lett.*, **1983**, 795.

[501] R. M. Adlington and S. J. Mantell, *Tetrahedron*, **48**, 6529 (1992).

[502] J. E. Baldwin and C. S. Li, *J. Chem. Soc., Chem. Commun.*, **1987**, 166.

[503] M. Newcomb, D. J. Marquardt, M. U. Kumar, and M. Udaya, *Tetrahedron*, **46**, 2345 (1990).

[504] C.-P. Chuang, *Tetrahedron*, **47**, 5425 (1991).

[505] S. Hatakeyama, K. Sugawara, and S. Takano, *J. Chem. Soc., Chem. Commun.*, **1993**, 125.

[506] P. N. Culshaw and J. C. Walton, *J. Chem. Soc., Perkin Trans. 2*, **1991**, 1201.

[507] C. Dupuy, M. P. Crozet, and J. M. Surzur, *Bull. Soc. Chim. Fr.*, **1980**, 361.

[508] M. O. Funk, R. Isaac, and N. A. Porter, *J. Am. Chem. Soc.*, **97**, 1281 (1975).

[509] M. P. Crozet, J. M. Surzur, and C. Dupuy, *Tetrahedron Lett.*, **23**, 2031 (1971).

[510] W. B. Motherwell, A. M. K. Pennell, and F. Ujjainwalla, *J. Chem. Soc., Chem. Commun.*, **1992**, 1067.

[511] S. P. Munt and E. J. Thomas, *J. Chem. Soc., Chem. Commun.*, **1989**, 480.

[512] R. D. Walkup, N. U. Obeyekesere, and R. R. Kane, *Chem. Lett.*, **1990**, 1055.

[513] K. S. Feldmann and H. M. Berven, *Synlett*, **1993**, 827.

[514] I. Lakomy and R. Scheffold, *Helv. Chim. Acta*, **76**, 804 (1993).

[515] S. Hatakeyama, N. Ochi, H. Numata, and S. Takano, *J. Chem. Soc., Chem. Commun.*, **1988**, 1202.

[516] N. A. Porter, N. A. Roe, and A. T. McPhail, *J. Am. Chem. Soc.*, **102**, 7574 (1980).

[517] M. D. Bachi and D. Denenmark, *J. Org. Chem.*, **55**, 3442 (1990).

[518] W. Xu, T. Y. Jeon, E. Hasegawa, U. C. Yoon, and S. P. Mariano, *J. Am. Chem. Soc.*, **111**, 406 (1989).

[519] M. Ihara, K. Yasai, N. Taniguchi, K. Fukumoto, and T. Kametani, *Tetrahedron Lett.*, **29**, 4963 (1988).

[520] G. Pandey, G. Kumaraswamy, and U. T. Bhalerao, *Tetrahedron Lett.*, **30**, 6059 (1989).

[521] L. Stella, B. Raynier, and J. M. Surzur, *Tetrahedron Lett.*, **31**, 2721 (1977).

[522] M. Ihara, N. Taniguchi, K. Kukumoto, and T. Kametani, *J. Chem. Soc., Chem. Commun.*, **1987**, 1438.

[523] S. Yoo, K. Y. Yi, S.-H. Li, and N. Jeong, *Synlett*, **1990**, 575.

[524] T. Yamada, Y. Iwahara, H. Nishino, and K. Kurosawa, *J. Chem. Soc., Perkin Trans. 1*, **1993**, 609.

[525] M. Kaafarani, M. P. Crozet, and J. M. Surzur, *Bull. Soc. Chim. Fr.*, **1987**, 885.

[526] J. H. Hutchinson, T. S. Daynard, and J. W. Gillard, *Tetrahedron Lett.*, **32**, 573 (1991).

[527] A. G. Myers, D. Y. Gin, and K. L. Widdowson, *J. Am. Chem. Soc.*, **113**, 9661 (1991).

[528] G. A. Kraus and Y. Wu, *J. Am. Chem. Soc.*, **114**, 8705 (1992).

[529] E. I. Troyansky, M. I. Lazareva, D. V. Demchuk, V. V. Samoshiu, Y. A. Strelenko, and G. I. Nikishiu, *Synlett*, **1992**, 233.

[530] D. L. Boger and R. J. Mathvink, *J. Am. Chem. Soc.*, **112**, 4008 (1990).

[531] J. E. Baldwin, R. M. Adlington, M. B. Mitchell, and J. Robertson, *J. Chem. Soc., Chem. Commun.*, **1990**, 1574.

[532] K. J. Shea, R. O'Dell, and D. Y. Sasaki, *Tetrahedron Lett.*, **33**, 4699 (1992).

[533] C. Descoins, M. Julia, and H. van Sang, *Bull. Soc. Chim. Fr.*, **1971**, 4087.

[534] S. Wolf and W. C. Agosta, *J. Org. Chem.*, **45**, 3139 (1980).

[535] P. Dowd and W. Zhang, *J. Am. Chem. Soc.*, **114**, 10084 (1992).

[536] T. Shono and N. Kise, *Tetrahedron Lett.*, **31**, 1303 (1990).

[537] C. K. Sha, T. S. Jean, and D. C. Wang, *Tetrahedron Lett.*, **31**, 3745 (1990).

[538] S. Wolf and W. C. Agosta, *J. Chem. Res. (S)*, **1981**, 78.

[539] M. Nagai, J. Lazor, and C. S. Wilcox, *J. Org. Chem.*, **55**, 3440 (1990).

[540] E. J. Corey and S. G. Pyne, *Tetrahedron Lett.*, **24**, 2821 (1983).

[541] V. H. Rawal and V. Krishnamurthy, *Tetrahedron Lett.*, **33**, 3439 (1992).

[542] F. H. Wartenberg, H. Junga, and S. Blechert, *Tetrahedron Lett.* **34**, 5251 (1993).

[543] B. B. Snider and B. O. Buckman, *J. Org. Chem.*, **57**, 322 (1992).

[544] E. J. Eukden and J. A. Burroff, *Tetrahedron Lett.*, **33**, 1835 (1992).

[545] J. D. Kilburn, *Tetrahedron Lett.*, **31**, 2193 (1990).

[546] C. P. Jasperse and D. P. Curran, *J. Am. Chem. Soc.*, **112**, 5601 (1990).

[547] A. J. Bloodworth, D. Crich, and T. Melvin, *J. Chem. Soc., Chem. Commun.*, **1987**, 786.

[548] D. L. J. Clive and H. W. Manning, *J. Chem. Soc., Chem. Commun.*, **1993**, 666.

[549] D. L. J. Clive, D. R. Cheshire, and L. Set, *J. Chem. Soc., Chem. Commun.*, **1987**, 353.

[549a] D. P. Curran, A. C. Abraham, and H. Liu, *J. Org. Chem.*, **56**, 4335 (1991).

[550] M. Journet, W. Smadja, and M. Malacria, *Synlett*, **1990**, 320.

[551] S. Kim and J. S. Kee, *Tetrahedron Lett.*, **34**, 4213 (1993).

[552] G. V. M. Sharma and S. R. Vepachedu, *Tetrahedron Lett.*, **31**, 4931 (1990).

[553] W. R. Leonard and T. Livinghouse, *Tetrahedron Lett.*, **26**, 6431 (1985).

[554] D. H. R. Barton, E. da Silva, and S. Z. Zard, *J. Chem. Soc., Chem. Commun.*, **1988**, 285.

[555] H. Hemmerle and H. J. Gais, *Angew. Chem.*, **101**, 362 (1989); *Angew. Chem., Int. Ed. Engl.*, **28**, 349 (1989).

[556] M. Suzuki, H. Koyano, and R. Noyori, *J. Org. Chem.*, **52**, 5583 (1987).

[557] F. MacCorquodale and J. C. Walton, *J. Chem. Soc., Chem. Commun.*, **1987**, 1456.

[558] F. MacCorquodale and J. C. Walton, *J. Chem. Soc., Perkin Trans. 1*, **1989**, 347.

[559] K. Weinges, H. Reichert, U. Huber–Patz, and H. Irngartinger, *Liebigs Ann. Chem.*, **1993**, 403.

[560] S. B. Booth, P. R. Jenkins, C. J. Swain, and J. B. Sweeney, *J. Chem. Soc., Chem. Commun.*, **1991**, 1656.

[561] S. B. Booth, P. R. Jenkins, and C. J. Swain, *J. Chem. Soc., Chem. Commun.*, **1991**, 1248.

[562] N. N. Marinovic and H. Ramanathan, *Tetrahedron Lett.*, **24**, 1872 (1983).

[563] J. K. Mcleod and L. C. Monahan, *Tetrahedron Lett.*, **29**, 391 (1988).

[564] E. W. Della, A. M. Knill, and P. E. Pigou, *J. Org. Chem.*, **58**, 2110 (1993).

[565] A. L. J. Beckwith and S. Gerba, *Aust. J. Chem.*, **45**, 289 (1992).

[566] J. K. MacLeod and L. C. Monahan, *Aust. J. Chem.*, **43**, 329 (1990).

[567] A. L. J. Beckwith, D. M. O'Shea and D. H. Roberts, *J. Chem. Soc., Chem. Commun.*, **1983**, 1445.

[568] G. A. Kraus and S. Liras, *Tetrahedron Lett.*, **31**, 5265 (1990).

[569] M. A. Dombroski, S. A. Kates, and B. B. Snider, *J. Am. Chem. Soc.*, **112**, 2759 (1990).

[570] A. L. J. Beckwith and D. H. Roberts, *J. Am. Chem. Soc.*, **108**, 5893 (1986).

[571] V. H. Rawal, V. Krishnamurthy, and A. Fabre, *Tetrahedron Lett.*, **34**, 2899 (1993).

[572] S. Kim and J. S. Koh, *J. Chem. Soc., Chem. Commun.*, **1992**, 1377.

[573] J. Cossy, J. P. Pete, and C. Portella, *Tetrahedron Lett.*, **30**, 7361 (1989).

[574] B. B. Snider and M. A. Dombroski, *J. Org. Chem.*, **52**, 5487 (1987).

[575] W. F. Berkowitz and P. J. Wilson, *J. Org. Chem.*, **56**, 3097 (1991).

[576] M. Julia and F. LeGoffic, *Bull. Soc. Chim. Fr.*, **1965**, 1555.

[577] D. F. Taber, Y. Wang, and S. J. Stachel, *Tetrahedron Lett.*, **34**, 6209 (1993).

[578] Y. Hanzawa, H. Ito, N. Kohara, H. Sasaki, H. Fukuda, T. Morikawa, and T. Taguchi, *Tetrahedron Lett.*, **32**, 4143 (1991).

[579] P. A. Zoretic, M. Ramchandani, and M. C. Caspar, *Synth. Commun.*, **21**, 923 (1991).

[580] T. V. RajanBabu and T. Fukunaga, *J. Am. Chem. Soc.*, **111**, 296 (1989).

[581] J. E. Einhorn, C. Einhorn, and J. L. Luche, *Tetrahedron Lett.*, **29**, 2183 (1988).

[582] B. B. Snider and Q. Zhang, *Tetrahedron Lett.*, **33**, 5921 (1993).

[583] B. B. Snider, B. Y.–F. Wan, B. O. Buckman, and B. M. Fox, *J. Org. Chem.*, **56**, 328 (1991).

[584] B. B Snider and Q. Zhang, *Tetrahedron Lett.*, **33**, 5921 (1992).

[585] D. L. J. Clive, *Pure Appl. Chem.*, **60**, 1645 (1988).

[586] J. W. Grissom, T. L. Calkins, and H. A. McMillen, *J. Org. Chem.*, **58**, 6556 (1993).

[587] P. J. Wagner and B. S. Park, *Tetrahedron Lett.*, **32**, 165 (1991).

[588] S. Kim and J. S. Koh, *Tetrahedron Lett.*, **33**, 7391 (1992).

[589] D. L. Boger and R. J. Mathvink, *J. Org. Chem.*, **55**, 5442 (1990).

[590] A. Srikrishna and P. Hemamalini, *Tetrahedron*, **48**, 9337 (1992).

[591] E. J. Corey, C. Shih, N. Y. Shih, and K. Shimaji, *Tetrahedron Lett.*, **25**, 5013 (1984).

[592] E. J. Corey, K. Shimaji, and C. Shih, *J. Am. Chem. Soc.*, **106**, 6425 (1984).

[593] D. L. J. Clive, H. W. Manning, T. L. B. Boivin, and M. H. D. Postema, *J. Org. Chem.*, **58**, 6857 (1993).

[594] D. L. J. Clive and M. H. D. Postema, *J. Chem. Soc., Chem. Commun.*, **5**, 429 (1993).

[595] S. Satoh, M. Sodeoka, H. Sasai, and M. Shibasaki, *J. Org. Chem.*, **57**, 2278 (1991).

[596] J. D. Harling and W. B. Motherwell, *J. Chem. Soc., Chem. Commun.*, **1988**, 1380.

[597] D. L. J. Clive, H. W. Manning, and T. L. B. Boivin, *J. Chem. Soc., Chem. Commun.*, **1990**, 972.

[598] P. A. Zoretic, X. Weng, C. K. Biggers, M. S. Biggers, M. L. Caspar, and G. D. Davis, *Tetrahedron Lett.*, **33**, 2637 (1992).

[599] D. Colclough, J. B. White, W. B. Smith, and Y. Chu, *J. Org. Chem.*, **58**, 6303 (1993).

[600] P. Dowd and W. Zhang, *J. Am. Chem. Soc.*, **113**, 9875 (1991).

[601] P. Dowd and W. Zhang, *J. Org. Chem.*, **57**, 7163 (1992).

[602] K. Weinges, W. Maurer, H. Reichert, G. Schilling, T. Oeser, and H. Irngartinger, *Chem. Ber.*, **123**, 901 (1990).

[603] C. Ellwood and G. Pattenden, *Tetrahedron Lett.*, **32**, 1591 (1991).

[604] T. Sugimura, T. Futagawa, and A. Tai, *Chem. Lett.*, **1990**, 2295.

[605] T. Satoh, M. Itoh, and K. Yamakawa, *Chem. Lett.*, **1987**, 1949.

[606] W. Fan and J. B. White, *Tetrahedron Lett.*, **34**, 957 (1993).

[607] H. Nishida, H. Takahashi, H. Takeda, N. Takada, and O. Yonemitsu, *J. Am. Chem. Soc.*, **112**, 902 (1990).

[608] F. Lombardo, R. A. Newmark, and E. Kariv–Miller, *J. Org. Chem.*, **56**, 2422 (1991).

[609] T. Ikeda, S. Yue, and C. R. Hutchinson, *J. Org. Chem.*, **50**, 5193 (1985).

[610] T. Shono, N. Kise, T. Suzumoto, and T. Morimoto, *J. Am. Chem. Soc.*, **108**, 4676 (1986).

[611] A. Srikrishna and P. Hemamalini, *J. Chem. Soc., Perkin Trans. 1*, **1989**, 2511.

[612] J. C. Chottard, M. Julia, and J. M. Salard, *Tetrahedron*, **25**, 4967 (1969).

[613] M. Kawaguchi, S. Satoh, M. Mori, and M. Shibasaki, *Chem. Lett.*, **1992**, 395.

[614] P. A. Zoretic, M. Ramchandani, and M. L. Caspar, *Synth. Commun.*, **21**, 915 (1991).

[615] G. H. Posner, E. Asirvatham, T. G. Hamill, and K. S. Webb, *J. Org. Chem.*, **55**, 3132 (1989).

[616] J. D. Winkler, V. Sridar, and M. G. Siegel, *Tetrahedron Lett.*, **30**, 4943 (1989).

[617] J. D. White, T. C. Somers, and K. M. Yager, *Tetrahedron Lett.*, **31**, 59 (1990).

[618] M. I. Colombo, S. Signorella, M. P. Mischue, M. Gonzales–Sierra, and E. A. Ruveda, *Tetrahedron*, **46**, 4149 (1990).

[619] J. L. Stein and M. P. Crozet, *C. R. Acad. Sci., Ser. 2*, **300**, 59 (1985).

[620] J. L. Stein, L. Stella, and J. M. Surzur, *Tetrahedron Lett.*, **21**, 287 (1980).

[621] W.–W. Weng and T.–Y. Luh, *J. Org. Chem.*, **58**, 5574 (1993).

[622] J. Knight and P. J. Parsons, *J. Chem. Soc, Perkin Trans. 1*, **1987**, 1237.

[623] J. Knight, P. J. Parsons, and R. Southgate, *J. Chem. Soc., Chem. Commun.*, **1986**, 78.

[624] M. D. Bachi, A. DeMesmaeker, and N. Stevenart–DeMesmaeker, *Tetrahedron Lett.*, **28**, 2887 (1987).

[625] M. D. Bachi, A. DeMesmaeker, and N. Stevenart–DeMesmaeker, *Tetrahedron Lett.*, **28**, 2637 (1987).

[626] J. Anaya, D. H. R. Barton, S. D. Gero, M. Grande, N. Martin, and C. Tachdijian, *Angew. Chem. Int. Ed. Engl.*, **32**, 867 (1993).

[627] G. I. Nikishin, E. I. Troyanskii, and M. I. Lazareva, *Tetrahedron*, **41**, 4279 (1985).

[628] T. Kametani and T. Honda, *Heterocycles*, **19**, 1861 (1982).

[629] T. Kametani, D. C. Shih, A. Itoh, S. Maeda, and T. Honda, *Heterocycles*, **27**, 875 (1988).

[630] A. L. J. Beckwith and D. R. Boate, *Tetrahedron Lett.*, **26**, 1761 (1985).

[631] T. Kametani, S. D. Chu, A. Itoh, S. Maeda, and T. Honda, *J. Org. Chem.*, **53**, 2683 (1988).

[632] M. D. Bachi and C. Hoornaert, *Tetrahedron Lett.*, **1982**, 2505.

[633] W. Cabri, I. Candiani, A. Bedeschi, and R. Santi, *Tetrahedron Lett.*, **33**, 4783 (1992).

[634] W. Cabri, D. Borghi, E. Arlandini, P. Sbraletta, and A. Bedeschi, *Tetrahedron*, **49**, 6837 (1993).

[635] V. M. Girijavallabhan and A. K. Ganguly, *Heterocycles*, **28**, 47 (1989).

[636] M. D. Bachi, F. Frolow, and C. Hoornaert, *J. Org. Chem.*, **48**, 1841 (1983).

[637] M. D. Bachi and C. Hoornaert, *Tetrahedron Lett.*, **1981**, 2693.

[638] M. D. Bachi and C. Hoornaert, *Tetrahedron Lett.*, **1981**, 2689.

[639] K. Miura, K. Oshima, and K. Utimoto, *Chem. Lett.*, **1992**, 2477.

[640] M. Pezechk, A. P. Brunetiere, and J. Y. Lallemand, *Tetrahedron Lett.*, **27**, 3715 (1986).

[641] H. Nagashima, K. Ara, H. Wakamatsu, and K. Itoh, *J. Chem. Soc., Chem. Commun.*, **1985**, 518.

[642] S. Knapp, F. S. Gibson, and Y. H. Choe, *Tetrahedron Lett.*, **31**, 5397 (1990).

[643] T. Sato, K. Tsujimoto, K. Matsubayashi, H. Ishibashi, and M. Ikeda, *Chem. Pharm. Bull.*, **40**, 2308 (1992).

[644] P. F. Keusenkothen and M. B. Smith, *Tetrahedron*, **48**, 2977 (1992).

[645] P. F. Keusenkothen and M. B. Smith, *Tetrahedron Lett.*, **30**, 3369 (1989).

[646] R. S. Jolly and T. Livinghouse, *J. Am. Chem. Soc.*, **110**, 7536 (1988).

[647] J. M. Surzur and L. Stella, *Tetrahedron Lett.*, **1974**, 2191.

[648] R. J. Kolt, D. Griller, and D. D. M. Wayner, *Tetrahedron Lett.*, **31**, 7539 (1990).

[649] S. Torii, T. Inokuchi, and T. Yukawa, *J. Org. Chem.*, **50**, 5875 (1985).

[650] H. Ishibashi, T. Sato, M. Irie, S. Harada, and M. Ikeda, *Chem. Lett.*, **1987**, 795.

[651] C. D. S. Brown, N. S. Simpkins, and K. Clinch, *Tetrahedron Lett.*, **34**, 131 (1993).

[652] S. Iwasa, M. Yamamoto, S. Kohmoto, and K. Yamada, *J. Org. Chem.*, **56**, 2849 (1991).

[653] A. Srikrishna, S. Nagaraju, and G. V. R. Sharma, *J. Chem. Soc., Chem. Commun.*, **1993**, 285.

[654] A. Nishida, M. Nishida, and O. Yonemitsu, *Tetrahedron Lett.*, **31**, 7035 (1990).

[655] G. Pandey and G. D. Reddy, *Tetrahedron Lett.*, **33**, 6533 (1992).

[656] G. Stork and O. Ouertelli, *New J. Chem.*, **16**, 95 (1992).

[657] C. S. Wilcox and L. M. Thomasco, *J. Org. Chem.*, **50**, 546 (1985).

[658] M. F. Jones and S. M. Roberts, *J. Chem. Soc., Perkin Trans. 1*, **1988**, 2927.

[659] C. Jenny, P. Wipf, and H. Heimgartner, *Helv. Chim. Acta*, **72**, 838 (1989).

[660] S. Takano, K. Ohashi, T. Sugihara, and K. Ogasawara, *Chem. Lett.*, **1991**, 203.

[661] M. J. Begley, R. J. Fletcher, J. A. Murphy, and M. S. Sherburn, *J. Chem. Soc., Chem. Commun.*, **1993**, 1723.

[662] C. Gennari, G. Poli, C. Scolastico, and M. Vassallo, *Tetrahedron Asymmetry*, **2**, 793 (1991).

[663] L. Belsivi, C. Gennari, G. Poli, C. Scolastico, B. Salom, and M. Vassallo, *Tetrahedron*, **48**, 3945 (1992).

[664] S. F. Martin, C. P. Yang, W. L. Laswell, and H. Rueger, *Tetrahedron Lett.*, **29**, 6685 (1988).

[665] D. A. Burnett, J. K. Choi, D. J. Hart, and Y. M. Tsai, *J. Am. Chem. Soc.*, **106**, 8201 (1984).

[666] D. J. Hart and Y. M. Tsai, *J. Am. Chem. Soc.*, **104**, 1430 (1982).

[667] A. Srikrishna, G. V. R. Sharma, and S. Nagaraju, *Synth. Commun.*, **22**, 1221 (1992).

[668] J.-K. Choi and D. J. Hart, *Tetrahedron*, **41**, 3959 (1985).

[669] D. J. Hart and Y. M. Tsai, *J. Am. Chem. Soc.*, **106**, 8209 (1984).

[670] A. G. H. Wee, *Tetrahedron*, **46**, 5065 (1990).

[671] J. Marco-Contelles, P. Ruiz, L. Martinez, and A. Martinez-Grau, *Tetrahedron*, **49**, 6669 (1993).

[672] J. E. Baldwin, M. G. Molonay, and A. F. Passons, *Tetrahedron*, **48**, 9373 (1992).

[673] D. S. Middleton, N. S. Simpkins, and N. K. Terrett, *Tetrahedron Lett.*, **30**, 3865 (1989).

[674] J. Cossy and A. Bouzide, *J. Chem. Soc., Chem. Commun.*, **1993**, 1218.

[675] H. Urbach and R. Henning, *Heterocycles*, **28**, 957 (1989).

[676] K. Matsumoto, K. Miura, K. Oshima, and K. Utimoto, *Tetrahedron Lett.*, **33**, 7031 (1992).

[677] Y. M. Tsai, B. W. Ben, C. T. Yang, and C. H. Lin, *Tetrahedron Lett.*, **33**, 7895 (1992).

[678] S. Kano, Y. Yuasa, K. Asami, and S. Shibuya, *Chem. Lett.*, **1986**, 735.

[679] C. L. Tu and P. S. Mariano, *J. Am. Chem. Soc.*, **109**, 5287 (1987).

[680] J. M. Contelles, P. Ruiz, B. Snchez, and M. L. Jimeno, *Tetrahedron Lett.*, **33**, 5261 (1993).

[681] J. K. Choi, D. J. Hart, and Y. M. Tsai, *Tetrahedron Lett.*, **23**, 4765 (1982).

[682] D. P. Curran, W.-T. Jiaang, M. Palovich, and Y.-M. Tsai, *Synlett*, **1993**, 403.

[683] R. Mahler and J. O. Metzger, *Fat. Sci. Technol.*, **95**, 337 (1993).

[684] C. E. McDonald and R. W. Dugger, *Tetrahedron Lett.*, **29**, 2413 (1988).

[685] J. Cossy, A. Bouzide, and C. Leblanc, *Synlett*, **1993**, 202.

[686] W. R. Bowman, D. N. Clark, and R. J. Marmon, *Tetrahedron Lett.*, **34**, 4993 (1992).

[687] D. R. Artis, J. S. Cho, and J. M. Muchowski, *Can. J. Chem.*, **70**, 1838 (1992).

[688] D. L. Flynn, D. L. Zabrowski, and R. Nosal, *Tetrahedron Lett.*, **33**, 7281 (1992).

[689] J. E. Baldwin and C. S. Li, *J. Chem. Soc., Chem. Commun.*, **1988**, 261.

[690] J. P. Marino, E. Laborde, and R. S. Paley, *J. Am. Chem. Soc.*, **110**, 966 (1988).

[691] J. J. Gaudino and C. S. Wilcox, *J. Am. Chem Soc.*, **112**, 4374 (1990).

[692] W. Koof, R. VanGinkel, M. Kranenburg, H. Hiemstra, S. Louwrier, M. J. Molenaar, and W. N. Speckamp, *Tetrahedron Lett.*, **32**, 401 (1991).

[693] J. M. Dener and D. J. Hart, *Tetrahedron*, **44**, 7037 (1988).

[694] S. Velázquez, S. Huss, and M.–J. Camarasa, *J. Chem. Soc., Chem. Commun.*, **1991**, 1263.

[695] G. E. Keck, E. N. K. Cressman, and E. J. Enholm, *J. Org. Chem.*, **54**, 4345 (1989).

[696] G. Stork, H. S. Suh, and G. Kim, *J. Am. Chem. Soc.*, **113**, 7054 (1991).

[697] M. Kaafarani, M. P. Crozet, and J. M. Surzur, *Bull. Soc. Chim. Fr.*, **1989**, 114.

[698] H. Nagashima, H. Wakamatsu, N. Ozaki, T. Ishii, M. Watanabe, T. Tajima, and U. Itoh, *J. Org. Chem.*, **57**, 1682 (1992).

[699] J. A. Murphy and M. S. Sherburn, *Tetrahedron*, **47**, 4077 (1991).

[700] J. A. Murphy and M. S. Sherburn, *Tetrahedron Lett.*, **31**, 3495 (1990).

[701] L. Benati, P. C. Montevecchi, and P. Sagnolo, *J. Chem. Soc., Perkin Trans. 1*, **1991**, 2103.

[702] S. Ozaki, H. Matsushita, and H. Ohmori, *J. Chem. Soc., Chem. Commun.*, **1992**, 1120.

[703] G. Boisvert and R. Giasson, *Tetrahedron Lett.*, **33**, 6587 (1992).

[704] A. L. J. Beckwith and G. F. Meijs, *J. Chem. Soc., Chem. Commun.*, **1981**, 595.

[705] G. F. Meijs and A. L. J. Beckwith, *J. Am. Chem. Soc.*, **108**, 5890 (1986).

[706] G. F. Meijs and A. L. J. Beckwith, *J. Chem. Soc., Chem. Commun.*, **1981**, 136.

[707] A. L. J. Beckwith and G. F. Meijs, *J. Org. Chem.*, **52**, 1922 (1987).

[708] Y. Yuasa, S. Kano, and S. Shibuya, *Heterocycles*, **32**, 2311 (1991).

[709] K. Nozaki, K. Oshima, and K. Utimoto, *Tetrahedron*, **45**, 923 (1989).

[710] C. Audin, J.–M. Lancelin, and J.–M. Beau, *Tetrahedron Lett.*, **29**, 3691 (1988).

[711] H. Ishibashi, T. S. So, T. Sato, K. Kuroda, and M. Ikeda, *J. Chem. Soc., Chem. Commun.*, **1989**, 762.

[712] H. Ishibashi, T. S. So, K. Okochi, T. Sato, N. Nakamura, H. Nakatani, and M. Ikeda, *J. Org. Chem.*, **56**, 95 (1991).

[713] A. Srikrishna, *J. Chem. Soc., Chem. Commun.*, **1987**, 587.

[714] H. Ishibashi, N. Nakamura, T. Sato, M. Takeuchi, and M. Ikeda, *Tetrahedron Lett.*, **32**, 1725 (1991).

[715] T. Sato, N. Nakamura, K. Ikeda, M. Okada, H. Ishibashi, and M. Ikeda, *J. Chem. Soc., Perkin Trans. 1*, **1992**, 2399.

[716] C. Lampard, J. A. Murphy, and N. Lewis, *J. Chem. Soc., Chem. Commun.*, **1993**, 295.

[717] Y. Watanabe, Y. Ueno, L. Tanaka, M. Okawara, and T. Endo, *Tetrahedron Lett.*, **28**, 3953 (1987).

[718] K. Shankaran, C. P. Sloan, and V. Snieckus, *Tetrahedron Lett.*, **26**, 6001 (1985).

[719] K. Jones, M. Thompson, and C. Wright, *J. Chem. Soc., Chem. Commun.*, **1986**, 115.

[720] K. Kobayashi, M. Itoh, and H. Suginone, *Tetrahedron Lett.*, **28**, 3369 (1987).

[721] C. P. A. Kunka and J. Warkentin, *Can. J. Chem.*, **68**, 575 (1990).

[722] K. R. Biggs, P. J. Parsons, D. J. Tapolzcay, and J. M. Underwood, *Tetrahedron Lett.*, **30**, 7115 (1989).

[723] A. L. J. Beckwith and S. W. Westwood, *Tetrahedron*, **45**, 5269 (1989).

[724] A. Srikrishna and K. C. Pullaiah, *Tetrahedron Lett.*, **28**, 5203 (1987).

[725] B. Roudot, T. Durand, J. P. Girard, J. C. Rossi, L. Schio, S. P. Khanapure, and J. Rokach, *Tetrahedron Lett.*, **34**, 8245 (1993).

[726] S. Fukuzawa, A. Nakanishi. T. Fujinami, and S. Sakai, *J. Chem. Soc., Perkin Trans. 1*, **1988**, 1669.

[727] R. Kiesewetter and P. Margaretha, *Helv. Chim. Acta*, **70**, 121 (1987).

[728] C. Wright, M. Shulkind, K. Jones, and M. Thompson, *Tetrahedron Lett.*, **28**, 6389 (1987).

[729] J. P. Dittani and H. Ramanathan, *Tetrahedron Lett.*, **29**, 45 (1988).

[730] M. Tsukazaki and V. Snieckus, *Can. J. Chem.*, **70**, 1486 (1992).

[731] U. Albrecht, R. Wartchow, and H. M. R. Hoffmann, *Angew. Chem., Int. Ed. Engl.*, **31**, 910 (1992).

[732] A. Srikishina and K. Kroshnan, *Tetrahedron Lett.*, **29**, 4995 (1988).

[733] E. R. Lee, I. Lakomy, P. Bigler, and R. Scheffold, *Helv. Chim. Acta*, **74**, 146 (1991).

[734] H. Togo and O. Kikuchi, *Hetereocycles*, **28**, 373 (1989).

735 J. C. Lopez, A. M. Gomez, and B. Fraser–Reid, *J. Chem. Soc., Chem. Commun.*, **1993**, 762.

736 W. Xu and P. S. Mariano, *J. Am. Chem. Soc.*, **113**, 1431 (1991).

737 K. Jones and J. M. D. Storey, *J. Chem. Soc., Chem. Commun.*, **1992**, 1766.

738 Y. Chapleur and N. Moufid, *J. Chem. Soc., Chem. Commun.*, **1989**, 39.

739 R. J. Ferrier and P. M. Petersen, *Tetrahedron*, **46**, 1 (1990).

740 A. DeMesmaeker, P. Hoffmann, and B. Ernst, *Tetrahedron Lett.*, **29**, 6585 (1988).

741 A. J. Clark and K. Jones, *Tetrahedron Lett.*, **30**, 5485 (1989).

742 A. J. Clark and K. Jones, *Tetrahedron*, **48**, 6875 (1992).

743 M. J. Begley, M. Ladlow, and G. Pattenden, *J. Chem. Soc., Perkin Trans. 1*, **1988**, 1095.

744 M. Koreeda and D. C. Visger, *Tetrahedron Lett.*, **33**, 6603 (1992).

745 A. N. Abeywickrema, A. L. J. Beckwith, and S. Gerba, *J. Org. Chem.*, **52**, 4072 (1987).

746 L. K. Dyall, *Aust. J. Chem.*, **32**, 643 (1979).

747 J. Yoshida, S. Nakatani, K. Sakaguchi, and S. Isoe, *J. Org. Chem.*, **54**, 3383 (1989).

748 A. Srikrishna and G. Veera Raghava Sharma, *Tetrahedron Lett.*, **29**, 6487 (1988).

749 R. J. Ferrier, P. M. Petersen, and M. A. Taylor, *J. Chem. Soc., Chem. Commun.*, **1989**, 1247.

750 N. Moufid and Y. Chapleur, *Tetrahedron Lett.*, **32**, 1799 (1991).

751 R. Nouguier, C. Lesueur, E. DeRiggi, M. P. Bertrand, and A. Virgili, *Tetrahedron Lett.*, **31**, 3541 (1990).

752 J. Cossy and D. Belotti, *Tetrahedron Lett.*, **29**, 6113 (1988).

753 K. Tamao, K. Maeda, T. Yamaguchi, and Y. Ito, *J. Am. Chem. Soc.*, **111**, 4984 (1989).

754 S. Knapp and F. S. Gibson, *J. Org. Chem.*, **57**, 4802 (1992).

755 J. Ardisson, J. P. Ferezou, M. Julia, Y. Li, L. W. Liu, and A. Pancrazi, *Bull. Soc. Chim. Fr.*, **129**, 387 (1992).

756 J. Arolisson, J. P. Frzou, and M. Julia, *Tetrahedron Lett.*, **28**, 2001 (1987).

757 A. L. J. Beckwith, S. P. Joseph, and R. T. A. Mayadunne, *J. Org. Chem.* **58**, 4198 (1993).

758 D. L. Boger and R. S. Coleman, *J. Am. Chem. Soc.*, **110**, 4796 (1988).

759 J. E. Lyons, C. H. Schiesser, and K. Sutej, *J. Org. Chem.*, **58**, 5632 (1993).

760 D. S. Middleton, N. S. Simpkins, and N. K. Terret, *Tetrahedron*, **46**, 545 (1990).

761 A. DeMesmaeker, P. Hoffmann, and B. Ernst, *Tetrahedron Lett.*, **30**, 57 (1989).

762 K. S. Grкninger, K. F. Jger, and B. Giese, *Liebigs Ann. Chem.*, **1987**, 731.

763 Y. Ueno, K. Chino, and M. Okawara, *Tetrahedron Lett.*, **23**, 2575 (1982).

764 J. C. Lopez, A. M. Gomez, and S. Valverde, *J. Chem. Soc., Chem. Commun.*, **1992**, 613.

765 D. Batty and D. Crich, *Tetrahedron Lett.*, **33**, 875 (1992).

766 D. Batty and D. Crich, *J. Chem. Soc., Perkin Trans. 1*, **1992**, 3193.

767 S. Hanessian, P. Beaulieu, and D. Dub, *Tetrahedron Lett.*, **27**, 5071 (1986).

768 D. L. J. Clive and A. Y. Mohammed, *Heterocycles*, **28**, 1157 (1989).

769 K. Jones and C. McCarthy, *Tetrahedron Lett.*, **30**, 2657 (1989).

770 H. McNab, *J. Chem. Soc., Perkin Trans. 1*, **1984**, 377.

771 H. McNab and G. S. Smith, *J. Chem. Soc., Chem. Commun.*, **1982**, 996.

772 T. A. K. Smith and G. H. Whitham, *J. Chem. Soc., Chem. Commun.*, **1989**, 313.

773 T. Honda, M. Satoh, and Y. Kobayashi, *J. Chem. Soc., Perkin Trans. 1*, **1992**, 1557.

774 M. Carda and J. A. Alberto, *Tetrahedron*, **48**, 9789 (1992).

775 E. J. Enholm and A. Trivellas, *J. Am. Chem. Soc.*, **111**, 6463 (1989).

776 R. Leardini, D. Nanni, M. Santori, and G. Zanardi, *Tetrahedron*, **48**, 3961 (1992).

777 D. Batty and D. Crich, *J. Chem. Soc., Perkin Trans. 1*, **1992**, 3205.

778 M. J. Begley, H. Bhandal, J. H. Hutchinson, and G. Pattenden, *Tetrahedron Lett.*, **28**, 1317 (1987).

779 M. J. Begley, D. R. Cheshire, T. Harrison, J. H. Hutchinson, P. L. Myers, and G. Pattenden, *Tetrahedron*, **45**, 5215 (1989).

780 M. Carda and J. A. Marco, *Tetrahedron Lett.*, **32**, 5191 (1991).

781 E. J. Enholm, H. Satici, and A. Trivellas, *J. Org. Chem.* **54**, 5841 (1989).

782 V. Pedretti, J.–M. Mallet, and P. Sina, *Carbohydr. Res.*, **244**, 247 (1993).

783 P. J. Parsons, P. Willis, and S. C. Eyley, *Tetrahedron*, **48**, 9461 (1992).

784 D. L. J. Clive and J. R. Bergstra, *J. Org. Chem.*, **56**, 4976 (1991).

[785] A. DeMaesmaeker, P. Hoffmann, T. Winkler, and A. Waldner, *Synlett*, **1990**, 201.

[786] T. Honda, M. Hoshi, and M. Tsubuki, *Heterocycles*, *34*, 1515 (1992).

[787] H. Nagano, Y. Seko, and K. Nakai, *J. Chem. Soc., Perkin Trans. 1*, **1990**, 2153.

[788] S. Kano, Y. Yuasa, T. Yokomatsu, K. Asami, and S. Shibuya, *J. Chem. Soc., Chem. Commun.*, **1986**, 1717.

[789] M. C. Fong and C. H. Schiesser, *Tetrahedron Lett.*, *34*, 4347 (1993).

[790] S. Kano, Y. Yuasa, K. Asami, and S. Shibuya, *Heterocycles*, *27*, 1437 (1988).

[791] W. R. Bowman, H. Heaney, and B. M. Jordan, *Tetrahedron Lett.*, *29*, 6657 (1988).

[792] A. DeMesmaeker, A. Waldner, P. Hoffmann, T. Mindt, P. Hug, and T. Winkler, *Synlett*, **1990**, 687.

[793] H. Ishibashi, H. Nakatani, S. Iwaml, T. Sato, N. Nakamura, and M. Ikeda, *J. Chem. Soc., Chem. Commun.*, **1989**, 1767.

[794] J. M. Dener, D. J. Hart, and S. Ramesh, *J. Org. Chem.*, *53*, 6022 (1988).

[795] R. A. Alonso, G. D. Vite, R. E. McDevitt, and B. Fraser–Reid, *J. Org. Chem.*, *57*, 573 (1992).

[796] G. D. Vite, R. A. Alonso, and B. Fraser–Reid, *J. Org. Chem.*, *54*, 2268 (1989).

[797] P. J. Parsons, P. A. Willis, and S. C. Eyley, *J. Chem. Soc., Chem. Commun.*, **1988**, 283.

[798] H. Hashimoto, K. Furuichi, and T. Miwa, *J. Chem. Soc., Chem. Commun.*, **1987**, 1002.

[799] T. V. RajanBabu, *J. Org. Chem.*, *53*, 4522 (1988).

[800] D. L. Boger, R. J. Wysocki, Jr., and T. Ishizaki, *J. Am. Chem. Soc.*, *112*, 5230 (1990).

[801] D. L. Boger and R. J. Wysocki, Jr., *J. Org. Chem.*, *54*, 1239 (1989).

[802] A. DeMesmaeker, P. Hoffmann, B. Ernst, P. Hug, and T. Winkler, *Tetrahedron Lett.*, *30*, 6307 (1989).

[802a] A. DeMesmaeker, P. Hoffmann, B. Ernst, P. Hug, and T. Winkler, *Tetrahedron Lett.*, *30*, 6311 (1989).

[803] A. Haudrechy and P. Sina, *Carbohydr. Res.*, *216*, 375 (1991).

[804] A. J. Bloodworth and M. D. Spencer, *Tetrahedron Lett.*, *31*, 5513 (1990).

[805] E. Lee, C. H. Yoon, and T. H. Lee, *J. Am. Chem. Soc.*, *114*, 10981 (1992).

[806] K. S. Feldman, H. M. Berven, and A. L. Romanelli, *J. Org. Chem.*, *58*, 6851 (1993).

[807] A. Gossen, C. W. McCleland, and F. C. Rinaldi, *J. Chem. Soc., Perkin Trans. 2*, **1993**, 279.

[808] T. Sato, S. Ishida, H. Ishibashi, and M. Ikeda, *J. Chem. Soc., Perkin Trans. 1*, **1991**, 353.

[809] A. J. Clark, K. Jones, C. McCarthy, and J. M. D. Storey, *Tetrahedron Lett.*, *23*, 2829 (1991).

[810] M. J. Tomaszewski and J. Warkentin, *Tetrahedron Lett.*, *33*, 2123 (1992).

[811] R. Leardini, D. Nanni, A. Tundo, G. Zanardi, and F. Ruggieri, *J. Org. Chem.*, *57*, 1842 (1992).

[812] K. Kobayashi, A. Kinishi, Y. Kanno, and H. Suginome, *J. Chem. Soc., Perkin Trans. 1*, **1993**, 111.

[813] T. Hamada, M. Ohmori, and O. Yoremitsu, *Tetrahedron Lett.*, **1977**, 1519.

[814] D. Crich, K. A. Eustace, and T. J. Ritchie, *Heterocycles*, *28*, 67 (1989).

[815] C. Chen and D. Crich, *Tetrahedron Lett.*, *34*, 1545 (1993).

[816] R. Leardini, D. Nanni, A. Tundo, and G. Zanardi, *Gazz. Chim. Ital.*, *119*, 637 (1989).

[817] R. Leardini, G. F. Pedulli, A. Tundo, and G. Zanardi, *J. Chem. Soc., Chem. Commun.*, **1984**, 1320.

[818] A. R. Forrester, M. Gill, and R. H. Thomson, *J. Chem. Soc., Chem. Commun.*, **1976**, 677.

[819] F. D. Lewis and G. D. Reddy, *Tetrahedron Lett.*, *33*, 4249 (1992).

[820] R. Leardini, D. Nanni, A. Tundo, and G. Zanardi, *J. Chem. Soc., Chem. Commun.*, **1989**, 757.

[821] S. Takano, M. Suzuki, A. Kijiwa, and K. Ogasawara, *Chem. Lett.*, **1990**, 315.

[822] F. A. Neugebauer and I. Umminger, *Chem. Ber.*, *113*, 1205 (1980).

[823] S. A. Hitchcock and G. Pattenden, *Tetrahedron Lett.*, *31*, 3641 (1990).

[824] D. J. Hart and J. A. McKinney, *Tetrahedron Lett.*, *30*, 2611 (1989).

[825] E. C. Ashby and N. P. Tung, *Tetrahedron Lett.*, **1984**, 4333.

[826] W. Zhang and P. Dowd, *Tetrahedron Lett.*, *33*, 7307 (1992).

[827] T. Rajamannar and K. K. Balasubramanian, *Tetrahedron Lett.*, *29*, 5789 (1988).

[828] D. P. Curran and S. C. Kuo, *Tetrahedron*, *43*, 5653 (1987).

[829] D. Hobbs–Mallyou and D. A. Whiting, *J. Chem. Soc., Chem. Commun.*, **1991**, 1324.

[830] J. S. Yadav, T. K. Praveen Kumar, and V. R. Gadgil, *Tetrahedron Lett.*, *33*, 3687 (1992).

[831] P. Bakuzis, O. O. S. Campos, and M. L. F. Bakuzis, *J. Org. Chem.*, **41**, 3261 (1976).

[832] S. Pal, M. Mukherjee, D. Podder, A. U. Mukherjee, and U. R. Ghahk, *J. Chem. Soc., Chem. Commun.*, **1991**, 1591.

[833] A. P. Neary and P. J. Parsons, *J. Chem. Soc., Chem. Commun.*, **1989**, 1090.

[834] D. P. Curran and D. M. Rakiewicz, *J. Am. Chem. Soc.*, **107**, 1448 (1985).

[835] J. Cossy, D. Belotti, and J. P. Pete, *Tetrahedron Lett.*, **28**, 4547 (1987).

[836] J. Cossy, D. Belotti, and J. P. Pete, *Tetrahedron*, **46**, 1859 (1990).

[837] C. E. Mowbray and G. Pattenden, *Tetrahedron Lett.*, **34**, 127 (1993).

[838] Y.-J. Chen and W.-Y. Lin, *Tetrahedron Lett.*, **33**, 1749 (1992).

[839] J. M. Fang and M. Y. Chen, *Tetrahedron Lett.*, **28**, 2853 (1987).

[840] A. K. Ghosh, K. Ghosh, P. Sitaram, and U. R. Chatak, *J. Chem. Soc., Chem. Commun.*, **1993**, 809.

[841] D. P. Curran and M. H. Chen, *Tetrahedron Lett.*, **26**, 4991 (1985).

[842] G. Pattenden and S. J. Teague, *J. Chem. Soc., Perkin Trans. 1*, **1988**, 1077.

[843] L. M. Van der Linde and A. J. A. Van der Weerdt, *Tetrahedron Lett.*, **25**, 1201 (1984).

[844] J. W. Grissom and T. L. Calkins, *J. Org. Chem.*, **58**, 5422 (1993).

[845] G. A. Kraus and Y. S. Hon, *J. Org. Chem.*, **50**, 4605 (1985)

[846] A. Srikrishna, G. V. R. Sharma, and P. Hemamalini, *J. Chem. Soc., Chem. Commun.*, **1990**, 1681.

[847] J. W. Grissom and T. L. Calkins, *Tetrahedron Lett.*, **33**, 2315 (1992).

[848] K. Vijaya Bhaskar and G. S. R. Subba Rao, *Tetrahedron Lett.*, **30**, 225 (1989).

[849] T. L. Fevig, R. L. Elliott, and D. P. Curran, *J. Am. Chem. Soc.*, **110**, 5064 (1988).

[850] J. C. Chottard and M. Julia, *Bull. Soc. Chim. Fr.*, **1968**, 3700.

[851] Y. S. Kulkarni, M. Niwa, E. Ron, and B. B. Snider, *J. Org. Chem.*, **52**, 1568 (1987).

[852] D. P. Curran and W. Shen, *Tetrahedron*, **49**, 755 (1993).

[853] M. J. Begley, G. Pattenden, and G. M. Robertson, *J. Chem. Soc., Perkin Trans. 1*, **1988**, 1085.

[854] G. Pattenden and G. M. Robertson, *Tetrahedron Lett.*, **27**, 399 (1986).

[855] Y.-J. Chen, C.-M. Chen, and W.-Y. Lin, *Tetrahedron Lett.*, **34**, 2962 (1993).

[856] A. Srikrishna, P. Hemamalini, and G. Veera Raghava Sharma, *Tetrahedron Lett.*, **32**, 6609 (1991).

[857] A. Srikrishna, P. Hemamalini, and G. Veera Raghava Sharma, *J. Org. Chem.*, **58**, 2509 (1993).

[858] Y. K. Rao and M. Nagarajan, *J. Org. Chem.*, **54**, 5678 (1989).

[859] Y. Koteswar Rao and M. Nagarajan, *Tetrahedron Lett.*, **29**, 107 (1988).

[860] K. Shishido, G. Kiyoto, A. Tsuda, Y. Takaishi, and M. Shibuya, *J. Chem. Soc., Chem. Commun.*, **1993**, 793.

[861] E. Wenkert, B. C. Bookser, and T. S. Arrhenius, *J. Am. Chem. Soc.*, **114**, 644 (1992).

[862] R. P. Quirk and F. H. Murphy, *Tetrahedron Lett.*, **1979**, 301.

[863] A. G. Myers and K. R. Condroski, *J. Am. Chem. Soc.*, **115**, 7926 (1993).

[864] L. A. Paquette, C. S. Ra, and T. W. Silvestri, *Tetrahedron*, **45**, 3099 (1989).

[865] A. Osuka, H. Shimizu, H. M. Chiba, H. Suzuki, and K. Maruyama, *J. Chem. Soc., Perkin Trans. 1*, **1983**, 2073.

[866] E. Hobloth and H. Lund, *Acta Chem. Scand., Ser. B*, **B31**, 395 (1977).

[867] A. S. Kende, F. H. Ebetino, and T. Ohta, *Tetrahedron Lett.*, **26**, 3063 (1985).

[868] A. S. Kende, K. Koch, and C. A. Smith, *J. Am. Chem. Soc.*, **110**, 2210 (1988).

[869] L. A. Paquette, I. A. Colapret, and D. R. Andrews, *J. Org. Chem.*, **50**, 201 (1985).

[870] S. K. Pradhan, T. V. Radhakrishnan, and R. Subramanian, *J. Org. Chem.*, **41**, 1943 (1976).

[871] B. Arreguy San Miguel, B. Maillard, and B. Delmond, *Tetrahedron Lett.*, **28**, 2127 (1987).

[872] P. A. Zoretic, X. Weng, and M. L. Caspar, *Tetrahedron Lett.*, **32**, 4819 (1991).

[873] A. V. R. Rao, B. V. Rao, D. R. Reddy, and A. K. Singh, *J. Chem. Soc., Chem. Commun.*, **1989**, 400.

[874] D. L. J. Clive, A. G. Angoh, and S. M. Bennet, *J. Org. Chem.*, **52**, 1339 (1987).

[875] S. Wolff and H. M. R. Hoffmann, *Synthesis*, **1988**, 760.

[876] L. Benati and P. C. Montevecchi, *J. Org. Chem.*, **46**, 4570 (1981).

[877] K. Last and H. M. R. Hoffmann, *Synthesis*, **1989**, 901.

[878] C. P. Chuang, J. C. Gallucci, D. J. Hart, and C. Hoffman, *J. Org. Chem.*, **53**, 3218 (1988).

[879] C. P. Chuang, J. C. Gallucci, and D. J. Hart, *J. Org. Chem.*, **53**, 3210 (1988).

[880] B. W. A. Yenng, J. L. M. Contelles, and B. Fraser–Reid, *J. Chem. Soc., Chem. Commun.*, **1989**, 1160.

[881] W. B. Motherwell and A. M. K. Pennell, *J. Chem. Soc., Chem. Commun.*, **1991**, 877.

[882] P. Bhattacharyya, S. S. Jash, and A. K. Dey, *J. Chem. Soc., Chem. Commun.*, **1984**, 1668.

[883] G. Ariamala and U. U. Balasubramanian, *Tetrahedron Lett.*, **29**, 3335 (1988).

[884] R. Yamaguchi, T. Hamasaki, and K. Utimoto, *Chem. Lett.*, **1988**, 913.

[885] C. P. Sloan, J. C. Cuevas, C. Quesnelle, and V. Snieckus, *Tetrahedron Lett.*, **29**, 4685 (1988).

[886] T. Sugawara, B. A. Otter, and T. Ueda, *Tetrahedron Lett.*, **29**, 75 (1988).

[887] T. Ghosh and H. Hart, *J. Org. Chem.*, **54**, 5073 (1989).

[888] C. P. Chuang and D. J. Hart, *J. Org. Chem.*, **48**, 1782 (1983).

[889] J. Marco–Contelles, A. Martinez–Grau, M. Bernabe, N. Martin, and C. Seoane, *Synlett*, **1991**, 165.

[890] M. P. Crozet and W. Kassar, *C. R. Acad. Sci. Ser. 2*, **300**, 99 (1985).

[891] K. S. Kim, J. H. Kim, Y. K. Kim, Y. S. Park, and C. S. Hahn, *Carbohydr. Res.*, **194**, C1 (1989).

[892] R. Kiesewetter, A. Grott, and P. Margharetha, *Helv. Chim. Acta*, **71**, 502 (1988).

[893] S. Danishefsky and E. Taniyama, *Tetrahedron Lett.*, **24**, 15 (1983).

[894] R. Leardini, A. Tundo, and G. Zanardi, *J. Chem. Soc., Perkin Trans. 1*, **1981**, 3164.

[895] D. P. Curran and H. Liu, *J. Am. Chem. Soc.*, **113**, 2127 (1991).

[896] H. M. R. Hoffmann, B. Schmidt, and S. Wolff, *Tetrahedron*, **45**, 6113 (1989).

[897] Y. D. Vankar and N. C. Chaudhuri, *Synth. Commun.*, **21**, 885 (1991).

[898] J. J. Koehler and W. N. Speckamp, *Tetrahedron Lett.*, **1977**, 631.

[899] S. Hanessian, Y. L. Bennani, and R. di Fabio, *Acta Cryst.*, **C46**, 934 (1990).

[900] A. V. Rama Rao, J. S. Yadav, C. S. Rao, and S. Chandrasekhar, *J. Chem. Soc., Perkin Trans. 1*, **1990**, 1211.

[901] M. E. Jung and G. L. Hatfield, *Tetrahedron Lett.*, **24**, 3175 (1983).

[902] A. Alonso, C. S. Burgey, B. V. Rao, G. D. Vite, R. Vollerthun, M. A. Zotolla, and B. Fraser–Reid, *J. Am. Chem. Soc.*, **115**, 6666 (1993).

[903] K. C. Nicolaou, D. G. McGarry, P. K. Somers, C. A. Veale, and G. T. Furst, *J. Am. Chem. Soc.*, **109**, 2504 (1987).

[904] J. Lejeune and J. Y. Lallemand, *Tetrahedron Lett.*, **33**, 2977 (1992).

[905] J. C. Lopez and B. Fraser–Reid, *J. Am. Chem. Soc.*, **111**, 3450 (1989).

[906] S. A. Gloves, S. L. Golding, A. Goosen, and C. W. McCleland, *J. Chem. Soc., Perkin Trans. 1*, **1981**, 842.

[907] C. C. Yang, H. T. Chang, and J. M. Fang, *J. Org. Chem.*, **58**, 3100 (1993).

[908] A. L. Beck, M. Mascal, C. J. Moody, A. M. Z. Slawin, D. J. Williams, and W. J. Coates, *J. Chem. Soc., Perkin Trans. 1*, **1992**, 797.

[909] A. L. J. Beckwith and D. R. Boate, *J. Org. Chem.*, **53**, 4339 (1988).

[910] S. A. Glover and J. Warkentin, *J. Org. Chem.*, **58**, 2115 (1993).

[911] P. Renaud, J.–P. Vionnet, and P. Vogel, *Tetrahedron Lett.*, **32**, 3491 (1991).

[912] D. J. Hart, H. C. Huang, R. Krishnamurthy, and T. Schwartz, *J. Am. Chem. Soc.*, **111**, 7507 (1989).

[913] J. Marco–Contelles, P. Ruiz–Fernández, and B. Sánchez, *J. Org. Chem.*, **58**, 2894 (1993).

[914] M. Ladlow and G. Pattenden, *J. Chem. Soc., Perkin Trans. 1*, **1988**, 1107.

[915] H. E. Bode, C. G. Sowell, and R. D. Little, *Tetrahedron Lett.*, **31**, 2525 (1990).

[916] A. R. Forrester, M. Gill, J. S. Sadd, and R. H. Thomson, *J. Chem. Soc., Chem. Commun.*, **1975**, 291.

[917] D. J. Hart and W. C. Wu, *Tetrahedron Lett.*, **32**, 4099 (1991).

[918] M. P. Bertrand, J.–M. Surzur, H. Oumar–Mahamat, and C. Moustrou, *J. Org. Chem.*, **56**, 3089 (1991).

[919] B. Chenera, C.–P. Chuang, D. J. Hart, and C.–S. Lai, *J. Org. Chem.*, **57**, 2018 (1992).

[920] D. L. J. Clive and S. Daigneault, *J. Org. Chem.*, **56**, 5285 (1991).

[921] B. Vacher, A. Samat, A. Allouche, A. Laknifly, A. Baldy, and M. Chanon, *Tetrahedron*, **44**, 2925 (1988).

[922] M. Ladlow and G. Pattenden, *Tetrahedron Lett.*, **26**, 4413 (1985).

[923] M. Ishizaki, K. Ozaki, A. Kaematsu, T. Isoda, and O. Hoshino, *J. Org. Chem.*, **58**, 3877 (1993).

[924] T. Ueda and S. Shuto, *Nucleosides Nucleotides*, **3**, 295 (1984).

[925] C. Lamas, C. Saa, L. Castedo, and D. Dominguez, *Tetrahedron Lett.*, **33**, 5663 (1992).

[926] F. E. Ziegler and P. G. Harran, *Tetrahedron Lett.*, **34**, 4505 (1993).

[927] L. A. Paquette, A. G. Schaefer, and J. P. Springer, *Tetrahedron*, **43**, 5567 (1987).

[928] T. Yamamoto, S. Ishibuchi, T. Ishizuka, M. Haratake, and T. Kunieda, *J. Org. Chem*, **58**, 1997 (1993).

[929] L. Benati, P. Spagnolo, A. Tundo, and G. Zanardi, *J. Chem. Soc., Perkin Trans. 1*, **1979**, 1536.

[930] L. Benati, G. Placucci, P. Spagnolo, A. Tundo, and G. Zanardi, *J. Chem. Soc., Perkin Trans. 1*, **1977**, 1694.

[931] S. Caddick, K. Aboutayab, and R. West, *Synlett*, **1993**, 231.

[932] R. Ahmed–Schofield and P. S. Mariano, *J. Org. Chem.*, **50**, 5667 (1985).

[933] R. Leardini, A. Tundo, G. Zanardi, and G. F. Pedulli, *Synthesis*, **1985**, 107.

[934] A. R. Forrester, A. S. Ingram, and R. H. Thomson, *J. Chem. Soc., Perkin Trans, 1*, **1972**, 2847.

[935] Y. Hirai, A. Hagiwara, T. Terada, and T. Yamazaki, *Chem. Lett.*, **1987**, 2417.

[936] L. Belvisi, C. Gennari, G. Poli, C. Scolastico, and B. Salom, *Tetrahedron: Asymmetry*, **4**, 273 (1993).

[937] D. C. Lathbury, P. J. Parsons, and I. Pinto, *J. Chem. Soc., Chem. Commun.*, **1988**, 81.

[938] R. Leardini, M. Lucarini, A. Nanni, D. Nanni, G. F. Pedulli, A. Tundo, and G. Zanardi, *J. Org. Chem.*, **58**, 2419 (1993).

[939] R. Ahmed–Schofield and P. S. Mariano, *J. Org. Chem.*, **52**, 1478 (1987).

[940] A. J. Walkington and D. A. Whiting, *Tetrahedron Lett.*, **30**, 4731 (1989).

[941] K. S. Feldman and C. J. Burns, *J. Org. Chem.*, **56**, 4601 (1991).

[942] J. K. Choi, D. C. Ha, D. J. Hart, C. S. Lee, S. Ramesh, and S. Wu, *J. Org. Chem.*, **54**, 279 (1989).

[943] J. K. Dickson, Jr., R. Tsang, J. M. Llera, and B. Fraser–Reid, *J. Org. Chem.*, **54**, 5350 (1989).

[944] S. Kano, Y. Ynasa, T. Yokomatsu, K. Asami, and S. Shibuya, *Chem. Pharm. Bull.*, **36**, 2934 (1988).

[945] D. L. J. Clive and A. C. Joussef, *J. Chem. Soc., Perkin Trans. 1*, **1991**, 2797.

[946] D. L. Boger, T. Ishizaki, R. J. Wysocki, Jr., S. A. Munk, P. A. Kitos, and O. Suntornwat, *J. Am. Chem. Soc.*, **111**, 6461 (1989).

[947] K. Jones and J. Wilkinson, *J. Chem. Soc., Chem. Commun.*, **1992**, 1767.

[948] G. Just and G. Sacripante, *Can. J. Chem.*, **65**, 104 (1987).

[949] R. C. Larock and N. H. Lee, *J. Org. Chem.*, **56**, 6253 (1991).

[950] V. I. Minkin, E. P. Ivakhnenko, A. I. Shif, L. P. Olekhnovich, O. E. Kompan, A. I. Yanovskii, and Y. T. Struchkov, *J. Chem. Soc., Chem. Commun.* **1988**, 990.

[951] Y. C. Xin, J.–M. Mallet, and P. Sinaÿ, *J. Chem. Soc., Chem. Commun.*, **1993**, 864.

[952] B. Vauzeilles, D. Cravo, J.–M. Mallet, and P. Sinaÿ, *Synlett*, **1993**, 522.

[953] A. G. Myers, D. Y. Gin, and D. H. Rogers, *J. Am. Chem. Soc.*, **115**, 2036 (1993).

[954] T. Sano, H. Inoue, and T. Ueda, *Chem. Pharm. Bull.*, **33**, 1856 (1985).

[955] A. M. Rosa, S. Prabhahav, and A. M. Lobo, *Tetrahedron Lett.*, **31**, 1881 (1990).

[956] G. Stork, R. K. Boeckmann, Jr., D. F. Taber, W. C. Still, and J. Singh, *J. Am. Chem. Soc.*, **101**, 7107 (1979).

[957] D. P. Curran and H. Liu, *J. Am. Chem. Soc.*, **114**, 5863 (1992).

[958] A. Gopalsamy and K. K. Balasubramanian, *J. Chem. Soc., Chem. Commun.*, **1988**, 28.

[959] G. A. Kraus and H. Kim, *Synth. Commun.*, **23**, 55 (1993).

[960] J. J. Koehler and W. N. Speckamp, *Tetrahedron Lett.*, **1977**, 635.

[961] K. A. Parker, D. M. Spero, and J. van Epp, *J. Org. Chem.*, **53**, 4628 (1988).

[962] T. Ghosh and H. Hart, *J. Org. Chem.*, **53**, 2396 (1988).

[963] G. Stork and R. Mah, *Tetrahedron Lett.*, **30**, 3609 (1989).

[964] S. Karady, E. G. Corley, N. L. Abrahamson, and L. M. Weinstock, *Tetrahedron Lett.*, **30**, 2191, (1989).

[965] A. I. Meyers and B. A. Lefka, *Tetrahedron*, **43**, 5663 (1987).

[966] S. A. Ahmad–Junan and D. A. Whiting, *J. Chem. Soc., Chem. Commun.*, **1988**, 1160.

[967] N. S. Narasimhan and I. S. Aidhen, *Tetrahedron Lett.*, **29**, 2987 (1988).

[968] S. Takauo, M. Suzuki, A. Kijima, and K. Ogosawara, *Tetrahedron Lett.*, **31**, 2315 (1990).

[969] Y. Ozlu, D. E. Cladingboel, and P. J. Parsons, *Synlett*, **1993**, 357.

[970] R. Tsang and B. Fraser–Reid, *J. Am. Chem. Soc.*, **108**, 2116 (1986).

[971] J. C. Estevez, M. C. Villaverde, R. J. Estevez, and L. Castedo, *Tetrahedron*, **49**, 2783 (1993).

[972] J. C. Estevez, M. C. Villaverde, R. J. Estevez, and L. Castedo, *Tetrahedron Lett.*, **32**, 529 (1991).

[973] H. Pak, J. K. Dickson, Jr., and B. Fraser–Reid, *J. Org. Chem.*, **54**, 5337 (1989).

[974] J. Grimshaw and A. Prasanna de Silva, *Can. J. Chem.*, **58**, 1880 (1980).

[975] J. Grimshaw and A. Prasanna de Silva, *J. Chem. Soc., Chem. Commun.*, **1979**, 193.

[976] Y. Kubo, N. Asai, and T. Araki, *J. Org. Chem.*, **50**, 5484 (1985).

[977] H. Pak, I. I. Canalda, and B. Fraser–Reid, *J. Org. Chem.*, **55**, 3009 (1990).

[978] R. V. Bonnert, M. J. Davies, J. Howarth, and P. R. Jenkis, *J. Chem. Soc., Chem. Commun.*, **1990**, 148.

[979] G. Majetich, J. S. Song, C. Ringold, and G. A. Nemeth, *Tetrahedron Lett.*, **31**, 2239 (1990).

[980] T. Ueda, S. Shuto, M. Satoh, and H. Iuone, *Nucleosides Nucleotides*, **4**, 401 (1985).

[981] M. D. Bachi and D. Denenmark, *J. Am. Chem. Soc.*, **111**, 1886 (1989).

[982] J. J. Jenkinson, J. P. Parsons, and S. C. Eyley, *Synlett*, **1992**, 679.

[983] S. Takano, M. Suzuki, and K. Ogasawara, *Heterocycles*, **31**, 1151 (1990).

[984] D. E. Cladingboel and P. J. Parsons, *J. Chem. Soc., Chem. Commun.* **1990**, 1543.

[985] R. A. Holton and A. D. Williams, *J. Org. Chem.*, **53**, 5983 (1988).

[986] K. A. Parker and D. Fokas, *J. Am. Chem. Soc.*, **114**, 9688 (1992).

[987] M. Kim, K. Kawada, R. S. Gross, and D. S. Watt, *J. Org. Chem.*, **55**, 504 (1990).

[988] K. Kawada, M. Kim, and D. S. Watt, *Tetrahedron Lett.*, **30**, 5989 (1989).

[989] H. Finch, L. M. Harwood, G. M. Robertson, and R. C. Sewell, *Tetrahedron Lett.*, **30**, 2585 (1989).

[990] M. Ishizaki, K. Kurihara, E. Tanazawa, and O. Hoshino, *J. Chem. Soc., Perkin Trans. 1*, **1993**, 101.

[991] A. Kurek–Tyrlik, J. Wicha, A. Zarecki, and G. Snatzke, *J. Org. Chem.*, **55**, 3484 (1990).

[992] S. Kim and P. L. Fuchs, *J. Am. Chem. Soc.*, **113**, 9864 (1991).

[993] K. N. V. Duong, A. Gaudemer, M. O. Johnson, R. Quilivic, and J. Zylber, *Tetrahedron Lett.*, **34**, 2997 (1975).

[994] S. A. Ahmad–Junan, A. J. Walkington, and D. A. Whiting, *J. Chem. Soc., Chem. Commun.*, **1989**, 1613.

[995] E. Ottow, G. Neef, and R. Wiechert, *Angew. Chem.*, **101**, 776 (1989).

[996] A. K. Singh, R. K. Bakshi, and E. J. Corey, *J. Am. Chem. Soc.*, **109**, 6187 (1987).

[997] M. Koreeda and I. A. George, *Chem. Lett.*, **1990**, 83.

CUMULATIVE CHAPTER TITLES
BY VOLUME

6. **The Preparation of Unsymmetrical Biaryls by the Diazo Reaction and the Nitrosoacetylamine Reaction**: Werner E. Bachmann and Roger A. Hoffman

7. **Replacement of the Aromatic Primary Amino Group by Hydrogen**: Nathan Kornblum

8. **Periodic Acid Oxidation**: Ernest L. Jackson

9. **The Resolution of Alcohols**: A. W. Ingersoll

10. **The Preparation of Aromatic Arsonic and Arsinic Acids by the Bart, Béchamp, and Rosenmund Reactions**: Cliff S. Hamilton and Jack F. Morgan

Volume 3 (1946)

1. **The Alkylation of Aromatic Compounds by the Friedel-Crafts Method**: Charles C. Price

2. **The Willgerodt Reaction**: Marvin Carmack and M. A. Spielman

3. **Preparation of Ketenes and Ketene Dimers**: W. E. Hanford and John C. Sauer

4. **Direct Sulfonation of Aromatic Hydrocarbons and Their Halogen Derivatives**: C. M. Suter and Arthur W. Weston

5. **Azlactones**: H. E. Carter

6. **Substitution and Addition Reactions of Thiocyanogen**: John L. Wood

7. **The Hofmann Reaction**: Everett L. Wallis and John F. Lane

8. **The Schmidt Reaction**: Hans Wolff

9. **The Curtius Reaction**: Peter A. S. Smith

Volume 4 (1948)

1. **The Diels-Alder Reaction with Maleic Anhydride**: Milton C. Kloetzel

2. **The Diels-Alder Reaction: Ethylenic and Acetylenic Dienophiles**: H. L. Holmes

3. **The Preparation of Amines by Reductive Alkylation**: William S. Emerson

4. **The Acyloins**: S. M. McElvain

5. **The Synthesis of Benzoins**: Walter S. Ide and Johannes S. Buck

6. **Synthesis of Benzoquinones by Oxidation**: James Cason

7. **The Rosenmund Reduction of Acid Chlorides to Aldehydes**: Erich Mosettig and Ralph Mozingo

8. **The Wolff-Kishner Reduction**: David Todd

2. **Halocyclopropanes from Halocarbenes**:   William E. Parham and Edward E. Schweizer

3. **Free Radical Addition to Olefins to Form Carbon-Carbon Bonds**:   Cheves Walling and Earl S. Huyser

4. **Formation of Carbon-Heteroatom Bonds by Free Radical Chain Additions to Carbon-Carbon Multiple Bonds**:   F. W. Stacey and J. F. Harris, Jr.

Volume 14 (1965)

1. **The Chapman Rearrangement**:   J. W. Schulenberg and S. Archer

2. **α-Amidoalkylations at Carbon**:   Harold E. Zaugg and William B. Martin

3. **The Wittig Reaction**:   Adalbert Maercker

Volume 15 (1967)

1. **The Dieckmann Condensation**:   John P. Schaefer and Jordan J. Bloomfield

2. **The Knoevenagel Condensation**:   G. Jones

Volume 16 (1968)

1. **The Aldol Condensation**:   Arnold T. Nielsen and William J. Houlihan

Volume 17 (1969)

1. **The Synthesis of Substituted Ferrocenes and Other π-Cyclopentadienyl-Transition Metal Compounds**:   Donald E. Bublitz and Kenneth L. Rinehart, Jr.

2. **The γ-Alkylation and γ-Arylation of Dianions of β-Dicarbonyl Compounds**:   Thomas M. Harris and Constance M. Harris

3. **The Ritter Reaction**:   L. I. Krimen and Donald J. Cota

Volume 18 (1970)

1. **Preparation of Ketones from the Reaction of Organolithium Reagents with Carboxylic Acids**:   Margaret J. Jorgenson

2. **The Smiles and Related Rearrangements of Aromatic Systems**:   W. E. Truce, Eunice M. Kreider, and William W. Brand

3. **The Reactions of Diazoacetic Esters with Alkenes, Alkynes, Heterocyclic, and Aromatic Compounds**:   Vinod Dave and E. W. Warnhoff

4. **The Base-Promoted Rearrangements of Quaternary Ammonium Salts**:   Stanley H. Pine

Volume 24 (1976)

1. **Homogeneous Hydrogenation Catalysts in Organic Solvents**:   Arthur J. Birch and David H. Williamson

2. **Ester Cleavages via S_N2-Type Dealkylation**:   John E. McMurry

3. **Arylation of Unsaturated Compounds by Diazonium Salts (The Meerwein Arylation Reaction)**:   Christian S. Rondestvedt, Jr.

4. **Selenium Dioxide Oxidation**:   Norman Rabjohn

Volume 25 (1977)

1. **The Ramberg-Bäcklund Rearrangement**:   Leo A. Paquette

2. **Synthetic Applications of Phosphoryl-Stabilized Anions**: William S. Wadsworth, Jr.

3. **Hydrocyanation of Conjugated Carbonyl Compounds**:   Wataru Nagata and Mitsuru Yoshioka

Volume 26 (1979)

1. **Heteroatom-Facilitated Lithiations**:   Heinz W. Gschwend and Herman R. Rodriguez

2. **Intramolecular Reactions of Diazocarbonyl Compounds**:   Steven D. Burke and Paul A. Grieco

Volume 27 (1982)

1. **Allylic and Benzylic Carbanions Substituted by Heteroatoms**: Jean-François Biellmann and Jean-Bernard Ducep

2. **Palladium-Catalyzed Vinylation of Organic Halides**:   Richard F. Heck

Volume 28 (1982)

1. **The Reimer-Tiemann Reaction**:   Hans Wynberg and Egbert W. Meijer

2. **The Friedländer Synthesis of Quinolines**:   Chia-Chung Cheng and Shou-Jen Yan

3. **The Directed Aldol Reaction**:   Teruaki Mukaiyama

Volume 29 (1983)

1. **Replacement of Alcoholic Hydroxy Groups by Halogens and Other Nucleophiles via Oxyphosphonium Intermediates**:   Bertrand R. Castro

Volume 36 (1988)

1. **The [3 + 2] Nitrone-Olefin Cycloaddition Reaction**:   Pat N. Confalone and Edward M. Huie

2. **Phosphorus Addition at sp^2 Carbon**:   Robert Engel

3. **Reduction by Metal Alkoxyaluminum Hydrides. Part II. Carboxylic Acids and Derivatives, Nitrogen Compounds, and Sulfur Compounds**:   Jaroslav Málek

Volume 37 (1989)

1. **Chiral Synthons by Ester Hydrolysis Catalyzed by Pig Liver Esterase**:   Masaji Ohno and Masami Otsuka

2. **The Electrophilic Substitution of Allylsilanes and Vinylsilanes**:   Ian Fleming, Jacques Dunoguès, and Roger Smithers

Volume 38 (1990)

1. **The Peterson Olefination Reaction**:   David J. Ager

2. **Tandem Vicinal Difunctionalization: β-Addition to α,β-Unsaturated Carbonyl Substrates Followed by α-Functionalization**:   Marc J. Chapdelaine and Martin Hulce

3. **The Nef Reaction**:   Harold W. Pinnick

Volume 39 (1990)

1. **Lithioalkenes from Arenesulfonylhydrazones**:   A. Richard Chamberlin and Steven H. Bloom

2. **The Polonovski Reaction**:   David Grierson

3. **Oxidation of Alcohols to Carbonyl Compounds via Alkoxysulfonium Ylides: The Moffatt, Swern, and Related Oxidations**:   Thomas T. Tidwell

Volume 40 (1991)

1. **The Pauson-Khand Cycloaddition Reaction for Synthesis of Cyclopentenones**:   Neil E. Schore

2. **Reduction with Diimide**:   Daniel J. Pasto and Richard T. Taylor

3. **The Pummerer Reaction of Sulfinyl Compounds**:   Ottorino DeLucchi, Umberto Miotti, and Giorgio Modena

4. **The Catalyzed Nucleophilic Addition of Aldehydes to Electrophilic Double Bonds**:   Hermann Stetter and Heinrich Kuhlmann

AUTHOR INDEX, VOLUMES 1–48

Volume number only is designated in this index.

CHAPTER AND TOPIC INDEX, VOLUMES 1-48

Many chapters contain brief discussions of reactions and comparisons of alternative synthetic methods related to the reaction that is the subject of the chapter. These related reactions and alternative methods are not usually listed in this index. In this index, the volume number is in **boldface**, the chapter number is in ordinary type.